国家自然科学基金项目（61571312） 国家重点研发项目（2018YFC0830300）

Fractional Calculus Principles of Signal Processing

信号处理的分数阶微积分原理

蒲亦非　张　妮　周激流 ◎ 著

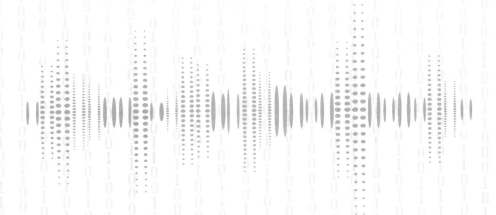

四川大学出版社

项目策划：毕　潜
责任编辑：毕　潜
责任校对：周维彬
封面设计：墨创文化
责任印制：王　炜

图书在版编目（CIP）数据

信号处理的分数阶微积分原理 / 蒲亦非，张妮，周激流著 . — 成都：四川大学出版社，2020.1
ISBN 978-7-5690-3697-8

Ⅰ . ①信… Ⅱ . ①蒲… ②张… ③周… Ⅲ . ①信号处理—研究 Ⅳ . ① TN911.7

中国版本图书馆 CIP 数据核字（2020）第 015245 号

书名　信号处理的分数阶微积分原理

著　　者	蒲亦非　张　妮　周激流
出　　版	四川大学出版社
地　　址	成都市一环路南一段 24 号（610065）
发　　行	四川大学出版社
书　　号	ISBN 978-7-5690-3697-8
印前制作	四川胜翔数码印务设计有限公司
印　　刷	四川彩美印务有限公司
成品尺寸	185mm×260mm
印　　张	37.75
字　　数	1157 千字
版　　次	2020 年 7 月第 1 版
印　　次	2020 年 7 月第 1 次印刷
定　　价	160.00 元

版权所有 ◆ 侵权必究

◆ 读者邮购本书，请与本社发行科联系。
　电话：(028)85408408/(028)85401670/
　(028)86408023　邮政编码：610065
◆ 本社图书如有印装质量问题，请寄回出版社调换。
◆ 网址：http://press.scu.edu.cn

四川大学出版社
微信公众号

前 言

　　信号处理是信息科学的一个核心基础组成部分。信号处理经历了从模拟到数字，从确知到随机的历程，已发展成以非线性和非确定性为主要特征的现代信号处理。现代信号处理经常会遇到许多关于非线性、非确定性、非因果、非高斯、非平稳、非最小相位、非白噪声、非整数维和非整数阶的难题，而传统整数阶信号处理方法已经不能有效解决上述"非"的难题。为了攻克这些"非"的难题，我们迫切需要寻求新的数学方法进行学科交叉。

　　近三百年以降，分数阶微积分（fractional calculus）业已成为数学分析的一个重要分支。然而，对于世界上大多数数学家、物理学家和工程技术学者而言，分数阶微积分都还比较陌生，它是一个新颖且有前景的研究方向。分数阶微积分是一门和整数阶微积分同样古老的学科，直至近期它的相关研究一直专注于纯数学领域。我们所熟知的是一阶微积分、二阶微积分，一直到 n 阶微积分，那么一个有趣的问题是 0.5 阶微积分是什么呢？其实，分数阶微积分的概念不局限于常用的有理分数阶，它还可以是无理分数阶，甚至是复数阶和四元数阶。因此，分数阶微积分最科学的名称应该是"非整数阶微积分"。但由于历史的原因，本书仍按习惯称其为"分数阶微积分"。分数阶微积分的物理意义是分数阶速度或分数流，几何意义是整数阶微积分的连续内插。分数阶微积分是对经典整数阶微积分理论的推广，扩展了整数阶差分和 Riemann 求和的概念，具有许多独特的性质。事实上，大多数函数都是可微积函数（differintegrable function）。众所周知，物理过程中的随机变量可以被视为粒子随机运动的位移，于是，分数阶微积分能够被应用于分析和处理许多特定的物理问题及其信号处理之中。然而不幸的是，目前它的主要应用仍集中于对物理变化的瞬态描述，较少涉猎系统的演化过程。科学研究表明，分数阶以及分数维的方法是对许多自然现象进行描述的最佳方法。目前，分数阶微积分已经被成功应用于诸如分数阶扩散过程、分数阶粘弹性理论、分形动力学、分数阶控制、分数阶生物工程、分数阶材料学、分数阶电磁场理论、分数阶信号处理、分数阶图像处理、分数阶神经网络、分数阶进化算法、分数阶电路与系统等许多研究领域，取得了一些令人满意的实验结果和进一步研究的科学启示。与传统整数阶方法相比，分数阶方法是"比最好还更好"的方法。当分数阶取为整数阶时，分数阶方法就退化为经典整数阶方法，可取得该整数阶方法所能达到的最好效果；当分数阶取为非整数阶时，就有可能取得比传统整数阶方法更好的效果。

如何将分数阶微积分应用于现代信号分析与处理之中，目前在国际上还是一个研究甚少的新兴学科分支。分数阶微积分被引入用于现代信号处理研究领域，主要是因为其具有常时记忆性（long-term memory）、非局域性（non-locality）和弱奇异性（weak singularity）。信号的分数阶微积分特性与其整数阶微积分特性有显著差别。随着基于分数阶微积分的现代信号处理的深入研究，人们很自然地会提出一些值得探索的有趣科学问题。譬如，除 Caputo 定义以外，赫维赛德函数（Heaviside function）的常数部分的分数阶微分不等于零，而其常数部分的整数阶微分却恒等于零。因此，分数阶微分可以用于非线性增强图像的复杂纹理细节特征。图像的分数阶微分具有特殊的马赫效应和拮抗特性，以致拥有一种分数阶仿生视觉感受野模型。分数阶微分不仅能非线性保持平滑图像区域的低频轮廓信息，而且能同时分数阶多尺度增强图像的中频和高频的边缘和纹理细节。又如，我们能否改造传统的整数阶变分法，将分数阶微积分的这一特性用于图像复原模型（图像去噪、图像增强、图像分割）之中呢？在分数阶自适应信号处理和分数阶自适应控制之中，分数阶极值点和传统的整数阶极值点、一阶驻点具有显著的不同。为了搜索能量泛函的分数阶极值点，可将整数阶最速下降法推广到分数阶，提出了分数阶最速下降法。我们能否利用分数阶最速下降法的分数阶极值点的独特性质，改造已被广泛应用的经典整数阶 BP 算法（Back Propagation algorithm），从而让其更容易收敛到全局最优点呢？再如，随着人们以模拟电路形式成功构造出分数阶微分器和分数阶积分器，出现了一种被称为"分抗元"（fractor）的崭新电路元器件。分抗元实质上是一种完成分数阶微积分功能的信号处理滤波器。以自然形式实现的理想分抗元是由普通的电阻和电容元或电感元以树枝型、双电路型、H 型、网格型等其他无限递归结构的模拟电路形式构成的，具有高度自相似的分形结构。分抗值（fractance）是指分抗元的分数阶阻抗（fractional-order impedance）。一个分抗元的驱动点阻抗函数（driving-point impedance function）即是其分数阶电抗（fractional-order reactance）。分抗元有两种类型：容性分抗元和感性分抗元。在二端电路元件的蔡氏（蔡少棠，L. O. Chua）周期表中，容性分抗元处于电容元和电阻之间的线段内。因此，容性分抗元的电气特性处于电容元和电阻的电气特性之间。同样地，在关于所有二端电路元件的蔡氏周期表中，感性分抗元处于电感元和电阻之间的线段内。因此，感性分抗元的电气特性处于电感元和电阻的电气特性之间。有趣的问题是：分抗值的度量单位和物理量纲是什么？分抗元的串并联规则是什么？根据蔡氏电路元件周期表、逻辑相容性、公理完备性与形式对称性，相应于容性分抗元和感性分抗元，在电容元和忆阻元（memristor）之间以及在电感元和忆阻元之间，还应该分别存在两种新兴的分数阶忆阻元（fractional order memristor），即容性分忆抗元（fracmemristor）和感性分忆抗元。那么，分忆抗元的指纹特征（fingerprints）是什么？分忆抗元的模拟电路实现形式又是什么？

笔者长期从事将分数阶微积分应用于现代信号处理，特别是图像处理、电路与系统和人工智能这一新兴学科分支的尝试性探索和系统研究。本书内容主要取材于国家重点研发项目（2018YFC0830300）、国家自然科学基金项目（61571312和60972131）、中国博士后科学基金项目（20060401016）、法中科学与应用基金博士后项目（Fondation Franco-Chinoise pour la Science et ses Applications）的部分研究成果，源自作者与合作者近年来的一些探索和思考。感谢中国自动化学会分数阶系统与控制专业委员会对本书的出版的大力支持。

各种函数的分数阶微积分具有一个显著的特征：大多数函数的分数阶微积分等于幂级数，或等于特定函数和幂函数的乘积或叠加。这一可贵的特征是否暗示着自然界的某种实质性变化规律呢？现代信号处理的分数阶微积分原理博大精深，笔者不自量力，拢拙成著，抛砖引玉，以期让读者了解并使用分数阶微积分这一前途无量的数学方法。由于本书编写仓促，加之作者才疏学浅，难免会有错漏遗误，恳请同行专家和广大读者批评斧正。

著　者

谨识于四川大学天工斋

2020年2月23日

目 录

第1章 分数阶微积分理论研究及其主要应用 ………………………………………（ 1 ）
　1.1 分数阶微积分理论研究 ………………………………………………………（ 2 ）
　1.2 分数阶微积分应用于描述各种物理系统和材料的动力学行为 ……………（ 3 ）
　1.3 分数阶微积分应用于生物工程 ………………………………………………（ 4 ）
　1.4 分数阶微积分应用于动力学系统 ……………………………………………（ 5 ）
　1.5 分数阶微积分应用于控制系统 ………………………………………………（ 5 ）
　1.6 分数阶微积分应用于信号处理 ………………………………………………（ 6 ）

第2章 连续子波变换数值实现中起始尺度的确定以及信号时间和扫描时间之间的几何关系 ………………………………………………………………………（ 9 ）
　2.1 问题提出 ………………………………………………………………………（ 9 ）
　2.2 连续子波的选择 ………………………………………………………………（ 10 ）
　2.3 （复）解析母波的尺度采样间隔的推导 ……………………………………（ 11 ）
　2.4 （实）偶母波的尺度采样间隔的推导 ………………………………………（ 13 ）
　2.5 （实）奇母波的尺度采样间隔的推导 ………………………………………（ 17 ）
　2.6 二进点格采样及二进抽取采样时起始尺度的确定 …………………………（ 20 ）
　2.7 推导连续子波变换中信号时间和扫描时间之间的几何关系 ………………（ 20 ）
　2.8 小结 ……………………………………………………………………………（ 21 ）

第3章 现代信号分析与处理中分数阶微积分的数值实现 ……………………（ 23 ）
　3.1 问题提出 ………………………………………………………………………（ 23 ）
　3.2 分数阶微积分四种常用的时域定义 …………………………………………（ 25 ）
　3.3 分数阶微积分三种常用的频域定义 …………………………………………（ 27 ）
　3.4 信号分数阶微积分的幂级数算法 ……………………………………………（ 30 ）
　3.5 信号分数阶微积分的 Fourier 级数算法 ……………………………………（ 32 ）
　3.6 信号分数阶微分基于 Grümwald-Letnikov 定义算法 ………………………（ 37 ）
　3.7 信号分数阶微分基于子波变换的算法 ………………………………………（ 40 ）
　3.8 信号分数阶微分基于子波变换的快速工程算法 ……………………………（ 44 ）
　3.9 小结 ……………………………………………………………………………（ 48 ）

第4章 分数阶微积分的模拟分抗电路实现 ……………………………………（ 49 ）
　4.1 问题提出 ………………………………………………………………………（ 49 ）
　4.2 模拟分抗电路的阻抗特性 ……………………………………………………（ 50 ）
　4.3 构造 $\frac{1}{2}$ 阶分数阶微积分的模拟分抗电路 ………………………………（ 53 ）

· 1 ·

4.4 构造 $\frac{1}{2^n}$ 阶模拟分抗电路 ……………………………………………… (60)
4.5 构造任意分数阶模拟分抗电路 ……………………………………………… (62)
4.6 实验仿真及结果分析 ………………………………………………………… (64)
4.7 小结 …………………………………………………………………………… (71)

第5章 分抗的单位和物理量纲 …………………………………………………………… (73)
5.1 Measurement Units and Physical Dimensions of Fractance-Part Ⅰ：Position of Purely Ideal Fractor in Chua's Axiomatic Circuit Element System and Fractional-Order Reactance of Fractor in Its Natural Implementation …… (73)
5.2 Measurement Units and Physical Dimensions of Fractance-Part Ⅱ：Fractional-Order Measurement Units and Physical Dimensions of Fractance and Rules for Fractors in Series and Parallel ……………………………… (108)

第6章 分抗元的阶—频特性 ……………………………………………………………… (141)
6.1 Introduction …………………………………………………………………… (141)
6.2 Related Work ………………………………………………………………… (143)
6.3 Order-Frequency Characteristics of Fractor ……………………………… (144)
6.4 Experiment and Analysis …………………………………………………… (151)
6.5 Conclusions …………………………………………………………………… (155)

第7章 分忆抗元：分数阶忆阻元 ………………………………………………………… (157)
7.1 Introduction …………………………………………………………………… (157)
7.2 Related Work ………………………………………………………………… (160)
7.3 Fracmemristor ………………………………………………………………… (163)
7.4 Analog Circuit Implementation of Fractional-Order Memristor：Arbitrary-Order Lattice Scaling Fracmemristor ……………………………………… (175)
7.5 Experiment and Analysis …………………………………………………… (188)
7.6 Conclusions and Discussion ………………………………………………… (206)

第8章 任意分数阶神经型脉冲振荡器 …………………………………………………… (209)
8.1 问题提出 ……………………………………………………………………… (209)
8.2 1/2阶网格型模拟分抗的等效实现及其电路特性 ………………………… (210)
8.3 基于分数阶演算的分数阶神经型振荡器 …………………………………… (214)
8.4 实验仿真及结果分析 ………………………………………………………… (215)
8.5 小结 …………………………………………………………………………… (217)

第9章 分数阶Hopfield神经网络：分数阶动态联想递归神经网络 …………………… (218)
9.1 问题提出 ……………………………………………………………………… (218)
9.2 多层动态联想神经网络 ……………………………………………………… (219)
9.3 基于广义Hebb规则的多层动态联想神经网络学习算法 ………………… (220)
9.4 分数阶Hopfield神经网络：分数阶多层动态联想神经网络 ……………… (225)

目 录

第10章 分数阶硬件安全：基于分数阶 Hopfield 神经网络的抗芯片克隆 ………… (262)
  10.1 Defense Against Chip Cloning Attacks Based on Fractional Hopfield Neural Networks ……………………………………………………………… (262)
  10.2 Analog Circuit Realization of Arbitrary-Order Fractional Hopfield Neural Networks：A Novel Application of Fractor to Defense against Chip Cloning Attacks ……………………………………………………… (303)

第11章 图像处理的分数阶微分算子：分数阶微分掩模 ……………………… (332)
  11.1 二维数字图像信号分数阶微分的数值实现 ……………………………… (332)
  11.2 Fractional Differential Mask：A Fractional Differential-Based Approach for Multiscale Texture Enhancement ……………………………… (371)

第12章 分数阶最速下降法：分数阶自适应信号处理 ………………………… (414)
  12.1 Introduction …………………………………………………………… (414)
  12.2 Related Works ………………………………………………………… (416)
  12.3 Experiments and Analysis …………………………………………… (421)
  12.4 Conclusions …………………………………………………………… (428)

第13章 分数阶变分法中的分数阶欧拉－拉格朗日方程 ……………………… (430)
  13.1 Introduction …………………………………………………………… (430)
  13.2 Mathematical Background …………………………………………… (432)
  13.3 Fractional-Order Euler-Lagrange Equation for Fractional-Order Variational Method ……………………………………………………………… (433)
  13.4 Experiment and Analysis …………………………………………… (450)
  13.5 Conclusions …………………………………………………………… (466)

第14章 基于分数阶偏微分方程的退化图像逆处理 …………………………… (468)
  14.1 引言 …………………………………………………………………… (468)
  14.2 相关工作 ……………………………………………………………… (470)
  14.3 构造分数阶偏微分方程：一类基于分数阶超全变差和分数阶最速下降法的纹理图像多尺度去噪模型的理论推导和分析 ……………………………… (471)
  14.4 实验仿真及其理论分析 ……………………………………………… (489)
  14.5 结论 …………………………………………………………………… (521)

第15章 图像的分数阶对比度增强：一种 Retinex 的分数阶变分框架 ……… (523)
  15.1 Introduction …………………………………………………………… (523)
  15.2 Mathematical Background …………………………………………… (526)
  15.3 Fractional-Order Variational Framework for Retinex Based on FPDE … (527)
  15.4 Experiment and Analysis …………………………………………… (532)
  15.5 Conclusions …………………………………………………………… (548)

参考文献 ……………………………………………………………………… (549)

第1章 分数阶微积分理论研究及其主要应用

分数阶微积分将通常的微分和积分从整数阶推广到任意阶，它和微分学一起诞生，可以追溯到 Leibniz 和 Newton 创造微分学的时代。1695 年，Leibniz 在写给 L'Hôpital 的信中提出了一个关于将微分阶数从整数阶推广到非整数阶的问题，他的原话是："Can the meaning of derivatives with integer order be generalized to derivatives with non-integer orders?"要想回答这个问题不是一件易事，即使对于大数学家 L'Hôpital 而言。L'Hôpital 感到有些好奇，于是以问代答，转而问了 Leibniz 另一个关于 1/2 阶微积分的问题，他的原话是："What if the order will be 1/2?" Leibniz 在署名日期为 1695 年 9 月 30 日（分数阶微积分的准确诞生日）的回信中写道：这将导出一个是非而是的论点，总有一天人们可以由它推导出一些有用的结论。他的原话是："It will lead to a paradox, from which one day useful consequences will be drawn."由 Leibniz 所提出的问题开创了一门蓬勃发展了 300 多年的关于分数阶微积分的学说。历史上许多数学家，包括 Liouville，Riemann，Weyl，Fourier，Abel，Lacroix，Leibniz，Grunwald 和 Letnikov 这样的数学巨匠，花费多年的时间去完善和发展了它的理论。他们的理论与观点在 K. B. Oldham，J. Spanier 或 K. S. Miller，B. Ross 所著的相关书籍中得到了比较详细的总结。

从数学观点来看，分数阶微积分或分数阶演算（fractional calculus）是数学分析的一个分支，它研究微分算子 $D = \dfrac{\mathrm{d}}{\mathrm{d}x}$ 和积分算子 J（或用 I 表示）实数次幂（即可以扩展为非整数，包括分数）的理论及应用。定义在区间 $[a,t]$ 上的函数或信号 $s(t)$ 的 v 阶分数阶微积分最常用的表示符号是 ${}_aD_t^v s(t)$ 或 $s^{(v)}(t)$。当阶数 v 取负数时，即进行 v 阶分数阶积分。分数阶微积分的幂数与 $G^2(x) = G(G(x))$ 中的平方指数意义相同，其中，G 表示某种算子。因此，可将微分算子 D 的非整数次幂表示为

$$\sqrt[p]{D} = D^{\frac{1}{p}} = D^v \tag{1.1}$$

特别地，作为微分算子的平方根（半次操作）$\sqrt[2]{D}$，即对 $D^{\frac{1}{2}}$，操作两次以后有一阶微分的效果 $D = D^{\frac{1}{2}} D^{\frac{1}{2}}$。令 $\mathbf{Z}$ 表示整数集，当 v 为实整数时，对于算子 $D^v = D^n$，$n \in \mathbf{Z}$ 而言，若 $n>0$，则等同于通常的幂 n 次操作；若 $n<0$，则等同于幂 n 次积分 J。于是，更一般地，可以将幂 D^n 组成的半群视为在一个连续半群中取离散值的部分。连续半群的理论在数学上已经有系统的研究。

值得注意的是，将上述这种非整数阶的微积分称为"分数阶微积分"或"分数阶演算"是一种错误的命名，其中的"分数"是一个错误的归类。因为指数 v 可以被推广到有理分数、无理数甚至复数，所以从严格意义上讲，它应该被称为"非整数阶微积分"或"非整数阶演算"。但是由于历史的原因，"分数阶微积分"或"分数阶演算"的命名已经成为习惯用法。为了不造成混淆，本书仍然沿用这种习惯命名。另一个值得注意的是，分

数阶微分算子是一个线性算子。

分数阶微积分是一种关于整数阶标准微积分的自然的数学推广。在数学中，与传统的标准算子相区别，分数阶微积分算子属于非标准算子。近年来，诸如分数阶积分—微分算子这样的非标准算子受到了越来越普遍的应用，并且它们正逐步成为应用科学领域内一个新的学科分支[9-24,50]。适合描述它们的数学理论框架是若干年前被确切定义的伪微分算子(Pseudo-Differential Operators，PDO)[1-2]。伪微分算子可以被视为对具有 $H(d/dt)$ 形式的微分算子的推广。$H(d/dt)$ 不再是一个多项式，而是一个具有条件限制的规则函数。关于这些非标准算子的理论和应用研究逐步表明，非标准算子是目前关于诸如流变学(rheology)、控制系统、信号处理、声波和电磁的传播(propagation)以及计量经济学(econometry)等许多复杂自然现象的已有描述中的唯一最佳描述。分数阶微积分已经在诸如电化学、扩散、概率、黏弹性、遗传结构、控制论和信号处理等方面进行了一些初步的应用，并取得了一些良好的效果，本章对其中部分研究成果进行陈述。由于作者知识水平有限，对许多学科相关知识的翻译和理解可能有误，也可能挂一漏万。

1.1 分数阶微积分理论研究

分数阶微积分和分形之间存在某种联系，分数阶微积分是分形的数学基础之一[4-5]。近几年来，分形在许多学科中体现出它的重要性。许多学者对分形的数学基础做出了出色的研究。有一类 Weierstrass 函数是一种具有特殊分形性质的分形函数。有学者以这类 Weierstrass 函数为例，试图将分数阶微积分和分形函数进行合并[6]。有学者对已知的复分数指数幂在时域中的分数阶积分的几何以及物理意义进行了探讨。他们研究发现分数阶积分虚数部分(the imaginary part of the fractional integral)和离散尺度恒定性(discrete scale invariance)现象是相关的，并且只能以实规则离散分形的形式呈现[7]。分数阶微分的 Grümwald 定义通常被用于数值估计分数阶微积分值，它是对整数阶微积分的有限差分公式的推广。有学者将向量分数阶微分的 Grümwald 定义式进行了推广。对空间变量而言，该推广更有利于求解分数阶的偏微分方程[8]。尽管分数阶微积分在理论和应用上有很多优势，但是其中还有一些不太清楚的地方使得它不能系统地运用。有学者从系统的观点研究了分数阶微积分的定义问题。他们不仅研究了局域(Grümwald-Letnikov)定义，还考虑了其全局定义。他们的研究表明，应该采用 Cauchy 公式，因为它在信号处理和控制方面的应用保持了一致性[16]。有学者论述了用功能微积分(functional-calculus)定义的线性分数阶(伪)微分算子(推广 Fourier 变换所得)的功率[11]。有学者研究了分数阶微积分数字化实现的关键步长。他们提出了两种方法：一种是递归离散化 Tustin 算子(Tustin operator)，另一种是用 Al-Alaoui 算子(Al-Alaoui operator)通过连续分数扩充来直接进行离散化。他们对其最小相位和状态进行近似离散化[15]。有学者将算子理论与归一化的分数阶算子相结合，以一种系统和完备的方式推演出了关于分数阶卷积和相关性算子的显式表达。通过算子运算，他们还提出了关于这些分数阶算子可以有效离散实现的可选公式。在此基础上，他们还将快速分数阶自相关应用于对线性 FM 信号的侦测与参数估计[22]。

1.2 分数阶微积分应用于描述各种物理系统和材料的动力学行为

在过去的几十年里，分数阶微积分常被用来描述各种物理系统和材料的动力学行为。散布性传送（Diffusive Transport）是自然界中一种最重要的传送机械原理。在显微镜可见的级别，扩散是每个粒子随机运动造成的结果。基于这种随机运动的是一种随机高斯过程（a stochastic Gaussian process）的关键假设，并运用拉普拉斯算子（Laplacian operators）去构建它的模型。然而，功的均方根在超扩散（superdiffusion）中比在高斯扩散（Gaussian diffusion）中增加得更快，在次扩散（subdiffusion）中比在高斯扩散中增加得更慢。不断出现的这种现象使得反常扩散（anomalous diffusion）随处可见，可以利用分数阶微分算子来构建非高斯反常散布性传送模型。分数阶扩散理论以某种方式涵盖了反常扩散。同时，如果用某种宽的（代数衰减的）传播子［broad (algebraic decaying) propagator］去审视积分算子，则分数阶扩散提供了一种关于非局域传送（non-local transport）的自然描述。国际上一些值得参考的关于分数阶扩散的项目是：①Numerical methods for the solution of partial differential equations of fractional order；②Front dynamics in reaction-diffusion systems with Levy flights: a fractional diffusion approach；③Non-diffusive transport in plasma turbulence: a fractional diffusion approach。在时间域内被非线性的瞬时力矩（second moment）所描述的反常扩散，不像用线性力矩描述的常规扩散，可以用关于时间的分数阶微分方程来刻画。求解这些分数阶扩散方程的基本解（fundamental solution）依赖于 H 函数（H-function）。然而不幸的是，没有关于分数阶扩散的一个概括性的物理解释。众所周知，普通扩散的解是高斯的，它可以用随机行走（random walks）来表现。由高斯函数导出的标准正交原理以及任何函数能够按照这一原则来进行推广。分数阶扩散方程的 H 函数解是用一个截短的高斯函数的线性合成来进行描述的。有学者用一套随机行走模拟了分数阶扩散，并且给出了一种可能的物理解释。有学者研究和分析了关于赫斯特参数（Hurst Parameter）大于二分之一的分数阶布朗运动（Brownian Motions）的 Clark-Ocone 定理。他们通过示例的方式分析了具有指标函数（indicator function）的乘法会按照相应的规则进行增加。他们讨论了该结论对于由 Hu 和 Ksendal（2000）引入的准条件下的期望值以及分数阶 Clark-Ocone 微分的重要意义[25]。分数阶 Cauchy 问题有利于构建不规则发散现象的物理模型。有学者研究了分数阶 Cauchy 问题的随机解。每个用于解决某抽象 Cauchy 问题的无限可分法则都定义了一个回旋半群（convolution semigroup）。在分数阶 Cauchy 问题中，他们用分数阶微分（fractional derivative）替换了关于时间的一阶微分。他们关于分数阶 Cauchy 问题的解决方案结合了传统的 Cauchy 问题解决方案[26]。

有学者研究了关于分形多空渗水介质的一种针对连续介质的改进 Fokker-Planck 方程。他们采用一个以时间和尺度为变量的基于分数阶微分和散布系数作为控制方程的散布条件。他们提出了一种改进 Fokker-Planck 方程的数值解法，同时用数值解法探讨了依赖时间和距离的分数阶微分和分数阶幂函数的差别。他们的研究表明，对于长尾边缘分布（heavy tailed marginal distribution）的运动可以用基于分数阶微分或以依赖时间和尺度为变量的散度方程来进行建模[27]。一个一维分数阶的水平对流差量方程（advection-dispersion equation）一般可以对氚羽（tritium plume）核在同位素异构点（heterogeneous MADE site）的运动状态进行精确建模。有学者提出了一种对这种方程参数（包括分数阶散布微分的阶数）的估计方法。这种方法假设液压传导场（hydraulic conductivity field）的长尾（heavy tail）功率将引起一个具有相似分布的流速场（velocity field）[28]。分数阶水平对流散布方程（Fractional advection-dispersion equations）可被应用于地下水文学，给在多空渗水媒介中被水流带动的追踪器的传输过程和状态进行建模。有学者提出了一种求解有限域内具有可变系数的一维分数阶水平对流散布方程可实现的数值算法，他们通过给一个散射流问题建模的方式对推导结论进行了演示。应用分数阶微分可使模型方程追踪在观测地点先到达的追踪器[29]。分数阶 diffusion-wave 方程是被广义化了的扩散和波动方程（通过对时间和空间进行分数阶微分）。分数阶 diffusion-wave 方程以 Levy 随机行走以及分数阶 Brownian 运动为理论基础，它对于诸如金融学、计算生物学、声学这样的数理学是至关重要的，但是这一点过去几乎没被提及。虽然分数阶 diffusion-wave 方程是被用来反映反常能量耗散的，但是该方程的物理意义仍没有被很好的解释。有学者试图通过一种新的时间—空间分数阶微分波动方程来解释分数阶 diffusion-wave 方程。当声波在人类身体组织、沉积物和岩层中传播时，该分数阶微分波动方程对被表现为某种复现象的关于频率的耗散进行了建模。与此同时，他们发现了在分数阶 diffusion-wave 方程中关于时间和空间微分阶次的一种新的限定条件。这种新的限定条件表明所谓的亚扩散过程（sub-diffusion process）与现实世界中的频域功率耗散律（frequency power law dissipation）相矛盾[30]。有学者提出一种用分数阶微积分对相对复介电常数的等时线描述，他们注意到了三种张弛现象。传统的微积分对这些张弛现象不能很好地描述，而分数阶微积分却可以很清晰地显示其相对复介电常数的等时线图表的三种电介质张弛现象[31]。有学者提出了一种关于电荷密度的分数阶多级性（"fractional-order" multipoles of electric-charge densities）定义，它们的标量电势（potentials）分布是相关的。分数阶多极子（multipoles）起到了有效联系诸如点磁单极子（point monopoles）的整数阶点多极子之间间隙的桥梁媒介的作用。这项技术包含关于 Dirac delta 函数的分数阶微分或积分，它提供了一种用公式表示电极分布（electric source distribution）的手段。这种电极分布的电势函数能够通过对整数阶点磁单极子的电势进行更高或更低的分数阶微分或积分得到[33]。

1.3 分数阶微积分应用于生物工程

分数阶微积分可以被定义为卷积形式。有学者研究了与头不同旋转位置相对应的前庭第一传入神经元所呈现出来的分数阶动力学系统[32]。有学者在时间—频率域内分析和比较了各种心脏微电压能量的时—频分布。他们的分析是以心脏不停周期脉搏为基础的。由于信号的平滑处理仍然广泛应用于临床医疗设备之中，所以这项研究是对经过平滑处理之后的信号进行比较的。时—频分布包括 Born-Jordan, Zhao-Atlas-Marks 和 Margenau-Hill 等。时间—尺度表示是基于 Morlet 子波变换（Morlet Wavelet）的。心脏微电压的研究既包括对心室后期电压（Ventricular Late Potentials）的研究，也包括对与心脏搏动早期电压相联系的 P 波（P wave）的研究。他们的主要研究目的是在时间和频率域内刻画这些心脏搏动微信号,从而提出对心脏脉搏规律的某种预测。不同的模型会有不同的参数。对心脏脉搏信号的研究主要得益于用以提高信噪比的信号平滑技术的完善。他们利用分数阶微分试图去构建一个较好的模型[34]。有学者将分数阶激光(Fractional Laser)用于治疗妊娠斑 [melasma (the "mask of pregnancy")]。

1.4 分数阶微积分应用于动力学系统

分数阶微积分应用于动力学系统的连续模型为

$$a_n D^{v_n} y(t) + a_{n-1} D^{v_{n-1}} y(t) + \cdots + a_0 D^{v_0} y(t)$$
$$= b_m D^{p_m} u(t) + b_{m-1} D^{p_{m-1}} u(t) + \cdots + b_0 D^{p_0} u(t) \quad (1.2)$$

其离散模型为

$$a_n \Delta_h^{v_n} y(t) + a_{n-1} \Delta_h^{v_{n-1}} y(t) + \cdots + a_0 \Delta_h^{v_0} y(t)$$
$$= b_m \Delta_h^{p_m} u(t) + b_{m-1} \Delta_h^{p_{m-1}} u(t) + \cdots + b_0 \Delta_h^{p_0} u(t) \quad (1.3)$$

目前,在系统论意义上,所谓的散布性表征(diffusive representations)使得许多伪微分算子通过具有耗散性(dissipative nature)的输入—输出动力学系统用一种具体的形式来实现成为可能。因此,那些用传统理论很难回答的关于分解和逼近(analysis and approximation)的问题,能够以一种更加一致和便利的方式,就如同传统的公式表达法一样来加以解决。将分数阶微分算子理论应用于动力学系统之中取得了不少的研究成果[1,38-39]。有学者将分数阶微分应用于只有一个解的动力学系统的相位空间的重构之中,其中,关于时间序列信号分数阶微分是由频域数据变换得到的。他们将这种方法应用于Duffing振荡器和Lorenz系统取得了良好的效果[35]。有学者提出了关于分数阶微分的离散逼近方法以及它们在由 FO 微分方程描述的 FO 动力学系统中一些可选类型的应用[36]。

1.5 分数阶微积分应用于控制系统

在控制方面,基于分数阶微分的自适应控制是目前一个较新的研究方向。如果令 e 表示误差,p 表示待调整的参数,其评价(criterion)函数为

$$J(p) = \frac{1}{2} e^2 \quad (1.4)$$

其学习训练的准则为

$$\frac{\mathrm{d}p}{\mathrm{d}t} = -\gamma \frac{\partial J}{\partial p} = -\gamma e \frac{\partial e}{\partial p} \quad (1.5)$$

引入分数阶微分后,其调整率的大小依赖于 γ 和分数阶阶数。

有学者在运动控制中输入命令信号的整形或预处理用于减少系统振动。他们对系统的输入进行了必要的变换,以使得系统在完成规定动作的同时没有残余振动。他们对在航空领域内专门用于柔性控制的 N. C. Singer 和 W. P. Seering 系统进行了改进,提出了直接以分数阶微分系统来进行 ZV(Zero Vibration)和 ZVD(Zero Vibration Derivative)造型合成[37]。由于分数阶模型可以很好地描述了许多物理系统,同时分数阶控制解决了阻断问题(这是传统整数阶控制的运行弱点),所以分数阶系统得到了相当大的关注。由于缺少

对线性反馈式分数阶系统在时域内的有效分析方法，有学者提出了两种可靠、准确转换分数阶 Laplace 变换的数值方法[40]。

1.6 分数阶微积分应用于信号处理

目前，将分数阶微积分应用于信号处理主要有以下四个方面。

(1) 分数阶系统辨识。

在 Fourier 变换域内，令系统输入—输出数据为 $U(\omega_m)$，$Y(\omega_m)$，其频率响应为 $H(\omega_m)$，$m=1, 2, \cdots, M$。

其候选模型结构一般为

$$H(s,\theta) = \frac{N(s,\theta)}{D(s,\theta)} = \frac{\sum_{k=0}^{n} a_k s^{kv}}{\sum_{k=0}^{m} b_k s^{kv}}, \quad v \in \mathbf{R}^+ \tag{1.6}$$

它从数据中选择模型的标准一般采用：

$$\frac{\int_0^\infty [y(t) - \tilde{y}(t)]^2 dt}{\int_0^\infty [h(t)]^2 dt} = \frac{\int_{-\infty}^{\infty} [H(\omega) - \widetilde{H}(\omega)][H(\omega) - \widetilde{H}(\omega)]^* \frac{d\omega}{\omega^2}}{\int_{-\infty}^{\infty} H(\omega) H^*(\omega) d\omega} \tag{1.7}$$

式中，$y(t)$ 和 $h(t)$ 为实际的输出值和冲击响应，$\tilde{y}(t)$ 和 $\widetilde{H}(\omega)$ 为系统模拟辨识的输出值和传输函数。

有学者将经典的离散线性系统进行了扩展，使之包括离散时间分数阶线性系统[41]。

(2) 分数阶内插（Splines）。

其 Piecewise 能量函数为

$$\beta_+^v(x) = \frac{1}{\Gamma(v+1)} \Delta_+^{v+1} x_+^v$$

$$= \frac{1}{\Gamma(v+1)} \sum_{k \geqslant 0} (-1)^k \binom{v+1}{k} (x-k)_+^v \tag{1.8}$$

$$\beta_+^v(\omega) = \left(\frac{1-e^{-j\omega}}{j\omega}\right)^{v+1} \tag{1.9}$$

$$\beta_+^n(x) = \underbrace{\beta_+^0(x) * \beta_+^0(x) * \cdots * \beta_+^0(x)}_{n} \tag{1.10}$$

它可以被应用于信号处理的滤波器设计、子波变换、模拟数字变换以及图像处理的放大和增强、图像的几何转换、图像压缩等。

有学者研究了分数阶微分、spline 和 X 线断层摄影技术。为了处理分数阶微分，他们发展了 spline 微积分学。他们在研究分数阶 spline 的基础上提出了关于其基本函数（the underlying basis functions）分数阶微分的主要计算公式。特别地，他们将一个 α 次幂的 B-spline 的第 γ 阶分数阶微分（不需要整数阶）假定为一个 $\alpha-\gamma$ 次幂的 B-spline 的第 γ 阶分数阶微分。他们用这些结论推导出一种改进的关于 X 线断层摄影技术的背景放映算法。他们的实验数据用 splines 进行一阶内插值替换，于是其连续模型就可以用来精确执行滤波和背景放映功能[42]。

（3）模拟分数阶分抗和滤波器。

目前如何用开关电容（switched-capacitors）构建可编程阶数和带宽的分抗是一个研究热点。有学者对基于宽带频率吸收的模型Ⅲ的分数阶微分模型进行了研究和分析[43]。有学者提出了由一小组整数阶微分来很好地逼近一维或二维的关于离散时间的分数阶微分滤波器，这样关于信号的任意分数阶微分（二维时是任意方向）可由整数阶微分的线性组合来逼近[44]。可以用分数阶延迟的方法来设计和调整基于多项式的 FIR 滤波器。通过最优化通频带范围、滤波器参数式，使在通频带中的最坏相位延迟误差最小。滤波器采用分数阶延迟相当于对幅度失真的下限取 $1/2$[52]。

（4）其他分数阶信号处理。

可以构建分数阶延迟器，分数阶维持器（Fractional order holds）理想的分数阶延迟器如图 1.1 所示。

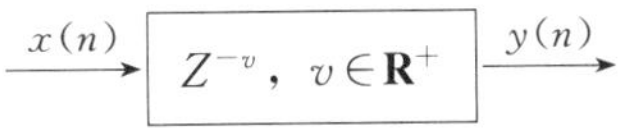

$$x(n) \longrightarrow \boxed{Z^{-v}, v \in \mathbf{R}^+} \longrightarrow y(n)$$

图 1.1　理想的分数阶延迟器

分数阶延迟器主要应用于信号的微量延迟、对数字调制方式的时间校正、语音编码和分析以及为数字波导的建模。有学者提出了用分数阶微分和分数阶采样延迟来设计数字分数阶微分滤波器的方法。为了提高用传统的分数阶微分设计方法在高频端的精确率，运用 well-documented FIR Lagrange 和 IIR allpass 分数阶延迟滤波器，他们提出分数阶微分甚至可以在进行分数阶采样延迟之前来完成[45]。有学者从分数 Hilbert 变换出发来考察实信号的复数形式，特别是新的解析信号的构造问题。他们简述了解析信号和分数 Hilbert 变换的概念，从经典的解析信号入手，提出了基于分数 Hilbert 变换的解析信号的构造方法，从而得到两类新的解析信号[46]。有学者研究了一类由 Levy 噪声激励的分数阶微分方程。由这种微分方程可以得到单一谱（singularity spectrum）[47]。有学者提出了一种关于分数阶积分过程的微分参数的调和 OLS 估计。他们采用了基于滤波核函数的连续子波变换理论，论述了子波系数变化与分数阶积分过程的平滑程度之间存在 Log-Log 线性关系[48]。有学者研究了用分数阶微积分通过 α-features 来对声表面波（surface waves）的空间和声学特性[49]。有学者将分数阶微积分应用于处理电磁学、辐射、分形辐射体的相关问题之中，辐射等势线轮廓符合分形结构[51]。有学者提出了计算机模拟电离辐射分光计的电子噪声，他们的技术以分数阶微积分为基础[53]。有学者提出了一种通过计算机模拟在无线电前置放大器或抗混滤波器（the anti-aliasing filter）的输出端观测到的电离—辐射分光计（ionizing-radiation spectrometers）电噪声的数学过程[54]。有学者提出了一种通过可能的无限平均延迟来将计算机网络传输描述为随机过程的模型，这个模型可以被用来解释依赖远程传输和具有分形特征的网络数据流特征。他们论述了存在于长尾（heavy-tailed）延迟分布、数据包延迟协方差函数的双曲线（hyperbolically）衰减以及分数阶微分方程之间的关系。他们用分数阶微分理论解释了数据包传输局域和广域特征[55]。因为基于信息包交换，局域网和广域网（LAN and WAN）的数据传输是一种触发式和不稳定的操作。具有统计学和动力学特征的数据包信息流存在较宽的电压值范围以适应 QoS 管理（QoS management）目的。有学者研究了可以用来分析网络传输性质和预测数据包级别的分数阶微积分形式。他们将这个理论用到远程网络模型设计之中，并讨论了用这种方法来描述网络数据包动力学系统的微分和积分特性的前景和局限性[56]。

第 2 章 连续子波变换数值实现中起始尺度的确定以及信号时间和扫描时间之间的几何关系

连续子波变换的实现除了少数能被解析表达的信号可进行解析计算外，大多数情况只能通过计算机做近似的数值积分。连续子波变换的数值实现首先必须解决的关键问题是起始尺度的确定以及对信号时间和扫描时间之间几何关系的正确分析，它们是信号的子波分析的工程实现和理论研究的前提和基础。连续子波变换数值实现中，起始尺度即起始尺度采样间隔，在不引起歧义的情况下，本书简称其为尺度采样间隔。

本章系统地论述了连续子波变换数值实现中尺度采样间隔确定的基本理论。本章按照信号最高数字频率等于或小于 π 的两种情况，论证了均匀点格采样时连续子波变换数值实现中 Morlet 母波以及偶对称或奇对称的各阶高斯函数导数解析母波的尺度采样间隔的最佳取值，并且特别分析了著名的墨西哥帽母波的尺度采样间隔的最佳取值；讨论了奇对称母波数值实现中同时所需的时间平移量；研究了偶对称或奇对称的各阶高斯函数导数解析母波的相应数字滤波器的波动情况，对其波动性进行了研究；将连续子波变换数值实现中均匀点格采样的研究结论推广到二进点格采样和二进抽取采样两种情况。

本章还系统地论述了连续子波变换数值实现中信号时间和扫描时间之间的几何关系，论述了连续子波变换数值实现中起始扫描时间的最佳取值范围。

本章推导出的结论不仅解决了信号子波分析理论研究和工程实现的基础性问题，而且解决了信号分数阶微积分的子波变换数值实现的基础性问题。

2.1 问题提出

众所周知，连续子波变换如式（2.1）、（2.2），其中 τ，a，ψ，$\psi_{\tau,a}$ 分别为扫描时间、尺度因子、母波、子波基函数[57]。本书以能量恒等化子波为例进行讨论显然不失对问题研究的一般性。

$$\psi_{\tau,a}(t) = 1/\sqrt{a}\,\psi\left(\frac{t-\tau}{a}\right) \Leftrightarrow \hat{\psi}_{\tau,a}(\Omega) = \sqrt{a}\,\psi(a\Omega)\mathrm{e}^{-\mathrm{j}\Omega\tau} \quad (2.1)$$

$$s(t) \Leftrightarrow W_s(\tau,a) = \int_t s(t)\psi^*_{\tau,a}(t)\mathrm{d}t = 1/\sqrt{a}\int_t s(t)\psi^*\left(\frac{t-\tau}{a}\right)\mathrm{d}t \quad (2.2)$$

要实现连续子波变换 $W_s(\tau,a)$，除了少数能被解析表达的信号可进行解析计算外，大多数信号只能通过计算机做近似的数值积分[58−59]；同时，我们获得的信号通常也是数字信号 $s[k]$，$k \in \mathbf{Z}$，或是用周期 $\delta_{T_s}(t)$ 信号对连续时间信号 $s(t)$，$t \in \mathbf{R}$ 采样所得的序列，所以 $W_s(\tau,a)$ 的数值近似计算式为

$$W_s(\tau,a) \approx \frac{T_s}{\sqrt{a}}\sum_k s[kT_s]\psi^*\left(\frac{kT_s - \tau}{a}\right) \quad (2.3)$$

$W_s(\tau,a)$ 是高度冗余的,去掉冗余的最简单的办法就是离散化[69]。因此,对于扫描时间 τ 和尺度 a,在具体数值实现时也必须离散化。时频相平面最简单的离散化就是均匀点格采样,即

$$\begin{cases} a = na_0 \\ \tau = m\tau_0 \end{cases}, \quad m \in \mathbf{Z}, \ n \in \mathbf{N} \tag{2.4}$$

在连续子波变换数值实现中,由于二进点格采样和二进抽取采样时起始尺度的确定方法与均匀点格采样时的确定原理可以类推,因此,本书仅详细推导均匀点格采样时连续子波变换数值实现中尺度采样间隔 a_0 的确定方法。对于二进点格采样和二进抽取采样时起始尺度的确定,本书仅给出推导结论。

把式(2.4)代入式(2.3),并按 $\tau_0 = T_s = 1$ 对其进行归一化,则得

$$\widetilde{\widetilde{W}}_s(m,n) = \frac{1}{\sqrt{na_0}} \sum_k s[k] \psi^* \left(\frac{k-m}{na_0} \right) \tag{2.5}$$

式(2.5)简记为式(2.6)。式(2.6)中 g_n 如式(2.7)所示,称为尺度 n 的(数字)子波滤波器,其中 n,m,a_0 分别为尺度、扫描时间、尺度采样间隔。

$$\widetilde{\widetilde{W}}_s(m,n) = s[m] * g_n[m] = \sum_l s[l] g_n[m-l] \tag{2.6}$$

$$g_n: g_n[m] = \frac{1}{\sqrt{na_0}} \psi^* \left(-\frac{m}{na_0} \right), \quad m \in \mathbf{Z}, n = 1,2,3,\cdots \tag{2.7}$$

从理论上讲,只要已知母波 $\psi(t)$,就可求得尺度 n 下相应的数字滤波器系数 g_n 的取值。显然,g_n 和母波 $\psi(t)$ 一样也必须满足波动性 $\sum_m g_n[m] = 0 (n=1,2,3,\cdots)$ 和能量归一性 $\sum_m |g_n[m]|^2 = 1 (n=1,2,3,\cdots)$[59]。同时,$g_n$ 与母波 $\psi(t)$ 具有相同的时间局域化特征[61,67]。对于时域紧支撑子波不存在截短问题,而对于非紧支撑的速降子波,相应的 g_n 也是非紧支撑的速降序列[62],在数值实现时,必须截短 g_n。用数值法近似计算 $\widetilde{\widetilde{W}}_s(m,n)$ 时,尺度采样间隔的确定是一个首先必须解决的关键问题。如果这个问题不解决,工程上就无从谈论对信号进行子波分析和研究了[73-76]。

在连续子波变换中,尺度 a 是连续的,且 $a>0$;扫描时间 $\tau \in \mathbf{R}$,令 $\tau_1 \leqslant \tau \leqslant \tau_2$;信号时间 $t \in \mathbf{R}$,令 $t_1 \leqslant t \leqslant t_2$。于是,可以很自然想到:$\tau_1 \stackrel{?}{=} t_1, \tau_2 \stackrel{?}{=} t_2$,即待论证序列信号的连续子波变换中信号时间 t 和扫描时间 τ 之间的几何关系。连续子波变换数值实现中信号时间和扫描时间之间几何关系的确定是另一个首先必须解决的关键问题。如果这个问题不解决,工程上对信号进行子波分析和研究同样也无从谈论[157-159]。

2.2 连续子波的选择

按照母波的对称状况可分为(复)解析母波、(实)偶母波、(实)奇母波三种情况[57,59]。对于不同的母波 ψ,本书采取"简单性策略",从考察最简单的情况,即尺度 $n=1$ 开始,亦即确定式(2.5)中的尺度采样间隔 a_0。

显然,离散有限信号 $s(k)$ 的频谱密度 $S(e^{j\omega})$ 是 ω 的连续且周期的函数,$S(e^{j\omega})$ 的周

期 $\omega_T = \dfrac{2\pi}{\tau_0}$, τ_0 是离散间隔。当 $\tau_0 = 1$ 时, $\omega_T = 2\pi$[60]; 另外, 由于高斯信号是不确定性原理(测不准原理)下的最佳信号[57,61], 于是我们构造母波为 $\psi(t) = g(t)e^{j\Omega_0 t} \Leftrightarrow \hat{\psi}(\Omega) = g(\Omega - \Omega_0)$, 其中 $g(t)$ 为 Gabor 变换的窗口函数[57,65]; 最后, 根据子波基函数的定义, 可知 $\psi_a(t) = \dfrac{1}{\sqrt{a}} \psi\left(\dfrac{t}{a}\right) \Leftrightarrow \hat{\psi}_a^*(\Omega) = \sqrt{a}\hat{\psi}^*(a\Omega)$[57]。因此, 为了精确计算连续子波变换 $W_s(\tau, a)$ 在 $(m\tau_0, na_0)|_{\tau_0=1, n=1} \Rightarrow (m, a_0) \Rightarrow (m, 1)$ 点的数值 $\widetilde{\widetilde{W}}_s(m, 1)$, 必须使相应的数字滤波器 g_1 的滤波函数 $G_1(\omega) = \sum_m g_1[m]e^{-j\omega m}$ 与尺度 $n = 1 \Rightarrow a_{n=1} = a_0$ 时的子波滤波函数 $\hat{\psi}_{a_0}^*(\Omega) = \sqrt{a_0}\hat{\psi}^*(a_0\Omega) = \sqrt{a_0}\hat{g}^*(a_0\Omega - \Omega_0)$, $a_0 > 0$ 的 2π 周期重复基本一致[63], 即

$$G_1(\omega) = \sum_l \sqrt{a_0}\hat{\psi}^*[a_0(\omega - 2\pi l)] = \sum_l \sqrt{a_0}\hat{g}^*[a_0(\omega - 2\pi l) - \Omega_0] \quad (2.8)$$

2.3 (复)解析母波的尺度采样间隔的推导

2.3.1 理论分析

(复)解析母波 $\psi_A(t)$ 的频谱 $\hat{\psi}_A(\Omega)$ 一般应当是速降的单峰实函数[65-66], 如图 2.1 所示, 其中 Ω_c 是峰值频率, Ω_m 是最高频率。

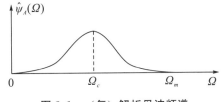

图 2.1 (复)解析母波频谱

在尺度 $a = a_0$, 即 $n = 1$ 时, 对应的数字滤波函数 $G_1(\omega)$ 如图 2.2 所示。

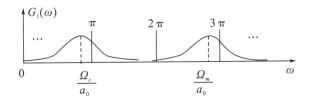

图 2.2 (复)解析母波相应的数字滤波器的频谱

由图 2.2 可知, $G_1(\omega)$ 的峰值频率 $\omega_c = \dfrac{\Omega_c}{a_0}$, 最高频率 $\omega_m = \dfrac{\Omega_m}{a_0}$[59,65]。众所周知, 数字信号的最高数字频率是 $\omega = \pi$[58]。因此, 为了能够分析数字信号 $s[k]$ 中的最高频率成分, 一般是用数字滤波器的频谱的峰值去考察, 即满足 $\omega_c \geqslant \pi$, 则有 $a_0 \leqslant \Omega_c / \pi$。但是, 为了避免过分减小 a_0 以致使 $G_1(\omega)$ 的最高数字频率 ω_m 增大而超过 2π, 造成频谱混叠的

情况发生，还必须使 $\omega_m \leqslant 2\pi$，则得 $a_0 \geqslant \frac{\Omega_m}{2\pi}$，于是有

$$\frac{\Omega_m}{2\pi} \leqslant a_0 \leqslant \frac{\Omega_c}{\pi} \qquad (2.9)$$

显然，只有当 $\Omega_m \leqslant 2\Omega_c$ 时，式（2.9）才成立。因为 $\hat{\psi}_A(\Omega)$ 是单峰速降的，只有当 Ω_m 在其取值范围内取到最大值时，才能保证 $\frac{\hat{\psi}_A(\Omega_m)}{\hat{\psi}_A(\Omega_c)}$ 的比值达到最小[64]，这样可在具体数值实现中尽可能地减小频谱混叠[57]。因此，一种简单的考虑是取 $\Omega_m = 2\Omega_c$，于是 a_0 的最佳取值为

$$a_0 = \frac{\Omega_c}{\pi} \qquad (2.10)$$

另外，若数字信号 $s[k]$，$k \in \mathbf{Z}$ 是连续时间信号 $s(t)$，$t \in \mathbf{R}$ 的采样序列，设 Ω_{smax} 是 $s(t)$ 的最高频率，信号的采样频率为 f_s，则 $s[k]$ 的最高数字频率 $\omega_{sm} = \Omega_{smax}/f_s$[58]。显然，可能存在 $\omega_{sm} < \pi$ 的情况[58]。此时只有当满足 $\omega_c \geqslant \omega_{sm}$ 时，才能用数字滤波器频谱的峰值去分析 $s[k]$ 的最高数字频率成分，于是有 $a_0 \leqslant \Omega_c f_s/\Omega_{smax}$，又因为 $a_0 \geqslant \frac{\Omega_m}{2\pi}$，故有

$$\frac{\Omega_m}{2\pi} \leqslant a_0 \leqslant \frac{\Omega_c f_s}{\Omega_{smax}} \qquad (2.11)$$

显然，只有当 $\Omega_m \leqslant 2\pi \Omega_c f_s/\Omega_{smax}$ 时，式（2.11）才成立。一种简单的考虑是取 $\Omega_m = 2\pi \Omega_c f_s/\Omega_{smax}$，于是 a_0 的最佳取值为

$$a_0 = \frac{\Omega_c f_s}{\Omega_{smax}} \qquad (2.12)$$

2.3.2 Morlet 母波的尺度采样间隔的确定

Morlet 母波为 $\psi_m(t) = \frac{1}{\sqrt{2\pi}} e^{-\frac{1}{2}t^2} e^{j\Omega_0 t} \Leftrightarrow \hat{\psi}_m(\Omega) = e^{-\frac{1}{2}(\Omega-\Omega_0)^2}$，当 $\Omega_0 \geqslant 5$ 时，Morlet 母波近似满足波动性。$\hat{\psi}_m(\Omega)$ 是单峰速降的，令 $\frac{d}{d\Omega}\hat{\psi}_m(\Omega) = 0$，即可求得其峰值频率 $\Omega_c = \Omega_0$。当数字信号的最高数字频率 $\omega = \pi$ 时，由式（2.10），取 $a_0 = \frac{\Omega_c}{\pi} = \frac{\Omega_0}{\pi}$。另外，当数字采样序列的最高数字频率 $\omega_{sm} < \pi$ 时，由式（2.12），取 $a_0 = \frac{\Omega_c f_s}{\Omega_{smax}} = \frac{\Omega_0 f_s}{\Omega_{smax}}$，见表 2.1。

表 2.1 Morlet 母波 $\psi_m(t)$ 的尺度采样间隔 a_0 的最佳取值

Morlet 母波 $\psi_m(t)$	信号的最高数字频率等于 π 时的尺度采样间隔 a_0	信号的最高数字频率小于 π 时的尺度采样间隔 a_0
$1/\sqrt{2\pi} e^{-\frac{1}{2}t^2} e^{j\Omega_0 t}$	$a_0 = \frac{\Omega_0}{\pi}$	$a_0 = \frac{\Omega_0 f_s}{\Omega_{smax}}$

2.4 （实）偶母波的尺度采样间隔的推导

2.4.1 理论分析

实偶对称母波的时域形式是 $\psi_E(-t)=\psi_E(t)$，其频谱 $\hat{\psi}_E(\Omega)$ 是速降的双峰实偶函数，如图 2.3 所示，其中 Ω_c 是峰值频率，Ω_m 是最高频率。当尺度 $n=1$ 时，$a=a_0$。

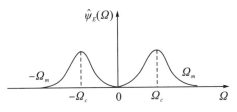

图 2.3 实偶对称母波频谱

当信号的最高数字频率 $\omega=\pi$ 时，如果满足关系 $\omega_m=\dfrac{\Omega_m}{a_0}\leqslant\pi\Rightarrow a_0\geqslant\dfrac{\Omega_m}{\pi}$，则相应的数字滤波函数 $G_1(\omega)$ 如图 2.4 所示，不发生频谱重叠，此时滤波器 g_1 系数是子波 ψ_{Ea_0} 的超采样和临界采样情形：g_1：$g_1[m]=\psi_E^*(-m/a_0)/\sqrt{a_0}=\psi_E(m/a_0)/\sqrt{a_0}$。

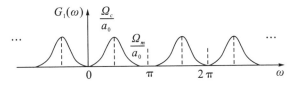

图 2.4 a_0 较大时的实偶对称母波的数字滤波器频谱

当满足 $a_0\geqslant\dfrac{\Omega_m}{\pi}$ 时，显然 ω_m 增至 π 时 $G_1(\omega)$ 即截止了，此时无论如何也不可能分析序列信号的最高频率成分。所以只有减小 a_0，使 $a_0<\dfrac{\Omega_m}{\pi}$，实现对实偶子波的欠采样，这样才能分析序列的最高频率成分，但此时必定造成频谱重叠。

由图 2.4 可知，当 $\omega_c=\dfrac{\Omega_c}{a_0}\geqslant\pi\Rightarrow a_0\leqslant\dfrac{\Omega_c}{\pi}$ 时，就能分析序列信号的最高频率成分，故 $a_0<\dfrac{\Omega_m}{\pi}$ 且 $a_0\leqslant\dfrac{\Omega_c}{\pi}$。因为 $\Omega_c<\Omega_m$，所以 $a_0\leqslant\dfrac{\Omega_c}{\pi}$。另外，因为采样序列的频谱会以 2π 周期重复，所以最高数字频率为 2π[58]。因此必须使 $\omega_m\leqslant 2\pi$，则得 $a_0\geqslant\dfrac{\Omega_m}{2\pi}$。于是有

$$\dfrac{\Omega_m}{2\pi}\leqslant a_0\leqslant\dfrac{\Omega_c}{\pi} \tag{2.13}$$

显然，只有当 $\Omega_m\leqslant 2\Omega_c$ 时，式（2.13）才成立。一种简单的考虑是取 $\Omega_m=2\Omega_c$，于是 a_0 的最佳取值为

$$a_0=\dfrac{\Omega_c}{\pi} \tag{2.14}$$

此时数字滤波器的频谱 $G_1(\omega)$ 在其峰值处正好完全重合，虽然发生了频谱混叠，但它却能分析信号的最高频率成分，如图 2.5 所示。

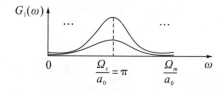

图 2.5 a_0 较小时的实偶对称母波的数字滤波器频谱

若数字信号 $s[k]$ 的最高数字频率 $\omega_{sm}=\Omega_{smax}/f_s<\pi$，此时只有当满足 $\omega_c \geqslant \omega_{sm}$ 时，才能用数字滤波器频谱的峰值去分析 $s[k]$ 的最高数字频率成分，于是有 $a_0 \leqslant \Omega_c f_s/\Omega_{smax}$，又因为 $a_0 \geqslant \dfrac{\Omega_m}{2\pi}$，故有

$$\frac{\Omega_m}{2\pi} \leqslant a_0 \leqslant \frac{\Omega_c f_s}{\Omega_{smax}} \tag{2.15}$$

显然，只有当 $\Omega_m \leqslant 2\pi\Omega_c f_s/\Omega_{smax}$ 时，式（2.15）才成立。一种简单的考虑是取 $\Omega_m=2\pi\Omega_c f_s/\Omega_{smax}$，于是 a_0 的最佳取值为

$$a_0 = \frac{\Omega_c f_s}{\Omega_{smax}} \tag{2.16}$$

2.4.2 （实）偶高斯函数各阶导数解析母波的尺度采样间隔的确定

因为高斯信号是不确定性原理（测不准原理）下的最佳信号[59]，所以推导高斯函数导数解析母波的尺度采样间隔 a_0 有极其广泛的工程应用需要。由高斯函数的各阶导数构造得到的（复）解析母波 $_k\psi$，其频谱函数为

$$_k\hat{\psi}(\Omega) = \begin{cases} A(k)\Omega^k \exp\left(-\dfrac{1}{2}\Omega^2\right), & \Omega \geqslant 0 \\ 0, & \Omega < 0 \end{cases} \tag{2.17}$$

式中，$A(k)$ 是能量归一化常数，它依赖于参量 k[66,68]。

显然 $_k\hat{\psi}$ 是单峰速降的，令 $\dfrac{\mathrm{d}}{\mathrm{d}\Omega}{_k\hat{\psi}}(\Omega)=0$，即可求得其峰值频率 $\Omega_c=\sqrt{k}$。显然 $\sqrt{k}=\Omega_c<\Omega_m\leqslant 2\Omega_c=2\sqrt{k}$，故 $\dfrac{_k\hat{\psi}(\Omega_m)}{_k\hat{\psi}(\Omega_c)}=\left(\dfrac{\mathrm{e}}{k}\right)^{\frac{k}{2}}\cdot\dfrac{\Omega_m^k}{\mathrm{e}^{\frac{1}{2}\Omega_m^2}}$，若令 $\dfrac{_k\hat{\psi}(\Omega_m)}{_k\hat{\psi}(\Omega_c)}=\left(\dfrac{\mathrm{e}}{k}\right)^{\frac{k}{2}}\cdot\dfrac{\Omega_m^k}{\mathrm{e}^{\frac{1}{2}\Omega_m^2}}=\dfrac{1}{m}$，$m$ 必为模值大于 1 的常数，显然这是一个只能用数值法（可用牛顿法或最速下降法[71]）或图像法求解的方程，因为方程中既包含线性项也包含指数项[70]，但这样做会使解得的 Ω_m 随 m 的取值而变化，因此就不一定满足前面论证的 Ω_m 的取值范围。一种简单的考虑是取 $\Omega_m=2\Omega_c=2\sqrt{k}$，这样可以保证 $\dfrac{_k\hat{\psi}(\Omega_m)}{_k\hat{\psi}(\Omega_c)}$ 的比值达到最小[64]，尽可能地在具体数值实现时减小频谱混叠[57]。当信号的最高数字频率 $\omega=\pi$ 时，由式（2.14）得

$$a_0 = \frac{\Omega_c}{\pi} = \frac{\sqrt{k}}{\pi}, \quad k=2q, q\in\mathbf{N} \tag{2.18}$$

因为高斯信号的傅立叶变换仍是高斯信号[58]，则得母波 $\psi(t)$ 的时域表示形式为

$$\psi(t) = (-1)^{k-1}\mathrm{j}^k (t\mathrm{e}^{-\frac{1}{2}t^2})^{k-1} \tag{2.19}$$

式中，j 为虚数单位，k 为偶数。

若数字信号 $s[k]$ 的最高数字频率 $\omega_{sm} = \Omega_{smax}/f_s < \pi$，由式（2.16）得

$$a_0 = \frac{\Omega_c f_s}{\Omega_{smax}} = \frac{\sqrt{k} f_s}{\Omega_{smax}}, \quad k = 2q, \quad q \in \mathbf{N} \tag{2.20}$$

（实）偶高斯函数各阶导数解析母波的尺度采样间隔的最佳取值见表 2.2。

表 2.2 （实）偶高斯函数各阶导数解析母波的尺度采样间隔的最佳取值

k	（实）偶母波 $\psi_E(t)$	信号的最高数字频率等于 π 时的尺度采样间隔 a_0	信号的最高数字频率小于 π 时的尺度采样间隔 a_0
2	$(1-t^2)\mathrm{e}^{-\frac{1}{2}t^2}$	$\dfrac{\sqrt{2}}{\pi}$	$\sqrt{2} f_s/\Omega_{smax}$
4	$(3-6t^2+t^4)\mathrm{e}^{-\frac{1}{2}t^2}$	$\dfrac{2}{\pi}$	$2 f_s/\Omega_{smax}$
6	$(15-45t^2+15t^4-t^6)\mathrm{e}^{-\frac{1}{2}t^2}$	$\dfrac{\sqrt{6}}{\pi}$	$\sqrt{6} f_s/\Omega_{smax}$
偶数	$(-1)^{k-1}\mathrm{j}^k (t\mathrm{e}^{-\frac{1}{2}t^2})^{k-1}$	$\dfrac{\sqrt{k}}{\pi}$	$\sqrt{k} f_s/\Omega_{smax}$

下面介绍墨西哥帽母波的尺度采样间隔的确定。

当 $k=2$ 时，此时的母波就是著名的墨西哥帽母波。令 $\dfrac{\mathrm{d}}{\mathrm{d}\Omega}\hat{\psi}_{\mathrm{hat}}(\Omega) = 0$，得 $\Omega_c = \sqrt{k} = \sqrt{2}$，由式（2.18），得 $a_0 = \dfrac{\sqrt{2}}{\pi}$。由式（2.19），得母波 $\psi_{\mathrm{hat}}(t)$ 的时域表示形式为

$$\psi_{\mathrm{hat}}(t) = (1-t^2)\mathrm{e}^{-\frac{1}{2}t^2} \Leftrightarrow \hat{\psi}_{\mathrm{hat}}(\Omega) = \Omega^2 \mathrm{e}^{-\frac{1}{2}\Omega^2} \tag{2.21}$$

在该尺度下，相应的子波 $\psi_{\mathrm{hat}a_0}(t)$ 与数字滤波器 g_1 的时域波形如图 2.6 所示。

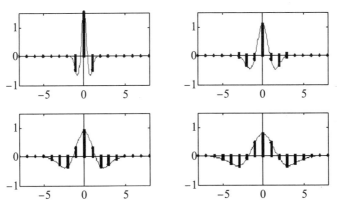

图 2.6 墨西哥帽子波与数字滤波器（$a_0 = \sqrt{2}/\pi$）

对 g_1 截短并能量归一化，则得

$$g_1 = \left\{ -\frac{\sqrt{2}}{4}, \frac{\sqrt{2}}{2}, -\frac{\sqrt{2}}{4} \right\} \tag{2.22}$$

由式（2.7）可得相应的数字滤波器为

$$g_n : g_n[m] = \frac{1}{\sqrt{n\sqrt{2}/\pi}} \left[1 - \left(\frac{m}{n\sqrt{2}/\pi}\right)^2\right] \exp\left[-\frac{1}{2}\left(\frac{m}{n\sqrt{2}/\pi}\right)\right], \quad n = 2,3,4,\cdots \tag{2.23}$$

2.4.3 （实）偶高斯函数各阶导数解析母波相应的数字滤波器的波纹系数

一般我们用波纹系数 $\rho = G_{1\min}(\omega)/G_{1\max}(\omega)$ 来量度函数波动状况。将式（2.17）代入式（2.8），可得

$$G_1(\omega) = A(k) a_0^{k+\frac{1}{2}} \sum_l (\omega - 2\pi l)^k e^{-\frac{1}{2}a_0^2(\omega - 2\pi l)^2} \tag{2.24}$$

令 $\dfrac{\mathrm{d}}{\mathrm{d}\omega} G_1(\omega) = 0$，可求得极小值点 $\omega_{\min} = 2\pi l$ 和极大值点 $\omega_{\max} = 2\pi l \pm \dfrac{\sqrt{k}}{a_0}$。由于 $\omega_{\max}$ 的两个解是关于 2π 对称的，两者具有相同的极大值特性，故我们只取 $\omega_{\max} = 2\pi l + \dfrac{\sqrt{k}}{a_0}$。另外，由于数字滤波器的极大值和极小值是周期交替出现的，故任取其中相邻两个不同极值特性的极值点就可求出其波纹系数。于是令 $l = 0$，即取 $\omega_{\min} = 0$ 和 $\omega_{\max} = \dfrac{\sqrt{k}}{a_0}$，将其分别代入式（2.24），则得

$$G_{1\min} = A(k) a_0^{k+\frac{1}{2}} (-2\pi)^k \sum_l l^k e^{-2\pi^2 a_0^2 l^2} \tag{2.25}$$

$$G_{1\max} = A(k) a_0^{k+\frac{1}{2}} \sum_l \left(\frac{\sqrt{k}}{a_0} - 2\pi l\right)^k e^{-\frac{1}{2}a_0^2(\frac{\sqrt{k}}{a_0} - 2\pi l)^2} \tag{2.26}$$

由式（2.25）和式（2.26），并代入式（2.18），则得波纹系数为

$$\rho = G_{1\min}(\omega)/G_{1\max}(\omega) = \frac{\sum_l l^k e^{-2kl^2}}{\sum_l \left(l - \frac{1}{2}\right)^k e^{-\frac{k}{2}(1-2l)^2}}, \quad k = 2q, q \in \mathbf{N} \tag{2.27}$$

由式（2.8）易知 $G_1(\omega)$ 的极值主要取决于相邻的两个波的叠加，故取 $l = 0, 1$ 来简化式（2.27），则得

$$\rho = G_{1\min}(\omega)/G_{1\max}(\omega) \approx \frac{1}{\left[\left(-\frac{1}{2}\right)^k + \left(\frac{1}{2}\right)^k\right] e^{\frac{3k}{2}}}, \quad k = 2q, q \in \mathbf{N} \tag{2.28}$$

可见，当 k 越大时，（实）偶高斯函数各阶导数解析母波相应的数字滤波器 g_1 的波动也会愈加厉害，见表2.3。

表 2.3 （实）偶高斯函数各阶导数解析母波相应的数字滤波器 g_1 的波纹系数

k	（实）偶母波 $\psi_E(t)$	数字滤波器 g_1 的波纹系数 ρ 的近似值
2	$(1-t^2)e^{-\frac{1}{2}t^2}$	$2/e^3$
4	$(3-6t^2+t^4)e^{-\frac{1}{2}t^2}$	$8/e^6$
6	$(15-45t^2+15t^4-t^6)e^{-\frac{1}{2}t^2}$	$32/e^9$
偶数	$(-1)^{k-1}j^k(te^{-\frac{1}{2}t^2})^{k-1}$	$\dfrac{1}{[(-\frac{1}{2})^k+(\frac{1}{2})^k]e^{\frac{3k}{2}}}$

2.5 （实）奇母波的尺度采样间隔的推导

2.5.1 理论分析

实奇对称母波的时域形式是 $\psi_O(-t)=-\psi_O(t)$，由于信号的时域形式是实奇对称的，则其频域形式必为虚奇对称的函数[58,60]，故其频谱 $\hat{\psi}_O(\Omega)$ 是速降的双峰虚奇函数，如图 2.7 所示。

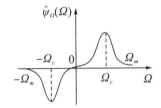

图 2.7 实奇对称母波频谱

当 $\omega_m=\dfrac{\Omega_m}{a_0}\leqslant\pi\Rightarrow a_0\geqslant\dfrac{\Omega_m}{\pi}$ 时，对应的数字滤波函数 $G_1(\omega)$ 不发生频谱重叠，但不能分析序列信号的最高频率成分。减小 a_0，实现对实奇子波的欠采样，使 $a_0<\Omega_m/\pi$，虽然造成频谱重叠，但由于重叠部分正负抵消，如图 2.8 所示，仍然不能分析序列信号的最高频率成分。

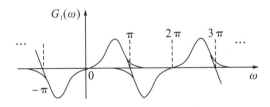

图 2.8 实奇对称母波的数字滤波器频谱

本章从分析一个具体的实奇母波入手。最简单的高斯一阶（微分）母波：$_1\psi(t)=-te^{-\frac{1}{2}t^2}\Leftrightarrow{}_1\hat{\psi}(\Omega)=j\Omega e^{-\frac{1}{2}\Omega^2}$，令 $\dfrac{d}{d\Omega}{}_1\hat{\psi}(\Omega)=0$，即得其峰值频率 $\Omega_c=1$。先按照（复）解

析母波和实偶对称母波确定尺度采样间隔 a_0 的方法,如式(2.10)和(2.18),则 $a_0=\frac{\Omega_c}{\pi}$,得到式(2.29)。对 g_1 截短并能量归一化,$_1\psi_{a_0}(t)$ 与 g_1 时域波形如图2.9所示。得到的 g_1 属于数字序列 $+1,0,-1,\cdots$ 类型,显然它的变化频率小于 π,所以 g_1 不可能用来分析序列的最高数字频率成分。

$$a_0=1/\pi \Rightarrow _1\psi_{a_0}(t)=-\pi^{\frac{3}{2}}te^{-\frac{1}{2}(\pi t)^2}\Rightarrow h_1:h_1[i]=-\pi^{\frac{3}{2}}ie^{-\frac{1}{2}(\pi i)^2} \tag{2.29}$$

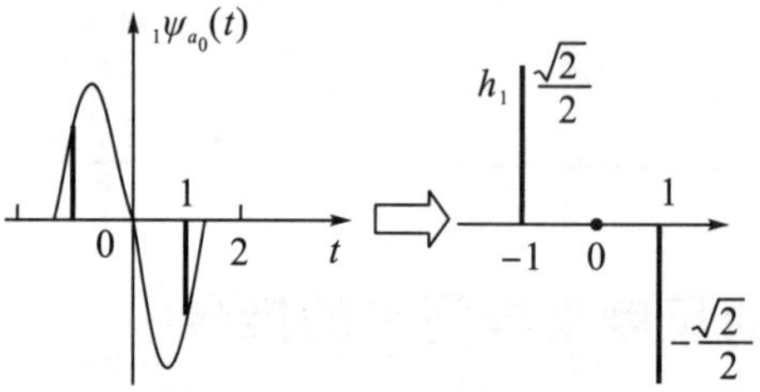

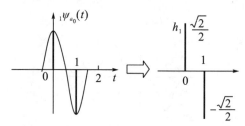

图 2.9　直接对实奇母波进行抽样

于是,本章仿照 haar 母波的形式,为了使 g_1 属于相位变化最快的数字序列 $+1$,$-1,\cdots$ 类型,我们只需要将母波 $_1\psi_{a_0}(t)$ 向右或左平移 $\Omega_c/2=1/2$ 个时间单位后所得的 $_1\psi_{a_0}(t\mp 1/2)$ 再均匀点格抽样即得,如图2.10所示,显然此时它的变化频率等于 π,所以此时的 g_1 能够用来分析序列信号的最高数字频率($\omega=\pi$)成分。

图 2.10　对平移 1/2 后的实奇母波进行抽样

用数学归纳法把上面的结论推而广之,对于任意实奇对称母波 $\psi_O(t)$,为了使 g_1 的最高数字频率为 π,我们只需先将母波 $\psi_{Oa_0}(t)$ 向右或左平移 $\Omega_c/2$ 个时间单位后所得的 $\psi_{Oa_0}(t\mp\Omega_c/2)$,再用与(复)解析母波和实偶对称母波确定尺度采样间隔 a_0 同样的方法对其进行均匀点格采样,则得

$$a_0=\frac{\Omega_c}{\pi} \tag{2.30}$$

$$a_0=\frac{\Omega_c f_s}{\Omega_{s\max}} \tag{2.31}$$

$$\tau=\frac{\Omega_c}{2} \tag{2.32}$$

式中,τ 为时间平移量。

2.5.2　(实)奇高斯函数各阶导数解析母波尺度采样间隔和时间平移量的确定

本章仍采用式(2.17)定义的高斯函数的各阶导数构造得到的(复)解析母波 $_k\psi$,由式(2.30)、(2.31)、(2.32),得式(2.33)、(2.34)、(2.35)。总结则得表2.4,其中母

第 2 章　连续子波变换数值实现中起始尺度的确定以及信号时间和扫描时间之间的几何关系

波 $\psi_O(t)$ 去掉了傅立叶反变换得到的虚数单位 j，以保证它是一个实函数。

$$a_0 = \frac{\Omega_c}{\pi} = \frac{\sqrt{k}}{\pi}, \quad k = 2q-1, q \in \mathbf{N} \tag{2.33}$$

$$a_0 = \frac{\Omega_c f_s}{\Omega_{s\max}} = \frac{\sqrt{k} f_s}{\Omega_{s\max}}, \quad k = 2q-1, q \in \mathbf{N} \tag{2.34}$$

$$\tau = \frac{\Omega_c}{2} = \frac{\sqrt{k}}{2}, \quad k = 2q-1, q \in \mathbf{N} \tag{2.35}$$

表 2.4　（实）奇高斯函数各阶导数解析母波的尺度采样间隔和时间平移量

k	（实）奇母波 $\psi_O(t)$	最高数字频率等于 π 时的 a_0	最高数字频率小于 π 时的 a_0	母波的时间平移量 τ
1	$t\mathrm{e}^{-\frac{1}{2}t^2}$	$\frac{1}{\pi}$	$f_s/\Omega_{s\max}$	$\frac{1}{2}$
3	$(3-t^2)t\mathrm{e}^{-\frac{1}{2}t^2}$	$\frac{\sqrt{3}}{\pi}$	$\sqrt{3}f_s/\Omega_{s\max}$	$\frac{\sqrt{3}}{2}$
5	$(15-10t^2+t^4)t\mathrm{e}^{-\frac{1}{2}t^2}$	$\frac{\sqrt{5}}{\pi}$	$\sqrt{5}f_s/\Omega_{s\max}$	$\frac{\sqrt{5}}{2}$
奇数	$(-1)^{k-1}\mathrm{j}^{k-1}(t\mathrm{e}^{-\frac{1}{2}t^2})^{k-1}$	$\frac{\sqrt{k}}{\pi}$	$\sqrt{k}f_s/\Omega_{s\max}$	$\frac{\sqrt{k}}{2}$

2.5.3　（实）奇高斯函数各阶导数解析母波相应的数字滤波器的波动性

当 k 为奇数，即 $k=2q-1$，$q \in \mathbf{N}$ 时，由式（2.26）可求得 $G_{1\max} \equiv 0$。因此，就不能再用波纹系数来度量奇对称高斯函数各阶导数解析母波相应的数字滤波器 g_1 的波动性了。我们换用 $G_1(\Omega)_{\min}$ 来度量 g_1 的波动性。由式（2.25）并取 $l=0$，1，可得

$$G_{1\min} \approx \frac{(-2)^k}{\sqrt{\pi}} A(k) k^{\frac{2k+1}{4}} \mathrm{e}^{-2k} \tag{2.36}$$

运用 parseval 定理，由式（2.17）可求得式（2.36）中的 $A(k)$，即

$$A(k) = -2\pi/\int \Omega^{k-1} \mathrm{d}\mathrm{e}^{-\frac{1}{2}\Omega^2} \tag{2.37}$$

可见，当 k 增大时，相应数字滤波器的波动愈厉害，见表 2.5。

表 2.5　（实）奇高斯函数各阶导数解析母波相应的数字滤波器 $G_1(\Omega)_{\min}$

k	（实）奇母波 $\psi(t)$	相应数字滤波器 $G_1(\Omega)_{\min}$ 的近似值
1	$t\mathrm{e}^{-\frac{1}{2}t^2}$	$\dfrac{-2A(1)}{\mathrm{e}^2\sqrt{\pi}}$
3	$(3-t^2)t\mathrm{e}^{-\frac{1}{2}t^2}$	$\dfrac{-8A(3)3^{7/4}}{\mathrm{e}^6\sqrt{\pi}}$
5	$(15-10t^2+t^4)t\mathrm{e}^{-\frac{1}{2}t^2}$	$\dfrac{-32A(5)5^{11/4}}{\mathrm{e}^{10}\sqrt{\pi}}$
奇数	$(-1)^{k-1}\mathrm{j}^{k-1}(t\mathrm{e}^{-\frac{1}{2}t^2})^{k-1}$	$\dfrac{(-2)^k A(k) k^{\frac{2k+1}{4}}}{\mathrm{e}^{2k}\sqrt{\pi}}$

2.6 二进点格采样及二进抽取采样时起始尺度的确定

在对信号的连续子波变换进行数值实现时,对于时频相平面的另外两种离散化就是二进点格采样和二进抽取采样。

二进抽取采样:

$$\begin{cases} a = 2^n a_0 \\ \tau = 2^n m \tau_0 \end{cases}, \quad m \in \mathbf{Z}, n \in \mathbf{N} \tag{2.38}$$

众所周知,数字信号的最高数字频率$\omega = \pi$,故为了能够分析数字信号$s[k]$中的最高频率成分,一般是用数字滤波器频谱的峰值去考察。此时a_0的最佳取值为

$$a_0^{opt} = \frac{\omega_c}{\pi} = \frac{\sqrt{v}}{\pi} \tag{2.39}$$

另外,若数字信号$s[k]$($k \in \mathbf{Z}$)是连续时间信号$s(t)$($t \in \mathbf{R}$)的采样序列,设Ω_{smax}是$s(t)$的最高频率,信号的采样频率为f_s,则$s[k]$的最高数字频率$\omega_{sm} = \Omega_{smax}/f_s$。显然,可能存在$\omega_{sm} < \pi$的情况。此时$a_0$的最佳取值为

$$a_0^{opt} = \frac{\omega_c f_s}{\Omega_{smax}} = \frac{\sqrt{v} \cdot f_s}{\Omega_{smax}} \tag{2.40}$$

二进点格采样:

$$\begin{cases} a = 2^n a_0 \\ \tau = m \tau_0 \end{cases}, \quad m \in \mathbf{Z}, n \in \mathbf{N} \tag{2.41}$$

当数字信号的最高数字频率$\omega = \pi$时,a_0的最佳取值为

$$a_0^{opt} = \frac{\omega_c}{2\pi} = \frac{\sqrt{v}}{2\pi} \tag{2.42}$$

当数字信号的最高数字频率$\omega_{sm} < \pi$时,a_0的最佳取值为

$$a_0^{opt} = \frac{\omega_c f_s}{2\Omega_{smax}} = \frac{\sqrt{v} \cdot f_s}{2\Omega_{smax}} \tag{2.43}$$

2.7 推导连续子波变换中信号时间和扫描时间之间的几何关系

在连续子波变换中,尺度a是连续的,且$a > 0$;扫描时间$\tau \in \mathbf{R}$,令$\tau_1 \leq \tau \leq \tau_2$;信号时间$t \in \mathbf{R}$,令$t_1 \leq t \leq t_2$。于是,可以很自然想到:$\tau_1 \stackrel{?}{=} t_1$,$\tau_2 \stackrel{?}{=} t_2$,即待论证序列信号的连续子波变换中信号时间$t$和扫描时间$\tau$之间的几何关系。

众所周知,信号$s(t)$的连续子波变换为:$W_s(\tau, a) = \frac{1}{\sqrt{a}} \int_{t_1}^{t_2} s(t) \psi^* (\frac{t-\tau}{a}) dt$。其中,母波的持续时间为

$$t' = \frac{t - \tau}{a} \tag{2.44}$$

令$t_1' \leq t' \leq t_2'$。显然,只要在$t_1 \leq t \leq t_2$内,若有$\psi^*(\frac{t-\tau}{a}) \neq 0$成立,即$\psi^*(t') \neq 0$,

$t'_1 \leqslant t' \leqslant t'_2$ 成立,便可得到 $W_s(\tau, a) \neq 0$。由 $t' = \dfrac{t-\tau}{a}$,得 $\tau = t - at'$。再根据不等式法则可以方便地推证得

$$t_1 - at'_2 \leqslant \tau \leqslant t_2 - at'_1 \tag{2.45}$$

信号时间和扫描时间之间的不等式关系如图 2.11 所示。

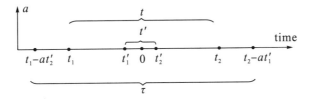

图 2.11　信号时间和扫描时间之间的不等式关系

因此,在信号的连续子波变换中,当尺度 a 连续变化时,信号时间 t 和扫描时间 τ 之间的几何关系如图 2.12 所示。

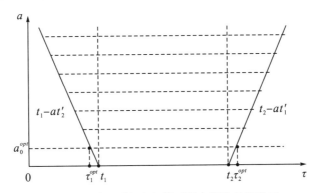

图 2.12　信号时间和扫描时间之间的几何关系

显然,当信号的连续子波变换数值实现时,若起始尺度 a_0 的最佳取值为 a_0^{opt},对应起始扫描时间 τ_0 的最佳范围为

$$t_1 - a_0^{opt} t'_2 \leqslant \tau_0^{opt} \leqslant t_2 - a_0^{opt} t'_1 \tag{2.46}$$

2.8　小结

本章系统地论述了连续子波变换数值实现中确定尺度采样间隔(起始尺度)的基本理论,均匀点格采样、二进点格采样和二进抽取采样时的起始尺度确定方法以及各自的最佳起始尺度;论述了连续子波变换数值实现中起始扫描时间的最佳取值范围;研究了偶对称或奇对称的各阶高斯函数导数解析母波的相应数字滤波器的波动情况;系统地论述了连续子波变换数值实现中信号时间和扫描时间之间的几何关系;论述了连续子波变换数值实现中起始扫描时间的最佳取值范围。本章推导出的结论不仅解决了信号子波分析理论研究和工程实现两个基础性问题,而且解决了信号分数阶微积分的子波变换数值实现的两个基础性问题。

对于用传统的整数阶微积分来构造子波变换中的母波而言,本章已经进行了系统的研

究工作，并得出了一些结论。但是，当我们把传统的整数阶微积分推广到分数阶微积分，进而用分数阶微积分来构造子波变换总的母波时，以上推导得出的结论又会是怎样的情况？换言之，在式（2.17）中，当由高斯函数的各阶导数构造得到（复）解析母波$_k\psi$时，其中各阶导数的阶数 k 不是整数，而是分数 v，当由高斯函数的各分数阶导数构造得到（复）解析母波$_v\psi$时，连续子波变换数值实现中均匀点格采样、二进点格采样和二进抽取采样时的起始尺度确定方法以及各自的最佳起始尺度以及信号时间和扫描时间之间的几何关系又会是怎样的结论？对于以上问题，笔者还将进一步进行系统的研究。

第3章 现代信号分析与处理中分数阶微积分的数值实现

本章研究的目的是在计算机上数值实现现代信号分析与处理中信号的分数阶微积分，特别是信号的分数阶微积分快速子波数值实现算法。首先，本章分析分数阶微积分在数学中常用的 Grümwald-Letnikov 分数阶微积分定义、Riemann-Liouville 分数阶微积分公式、Caputo 分数阶微分定义、分数阶 Cauchy 积分公式四种常用的时域定义，以及其在 Fourier 变换域、Laplace 变换域和 Wavelet 变换域中三种常用的频域定义。其次，在现代信号分析与处理中，推导并比较信号分数阶微分的幂级数数值算法、Fourier 级数数值算法，并将这两种算法与经典的基于 Grümwald-Letnikov 定义的数值算法相比较，在此基础上推导具有较高精度和计算速度的基于子波变换的分数阶微积分快速数值算法。最后，以牺牲计算精度为代价进一步提高计算速度，推导基于子波变换和连续内插的快速工程算法。理论分析和仿真实验均表明，本章研究的五种数值算法中，基于子波变换的数值算法具有较高的精度和运算速度，改进的快速工程算法运算速度最高，但精度下降，这两种算法都具有较强的实用价值。

3.1 问题提出

分数阶微积分的概念的产生最早可追溯到三百多年前法国著名数学家 Guillaume François Antoine L'Hôpital 问过微积分学的奠基人之一 Gattfried Wilhelm Leibnitz 的一个关于当微分阶次是分数时其数学含义的问题[77]。近三百年以降，分数阶微积分主要侧重于纯数学理论研究，它作为整数阶微积分一种重要的推广，成为非线性数学的一个分支，其相关的数学理论研究已逐渐完善，至今相关的新数学理论与概念仍层出不穷。

不幸的是，直至近年，对于大多数工程技术界的学者，甚至一些数学家而言，分数阶微积分都还鲜为人知。这种束之高阁、藏在深闺中的现象是什么原因导致的呢？我认为其中关键的因素是，各应用学科的工程实践中没有能找到很好的应用前景，因此不能激起工程技术界学者对它进行研究的广泛热情。分数阶微积分与分数阶 Fourier 变换在理论上都是一种由整数阶到分数阶的拓展，而且它们都要逾越人们关于整数阶和分数阶传统理念束缚的门槛，因此它们产生、发展的历程也有相似之处。众所周知，将 Wigner 分布应用于处理非平稳随机信号或解析信号时，就是 Wigner-Ville 分布。p 阶分数阶 Fourier 变换可由信号的 Wigner 分布在时—频平面上逆时针旋转 $\alpha = p\pi/2$ 得到[78-79]。诺贝尔物理学奖得主 Wigner 在 1932 年创立 Wigner 分布理论，1948 年 Ville 将 Wigner 分布引入信号分析，但是很长一段时间内 Wigner 分布都没能引起足够的重视。直至 Wigner 分布在光学领域，包括经典的 Fourier 光学和量子光学中找到了很好的应用前景，大家才开始对它投入了足够的重视和广泛的研究,开始领悟到它在信号时—频分析中对非平稳信号分析的独到优势。与之相应,图灵奖得主 Mandelbrot 首先将 Riemann-Liouville 定义的分数阶微积分应用于分析和研究分形媒介中的布朗运动,提出分形学说。由于混沌在本质上是一种分形,混沌产生于非线性方程的数值计算结果,但描述混沌吸引子的分形维数却不依赖于方程。他将寻找混沌吸引子普适常数的物理内容转化为研究分形维数的物理意义。他的研究表明,整数阶微积分有力地描述了 Euclid 空间;相应地,分数阶微积分有力地描述了分数维空间[80]。正是由于分数阶微积分理论的这一次成功实践,近年来才激起了工程技术界学者对分数阶微积分理论和应用的关注。近年来,将分数阶微积分应用于自动控制领域,出现了分数阶控制理论这一新兴的学科分支[81−93,143];将分数阶微积分应用于生物工程领域,出现了基于分数阶的生物工程理论这一新兴的学科分支[94];分数阶微积分还在一些学科的现代工程分析与计算中,特别是在化学、电磁学、材料科学、力学、交通管制、金融、地质探矿等学科中得到关注和应用,并取得了一些可喜的研究成果[95−118]。

然而,如何将分数阶微积分应用于通信和信息处理,特别是应用于现代信号分析与处理中,在国内外目前还是一个研究甚少的领域。一方面,整数阶微积分已经广泛地应用于传统的信号分析与处理领域,特别是对信号的奇异性检测和提取[119];另一方面,现代信号分析与处理本质上是研究非线性、非因果、非最小相位系统、非高斯、非平稳、非整数维(分形)信号、非白色的加性噪声等[120−121]。分数阶微积分是否和分数阶傅立叶变换一样,也是对以上这些"非"问题进行分析和研究的一个有力工具呢?作者对它作了一些初步的尝试,将在本书后续章节中系统论述。

将分数阶微积分应用于现代信号分析与处理中遇到的两个首要基础性问题:一是怎样在计算机上快速数值实现信号的任意分数阶微积分,二是怎样构造模拟分抗电路实现信号的任意分数阶微积分。这两个基础性问题是将分数阶微积分应用于现代信号分析与处理这项研究的工程实现和进一步理论深究的前提和基础。本章将详细论述信号任意分数阶微积分的快速数值实现,第 4 章将详细论述信号任意分数阶微积分的模拟分抗电路实现。

实现分数阶微积分分为解析算法和数值算法两种。解析算法有拉普拉斯变换法和傅立叶变换法[121],数值算法有 Zhang and Shimizu 法[122]、L−1 法[123]和池田法[124]等[127−145]。另外,在信号处理方面可以用数字滤波器[128−134]和 FFT[135]的方法来实现分数阶微积分。普遍的解析法虽然精度高,但计算中常常伴有发散不收敛的问题;普遍的数值算法需要存储全部信号数据,若信号的时间很大,计算量随之有大的增加;用数字滤波器来实现的算法也无法回避甚低阶零频归零问题,这是在甚低阶 FIR 数字微分滤波器设计中,零频归零问题与零频附近(甚低频)逼近精度问题所构成的一对矛盾。虽然用 FFT 的算法对于低阶分数阶微积分的相对误差较小,但是对于高阶分数阶微积分,它的相对误差与阶数成正比。

针对快速数值实现信号的分数阶微积分这一问题,本章的主要研究内容有:①分析比较分数阶微积分在数学中常用的 Grünwald-Letnikov 分数阶微积分定义、Riemann-Liouville 分数阶微积分公式、Caputo 分数阶微分定义、分数阶 Cauchy 积分公式四种常用的时域定义,以及其在 Fourier 变换域、Laplace 变换域和 Wavelet 变换域中三种常用的频域定义。②推导并研究信号分数阶微积分的五种数值算法实现算法。首先推导并比较信号分数阶微分的幂级数数值算法、Fourier 级数数值算法,并将这两种算法与经典的基于

Grümwald-Letnikov 定义的数值算法相比较；其次，推导具有较高精度和计算速度的基于子波变换的分数阶微积分快速数值算法；最后，以牺牲计算精度为代价进一步提高计算速度，推导基于子波变换的快速工程算法。

3.2 分数阶微积分四种常用的时域定义

因为从不同角度去考察分数阶微积分我们可以得到不同的定义，所以至今分数阶微积分在数学上仍旧没有一个统一的时域定义表达式。虽然这是对同一事物殊途同归的处理方法，但是它也给进一步研究分数阶微积分带来了一些难度。因此，本节有必要介绍几种经典的分数阶微积分定义及其相互关系。

3.2.1 Grümwald-Letnikov 定义

Grümwald-Letnikov 定义将连续函数经典的整数阶微分的阶数与因次从整数推广到分数，是通过对原整数阶微分的差分近似递推式求极限推衍而来的[136-151]。若用 $\mathbf{R}$ 表示实数域，$\forall v \in \mathbf{R}$，用 $[v]$ 表示 v 的整数部分，设信号 $s(t)$ 满足 $s(t) \in [a, t]$ ($a<t$, $a \in \mathbf{R}$, $t \in \mathbf{R}$)；若用 $\mathbf{Z}$ 表示整数域，设 $p \in \mathbf{Z}$，令信号 $s(t)$ 存在 $p+1$ 阶连续导数；当 $v>0$ 时，p 至少取到 $[v]$，则信号 $s(t)$ 的 v 阶导数定义为

$$_a^G D_t^v s(t) \overset{\Delta}{=} \lim_{h \to 0} s_h^{(v)}(t) \overset{\Delta}{=} \lim_{\substack{h \to 0 \\ nh = t-a}} h^{-v} \sum_{r=0}^{n} \begin{bmatrix} -v \\ r \end{bmatrix} s(t-rh) \qquad (3.1)$$

式中，$_a^G D_t^v$ 的左上标 G 表示 Grümwald-Letnikov 定义，右上标 v 表示求 v 阶微分，下标 a 和 t 表示积分式的下界和上界，a 为时间 t 的初值。在实际的因果物理系统中，常常令 $a=0$。使 $s_h^{(-v)}(t)$ 达到非零极限，必须满足当 $h \to 0$ 时，$n \to \infty$。于是令 $h = \dfrac{t-a}{n}$，可得 $n = \left[\dfrac{t-a}{h}\right]$。如果将组合数 $\binom{b}{r} = \dfrac{(b)(b-1)\cdots(b+r-1)}{r!}$ 中的 b 扩展为任意实数，再定义 $\begin{bmatrix} -v \\ r \end{bmatrix} = \dfrac{(-v)(-v+1)\cdots(-v+r-1)}{r!}$，则可以得到 $\binom{-b}{r} = (-1)^r \begin{bmatrix} b \\ r \end{bmatrix}$ 成立。与组合数的性质相似，我们可以令 $\begin{bmatrix} 0 \\ r \end{bmatrix} = 0$，$\binom{0}{0} = \begin{bmatrix} 0 \\ 0 \end{bmatrix} = 1$。由式（3.1）可以推得：①$s^{(0)}(t) = s(t)$，当分数阶微积分的阶数 $v=0$ 时，表示既不微分也不积分；②对于 $0<v \leqslant n$ ($v \in \mathbf{Z}$)，如果存在 $q>v$，$\binom{v}{q} = \dfrac{v(v-1)\cdots1 \cdot 0 \cdots (v-q+1)}{q!} = 0 = \begin{bmatrix} -v \\ q \end{bmatrix}$；③根据算子理论，当 $v<0$，即 v 为负实数时，$|v|$ 阶微分转化为 $|v|$ 阶积分。我们可以运用数学归纳法和分部积分法，由式（3.1）可推知

$$_a^G D_t^v s(t) = \sum_{k=0}^{m} \dfrac{s^{(k)}(a)(t-a)^{-v+k}}{\Gamma(-v+k+1)} + \dfrac{1}{\Gamma(-v+m+1)} \int_a^t (t-\tau)^{-v+m} s^{(m+1)}(\tau) \mathrm{d}\tau \qquad (3.2)$$

$\Gamma(\alpha) = \int_0^\infty \mathrm{e}^{-x} x^{\alpha-1} \mathrm{d}x = (\alpha-1)!$ 即是 Gamma 函数。

如果我们令 v 阶微积分算子 $D^v=D_v$，那么信号 $s(t)$ 的 v 阶导数存在以下六个性质[143-144]：

(1) 解析信号 $s(t)$ 的分数阶微分 $D^v s(t)=D_v s(t)$ 不仅对于时间 t 而且对于阶数 v 都是解析的。

(2) 分数阶微分值是有界的，即 $|D^v s(t)|=|D_v s(t)|=|s^{(v)}(t)|<\infty$ 成立。

(3) 分数阶微分值是连续内插的，即 $\lim_{v_1 \to v_2} D^{v_1} s(t)=D^{v_2} s(t)$，$v_1$，$v_2 \in \mathbf{R}$ 成立。

(4) 分数阶微分值是实值的，即 $D^v s(t) \in \mathbf{R}$。当 $v \in \mathbf{Z}^+$ 时，分数阶微分退化为整数阶微分；当 $v \in \mathbf{Z}^-$ 时，分数阶积分退化为整数阶积分；当 $v=0$ 时，分数阶微积分既不作微分运算也不作积分运算，保持原信号不变。

(5) 分数阶微积分算子满足线性可加性。设存在任意常数 i 和 j，有 $D^v[is_1(t)+js_2(t)]=iD^v s_1(t)+jD^v s_2(t)$ 成立。

(6) 与整数阶微分算子一样，分数阶微分算子也满足交换率和算子叠加准则，即 $D^{v_1} D^{v_2} s(t)=D^{v_2} D^{v_1} s(t)=D^{v_1+v_2} s(t)$。

3.2.2 Riemann-Liouville 定义

为了使分数阶微积分的计算简化，对照分数阶微积分应满足的以上性质，Riemann-Liouville 定义对 Grümwald-Letnikov 定义进行了改进[136-151]。Riemann-Liouville 定义是目前最常用的分数阶微积分定义[150]。Riemann-Liouville 定义的分数阶微分为

$$_a^R D_t^v s(t) = \begin{cases} \dfrac{\mathrm{d}^n s}{\mathrm{d} t^n}, & v=n \in \mathbf{N} \\ \dfrac{\mathrm{d}^n}{\mathrm{d} t^n} \dfrac{1}{\Gamma(n-v)} \int_a^t \dfrac{s(\tau)}{(t-\tau)^{v-n+1}} \mathrm{d}\tau, & 0 \leqslant n-1 < v < n \end{cases} \quad (3.3)$$

Riemann-Liouville 定义的分数阶微分是先进行 $n-v$ 阶积分，然后进行 n 阶微分。Riemann-Liouville 定义的分数阶积分为

$$_a^R D_t^v s(t) = \frac{1}{\Gamma(-v)} \int_a^t (t-\tau)^{-v-1} s(\tau) \mathrm{d}\tau \quad (v<0) \quad (3.4)$$

3.2.3 Caputo 定义

Caputo 定义是对 Grümwald-Letnikov 定义的另一种改进，其目的是让拉普拉斯变换更加简洁，从而便于解分数阶微分方程。Caputo 定义更适合于对分数阶微分方程的初值问题进行描述[136-151]。Caputo 定义的分数阶微分（$0 \leqslant n-1 < v < n$，$n \in \mathbf{R}$）为

$$_a^{Caputo} D_t^v s(t) \stackrel{\Delta}{=} \frac{1}{\Gamma(n-v)} \int_a^t (t-\tau)^{n-v-1} s^{(n)}(\tau) \mathrm{d}\tau \quad (3.5)$$

本定义先进行 n 阶微分，再进行 $n-v$ 阶积分。

3.2.4 分数阶 Cauchy 积分公式

分数阶 Cauchy 积分公式由整数阶微积分公式直接扩展得到[136-151]，即

$$^{Cauchy} D^v s(t) \stackrel{\Delta}{=} \frac{\Gamma(v-n+1)}{2\pi j} \int_l \frac{s(\tau)}{(\tau-t)^{v-n+1}} \mathrm{d}\tau \quad (3.6)$$

式中，积分曲线 l 是包围信号 $s(t)$ 单值与解析区域的光滑曲线。

3.2.5　各分数阶微积分定义的关系

当信号 $s(t)$ 存在 $(m+1)$ 阶连续导数，且 m 至少取得 $[v]=n-1$ 时，Grümwald-Letnikov 定义和 Riemann-Liouville 定义是等价的。但若不满足上述条件，Riemann-Liouville 定义是 Grümwald-Letnikov 定义的拓展，具有更广泛的应用[138-151]。可见，对于大多数信号的分数阶微积分，其 Grümwald-Letnikov 定义和 Riemann-Liouville 定义是可以互换的。当 $s(t)$ 满足 $s^{(k)}(C)=0$，$k=0,1,\cdots,n-1$，其中 C 是常数，且当 $s(t)$ 存在 $m+1$ 阶连续导数，m 至少可以取到 $[v]=n-1$ 时，Caputo 定义和 Grümwald-Letnikov 定义等价；否则，两个定义不等价[138-151]。令 $\mathbf{N}$ 表示自然数，如果我们对 Riemann-Liouville 和 Caputo 定义的分数阶微分分别求 $v \to n$ 极限，可以推导得 ${}_a^{Caputo}D_t^v s(t) = {}_a^G D_t^n s(t) = s^{(n)}(t)$，$v=n\in\mathbf{N}$。Caputo 定义的分数阶积分和 Riemann-Liouville 定义的分数阶积分相同。当 v 为正整数和负实数时，Caputo 定义和 Riemann-Liouville 定义是等价的，两个定义的关系式为 ${}_a^R D_t^v s(t) = \sum_{k=0}^{n-1} \frac{s^{(k)}(a)(t-a)^{k-v}}{\Gamma(k-p+1)} + {}_a^{Caputo}D_t^v s(t)$。

Riemann-Liouville 定义和 Caputo 定义的区别主要表现在对常数的求导数的定义上。Riemann-Liouville 定义的分数阶微积分对常数求导数是有界的，其值为零；Caputo 定义的分数阶微积分对常数求导数，其值是无界的[138-151]。

3.2.6　分数阶微分和积分的关系

根据 Riemann-Liouville 定义，令信号 $s(t)$ 是定义在非负实数域内的全部绝对连续的实函数，$s(t)$ 在零点处存在右极限，$s'(t)$ 在正实数域内的任意有界子区间上 Lebesgue 可积，$0<v<1$，于是由式（3.3）可以推导 ${}_0^R D_t^v s(t) = \frac{1}{\Gamma(1-v)}\left[\frac{s(0)}{t^v} + \int_0^t \frac{s'(\tau)}{(t-\tau)^v}d\tau\right]$，$0<v<1$ 成立。如果 ${}_0^R I_t^v$ 为 v 阶 Riemann-Liouville 定义的分数阶积分算子，D 为传统的一阶微分算子，于是由式（3.4），我们可以推导得到 Riemann-Liouville 定义的分数阶微分和分数阶积分存在 ${}_0^R D_t^v s(t) = {}_0^R I_t^{1-v} D s(t) + \frac{1}{\Gamma(1-v)}\frac{s(0)}{t^v}$ 的关系。因此，如果 $s(0)=0$，则有 ${}_0^R D_t^v s(t) = {}_0^R I_t^{1-v} D s(t)$ 成立。

这就是分数阶微分和分数阶积分的内在关系。根据它，由分数阶微分的计算结果，就可以推导得到分数阶积分的计算结果[138-151]。

3.3　分数阶微积分三种常用的频域定义

3.3.1　Fourier 变换域定义

如果信号 $s(t) \in L^2(R)$ 是平方可积的能量型信号，由信号与信息系统的理论知，它的 Fourier 变换为 $\hat{s}(\omega) = \int_R s(t)\exp(-i\omega t)dt$。令信号 $s(t)$ 的 n 阶导数为 $s^{(n)}(t)$，根据 Fourier 变换的性质[58,60]，$s^{(n)}(t)$ 在频域的等价形式为 $D^n s(t) \overset{FT}{\Longleftrightarrow} (D\hat{s})^n(\omega) = (i\omega)^n \cdot \hat{s}(\omega) =$

$d^n(\hat{\omega})\hat{s}(\omega)$。令信号 $s(t)$ 的 v 阶分数阶导数为 $s^{(v)}(t)$，同样根据 Fourier 变换的性质，将整数求导阶数 n 推广为分数求导阶数 v，可以推导得到信号 $s(t)$ 的 v 阶分数阶微分在 Fourier 变换域的等价形式为

$$\begin{cases} D_v s(t) = \dfrac{\mathrm{d}^v s(t)}{\mathrm{d}t^v}, & v \in \mathbf{R}^+ \\ \Updownarrow FT \\ (D_v \hat{s})(\omega) = (i\omega)^v \cdot \hat{s}(\omega) = \hat{d}_v(\omega) \cdot \hat{s}(\omega), & v \in \mathbf{R}^+ \end{cases} \quad (3.7)$$

令 v 阶微分算子 $D_v = D^v$，显然，D^v 是 v 阶微分乘子函数 $\hat{d}_v(\omega) = (i\omega)^v$ 的乘性算子。在复数域中 $\hat{d}_v(\omega)$ 指数形式为

$$\begin{cases} \hat{d}_v(\omega) = (i\omega)^v = \hat{a}_v(\omega) \cdot \exp(i\theta_v(\omega)) = \hat{a}(\omega) \cdot \hat{p}_v(\omega) \\ \hat{a}_v(\omega) = |\omega|^v, \quad \hat{\theta}_v(\omega) = \dfrac{v\pi}{2} \mathrm{sgn}(\omega) \end{cases} \quad (3.8)$$

从信号处理角度看，v 阶微积分运算其实是对信号的一个线性时不变滤波系统[135]，其滤波函数为 $\hat{d}_v(\omega) = (i\omega)^v = |\omega|^v \cdot \exp(i\theta_v(\omega))$。当 $v>0$ 时，它是微分器；当 $v=0$ 时，它是全通滤波器；当 $v<0$ 时，它是积分器。由式（3.8），从通信调制角度看，笔者认为信号的分数阶微分的物理意义可以理解为广义的调幅调相，其振幅随频率是分数阶幂指数变化，其相位是频率的广义 Hilbert 变换。

将式（3.8）写成对应的时域形式为

$$d_v(t) = a_v(t) * p_v(t) = \frac{1}{2\pi} \int_{-\infty}^{+\infty} (i\omega)^v \cdot \mathrm{e}^{i\omega t} \mathrm{d}\omega \quad (3.9)$$

$$a_v(t) = \int_{-\infty}^{+\infty} \hat{a}(\omega) \cdot \mathrm{e}^{i\omega t} \mathrm{d}\omega = \frac{1}{\pi} \int_{0}^{+\infty} |\omega|^v \cdot \cos(i\omega t) \mathrm{d}\omega \quad (3.10)$$

$$p_v(t) = \frac{1}{2\pi} \int_{-\infty}^{+\infty} \hat{p}_v(\omega) \cdot \mathrm{e}^{i\omega t} \mathrm{d}\omega = \cos\frac{v\pi}{2} \cdot \delta(t) - \sin\frac{v\pi}{2} \cdot \frac{1}{\pi t} \quad (3.11)$$

从信号处理的角度来看，$d_v(t)$ 是分数阶微积分滤波器的冲击响应。由式（3.10）、（3.11）知，分数阶微积分滤波器的振幅特性是偶函数，其相位特性是奇函数。于是，本章仅需讨论当 $\omega>0$ 时分数阶微积分滤波器的振幅特性和相位特性即可，当 $\omega<0$ 时的分数阶微积分滤波器特性可由 $\omega>0$ 时的分数阶微积分滤波器特性推导得到。由信号与信息系统理论可知，信号 $s(t)$ 的分数阶微积分在时域中的卷积形式为

$$\begin{aligned} D_v s(t) &= d_v(t) * s(t) = a_v(t) * p_v(t) * s(t) \\ &= \left[\cos\frac{v\pi}{2} \cdot \delta(t) - \sin\frac{v\pi}{2} \cdot \frac{1}{\pi t}\right] * \int_{-\infty}^{+\infty} a_v(t-\tau) s(\tau) \mathrm{d}\tau \end{aligned} \quad (3.12)$$

3.3.2　Laplace 变换域定义

与信号 $s(t)$ 分数阶微积分的 Fourier 变换域定义的推导类似，有因果实信号 $s(t)$，$t \in [0, t]$，如果将 $s(t)$ 的 Laplace 变换表示为 $s(t) \overset{LT}{\Leftrightarrow} \hat{s}(S)$，令 $v>0$，那么信号 $s(t)$ 的 v 阶分数阶积分的 Laplace 变换域定义式为[135,138]

$$D^{-v} s(t) \overset{LT}{\Leftrightarrow} S^{-v} \hat{s}(S) \quad (3.13)$$

与此相应，信号 $s(t)$ 的 v 阶分数阶微分的 Laplace 变换域的 Grünwald-Letnikov 定义式为[135,138]

$$D^v s(t) \overset{LT}{\Leftrightarrow} S^v \hat{s}(S) \tag{3.14a}$$

信号 $s(t)$ 的 v 阶分数阶微分的 Laplace 变换域 Caputo 定义式为[135,138]

$$D^v s(t) \overset{LT}{\Leftrightarrow} S^v \hat{s}(S) - \sum_{m=0}^{n-1} S^{v-m-1} [D^m s(t)]_{t=0}, \quad n-1 \leqslant v < n \tag{3.14b}$$

信号 $s(t)$ 的 v 阶分数阶微分的 Laplace 变换域 Riemann-Liouville 定义式为[135,138]

$$D^v s(t) \overset{LT}{\Leftrightarrow} S^v \hat{s}(S) - \sum_{m=0}^{n-1} S^m [D^{v-m-1} s(t)]_{t=0}, \quad n-1 \leqslant v < n \tag{3.14c}$$

如果信号 $s(t)$ 及其各阶导数在 $t=0$ 时其初值均为零,则有

$$D^v s(t) \overset{LT}{\Leftrightarrow} S^v \hat{s}(S) \tag{3.15}$$

3.3.3 Wavelet 变换域定义

众所周知,子波变换是信号时—频分析的有力手段,那么信号的分数阶微积分与其子波变换有什么样的关系呢?为了利用业已成熟的子波变换来实现信号的分数阶微分,从而使工程计算更简单,有学者对信号的分数阶微分与其子波变换的关系进行了研究[128-129]。

如果 $\varphi(t) \overset{FT}{\Leftrightarrow} \hat{\varphi}(\omega)$ 是一个光滑实偶信号,为 $\int_{-\infty}^{+\infty} \varphi(t) \mathrm{d}t = 1$,即 $\varphi(t)$ 满足低通条件时,对 $\varphi(t)$ 进行 v 阶分数阶微分 $\varphi^{(v)}(t)$,再用 $\varphi^{(v)}(t)$ 来构造出能量规范化母波 $\psi_v(t)$。其运算过程为

$$\psi_v(t) = E^{-1}(v) D_v [\varphi(t)] \overset{FT}{\Leftrightarrow} \hat{\psi}_v(\omega) = E^{-1}(v)(i\omega)^v \hat{\varphi}(\omega) \tag{3.16}$$

式中,$E(v) = \{\int_{-\infty}^{\infty} | D_v[\varphi(t)] |^2 \mathrm{d}t\}^{\frac{1}{2}}$,它是能量归一化因子。令子波变换的伸缩系数或尺度因子 $a > 0$,则实信号 $s(t)$ 的子波变换为

$$Ws(\tau, a) = s(\tau) * \psi_{va}(t) = \int_{-\infty}^{+\infty} s(\tau - t) \cdot \frac{1}{a} \psi_v(\frac{t}{a}) \mathrm{d}t \tag{3.17}$$

式中,τ 是扫描时间,能量恒等化子波 $\psi_{va}(t) = \frac{1}{a} \psi_v(\frac{t}{a})$。由式(3.16)、(3.17)以及卷积性质[152]可得

$$D_v[s(\tau) * \varphi_a(t)] = \frac{E(v) Ws(\tau, a)}{a^v} \tag{3.18}$$

式中,$\varphi_a(t) = \frac{1}{a}\varphi(\frac{t}{a})$,根据实分析理论可知 $\varphi_a(t)$ 是一个恒等逼近核[152],可以推知 $\lim_{a \to 0} [s(\tau) * \varphi_a(t)] = s(\tau)$ 成立。代入式(3.18),可得

$$D_v[s(\tau)] = \lim_{a \to 0} \frac{E(v) Ws(\tau, a)}{a^v} \tag{3.19}$$

这便在数学上找到了分数阶微积分与其子波变换的转换公式。

那么,信号的分数阶微积分与其子波变换的转换公式在信号与信息系统理论中的物理意义是什么呢?笔者认为,Heisenberg 不确定性原理是我们对信号时—频分析的极限制约[61]。对于同一子波而言,信号子波变换中的时—频胞元面积是一定的,如图 3.1 所示。

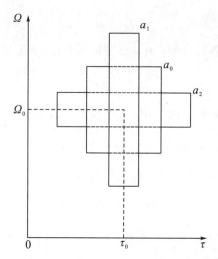

图 3.1 信号子波变换的时—频胞元和时间尺度因子的关系

图 3.1 中，$a_2 > a_0 > a_1$。可见，当信号子波变换的时间伸缩系数或尺度因子越小时，其时—频胞元的时间宽度越小，越有利于分析细节；当时间尺度因子越大时，其时—频胞元的带宽越小，越有利于分析轮廓，从而实现对信号的时—频分析。显然，当信号的分数阶微积分与其子波变换的转换公式（3.19）中的时间尺度因子 $a \to 0$ 时，其时—频胞元的带宽趋于无穷大，此时信号的子波变换就失去了对信号频率的分析能力，而不是完整意义上的时—频分析，只具有对信号时间细节变化的分析能力。这恰好和信号的分数阶微积分只是对信号做时间细节分析的物理含义相一致。换言之，一方面，当子波变换的时间尺度因子 $a \to 0$ 时，信号的子波变换从对信号的时—频分析褪化为只对信号进行时间分析的变换，这一点恰好与信号的分数阶微积分只是对信号进行时间分析的物理意义相一致；另一方面，当时间尺度因子 $a \to 0$ 时，信号的子波变换对信号具有极强的分析时间细节的能力，这一点恰好与信号的分数阶微积分是对信号进行高通滤波，加强高频信息的物理意义相吻合。这就是式（3.19）的物理含义。

3.4 信号分数阶微积分的幂级数算法

一些学者研究并推导了幂函数 x^n（$n \in \mathbf{Z}$）的分数阶微分形式[145-147]，本节将其结论推广到一般信号分数阶微积分的求解中。

3.4.1 理论分析

由于实数域具有完备性，所以可以直接把阶乘 $x!$ 推广到实数集中，即 $x \in \mathbf{R}$（包括分数）。为保证这种推广的合理性和唯一性，要求推广后的阶乘具有连续性和光滑性。该广义阶乘实质上就是 Gamma 函数，$x! = \int_0^\infty t^x e^{-t} dt = \Gamma(x+1)$。进而，由 Grümwald-Letnikov 分数阶微积分定义式（3.1）和（3.2），可以推导出幂级数 t^n 的 v（$v \in \mathbf{R}$）阶微分为

$$(t^n)^{(v)} = \frac{n!}{(n-v)!} t^{n-v} = \frac{\Gamma(n+1)}{\Gamma(n-v+1)} t^{n-v} \qquad (3.20)$$

为保证式（3.20）在 $t=0$ 处有意义，必须满足 $v \leqslant n$。

若信号 $s(t)$ 在区间（$-\mathbf{R}$，$\mathbf{R}$）存在任意阶导数，则它总能够展开成幂级数的形式[70,153]。若 $s(t)$ 在 $t=0$ 的某邻域内存在任意阶导数，将 $s(t)$ 的泰勒级数在原点展开，即得到它的麦克劳林展开式：

$$s(t) = \sum_{k=0}^{n} \frac{s^{(k)}(0)}{k!} t^k + R_n(t) \tag{3.21}$$

$$R_n(t) = \frac{s^{(n+1)}(\xi)}{(n+1)!} t^{n+1}, \quad 0 < \xi < t \tag{3.22}$$

要使该展开式收敛，其中拉格朗日余项 $R_n(t)$ 须满足 $\lim_{n \to \infty} R_n(t) = 0$，$t \in (-\mathbf{R}, \mathbf{R})$，即 t 必须在其收敛域内。

众所周知，从函数论的角度来看，求导数是函数的一个线性变换，因此推广后的导数必具有线性性。故信号 $s(t)$ 的 v 阶微分等价于其麦克劳林级数各项的 v 阶微分之和。

如前所述，当且仅当时间 t 在其收敛域内时，信号 $s(t)$ 的麦克劳林级数才收敛。然而，一般仅有纯频率信号 $\sin\omega t$，$\cos\omega t$，$e^{\omega t}$ 的 t 收敛域为整个实数空间，它们的麦克劳林级数及其收敛域分别为

$$\sin\omega t = \sum_{n=0}^{\infty} (-1)^n \frac{(\omega t)^{2n+1}}{(2n+1)!}, \quad -\infty < t < \infty \tag{3.23}$$

$$\cos\omega t = \sum_{n=0}^{\infty} (-1)^n \frac{(\omega t)^{2n}}{(2n)!}, \quad -\infty < t < \infty \tag{3.24}$$

$$e^{\omega t} = \sum_{n=0}^{\infty} \frac{(\omega t)^n}{n!}, \quad -\infty < t < \infty \tag{3.25}$$

由式（3.20），分别求得其分数阶微分的闭式表达式为

$$(\sin\omega t)^{(v)} = \sum_{n=0}^{\infty} (-1)^n \frac{1}{\Gamma(2n-v+2)} (\omega t)^{2n-v+1}, \quad -\infty < t < \infty \tag{3.26}$$

$$(\cos\omega t)^{(v)} = \sum_{n=0}^{\infty} (-1)^n \frac{1}{\Gamma(2n-v+1)} (\omega t)^{2n-v}, \quad -\infty < t < \infty \tag{3.27}$$

$$(e^{\omega t})^{(v)} = \sum_{n=0}^{\infty} \frac{1}{\Gamma(n-v+1)} (\omega t)^{n-v}, \quad -\infty < t < \infty \tag{3.28}$$

但是，因为满足级数收敛 $\lim_{n \to \infty} R_n(t) = 0$ 的时间 t 不一定遍历整个实数域，故本算法并非对所有信号在整个时间域内都行之有效。信号 $\ln(1+t)$，$\arctan t$，$\ln t$ 的麦克劳林级数及其收敛域分别为

$$\ln(1+t) = \sum_{n=0}^{\infty} (-1)^n \frac{t^{n+1}}{n+1}, \quad -1 < t \leqslant 1 \tag{3.29}$$

$$\arctan t = \sum_{n=0}^{\infty} (-1)^n \frac{t^{2n+1}}{2n+1}, \quad -1 < t < 1 \tag{3.30}$$

$$\ln t = 2 \sum_{n=0}^{\infty} \frac{1}{2n+1} \cdot \left(\frac{t-1}{t+1}\right)^{2n+1}, \quad 0 < t \tag{3.31}$$

由式（3.20）分别求得其分数阶微分的闭式表达式为

$$(\ln(1+t))^{(v)} = \sum_{n=0}^{\infty} (-1)^n \frac{\Gamma(n+1)}{\Gamma(n-v+2)} t^{n-v+1}, \quad -1 < t \leqslant 1 \tag{3.32}$$

$$(\arctan t)^{(v)} = \sum_{n=0}^{\infty} (-1)^n \frac{\Gamma(2n+1)}{\Gamma(2n-v+2)} t^{2n-v+1}, \quad -1 < t < 1 \quad (3.33)$$

$$\ln t^{(v)} = 2 \sum_{n=0}^{\infty} \frac{\Gamma(2n+1)}{\Gamma(2n-v+2)} \cdot \left(\frac{t-1}{t+1}\right)^{2n-v+1}, \quad 0 < t \quad (3.34)$$

3.4.2 实验仿真及结果分析

不失一般性，下面以信号麦克劳林级数的收敛域为整个实数域的 $s(t) = \sin t$ 进行实验。特别地，在数值实现时必须注意，为保证式（3.26）在 $t=0$ 处有意义，必须满足 $v \leqslant 2n+1$，故 n 的初始值为 $\left|\left[\frac{v-1}{2}\right]\right| \geqslant 0$，而不一定为零。取 $v=0.75$，$n \leqslant 100$，仅观测 $-1.5\pi < t < 1.5\pi$，实验结果如图 3.2 所示。

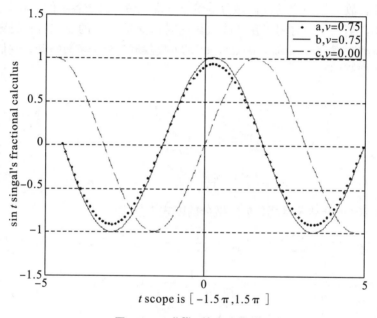

图 3.2　$\sin^{(0.75)} t$ 的实验结果

在图 3.2 中，曲线 c 是 $s(t) = \sin t$，曲线 b 是 $\sin^{(0.75)} t$ 理想的解析解，曲线 a 是用本幂级数算法求得的 $\sin^{(0.75)} t$ 的近似解。可见，用本幂级数算法求得的近似解在幅度上比理想的解析解小，这是由于 n 的取值上限有限，当 n 的取值上限逐渐增大时，这种相对误差会随之减小。另外，用本幂级数算法求得的近似解在相位上和理想的解析解一致。这就证明了本算法的可行性。

3.5　信号分数阶微积分的 Fourier 级数算法

由式（3.9）、（3.10）、（3.11），只需求得积分核 $d_v(t)$，$a_v(t)$ 以及 $p_v(t)$，通过积分运算就能求得信号的分数阶微分。由式（3.8）、（3.12），可求得纯频率信号[61] $\sin \omega_0 t$，$\cos \omega_0 t$ 和 $e^{i\omega_0 t}$（频率 $\omega_0 > 0$）的分数阶微分的解析解为

$$(\sin\omega_0 t)^{(v)} = \omega_0^v \sin(\omega_0 t + \frac{\pi v}{2}) \tag{3.35}$$

$$(\cos\omega_0 t)^{(v)} = \omega_0^v \cos(\omega_0 t + \frac{\pi v}{2}) \tag{3.36}$$

$$(e^{i\omega_0 t})^{(v)} = \omega_0^v e^{i(\omega_0 t + \frac{\pi v}{2})} = \omega_0^v e^{i\omega_0 t} e^{i\frac{\pi v}{2}} \tag{3.37}$$

当 $v=k$ （$k \in \mathbf{Z}^+$）时就是传统的整数阶微分。显然，常用周期信号可以展开为傅立叶级数，若用式（3.35）、（3.36）、（3.37）分别求得其傅立叶级数各分量的分数阶微分值，再求和，即该周期信号的分数阶微分值[154]。本节在深入研究傅立叶级数性质的基础上推导出一般能量性信号的分数阶微分数值实现算法。

3.5.1 理论分析

首先，通过截断的办法，可由一个时域无限的能量性信号得到分布在其主要能量分布区间内时域有限的能量性信号。其次，任何一个时域有限的能量性信号都可以视为一个周期信号的主值区间，若把该主值区间向时域两极进行周期延拓，则得到相应的周期信号[61,153]。因此，不仅周期性信号可进行傅立叶级数展开，而且一般能量性信号也可用傅立叶级数表示或逼近。由于对信号求导数是一个线性变换，因此分数阶微分必具有线性性。故可用对信号 $s(t)$ 傅立叶级数的相应分数阶微分之和来代替 $s(t)$ 的分数阶微分。

3.5.1.1 一般能量性信号的分数阶微分

当信号时间 $t \in [-\pi, \pi]$ 或 $t \in [a, a+2\pi]$ 时，能量性信号 $s(t)$ 经周期延拓后的傅立叶级数展开式为

$$s(t) = \frac{a_0}{2} + \sum_{k=1}^{\infty}(a_k \cos kt + b_k \sin kt) \tag{3.38}$$

式中，$a_0 = \frac{1}{\pi}\int_{-\pi}^{\pi} s(t) dt$，$a_k = \frac{1}{\pi}\int_{-\pi}^{\pi} s(t) \cos kt \, dt$，$b_k = \frac{1}{\pi}\int_{-\pi}^{\pi} s(t) \sin kt \, dt$。由式（3.35）、（3.36）可求得信号 $s^{(v)}(t)$ 的闭式表达式为

$$s^{(v)}(t) = \sum_{k=1}^{\infty}\left[a_k k^v \cos(kt + \frac{\pi v}{2}) + b_k k^v \sin(kt + \frac{\pi v}{2})\right] \tag{3.39}$$

当信号的持续时间为任意长度时，即信号时间 $t \in [-l, l]$（$l \in \mathbf{R}^+$），该能量性信号 $s(t)$ 的傅立叶级数展开式为

$$s(t) = \frac{a_0}{2} + \sum_{k=1}^{\infty}\left(a_k \cos\frac{k\pi}{l}t + b_k \sin\frac{k\pi}{l}t\right)$$

式中，$a_0 = \frac{1}{l}\int_{-l}^{l} s(t) dt$，$a_k = \frac{1}{l}\int_{-l}^{l} s(t) \cos\frac{k\pi}{l}t \, dt$，$b_k = \frac{1}{l}\int_{-l}^{l} s(t) \sin\frac{k\pi}{l}t \, dt$。同理，可求得信号 $s^{(v)}(t)$ 的闭式表达式为

$$s^{(v)}(t) = \sum_{k=1}^{\infty}\left(\frac{k\pi}{l}\right)^v \left[a_k \cos(\frac{k\pi}{l}t + \frac{\pi v}{2}) + b_k \sin(\frac{k\pi}{l}t + \frac{\pi v}{2})\right] \tag{3.40}$$

特别地，当 $t \in [-l, l]$ 时，只要令 $t' = \frac{\pi t}{l}$，即可使得 $t' \in [-\pi, \pi]$。当 $t \in [-a, b]$（$a < b$，$a \in \mathbf{R}$，$b \in \mathbf{R}$）时，只要令 $t' = \frac{\pi(t - \frac{a+b}{2})}{\frac{b-a}{2}}$，即可使得 $t' \in [-\pi, \pi]$。

3.5.1.2 能量奇信号的分数阶微分

当 $t \in [-\pi, \pi]$ 时，$s(t)$ 周期延拓后的傅立叶级数为 $s(t) = \sum_{k=1}^{\infty} b_k \sin kt$，其中，$b_k = \frac{2}{\pi} \int_0^{\pi} s(t) \sin kt \, dt$。同理，可求得信号 $s^{(v)}(t)$ 的闭式表达式为

$$s^{(v)}(t) = \sum_{k=1}^{\infty} b_k k^v \sin(kt + \frac{\pi v}{2}) \tag{3.41}$$

当 $t \in [-l, l]$ 时，$s(t)$ 周期延拓后的傅立叶级数为 $s(t) = \sum_{k=1}^{\infty} b_k \sin \frac{k\pi}{l} t$，其中，$b_k = \frac{2}{l} \int_0^{l} s(t) \sin \frac{k\pi}{l} t \, dt$。同理，可求得信号 $s^{(v)}(t)$ 的闭式表达式为

$$s^{(v)}(t) = \sum_{k=1}^{\infty} b_k \left(\frac{k\pi}{l}\right)^v \sin(\frac{k\pi}{l} t + \frac{\pi v}{2}) \tag{3.42}$$

3.5.1.3 能量偶信号的分数阶微分

当 $t \in [-\pi, \pi]$ 时，$s(t) = \frac{a_0}{2} + \sum_{k=1}^{\infty} (a_k \cos kt)$，其中，$a_0 = \frac{2}{\pi} \int_0^{\pi} s(t) \, dt$，$a_k = \frac{2}{\pi} \int_0^{\pi} s(t) \cos kt \, dt$。同理，可求得信号 $s^{(v)}(t)$ 的闭式表达式为

$$s^{(v)}(t) = \sum_{k=1}^{\infty} [a_k k^v \cos(kt + \frac{\pi v}{2})] \tag{3.43}$$

当 $t \in [-l, l]$ 时，$s(t) = \frac{a_0}{2} + \sum_{k=1}^{\infty} (a_k \cos \frac{k\pi}{l} t)$，其中，$a_0 = \frac{2}{l} \int_0^{l} s(t) \, dt$，$a_k = \frac{2}{l} \int_0^{l} s(t) \cos \frac{k\pi}{l} t \, dt$。同理，可求得信号 $s^{(v)}(t)$ 的闭式表达式为

$$s^{(v)}(t) = \sum_{k=1}^{\infty} \left[a_k \left(\frac{k\pi}{l}\right)^v \cos(\frac{k\pi}{l} t + \frac{\pi v}{2})\right] \tag{3.44}$$

3.5.1.4 对一般能量性信号奇式开拓求其分数阶微分

在实际问题中，一般能量性信号并非奇函数或偶函数，而且 $t \in [0, l]$，我们采用前面的开拓法，即先补充 $t \in [-l, 0]$ 上的信号部分，再开拓以 $2l$ 为周期的信号。若在 $t \in [-l, 0]$ 时把信号补充为奇函数，即是奇式开拓，则有 $S(t) = \begin{cases} -s(-t), & -l \leq t < 0 \\ s(t), & 0 \leq t \leq l \end{cases}$。

这样，信号 $s(t)$ 奇式延拓后的傅立叶级数展开式为 $s(t) = \sum_{k=1}^{\infty} b_k \sin kt$，其中，$b_k = \frac{2}{l} \int_0^{l} s(t) \sin kt \, dt$。同理，可求得信号 $s^{(v)}(t)$ 的闭式表达式为

$$s^{(v)}(t) = \sum_{k=1}^{\infty} b_k k^v \sin(kt + \frac{\pi v}{2}) \tag{3.45}$$

3.5.1.5 对一般能量性信号偶式开拓求其分数阶微分

若在 $t \in [-l, 0]$ 时把信号补充为偶函数，即是偶式开拓，则有 $S(t) = \begin{cases} s(-t), & -l \leq t < 0 \\ s(t), & 0 \leq t \leq l \end{cases}$。这样，信号 $s(t)$ 偶式延拓后的傅立叶级数展开式为 $s(t) = \frac{a_0}{2} +$

$\sum_{k=1}^{\infty}(a_k\cos kt)$,其中,$a_0 = \frac{2}{l}\int_0^l s(t)\mathrm{d}t$,$a_k = \frac{2}{l}\int_0^l s(t)\cos kt\,\mathrm{d}t$。同理,可求得信号 $s^{(v)}(t)$ 的闭式表达式为

$$s^{(v)}(t) = \sum_{k=1}^{\infty}\left[a_k k^v \cos\left(kt + \frac{\pi v}{2}\right)\right] \tag{3.46}$$

3.5.1.6 本算法的均方误差

在物理、统计学和工程计算中,经常把均方误差(平均平方误差)用来衡量观测所得数据的准确性。不失一般性,这里以信号时间 $t \in [-\pi, \pi]$ 为例进行分析,用均方误差来衡量三角多项式(3.38)对信号 $s(t)$ 的近似程度。所不同的是,这里不是对有限多个值来求误差,而是对连续分布在整个区间 $[-\pi, \pi]$ 上的无穷多个值来求误差。显然,考察近似式的优劣,只须在一个周期中进行即可,当式(3.38)中 $1 \leqslant k \leqslant n$ 时,其均方误差为 $\delta_n^2 = \frac{1}{2\pi}\int_{-\pi}^{\pi}[s(t)-s_n(t)]^2\mathrm{d}t$。可以证明用 n 阶三角多项式作可积信号 $s(t)$ 的近似表达式时,如果该多项式的系数是 $s(t)$ 的傅立叶系数,其均方误差最小[153],即

$$\delta_n^2 = \frac{1}{2\pi}\int_{-\pi}^{\pi} s(t)^2 \mathrm{d}t - \frac{a_0^2}{4} - \frac{1}{2}\sum_{k=1}^{n}(a_k^2 + b_k^2) \tag{3.47}$$

因为 $\delta_n^2 \geqslant 0$,由式(3.40)得 $\frac{1}{\pi}\int_{-\pi}^{\pi} s(t)^2 \mathrm{d}t \geqslant \frac{a_0^2}{2} + \sum_{k=1}^{n}(a_k^2 + b_k^2)$。在此不等式中,$n$ 是任意的,右边的和式随 n 的增大而增大,且以左边的积分值为上界。当 $n \to \infty$ 时,即得 Bessel 不等式;当均方误差趋于零时,即得 Parseval 等式 $\frac{1}{\pi}\int_{-\pi}^{\pi} f(t)^2 \mathrm{d}t = \frac{a_0^2}{2} + \sum_{k=1}^{\infty}(a_k^2 + b_k^2)$。由此,当工程计算的均方误差 δ_n^2 确定时,式(3.38)中 k 的上限 n 亦相应确定。

3.5.2 实验仿真及结果分析

首先,取 $k \leqslant 100$,以信号 $s(t) = \sin t$ 进行实验,实验结果如图 3.3 所示。

图 3.3 中,曲线 a 是 $s(t) = \sin t$,曲线 b 是 $\sin^{(0.75)}t$ 的解析解,曲线 c 是用本数值算法求得的 $\sin^{(0.75)}t$ 的解。可见曲线 c 和 b 基本重合,但在时间 t 区域的两端计算结果的收敛性较差。同时,由于 $1 \leqslant k \leqslant n$,上限 n 有限,故实验结果必具有均方误差,如图 3.4 所示。但是,均方误差会随 n 的增大而减小。

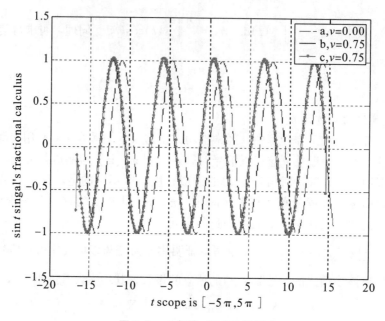

图 3.3 $\sin^{(0.75)}t$ 的实验结果

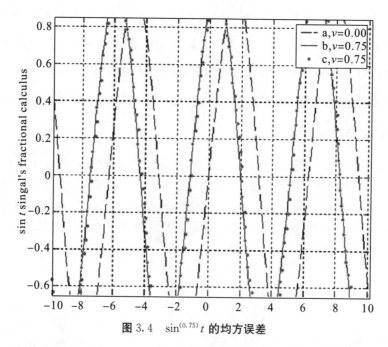

图 3.4 $\sin^{(0.75)}t$ 的均方误差

对比图 3.4 和图 3.2 可知,在相同级数展开项的情况下,本算法的相对误差要小于用幂级数求分数阶微分的算法。由于用傅立叶级数求分数阶微分的过程中涉及许多积分运算,故用幂级数的算法比本算法运算速度明显要快,实验证实了这一点。

为进一步验证本算法计算分数阶微积分的正确性,令整数 n 为 L 个分数 v_i 之和,即 $n=\sum_{i=1}^{L}v_i$。用本算法对信号 $s(t)=\cos t$ 进行 4 次 $v_i=0.25$ 阶微分与对 $s(t)$ 进行一次 1 阶微分进行比较,实验结果如图 3.5 所示。

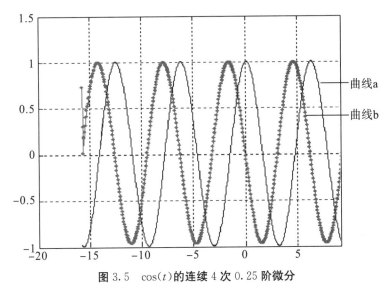

图 3.5 $\cos(t)$ 的连续 4 次 0.25 阶微分

图 3.5 中，曲线 a 是 $s(t)=\cos t$，曲线 b 是用本算法连续进行 4 次 0.25 阶微分所得的结果，它和解析解 $s^{(1)}t=\cos^{(1)}t=-\sin t$ 基本一致，具有可比性，从而证明了本算法计算分数阶微积分的能力。

3.6 信号分数阶微分基于 Grümwald-Letnikov 定义算法

3.6.1 理论分析

由 Grümwald-Letnikov 定义，利用组合数以及 Gamma 函数的性质对式（3.1）做如下变形：

$$_a^G D_t^v s(t) \stackrel{\Delta}{=} \lim_{h \to 0} s_h^{(v)}(t) \stackrel{\Delta}{=} \lim_{h \to 0} \frac{_a\Delta_h^v s(t)}{h^v} \tag{3.48}$$

式中，$_a\Delta_h^v s(t) = \sum_{k=0}^{\left[\frac{t-a}{h}\right]} w_k^v s(t-kh)$，$w_k^v = \frac{(-1)^k \Gamma(v+1)}{k!\Gamma(v-k+1)}$。由于 $\Gamma(x+1)=x\Gamma(x)=x!$，故 $\Gamma(k+1)=k\Gamma(k)=k!$，由式（3.48）可得 Grümwald-Letnikov 定义的分数阶导数的近似表达式为 $_a^G D_t^v s(t) \approx _a^G \widetilde{D}_t^v = \frac{_a\Delta_h^v s(t)}{h^v} = \frac{1}{h^v}\sum_{k=0}^{n} w_k^v s(t-kh)$。Podlubny 已经证明了 $_a^G \widetilde{D}_t^v$ 以同阶无穷小 $o(h)$ 的精度收敛于 $_a^G D_t^v s(t)$ [121]。

对于 Grümwald-Letnikov 定义的分数阶导数的一阶近似而言，权系数 w_k^v 恰好等于幂级数 $(1-x)^v$ 在原点的泰勒展开式的系数。在经典的导数定义中，信号 $s(t)$ 的一阶导数可用两点的后向差分方程近似表达[70,153]。对于 Grümwald-Letnikov 定义的分数阶微分的 2—6 阶高阶近似而言，Lubich 证明了下面的结论[155]：一阶导数的 n 阶近似表达式可用 $n+1$ 个点的后向差分表示，同样，$(n+1)$ 个点的后向差分表达式的 v 次幂也可表示 v 阶导数的 n 阶近似。由式（3.48）可得信号 $s(t)$ 的 v 阶微分为

$$s^v(t) = \frac{d^v}{dt^v} = \frac{n^v t^v}{\Gamma(-v)} \sum_{k=0}^{n-1} \frac{\Gamma(k-v)}{\Gamma(k+1)} s\left(t - \frac{kt}{n}\right) \tag{3.49}$$

式中，$n=\dfrac{t-a}{h}$ 是信号数据长度。

3.6.2 实验仿真及结果分析

为方便与上面两种算法比较，本实验仍以信号 $s(t)=\sin t$ 进行实验，取 $k=100$，$v=0.75$，实验结果如图 3.6 所示。

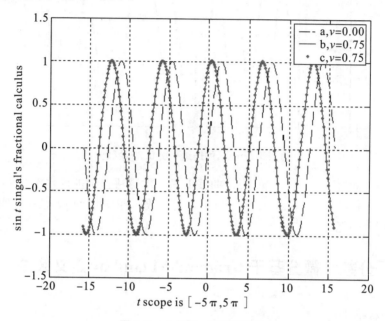

图 3.6 $\sin^{(0.75)}t$ 的实验结果

图 3.6 中，曲线 a 是 $s(t)=\sin t$，曲线 b 是 $s^{(0.75)}(t)=\sin^{(0.75)}t$ 的解析解，曲线 c 是用本算法求得的 $s^{(0.75)}(t)=\sin^{(0.75)}t$。可见，曲线 c 和 b 基本重合。与傅立叶级数法所得结果（见图 3.3）对比可知，本算法在时间 t 区域的两端计算结果的收敛性较好，没有发散的情况。虽然本算法仍需若干次迭代，计算速度因而受限，但明显快于前两种算法。

为了便于观测本算法的相对误差，对图 3.6 进行局部放大，如图 3.7 所示。

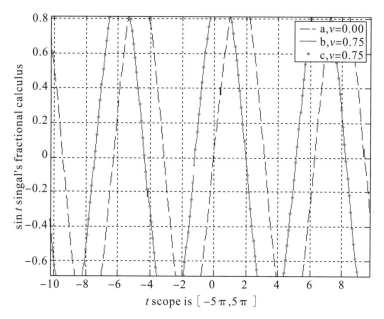

图 3.7　$\sin^{(0.75)}(t)$ 的均方误差

对比图 3.2、图 3.4 和图 3.6、图 3.7 可知，在同样的运算次数的情况下，本算法的相对误差明显小于前两种算法，本算法的收敛性较好。用本算法对锯齿波信号和方波信号进行分数阶微分，其计算结果分别如图 3.8、图 3.9 所示。

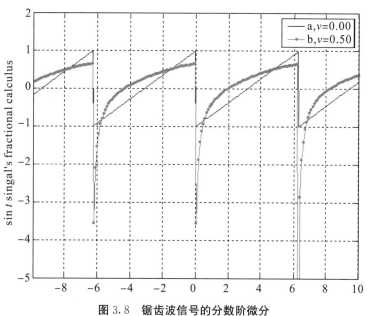

图 3.8　锯齿波信号的分数阶微分

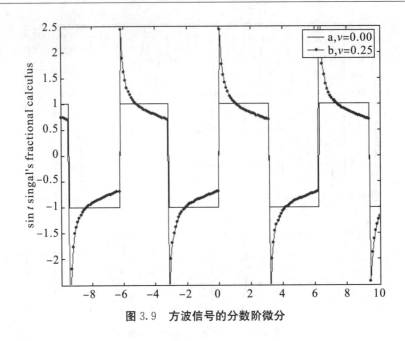

图 3.9 方波信号的分数阶微分

3.7 信号分数阶微分基于子波变换的算法

3.7.1 理论分析

如果 $\varphi(t) \in L^2(\mathbf{R}) \overset{FT}{\Longleftrightarrow} \hat{\varphi}(\omega)$ 是一个任意阶可微的低通信号,那么它的 v 阶分数阶微分为 $\varphi^{(v)}(t) = D_v[\varphi(t)]$,均能得到一个满足允许性条件的母波。由于需要计算奇异积分,因此在时域中计算 $\varphi^{(v)}(t) = D_v[\varphi(t)] = A_v P_v[\varphi(t)]$ 比较困难。即使是一个不光滑函数,对于那些具有良好时间局域化特征、低通紧支撑的信号 $\varphi(t) \in L^2(\mathbf{R})$,得到其频谱 $\hat{\varphi}(\omega)$ 是较容易的。对 $\hat{\varphi}_v(\omega) = (i\omega)^v \hat{\varphi}(\omega)$ 作 Fourier 反变换就可以得到 $\varphi^{(v)}(t)$。分数阶微积分的 Wavelet 变换域定义取 $\varphi(t)$ 为低通实偶信号。显然,其频谱 $\hat{\varphi}(\omega)$ 也是实偶函数,于是可得

$$\begin{aligned}\varphi^{(v)}(t) &= \frac{1}{2\pi}\int_{-\infty}^{\infty}(i\omega)^v\hat{\varphi}(\omega)\exp(i\omega t)\mathrm{d}\omega \\ &= \frac{1}{\pi}\int_0^{\infty}|\omega|^v\hat{\varphi}(\omega)\cos(\omega t + \frac{\pi v}{2})\mathrm{d}\omega\end{aligned} \quad (3.50)$$

只要 $\varphi(t)$ 在频域内也同样存在良好的频率局域化特性,便可使 $|\omega|^v \hat{\varphi}(\omega)$, $v \in \mathbf{R}^+$ 可积。将 $\varphi^{(v)}(t)$ 代入式 (3.16),求得能量规范化母波 $\psi_v(t)$。

为了构造出具有良好时间、频率局域化特性的能量规范化母波 $\psi_v(t)$,显然应要求 $\varphi(t)$ 也具有良好的时间、频率局域化特性。本节把低通光滑信号 $\varphi(t)$ 选取为测不准原理下具有最佳时—频局域化特征的高斯信号。

在式 (3.50) 中,$\varphi^{(v)}(t)$ 的积分区间为 $t \in [0, \infty]$,因此,在该无穷区间用极小微元代替积分自变量的微分,用离散积和来近似逼近连续积分的方法是行不通的。

本节以服从高斯分布的概率密度函数[156-159]来构造低通光滑信号 $\varphi(t)$。如果 $T \sim N$

(μ，σ^2)，令 $\varphi(t)$ 为连续型随机变量 T 的概率密度函数。令数学期望 $\mu=0$，即可使 $\varphi(t)$ 是一个实偶信号。另外，为使计算简便，取方差 $\sigma=1$，这样信号 $\varphi(t)$ 的曲线是标准正态分布 $T \sim N(0,1)$ 的曲线，即

$$\varphi(t) = \frac{1}{\sqrt{2\pi}} e^{-\frac{t^2}{2}}, \quad -\infty < t < \infty \tag{3.51}$$

$t=\pm 1$ 是信号 $\varphi(t)$ 的拐点。信号 $\varphi(t)$ 是以时间轴为渐进线的单峰速降实偶信号。显然，$\varphi(t)$ 满足 $\int_{-\infty}^{\infty} \varphi(t) \mathrm{d}t = 1$。

如前所述，分数阶微积分基于 Grümwald-Letnikov 定义的数值算法不论是在算法精度上还是在运算速度上都优于基于幂级数和傅立叶级数的算法。故本节选取基于 Grümwald-Letnikov 定义的数值算法来求解 $\varphi^{(v)}(t)$，如图 3.10 所示。

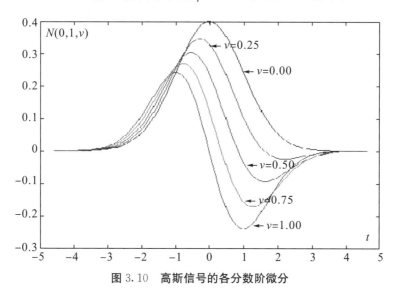

图 3.10 高斯信号的各分数阶微分

可见，高斯信号的分数阶微分结果类似于一个具有急速变化的高耸"正峰头"，然后拖着一个缓慢衰减的细长的"负尾巴"的"孤波"。高斯信号的分数阶微分结果在整个时域波形的积分为零（由于数值实验的叠代次数仅为 80 次，故图中高斯信号分数阶微分结果的"负尾巴"拖得还不够长，其波形的积分看似不为零），具有波动性。实验结果还表明，当 $v > \frac{1}{2}$ 时，随着 v 的增大，该"负尾巴"迅速缩短并逐渐消失。

由式（3.16）知，要得到能量规范化母波 $\psi_v(t)$，需要先求得其能量归一化因子 $E(v)$。由广义 Parseval 定理（$\langle x(t), y(t) \rangle = \frac{1}{2\pi} \langle \hat{X}(\Omega), \hat{Y}(\Omega) \rangle$），$\varphi(t) = \frac{1}{\sqrt{2\pi}} e^{-\frac{t^2}{2}} \overset{FT}{\Longleftrightarrow}$

$\hat{\varphi}(\omega) = e^{-\frac{\omega^2}{2}}$，$\varphi^{(v)}(t) \overset{FT}{\Longleftrightarrow} \hat{\varphi}_v(\omega) = (i\omega)^v \cdot e^{-\frac{\omega^2}{2}}$，可求得

$$E(v) = \left\{ \int_{-\infty}^{+\infty} \left| D_v[\varphi(t)] \right|^2 \mathrm{d}t \right\}^{\frac{1}{2}}$$

$$= \left\{ \frac{1}{2\pi} \int_{-\infty}^{+\infty} \left| \hat{\psi}_v(\omega) \right|^2 \mathrm{d}\omega \right\}^{\frac{1}{2}}$$

$$= \left\{ \frac{1}{2\pi} \int_{-\infty}^{\infty} |\omega|^{2v} \cdot e^{-\omega^2} d\omega \right\}^{\frac{1}{2}}$$

$$= \left\{ \frac{1}{\pi} \int_{0}^{\infty} \omega^{2v} \cdot e^{-\omega^2} d\omega \right\}^{\frac{1}{2}} \tag{3.52}$$

令 $\omega^2 = x$，故 $E(v) = \left\{ \frac{1}{2\pi} \int_{0}^{\infty} x^{v-\frac{1}{2}} e^{-x} dx \right\}^{\frac{1}{2}}$。该式中既包含线性项又包含指数项，所以只能用数值法（牛顿法或最速下降法）或图像法对其进行求解[70]。如果 $v = \frac{1}{2}$，可求得 $E(v) = \sqrt{\frac{1}{2\pi}} \approx 0.39894$，这与解析解基本一致。于是将能量规范化母波表示为 $\psi_v(t) = E^{-1}(v)\varphi_v(t) \overset{FT}{\Leftrightarrow} \hat{\psi}_v(\omega) = E^{-1}(v)(i\omega)^v \cdot e^{-\frac{\omega^2}{2}}$。

如第 2 章所述，计算机只能用离散的数值计算近似求解连续解析函数的计算式，为了最大限度地避免产生频谱混叠，子波变换在数值实现时其起始时间尺度因子 a_0 不可能趋近于零[72-76]。用计算机数值实现式（3.19）时，子波变换的起始尺度因子只能取到一个最佳的最小值 $a_0^{opt} > 0$，不能取 $a \to 0$。a_0^{opt} 的取值在第 2 章中已有系统的论述。

将 a_0^{opt} 代入式（2.3）得

$$\widetilde{\widetilde{W}}^{opt} s(m,n) = \frac{1}{\sqrt{na_0^{opt}}} \sum_k s[k] \psi_{v,a_0^{opt}} \left(\frac{k-m}{na_0^{opt}} \right) \tag{3.53}$$

式（3.19）不是 $\frac{0}{0}$ 型，将式（3.53）代入式（3.19）得

$$D_v[s(\tau)] \cong \lim_{a \to a_0^{opt}} \frac{E(v)Ws(\tau,a)}{a^v} = \frac{E(v)Ws(\tau,a_0^{opt})}{a_0^{vopt}} \tag{3.54}$$

式中，$Ws(\tau, a_0^{opt}) = \widetilde{\widetilde{W}}^{opt} s(m,n)$，$\psi_{v,a_0^{opt}} = \frac{1}{a_0^{opt}} \psi_v \left(\frac{t}{a_0^{opt}} \right) \overset{FT}{\Leftrightarrow} \hat{\psi}_{v,a_0^{opt}}(\omega) = E^{-1}(v)(i\omega)^v \cdot e^{-\frac{\omega^2}{2}}$。式（3.54）的计算结果与信号的分数阶微分的解析表达式（3.19）的值存在偏差，但这是用子波变换法在计算机上数值计算信号分数阶导数的最佳逼近值。

用 ω_{max} 表示 $\hat{\psi}_v(\omega)$ 的峰值频率，将 $\left| \frac{\hat{\psi}_v(\omega_b)}{\hat{\psi}_v(\omega_{max})} \right| = \frac{1}{\sqrt{2}}$ 的圆频率 ω_b 称为通带的截止圆频率。当 $|\omega| > \omega_b$ 时，信号幅度衰减很大，幅频特性逐渐下降，相频特性愈趋饱和[58,60]。由信号的时—频关系[59]可求得所取的母波 $\psi_v(t)$ 持续时间的近似范围为 $-4 \leqslant t' \leqslant 4$。由图 3.10 可知，其值是合理和可接受的。

3.7.2 实验仿真及结果分析

用本算法对 $s(t) = \text{sinc}(t)$ 进行分数阶微分，其计算结果如图 3.11 所示。

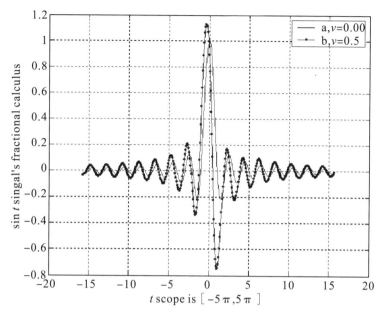

图 3.11 $\sin c^{(0.5)}t$ 的实验结果

对被加性高斯噪声污染了的余弦信号进行 $v=0.75$ 的分数阶微分,其计算结果如图 3.12 所示。

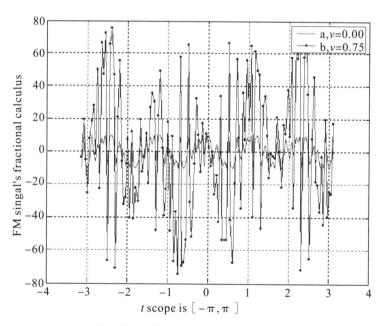

图 3.12 叠加高斯噪声的余弦信号的分数阶微分

由图 3.12 可知,分数阶微分和整数阶微分一样,具有加强信号的高频信息的作用。另外,分数阶微分对高频成分的加强程度没有整数阶微分那么强,在一定程度上保留了信号的包络,这就为信号的去噪提供了一种新的方法。关于这个问题笔者还在做进一步的研究。

通过以上的分析和实验可知,本算法的最大优点在于其运算速度远快于前面三种算法。理论分析发现原因在于,前面三种算法需要若干次迭代以计算信号的级数展开式各项,并且对于信号级数展开式各项都需要求其分数阶微分值,这又需要若干次迭代。而基于子波变换的分数阶微分的数值算法只需在计算能量规范化母波 $\psi_v(t)$ 时迭代计算 $\varphi(t)$ 的分数阶微分 $\varphi^{(v)}(t)$,而不涉及迭代求解信号 $s(t)$ 的级数各分量的分数阶微分。受此启发,只要我们能进一步减少求解分数阶微分运算的次数,就能进一步提高本算法的速度。

3.8 信号分数阶微分基于子波变换的快速工程算法

3.8.1 理论分析

如前所述,信号 $s(t)$ 的分数阶微分 $s^{(v)}(t)$ 具有连续性,即 $\lim_{v_1 \to v_2} D^{v_1} s(t) = D^{v_2} s(t)$,$v_1, v_2 \in \mathbf{R}$。换言之,$s^{(v)}(t)$ 是 v 在定义域内的连续内插。显然,只要我们利用保存的已事先求得的 $\varphi^{(v)}(t)$ 的先验知识表,在 $\varphi^{(v_1)}(t)$ 和 $\varphi^{(v_3)}(t)$($v_1 < v_2 < v_3$)之间就可连续内插得到 $\varphi^{(v_2)}(t)$。当信号的工程计算精度要求不高,并且 v_1 和 v_3 相差不大时,最简单的内插方法是取 $\varphi^{(v_2)}(t) = \frac{1}{2}[\varphi^{(v_1)}(t) + \varphi^{(v_3)}(t)]$,其中,$\varphi^{(v_1)}(t)$ 和 $\varphi^{(v_3)}(t)$ 的值查自预先求得的先验知识表。虽然在制作这张关于 $\varphi^{(v)}(t)$ 值的先验知识表时比较费力,但当用它来作连续内插时却避免了迭代求解 $\varphi^{(v)}(t)$。因此,本改进算法可进一步提高求解信号分数阶微积分的速度。

为适应计算机的数字运算,在作连续内插时,处理的应是 $\varphi^{(v)}(t)$ 的离散采样值 $\varphi^{(v)}(n)$,$n \in \mathbf{Z}$,然后对 $\varphi^{(v)}(n)$ 通过低通滤波来复原连续值 $\varphi^{(v)}(t)$。由式(3.51)知,$\varphi(t)$ 是标准正态高斯信号,其具有良好的时间、频率局域化特性。故 $\varphi^{(v)}(t)$ 也应该具有良好的时间、频率局域化特性[61]。因此,$\varphi^{(v)}(t)$ 满足抽样定理[58,60]。如前所述,$\hat{\varphi}_v(\omega) = (i\omega)^v \cdot e^{-\frac{\omega^2}{2}}$。若 $\omega_{\max}$ 是 $\hat{\varphi}_v(\omega)$ 的峰值频率,则把 $\left|\frac{\hat{\varphi}_v(\omega_b)}{\hat{\varphi}_v(\omega_{\max})}\right| = \frac{1}{\sqrt{2}}$ 的圆频率 ω_b 称为通带的截止圆频率。当 $\omega > 0$ 时,$\hat{\varphi}_v(\omega)$ 是单峰速降的,令 $\frac{\mathrm{d}}{\mathrm{d}\omega} \hat{\varphi}_v(\omega) = 0$,求得其峰值频率 $\omega_{\max} = \sqrt{v}$。故 $\omega_b^v \cdot e^{-\frac{\omega_b^2}{2}} = \frac{1}{\sqrt{2}} (\sqrt{v})^v e^{-\frac{v}{2}}$,显然,这是一个只能用数值法或图像法求解的方程,因为方程中既包含非线性项也包含指数项[70]。采用图像法求解,实验结果如图3.13所示。

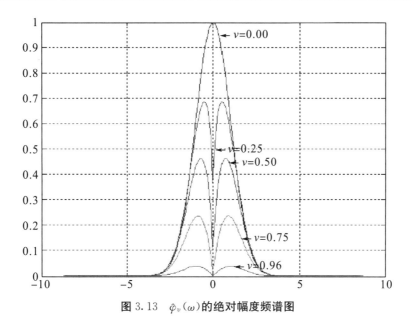

图 3.13 $\hat{\varphi}_v(\omega)$ 的绝对幅度频谱图

由图 3.13 可见，$\hat{\varphi}_v(\omega)$ 的截止圆频率 $\omega_b \leqslant 4$，取 $\omega_b = 4$。根据奈奎斯特抽样定理，对 $\varphi(t)$ 的抽样间隔为 $T_s \leqslant \dfrac{\pi}{\omega_b} = \dfrac{\pi}{4}$。为保证计算精度，采用饱和抽样。若取 $T_s = \dfrac{\pi}{20}$，即抽样频率为 $f_s = 2\pi \dfrac{1}{T_s} = 40$，此时，连续信号 $\varphi^{(v)}(t)$ 就可表示为

$$\varphi^{(v)}(t) = \sum_{n=0}^{N} \varphi^{(v)}(nT_s) \operatorname{sinc}(\omega_b t - n\pi)$$
$$= \sum_{n=0}^{N} \varphi^{(v)}\left(n\dfrac{\pi}{20}\right) \operatorname{sinc}(4t - n\pi) \tag{3.55}$$

式中，$\varphi^{(v_2)}(nT_s) = \dfrac{1}{2}[\varphi^{(v_1)}(nT_s) + \varphi^{(v_3)}(nT_s)]$。实际上不存在理想的低通滤波器，由于混叠效应和泄漏效应的影响，经低通滤波所得的信号只是 $\varphi^{(v)}(t)$ 的近似。可以根据需要将误差限制在工程精度许可的范围内。本改进的工程算法的其余各步均与未改进的基于子波变换的算法相同。

3.8.2 实验仿真及结果分析

若已知 $\varphi^{(0.4)}\left(n\dfrac{\pi}{20}\right)$ 和 $\varphi^{(0.6)}\left(n\dfrac{\pi}{20}\right)$ 的先验值，采用最简单的插值法 $\varphi^{(0.5)}\left(n\dfrac{\pi}{20}\right) = \dfrac{1}{2}\left[\varphi^{(0.4)}\left(n\dfrac{\pi}{20}\right) + \varphi^{(0.6)}\left(n\dfrac{\pi}{20}\right)\right]$，实验结果如图 3.14 所示。

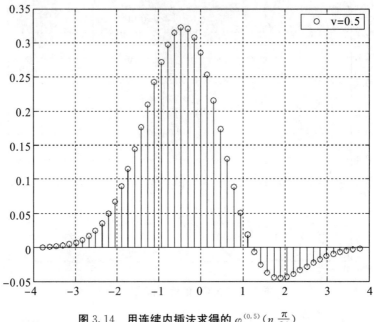

图 3.14 用连续内插法求得的 $\varphi^{(0.5)}(n\frac{\pi}{20})$

然后根据式（3.55）进行低通滤波恢复得 $\varphi^{(0.5)}(t)$，实验结果如图 3.15 所示。

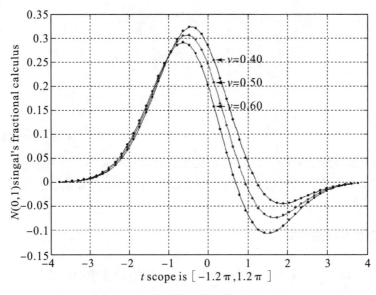

图 3.15 低通滤波恢复得 $\varphi^{(0.5)}(t)$

图 3.15 中，圆点表示采样点。与图 3.10 对比可知，低通滤波恢复所得的 $\varphi^{(0.5)}(t)$ 只是 $\varphi^{(0.5)}(t)$ 解析解的近似。

用本改进算法对 $\sin t$ 进行分数阶微分，如图 3.16 所示。曲线 a 是 $s(t)=\sin t$，曲线 b 是本改进算法求得的 $s(t)^{(0.75)}$，曲线 c 是解析法求得的 $s(t)^{(0.75)}$。图中，曲线 b 和 c 基本一致，存在较小相对误差的原因是子波变换的尺度最小只能取起始尺度的最佳取值 a_0^{opt} >0，而不可能无限逼近于零。在具体实现时，还需注意子波变换的最佳扫描时间 τ（即最佳观测时间）的范围。

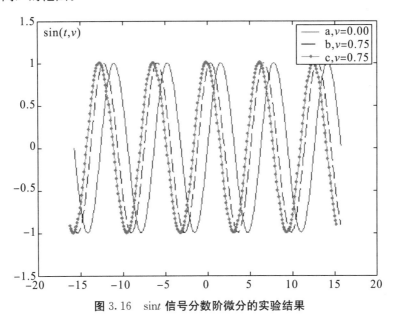

图 3.16 sin$t$ 信号分数阶微分的实验结果

用本改进算法分别对对称三角波、Dirichlet 信号进行了分数阶微分计算，其结果如图 3.17、图 3.18 所示。

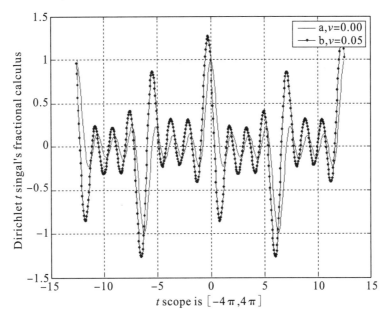

图 3.17 Dirichlet 信号分数阶微分的实验结果

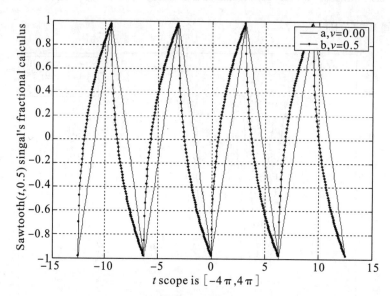

图 3.18　对称三角波信号分数阶微分的实验结果

实验证明，本改进的基于子波变换的快速工程算法不仅具有很高的运算速度，而且具有一定的计算准确性。

3.9　小结

分数阶微积分在信息科学领域中，特别是在现代信号处理方面的应用研究国内外都才刚刚兴起。本章的工作只是对现代信号分析与处理中分数阶微积分快速数值实现作了一些初步尝试，还有许多问题有待进一步研究。目前，笔者正在对分数阶微分算子序贯法等其他数值算法进行研究。

第4章　分数阶微积分的模拟分抗电路实现

本章的研究目的是用普通的模拟无源电路元件构造出实现任意分数阶微积分的模拟分抗电路。首先，本章分析了用普通的无源电路元件构造出实现任意分数阶微积分的模拟分抗电路的电路特性及其可能性；分析了一种目前国际上经典的1/2分数阶树型模拟分抗电路。其次，提出并推导了三种1/2分数阶演算的模拟分抗电路；分析比较了作者提出的三种1/2阶模拟分抗电路与国际上经典的1/2阶树型模拟分抗电路之间的优劣，论述了网格型1/2阶递归模拟分抗电路在电路结构上是这四种1/2阶分抗电路中最优的一种；在此基础上，提出并论述了一种实现任意分数阶演算的递归模拟分抗电路模型，并以任意分数阶递归网格型模拟分抗电路模型为例进行了分析。最后，通过仿真实验探讨了分数阶演算在通信与信息处理应用中有待进一步研究的热点问题。本章的研究结论是进一步理论研究和工程实现构造任意分数阶模拟分抗电路的基础。

4.1　问题提出

正如第3章的问题提出部分所述，怎样构造模拟分抗电路用以实现信号的任意分数阶微积分是将分数阶微积分应用于现代信号分析与处理中遇到的首要基础性问题之一。目前，构造模拟分抗电路的方法大体上可分为用无源电路元件和用有源电路元件来构造[77,86−93,123,130−143,160−172]。其中，如何用普通的模拟无源电路元件构造出任意分数阶模拟分抗电路是将来制造集成化的模拟分抗电路元件芯片的基础。本章的主要研究内容就围绕这个课题展开。目前，用普通的模拟电路元件（如电阻、电容）来构造任意阶的模拟的方法主要有两种：一种是采用诸如牛顿法来进行逼近；另一种是构造具有高度自相似性的递归电路。逼近的方法一般采用链式电路来达到逼近任意分数阶微积分的效果。用这种方法求得其中的各电路元件一般都是4位以上的小数，这在实际的物理电路中是难以实现的，这就是逼近法最大的缺点[138,169−170]。构造高度自相似递归电路的方法通常使得构造出的电路的元件数与电路递归级数成指数关系增加，这样就极大地增加了制造成本。另外，由于电路具有高度自相似性，这也给其电路制造带来了一定的难度[138,171−174]。

对于用无源模拟电路元件来构造模拟分抗电路而言，怎样构造出结构简单易于制造的模拟分抗电路模型，国内外的学者都还研究不多。本章的研究内容和目的主要围绕怎样构造出结构简单易于制造的无源模拟分抗电路模型展开。笔者的研究表明[171−174]，除了国际上经典的1/2分数阶树型模拟分抗电路模型外，还可以找到其他类型的1/2分数阶模拟分抗电路模型。在保持同样的电路特性的情况下，笔者提出的网格型链式1/2分数阶模拟分抗电路模型在电路结构上远优于国际上经典的1/2分数阶树型模拟分抗电路模型。另外，可以用1/2分数阶模拟分抗电路通过递归嵌套的方式构造出任意分数阶模拟分抗电路。通过以上的研究可以发现，既然用普通的电路元件电阻、电容或电感按照某种高度自相似的分形结构就能够找出分抗电路，那么从这个意义上讲，分抗（fractance）可以理解为分数阶的阻抗。由于目前在国际上还没有一个用以表示分抗电路元件的符号，因此，笔者建议把电容和电阻的符号相结合，将分抗电路元件的符号表示成 ⊣⊢ 或 ⌐W⌐，其中，F 是 fractance 的缩写。

4.2 模拟分抗电路的阻抗特性

本章将模拟分抗电路的阻抗特性简称为分抗特性[171-174]。

4.2.1 一阶 $R-C$ 电路的微分特性

若 $\forall v \in \mathbf{R}^+$（包括分数），令其整数部分为 $[v]$，其小数部分为 $v-[v]$，根据算子理论，$D^v = D^{[v]} D^{v-[v]}$。由于整数阶微积分的理论和实现已经非常成熟，故本章只需要讨论 $|v| < 1$ 的情况。

由于 $R-C$ 电路具有一阶微积分的功能，于是很自然地联想到能否用电阻 R 和电容 C 的组合来构造出模拟分抗电路元件。一阶 $R-C$ 串联电路如图 4.1 所示。

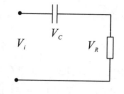

图 4.1 一阶 $R-C$ 串联电路

显然，$V_R = R \cdot i_C = RC \dfrac{\mathrm{d}}{\mathrm{d}t} V_C(t) = RC \dfrac{\mathrm{d}}{\mathrm{d}t} [V_i - V_R]$。当激励和响应为复简谐信号或任意信号的频谱函数时，正弦稳态电路产生的输入正弦型信号的角频率为 $\bar{\omega}$，根据分压原理得

$$V_R(t) = \frac{R}{\dfrac{1}{i\bar{\omega}C} + R} \cdot V_i(t)$$

$$= \frac{\bar{\omega}RC}{\sqrt{1+\bar{\omega}^2 R^2 C^2}} \cdot \mathrm{e}^{\mathrm{i}\arcsin\frac{1}{\sqrt{1+\bar{\omega}^2 R^2 C^2}}} \cdot V_i(t) \quad (4.1)$$

式中，i 为虚数单位。令 $\varphi = \arcsin \dfrac{1}{\sqrt{1+\bar{\omega}^2 R^2 C^2}}$，得 $V_R(t) = \cos\varphi \cdot \mathrm{e}^{\mathrm{i}\varphi} \cdot V_i(t)$。对于时间 t 而言，$\cos\varphi \cdot \mathrm{e}^{\mathrm{i}\varphi}$ 是常数，所以其傅立叶变换为

$$\hat{V}_R(\omega) = \frac{1}{2\pi} \cdot 2\pi \cos\varphi \cdot \mathrm{e}^{\mathrm{i}\varphi} \delta(\omega) * \hat{V}_i(\omega)$$

$$= \cos\varphi \cdot \mathrm{e}^{\mathrm{i}\varphi} \cdot \hat{V}_i(\omega) \quad (4.2)$$

显然，$\cos\varphi \cdot \mathrm{e}^{\mathrm{i}\varphi}$ 为该电路系统的传递函数。由式（3.7）、（3.8）知，若欲用一阶 $R-C$ 电路来实现分数阶演算功能，则必须满足 $\hat{d}(\omega) = (\mathrm{i}\omega)^v = \cos\varphi \cdot \mathrm{e}^{\mathrm{i}\varphi}$，故有

$$\begin{cases} |\omega|^v = \cos\varphi \\ \theta(\omega) = \dfrac{\pi v}{2} = \varphi \end{cases} \quad (4.3)$$

该方程含有两个变量 ω 和 v。根据线性方程组的性质，若适当调整一阶 $R-C$ 电路中的 R 和 C，可使该方程满秩，此时有且仅有一解，而且这个解只可能在 $0 \leqslant \omega \leqslant 1$ 的低频段出现，如图 4.2 所示。

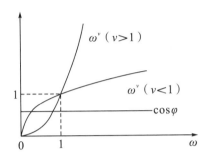

图 4.2　一阶 $R-C$ 电路传递函数和 $(i\omega)^v$ 的关系

无论分数阶微分还是整数阶微分，都是针对频域中信号频谱覆盖的所有复频率 ω 而言的。因此，如果用电路来实现分数阶演算，就必须要求该电路对信号频谱覆盖的所有复频率 ω 都有效。由以上分析可知，用单纯的一阶 $R-C$ 电路来实现分数阶演算是不可能的。

那么，能否用 $R-C$ 一阶微积分电路的组合构造来逼近分数阶微积分呢？由第 3 章可知，在区间 $[-l, l]$ 上分数阶微分的 $R-C$ 电路实现显然具有普适性。由式（3.40）可知，用 $R-C$ 电路来逼近分数阶微分，首先就需用 $R-C$ 电路来求出函数或信号的傅立叶级数的前 n 项的系数，这样可以得到函数或信号用纯频率信号的近似逼近。由式（3.35）、（3.36）、（3.37）可知，纯频率信号的分数阶微分实质上是其自身的广义希尔伯特变换。因此，信号的分数阶微分就是对傅立叶级数各项的广义希尔伯特变换求和。广义希尔伯特变换已有许多学者研究了它的实现，这里不再详述，那么问题的关键就集中到了如何用 $R-C$ 电路的并联来实现傅立叶级数的系数的运算。由于傅立叶系数 a_0 是常数，用 Riemann-Liouville 定义的分数阶微分对 a_0 求导后其值为零，所以在实现时不予考虑它。于是，我们似乎可以实现如图 4.3 所示的任意分数阶微积分的 $R-C$ 电路模型。

在图 4.3 所示的模型中，n 阶近似总共需要 $4n+1$ 个乘法器，$2n$ 个一阶 $R-C$ 积分电路，$n-1$ 个分数倍频器，n 个 $\dfrac{\pi}{2}$ 频移器，$2n$ 个 $\dfrac{\pi v}{2}$ 频移器。该电路中的积分电路是由 $R-C$ 积分电路构成的。虽然本算法是以高阶无穷小 $o(h)$ 收敛的，但该电路中的分数倍频器和 $\dfrac{\pi v}{2}$ 的广义希尔伯特变换器在实际电路中是难以实现的。

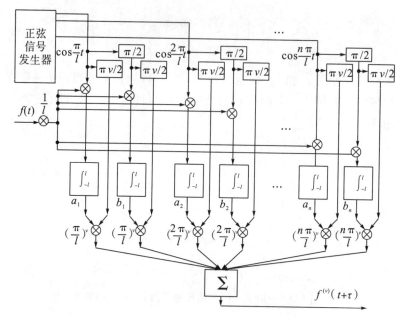

图 4.3 任意分数阶微积分的 $R-C$ 实现电路模型假设一

另外，由 Γ 函数的定义式显然可得

$$\int_0^1 \left(\log \frac{1}{x}\right)^{p-1} \mathrm{d}x = \Gamma(p) \tag{4.4}$$

因此，我们可用 $R-C$ 积分电路来得到 Γ 函数的值。令 $n = \left[\dfrac{t-a}{h}\right]$，由式（3.49），于是我们似乎可以实现如图 4.4 所示的任意分数阶微积分的 $R-C$ 电路模型。

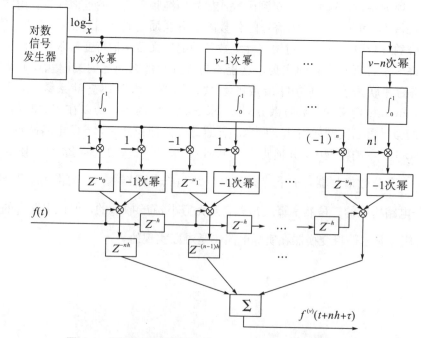

图 4.4 任意分数阶微积分的 $R-C$ 实现电路模型假设二

在图 4.4 所示的模型中，分数阶微积分的 n 阶近似总共需要 $3n$ 个乘法器，n 个一阶 $R-C$ 积分电路，$2n-1$ 个延时器，n 个 $v-i$（$0<i<n$）次幂运算器，n 个 -1 次幂运算器。其中，延时 $Z^{-u_{k-1}}$，$k \in [0, n]$ 的目的是抵消对数信号 $\log \frac{1}{x}$ 的 v 次幂比较 $v-k$ 次幂与 -1 次幂运算所需的时差。虽然本算法理论上以同阶无穷小 $o(h)$ 收敛，但是该电路中对数信号发生器和幂次运算器在实际电路中是难以实现的。

那么，还能否用普通的电阻 R 和电容 C 来构造模拟分抗电路，以近似实现分数阶微积分呢？进一步研究发现，这类分抗电路都具有高度自相似性的分形结构[138,171-174]。

4.2.2 模拟分抗电路的阻抗特性

电容、电阻和电感都是我们普遍使用的电子器件。这类电路元件的阻抗通常可表示为 $Z \propto (i\omega)^m$，其中 ω 是角频率。一般而言，$m \in \{-1, 0, +1\}$，当 m 分别取 -1，0，-1 时，Z 依次表示电容、电阻和电感。但是，模拟分抗电路的输入阻抗 Z 中的 m 就不一定属于集合 $\{-1, 0, +1\}$，它应该可以为分数。这类模拟分抗电路的阻抗特性可以表示为

$$Z \propto (i\omega)^\alpha \exp(i\theta) \tag{4.5}$$

式中，α 一般为分数，θ 为相位常数，θ 是一个和频率相互独立的量。

4.3 构造 $\frac{1}{2}$ 阶分数阶微积分的模拟分抗电路

4.3.1 经典的树型 $\frac{1}{2}$ 阶模拟分抗电路

一种经典的树型 $\frac{1}{2}$ 阶模拟分抗电路模型[138,171-174]如图 4.5 所示。

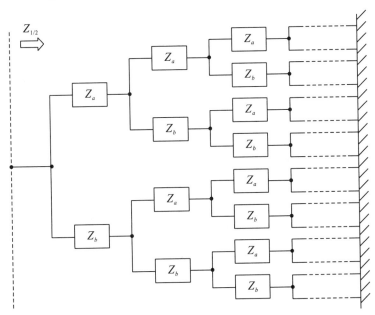

图 4.5 树型 $\frac{1}{2}$ 阶模拟分抗电路

该分抗电路由阻抗 Z_a 和 Z_b 构成树型无限递归结构，该电路具有高度自相似性，其等效电路如图 4.6 所示。

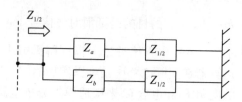

图 4.6 树型 $\frac{1}{2}$ 阶模拟分抗等效电路

由电路的串并联法则可得

$$Z_{1/2} = (Z_a Z_b)^{\frac{1}{2}} \tag{4.6}$$

假设电路元件中的初始储能为零，且在 Laplace 变换域中表示阻抗时，取 $Z_a = r$ 为电阻，$Z_b = 1/Sc$ 为电容，其中，S 为 Laplace 算子。此时，该 $\frac{1}{2}$ 阶模拟分抗电路的输入阻抗为

$$Z_{1/2} = \xi^{\frac{1}{2}} S^{-\frac{1}{2}} \tag{4.7}$$

式中，$\xi = \dfrac{r}{c}$。由式（4.7）可得输出电压 $V(S)$ 和输入电流 $I(S)$ 的关系为

$$V(S) = \xi^{\frac{1}{2}} S^{-\frac{1}{2}} I(S) \tag{4.8}$$

对式（4.8）作 Laplace 反变换，可得

$$v(t) = \frac{\xi^{\frac{1}{2}}}{\Gamma(1/2)} \int_{-\infty}^{t} \frac{i(\tau)}{(t-\tau)^{1/2}} \mathrm{d}\tau \tag{4.9}$$

式中，Gamma 函数 $\Gamma(\alpha) = \int_0^\infty \mathrm{e}^{-x} x^{\alpha-1} \mathrm{d}x = (\alpha - 1)!$。由分数阶积分的 Caputo 定义式（3.5）可知，式（4.9）表明 $v(t)$ 正比于 $i(t)$ 的 $\frac{1}{2}$ 阶积分；反之，可得

$$I(S) = \xi^{-\frac{1}{2}} S^{\frac{1}{2}} V(S) \tag{4.10}$$

对式（4.10）作 Laplace 反变换，可得

$$i(t) = \frac{1}{\xi^{\frac{1}{2}} \Gamma(1/2)} \int_{-\infty}^{t} \frac{v'(\tau)}{(t-\tau)^{1/2}} \mathrm{d}\tau \tag{4.11}$$

由分数阶微积分的 Caputo 定义式（3.5）可知，式（4.11）表明 $i(t)$ 正比于 $v(t)$ 的 $\frac{1}{2}$ 阶微分。

4.3.2 两回路串联的 $\frac{1}{2}$ 阶模拟分抗电路

笔者经研究提出了一种两回路串联的 $\frac{1}{2}$ 阶模拟分抗电路模型[171-174]，该分抗电路由阻抗 Z_a 和 Z_b 构成无限递归两回路串联连接结构，它具有高度自相似性，如图 4.7 所示。

第 4 章 分数阶微积分的模拟分抗电路实现

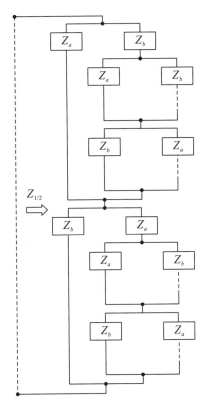

图 4.7 两回路串联的 $\frac{1}{2}$ 阶模拟分抗电路

其等效电路如图 4.8 所示。

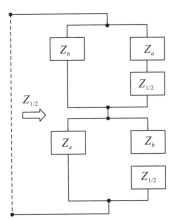

图 4.8 两回路串联的 $\frac{1}{2}$ 阶模拟分抗等效电路

由电路的串并联法则可得

$$Z_{1/2} = \sqrt{2}\,(Z_a Z_b)^{\frac{1}{2}} \tag{4.12}$$

假设电路元件中的初始储能为零,且在 Laplace 变换域中表示阻抗时,取 $Z_a = r$ 为电阻,$Z_b = 1/Sc$ 为电容,其中,S 为 Laplace 算子。此时,该 $\frac{1}{2}$ 阶模拟分抗电路的输入阻

抗为

$$Z_{1/2} = \sqrt{2}\xi^{\frac{1}{2}}S^{-\frac{1}{2}} \tag{4.13}$$

式中，$\xi = \dfrac{r}{c}$。由式（4.13）可得输出电压 $V(S)$ 和输入电流 $I(S)$ 的关系为

$$V(S) = \sqrt{2}\xi^{\frac{1}{2}}S^{-\frac{1}{2}}I(S) \tag{4.14}$$

对式（4.14）作 Laplace 反变换，可得

$$v(t) = \frac{\sqrt{2}\xi^{\frac{1}{2}}}{\Gamma(1/2)}\int_{-\infty}^{t}\frac{i(\tau)}{(t-\tau)^{1/2}}\mathrm{d}\tau \tag{4.15}$$

式中，Gamma 函数 $\Gamma(\alpha) = \int_0^{\infty} e^{-x} x^{\alpha-1} \mathrm{d}x = (\alpha-1)!$。由分数阶积分的 Caputo 定义式可知，式（4.15）表明 $v(t)$ 正比于 $i(t)$ 的 $\dfrac{1}{2}$ 阶积分；反之，可得

$$I(S) = \frac{1}{\sqrt{2}}\xi^{-\frac{1}{2}}S^{\frac{1}{2}}V(S) \tag{4.16}$$

对式（4.16）作 Laplace 反变换，可得

$$i(t) = \frac{1}{\sqrt{2}\xi^{\frac{1}{2}}\Gamma(1/2)}\int_{-\infty}^{t}\frac{v'(\tau)}{(t-\tau)^{1/2}}\mathrm{d}\tau \tag{4.17}$$

由分数阶微分的 Caputo 定义式可知，式（4.17）表明 $i(t)$ 正比于 $v(t)$ 的 $\dfrac{1}{2}$ 阶微分。

4.3.3　H 型 $\dfrac{1}{2}$ 阶模拟分抗电路

笔者经研究提出了一种 H 型的 $\dfrac{1}{2}$ 阶模拟分抗电路模型[171-174]，该分抗电路由阻抗 Z_a 和 Z_b 构成 H 型无限递归结构，它具有高度自相似性，如图 4.9 所示。

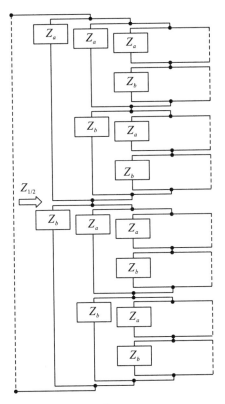

图 4.9　H 型 $\frac{1}{2}$ 阶模拟分抗电路

其等效电路如图 4.10 所示。

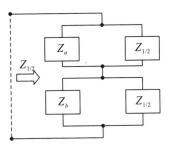

图 4.10　H 型 $\frac{1}{2}$ 阶模拟分抗等效电路

由电路的串并联法则可得

$$Z_{1/2} = (Z_a Z_b)^{\frac{1}{2}} \tag{4.18}$$

与前述同理，可以证明，输出电压 $v(t)$ 正比于输入电流 $i(t)$ 的 $\frac{1}{2}$ 阶积分；反之，输入电流 $i(t)$ 正比于输出电压 $v(t)$ 的 $\frac{1}{2}$ 阶微分。

4.3.4 网格型 $\frac{1}{2}$ 阶模拟分抗电路

笔者经研究提出了一种网格型 $\frac{1}{2}$ 阶模拟分抗电路模型[171-174]，该分抗电路由阻抗 Z_a 和 Z_b 构成无限重复的网格结构，它具有高度自相似性，如图 4.11 所示。

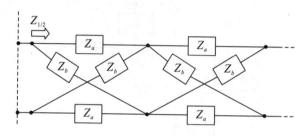

图 4.11 网格型 $\frac{1}{2}$ 阶模拟分抗电路

其等效电路如图 4.12 所示。

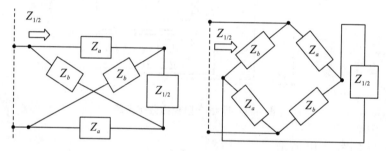

图 4.12 网格型 $\frac{1}{2}$ 阶模拟分抗等效电路

令阻抗 Z_a，Z_b 上流过的电流分别为 i_a，i_b，输入电压为 V_i。根据基尔霍夫电流定律（KCL）和基尔霍夫电压定律（KVL），可得

$$\begin{cases} Z_a i_a + Z_b i_b = V_i \\ (Z_a + Z_{1/2})i_a - (Z_{1/2} + Z_b)i_b = 0 \end{cases} \quad (4.19)$$

由线性代数中的克莱姆法则，可得

$$\begin{cases} i_a = \dfrac{(Z_b + Z_{1/2})V_i}{\begin{vmatrix} Z_a + Z_{1/2} & -(Z_{1/2} + Z_b) \\ Z_a & Z_b \end{vmatrix}} \\ i_b = \dfrac{(Z_a + Z_{1/2})V_i}{\begin{vmatrix} Z_a + Z_{1/2} & -(Z_{1/2} + Z_b) \\ Z_a & Z_b \end{vmatrix}} \end{cases} \quad (4.20)$$

因此，端口总电阻 $Z_{1/2}$ 为

$$Z_{1/2} = \frac{V_i}{i_a + i_b} = \frac{2Z_a Z_b + Z_{1/2}(Z_a + Z_b)}{2Z_{1/2} + Z_a + Z_b} \quad (4.21)$$

于是解得

$$Z_{1/2} = (Z_a Z_b)^{\frac{1}{2}} \quad (4.22)$$

若电路元件中的初始储能为零,且在 Laplace 变换域中表示阻抗时,取 $Z_a=r$ 为电阻,$Z_b=1/Sc$ 为电容,其中,S 为 Laplace 算子。此时,该 $\frac{1}{2}$ 阶模拟分抗电路的输入阻抗约为

$$Z_{1/2} = \xi^{\frac{1}{2}} S^{-\frac{1}{2}} \tag{4.23}$$

式中,$\xi=\frac{r}{c}$。由式(4.23)可得输出电压 $V(S)$ 和输入电流 $I(S)$ 的关系为

$$V(S) = \xi^{\frac{1}{2}} S^{-\frac{1}{2}} I(S) \tag{4.24}$$

对式(4.24)作 Laplace 反变换,可得

$$v(t) = \frac{\xi^{\frac{1}{2}}}{\Gamma(1/2)} \int_{-\infty}^{t} \frac{i(\tau)}{(t-\tau)^{1/2}} \mathrm{d}\tau \tag{4.25}$$

由分数阶积分的 Caputo 定义可知,式(4.25)表明 $v(t)$ 正比于 $i(t)$ 的 $\frac{1}{2}$ 阶积分;反之,可得

$$I(S) = \xi^{-\frac{1}{2}} S^{\frac{1}{2}} V(S) \tag{4.26}$$

对式(4.26)作 Laplace 反变换,可得

$$i(t) = \frac{1}{\xi^{\frac{1}{2}} \Gamma(1/2)} \int_{-\infty}^{t} \frac{v'(\tau)}{(t-\tau)^{1/2}} \mathrm{d}\tau \tag{4.27}$$

由分数阶微分的 Caputo 定义可知,式(4.27)表明 $i(t)$ 正比于 $v(t)$ 的 $\frac{1}{2}$ 阶微分。

4.3.5 分析比较四种 $\frac{1}{2}$ 阶模拟分抗电路

首先,比较图 4.5、图 4.6、图 4.7、图 4.8 可知,两回路串联的 $\frac{1}{2}$ 阶模拟分抗电路在结构上与经典的树型 $\frac{1}{2}$ 阶模拟分抗电路复杂程度相同,两种模型同样具有无限递归结构。但是,两回路串联的 $\frac{1}{2}$ 阶模拟分抗电路模型所用的电路元件(Z_a 或 Z_b)却是树型 $\frac{1}{2}$ 阶模拟分抗电路模型的两倍。

其次,比较图 4.5、图 4.6、图 4.9、图 4.10 可知,H 型 $\frac{1}{2}$ 阶模拟分抗电路在结构上与经典的树型 $\frac{1}{2}$ 阶模拟分抗电路复杂程度相同,两种模型同样具有无限递归结构。同时,H 型的 $\frac{1}{2}$ 阶模拟分抗电路模型所用的电路元件(Z_a 或 Z_b)和树型 $\frac{1}{2}$ 阶模拟分抗电路模型相同。

最后,比较图 4.5、图 4.6、图 4.11、图 4.12 可知,由于网格型 $\frac{1}{2}$ 阶模拟分抗电路只是无限重复同样的交叉网格结构,故该电路模型中所用阻抗 Z_a 和 Z_b 的个数与电路层数只是成一个 2 倍数关系,然而经典的树型 $\frac{1}{2}$ 阶模拟分抗电路模型所用的电路元件是按几

何级数增加的。同时，两者完成的功能是相同的。显然，网格型 $\frac{1}{2}$ 阶模拟分抗电路模型比树型 $\frac{1}{2}$ 阶模拟分抗电路模型简洁易行得多。可见，$\frac{1}{2}$ 阶网格型模拟分抗电路是这四种 $\frac{1}{2}$ 阶分抗电路模型中最优的。

4.4 构造 $\frac{1}{2^n}$ 阶模拟分抗电路

不失一般性，本节以网格型模拟分抗电路为例进行推导[171-174]。

4.4.1 $\frac{1}{4}$ 阶模拟分抗电路

在前面所述的 $\frac{1}{2}$ 阶模拟分抗电路模型的基础上，我们研究构造得到一种 $\frac{1}{4}$ 阶模拟分抗电路模型。该分抗电路由阻抗 Z_a 和 $\frac{1}{2}$ 阶模拟分抗电路 $Z_{1/2}$ 构成无限重复的交叉平衡电桥结构，如图 4.13 所示。

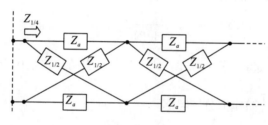

图 4.13 网格型 $\frac{1}{4}$ 阶模拟分抗电路

其等效电路如图 4.14 所示。

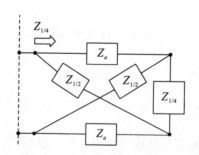

图 4.14 网格型 $\frac{1}{4}$ 阶模拟分抗等效电路

由电路的串并联法则可得

$$Z_{1/4} = (Z_a Z_{1/2})^{\frac{1}{2}} = (r\xi^{\frac{1}{2}} S^{-\frac{1}{2}})^{\frac{1}{2}} = r^{\frac{1}{2}} \xi^{\frac{1}{4}} S^{-\frac{1}{4}} \quad (4.28)$$

于是，可得输出电压 $V(S)$ 和输入电流 $I(S)$ 的关系为

$$V(S) = r^{\frac{1}{2}} \xi^{\frac{1}{4}} S^{-\frac{1}{4}} I(S) \quad (4.29)$$

对式（4.29）作 Laplace 反变换，可得

$$v(t) = \frac{r^{\frac{1}{2}}\xi^{\frac{1}{4}}}{\Gamma(1/4)}\int_{-\infty}^{t}\frac{i(\tau)}{(t-\tau)^{3/4}}d\tau \tag{4.30}$$

式（4.30）表明 $v(t)$ 正比于 $i(t)$ 的 $\frac{1}{4}$ 阶积分；反之，可得

$$I(S) = r^{-\frac{1}{2}}\xi^{-\frac{1}{4}}S^{\frac{1}{4}}V(S) \tag{4.31}$$

对式（4.31）作 Laplace 反变换，可得

$$i(t) = \frac{1}{r^{\frac{1}{2}}\xi^{\frac{1}{2}}\Gamma(3/4)}\int_{-\infty}^{t}\frac{v'(\tau)}{(t-\tau)^{1/4}}d\tau \tag{4.32}$$

式（4.32）表明 $i(t)$ 正比于 $v(t)$ 的 $\frac{1}{4}$ 阶微分。

4.4.2 $\frac{1}{2^n}$ 阶模拟分抗电路

下面用数学归纳法求证 $\frac{1}{2^n}$ 阶模拟分抗电路模型的分数阶微积分规律。该分抗电路由阻抗 Z_a 和 $\frac{1}{2^{n-1}}$ 阶模拟分抗电路 $Z_{1/2^{n-1}}$ 构成无限重复的交叉平衡电桥结构，其等效电路如图 4.15 所示。

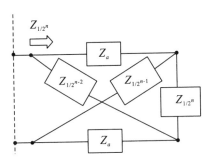

图 4.15 网格型 $\frac{1}{2^n}$ 阶模拟分抗等效电路

令 $Z_{1/2^{n-1}} = r^{\frac{2^{n-2}-1}{2^{n-2}}}\xi^{\frac{1}{2^{n-1}}}S^{-\frac{1}{2^{n-1}}}$，由电路的串并联法则可得

$$Z_{1/2^n} = (Z_a Z_{1/2^{n-1}})^{\frac{1}{2}} = r^{\frac{2^{n-1}-1}{2^{n-1}}}\xi^{\frac{1}{2^n}}S^{-\frac{1}{2^n}} \tag{4.33}$$

于是，可得输出电压 $V(S)$ 和输入电流 $I(S)$ 的关系为

$$V(S) = r^{\frac{2^{n-1}-1}{2^{n-1}}}\xi^{\frac{1}{2^n}}S^{-\frac{1}{2^n}}I(S) \tag{4.34}$$

对式（4.34）作 Laplace 反变换，可得

$$v(t) = \frac{r^{\frac{2^{n-1}-1}{2^{n-1}}}\xi^{\frac{1}{2^n}}}{\Gamma(1/2^n)}\int_{-\infty}^{t}\frac{i(\tau)}{(t-\tau)^{1/2^n+1}}d\tau \tag{4.35}$$

式（4.35）表明 $v(t)$ 正比于 $i(t)$ 的 $\frac{1}{2^n}$ 阶积分；反之，可得

$$I(S) = r^{\frac{1-2^{n-1}}{2^{n-1}}}\xi^{-\frac{1}{2^n}}S^{\frac{1}{2^n}}V(S) \tag{4.36}$$

对式（4.36）作 Laplace 反变换，可得

$$i(t) = \frac{1}{r^{\frac{2^{n-1}-1}{2^{n-1}}} \xi^{\frac{1}{2^n}} \Gamma(1-1/2^n)} \int_{-\infty}^{t} \frac{v'(\tau)}{(t-\tau)^{1/2^n}} d\tau \qquad (4.37)$$

式（4.37）表明 $i(t)$ 正比于 $v(t)$ 的 $\frac{1}{2^n}$ 阶微分。

4.4.3 分析 $\frac{1}{2^n}$ 阶模拟分抗电路

由以上分析可知，$\frac{1}{2^n}$ 阶模拟分抗电路其实是在 $\frac{1}{2}$ 阶模拟分抗电路中进行 $n-1$ 次递归嵌套 $\frac{1}{2}$ 阶模拟分抗电路模型。显然，该类型电路的局部和整体之间具有高度自相似性，这恰恰符合分形学说。这也正是分数阶微积分（亦称分数阶演算）是分形理论的数学基础的具体电路印证。

另外，根据微分算子理论，$\frac{1}{2^n}$ 阶模拟分抗电路完成的功能其实是 $D^{2^{-n}}$。这恰恰是数学中微分算子的幂在电路中的具体实现范例。可见，D^{v^n} 表示在 D^v 中 n 次递归嵌入 D^v，而不是 n 个 D^v 的联乘。

4.5 构造任意分数阶模拟分抗电路

根据代数理论和数论，任意有理小数都可以表示为一个有理分数。根据分数的运算法则，可以把一个有理分数表示为若干个有理分数项的和差。因此，根据微分算子联乘的序贯算法，若干个有理分数阶微分算子的联乘可以恒等于某分数阶的微分算子。显然，为了相应分抗电路构造的简便，要求参加联乘的有理分数阶微分算子的阶数必须符合 $\frac{1}{2^n}$ 规律。因此，实际上相应有理分数阶微分算子联乘的结果一般只能近似逼近于某分数阶的微分算子，而非恒等，逼近的精度与参加联乘的有理分数阶微分算子的项数直接有关，由要求的工程精度决定。

不失一般性，本节以如何构造 1/3 阶模拟分抗电路为例进行说明。

4.5.1 $\frac{2}{3}$ 阶模拟分抗电路

令 $\alpha = \frac{2}{3}$，$\beta = \frac{1}{2} + \frac{1}{2^3} + \frac{1}{2^5}$，则有 $\alpha - \beta = \frac{1}{96} \approx 0.01$。显然，按一般工程精度，可以用 β 代替 α。根据算子联乘的序贯算法，$D^{\frac{2}{3}} \approx D^{\frac{1}{2}} D^{\frac{1}{2^3}} D^{\frac{1}{2^5}}$。根据信号与电路系统理论，$\frac{2}{3}$ 阶模拟分抗电路结构如图 4.16 所示。

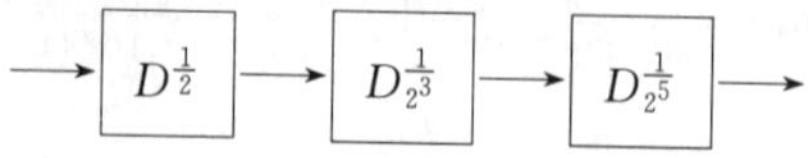

$$\longrightarrow \boxed{D^{\frac{1}{2}}} \longrightarrow \boxed{D^{\frac{1}{2^3}}} \longrightarrow \boxed{D^{\frac{1}{2^5}}} \longrightarrow$$

图 4.16　$\frac{2}{3}$ 阶模拟分抗电路结构

$D^{\frac{1}{2}}$，$D^{\frac{1}{2^3}}$，$D^{\frac{1}{2^5}}$ 阶模拟分抗电路都是前面已经实现了的电路。由式（4.33）可知 $\frac{2}{3}$ 阶

模拟分抗电路的输入阻抗（系统的传输函数）为

$$Z_{\frac{2}{3}} = Z_{\frac{1}{2}} Z_{\frac{1}{2^3}} Z_{\frac{1}{2^5}} \approx r^{\frac{27}{16}} \xi^{\frac{2}{3}} S^{-\frac{2}{3}} \tag{4.38}$$

同理可知，式（4.38）表明 $v(t)$ 正比于 $i(t)$ 的 $\frac{2}{3}$ 阶积分；反之，$i(t)$ 正比于 $v(t)$ 的 $\frac{2}{3}$ 阶微分。

4.5.2 $\frac{1}{3}$ 阶模拟分抗电路

$\frac{1}{3}$ 阶模拟分抗电路可以是 $\frac{2}{3}$ 阶模拟分抗电路和阻抗 Z_a 构成的无限递归结构，其等效电路如图 4.17 所示。

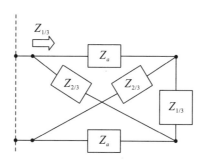

图 4.17　$\frac{1}{3}$ 阶模拟分抗等效电路

由电路的串并联法则以及式（4.38）可得

$$Z_{1/3} = (Z_a Z_{2/3})^{\frac{1}{2}} = r^{\frac{43}{32}} \xi^{\frac{1}{3}} S^{-\frac{1}{3}} \tag{4.39}$$

于是，可得输出电压 $V(S)$ 和输入电流 $I(S)$ 的关系为

$$V(S) = r^{\frac{43}{32}} \xi^{\frac{1}{3}} S^{-\frac{1}{3}} I(S) \tag{4.40}$$

对式（4.40）作 Laplace 反变换，可得

$$v(t) = \frac{r^{\frac{43}{32}} \xi^{\frac{1}{3}}}{\Gamma(1/3)} \int_{-\infty}^{t} \frac{i(\tau)}{(t-\tau)^{2/3}} \mathrm{d}\tau \tag{4.41}$$

式（4.41）表明 $v(t)$ 正比于 $i(t)$ 的 $\frac{1}{3}$ 阶积分；反之，可得

$$I(S) = r^{-\frac{43}{32}} \xi^{-\frac{1}{3}} S^{\frac{1}{3}} V(S) \tag{4.42}$$

对式（4.42）作 Laplace 反变换，可得

$$i(t) = \frac{1}{r^{\frac{43}{32}} \xi^{\frac{1}{3}} \Gamma(2/3)} \int_{-\infty}^{t} \frac{v'(\tau)}{(t-\tau)^{1/3}} \mathrm{d}\tau \tag{4.43}$$

式（4.43）表明 $i(t)$ 正比于 $v(t)$ 的 $\frac{1}{3}$ 阶微分。

同理，仿照上述方法便可以容易地构造任意分数阶模拟分抗电路。

4.6 实验仿真及结果分析

不失一般性，本节仅列举和分析网格型模拟分抗电路的实验结果。网格型 $\frac{1}{2}$ 阶模拟分抗电路的级数 n 和该模拟分抗电路输入阻抗 Z 之间存在关系，其幅度 $|Z(n)|$ 和相位 $\theta(n)$ 分别如图 4.18、图 4.19 所示。

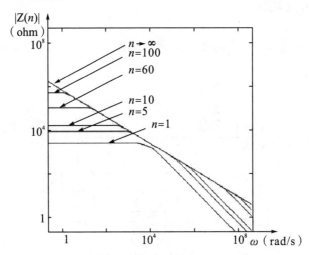

图 4.18　网格型 $\frac{1}{2}$ 阶模拟分抗电路的 $|Z(n)|$

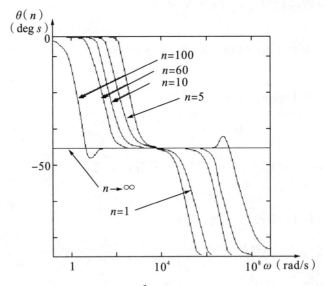

图 4.19　网格型 $\frac{1}{2}$ 阶模拟分抗电路的 $\theta(n)$

由图 4.18、图 4.19 可知，当电路级数 $n=5$ 时，网格型 $\frac{1}{2}$ 阶模拟分抗电路在一定的频带内已经能够较好地逼近理想状态（$n\to\infty$）。因此，本节构建了 5 级网格型 $\frac{1}{2}$ 阶模拟微分电路，如图 4.20 所示。

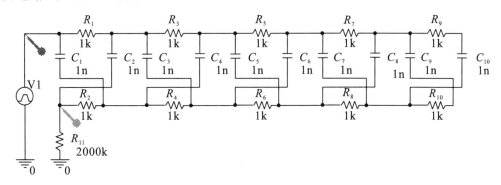

图 4.20　5 级网格型 $\frac{1}{2}$ 阶模拟分抗电路

用该电路对传统的正弦信号进行了 $\frac{1}{2}$ 阶微分，用 Orcad 电路软件进行仿真，在示波器上显示的实验结果与其解析解分别如图 4.21、图 4.22 所示。

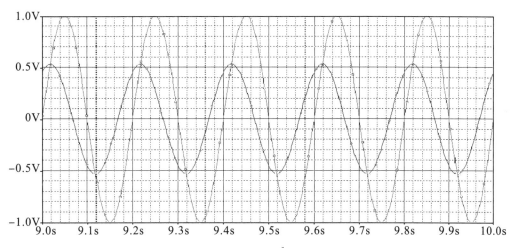

图 4.21　正弦信号的 $\frac{1}{2}$ 分数阶微分

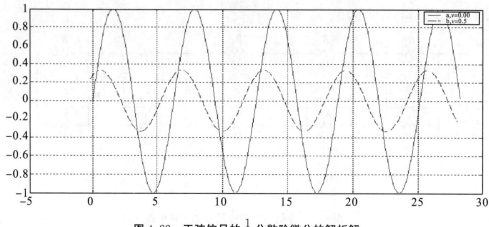

图 4.22 正弦信号的 $\frac{1}{2}$ 分数阶微分的解析解

对比图 4.21、图 4.22 可知，其结果具有可比性，说明了网格型模拟分抗电路模型的可行性。于是，可用上述递归嵌套的方法构建 5 级递归网格型 $\frac{1}{3}$ 阶模拟微分电路，如图 4.23 所示。

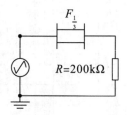

图 4.23　5 级递归网格型 $\frac{1}{3}$ 阶模拟分抗电路

用 5 级递归网格型 $\frac{1}{3}$ 阶模拟分抗电路分别对 δ 信号、锯齿波信号、方波信号、三角波信号、梯形方波信号、锯齿方波信号进行了 $\frac{1}{3}$ 阶微分，在示波器上显示的实验结果如图 4.24～图 4.29 所示。

第4章 分数阶微积分的模拟分抗电路实现

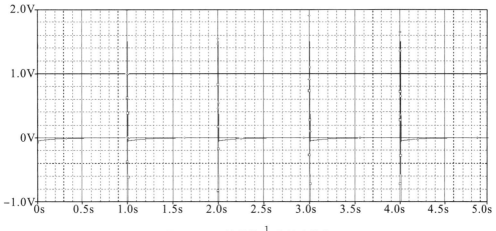

图 4.24 δ 信号的 $\frac{1}{3}$ 分数阶微分

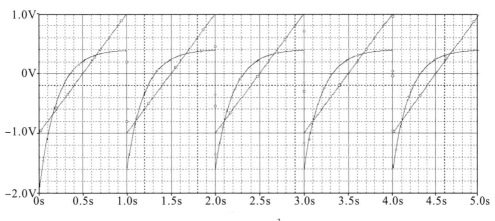

图 4.25 锯齿波信号的 $\frac{1}{3}$ 分数阶微分

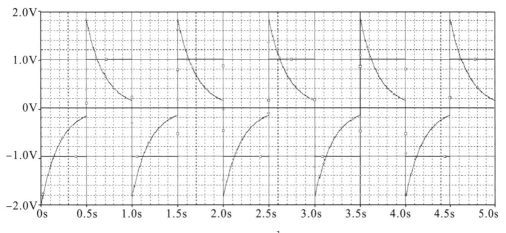

图 4.26 方波信号的 $\frac{1}{3}$ 分数阶微分

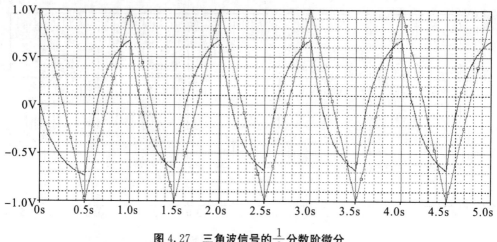

图 4.27　三角波信号的 $\frac{1}{3}$ 分数阶微分

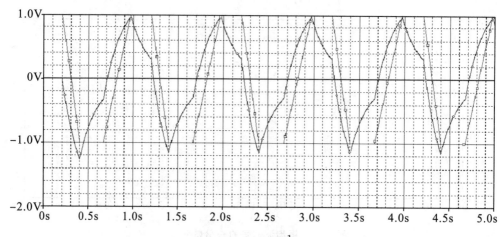

图 4.28　梯形方波信号的 $\frac{1}{3}$ 分数阶微分

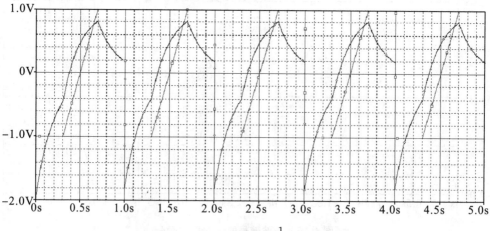

图 4.29　锯齿方波信号的 $\frac{1}{3}$ 分数阶微分

用 MATLAB 进行仿真实验可知，上述实验结果和该信号 $\frac{1}{3}$ 阶分数阶微分的解析解具有可比性,从而进一步证明了本方法的可行性。由图 4.24 的实验结果可知,δ 信号的分数阶微分拖着长长的负尾巴,这与拖着长长负尾巴的生物神经脉冲振荡信号具有相似的特点。于是很自然地联想到:是否可以对某些信号作分数阶微分来产生神经脉冲振荡信号?这将在下一章中进行详细论述。

为了研究分数阶微分在信号处理中的应用,分别对幂信号、指数信号用 5 级递归网格型 $\frac{1}{3}$ 阶模拟分抗电路进行了 $\frac{1}{3}$ 阶微分,其实验结果如图 4.30、图 4.31 所示。

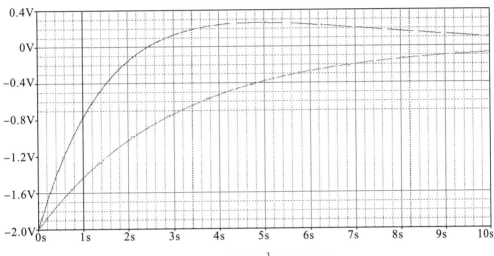

图 4.30　幂信号的 $\frac{1}{3}$ 分数阶微分

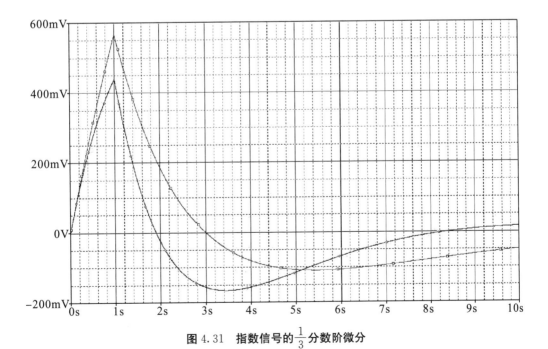

图 4.31　指数信号的 $\frac{1}{3}$ 分数阶微分

对照式(3.8)和图 4.30、图 4.31,从信号处理角度看,v 阶微积分运算其实是对信

号的一个线性时不变滤波器，其滤波函数为 $\hat{d}_v(\omega) = (i\omega)^v = |\omega|^v \cdot \exp(i\theta_v(\omega))$。当 $v >$ 0 时，它是微分器，$\lim\limits_{|\omega|\to\infty}|\hat{d}_v(\omega)| \to \infty$，$\hat{d}_v(\omega)$ 是奇异高通滤波器。v 越大，通频带越窄，高通特性越明显，相应加强了信号 $s(t)$ 的高频成分，相对压制了其低频成分，这有利于突出信号的细节，但抗高频干扰成分性能差。当 $v=0$ 时，它是全通滤波器，$\hat{d}_v(\omega) \equiv 1 \Leftrightarrow d_v(t) = \delta(t)$。当 $v<0$ 时，它是积分器，$\lim\limits_{|\omega|\to 0}|\hat{d}_v(\omega)| \to \infty$，$\hat{d}_v(\omega)$ 是奇异低通滤波。$|v|$ 越大，低通特性越明显，突出了信号的低频成分，压制了高频成分，所以积分后信号变得更加平滑，这有利于观测信号的总体变化趋势，但抹去了信号的变化细节。相对而言，微分滤波器振幅特性的通频带较宽，而积分滤波器振幅特性的通频带较窄。

对简单的调频信号、调频调幅信号用 5 级递归网格型 $\frac{1}{3}$ 阶模拟分抗电路进行了 $\frac{1}{3}$ 阶微分，其实验结果如图 4.32、图 4.33 所示。

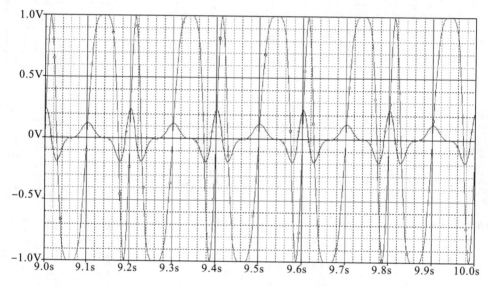

图 4.32　调频信号的 $\frac{1}{3}$ 分数阶微分

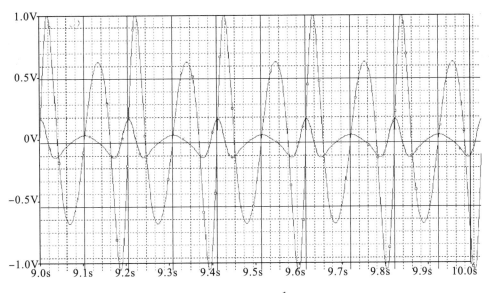

图 4.33 调频调幅信号的 $\frac{1}{3}$ 分数阶微分

由图 4.32、图 4.33 可知，对于调制信号的分数阶微分，可以强化和提取信号的调制信息和规律，然后再通过过零点分析等手段便可以提取这些信息。这就为通信中信号的调制解调找到了除传统的相干解调、非相干解调之外的另外一种调制解调方法。

4.7 小结

关于分数阶微积分（分数阶演算）在通信和现代信号处理中应用方面的研究方兴未艾，本章关于用无源电路元件构造模拟分抗的工作只是在分数阶微积分的模拟电路近似实现方面作了一些初步的尝试。倘若本章的工作能起到抛砖引玉的效果，吸引更多的学者一起来探究该领域，做出更好的工作，笔者将甚感慰藉。目前，笔者还在对一些相关课题做进一步的研究：

(1) 如何用模拟分抗电路来构造分数阶的混沌电路？分数阶混沌电路也是对整数阶混沌电路的推广和改造，它和整数阶混沌电路在性能上有什么异同？

(2) 如何把分数阶微积分运用于信号的调制解调之中，为通信中信号的调制解调找到一种新的方法？

(3) 如何用分数阶微积分来去除噪声，并将其应用于对扩频通信的侦测之中？

(4) 如何把分数阶微积分应用于模式识别之中？

(5) 如何把分数阶微积分应用于自适应信号处理中？换言之，如何构造基于分数阶梯度的自适应信号处理？

(6) 如何用分数阶微积分来迈过二次曲面的局部最小值？

(7) 如何构造分数阶子波？

(8) 模拟分抗电路的分数维怎样计算？能否利用分形学中的分形因子找出更多的模拟分抗电路？

（9）模拟分抗电路与分析化学以及纳米材料科学之间的联系和应用有哪些？分形结构的高度自相似性是自然界物质结构的普遍形式，模拟分抗电路模型的分形结构和某些自然界物质的结构具有相似性。能否找出这些自然物质？这个工作会对分抗电路元件的集成化实现产生深远影响。

（10）如何在光学领域实现分抗元件？笔者已经在这方面取得了一些初步的研究成果。

第5章 分抗的单位和物理量纲[①]

本章主要讨论了分抗值的测量单位和物理意义以及分抗的串联和并联规则。"分抗"是在模拟电路中分数阶电容或分数阶电感之后出现的新概念。分抗是一种新兴的分数阶电路元件,是分数阶电路和系统的硬件实现的核心部件。"分抗值"是"分数阶阻抗"的缩写,是指一个分抗元的分数阶阻抗。迄今为止,分抗的测量单位和物理意义尚未被提出,这是一个具有挑战性的理论问题。为突破该理论难题,我们研究了蔡氏电路系统中理想状态下电路元件的位置及其在自然实现中的分数阶阻抗,将一种新颖的数学方法分数阶微积分用于分析该概念框架。第一,介绍了理想状态下分抗元在蔡氏电路元件系统中的位置。第二,通过数学和电路分析,研究了1/2阶分抗元在自然实现中的1/2阶阻抗,并提出了任意阶分抗元在自然实现中的分数阶阻抗。第三,分析了任意阶分抗元在自然实现中的支路电流。第四,提出了分抗元的阶敏特性。第五,通过实验详细地分析了分抗元的温度影响和任意阶理想状态下分抗元的分数阶次对其固有电特性的影响。第一部分讨论分抗的基本问题,而第二部分则讨论分抗的测量单位和物理尺寸以及分抗元在串联和并联中的规则。

5.1 Measurement Units and Physical Dimensions of Fractance-Part Ⅰ: Position of Purely Ideal Fractor in Chua's Axiomatic Circuit Element System and Fractional-Order Reactance of Fractor in Its Natural Implementation

5.1.1 INTRODUCTION

Nowadays, fractional calculus has evolved as an important branch of mathematical analyses [142], [175] − [179]. The origin of fractional calculus was concurrent with that of integer-order calculus; however, until recently, the applications of fractional

---

[①] PU Yifei. Measurement Units and Physical Dimensions of Fractance-Part Ⅰ: Position of Purely Ideal Fractor in Chua's Axiomatic Circuit Element System and Fractional-Order Reactance of Fractor in Its Natural Implementation [J]. IEEE Access, 2016 (4): 3379−3397.

PU Yifei. Measurement Units and Physical Dimensions of Fractance-Part Ⅱ: Fractional-Order Measurement Units and Physical Dimensions Fractance and Rules for Fractors in Series and Parallel [J]. IEEE Access, 2016 (4): 3398−3416.

calculus were limited to mathematics. For physical scientists and engineering technicians, fractional calculus is a novel promising mathematical method [11] [94] [142], [179] − [180]. Fractional calculus extends the concepts of the integer-order difference and Riemann sum. The fractional differential, except for Caputo defined fractional calculus, of a Heaviside function is non-zero, whereas its integer-order differential must be zero [142], [175] − [179]. It has been shown that a fractional-order or fractional dimensional approach is now the most appropriate description for many natural phenomena. Many scientific fields such as the fractional-order diffusion processes [181] − [183], fractional-order viscoelasticity theory [126], fractal dynamics [184], fractional-order control [89], fractional-order image processing [185] − [191], and fractional-order adaptive signal processing [192], presently use fractional calculus.

The application of fractional calculus to circuits and systems, especially to a promising fractional-order circuit element called fractor, is an emerging and interesting discipline, in which insufficient studies have been implemented. Fractional calculus has been applied to circuits and systems mainly because of its intrinsic strengths of long-term memory, non-locality, and weak singularity. The term "fractor" arose following the successful synthesis of a fractional-order capacitor or a fractional-order inductor in an analog circuit. The term "fractance", as a portmanteau of "the fractional-order impedance", refers to the fractional-order impedance of a fractor [130], [171] − [173], [175], [185], [193] − [215]. Further, the driving-point impedance function of fractor is its fractional-order reactance. There are two types of fractor in nature: capacitive fractor and inductive fractor. On the one hand, a capacitive fractor is a fractional-order capacitor, which involves a negative-order fractional differential filter. The capacitive fractance is the fractional-order impedance of a capacitive fractor. Since the position of purely ideal capacitive fractor in Chua's circuit axiomatic element system is between that of a capacitor and that of a resistor, the electrical properties of a purely ideal capacitive fractor should fall in between those of a capacitor and those of a resistor. On the other hand, an inductive fractor is a fractional-order inductor, which involves a positive-order fractional differential filter. The inductive fractance is the fractional-order impedance of an inductive fractor. Since the position of purely ideal inductive fractor in Chua's circuit axiomatic element system is between that of an inductor and that of a resistor, the electrical properties of a purely ideal inductive fractor should fall in between those of an inductor and those of a resistor.

In particular, the tree type [175], [198], two-circuit type [172], [185], H type [172], [185], and net-grid type [185], [171], [193] − [194] should be four discovered natural implementations of purely ideal fractor. On one hand, a purely ideal capacitive fractor consists of ordinary capacitors and resistors in an analog circuit of the tree type [175], [198], two-circuit type [173], [185], H type [172], [185], net-grid type [171], [185], [193] − [194], and some other undiscovered infinite recursive types with extreme self-similar fractal structures. On the other hand, a purely ideal inductive fractor also consists of a series of ordinary inductors and resistors with the same types of extreme self-similar fractal structures as a corresponding purely ideal capacitive fractor. What distinguishes the aforementioned four types of fractor [175], [171] − [173], [185], [193] − [194], [198] from the other approximate implementations of fractor [130], [195] − [203], [206] − [215] is that the floating point values of the capacitance, inductance, and resistance of these four natural fractal structure types of fractor [175], [171] − [173], [185], [193] − [194], [198] are never required in deed. In fact, there are zero errors between these four types of fractance [175], [171] − [173], [185], [193] − [194], [198] with infinite recursive extreme self-similar structures and a purely ideal fractance, whereas the corresponding devices manufactured utilizing the other approximate implementations of fractor [130], [195] − [203], [206] − [215] could never represent a purely ideal fractor.

In the field of circuits and systems, many remarkable progresses [130], [171] − [173], [175], [185], [193] − [215] in the study of fractor have been not only validated actually its role as a promising fractional-order circuit element, which is a core component for the hardware implementation of the fractional-order circuits and systems, but has also provided some interesting and practical suggestions for future research. For example, L. Dorčák et al. proposed the analogue electronic realization of the fractional-order (FO) systems, e. g. controlled objects and/or controllers whose mathematical models are FO differential equations. The electronic realization is based on FO differentiator and FO integrator where operational amplifiers are connected with appropriate impedance, i. e. the FO element or constant phase element (CPE) [206] − [208]. J. Valsa et al. described a possible realization of such a model that is quite simple and in spite of its simplicity makes it possible to simulate the properties of ideal CPEs [209]. E. A. Gonzalez and I. Petráš offered a comprehensive discussion on the applications of fractional calculus in the design and implementation of fractional-order systems in the form of electronic circuits that could be used for signal processing and control engineering applications [210]. E. A. Gonzalez et al. presented the mathematical properties of a generalized fractional-order two-port network represented as a symmetrical T-section through its hybrid parameters [211]. G. L. Abulencia et al. studied the analog realization of a selectable fractional-order differentiator in a microelectronics scale, whose order of differentiation can be selected between 0.25 and 0.50 [212]. A. Tepljakov et al. proposed a modification of Newton's method for approximating a first-order implicit fractional transfer function, which corresponds to a frequency-bounded fractional differentiator or integrator [213]. A. Tepljakov et al. investigated the possibilities of network generation from the fractional-order controller approximations derived using different methods proposed over the years [214]. E. A. Gonzalez et al. proposed the conceptual design of a variable fractional-order differentiator in which the order can be selected from 0 to 1 with an increment of 0.05 [215]. Thus, it is natural to ponder what the measurement units and physical dimensions of fractance are. Until now, however, no efficient measurement units and physical dimensions of fractance have been proposed, which is a challenging theoretical problem [204] — [205]. Motivated by this need, in this chapter, based on the aforementioned studies [130], [171] — [173], [175], [185], [193] — [215], here and in the companion paper (Part II), the measurement units and physical dimensions of fractance and rules for fractors in series and parallel are discussed in detail. A state-of-the-art mathematical method, fractional calculus, is used to analyze the proposed conceptual framework. In particular, part I proposed to introduce the position of purely ideal fractor in Chua's circuit axiomatic element system and the fractional-order reactance of fractor in its natural implementation. The chapter discusses fundamental issues, whereas Part II is devoted to the measurement units and physical dimensions of purely ideal fractance and the rules for fractors in series and parallel. These two companion papers are two elaborately expanded versions of the conference publications of my previous work [204] — [205].

The remainder of this chapter is organized as follows: Section 2 recalls the necessary mathematical background of fractional calculus. Section 3 introduces the position of purely ideal fractor in Chua's axiomatic circuit element system. Section 4, through mathematical and circuit analysis, discusses in detail the fractional-order reactance of the arbitrary-order purely ideal fractor in its natural implementation. In Section 4, first, the 1/2-order reactance of the 1/2-order purely ideal fractor in its natural implementation is proposed. Second, the fractional-order reactance of the arbitrary-order purely ideal fractor in its natural implementation is proposed. Third, the branch-current of the arbitrary-order fractor in its natural implementation is analyzed, following which the order-sensitivity characteristics of fractor are proposed. Section 5 presents the experiment results obtained and the associated analyses carried out. Here, we discuss the temperature effect of fractor and the influence of the fractional-order of an arbitrary-order purely ideal fractor on its intrinsic electrical characteristics. In Section 6, the conclusions of this manuscript are presented.

5.1.2 MATHEMATICAL BACKGROUND

In order to let this chapter self-contained for a reader being not familiar with fractional calculus, this section includes a brief necessary recall on three basic definitions of fractional calculus in the domain of Euclidean measure. With more than 90 years developing, until now, however, the computability based on the definition of fractional calculus in the domain of Hausdorff measure is not yet developed. Whereas, the computability based on the definition of fractional calculus in the domain of Euclidean measure is well-developed. The commonly used fractional calculus definitions in the domain of Euclidean measure are those of Grünwald-Letnikov, Riemann-Liouville, and Caputo [142], [175] — [178].

The Grünwald-Letnikov fractional calculus for causal signal $f(x)$ is defined by:

第 5 章 分抗的单位和物理量纲

$$
{}_a^{G-L}D_x^v f(x) = \lim_{N \to \infty} \left\{ \frac{\left(\frac{x-a}{N}\right)^{-v}}{\Gamma(-v)} \sum_{k=0}^{N-1} \frac{\Gamma(k-v)}{\Gamma(k+1)} f\left(x - k\left(\frac{x-a}{N}\right)\right) \right\}, \quad (5.1)
$$

where $f(x)$ is a differintegrable function under consideration [1] − [5], $[a,x]$ is the duration of $f(x)$, v is a real number (can be a fraction in nature), $\Gamma(a) = \int_0^\infty e^{-x} x^{a-1} dx$ is the Gamma function, and ${}_a^{G-L}D_x^v$ denotes the Grünwald-Letnikov defined fractional differential operator. The numerical computation of the Grünwald-Letnikov defined fractional calculus only requires the discrete sampling values, $f(x - k(\frac{x-a}{N}))$, of causal signal $f(x)$. It only needs to carry out the differential or integral operation for the signal $f(x)$.

The Riemann-Liouville defined v-order integral, for causal signal $f(x)$ is defined by:

$$
{}_a^{R-L}I_x^v f(x) = \frac{1}{\Gamma(v)} \int_a^x \frac{f(\tau)}{(x-\tau)^{1-v}} d\tau, \quad (5.2)
$$

where $v > 0$. ${}_a^{R-L}I_x^v$ denotes the Riemann-Liouville defined fractional integral operator. The Riemann-Liouville defined v-order derivative is defined by:

$$
{}_a^{R-L}D_x^v f(x) = \frac{1}{\Gamma(n-v)} \frac{d^n}{dx^n} \int_a^x \frac{f(\tau)}{(x-\tau)^{v-n+1}} d\tau, \quad (5.3)
$$

where $n-1 \leqslant v < n$. ${}_a^{R-L}D_x^v$ denotes the Riemann-Liouville defined fractional differential operator. The Laplace transform of the Riemann-Liouville defined v-order differential operator is $L[{}_0^{R-L}D_x^v f(x)] = s^v L[f(x)] - \sum_{k=0}^{n-1} s^k [{}_0^{R-L}D_x^{v-1-k} f(x)]_{x=0}$, where s denotes the Laplace operator. If the fractional primitives of $f(x)$ are zero, the Laplace transform of ${}_0^{R-L}D_x^v f(x)$ can be simplified as $L[{}_0^{R-L}D_x^v f(x)] = s^v L[f(x)]$.

The Caputo defined v-order derivative for causal signal $f(x)$ is defined by:

$$
{}_a^C D_x^v f(x) = \frac{1}{\Gamma(n-v)} \int_a^x (x-\tau)^{n-v-1} f^{(n)}(\tau) d\tau, \quad (5.4)
$$

where $0 \leqslant n-1 < v < n$, $n \in \mathbf{R}$. ${}_a^C D_x^v$ denotes a Caputo defined fractional differential operator. Equation (5.4) shows that ${}_a^C D_x^v$ is equivalent to a successive implementation of an n-order differential with an $(n-v)$-order integral of $f(x)$. The Laplace transform of the v-order Caputo differential operator is $L[{}_0^C D_x^v f(x)] = s^v L[f(x)] - \sum_{k=0}^{n-1} s^k f^{(k)}(x)|_{x=0}$. If the fractional primitives of $f(x)$ are zero, the Laplace transform of ${}_0^C D_x^v f(x)$ can simplified as $L[{}_0^C D_x^v f(x)] = s^v L[f(x)]$. In this case, the three aforementioned definitions of fractional derivatives are equivalent. In this chapter, let me employ the equivalent notations in an interchangeable manner as follows:

$$
D_x^v = {}^{G-L}D_x^v
$$
$$
= {}_0^{R-L}D_x^v
$$

$$= \S D_x^v. \tag{5.5}$$

In order to achieve the qualitative analysis of the electrical characteristics of an arbitrary-order purely ideal fractor, in this section, the position of purely ideal fractor in Chua's axiomatic circuit element system is studied firstly.

For the convenience of illustration, we use Chua's periodic table of all two-terminal circuit elements [216] – [223] to analyse the electrical characteristics of an arbitrary-order purely ideal fractor. In recent times, a broader definition of all 2-terminal non-volatile memory devices based on resistance switching is argued for by Chua [216] – [219]. The periodic table of all two-terminal circuit elements is presented in Figure 5.1 [216] – [223].

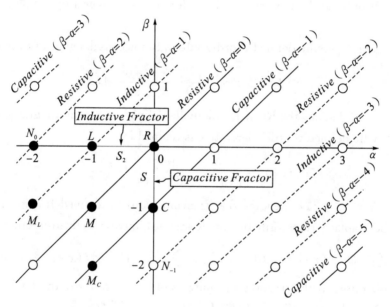

Figure 5.1 Chua's periodic table of all two-terminal circuit elements.

In Figure 5.1, the Chua's axiomatic element and its corresponding electrical characteristics is denoted by the symbol $C^{\langle \alpha, \beta \rangle}$, where the voltage exponent is denoted by α, the current exponent is denoted by β. α and β are the orders of the time differential of $v(t)$ and $i(t)$, respectively. $D_t^\alpha v(t)$ and $D_t^\beta i(t)$ are collectively the Chua's constitutive variables, where D is the differential operator. Thus, (α, β) is referred to the plane of the Chua's periodic table of all two-terminal circuit elements. C, R, L, M, M_L, and M_C denote capacitor, resistor, inductor, memristor, meminductor, and memcapacitor, respectively. The symbol O denotes the other postulated elements of the Chua's axiomatic element system. All Chua's axiomatic elements have the element independence property. Therefore, a corresponding constitutive relation is established by $(D_t^\alpha v(t), D_t^\beta i(t))$ as follows [218] – [221]:

$$D_t^\alpha v(t) - C^{\langle \alpha, \beta \rangle} D_t^\beta i(t) = 0, \tag{5.6}$$

where $\alpha \in \mathbf{R}$, and $\beta \in \mathbf{R}$. Thus, the following is true [216] – [223]:

$$C^{(\alpha,\beta)} = \begin{cases} C, & \text{if } \alpha = 0, \beta = -1 \\ R, & \text{if } \alpha = 0, \beta = 0 \\ L, & \text{if } \alpha = -1, \beta = 0 \\ M, & \text{if } \alpha = -1, \beta = -1 \end{cases}, \quad (5.7)$$

where for the convenience of illustration, C, R, L, and M also denote the electrical characteristics of capacitor, resistor, inductor, and memristor, respectively.

From Figure 5.1, we can see that with respect to ordinary two-terminal circuit elements such as capacitor C, resistor R, and inductor L, their first-order purely ideal reactances $Z(s)$ can be expressed in an unified formula:

$$Z(s) = \mathcal{R}_p s^p, \quad (5.8)$$

where the order of first-order purely ideal reactance, p, is an integer. Equation (5.8) shows that first, when $p = -1$, we have $\mathcal{R}_p = \mathcal{R}_{-1} = 1/c$ and $Z(s) = 1/(cs)$, where c denotes the capacitance of a purely ideal capacitor. The operator s^{-1} denotes the first-order integral operation. In this case, (5.8) represents the -1-order reactance of a purely ideal capacitor. Second, when $p = 0$, we have $\mathcal{R}_p = \mathcal{R}_0 = r$ and $Z(s) = rs^0 = r$, where r denotes the resistance of a purely ideal resistor. The operator s^0 denotes an identity operation, which neither implements differential nor achieves integral. In this case, (5.8) represents the 0-order reactance of a purely ideal resistor. Third, when $p = 1$, we have $\mathcal{R}_p = \mathcal{R}_1 = l$ and $Z(s) = ls$, where l denotes the inductance of a purely ideal inductor. The operator s^1 denotes the first-order differential operation. In this case, (5.8) represents the $+1$-order reactance of a purely ideal inductor. Note that the order of purely ideal reactance, p, represents the operational characteristics of the operator s^p.

Furthermore, fractor extends the concepts of the ordinary two-terminal circuit elements. The fractional-order reactance of a purely ideal fractor can be also expressed in the form of (5.8), where p is a real number (can be a fraction in nature) [130], [171] − [173], [175], [185], [193] − [215]. For the convenience of analysis and hardware implementation, we usually directly set $\mathcal{R}_p = 1$ for the fractional-order reactance of a purely ideal fractor [130], [171] − [173], [175], [185], [193] − [215]. Up to now, however, no exact expression of $\mathcal{R}_p$ of the fractional-order reactance of a purely ideal fractor has been proposed. Therefore, it is natural to ponder what the fractional-order reactance of fractor in its natural implementation is, and what the relationship between this fractional-order reactance of purely ideal fractor and the measurement units and physical dimensions of fractance is, which is a challenging theoretical problem.

In addition, from Figure 5.1, we can see that each dot with integer coordinates (α, β) represents a basic circuit element, $(D_t^\alpha v(t), D_t^\beta i(t))$. Let's suppose that at an operating point Q, $(D_t^\alpha v(t), D_t^\beta i(t))$ is characterized by $D_t^\alpha v(t) = f[D_t^\beta i(t)]$. Thus, the small-signal behavior of $(D_t^\alpha v(t), D_t^\beta i(t))$ about Q, the small-signal reactance of $(D_t^\alpha v(t), D_t^\beta i(t))$, can be defined by $Z_{(\alpha,\beta)}(s) = m_Q s^{(\beta-\alpha)} \stackrel{s=j\omega}{=} Z_{(\alpha,\beta)}(j\omega) = m_Q (j\omega)^{(\beta-\alpha)}$, where $m_Q > 0$ denotes the slope $D_t^1 f[D_t^\beta i(t)]$ at Q, $s = j\omega$, j denotes an imaginary unit, and ω denotes an angular frequency.

The four-periodic law of the Chua's periodic table of all two-terminal circuit elements [218] − [219] shows that first, if $\beta - \alpha = 4n - 1$, $Z_{\langle \alpha,\beta \rangle}(j\omega) = m_Q \omega^{4n}/j\omega$, where n is an integer. In this case, we can consider that $(D_t^\alpha v(t), D_t^\beta i(t))$ is a family of the nth-periodic frequency-dependent capacitors. When $n = 0$, $(D_t^0 v(t), D_t^{-1} i(t))$ is traditional capacitor, which achieves the first-order integral. Thus, traditional capacitor is the $0th$-periodic frequency-dependent capacitor.

Second, if $\beta - \alpha = 4n + 0$, $Z_{\langle \alpha,\beta \rangle}(j\omega) = m_Q \omega^{4n}$. In this case, we can consider that $(D_t^\alpha v(t), D_t^\beta i(t))$ is a family of the nth-periodic frequency-dependent positive resistors. When $n = 0$, $(D_t^0 v(t), D_t^0 i(t))$ is traditional resistor, and $(D_t^{-1} v(t), D_t^{-1} i(t))$ is traditional memristor. Thus, traditional resistor and memristor are the $0th$-periodic frequency-dependent positive resistors.

Third, if $\beta - \alpha = 4n + 1$, $Z_{\langle \alpha,\beta \rangle}(j\omega) = m_Q (j\omega) \omega^{4n}$. In this case, we can consider that $(D_t^\alpha v(t), D_t^\beta i(t))$ is a family of the nth-periodic frequency-dependent inductors. When $n = 0$, $(D_t^{-1} v(t), D_t^0 i(t))$ is traditional inductor. Thus, traditional inductor is the $0th$-periodic frequency-dependent inductor.

Fourth, if $\beta - \alpha = 4n + 2$, $Z_{\langle \alpha,\beta \rangle}(j\omega) = -m_Q \omega^{4n}$. In this case, we can consider that $(D_t^\alpha v(t), D_t^\beta i(t))$ is a family of the nth-periodic frequency-dependent negative resistors (FDNR)[54]. When $n = 0$, $(D_t^{-2} v(t), D_t^0 i(t))$ is the $0th$-periodic FDNR. $(D_t^{-2} v(t), D_t^0 i(t))$ is a recently introduced linear circuit element, which is now widely used as a basic building block in active circuit design. Let's denote $(D_t^{-2} v(t), D_t^0 i(t))$ as the symbol N_0, in which N is the abbreviation of FDNR, and subscript 0 denotes the $0th$-periodic one. In a similar way, when $n = -1$, $(D_t^0 v(t), D_t^{-2} i(t))$ is the $-1th$-periodic FDNR. Let's denote $(D_t^0 v(t), D_t^{-2} i(t))$ as the symbol N_{-1}, in which subscript -1 denotes the $-1th$-periodic one. The basic circuit devices of N_0 and N_{-1} have not been achieved until now, but N_0 or N_{-1} is certainly not a cascade system of the 2-stage first-order differentiators or integrators, respectively. N_0 or N_{-1} is not the second-order inductor or the second-order capacitor, respectively. Note that m_Q of $N_{n=q}$ and $N_{n=-q}$ remain unclear until now, where q is a non-negative integer.

Fifth, when α and β are integers, $(D_t^\alpha v(t), D_t^\beta i(t))$ has element independence property. The two-terminal elements with integer coordinates (α, β) in Figure 5.1 are all independent of each other in the sense that no basic circuit element defined by a $(D_t^\alpha v(t), D_t^\beta i(t))$ can be synthesized by a combination of any other basic circuit elements in the Chua's periodic table of all two-terminal circuit elements [53]. In addition, $(D_t^\alpha v(t), D_t^\beta i(t))$ has element closure property. Arbitrary interconnection of elements of the same type (i.e., same α and β) always results in another element of the same type [218].

Note that first, if the order of fractor is a non-integer, neither capacitive fractor nor inductive fractor is a basic circuit element, which can be synthesized by a combination of other basic circuit elements, such as resistor and capacitor or inductor. Meanwhile, if the order of fractor is an integer, resistor, capacitor, and inductor are three special cases of fractor. Second, in view of logical consistency, axiomatic completeness, formal symmetry, and constitutive relation, from Figure 5.1, we can see that the v-order purely ideal capacitive fractor, where $0 < v < 1$, should be lying on the line segment, S_1, between C and R. Meanwhile, the v-order purely ideal inductive fractor, where $0 < v < 1$, should be lying on the line segment, S_2, between L and R. Similarly, if the fractional-order v satisfies $n < v < n+1$, where n is a non-negative integer, the v-order purely ideal capacitive fractor and the v-order purely ideal inductive fractor should be still lying on the line segments S_1 and S_2, respectively.

Therefore, from Figure 5.1, we can further see that on the one hand, the fractional-order v, capacitance c, and resistance r are nonlinearly hybridized with the v-order purely ideal capacitive fractance in its natural implementations. The fractional-order capacitive reactance of an arbitrary-order purely ideal capacitive fractor should be closely related with the fractional-order v, capacitance c, and resistance r, simultaneously. On the other hand, the fractional-order v, inductance l, and resistance r are nonlinearly hybridized with the v-order purely ideal inductive fractance in its natural implementations. The fractional-order inductive reactance of an arbitrary-order purely ideal inductive fractor should be closely related with the fractional-order v, inductance l, and resistance r, simultaneously.

In general, the natural implementations of various materials could usually indicate their substantive characteristics. As aforementioned discussion, the tree type [175], [198], two-circuit type [173], [185], H type [172], [185], and net-grid type [171], [185], [193] − [194] should be four discovered natural implementations of purely ideal fractor. There are zero errors between the aforementioned four types of fractance [175], [171] − [173], [185], [193] − [194], [198] with infinite recursive extreme self-similar structures and a purely ideal fractance. What distinguishes the aforementioned four types of fractor [175], [171] − [173], [185], [193] − [194], [198] from the other approximate implementations of fractor [130], [195] − [203], [206] − [215] is that the floating point values of the capacitance, inductance, and resistance of these four natural fractal structure types of fractor [175], [171] − [173], [185], [193] − [194], [198] are never required in deed, whereas the corresponding devices manufactured utilizing the other approximate implementations of fractor [130], [195] − [203], [206] − [215] could never represent a purely ideal fractor. The electrical characteristics of the aforementioned four types of fractor in its natural implementations [175], [171] − [173], [185], [193] − [194], [198] inspire me to apply their infinite recursive extreme self-similar structures to the theoretical derivation of the fractional-order reactance of an arbitrary-order purely ideal fractor.

5.1.3 FRACTIONAL-ORDER REACTANCE OF FRACTOR IN ITS NATURAL IMPLEMENTATION

5.1.3.1 Fractional-order reactance of 1/2-order fractor in its natural implementation

In this subsection, the fractional-order reactance of the 1/2-order fractor in its natural implementation is studied.

As in the aforementioned discussion, fractance is the fractional-order impedance of a fractor. The driving-point impedance function of fractor is its fractional-order reactance. The tree type [175], [198], two-circuit type [173], [185], H type [172], [185], and net-grid type [171], [185], [193] − [194], [198] should be four discovered natural implementations of purely ideal fractor. What distinguishes the aforementioned four types of fractor [171] − [173], [175], [185], [193] − [194], [198] from the other approximate implementations of fractor [130], [195] − [203], [206] − [215] is that the floating point values of the capacitance, inductance, and resistance of these four natural fractal structure types of fractor [171] − [173], [175], [185], [193] − [194], [198] are never required in deed. In fact, there are zero errors between these four types of fractance [171] − [173], [175], [185], [193] − [194], [198] with infinite recursive extreme self-similar structures and a purely ideal fractance. Whereas, the corresponding devices manufactured utilizing the other approximate implementations of fractor [130], [195] − [203], [206] − [215] could never represent a purely ideal fractor.

With respect to the natural implementations of purely ideal fractor, on the one hand, the 1/2-order purely ideal capacitive fractor consists of a series of ordinary capacitors and resistors in the discovered form of an analog circuit on the tree type [175], [198], two-circuit type [173], [185], H type [172], [185], net-grid type [171], [185], [193] − [194] or other undiscovered infinite recursive structures with self-similarity. On the other hand, the 1/2-order purely ideal inductive fractor consists of a series of ordinary inductors and resistors with the same fractal structures as the 1/2-order purely ideal capacitive fractor in its natural implementation. Let's denote fractor as the symbol ⊣〰⊢, in which F is the abbreviation of fractor [171], [193] − [198], and denote the v-order reactance of the v-order fractor as the symbol F_v. For the convenience of illustration, F_v is also referred to as the v-order fractor in this chapter. The structural representation of the 1/2-order fractor in its natural implementation is shown in Figure 5.2.

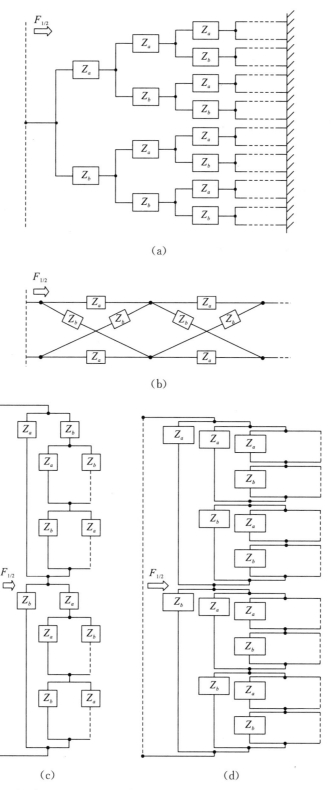

Figure 5.2 1/2-order fractor in its natural implementation: (a) 1/2-order tree type fractor; (b) 1/2-order net-grid type fractor; (c) 1/2-order two-circuit type fractor; (d) 1/2-order H type fractor.

In Figure 5.2, $F_{1/2}$ denotes the 1/2-order reactance of the purely ideal 1/2-order fractor. Figure 5.2 shows that the 1/2-order fractor in its natural implementation has an extreme self-similar fractal structure [171] – [173], [175], [185], [193] – [194], [198]. The 1/2-order purely ideal fractor of the net-grid type requires the minimum quantity of resistors and capacitors and is endowed with the simplest structure. Figure 5.2 (b) shows that the 1/2-order net-grid type purely ideal fractor has an extreme self-similar fractal structure with a series connection of infinitely repeated net-grid type structures, where Z_a and Z_b are the reactances of two ordinary passive circuit elements, resistor and capacitor/inductor. For the convenience of illustration, Z_a and Z_b are also referred to as two ordinary passive circuit elements in this chapter, respectively. The number of Z_a and Z_b is equal to twice the number of layers. The equivalent circuit of the 1/2-order net-grid type purely ideal fractor is shown in Figure 5.3.

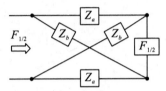

Figure 5.3 Equivalent circuit of the 1/2-order net-grid type fractor.

Suppose the currents of Z_a and Z_b are $i_a(s)$ and $i_b(s)$, respectively. $V_i(s)$ and $I_i(s)$ denote the input voltage and input current of $F_{1/2}$, respectively. According to Kirchoff's current law and Kirchoff's voltage law, we obtain from Figure 5.3 the following relationship:

$$\begin{cases} Z_a i_a + Z_b i_b = V_i \\ (Z_a + F_{1/2}) i_a - (F_{1/2} + Z_b) i_b = 0. \end{cases} \tag{5.9}$$

According to the Cramer's rule of linear algebra, from (5.9), we have:

$$\begin{cases} i_a = \dfrac{(Z_b + F_{1/2})V_i}{\begin{vmatrix} Z_a + F_{1/2} & -(F_{1/2} + Z_b) \\ Z_a & Z_b \end{vmatrix}} \\ i_b = \dfrac{(Z_a + F_{1/2})V_i}{\begin{vmatrix} Z_a + F_{1/2} & -(F_{1/2} + Z_b) \\ Z_a & Z_b \end{vmatrix}}. \end{cases} \tag{5.10}$$

Hence, from (5.10), we can derive the 1/2-order reactance $F_{1/2}$:

$$\begin{aligned} F_{1/2} &= \frac{V_i(s)}{I_i(s)} \\ &= \frac{V_i(s)}{i_a(s) + i_b(s)} \\ &= \frac{2Z_a Z_b + F_{1/2}(Z_a + Z_b)}{2F_{1/2} + Z_a + Z_b}. \end{aligned} \tag{5.11}$$

From (5.11), the following can be obtained:

$$F_{1/2} = V_i(s)/I_i(s) = (Z_a Z_b)^{\frac{1}{2}}. \tag{5.12}$$

On the one hand, with respect to the 1/2-order capacitive fractor in its natural implementation, let's suppose its 1/2-order capacitive reactance is $F^c_{-1/2}$, its resistance and capacitance are r and c, respectively, and the initial electrical energy of the 1/2-order capacitive fractor is equal to zero. Thus, in the Laplace transform domain, $Z_a = r$, and $Z_b = 1/(cs)$. From (5.12), the following is true:

$$F^c_{-1/2} = \xi^{-1/2} s^{-1/2}, \tag{5.13}$$

where $\xi = c/r$. Equation (5.13) shows that the 1/2-order capacitive fractor can be treated as a 1/2-order integrator. $F^c_{-1/2}$ denotes the fractional-order reactance of the 1/2-order capacitive fractor. From (5.13), the relationship between the input voltage $V_i(s)$ and input current $I_i(s)$ of the 1/2-order capacitive fractor can be derived as:

$$V_i(s) = \xi^{-1/2} s^{-1/2} I_i(s). \tag{5.14}$$

The inverse Laplace transform of (5.14) can be derived as:

$$V_i(t) = \frac{1}{\xi^{1/2} \Gamma(1/2)} \int_0^t \frac{I_i(\tau)}{(t-\tau)^{1/2}} d\tau. \tag{5.15}$$

Equations (5.14) and (5.15) show that $V_i(t)$ is in direct proportion to the 1/2-order fractional integral of $I_i(t)$. Whereas, from (5.14), the following can be obtained:

$$I_i(s) = \xi^{1/2} s^{1/2} V_i(s). \tag{5.16}$$

The inverse Laplace transform of (5.16) can be derived as:

$$I_i(t) = \frac{\xi^{1/2}}{\Gamma(1/2)} \int_0^t \frac{V_t^{(1)}(\tau)}{(t-\tau)^{1/2}} d\tau. \tag{5.17}$$

Equations (5.16) and (5.17) show that $I_i(t)$ is in direct proportion to the 1/2-order fractional derivative of $V_i(t)$.

On the other hand, with respect to the 1/2-order inductive fractor in its natural implementation, let's suppose its 1/2-order inductive reactance is $F^l_{1/2}$, its resistance and inductance are r and l, respectively, and the initial electrical energy of the 1/2-order inductive fractor is equal to zero. Therefore, in the Laplace transform domain, $Z_a = r$, and $Z_b = ls$. From (5.12), the following is true:

$$F^l_{1/2} = \xi^{1/2} s^{1/2}, \tag{5.18}$$

where $\xi = l/r$. Equation (5.18) shows that the 1/2-order inductive fractor can be treated as a 1/2-order differentiator. $F^l_{1/2}$ denotes the fractional-order reactance of the 1/2-order inductive fractor. Similarly, from (5.18), the relationship between input voltage $V_i(s)$ and input current $I_i(s)$ of the 1/2-order inductive fractor can be derived as follows, respectively:

$$V_i(s) = \xi^{1/2} s^{1/2} I_i(s), \tag{5.19}$$

$$I_i(s) = \xi^{-1/2} s^{-1/2} V_i(s). \tag{5.20}$$

Equations (5.19) and (5.20) show that $V_i(t)$ is in direct proportion to the 1/2-order fractional derivative of $I_i(t)$, whereas $I_i(t)$ is in direct ratio to the 1/2-order fractional integral of $V_i(t)$.

5.1.3.2 Fractional-order reactance of the arbitrary-order fractor in its natural implementation

In this subsection, the fractional-order reactance of the arbitrary-order fractor in its natural implementation is further proposed. The aforementioned four infinite recursive extreme self-similar structures of the 1/2-order fractor in its natural implementations [1, 16, 24−28, 32] can be applied to the derivation of the fractional-order reactance of the arbitrary-order fractor in its natural implementations.

On the one hand, by extending aforementioned derivation, the fractional-order capacitive reactance of the arbitrary-order capacitive fractor can be derived. Suppose that $Z_a = r$ and $Z_b = F^c_{-1/2}$ in Figure 5.2 (b). From (5.12) and (5.13), the following is true:

$$F^c_{-1/4} = r^{1/2} \xi^{-1/4} s^{-1/4}. \tag{5.21}$$

Suppose that $Z_a = r$ and $Z_b = F^c_{-1/2^{n-1}}$ in Figure 5.2(b), where n is a positive integer. From (5.12), the following is true:

$$F^c_{-1/2^n} = r^{\frac{2^{n-1}-1}{2^{n-1}}} \xi^{\frac{-1}{2^n}} s^{\frac{-1}{2^n}}. \tag{5.22}$$

Similarly, suppose that $Z_a = 1/(cs)$ and $Z_b = F^c_{-1/2}$ in Figure 5.2(b). From (5.12) and (5.13), the following can be obtained:

$$F^c_{-3/4} = c^{-1/2} \xi^{-1/4} s^{-3/4}. \tag{5.23}$$

Then, suppose that $Z_a = 1/(cs)$ and $Z_b = F^c_{-(2^{n-1}-1)/2^{n-1}}$ in Figure 5.2(b). It follows from (5.12) that:

$$F^c_{-(2^n-1)/2^n} = c^{\frac{-(2^{n-1}-1)}{2^{n-1}}} \xi^{\frac{-1}{2^n}} s^{\frac{-(2^n-1)}{2^n}}. \tag{5.24}$$

Similarly, suppose that $Z_a = 1/(cs)$ and $Z_b = F^c_{-1/2^n}$ in Figure 5.2(b). From (5.12) and (5.22), one can obtain:

$$F^c_{-(2^n+1)/2^{n+1}} = c^{\frac{-1}{2}} r^{\frac{2^{n-1}-1}{2^n}} \xi^{\frac{-1}{2^{n+1}}} s^{\frac{-(2^n+1)}{2^{n+1}}}. \tag{5.25}$$

Suppose that $Z_a = 1/(cs)$ and $Z_b = F^c_{-(2^n+1)/2^{n+1}}$ in Figure 5.2(b). From (5.12) and (5.25), one can obtain:

$$F^c_{-(2^{n+1}+2^n+1)/2^{n+2}} = c^{\frac{-3}{4}} r^{\frac{2^{n-1}-1}{2^{n+1}}} \xi^{\frac{-1}{2^{n+2}}} s^{\frac{-(2^{n+1}+2^n+1)}{2^{n+2}}}. \tag{5.26}$$

Then, suppose that $Z_a = 1/(cs)$ and $Z_b = F^c_{-(2^{n+k-1}-2^n+1)/2^{n+k-1}}$ in Figure 5.2(b), where k is a positive integer. It follows from (5.12) that:

$$F^c_{-(2^{n+k}-2^n+1)/2^{n+k}} = c^{\frac{-(2^k-1)}{2^k}} r^{\frac{2^{n-1}-1}{2^{n+k-1}}} \xi^{\frac{-1}{2^{n+k}}} s^{\frac{-(2^{n+k}-2^n+1)}{2^{n+k}}}. \tag{5.27}$$

Similarly, suppose that $Z_a = r$ and $Z_b = F^c_{-(2^n-1)/2^n}$ in Figure 5.2(b). From (5.12) and (5.24), one can obtain:

$$F^c_{-(2^n-1)/2^{n+1}} = r^{\frac{1}{2}} c^{\frac{-(2^{n-1}-1)}{2^n}} \xi^{\frac{-1}{2^{n+1}}} s^{\frac{-(2^n-1)}{2^{n+1}}}. \tag{5.28}$$

Suppose that $Z_a = r$ and $Z_b = F^c_{-(2^n-1)/2^{n+1}}$ in Figure 5.2(b). From (5.12) and (5.28), the following can be obtained:

$$F^c_{-(2^n-1)/2^{n+2}} = r^{\frac{3}{4}} c^{\frac{-(2^{n-1}-1)}{2^{n+1}}} \xi^{\frac{-1}{2^{n+2}}} s^{\frac{-(2^n-1)}{2^{n+2}}}. \tag{5.29}$$

Then, suppose that $Z_a = r$ and $Z_b = F^c_{-(2^n-1)/2^{n+k-1}}$ in Figure 5.2(b), where k is a positive integer. It follows from (5.12) that:

$$F^c_{-(2^n-1)/2^{n+k}} = r^{\frac{2^k-1}{2^k}} c^{\frac{-(2^{n-1}-1)}{2^{n+k-1}}} \xi^{\frac{-1}{2^{n+k}}} s^{\frac{-(2^n-1)}{2^{n+k}}}. \tag{5.30}$$

In addition, if n and k are non-negative integers, respectively, the following is true:

$$\frac{1}{2} \leqslant \frac{2^{n+k} - 2^n + 1}{2^{n+k}} = p \leqslant 1, \tag{5.31}$$

$$0 \leqslant \frac{2^n - 1}{2^{n+k}} = p \leqslant \frac{1}{2}. \tag{5.32}$$

From (5.27) and (5.31), when $1/2 \leqslant p \leqslant 1$, the following can be obtained:

$$\begin{aligned} F^c_{-p} &= c^{\frac{(2^n-1)(1-p)-(2^n-1)}{2^n-1}} r^{\frac{(2^n-1)(1-p)}{2^n-1}} s^{-p} \\ &= c^{-p} r^{1-p} s^{-p}. \end{aligned} \tag{5.33}$$

From (5.30) and (5.32), when $0 \leqslant p \leqslant 1/2$, the following can be obtained:

$$\begin{aligned} F^c_{-p} &= c^{\frac{-(2^n-1)p}{2^n-1}} r^{\frac{(2^n-1)(1-p)}{2^n-1}} s^{-p} \\ &= c^{-p} r^{1-p} s^{-p}. \end{aligned} \tag{5.34}$$

Equations (5.33) and (5.34) show that no matter whether $1/2 \leqslant p \leqslant 1$ or $0 \leqslant p \leqslant 1/2$, F^c_{-p} has the same analytical expression, $c^{-p} r^{1-p} s^{-p}$. Therefore, (5.33) and (5.34) are the joint analytical expression of the fractional-order capacitive reactance of the p-order capacitive fractor in its natural implementation, where $0 \leqslant p \leqslant 1$. Hence, from (5.33) and (5.34), if $0 \leqslant p \leqslant 1$, the following is true:

$$F^c_{-p} = c^{-p} r^{1-p} s^{-p}. \tag{5.35}$$

Further, (5.8) shows that the first-order capacitive reactance of a capacitor is as follows:

$$\begin{aligned} Z^c_{-1}(s) &= V_i(s)/I_i(s) \\ &= c^{-1} s^{-1}. \end{aligned} \tag{5.36}$$

Equation (5.36) shows that with respect to capacitor, $V_i(t)$ is in direct proportion to the first-order integral of $I_i(t)$, whereas $I_i(t)$ is in direct ratio to the first-order derivative of $V_i(t)$. A capacitor achieves the first-order integrator. Then, from (5.36), the capacitive reactance of a cascade system of the q-stage first-order integrators can be derived as:

$$\begin{aligned} Z^c_{-q}(s) &= V_i(s)/I_i(s) \\ &= c^{-q} s^{-q}, \end{aligned} \tag{5.37}$$

where q is a positive integer. Note that in order to synthesize the higher-order purely ideal capacitive fractor by a combination of two basic circuit elements (resistor and capacitor), we only intend to achieve a cascade system of the q-stage first-order integrators in (5.37), which is not the same as $N_{n=-q}$.

Then, we can easily achieve the arbitrary-order capacitive fractor in its natural implementation in a cascade manner. The fractional-order capacitive reactance of the arbitrary-order capacitive fractor in its natural implementation can be derived from (5.35) and (5.37) as:

$$\begin{aligned} F^c_{-v} &= F^c_{-(q+p)} \\ &= V_i(s)/I_i(s) \\ &= c^{-v} r^{1-p} s^{-v} \end{aligned}$$

$$= c^{-(q+p)} r^{1-p} s^{-(q+p)}, \tag{5.38}$$

where $v = q+p$ is a positive real number, and q is a positive integer, $0 \leqslant p \leqslant 1$. Equations (5.5) and (5.38) show that the v-order capacitive fractor implements the v-order fractional integrator. Equation (5.38) represents the fractional-order capacitive reactance of an arbitrary-order purely ideal capacitive fractor. For example, with regard to a 1.75-order purely ideal capacitive fractor, its 1.75-order capacitive reactance should be $F^c_{1.75} = c^{-1.75} r^{0.25} s^{-1.75}$. Furthermore, (5.11), (5.14), and (5.38), show that the purely ideal v-order capacitive fractor achieves a cascade system of a v-order fractional integral operator, a fractional power function of resistance, and a fractional power function of capacitance. Figure 5.1 shows that the v-order purely ideal capacitive fractor, where $0 < v < 1$, should be lying on the line segment, S_1, between C and R. If the fractional-order v satisfies $n < v < n+1$, where n is a non-negative integer, the v-order purely ideal capacitive fractor should be still lying on the line segment, S_1, between C and R, of Chua's axiomatic element system in Figure 5.1. Therefore, the electrical properties of the capacitive fractor should fall in between the electrical properties of the capacitor and those of the resistor. The capacitive fractor can be considered in a certain way as an interpolation of the resistor and capacitor. The fractional-order v, capacitance c, and resistance r are nonlinearly hybridized with the v-order purely ideal capacitive fractance in its natural implementations. The fractional-order capacitive reactance of an arbitrary-order purely ideal capacitive fractor should be closely related with the fractional-order v, capacitance c, and resistance r, simultaneously. Equation (5.38) represents exactly the nonlinear relationship of the fractional-order v, capacitance c, and resistance r for the fractional-order capacitive reactance of a purely ideal arbitrary-order capacitive fractor.

On the other hand, similarly, the fractional-order inductive reactance of the arbitrary-order inductive fractor can be derived. Suppose that $Z_a = r$, and $Z_b = F^l_{1/2}$ in Figure 5.2 (b). It follows from (5.12) and (5.18) that:

$$F^l_{1/4} = r^{1/2} \xi^{1/4} s^{1/4}. \tag{5.39}$$

Then, suppose that $Z_a = r$, and $Z_b = F^l_{1/2^{n-1}}$ in Figure 5.2(b), where n is a positive integer. It follows from (5.12) that:

$$F^l_{1/2^n} = r^{\frac{2^{n-1}-1}{2^{n-1}}} \xi^{\frac{1}{2^n}} s^{\frac{1}{2^n}}. \tag{5.40}$$

Similarly, let's suppose that $Z_a = ls$ and $Z_b = F^l_{1/2}$ in Figure 5.2 (b). From (5.12) and (5.18), one can obtain:

$$F^l_{3/4} = l^{1/2} \xi^{1/4} s^{3/4}. \tag{5.41}$$

Then, suppose that $Z_a = ls$ and $Z_b = F^l_{(2^{n-1}-1)/2^{n-1}}$ in Figure 5.2 (b). From (5.12), the following can be obtained:

$$F^l_{(2^n-1)/2^n} = l^{\frac{2^{n-1}-1}{2^{n-1}}} \xi^{\frac{1}{2^n}} s^{\frac{2^n-1}{2^n}}. \tag{5.42}$$

Similarly, let's suppose that $Z_a = ls$ and $Z_b = F^l_{1/2^n}$ in Figure 5.2(b). From (5.12) and (5.40), one can obtain:

$$F^l_{(2^n+1)/2^{n+1}} = l^{\frac{1}{2}} r^{\frac{2^{n-1}-1}{2^n}} \xi^{\frac{1}{2^{n+1}}} s^{\frac{2^n+1}{2^{n+1}}}. \tag{5.43}$$

Then, suppose that $Z_a = ls$ and $Z_b = F^l_{(2^{n+k-1}-2^n+1)/2^{n+k-1}}$ in Figure 5.2(b), where k is a positive integer. It follows from (5.12) that:

$$F^l_{(2^{n+k}-2^n+1)/2^{n+k}} = l^{\frac{2^k-1}{2^k}} r^{\frac{2^{n-1}-1}{2^{n+k-1}}} \xi^{\frac{1}{2^{n+k}}} s^{\frac{2^{n+k}-2^n+1}{2^{n+k}}}. \tag{5.44}$$

Similarly, suppose that $Z_a = r$ and $Z_b = F^l_{(2^n-1)/2^n}$ in Figure 5.2 (b). It follows from (5.12) and (5.42) that:

$$F^l_{(2^n-1)/2^{n+1}} = r^{\frac{1}{2}} l^{\frac{2^{n-1}-1}{2^n}} \xi^{\frac{1}{2^{n+1}}} s^{\frac{2^n-1}{2^{n+1}}}. \tag{5.45}$$

Then, suppose that $Z_a = r$ and $Z_b = F^l_{(2^n-1)/2^{n+k-1}}$ in Figure 5.2(b), where k is a positive integer. From (5.12), one can obtain:

$$F^l_{(2^n-1)/2^{n+k}} = r^{\frac{2^k-1}{2^k}} l^{\frac{2^{n-1}-1}{2^{n+k-1}}} \xi^{\frac{1}{2^{n+k}}} s^{\frac{2^n-1}{2^{n+k}}} \tag{5.46}$$

Therefore, similarly, the fractional-order inductive reactance of the p-order inductive fractor can be derived, where $0 \leqslant p \leqslant 1$. From (5.44) and (5.46), if $0 \leqslant p \leqslant 1$, the following is true:

$$F^l_p = l^p r^{1-p} s^p. \tag{5.47}$$

Further, (5.8) shows that the first-order inductive reactance of an inductor can be derived as:

$$Z^l_1(s) = V_i(s)/I_i(s) \\ = l^1 s^1. \tag{5.48}$$

Equation (5.48) shows that an inductor implements the first-order differentiator. Then, from (5.48), the inductive reactance of a cascade system of the q-stage first-order differentiators can be derived as:

$$Z^1_q(s) = V_i(s)/I_i(s) \\ = l^q s^q, \tag{5.49}$$

where q is a positive integer. Note that in order to synthesize the higher-order purely ideal inductive fractor by a combination of two basic circuit elements (resistor and inductor), we only intend to achieve a cascade system of the q-stage first-order differentiators in (5.49), which is not the same as $N_{n=q}$.

Thus, we can easily achieve the arbitrary-order inductive fractor in its natural implementation in a cascade manner. The fractional-order inductive reactance of the arbitrary-order inductive fractor in its natural implementation can be derived from (5.47) and (5.49) as:

$$F^l_v = F^l_{q+p} \\ = V^i(s)/I_i(s) \\ = l^v r^{1-p} s^v \\ = l^{q+p} r^{1-p} s^{q+p}, \tag{5.50}$$

where $v = q+p$ is a positive real number, and q is a positive integer, $0 \leqslant p \leqslant 1$. Equations (5.5) and (5.50) shows that the v-order inductive fractor implements the v-order fractional differentiator. Equation (5.50) represents the fractional-order inductive reactance of an arbitrary-order purely ideal inductive fractor. For example, with regard to a 1.75-order purely ideal inductive fractor, its 1.75-order inductive reactance should be $F^l_{1.75} = l^{1.75} r^{0.25} s^{1.75}$. Furthermore, (5.11), (5.19), and (5.50) show that the purely ideal v-order inductive fractor achieves a cascade system of a v-order fractional differential operator, a fractional power function of resistance, and a fractional power function of inductance. Figure 5.1 shows that the v-order purely ideal inductive fractor, where $0 < v < 1$, should be lying on the line segment, S_2, between L and R. If the fractional-order v satisfies $n < v < n+1$, where n is a non-negative integer, the v-order purely ideal inductive fractor should be still lying on the line segment, S_2, between L and R, of Chua's axiomatic element system in Figure 5.1. Therefore, the electrical properties of the inductive fractor should fall in between the electrical properties of the inductor and those of the resistor. The inductive fractor can be considered in a certain way as an interpolation of the resistor and inductor. The fractional-order v, inductance l, and resistance r are nonlinearly hybridized with the v-order purely ideal inductive fractance in its natural implementations. The fractional-order inductive reactance of an arbitrary-order purely ideal inductive fractor should be closely related with the fractional-order v, inductance l, and resistance r, simultaneously. Equation (5.50) represents exactly the nonlinear relationship of the fractional-order v, inductance l, and resistance r for the fractional-order inductive reactance of a purely ideal arbitrary-order inductive fractor.

5.1.3.3 Branch-current of fractor in its natural implementation

In this subsection, according to Kirchhoff's current law and Kirchhoff's voltage law, the branch-current of fractor in its natural implementation is analyzed.

At first, for the 1/2-order purely ideal fractor, from (5.10), it follows that:

$$\frac{i_a(s)}{i_b(s)} = \frac{Z_b + F_{1/2}}{Z_a + F_{1/2}}, \quad (5.51)$$

where $i_a(s)$ and $Z_a = r$ are the branch-current and the reactance of the resistor in a 1/2-order purely ideal fractor in the Laplace transform domain, respectively. $i_b(s)$ and Z_b are the branch-current and the capacitive or inductive reactance of the corresponding capacitor or inductor in a 1/2-order purely ideal fractor in the Laplace transform domain, respectively. Therefore, if $i_b(s)$ and Z_b denote the branch-current of the capacitor and its capacitive reactance, respectively, $F_{1/2}$ is a 1/2-order purely ideal capacitive fractor. If $i_b(s)$ and Z_b denote the branch-current of the inductor and the inductive reactance, respectively, $F_{1/2}$ is a 1/2-order purely ideal inductive fractor. Thus, from (5.11), it follows that:

$$I_i(s) = i_a(s) + i_b(s), \quad (5.52)$$

where $I_i(s)$ is the input current of the 1/2-order purely ideal fractor. Thus, from (5.51) and (5.52), it follows that:

$$i_a(s) = \frac{F_{1/2} + Z_b}{2F_{1/2} + Z_a + Z_b} I_i(s). \quad (5.53)$$

Substituting (5.12) into (5.53), one can obtain:
$$i_a(s) = \left[1 - \frac{Z_a + (Z_a Z_b)^{1/2}}{(Z_a^{1/2} + Z_b^{1/2})^2}\right] I_i(s). \tag{5.54}$$

Secondly, with regard to the 1/4-order purely ideal fractor, from (5.21) and (5.39), its equivalent circuit is shown in Figure 5.4.

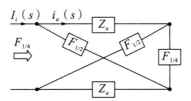

Figure 5.4　Equivalent circuit of 1/4-order net-grid-type purely ideal fractor.

In Figure 5.4, $I_i(s)$ is the input current of a 1/4-order purely ideal fractor. $i_a(s)$ and $Z_a = r$ are the branch-current and the reactance of the resistor in a 1/4-order purely ideal fractor in the Laplace transform domain, respectively. Therefore, if $F_{1/2}$ denotes a 1/2-order purely ideal capacitive fractor, $F_{1/4}$ is a corresponding 1/4-order purely ideal capacitive fractor. If $F_{1/2}$ denotes a 1/2-order purely ideal inductive fractor, $F_{1/4}$ is a corresponding 1/4-order purely ideal inductive fractor. Similarly, from (5.12) and Figure 5.4, the following can be obtained:

$$\begin{aligned} i_a(s) &= \left[1 - \frac{Z_a + (Z_a F_{1/2})^{1/2}}{(Z_a^{1/2} + F_{1/2}^{1/2})^2}\right] I_i(s) \\ &= \left[1 - \frac{Z_a + (Z_a^{3/2} Z_b^{1/2})^{1/2}}{(Z_a^{1/2} + Z_a^{1/2} Z_b^{1/2})^2}\right] I_i(s) \\ &= \left[1 - \frac{Z_a + Z_a^{3/4} Z_b^{1/4}}{(Z_a^{1/2} + Z_a^{1/2} Z_b^{1/2})^2}\right] I_i(s), \end{aligned} \tag{5.55}$$

where Z_b is the capacitive or inductive reactance of the corresponding capacitor or inductor in $F_{1/2}$.

Thirdly, for a p-order purely ideal fractor, from (5.35) and (5.47), the following can be obtained:
$$F_p = Z_a^{1-p} Z_b^p, \tag{5.56}$$

where $0 \leqslant p \leqslant 1$. $Z_a = r$ and Z_b are the reactance of the resistor and the capacitor or inductor in a p-order purely ideal fractor in the Laplace transform domain, respectively. From (5.56), the equivalent circuit of a purely ideal $p/2$-order fractor is shown in Figure 5.5.

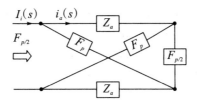

Figure 5.5　Equivalent circuit of $p/2$-order net-grid-type purely ideal fractor.

In Figure 5.5, $I_i(s)$ is the input current of a $p/2$-order purely ideal fractor. $i_a(s)$ and $Z_a = r$ are the branch-current and the reactance of the resistor in a $p/2$-order purely ideal fractor in the Laplace transform domain, respectively. Therefore, if F_p denotes a p-order purely ideal capacitive fractor, $F_{p/2}$ is a corresponding $p/2$-order purely ideal capacitive fractor. If F_p denotes a p-order purely ideal inductive fractor, $F_{p/2}$ is a corresponding $p/2$-order purely ideal inductive fractor. Similarly, from (5.56) and Figure 5.5, the following can be obtained:

$$i_a(s) = \left[1 - \frac{Z_a + (Z_a F_p)^{1/2}}{(Z_a^{1/2} + F_p^{1/2})^2}\right] I_i(s)$$

$$= \left[1 - \frac{Z_a + (Z_a^{2-p} Z_b^p)^{1/2}}{(Z_a^{1/2} + Z_a^{1-p} Z_b^p)^2}\right] I_i(s)$$

$$= \left[1 - \frac{Z_a + Z_a^{(2-p)/2} Z_b^{p/2}}{(Z_a^{1/2} + Z_a^{1-p} Z_b^p)^2}\right] I_i(s). \tag{5.57}$$

In addition, from (5.37) and (5.49), we can easily achieve an arbitrary-order purely ideal capacitive or inductive fractor by connecting a purely ideal $p/2$-order fractor with a m-stage first-order integrators or differentiators in a cascade manner. For a series circuit, the input current for each stage first-order integrator or differentiator is identical to the input current of a $p/2$-order purely ideal fractor, $I_i(s)$. Then, in the Laplace transform domain, (5.57) represents the branch-current of the resistor in the $(m + p/2)$-order, i.e. arbitrary-order, net-grid-type purely ideal fractor.

In particular, we would like to offer some additional remarks to supplement my previous article on fracmemristor [224]. Fractor and fracmemristor are different from each other but have a closed relationship between each other [59]. It should be noted that there is a negligent mistake in my previous article [224]. The branch-current of the memristor in a purely ideal $1/2$-order fracmemristor, a purely ideal $1/4$-order fracmemristor, and a purely ideal $p/2$-order fracmemristor in the Laplace transform domain should be $i_a(s) = \left[1 - \frac{Z_a + (Z_a Z_b)^{1/2}}{(Z_a^{1/2} + Z_b^{1/2})^2}\right] I_i(s)$, $i_a(s) = \left[1 - \frac{Z_a + Z_a^{3/4} Z_b^{1/4}}{(Z_a^{1/2} + Z_a^{1/2} Z_b^{1/2})^2}\right] I_i(s)$, and $i_a(s) = \left[1 - \frac{Z_a + Z_a^{(2-p)/2} Z_b^{p/2}}{(Z_a^{1/2} + Z_a^{1-p} Z_b^p)^2}\right] I_i(s)$, respectively.

5.1.3.4 Order-sensitivity characteristics of fractor

In this subsection, the order-sensitivity characteristics of a fractor are proposed.

Equaitons (5.5), (5.38), and (5.50) shows that the crux of the hardware implementation of an arbitrary-order fractor is to achive a fractional integral operator s^{-v} or a fractional differential operator s^v in the form of an analog circuit. For s^{-v} is the reciprocal of s^v, we only need analyze the issue of either s^{-v} or s^v. If v is an arbitrary positive rational number, v is equal to the summation of a positive integer and a positive rational fraction. Thus, the following can be obtained:

$$s^v = s^m s^{i/n}, \tag{5.58}$$

where m, n, and i are positive integers, $n \neq 1$, $i = 1, 2, \cdots, n-1$, and i/n is a positive rational fraction. s^m can be achieved by means of inductors or capacitors. Hence, the key point of the hardware implementation of s^v is to achieve $s^{i/n}$ in the form of an analog circuit. In view of circuits and systems, $s^{i/n}$ can be treated as the fractional-order transfer function of a i/n-order system that is equal to the quotient of dividing its input by its output. Thus, $s^{i/n}$ can be rewritten as:

$$\begin{aligned} s^{i/n} &= B(s)/A(s) \\ &= \frac{b_\alpha s^\alpha + b_{\alpha-1} s^{\alpha-1} + \cdots + b_1 s + b_0}{s^\beta + a_{\beta-1} s^{\beta-1} + \cdots + a_1 s + a_0} \\ &= \frac{\sum_{k=0}^{\alpha} b_k s^{k-\beta}}{1 + \sum_{l=0}^{\beta-1} a_l s^{l-\beta}}, \end{aligned} \quad (5.59)$$

where $B(s)$ and $A(s)$ denote the input and output of $s^{i/n}$, respectively, α, β, k, and l are positive integers, and a_l and b_k are constant. In light of systems theory, all physically realizable systems must satisfy the condition of $\alpha \leqslant \beta$ in (5.59). In addition, if $|\alpha - (\beta s)^{i/n}| < 1$, then from the binomial theorem, one can obtain $\lim_{j \to \infty}[\alpha - (\beta s)^{i/n}]^j = \lim_{j \to \infty} \sum_{k=0}^{j}(-1)^k \binom{j}{k} \alpha^{j-k} \beta^{ik/n} s^{quo(ik/n)} s^{rem(ik/n)} = 0$, where α and β are positive numbers, $\binom{j}{k} = \frac{\Gamma(1+j)}{\Gamma(j-k+1)\Gamma(k+1)}$, quo( ) denotes quotient calculation, and rem( ) denotes the remainder calculation. Therefore, if $j \to \infty$, we can derive the computational completeness of $s^{rem(ik/n)}$ as $\lim_{j \to \infty} s^{rem(ik/n)} = s^{i/n} = s^{1/n}$, $s^{2/n}$, $\cdots$, $s^{n-1/n}$ corresponding to $i = 1, 2, \cdots, n-1$. Thus, in light of the binomial theorem, when $i = 1, 2, \cdots, n-1$, the denominators of each $s^{i/n}$ in (5.59) are identical to each other, whereas the numerators of each $s^{i/n}$ in (5.59) are not equal to each other, respectively. For example, if $n = 6$, the denominators of $s^{1/6}$, $s^{2/6}$, $s^{3/6}$, $s^{4/6}$, and $s^{5/6}$ in (5.59) are identical to each other, whereas the numerators of $s^{1/6}$, $s^{2/6}$, $s^{3/6}$, $s^{4/6}$, and $s^{5/6}$ in (5.59) are not equal, respectively. Hence, the numerators of (5.59) determine the numerator i of the fractional-order i/n of $s^{i/n}$, whereas the denominator of (5.59) determines the denominator n of its fractional-order i/n. The variations of its numerators and denominator are both order-sensitive to $s^{i/n}$, but the variation of its denominator is relatively more order-sensitive than the variations of its numerators. According to Mason rules, the cascaded signal flow graph of $s^{i/n}$ can be derived from (5.59). It is shown in Figure 5.6.

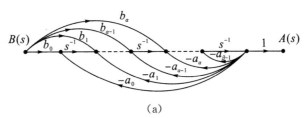

(a)

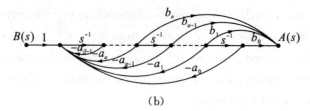

(b)

Figure 5.6 Cascaded signal flow graph of $s^{i/n}$: (a) Set rightmost s^{-1} as common part of $s^{i/n}$; (b) Set leftmost s^{-1} as common part of $s^{i/n}$.

Figure 5.6 shows that, according to the Mason rules, if all forward paths and feedback paths of $s^{i/n}$ have common parts, each numerator term of (5.59) constitutes an individual forward path, while each denominator term of (5.59) except for 1 constitutes an individual feedback path. Figures 5.6 (a) and 5.6 (b) are two equivalent cascaded signal flow graphs of $s^{i/n}$. We set the rightmost s^{-1} in Figure 5.6 (a) and the leftmost s^{-1} in Figure 5.6 (b) as a common part of $s^{i/n}$, respectively. Equation (5.59) and Figure 5.6 show that $s^{i/n}$ only has three fundamental operations, namely, s^{-1}, scalar product, and addition. Hence, we only need to implement the first-order integrator, multiplier, and summator to implement s^{-1}, scalar product, and addition, respectively. Figure 5.6 further shows that the forward path of the cascaded signal flow graphs of $s^{i/n}$ determines the numerator i of its fractional-order i/n, while its feedback path determines the denominator n of its fractional-order i/n. From aforementioned discussion, we can see that the variation of the forward path and feedback path of the cascaded signal flow graphs of $s^{i/n}$ are both order-sensitive to $s^{i/n}$, but the variation of its feedback path is relatively more order-sensitive to $s^{i/n}$ than the variation of its forward path.

In view of circuits and systems, the fractional differential operator $s^{i/n}$ can be treated as a driving-point impedance function of a i/n-order inductive fractor. Thus, $s^{i/n}=V_i(s)/I_i(s)$ denotes the fractional-order reactance, where $V_i(s)=B(s)$ is the input voltage and $I_i(s)=A(s)$ is the input current of a i/n-order inductive fractor. On the one hand, if $s^{i/n}$ has not multiple poles, the following can be obtained:

$$I_i(s) = (s+\vartheta_1)\cdots(s+\vartheta_k)\cdots(s+\vartheta_m), \tag{5.60}$$

where $-\vartheta_1, \cdots, -\vartheta_k, \cdots, -\vartheta_m$ are different simple roots of $I_i(s)=0$ for $k=1, 2, \cdots, m$, respectively. In light of systems theory, all stable systems must meet the condition of $\vartheta_k>0$. Thus, from (5.59) and (5.60), and in light of the binomial theorem, it follows that:

$$\begin{aligned} s^{i/n} &= V_i(s)/I_i(s) \\ &= C_0 + \frac{C_1}{s+\vartheta_1} + \cdots + \frac{C_k}{s+\vartheta_k} + \cdots + \frac{C_m}{s+\vartheta_m}, \end{aligned} \tag{5.61}$$

where $C_0 = \lim\limits_{s\to\infty} s^{i/n}$, and $C_k = \lim\limits_{s\to\vartheta_k}(s+\vartheta_k)s^{i/n}$. On the other hand, if $s^{i/n}$ has the r-order multiple poles, the following can be obtained:

$$s^{i/n} = V_i(s)/I_i(s)$$

$$= C_0 + \frac{C_1^1}{(s+\vartheta_1)} + \cdots + \frac{C_1^r}{(s+\vartheta_1)} + \frac{C_2}{(s+\vartheta_2)} + \cdots + \frac{C_m}{(s+\vartheta_m)}, \tag{5.62}$$

where $C_1^q = \lim_{s \to \vartheta_1}(s+\vartheta_1)^q s^{i/n}$, $q=1, 2, \cdots, r$. Equations (5.61) and (5.62) shows that the i/n-order inductive fractor $s^{i/n}$ is equivalent to the series circuit of the first-order integral analog circuit, i.e. fractional calculus is the result of the continuous interpolation of the integer-order calculus. $s^{i/n}$ is the result of the nonlinear interaction of the series circuit of the first-order integral analog circuit. $s^{i/n}$ is equal to a nonlinear expression of the first-order differentiator s. The kth series first-order integral analog circuit can be shown in Figure 5.7.

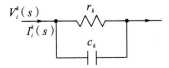

Figure 5.7 kth series first-order integral analog circuit.

Thus, from Figure 5.7, r_k and c_k can be derived as follows, respectively:

$$r_k = C_k/\vartheta_k \, \Omega, \tag{5.63}$$
$$c_k = 1/C_k \, \text{F}, \tag{5.64}$$

where the measurement units of resistance and capacitance are Ohm (Ω) and Farad (F), respectively. Thus, the input impedance of the kth series first-order integral analog circuit can be derived as:

$$Z^k(s) = V_i^k(s)/I_i^k(s)$$
$$= \frac{C_k}{s+\vartheta_k}. \tag{5.65}$$

Therefore, from (5.59) and (5.60), the corresponding analog cascade circuit of $s^{i/n}$ can be shown in Figure 5.8.

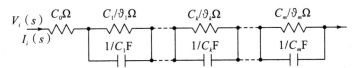

Figure 5.8 Analog cascade circuit of $s^{i/n}$.

Equations (5.61) − (5.65) and Figure 5.8 shows that the resistances of a i/n-order inductive fractor essentially depend on the numerator and poles of $s^{i/n}$ simultaneously, whereas the capacitances of a i/n-order inductive fractor only essentially depend on the numerator of $s^{i/n}$.

Note that the numerators of (5.59) determine the numerator i of the fractional-order i/n of $s^{i/n}$, while the denominator of (5.59) determines the denominator n of its fractional-order i/n. The variations of its numerators and denominator are both order-sensitive to $s^{i/n}$, but the variation of its denominator is relatively more order-sensitive than the variations of its numerators. Thus, from (5.61) − (5.65) and Figure 5.8, we can further see that the denominator n of its fractional-order i/n of $s^{i/n}$ is solely determined by the resistances of $s^{i/n}$, whereas the numerator i of its fractional-order i/n of $s^{i/n}$ is jointly determined by the resistances and capacitances of $s^{i/n}$ simultaneously. Hence, the variations of the resistances and capacitances of $s^{i/n}$ are both order-sensitive to $s^{i/n}$, but the variations of its resistances are relatively more order-sensitive than the variations of its capacitances.

5.1.4 EXPERIMENT AND ANALYSIS

In this section, through mathematical analysis and simulation results, some related issues of the fractional-order reactance of a fractor in its natural implementation are discussed in detail experimentally.

5.1.4.1 Temperature effect of fractor

In this subsection, in order to implement a robust fractor, the temperature effect of a fractor should be studied.

The resistor and capacitor in most fractors are subject to working conditions and environmental variations such as temperature, supply voltage, and electromagnetic interference, which can affect their performance. One crux of the implementation of a robust fractor is to keep its fractional-order being the same under different working conditions and environmental variations, especially across varying temperatures.

Example 1: With respect to a v-order purely ideal capacitive or inductive fractor in its natural implementation [171] − [173], [175], [185], [193] − [194], [198] from (5.38) and (5.50), we can see that the temperature effect can only change the magnitude of its v-order reactance, but cannot change its fractional-order. Suppose the temperature coefficient of resistance (TCR) and the temperature coefficient of capacitance (TCC) are equal to constants ζ_r and ζ_c, respectively. For a 2.5-order purely ideal capacitive fractor in its natural implementation, let's set $v=2.5$, $c=2$, and $r=2$ in (5.38), and set $\zeta_r=1$ and $\zeta_c=1$, $\zeta_r=2$ and $\zeta_c=3$, and $\zeta_r=5$ and $\zeta_c=18$, respectively. Substituting $j\omega$ for s in (5.38), where j denotes an imaginary unit, and ω denotes angular frequency, we can derive the Bode diagram of temperature effect of a purely ideal capacitive fractor $2^{-2.5} \cdot 2^{0.5} \cdot s^{-2.5}$ in its natural implementation, as shown in Figure 5.9.

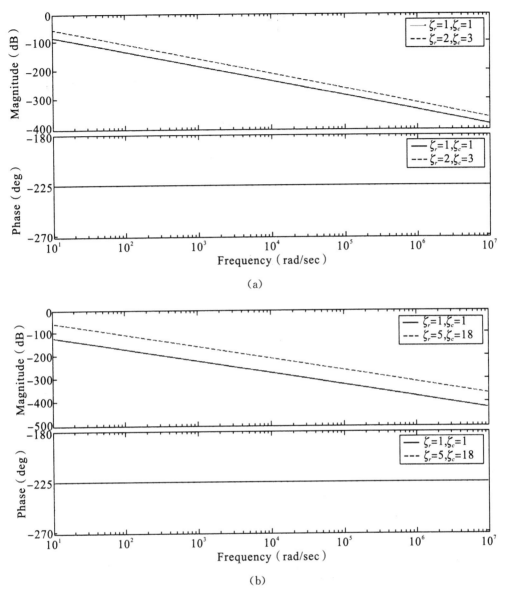

Figure 5.9 Bode diagram of temperature effect of $2^{-2.5} \cdot 2^{0.5} \cdot s^{-2.5}$ ($v=2.5, c=2, r=2$): (a) $\zeta_r=2$ and $\zeta_3=3$; (b) $\zeta_r=5$ and $\zeta_c=18$.

In Figure 5.9, the curve of $\zeta_r=1$ and $\zeta_c=1$ represents the Bode diagram of a 2.5-order purely ideal capacitive fractor $2^{-2.5} \cdot 2^{0.5} \cdot s^{-2.5}$ in its natural implementation, while the curve of $\zeta_r \neq 1$ and $\zeta_c \neq 1$ represents the Bode diagram of a temperature influenced capacitive fractor $s_T^{-2.5} = (\zeta_c)^{-2.5} (\zeta_r)^{0.5} \cdot 2^{-2.5} \cdot 2^{0.5} \cdot s^{-2.5}$. Figure 5.9 shows that the magnitude of a temperature influenced capacitive fractor $s_T^{-2.5}$ obviously deviates from that of a 2.5-order purely ideal capacitive fractor $2^{-2.5} \cdot 2^{0.5} \cdot s^{-2.5}$ in its natural implementation. Meanwhile, the phase of the former coincides exactly with that of the latter. In general, the temperature effect can only change the magnitude of the fractiona-order reactance of a purely ideal capacitive fractor in its natural implementation, but cannot change its fractional-order.

Example 2: With respect to an approximately implemented capacitive or inductive fractor, in order to eliminate the change of resistance or capacitance caused by temperature variation, the commonly used method is to connect in series a pair of resistors with opposite temperature coefficient, and to connect in parallel a pair of capacitors with opposite temperature coefficient in an approximately implemented fractor, whereas the temperature effect of an approximately implemented fractor can be also defensed against by employing a set of specific resistor and capacitor, whose temperature coefficients are the reciprocal of each other. We suggest setting $\zeta_c = 1/\zeta_r$. Thus, when temperature varies, from (5.63) and (5.64), one can obtain $r_k = \zeta_r C_k/\vartheta_k \Omega$, and $c_k = \zeta_c/C_k = 1/\zeta_r C_k \text{F}$ in Figure 5.7. From (5.65), the following can be obtained:

$$Z_T^k(s) = V_i^k(s)/I_i^k(s)$$
$$= \frac{\zeta_r C_k}{s + \vartheta_k}. \tag{5.66}$$

Substituting (5.66) into (5.61) or (5.62) and it follows that:

$$s_T^{i/n} = \zeta_r C_0 + \frac{\zeta_r C_1}{s + \vartheta_1} + \cdots \frac{\zeta_r C_k}{s + \vartheta_k} + \cdots + \frac{\zeta_r C_n}{s + \vartheta_n}$$
$$= \zeta_r s^{i/n}, \tag{5.67}$$

where $s_T^{i/n}$ denotes a temperature influenced fractor. Equation (5.67) shows that if $\zeta_c = 1/\zeta_r$ is satisfied, temperature effect produces an effect on $s_T^{i/n} = \zeta_r s^{i/n}$. If $\zeta_c = 1/\zeta_r$ is satisfied, temperature effect can only change the magnitude of the fractional-order reactance of fractor $s^{i/n}$, but cannot change its fractional-order i/n. Note that $\zeta_c = 1/\zeta_r$ is a very strictly restricted condition for a set of resistor and capacitor. But with progress made in prospective material science, some innovative materials maybe make it possible for us to find out the set of resistor and capacitor, whose temperature coefficients are the reciprocal of each other.

According to the binomial theorem, an approximate implementation of a band-pass filtering 1/2-order capacitive fractor $s^{-1/2}$ can be derived as:

$$s^{-1/2} \approx \frac{1.2 \cdot 10^{3/2} \cdot s^5 + 2.2 \cdot 10^{7/2} \cdot s^4 + 7.92 \cdot 10^{9/2} \cdot s^3 + 7.92 \cdot 10^{11/2} \cdot s^2 + 2.2 \cdot 10^{13/2} \cdot s + 1.2 \cdot 10^{13/2}}{s^6 + 6.6 \cdot 10^2 \cdot s^5 + 4.95 \cdot 10^4 \cdot s^4 + 9.24 \cdot 10^5 \cdot s^3 + 4.95 \cdot 10^6 \cdot s^2 + 6.6 \cdot 10^6 \cdot s + 10^6}$$
$$= \frac{3.09352345 \cdot 10}{s + 5.76954805 \cdot 10^2} + \frac{3.5988971}{s + 5.82842742 \cdot 10} + \frac{1.42217976}{s + 1.39839637 \cdot 10} + \frac{8.3736612263 \cdot 10^{-1}}{s + 5.88790706}$$
$$+ \frac{6.17473122 \cdot 10^{-1}}{s + 1.71572875} + \frac{5.3618124 \cdot 10^{-1}}{s + 1.73323801 \cdot 10^{-1}}. \tag{5.68}$$

Equation (5.68) shows that all of the poles of the 1/2-order approximately implemented capacitive fractor $s^{-1/2}$ are on the lift half plane. Thus, the 1/2-order approximately implemented capacitive fractor $s^{-1/2}$ is a stable system. From (5.63), (5.64), and Figure 5.7, we can derive the corresponding analog cascade circuit of the 1/2-order approximately implemented capacitive fractor $s^{-1/2}$ as shown in Figure 5.10.

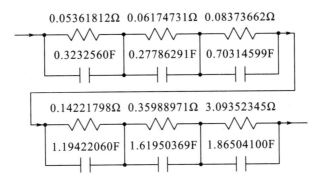

Figure 5.10 Analog cascade circuit of approximately implemented $s^{-1/2}$.

Substituting $j\omega$ for s in (5.68), we can derive the Bode diagram of the 1/2-order approximately implemented capacitive fractor $s^{-1/2}$ and that of the ideal filter $s^{-1/2}$, respectively, as shown in Figure 5.11.

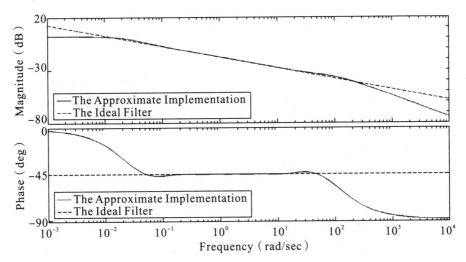

Figure 5.11 Bode diagram of approximately implemented $s^{-1/2}$ and that of the ideal filter $s^{-1/2}$.

Figure 5.11 shows that the frequency characteristic of the 1/2-order approximately implemented capacitive fractor $s^{-1/2}$ can approach that of the ideal filter $s^{-1/2}$ with a high degree of accuracy in the pass-band of [0.1rad/sec, 12rad/sec]. Both the magnitude and the phase of the fractional-order reactance of the approximately implemented capacitive fractor $s^{-1/2}$ can approach those of the ideal filter $s^{-1/2}$ in the pass-band of [0.1rad/sec, 12rad/sec].

As aforementioned discussion in (5.67), we suggest to set TCR and TCC be equal to constants ζ_r and ζ_c respectively, and to set $\zeta_c = 1/\zeta$. Without loss of generality, suppose that $\zeta_r = 2$. Thus, from (5.67) and Figure 5.10, we can get an analog cascade circuit of the capacitive fractor $s_T^{-1/2}$ of the temperature variation, as shown in Figure 5.12.

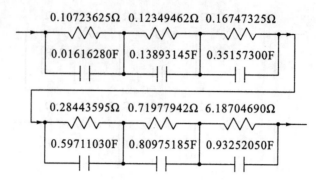

Figure 5.12 Analog cascade circuit of temperature influenced $s_T^{-1/2}$.

From (5.67) and (5.68), the fractional-order reactance of temperature influenced 1/2-order capacitance fractor $s_T^{-1/2}$ can be derived as:

$$s_T^{-1/2} \approx \frac{2.4 \cdot 10^{3/2} \cdot s^5 + 4.4 \cdot 10^{7/2} \cdot s^4 + 1.584 \cdot 10^{11/2} \cdot s^3 + 1.584 \cdot 10^{13/2} \cdot s^2 + 4.4 \cdot 10^{13/2} \cdot s + 2.4 \cdot 10^{13/2}}{s^6 + 6.6 \cdot 10^2 \cdot s^5 + 4.95 \cdot 10^4 \cdot s^4 + 9.24 \cdot 10^5 \cdot s^3 + 4.95 \cdot 10^6 \cdot s^2 + 6.6 \cdot 10^6 \cdot s + 10^6}$$

$$= \frac{6.18704690 \cdot 10}{s + 5.76954805 \cdot 10^2} + \frac{7.19779418}{s + 5.82842712 \cdot 10} + \frac{8.84435952}{s + 1.69839637 \cdot 10} + \frac{1.6747325}{s + 5.887909706} + \frac{1.23494624}{s + 1.71572875}$$

$$+ \frac{1.07236249}{s + 1.73323801 \cdot 10^{-1}}. \tag{5.69}$$

Substituting $j\omega$ for s in (5.69), we can derive the Bode diagram of the 1/2-order approximately implemented capacitive fractor $s^{-1/2}$ in (5.68) and the temperature influenced 1/2-order capacitive fractor $s_T^{-1/2}$ in (5.69), as shown in Figure 5.13.

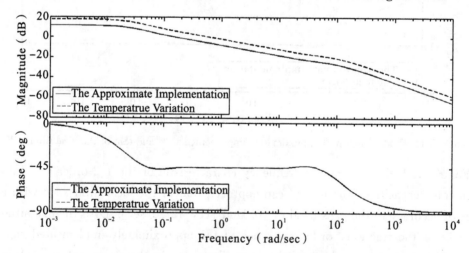

Figure 5.13 Bode diagram of approximately implemented $s^{-1/2}$ and temperature influenced $s_T^{-1/2}$.

Figure 5.13 shows that the magnitude of the 1/2-order reactance of the temperature influenced capacitive fractor $s_T^{-1/2}$ in (5.69) obviously deviates from that of the 1/2-order approximately implemented capacitive fractor $s^{-1/2}$ in (5.68). Meanwhile, the phase of the former coincides exactly with that of the latter. If $\zeta_c = 1/\zeta_r$ is satisfied, the temperature effect can only change the magnitude of the fractional-order reactance of an approximately implemented fractor, but cannot change its phase, i.e. if $\zeta_c = 1/\zeta_r$ is satisfied, the temperature effect cannot change the fractional-order of an approximately implemented fractor.

5.1.4.2　Influence of fractional-order on intrinsic electrical characteristics of a fractor

In this subsection, the influence of the fractional-order of an arbitrary-order purely ideal fractor on its intrinsic electrical characteristics is analyzed.

Example 3: The influence of fractional-order v on the intrinsic electrical characteristics of a v-order purely ideal capacitive fractor is analyzed.

Equation (5.38) represents the fractional-order capacitive reactance of an arbitrary-order purely ideal capacitive fractor. In (5.38), the fractional-order $v = q + p$ of a purely ideal capacitive fractor is a positive real number, where q is a positive integer, $0 \leqslant p \leqslant 1$. Without loss of generality, let's set $c = 2$ and $r = 2$ in (5.38). Substituting $j\omega$ for s in (5.38), we can derive the Bode diagram of a v-order purely ideal capacitive fractor $2^{-v} 2^{1-p} s^{-v}$, as shown in Figure 5.14.

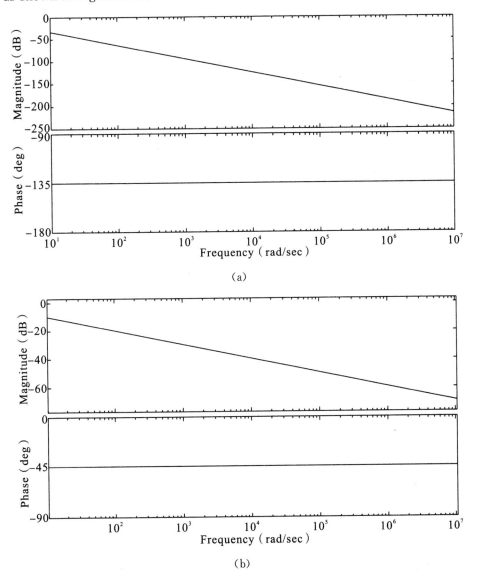

(a)

(b)

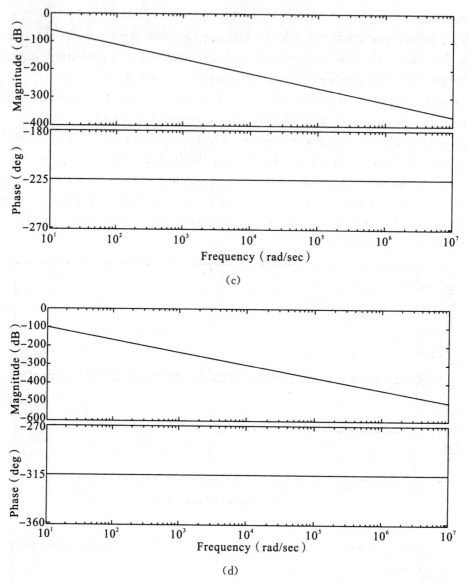

Figure 5.14 Bode diagram of a v-order purely ideal capacitive fractor $2^{-v}2^{1-p}s^{-v}$: (a) $v=0.5$; (b) $v=1.5$; (c) $v=2.5$; (d) $v=3.5$.

Figure 5.14 shows that with respect to a v-order purely ideal capacitive fractor $2^{-v}2^{1-p}s^{-v}$, the magnitude of its v-order capacitive reactance will decrease as the fractional-order v increases, whereas the phase of its v-order capacitive reactance will decrease 90deg as the fractional-order v increases every first-order. In addition, the magnitude of its v-order capacitive reactance will linearly decrease as frequency increases, whereas the phase of its v-order capacitive reactance will keep constant as frequency increases.

Equation (5.38) and Figure 5.14 further show that the fractional-order v essentially influences on the intrinsic electrical characteristics of a v-order purely ideal capacitive fractor $c^{-(q+p)}r^{1-p}s^{-(q+p)}$. Let's set the fractional-power-exponents, $-v=-(q+p)$ and

$(1-p)$, of capacitance c and resistance r in (5.38) as the electrical characteristics exponents of the capacitance c and resistance r of a v-order purely ideal capacitive fractor $c^{-(q+p)}r^{1-p}s^{-(q+p)}$, respectively. Thus, from (5.38), we can derive the influence of the fractional-order v on the electrical characteristics of capacitance c and resistance r of a v-order purely ideal capacitive fractor $c^{-(q+p)}r^{1-p}s^{-(q+p)}$, as shown in Figure 5.15.

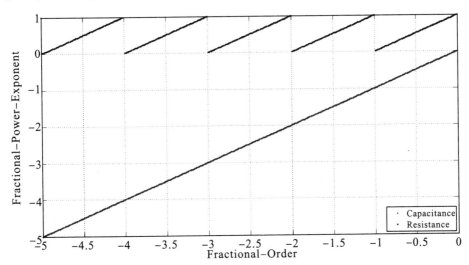

Figure 5.15 Influence of fractional-order on capacitance and resistance of a purely ideal capacitive fractor.

Figure 5.15 shows that with respect to a v-order purely ideal capacitive fractor $c^{-(q+p)}r^{1-p}s^{-(q+p)}$, the fractional-power-exponent of its capacitance c will linearly decrease as its fractional-order v increases, whereas the fractional-power-exponent of its resistance r is a piecewise continuously differentiable function that will linearly decrease as its fractional-order v increases in every sectional-continuous domain of definition.

In addition, (5.38), Figure 5.1, and Figure 5.15 furtherindicate some intrinsic electrical characteristics of a v-order purely ideal capacitive fractor $c^{-(q+p)}r^{1-p}s^{-(q+p)}$. First, when the fractional-order v of a v-order purely ideal capacitive fractor $c^{-(q+p)}r^{1-p}s^{-(q+p)}$ is equal to zero, the fractional-power-exponents of its capacitance c and resistance r are equal to zero and 1, respectively. Note that when $v=0$, the v-order purely ideal capacitive fractor $c^{-(q+p)}r^{1-p}s^{-(q+p)}$ converts to be an ordinary resistor, which is lying on the point R of Chua's periodic table of all two-terminal circuit elements in Figure 5.1.

Second, when the fractional-order $v=1$, the fractional-power-exponents of the capacitance c and resistance r of the v-order purely ideal capacitive fractor $c^{-(q+p)}r^{1-p}s^{-(q+p)}$ are equal to -1 and zero, respectively. Note that when $v=1$, the v-order purely ideal capacitive fractor $c^{-(q+p)}r^{1-p}s^{-(q+p)}$ converts to be an ordinary capacitor, which is lying on the point C of Chua's periodic table of all two-terminal circuit elements in Figure 5.1.

Third, when the fractional-order $v=n$, where n is a non-negative integer, the fractional-power-exponents of the capacitance c and resistance r of the v-order purely ideal capacitive fractor $c^{-(q+p)}r^{1-p}s^{-(q+p)}$ are equal to $-n$ and zero, respectively. Note that when $v=n$, the v-order purely ideal capacitive fractor $c^{-(q+p)}r^{1-p}s^{-(q+p)}$ converts to be a n-order integrator, which is still lying on the point C of Chua's periodic table of all two-terminal circuit elements in Figure 5.1.

Fourth, when the fractional-order v satisfies $n<v<n+1$, where n is a non-negative integer, the fractional-power-exponents of the capacitance c and resistance r of the v-order purely ideal capacitive fractor $c^{-(q+p)}r^{1-p}s^{-(q+p)}$ take values in the open intervals of $(-n-1, -n)$ and $(0, 1)$, respectively. Note that when $n<v<n+1$, the v-order purely ideal capacitive fractor $c^{-(q+p)}r^{1-p}s^{-(q+p)}$ should be still lying on lying on the line segment, S_1, between C and R, of Chua's periodic table of all two-terminal circuit elements in Figure 5.1.

Example 4: the influence of the fractional-order v on the intrinsic electrical characteristics of a v-order purely ideal inductive fractor is analyzed.

Equation (5.50) represents the fractional-order inductive reactance of an arbitrary-order purely ideal inductive fractor. In (5.50), the fractional-order $v=q+p$ of a inductive fractor is a positive real number, where q is a positive integer, $0 \leqslant p \leqslant 1$. Without loss of generality, let's set $l=2$ and $r=2$ in (5.50). Substituting $j\omega$ for s in (5.50), we can derive the Bode diagram of a v-order purely ideal inductive fractor $2^v 2^{1-p} s^v$, as shown in Figure 5.16.

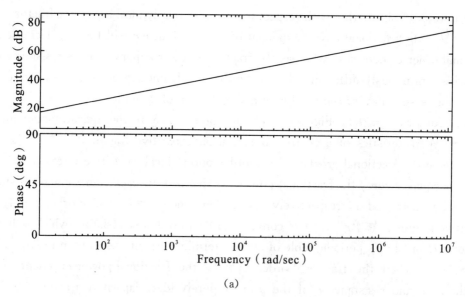

(a)

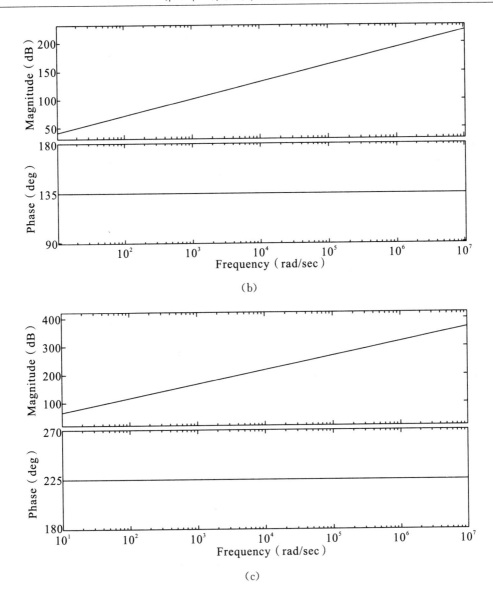

(b)

(c)

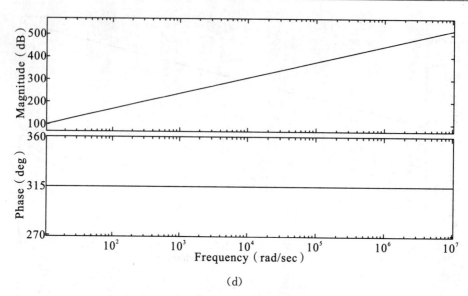

Figure 5.16 Bode diagram of a v-order purely ideal inductive fractor $2^v 2^{1-p} s^v$: (a) $v=0.5$; (b) $v=1.5$; (c) $v=2.5$; (d) $v=3.5$.

Figure 5.16 shows that with regard to a v-order purely ideal inductive fractor $2^v 2^{1-p} s^v$, the magnitude of its v-order inductive reactance will increase as the fractional-order v increases, whereas the phase of its v-order inductive reactance will increase 90deg as the fractional-order v increases every first-order. In addition, the magnitude of its v-order inductive reactance will linearly increase as frequency increases, whereas the phase of its v-order inductive reactance will keep constant as frequency increases.

Equation (5.50) and Figure 5.16 further show that the fractional-order v essentially influences on the intrinsic electrical characteristics of a v-order purely ideal inductive fractor $l^{q+p} r^{1-p} s^{q+p}$. Let's set the fractional-power-exponents, $v=(q+p)$ and $(1-p)$, of inductance l and resistance r in (5.50) as the electrical characteristics exponents of the inductance l and resistance r of a v-order purely ideal inductive fractor $l^{q+p} r^{1-p} s^{q+p}$, respectively. Thus, from (5.50), we can derive the influence of the fractional-order v on the electrical characteristics of inductance l and resistance r of a v-order inductive fractor $l^{q+p} r^{1-p} s^{q+p}$, as shown in Figure 5.17.

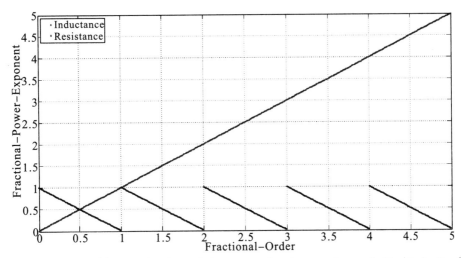

Figure 5.17 Influence of fractional-order on inductance and resistance of a purely ideal inductive fractor.

Figure 5.17 shows that with regard to a v-order purely ideal inductive fractor $l^{q+p}r^{1-p}s^{q+p}$, the fractional-power-exponent of its inductance l will linearly increase as its fractional-order v increases, whereas the fractional-power-exponent of its resistance r is a piecewise continuously differentiable function that will linearly decrease as its fractional-order v increases in every sectional-continuous domain of definition.

In addition, (5.50), Figure 5.1, and Figure 5.17 further indicate some intrinsic electrical characteristics of a v-order purely ideal inductive fractor $l^{q+p}r^{1-p}s^{q+p}$. First, when the fractional-order v of a v-order purely ideal inductive fractor $l^{q+p}r^{1-p}s^{q+p}$ is equal to zero, the fractional-power-exponents of its inductance l and resistance r are equal to zero and 1, respectively. Note that when $v=0$, the v-order inductive fractor $l^{q+p}r^{1-p}s^{q+p}$ converts to be an ordinary resistor, which is lying on the point R of Chua's periodic table of all two-terminal circuit elements in Figure 5.1. Second, when the fractional-order $v=1$, the fractional-power-exponents of the inductance l and resistance r of the v-order inductive fractor $l^{q+p}r^{1-p}s^{q+p}$ are equal to 1 and zero, respectively. Note that when $v=1$, the v-order inductive fractor $l^{q+p}r^{1-p}s^{q+p}$ converts to be an ordinary inductor, which is lying on the point L of Chua's periodic table of all two-terminal circuit elements in Figure 5.1. Third, when the fractional-order $v=n$, where n is a non-negative integer, the fractional-power-exponents of the inductance l and resistance r of the v-order inductive fractor $l^{q+p}r^{1-p}s^{q+p}$ are equal to n and zero, respectively. Note that when $v=n$, the v-order inductive fractor $l^{q+p}r^{1-p}s^{q+p}$ converts to be a n-order differentiator, which is still lying on the point L of Chua's periodic table of all two-terminal circuit elements in Figure 5.1. Fourth, when the fractional-order v satisfies $n<v<n+1$, where n is a non-negative integer, the fractional-power-exponents of the inductance l and resistance r of the v-order inductive fractor $l^{q+p}r^{1-p}s^{q+p}$ take values in the open intervals of $(n-1, n)$ and $(0, 1)$, respectively. Note that when $n<v<n+1$, the v-order inductive fractor $l^{q+p}r^{1-p}s^{q+p}$ should be still lying on the line segment, S_2, between L and R, of Chua's periodic table of

all two-terminal circuit elements in Figure 5.1.

5.1.5 CONCLUSIONS

Fractional calculus has evolved as an important, contemporary branch of mathematical analyses. The origin of fractional calculus was concurrent with that of integer-order calculus; however, until recently, the applications of fractional calculus were limited to mathematics. For physical scientists and engineering technicians, fractional calculus is a novel promising mathematical method. The application of fractional calculus to circuits and systems, especially to a promising fractional-order circuit element called fractor, is an emerging and interesting discipline, in which insufficient studies have been implemented. Fractional calculus has been applied to circuits and systems mainly because of its intrinsic strengths of long-term memory, non-locality, and weak singularity. The term "fractor" arose following the successful synthesis of a fractional-order capacitor or a fractional-order inductor in an analog circuit. The term "fractance", as a portmanteau of "the fractional-order impedance", refers to the fractional-order impedance of a fractor. Thus, it is natural to ponder what the measurement units and physical dimensions of fractance are. Until now, however, no efficient measurement units and physical dimensions of fractance have been proposed, which is a challenging theoretical problem. Motivated by this need, in this chapter, based on the aforementioned studies, here and in the companion paper (Part II), the measurement units and physical dimensions of fractance and rules for fractors in series and parallel are discussed in detail. A state-of-the-art mathematical method, fractional calculus, is used to analyze the proposed conceptual framework of fractor. In particular, part I proposed to introduce the position of purely ideal fractor in Chua's circuit axiomatic element system and the fractional-order reactance of fractor in its natural implementation. this chapter discusses fundamental issues, whereas Part II is devoted to the measurement units and physical dimensions of purely ideal fractance and the rules for fractors in series and parallel.

5.2 Measurement Units and Physical Dimensions of Fractance-Part II: Fractional-Order Measurement Units and Physical Dimensions of Fractance and Rules for Fractors in Series and Parallel

Here and in the companion paper (Part I), a novel conceptual framework on the measurement units and physical dimensions of fractance and rules for fractors in series and parallel is mainly discussed. The term "fractor" arose following the successful synthesis of a fractional-order capacitor or a fractional-order inductor in an analog circuit. Fractor is actually a promising fractional-order circuit element that is a core component for the hardware implementation of the fractional-order circuits and systems. The term "fractance", as a portmanteau of "the fractional-order impedance", refers to the fractional-order impedance of a fractor. Up to now, however, no effective measurement units and physical dimensions of fractance have been proposed for fractor, which is a challenging theoretical problem. Motivated by this need, in this part, we studied the fractional-order measurement units and physical dimensions of fractance and rules for fractors in series and parallel. We use a state-of-the-art mathematical method, fractional calculus, to analyze the proposed conceptual framework. In particular, the fundamental issues introduced in the companion paper (Part I) are combined with an analysis for the realistic requirement of the fractional-order measurement units of fractance, and a proposal for the fractional-order measurement units for capacitive fractance and inductive fractance, respectively, as well as the fractional-order physical dimensions of fractance, together with the rules for fractors in series and parallel, respectively. Finally, an arbitrary-order fractor in the form of an analog circuit is achieved by the binomial theorem based approximate implementation, the implementations of capacitance and resistance of arbitrary value are discussed, and approximately implemented fractors in series and parallel are analyzed in detail experimentally.

5.2.1 INTRODUCTION

In this second part of a two-part sequence, the fundamental issues introduced in the companion paper (Part I) [225] are combined with an analysis for the realistic requirement of the fractional-order measurement units of fractance, and a proposal for the fractional-order measurement units for capacitive fractance and inductive fractance, respectively, as well as the fractional-order physical dimensions of fractance, together with the rules for fractors in series and parallel, respectively. This part uses the same notation as Part I [225].

The application of fractional calculus to circuits and systems, especially to a promising fractional-order circuit element called fractor, is an emerging and interesting discipline, in which insufficient studies have been implemented. Fractional calculus has been applied to circuits and systems mainly because of its inherent strength of long-term memory, non-locality, and weak singularity. The term "fractor" arose following the successful synthesis of a fractional-order capacitor or a fractional-order inductor in an analog circuit. The term "fractance", as a portmanteau of "the fractional-order impedance," refers to the fractional-order impedance of a fractor [171] – [173], [175], [185], [193] – [194], [198], [226] Further, the driving-point impedance function of fractor is its fractional-order reactance. There are two types of fractor in nature: capacitive fractor and inductive fractor. Moreover, to seek the fractional-order extreme points of an energy norm, the integer-order steepest descent approach has been generalized to a fractional-order approach [192]. In light of the fundamental issues introduced in Part I [225], on the one hand, a capacitive fractor is a fractional-order capacitor, which involves a negative-order fractional differential filter. The capacitive fractance is the fractional-order impedance of a capacitive fractor. Since the position of purely ideal capacitive fractor in Chua's circuit axiomatic element system is between that of a capacitor and that of a resistor, the electrical properties of a purely ideal capacitive fractor should fall in between those of a capacitor and those of a resistor. On the other hand, an inductive fractor is a fractional-order inductor, which involves a positive-order fractional differential filter. The inductive fractance is the fractional-order impedance of an inductive fractor. Since the position of purely ideal inductive fractor in Chua's circuit axiomatic element system is between that of an inductor and that of a resistor, the electrical properties of a purely ideal inductive fractor should fall in between those of an inductor and those of a resistor.

In particular, with respect to the fundamental issues introduced in Part I [225], the tree type [175], [198], two-circuit type [173], [185], H type [172], [185], net-grid type [171], [185], [193] − [194] should be four discovered natural implementations of purely ideal fractor. What distinguishes the aforementioned four types of fractor [171] − [173], [175], [185], [198], [193] − [194] from the other approximate implementations of fractor [130], [196], [198] − [203], [206] − [215], [227] − [228] is that the floating point values of the capacitance, inductance, and resistance of these four natural fractal structure types of fractor [171] − [173], [175], [185], [198], [193] − [194] are never required in deed. In fact, there are zero errors between these four types of fractance [171] − [173], [175], [185], [198], [193] − [194] with infinite recursive extreme self-similar structures and a purely ideal fractance, whereas the corresponding devices manufactured utilizing the other approximate implementations of fractor [130], [196], [198] − [203], [206] − [215], [227] − [228] could never represent a purely ideal fractor. There are many remarkable progresses in the approximate implementations of fractor. For example, L. Dorčák et al. proposed the analogue electronic realization of the fractional-order (FO) systems, e. g. controlled objects and/or controllers whose mathematical models are FO differential equations. The electronic realization is based on FO differentiator and FO integrator where operational amplifiers are connected with appropriate impedance, i. e. the FO element or constant phase element (CPE) [206] − [208]. J. Valsa et al. described a possible realization of such a model that is quite simple and in spite of its simplicity makes it possible to simulate the properties of ideal CPEs [209]. E. A. Gonzalez and I. Petráš offered a comprehensive discussion on the applications of fractional calculus in the design and implementation of fractional-order systems in the form of electronic circuits that could be used for signal processing and control engineering applications [210]. E. A. Gonzalez et al. presented the mathematical properties of a generalized fractional-order two-port network represented as a symmetrical T-section through its hybrid parameters [211]. G. L. Abulencia et al. studied the analog realization of a selectable fractional-order differentiator in a microelectronics scale, whose order of differentiation can be selected between 0.25 and 0.50 [212]. A. Tepljakov et al. proposed a modification of Newton's method for approximating a first-order implicit fractional transfer function, which corresponds to a frequency-bounded fractional differentiator or integrator [213]. A. Tepljakov et al. investigated the possibilities of network generation from the fractional-order controller approximations derived using different methods proposed over the years [214]. E. A. Gonzalez et al. proposed the conceptual design of a variable fractional-order differentiator in which the order can be selected from 0 to 1 with an increment of 0.05 [215].

In general, the natural implementations of various materials could usually indicate their substantive characteristics. The electrical characteristics of the aforementioned four types of fractor in its natural implementations [171] − [173], [175], [185], [198], [193] − [194] inspire me to apply their infinite recursive extreme self-similar structures to the theoretical derivation of the fractional-order reactance of an arbitrary-order purely ideal fractor. In this part, based on the studies already mentioned [171] − [173], [175], [185], [198], [192] − [194], [225] − [226], we propose the fractional-order measurement unit of capacitive fractance and the fractional-order measurement unit of inductive fractance, respectively, as well as the fractional-order physical dimensions of fractance and the rules for fractors in series and parallel. These two companion papers are two elaborately expanded versions of the conference publications of my previous work [204], [205].

The remainder of this part is organized as follows: Section 2 includes a brief necessary recall on some intrinsic electrical characteristics of an arbitrary-order purely ideal fractor in its natural implementation. Section 3 discusses the fractional-order measurement units and physical dimensions of fractance. In Section 3, based on mathematical and physical analysis, the realistic requirement of the fractional-order measurement units of fractance is analyzed; the fractional-order measurement units and physical dimensions of fractance are proposed, followed by the rules for fractors in series and parallel. Section 4 presents the experimental results obtained and the associated analysis carried out in detail. Here, at first, an arbitrary-order fractor in the form of an analog circuit is achieved approximately. Second, the implementation of capacitance and resistance of arbitrary value by active elements is studied. Third, approximately implemented fractors in series and parallel are further analyzed in detail. In Section 5, the conclusions of this manuscript are presented.

5.2.2 RELATED WORK

This section includes a brief necessary recall on some intrinsic electrical characteristics of an arbitrary-order purely ideal fractor in its natural implementation.

In light of the fundamental issues introduced in Part I [225], we can see that first, the position of purely ideal fractor in Chua's axiomatic circuit element system can be shown as given in Figure 5.18.

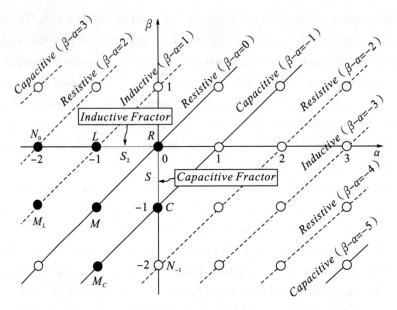

Figure 5.18 Position of purely ideal fractor in Chua's axiomatic circuit element system.

With respect to the fundamental issues introduced in Part I [225], in view of logical consistency, axiomatic completeness, formal symmetry, and constitutive relation, from Figure 5.1, we can see that in Chua's periodic table of all two-terminal circuit elements, the v-order purely ideal capacitive fractor, where $0 < v < 1$, should be lying on the line segment, S_1, between C and R. Meanwhile, the v-order purely ideal inductive fractor, where $0 < v < 1$, should be lying on the line segment, S_2, between L and R. Moreover, if the fractional-order v satisfies $n < v < n+1$, where v is a non-negative integer, the v-order purely ideal capacitive fractor and the v-order purely ideal inductive fractor should be still lying on the line segments S_1 and S_2 of Chua's periodic table of all two-terminal circuit elements in Figure 5.1, respectively. Thus, it is natural to consider that the electrical properties of purely ideal capacitive fractor should fall in between the electrical properties of capacitor and those of resistor. In a similar way, the electrical characteristics of inductive fractor should fall in between the electrical characteristics of inductor and those of resistor.

Second, the capacitive reactance of an ordinary capacitor in its natural implementation can be given as:

$$Z_{-1}^c(s) = V_i(s)/I_i(s)$$
$$= c^{-1}s^{-1}, \qquad (5.70)$$

where c is the capacitance of an ordinary capacitor, and s is the Laplace operator.

Third, the fractional-order v, capacitance c, and resistance r are nonlinearly hybridized with the v-order purely ideal capacitive fractance in its natural implementations. The fractional-order capacitive reactance of an arbitrary-order purely ideal capacitive fractor should be closely related with the fractional-order v, capacitance c, and resistance r, simultaneously. The fractional-order capacitive reactance of an arbitrary-order purely ideal capacitive fractor in its natural implementation can be derived as:

$$\begin{aligned} F_{-v}^c &= F_{-(q+p)}^c \\ &= V_i(s)/I_i(s) \\ &= c^{-v} r^{1-p} s^{-v} \\ &= c^{-(q+p)} r^{1-p} s^{-(q+p)}, \end{aligned} \quad (5.71)$$

where $v = q + p$ is a positive real number, q is a positive integer, and $0 \leqslant p \leqslant 1$.

Fourth, the inductive reactance of an ordinary inductor can be given as:

$$\begin{aligned} Z_1^l(s) &= V_i(s)/I_i(s) \\ &= l^1 s^1, \end{aligned} \quad (5.72)$$

where l is the inductance of an ordinary inductor.

Fifth, the fractional-order v, inductance l, and resistance r are nonlinearly hybridized with the v-order purely ideal inductive fractance in its natural implementations. The fractional-order inductive reactance of an arbitrary-order purely ideal inductive fractor should be closely related with the fractional-order v, inductance l, and resistance r, simultaneously. The fractional-order inductive reactance of an arbitrary-order purely ideal inductive fractor in its natural implementation can be derived as:

$$\begin{aligned} F_v^l &= F_{q+p}^l \\ &= V_i(s)/I_i(s) \\ &= l^v r^{1-p} s^v \\ &= l^{q+p} r^{1-p} s^{q+p}, \end{aligned} \quad (5.73)$$

where $v = q + p$ is a positive real number, q is a positive integer, and $0 \leqslant p \leqslant 1$.

5.2.3 FRACTIONAL-ORDER MEASUREMENT UNITS AND PHYSICAL DIMENSIONS OF FRACTANCE AND RULES FOR FRACTORS IN SERIES AND PARALLEL

5.2.3.1 Realistic requirement of fractional-order measurement units of fractance

In this subsection, to better understand the proposed theoretical framework of this part, the realistic requirement of the fractional-order measurement units of fractance is analyzed.

There is no doubt that we are now experiencing an unprecedented great innovation of circuits and systems theory. The ordinary integer-order two-terminal circuit elements such as capacitor, resistor, inductor, memristor, memcapacitor, and meminductor, belong to an integer-order space, whereas the emerging fractional-order two-terminal circuit elements such as fractor and fractional-order memristor belong to a fractional-order space, which jointly constitute our fascinating real world of circuits and systems. Figure 5.1 shows that in view of Chua's axiomatic circuit element system, the position of an ordinary integer-order two-terminal basic circuit element should be on a discrete point lattice position of Chua's periodic table of all two-terminal circuit elements, whereas the position of an emerging fractional-order two-terminal circuit element, such as fractor, should be lying on a line segment between two discrete point lattice positions of Chua's periodic table of all two-terminal circuit elements. The emerging fractional-order two-terminal circuit elements, such as fractors, extend the theoretical concepts of the ordinary integer-order two-terminal circuit elements from the integer-order to the fractional-order. The traditional integer-order measurement units are simple and logical ways for describing the ordinary integer-order two-terminal circuit elements, but they cannot efficiently describe the emerging fractional-order two-terminal circuit elements. This is a realistic requirement for us to use the fractional-order measurement units of fractance.

As we know, log-log plot is very useful for recognizing the physical relationship and estimating parameters of a power function. Therefore, it is natural to ponder how we can apply log-log plot to analyzing the electrical characteristics of fractance. With respect to an arbitrary-order purely ideal capacitive fractor and an arbitrary-order purely ideal inductive fractor, (5.71) and (5.73) can be rewritten as, respectively:

$$F^c_{-v}(s) = \frac{r^{1-p}}{(cs)^v}$$
$$= \frac{r^{1-p+v}}{(rcs)^v}$$
$$= \frac{K_{CF}}{(\tau_{CF}s)^v}, \qquad (5.74)$$

$$F^l_v(s) = r^{1-p}(ls)^v$$
$$= r^{1-p+v}\left(\frac{l}{r}s\right)^v$$
$$= K_{IF}(\tau_{IF}s)^v, \qquad (5.75)$$

where $K_{CF} = r^{1-p+v}$, $K_{IF} = r^{1-p+v}$, $\tau_{CF} = rc = 1/\omega_c$, and $\tau_{IF} = l/r = 1/\omega_c$. K_{CF} and τ_{CF} denote the reference impedance in magnitude and the reference time scale of a v-order purely ideal capacitive fractor, respectively. K_{IF} and τ_{IF} denote the reference impedance in magnitude and the reference time scale of a v-order purely ideal inductive fractor, respectively. ω_c denotes the reference angular frequency. Then, with the standard interpretation of $s = j\omega$, and $j = \sqrt{1}$, where ω denotes angular frequency, from (5.74) and (5.75), their charge storing v-order reactances can be also described by, respectively:

$$F^c_{-v}(j\omega) = \frac{r^{1-p+v}}{(jrc\omega)^v}$$
$$= \frac{K_{CF}}{(j\tau_{CF}\omega)^v}$$
$$= \frac{K_{CF}}{\left(j\dfrac{\omega}{\omega_c}\right)^v}, \qquad (5.76)$$

$$F^l_v(j\omega) = r^{1-p+v}\left(j\frac{l}{r}\omega\right)^v$$
$$= K_{IF}(j\tau_{IF}\omega)^v$$

$$= K_{IF}\left(j\frac{\omega}{\omega_c}\right)^v. \tag{5.77}$$

Note that (5.74) and (5.76) allow for the description of an ordinary resistor with $v = 0$, through a v-order purely ideal capacitive fractor to an ordinary first-order capacitor with $v = 1$, and through a v-order purely ideal capacitive fractor to an k-order RC network with $v = k$, where k is a nonzero integer. Likewise, (5.75) and (5.77) allow for the description of an ordinary resistor with $v = 0$, through a v-order purely ideal inductive fractor to an ordinary first-order inductor with $v = 1$, and through a v-order purely ideal inductive fractor to an k-order RL network with $v = k$. Therefore, from (5.76) and (5.77), the k-order reactance of a k-order RC network and a k-order RL network can be described by, respectively:

$$Z^{rc}_{-k}(j\omega) = \frac{r^k}{(jrc\omega)^k}$$
$$= \frac{K_{RC}}{(j\tau_{RC}\omega)^k}$$
$$= \frac{K_{RC}}{\left(j\dfrac{\omega}{\omega_c}\right)^k}, \tag{5.78}$$

$$Z^{rl}_k(j\omega) = r^k \left(j\frac{l}{r}\omega\right)^k$$
$$= K_{RL}(j\tau_{RL}\omega)^k$$
$$= K_{RL}\left(j\frac{\omega}{\omega_c}\right)^k, \tag{5.79}$$

where Z^{rc}_{-k} and Z^{rl}_k denote the k-order reactance of a k-order RC network and a k-order RL network, respectively. $K_{RC} = r^k$, $K_{RL} = r^k$, $\tau_{RC} = rc = 1/\omega_c$, and $\tau_{RL} = l/r = 1/\omega_c$. K_{RC} and τ_{RC} denote the reference impedance in magnitude and the reference time scale of a k-order RC network, respectively. K_{RL} and τ_{RL} denote the reference impedance in magnitude and the reference time scale of a k-order RL network, respectively. ω_c denotes the reference angular frequency.

Comparing the (5.74) and (5.76) with (5.78), we can see that with respect to a v-order purely ideal capacitive fractor, $K_{CF} = r^{1-p+v}$ is a $(1-p+v)$-order power function of both resistance r and the fractional-order v, whereas with regard to a k-order RC network, $K_{RC} = r^k$ is a k-order power function of both resistance r and the integer-order k. Likewise, comparing the (5.6) and (5.8) with (5.10), we can see that with respect to a v-order purely ideal inductive fractor, $K_{IF} = r^{1-p+v}$ is a $(1-p+v)$-order power function of both resistance r and the fractional-order v, whereas with regard to a k-order RL network, $K_{RL} = r^k$ is a k-order power function of both resistance r and the integer-order k.

In addition, Bode diagram is a well-known graph of the frequency response of a system, which is a kind of log-log plot. Let me apply Bode diagram to analyzing the electrical characteristics of fractance. Thus, from (5.76), (5.77), (5.78), and (5.79), it follows that, respectively:

$$\begin{cases} L^c_{-v}(\omega) = 20 \cdot \log \left| \dfrac{K_{CF}}{\left(j\dfrac{\omega}{\omega_c}\right)^v} \right| = 20 \cdot (1-p) \cdot \log r - 20 \cdot v \cdot \log c - 20 \cdot v \cdot \log \omega \, (dB) \\ \varphi^c_{-v}(\omega) = -v \cdot \arctan\left(\dfrac{\omega}{\omega_c}/0\right) = -\dfrac{v\pi}{2} \end{cases},$$

(5.80)

$$\begin{cases} L^l_v(\omega) = 20 \cdot \log \left| K_{IF}\left(j\dfrac{\omega}{\omega_c}\right)^v \right| = 20 \cdot (1-p) \cdot \log r + 20 \cdot v \cdot \log l + 20 \cdot v \cdot \log \omega \, (dB) \\ \varphi^l_v(\omega) = v \cdot \arctan\left(\dfrac{\omega}{\omega_c}/0\right) = \dfrac{v\pi}{2} \end{cases},$$

(5.81)

$$\begin{cases} L^{rc}_{-k}(\omega) = 20 \cdot \log \left| \dfrac{K_{RC}}{\left(j\dfrac{\omega}{\omega_c}\right)^k} \right| = 20 \cdot (k-1) \cdot \log r - 20 \cdot k \cdot \log c - 20 \cdot k \cdot \log \omega \, (dB) \\ \varphi^{rc}_{-k}(\omega) = -k \cdot \arctan\left(\dfrac{\omega}{\omega_c}/0\right) = -\dfrac{k\pi}{2} \end{cases},$$

(5.82)

$$\begin{cases} L^{rl}_k(\omega) = 20 \cdot \log \left| K_{RL}\left(j\dfrac{\omega}{\omega_c}\right)^k \right| = 20 \cdot (k-1) \cdot \log r + 20 \cdot k \cdot \log l + 20 \cdot k \cdot \log \omega \, (dB) \\ \varphi^{rl}_k(\omega) = k \cdot \arctan\left(\dfrac{\omega}{\omega_c}/0\right) = \dfrac{k\pi}{2} \end{cases},$$

(5.83)

where $L^c_{-v}(\omega)$, $L^l_v(\omega)$, $L^{rc}_{-k}(\omega)$, and $L^{rl}_k(\omega)$ are the reference magnitude at the reference angular frequency of the Bode diagram of an arbitrary-order purely ideal capacitive fractor, an arbitrary-order purely ideal inductive fractor, a k-order RC network, and a k-order RL network, respecively. $\varphi^c_{-v}(\omega)$, $\varphi^l_v(\omega)$, $\varphi^{rc}_{-k}(\omega)$, and $\varphi^{rl}_k(\omega)$ are the phase of the Bode diagram of an arbitrary-order purely ideal capacitive fractor, an arbitrary-order purely ideal inductive fractor, a k-order RC network, and a k-order RL network, respecively. ω_c denotes the reference angular frequency.

Equations (5.80) − (5.83) show that first, the physical impedances of a v-order purely ideal capacitive fractor, a v-order purely ideal inductive fractor, a k-order RC network, and a k-order RL network cannot be efficiently described in the traditional integer-order measurement units with their reference magnitudes at the reference angular frequency and phases on a Bode diagram, respectively. The reference magnitudes at the reference angular frequency in (5.80) - (5.83) can be plotted as a straight line on Bode diagram versus frequency, which can be described by three pieces of information: the horizontal coordinate and vertical coordinate of a reference point on the straight line, and the slope of the straight line, however, their reference magnitudes at the reference angular frequency nonlinearly contain the collective physical information of order (v or k), resistance (r), capacitance (c) or inductance (l), and the angular frequency (ω) at the same time. Furthermore, their inclusive physical information of resistance and capacitance or inductance present in the forms of logarithm: log$r$ and log$c$ or log$l$, respectively. Apparently, the reference magnitude of fractance in the form of logarithm on Bode diagram cannot be describe separately by order (v), resistance (r), capacitance (c) or inductance (l), and the angular frequency (ω). Then, if we attempt to describe the physical impedance of fractance in the traditional integer-order measurement units with magnitude and phase on Bode diagram, it leads to another theoretical problem: what physical meaning and of the measurement units of $(1-p) \cdot \log(\text{Ohm}) - v \cdot \log(\text{Farad})$ and $(1-p) \cdot \log(\text{Ohm}) + v \cdot \log(\text{Henry})$ are. This issue is a paradox. Thus, we can see that the traditional integer-order measurement units are simple and logical ways for describing the ordinary two-terminal circuit elements, but they cannot efficiently describe the emerging fractional-order two-terminal circuit elements.

Second, in view of signal analysis and signal processing, Bode diagram could be actually treated as a specific time-frequency analysis. In light of Heisenberg's uncertainty principle, just like other methods of time-frequency analysis such as Fourier transform, Gauss transform, Z transform, the fractional-order Fourier transform, Gabor transform, wavelet transform, and the fractional-order wavelet transform, Bode diagram is merely an analyzing method converting original physical problems from one time-frequency-space to another time-frequency-space domain. The nature of a realistic physical rule of circuits and systems in real world cannot be changed because of the change of its analysis method. Bode diagram can easily convert the physical variable combinations and regular form of a power function and simplify its analysis, but it actually cannot change the intrinsic fractional-order relationship among physical variables of fractance. The intrinsic physical problems of the measurement units of $(1-p) \cdot \log(\text{Ohm}) - v \cdot \log(\text{Farad})$ and $(1-p) \cdot \log(\text{Ohm}) + v \cdot \log(\text{Henry})$ are essentially still to deal with the natures of $(\text{Ohm})^{1-p}(\text{Farad})^{-v}$ and $(\text{Ohm})^{1-p}(\text{Henry})^{v}$, respectively. This is a realistic requirement for us to use the fractional-order measurement units of fractance. Common usage of reference impedance magnitude and phase in integer-order measurement units are inadequate to cover description of fractance behavior, and are unable to allow for straight forward derivation of the rules for fractors in series and parallel. It is well known that the rules for ordinary resistors in series and parallel can be given as, respectively:

$$Z^s(s) = \sum_{i=l}^{k} Z_i(s), \tag{5.84}$$

$$Z^p(s) = \frac{1}{\sum_{i=l}^{k} \frac{1}{Z_i(s)}}, \tag{5.85}$$

where $Z_i(s)$, $Z^s(s)$, and $Z^p(s)$ denote the resistance of resistor, resistors in series, and resistors in parallel in Laplace transform domain, respectively. These simple rules for ordinary resistors in series and parallel hold if and only if the integer-order measurement units and physical dimensions are maintained, which should be not suitable for fractors in series and parallel.

Third, the slope of the impedance straight lines of a k-order RC network, a k-order RL network, a v-order purely ideal capacitive fractor, and a v-order purely ideal inductive fractor can be given by corresponding exponents. Further, the phases of a k-order RC network, a k-order RL network, a v-order purely ideal capacitive fractor, and a v-order purely ideal inductive fractor can be given by $\varphi = -k\pi/2$, $\varphi = k\pi/2$, $\varphi = -v\pi/2$, and $\varphi = v\pi/2$, respectively. Their phases should be approximately constant over the same range of operating frequency, which is only determined by a single factor, namely, the order. Thus, we can further see that an arbitrary-order purely ideal capacitive fractor, an arbitrary-order purely ideal inductive fractor, a k-order RC network, and a k-order RL network are all essentially constant phase circuit elements.

Fourth, although log-log regression can be even used to estimate the fractional dimension of a naturally occurring fractal, going in the reverse direction, observing that data appears as an approximate line on a log-log scale and concluding that the data follows a power law, is invalid [229]. Therefore, with regard to a noised fractor, log-log regression is inefficient.

In particular, with regard to an actual analog circuit of a capacitive fractor, the schematic for a fractional-order integrator can be shown as given in Figure 5.19.

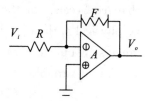

Figure 5.19 Schematic for a fractional-order integrator.

In Figure 5.19, F, R, A, V_i, and V_o represent a v-order capacitive fractor, a resistor, an amplifier, input voltage, and output voltage of a v-order integrator, respectively. In this instance, the v-order feedback reactance of a v-order capacitive fractor can be given by (5.5) and the input impedance is a resistance, R. Therefore, we can derive the transform function of the v-order integrator in Figure 5.2, the overall gain H_A of the amplifier A, from Kirchhoff's voltage law and Kirchhoff's current law. It follows that:

$$H_A(s) = \frac{V_o(s)}{V_i(s)}$$
$$= -\frac{r^{1-p}}{R} \frac{1}{(cs)^v}$$
$$= -\frac{K_{CF}}{R} \frac{1}{(\tau_{CF}s)^v}. \qquad (5.86)$$

It is apparent that (5.86) has the form of the Laplace transform of a fractional-order integrator of order v. Rewriting (5.86) in terms of the Laplace variable, $s = j\omega$, the following can be obtained:

$$H_A(j\omega) = \frac{V_o(j\omega)}{V_i(j\omega)}$$
$$= -\frac{r^{1-p}}{R}\frac{1}{(jc\omega)^v}$$
$$= -\frac{K_{CF}}{R}\frac{1}{(j\tau_{CF}\omega)^v}. \quad (5.87)$$

Further, swapping positions of the resistor and the v-order capacitive fractor in Figure 5.2 could result in a v-order differentiator. Note that the transform function of this v-order integrator is a dimensionless quantity that can be easily calculated from the generalized form of equation (5.74). Cascading such circuits will result in their transform functions multiplying and their phases adding in Laplace transform domain.

Using (5.86) and (5.87), from the Riemann-Liouville defined fractional integral, one can explicitly write out the fractional integral modeled by the analogue circuit of Figure 5.19 as follows:

$$V_o(t) = -\frac{r^{1-p}}{c^v}\frac{1}{R\Gamma(v)}\int_a^t \frac{V_i(t)}{(x-\tau)^{1-v}}d\tau$$
$$= -\frac{K_{CF}}{\tau_{CF}^v}\frac{1}{R\Gamma(v)}\int_a^t \frac{V_i(\tau)}{(x-\tau)^{1-v}}d\tau. \quad (5.88)$$

Equation (5.88) show that it now becomes clear how the reference impedance in magnitude $K_{CF} = r^{1-p+v}$ and the reference time constant $\tau_{CF} = rc = 1/\omega_c$ (derived from the reference frequency) collectively contribute to the Riemann-Liouville defined fractional integral, i.e. $K_{CF} \cdot \tau_{CF}^{-v} = c^{-v}r^{1-p}$. Furthermore, it could become even clearer when we use the Grünwald-Letnikov defined limit sum for fractional integral, given as:

$$V_o(t) = -\lim_{N\to\infty}\sum_{k=0}^{N-1}\left\{\frac{r^{1-p}}{c^v}\Delta t^v \frac{\Gamma(k+v)}{R\Gamma(v)\Gamma(k+1)}V_i(t-k\Delta t)\right\}$$
$$= -\lim_{N\to\infty}\sum_{k=0}^{N-1}\left\{\frac{K_{CF}}{\tau_{CF}^v}\Delta t^v \frac{\Gamma(k+v)}{R\Gamma(v)\Gamma(k+1)}V_i(t-k\Delta t)\right\}, \quad (5.89)$$

where $\Delta t = (t-a)/N$. Equation (5.89) show that a reasonable interpretation is that $K_{CF} \cdot \tau_{CF}^{-v} = c^{-v}r^{1-p}$ rescales time in the sum over the history of the input signal. Fractional calculus has been applied to circuits and systems mainly because of its inherent strength of long-term memory, non-locality, and weak singularity. Note especially that the v-order capacitive fractor can only be properly interpreted when $K_{CF} = r^{1-p+v}$ and $\tau_{CF} = rc = 1/\omega_c$ are considered collectively. Likewise, the v-order inductive fractor can only be properly interpreted when $K_{IF} = r^{1-p+v}$ and $\tau_{IF} = l/r = 1/\omega_c$ are considered collectively, i.e. $K_{IF} \cdot \tau_{IF}^v = l^v r^{1-p}$. Equations (5.80), (5.81), (5.88), and (5.89) show that how $K_{CF} = r^{1-p+v}$ and $\tau_{CF} = rc = 1/\omega_c$ or $K_{IF} = r^{1-p+v}$ and $\tau_{IF} = l/r = 1/\omega$ collectively affect overall aspects of the computation of a capacitive fractor or an inductive fractor, respectively, so separating $\log r$ and $\log c$ or $\log l$ respectively to consider the measurement units of fractance will cause a loss of much information. In fact, Figure 5.1 shows that the electrical properties of purely ideal capacitive fractor should fall in between the electrical properties of capacitor and those of resistor. Similarly, the electrical characteristics of inductive fractor should fall in between the electrical characteristics of inductor and those of resistor. Therefore, to collectively compute $K_{CF} \cdot \tau_{CF}^{-v} = c^{-v} r^{1-p}$ or $K_{IF} \cdot \tau_{IF}^{v} = l^{v} r^{1-p}$ of fractance, the traditional integer-order measurement units actually need be extended to the fractional-order measurement units.

5.2.3.2 Fractional-order measurement units and physical dimensions of fractance

In this subsection, to satisfy the realistic requirement of the fractional-order measurement units of fractance, the fractional-order measurement units and physical dimensions of fractance are proposed.

Equation (5.71) represents the fractional-order capacitive reactance of an arbitrary-order purely ideal capacitive fractor. Equation (5.73) represents the fractional-order inductive reactance of an arbitrary-order purely ideal inductive fractor. Thus, based on (5.71) and (5.73), in this subsection, the fractional-order measurement units for capacitive and inductive fractance are proposed, respectively, as well as the fractional-order physical dimensions of fractance. In next subsection, the rules for fractors in series and parallel are then introduced.

At first, with respect to capacitive fractance, from comparison of (5.70) and (5.71), we can see that the dimension of $c^{v} r^{1-p} = c^{q+p} r^{1-p}$ essentially determines the fractional-order measurement units and physical dimensions of capacitive fractance. Equation (5.71) reflects the nature of the electrical characteristics of an arbitrary-order capacitive fractor in its natural implementation. It is well known that the measurement units of capacitance, resistance, and time are Farad (F), Ohm (Ω), and Second (s), respectively. Consequently, it is natural to suggest defining (Yi-Fei PU)$_{-v}$ (abbreviated as P$_{-v}$) as the fractional-order measurement unit of the v-order capacitive fractance F_{-v}, where $v = q + p$ is a positive real number, q is a positive integer, and $0 \leqslant p \leqslant 1$, given as:

$$\begin{aligned} P_{-v} &= P_{-(q+p)} \\ &= F^{q+p} \Omega^{1-p} \\ &= F^{v} \Omega^{1-p} \\ &= sF^{q+p-1} \Omega^{-p} \\ &= A^{2q+4p-2} s^{2q+5p-3} kg^{1-q-2p} m^{2-2q-4p}. \end{aligned} \quad (5.90)$$

Thus, the fractional-order physical dimension of a v-order capacitive fractance F_{-v} are defined as:

$$dim F_{-v} = L^{2-2q-4p} T^{4q+7p-3} M^{1-q-2p} I^{2q+4p-2}. \quad (5.91)$$

Equation (5.90) shows that the v-order capacitive fractance F_{-v} of (5.71) is equal to:

$$F_{-v} = c^{q+p} r^{1-p} P_{-v}. \quad (5.92)$$

Equations (5.90) and (5.91) further show that the electrical characteristics of a capacitive fractor fall in between those of a capacitor and those of a resistor. As aforementioned discussion, a capacitive fractor in its natural implementation consists of a series of ordinary capacitors and resistors in the form of an analog circuit with an infinite recursive structure exhibiting self-similarity. Then, the electrical characteristics of a capacitive fractor must involve both capacitor-like and resistor-like electrical characteristics simultaneously, which indicates that a capacitive fractor is the result of a nonlinear fractional-order interaction among the capacitors and resistors of which it is composed.

In particular, if $q=0$ and $p=1$, from (5.90), one can obtain:

$$P_{-1} = P_{-(0+1)} = F. \tag{5.93}$$

Equation (5.93) shows that if $q=0$ and $p=1$, the v-order capacitive fractance $F_{-v} = F_{-(q+p)}$ converts into ordinary capacitance. In other words, an ordinary capacitor is a special case of an arbitrary-order capacitive fractor in its natural implementation. Further, if $q=0$ and $p=0$, from (5.90), the following can be obtained:

$$P_{-0} = P_{-(0+0)} = \Omega. \tag{5.94}$$

Equation (5.94) shows that if $q=0$ and $p=0$, the v-order capacitive fractance $F_{-v} = F_{-(q+p)}$ converts into ordinary resistance. In other words, an ordinary resistor is a special case of an arbitrary-order capacitive fractor in its natural implementation.

Secondly, with respect to inductive fractance, from comparison of (5.72) and (5.73), we can see that the dimension of $l^v r^{1-p} = l^{q+p} r^{1-p}$ essentially determines the fractional-order measurement units and physical dimensions of inductive fractance. Equation (5.73) reflects the nature of the electrical characteristics of an arbitrary-order inductive fractor in its natural implementation. It is well known that the measurement unit of inductor is Henry (H). Consequently, it is natural to suggest defining (Yi-Fei PU)$_v$ (abbreviated as P_v) as the fractional-order measurement unit of the v-order inductive fractance F_v, where $v=q+p$ is a positive real number, q is a positive integer, and $0 \leqslant p \leqslant 1$, given as:

$$\begin{aligned} P_v &= P_{(q+p)} \\ &= H^{q+p} \Omega^{1-p} \\ &= H^v \Omega^{1-p} \\ &= s^p H^q \Omega \\ &= A^{-2q-2} s^{-2q+p-3} kg^{q+1} m^{2q+2}. \end{aligned} \tag{5.95}$$

Thus, the fractional-order physical dimension of a v-order inductive fractance F_v are defined as:

$$dim F_v = L^{2q+2} T^{-2q+p-3} M^{q+1} I^{-2q-2}. \tag{5.96}$$

Equation (5.95) show that the v-order inductive fractance F_v of (5.73) is equal to:

$$F_v = l^{q+p} r^{1-p} P_v. \tag{5.97}$$

Equations (5.95) and (5.96) further show that the electrical characteristics of an inductive fractor fall in between those of an inductor and those of a resistor. As aforementioned discussion, an inductive fractor in its natural implementation consists of a series of ordinary inductors and resistors in the form of an analog circuit with an infinite recursive structures exhibiting self-similarity. Then, the electrical characteristics of an inductive fractor must involve both inductor-like and resistor-like electrical characteristics simultaneously, which indicates that an inductive fractor is the result of a nonlinear fractional-order interaction among the inductors and resistors of which it is composed.

In particular, if $q=0$ and $p=1$, from (5.95), one can obtain:

$$P_1 = P_{(0+1)} = H. \tag{5.98}$$

Equation (5.98) shows that if $q=0$ and $p=1$, the v-order inductive fractance $F_v = F_{q+p}$ converts into ordinary inductance. In other words, an ordinary inductor is a special case of an arbitrary-order inductive fractor in its natural implementation. Further, if $q=0$ and $p=0$, from (5.95), the following can be obtained:

$$P_0 = P_{(0+0)} = \Omega. \tag{5.99}$$

Equation (5.99) shows that if $q=0$ and $p=0$, the v-order inductive fractance $F_{-v} = F_{-(q+p)}$ converts into ordinary resistance. In other words, an ordinary resistor is a special case of an arbitrary-order inductive fractor in its natural implementation.

5.2.3.3 Rules for fractors in series and parallel

In this subsection, based on the proposed the fractional-order measurement units and physical dimensions of fractance, the rules for fractors in series and parallel are further introduced. For the convenience of illustration, the v-order capacitive fractance F_{-v} and the v-order inductive fractance F_v are discussed in turn.

At first, assume that the fractional primitives of a v-order capacitive fractor are zero, and F_{-v} denotes its v-order capacitive fractance. Thus, from (5.71), the following is true:

$$\begin{aligned} e &= \int_0^t I_i(t)\mathrm{d}t \\ &= \frac{c^{q+p}r^{-(1-p)}}{\Gamma(n-v)}\int_0^t\int_0^t (t-\tau)^{n-v-1}V_i^{(n)}(\tau)\mathrm{d}\tau\mathrm{d}t, \end{aligned} \tag{5.100}$$

where $0\leqslant n-1<v<n$, $n\in\mathbf{R}$, $v=q+p$ is a positive real number, q is a positive integer, $0\leqslant p\leqslant 1$, and e denotes the electric charge of a v-order capacitive fractor.

Secondly, assume that the jth v-order capacitive fractors $^jF_{-v}^c$ are connected in series, where $j=1, 2, 3, \cdots, k$, as shown in Figure 5.20.

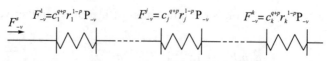

Figure 5.20　v-order capacitive fractors $^jF_{-v}^c$ connected in series.

See Figure 5.20, F_{-v}^j is the capacitive fractance of $^jF^{-v}$, F_{-v}^s is the total fractance of capacitive fractors $^jF_{-v}^c$ connected in series, $v=q+p$ is a positive real number, q is a positive integer, and $0\leqslant p\leqslant 1$. Thus, the fractance F_{-v}^j of each jth v-order capacitive fractor $^jF_{-v}^c$ can be obtained from (5.92):

$$F_{-v}^j = c_j^{q+p} r_j^{1-p} \mathrm{P}_{-v}. \tag{5.101}$$

As for a series circuit, the input current of each jth v-order capacitive fractor $^j F_{-v}^c$ is identical. Then, $I_j^i(s) = I_i(s)$, where $I_j^i(s)$ denotes the input current of each jth v-order capacitive fractor $^j F_{-v}^c$. From (5.71), one can obtain:

$$V_i^j(s) = c_j^{-(q+p)} r_j^{1-p} s^{-v} I_i(s), \tag{5.102}$$

where $V_j^i(s)$ denotes the input voltage of each of the jth v-order capacitive fractor $^j F_{-v}^c$. Then, the total input voltage $V^s(s)$ of the v-order capacitive fractors $^j F_{-v}^c$ connected in series can be derived as:

$$\begin{aligned} V^s(s) &= \sum_{j=1}^k V_i^j(s) \\ &= \Big(\sum_{j=1}^k c_j^{-(q+p)} r_j^{1-p}\Big) s^{-v} I_i(s) \\ &= \frac{\sum_{z=1}^k \Big(\frac{r_z^{1-p}}{c_z^{q+p}} \prod_{j=1}^k c_j^{q+p}\Big)}{\prod_{j=1}^k c_j^{q+p}} s^{-v} I_i(s) = h(s_{j;1 \to k}^{q+p}) g(r_{j;1 \to k}^{1-p}) s^{-v} I_i(s), \end{aligned} \tag{5.103}$$

where $h(c_{j;1\to k}^{q+p})$ is a nonlinear function of $(c_1^{q+p}, \cdots, c_j^{q+p}, \cdots, c_k^{q+p})$ while $g(r_{j;1\to k}^{1-p})$ is a nonlinear function of $(r_1^{1-p}, \cdots, r_j^{1-p}, \cdots, r_k^{1-p})$. Equations (5.71), (5.90), and (5.103) show that the total fractance F_{-v}^s of the v-order capacitive fractors $^j F_{-v}^c$ connected in series, where $j = 1, 2, 3, \cdots, k$, can be derived as:

$$F_{-v}^s = [h(c_{j;1\to k}^{q+p})]^{-1} g(r_{j;1\to k}^{1-p}) \mathrm{P}_{-v}. \tag{5.104}$$

In particular, first, if $c_{j;1\to k} \equiv c$ and $r_{j;1\to k}$ are not equal to each other, from (5.90), (5.103), and (5.104), the following can be obtained:

$$F_{-v}^s = c^{q+p} \sum_{j=1}^k r_j^{1-p} \mathrm{P}_{-v}. \tag{5.105}$$

Equation (5.105) shows that if $c_{j;1\to k} \equiv 0$, $p = 0$, and $q = 0$, one can obtain $F_{-v}^s = \sum_{j=1}^k r_j \Omega$. On this occasion, the v-order capacitive fractors $^j F_{-v}^c$ connected in series are actually equivalent to ordinary resistors r_j connected in series. In other words, the rule for resistors in series is a special case of the rule for capacitive fractors in series. Second, if $r_{j;1\to k} \equiv r$ and $c_{j;1\to k}$ are not equal to each other, one can obtain:

$$F_{-v}^s = r^{1-p} \Big(\sum_{j=1}^k c_j^{-(q+p)}\Big)^{-1} \mathrm{P}_{-v}. \tag{5.106}$$

Further, if $r_{j;1\to k} \equiv r$ and $p = 1$, the following can be obtained:

$$F_{-v}^s = \Big(\sum_{j=1}^k c_j^{-(q+1)}\Big)^{-1} \mathrm{F}^{q+1}. \tag{5.107}$$

Equation (5.107) shows that if $r_{j;1\to k} \equiv r$, $p = 1$, and $q = 0$, $F_{-v}^s = \Big(\sum_{j=1}^k c_j^{-1}\Big)^{-1} \mathrm{F}$ can be derived. In this case, the v-order capacitive fractors $^j F_{-v}^c$ connected in series are actually equivalent to ordinary capacitors c_j connected in series. In other words, the rule for capacitors in series is a special case of the rule for capacitive fractors in series.

Moreover, if the fractional-order of each v_j-order capacitive fractor $^jF^c_{-v_j}$ connected in series is different, the total input voltage $V^s(s)$ of the v_j-order capacitive fractors $^jF^c_{-v_j}$ connected in series can be obtained:

$$V^s(s) = \sum_{j=1}^{k} V_i^j(s)$$
$$= \Big(\sum_{j=1}^{k} c_j^{-(q_j+p_j)} r_j^{1-p_j} s^{-v_j}\Big) I_i(s), \qquad (5.108)$$

where $v_j = q_j + p_j$ is a positive real number, q_j is a positive integer, and $0 \leqslant p_j \leqslant 1$. Equation (5.39) shows that the total input voltage $V^s(s)$ is the result of the nonlinear interaction among the v_j-order capacitive fractors $^jF^c_{-v_j}$ connected in series. In this case, an analytical expression of the total fractance F^s_{-v} of the v_j-order capacitive fractors $^jF^c_{-v_j}$ connected in series is quite difficult to be derived directly.

Thirdly, assume that the jth v-order capacitive fractors $^jF^c_{-v}$ are connected in parallel, where $j = 1, 2, 3, \cdots, k$, as shown in Figure 5.21.

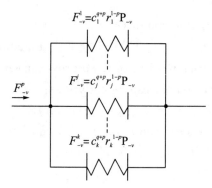

Figure 5.21 v-order capacitive fractors $^jF^c_{-v}$ connected in parallel.

In Figure 5.21, F^j_{-v} is the capacitive fractance of $^jF^c_{-v}$, F^p_{-v} is the total fractance of capacitive fractors $^jF^c_{-v}$ connected in parallel, $v = q + p$ is a positive real number, q is a positive integer, and $0 \leqslant p \leqslant 1$. As for a parallel circuit, the input voltage of each jth v-order capacitive fractor $^jF^c_{-v}$ is identical. Therefore, $V_i^j(s) = V_i(s)$, where $V_i^j(s)$ denotes the input voltage of each jth v-order capacitive fractor $^jF^c_{-v}$. From (5.71), one can obtain:

$$I_i^j(s) = c_j^{q+p} r_j^{-(1-p)} s^v V_i(s), \qquad (5.109)$$

where $I_i^j(s)$ denotes the input current of each jth v-order capacitive fractor $^jF^c_{-v}$. Thus, the total input current $I^p(s)$ of the v-order capacitive fractors $^jF^c_{-v}$ connected in parallel can be derived as:

$$I^p(s) = \sum_{j=1}^{k} I_i^j(s)$$
$$= \Big(\sum_{j=1}^{k} c_j^{q+p} r_j^{-(1-p)}\Big) s^v V_i(s)$$

$$=\frac{\sum_{z=1}^{k}\left(\frac{c_z^{q+p}}{r_z^{1-p}}\prod_{j=1}^{k}r_j^{1-p}\right)}{\prod_{j=1}^{k}r_j^{1-p}}s^v V_i(s) = \alpha(c_{j;1\to k}^{q+p})\beta(r_{j;1\to k}^{1-p})s^v V_i(s), \quad (5.110)$$

where $\alpha(c_{j;1\to k}^{q+p})$ is a nonlinear function of $(c_1^{q+p}, \cdots, c_j^{q+p}, \cdots, c_k^{q+p})$ while $\beta(r_{j;1\to k}^{1-p})$ is a nonlinear function of $(r_1^{1-p}, \cdots, r_j^{1-p}, \cdots, r_k^{1-p})$. Equations (5.71), (5.90), and (5.110) show that the total fractance F_{-v}^p of the v-order capacitive fractors ${}^j F_{-v}^c$ connected in parallel, where $j = 1, 2, 3, \cdots, k$, can be derived as:

$$F_{-v}^p = \alpha(c_{j;1\to k}^{q+p})[\beta(r_{j;1\to k}^{1-p})]^{-1}\mathrm{P}_{-v}. \quad (5.111)$$

In particular, first, if $c_{j;1\to k} \equiv c$ and $r_{j;1\to k}$ are not equal to each other, from (5.90), (5.110), and (5.111), the following can be obtained:

$$F_{-v}^p = c^{q+p}\left(\sum_{j=1}^{k} r_j^{-(1-p)}\right)^{-1}\mathrm{P}_{-v}. \quad (5.112)$$

Equation (5.112) shows that if $c_{j;1\to k} \equiv c$, $p = 0$, and $q = 0$, one can obtain $F_{-v}^p = \left(\sum_{j=1}^{k} r_j^{-1}\right)^{-1}\Omega$. On this occasion, the v-order capacitive fractors ${}^j F_{-v}^c$ connected in parallel are actually equivalent to ordinary resistors r_j connected in parallel. In other words, the rule for resistors in parallel is a special case of the rule for capacitive fractors in parallel. Second, if $r_{j;1\to k} \equiv r$ and $c_{j;1\to k}$ are not equal to each other, one can obtain:

$$F_{-v}^p = r^{1-p}\left(\sum_{j=1}^{k} c_j^{q+p}\right)\mathrm{P}_{-v}. \quad (5.113)$$

Further, if $r_{j;1\to k} \equiv r$ and $p = 1$, the following can be obtained:

$$F_{-v}^p = \left(\sum_{j=1}^{k} c_j^{q+1}\right)\mathrm{F}^{q+1}. \quad (5.114)$$

Equation (5.114) shows that if $r_{j;1\to k} \equiv r$, $p = 1$, and $q = 0$, $F_{-v}^p = \left(\sum_{j=1}^{k} c_j\right)\mathrm{F}$ can be derived. In this case, the v-order capacitive fractors ${}^j F_{-v}^c$ connected in parallel are actually equivalent to ordinary capacitors c_j connected in parallel. In other words, the rule for capacitors in parallel is a special case of the rule for capacitive fractors in parallel.

Moreover, if the fractional-order of each v_j-order capacitive fractor ${}^j F_{-v_j}^c$ connected in parallel is different, the total input current $I^p(s)$ of the v_j-order capacitive fractors ${}^j F_{-v_j}^c$ connected in parallel can be derived as:

$$I^p(s) = \sum_{j=1}^{k} I_i^j(s)$$
$$= \left(\sum_{j=1}^{k} c_j^{q_j+p_j} r_j^{-(1-p_j)} s^{v_j}\right) V_i(s), \quad (5.115)$$

where $v_j = q_j + p_j$ is a positive real number, q_j is a positive integer, and $0 \leqslant p_j \leqslant 1$. Equation (5.115) shows that the total input current $I^p(s)$ is the result of the nonlinear interaction among the v_j-order capacitive fractors ${}^j F_{-v_j}^c$ connected in parallel. In this case, an analytical expression of the total fractance F_{-v}^p of the v_j-order capacitive fractors ${}^j F_{-v_j}^c$

connected in parallel is quite difficult to be derived directly.

Fourthly, assume that the fractional primitives of a v-order inductive fractor is equal to zero, and F_v denotes its v-order inductive fractance. Thus, from (5.73), the following is true:

$$\begin{aligned} e &= \int_0^t I_i(t)\mathrm{d}t \\ &= \frac{l^{-(q+p)}r^{-(1-p)}}{\Gamma(v)}\int_0^t\int_0^t (t-\tau)^{v-1}V_i(\tau)\mathrm{d}\tau\mathrm{d}t, \end{aligned} \qquad (5.116)$$

where $v=q+p$ is a positive real number, q is a positive integer, $0 \leqslant p \leqslant 1$, and e denotes the electric charge of a v-order inductive fractor.

Fifthly, assume that the jth v-order inductive fractors ${}^jF_v^l$ are connected in series, where $j=1, 2, 3, \cdots, k$. Thus, the fractance F_v^j of each jth v-order inductive fractor ${}^jF_v^l$ can be obtained from (5.97):

$$F_v^j = l_j^{q+p}r_j^{1-p}P_v. \qquad (5.117)$$

Then, from (5.73), the following is true:

$$V_i^j(s) = l_j^{q+p}r_j^{1-p}s^v I_i(s), \qquad (5.118)$$

where $V_i^j(s)$ and $I_i^j(s)$ denote the input voltage and input current of each jth v-order inductive fractor ${}^jF_v^l$, respectively. As for a series circuit, $I_i^j(s) = I_i(s)$. Then, the total input voltage $V^s(s)$ of the v-order inductive fractors ${}^jF_v^l$ connected in series can be derived as:

$$\begin{aligned} V^s(s) &= \sum_{j=1}^k V_i^j(s) \\ &= (\sum_{j=1}^k l_j^{q+p}r_j^{1-p})s^v I_i(s) \\ &= \eta(l_{j,1\to k}^{q+p})\mu(r_{j,1\to k}^{1-p})s^v I_i(s), \end{aligned} \qquad (5.119)$$

where $\eta(l_{j,1\to k}^{q+p})$ is a nonlinear function of $(l_1^{q+p}, \cdots, l_j^{q+p}, \cdots, l_k^{q+p})$ while $\mu(r_{j,1\to k}^{1-p})$ is a nonlinear function of $(r_1^{1-p}, \cdots, r_j^{1-p}, \cdots, r_k^{1-p})$. Equations (5.73), (5.95), and (5.119) show that the total fractance F_v^s of the v-order inductive fractors ${}^jF_v^l$ connected in series, where $j=1, 2, 3, \cdots, k$, can be derived as:

$$F_v^s = \eta(l_{j,1\to k}^{q+p})\mu(r_{j,1\to k}^{1-p})P_v. \qquad (5.120)$$

In particular, first, if $l_{j,1\to k} \equiv l$ and $r_{j,1\to k}$ are not equal to each other, from (5.95), (5.119), and (5.120), the following can be obtained:

$$F_v^s = l^{q+p}\sum_{j=1}^k r_j^{1-p}P_v. \qquad (5.121)$$

Equation (5.121) shows that if $l_{j,1\to k} \equiv l$, $p=0$, and $q=0$, one can obtain $F_v^s = \sum_{j=1}^k r_j \Omega$. On this occasion, the v-order inductive fractors ${}^jF_v^l$ connected in series are actually equivalent to ordinary resistors r_j connected in series. In other words, the rule for resistors in series is a special case of the rule for inductive fractors in series. Second, if

$r_{j;1\to k} \equiv r$ and $l_{j;1\to k}$ are not equal to each other, one can obtain:

$$F_v^s = r^{1-p} \left(\sum_{j=1}^{k} l_j^{q+p} \right) \mathrm{P}_v. \tag{5.122}$$

Further, if $r_{j;1\to k} \equiv r$ and $p=1$, the following can be obtained:

$$F_v^s = \left(\sum_{j=1}^{k} l_j^{q+1} \right) \mathrm{H}^{q+1}. \tag{5.123}$$

Equation (5.123) shows that if $r_{j;1\to k} \equiv r$, $p=1$, and $q=0$, $F_v^s = \left(\sum_{j=1}^{k} l_j \right) \mathrm{H}$ can be derived. In this case, the v-order inductive fractors $^j F_v^l$ connected in series are actually equivalent to ordinary inductors l_j connected in series. In other words, the rule for inductors in series is a special case of the rule for inductive fractors in series.

Sixthly, assume that the jth v-order inductive fractors $^j F_v^l$ are connected in parallel, where $j=1, 2, 3, \cdots, k$. From (5.4), the following is true:

$$I_i^j(s) = l_j^{-(q+p)} r_j^{-(1-p)} s^{-v} V_i(s), \tag{5.124}$$

where $V_j^i(s)$ and $I_i^j(s)$ denote the input voltage and input current of each jth v-order inductive fractor $^j F_v^l$, respectively. As for a parallel circuit, $V_i^j(s) = V_i(s)$. Then, the total input current $I^p(s)$ of the v-order inductive fractors $^j F_v^l$ connected in parallel can be derived as:

$$\begin{aligned} I^p(s) &= \sum_{j=1}^{k} I_i^j(s) \\ &= \left(\sum_{j=1}^{k} l_j^{-(q+p)} r_j^{-(1-p)} \right) s^{-v} V_i(s) \\ &= \lambda(l_{j;1\to k}^{q+p}) \varphi(r_{j;1\to k}^{1-p}) s^{-v} V_i(s), \end{aligned} \tag{5.125}$$

where $\lambda(l_{j;1\to k}^{q+p})$ is a nonlinear function of $(l_1^{q+p}, \cdots, l_j^{q+p}, \cdots, l_k^{q+p})$ while $\varphi(r_{j;1\to k}^{1-p})$ is a nonlinear function of $(r_1^{1-p}, \cdots, r_j^{1-p}, \cdots, r_k^{1-p})$. Equations (5.73), (5.95), and (5.125) show that the total fractance F_v^p of the v-order inductive fractors $^j F_v^l$ connected in parallel, where $j=1, 2, 3, \cdots, k$, can be derived as:

$$F_v^p = [\lambda(l_{j;1\to k}^{q+p})]^{-1} [\varphi(r_{j;1\to k}^{1-p})]^{-1} \mathrm{P}_v. \tag{5.126}$$

In particular, first, if $l_{j;1\to k} \equiv l$ and $r_{j;1\to k}$ are not equal to each other, from (5.73), (5.125), and (5.126), the following can be obtained:

$$F_v^p = l^{q+p} \left(\sum_{j=1}^{k} r_j^{-(1-p)} \right)^{-1} \mathrm{P}_v. \tag{5.127}$$

Equation (5.127) shows that if $l_{j;1\to k} \equiv c$, $p=0$, and $q=0$, one can obtain $F_v^p = \left(\sum_{j=1}^{k} r_j^{-1} \right)^{-1} \Omega$. On this occasion, the v-order inductive fractors $^j F_v^l$ connected in parallel are actually equivalent to ordinary resistors r_j connected in parallel. In other words, the rule for resistors in parallel is a special case of the rule for inductive fractors in parallel. Second, if $r_{j;1\to k} \equiv r$ and $l_{j;1\to k}$ are not equal to each other, one can obtain:

$$F_v^p = r^{1-p} \left(\sum_{j=1}^{k} l_j^{-(q+p)} \right)^{-1} \mathrm{P}_v. \tag{5.128}$$

Further, if $r_{j,1 \to k} \equiv r$, and $p=1$, the following can be obtained:

$$F_v^p = \Big(\sum_{j=1}^{k} l_j^{-(q+1)}\Big)^{-1} H^{q+1}. \qquad (5.129)$$

Equation (5.129) shows that if $r_{j,1 \to k} \equiv r$, $p=1$, and $q=0$, $F_v^p = \Big(\sum_{j=1}^{k} l_j^{-1}\Big)^{-1} H$ can be derived. In this case, the v-order inductive fractors $^jF_{-v}^c$ connected in parallel are actually equivalent to ordinary inductors l_j connected in parallel. In other words, the rule for inductors in parallel is a special case of the rule for inductive fractors in parallel.

5.2.4 EXPERIMENT AND ANALYSIS

In this section, through mathematical analysis and simulation results, some issues of the rules for fractors in series and parallel are discussed in detail.

5.2.4.1 Implementation of arbitrary-order fractor

In this subsection, before discussing the rules for fractors in series and parallel, an arbitrary-order fractaor in the form of an analog circuit need to be achieved.

As previously discussed, the tree type [175], [198], two-circuit type [173], [185], H type [172], [185], net-grid type [171], [185], [193] − [194] should be four discovered natural implementations of fractor. In fact, there are zero errors between these four types of fractance [171] − [173], [175], [185], [193] − [194], [198], [226] with infinite recursive extreme self-similar structures and an ideal fractance. The corresponding devices manufactured using the other approximate implementations of fractor [130], [196], [198] − [203], [206] − [215], [227] − [228] can never represent a purely ideal fractor. Note that for the four discovered natural implementations of fractor [171] − [173], [175], [185], [193] − [194], [198], [226] must be of infinite recursive extreme self-similar structures, theoretically, they are usually applied to the analysis of the electrical characteristics of a purely ideal fractor. Actually, the other approximate implementations of fractor [130], [196], [198] − [203], [206] − [215], [227] − [228] are usually applied to the approximately achieved fractors. Therefore, here, the rules for approximately implemented fractors in series and parallel are mainly discussed.

Equations (5.70) − (5.73) further that a capacitor implements the first-order integral, an inductor implements the first-order differential, a v-order capacitive fractor F_{-v}^c implements the v-order fractional integral, and a v-order inductive fractor F_v^l implements the v-order fractional integral. Note that one crux of the approximate implementations of an arbitrary-order fractor is to perform a fractional integral operator s^{-v} or a fractional differential operator s^v in the form of an analog circuit. s^{-v} is the reciprocal of s^v. Let's suppose that v is a positive rational number. Thus, v is equal to a positive integer plus a positive rational fraction, which can be given as $s^v = s^m s^{i/n}$, where m, n, and i are positive integers, respectively, $n \neq 1$, $i = 1, 2, \cdots, n-1$, i/n is a positive rational fraction, and n is an arbitrary value corresponding to v. Equation (5.129) shows that for the integer-order differential operator s^m can be by means of inductors or capacitors, approximate achievement of fractional differential operator $s^{i/n}$ in the form of an analog circuit is a key problem of the approximate implementation of fractional differential operator s^v. In addition, the binomial theorem can be used to implement an arbitrary-order fractional differential operator $s^{i/n}$. if $|\alpha - (\beta s)^{i/n}| < 1$, then from the binomial theorem, one can obtain $\lim_{j \to \infty}[\alpha - (\beta s)^{i/n}]^j = \lim_{j \to \infty} \sum_{k=0}^{j} (-1)^k \binom{j}{k} \alpha^{j-k} \beta^{ik/n} s^{quo(ik/n)} s^{rem(ik/n)} = 0$, where α and β are positive numbers, $\binom{j}{k} = \frac{\Gamma(1+j)}{\Gamma(j-k+1)\Gamma(k+1)}$, $quo()$ denotes quotient calculation, and $rem()$ denotes the remainder calculation. Thus, if $j \to \infty$, we can derive the computational completeness of $s^{rem(ik/n)}$ as $\lim_{j \to \infty} s^{rem(ik/n)} = s^{i/n} = s^{1/n}$, $s^{2/n}$, ..., $s^{n-1/n}$ corresponding to $i = 1, 2, \cdots, n-1$. $|\alpha - (\beta s)^{i/n}| < 1$ is the restricted condition of convergence of the binomial theorem based approximate implementation of an arbitrary-order fractor. Hence, in order to get desired pass-band width for a fractor, α and β be optimal values should be set according to the actual conditions.

5.2.4.2 Implementations of capacitance and resistance of arbitrary value

In this subsection, to approximately achieve an arbitrary-order fractor, the implementation of capacitance and resistance of arbitrary value by active elements need to be studied.

Equations (5.71) and (5.73) further show that one essential requirement of all approximate implementations of an arbitrary-order fractor is able to perform capacitance and resistance of arbitrary value in the form of an analog circuit. For a capacitive fractor, it is constituted by a set of capacitors and resistors in a certain structure; for an inductive fractor, it is constituted by a set of inductors and resistors in a certain structure. Since an inductor can be implemented by a capacitor using gyrator, only the achievement of capacitance and resistance of arbitrary value by active elements are discussed following. In order to easily achieve an integrated circuit, gain a wider pass-band, and gain better high-frequency characteristics, an operational transconductance amplifier (OTA) is usually emploied in actual circuits to implement capacitance and resistance of arbitrary value. We propose to implement two equivalent OTA circuits for an analog resistor and an analog capacitor. They are shown in Figure 5.22.

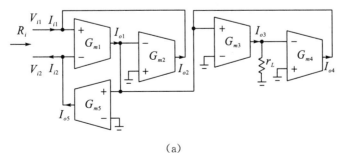

(a)

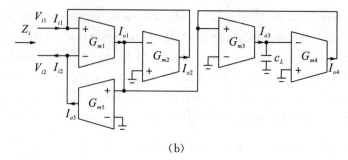

(b)

Figure 5.22　Two equivalent OTA circuits for an analog resistor and an analog capacitor: (a) Resistor; (b) Capacitor.

In Figure 5.22, G_m is the amplifier gain of OTA. From Figure 5.22, according to circuits and systems theory, we can derive that the input impedance R_i of a resistor and the input capacitance Z_i of a capacitor in Laplace transform domain. It follows that, respectively:

$$R_i(s) = \frac{V_{i1} - V_{i2}}{I_{i1}} = \frac{G_{m1}G_{m2}}{G_{m3}G_{m4}} r_L s, \qquad (5.130)$$

$$Z_i(s) = \frac{V_{i1} - V_{i2}}{I_{i1}} = 1 \bigg/ \left(\frac{G_{m1}G_{m2}}{G_{m3}G_{m4}} c_L s \right). \qquad (5.131)$$

Thus, we can derive the input resistance r_i and the input capacitance c_i in the time domain. It follows that, respectively:

$$r_i = \frac{G_{m1}G_{m2}}{G_{m3}G_{m4}} r_L, \qquad (5.132)$$

$$c_i = \frac{G_{m1}G_{m2}}{G_{m3}G_{m4}} c_L. \qquad (5.133)$$

According to Kirchhoff's current law, the input current and output current of a two-terminal analog network are equal to each other. Thus, it follows that:

$$G_{m5} = G_{m2}. \qquad (5.134)$$

If $G_{m5} \neq G_{m2}$, it yields an interference voltage-controlled current source. It follows that, respectively:

$$I_{i1} - I_{i2} = \frac{G_{m1}(V_{i1} - V_{i2})}{G_{m3}G_{m4}r_L s}(G_{m2} - G_{m5}). \qquad (5.135)$$

$$I_{i1} - I_{i2} = \frac{G_{m1}c_L(V_{i1} - V_{i2})s}{G_{m3}G_{m4}}(G_{m2} - G_{m5}). \qquad (5.136)$$

Equations (5.132) and (5.133) show that we can adjust G_{m1}, G_{m2}, G_{m3}, and G_{m4} to achieve capacitance and resistance of arbitrary value. Further, temperature variation has effects on the G_m of OTA, but the effects of temperature variation on r_i and c_i are restrained because the power of G_m is equal in (5.132) and (5.133).

In addition, we propose to implement $s^{-m}s^{-i/n}$ by OTA and voltage mode operational amplifier (VOA), where m, n, and i are positive integers, $n \neq 1$, $i = 1, 2, \cdots, n-1$, and i/n is a positive rational fraction. v is an arbitrary-order, and n is corresponding to v. The analogue circuit implementation of $s^{-1}s^{-i/n}$ is shown in Figure 5.23.

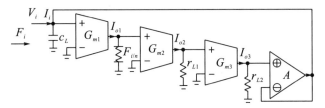

Figure 5.23 Analogue circuit implementation of $s^{-1}s^{-i/n}$ by OTA and VOA.

In Figure 5.23, where $F_{i/n}=s^{-i/n}$. From Figure 5.6, we can derive that the $(-1-i/n)$-order input capacitive reactance of a $(-1-i/n)$-order capacitive fractor $s^{-1}s^{-1/n}$. It follows that:

$$F_i(s) = \frac{V_i(s)}{I_i(s)} = \frac{G_{m1}G_{m2}G_{m3}r_{L1}r_{L2}}{c_L}s^{-1}s^{-i/n}. \tag{5.137}$$

5.2.4.3 Approximately implemented fractors in series and parallel

In this subsection, approximately implemented fractors in series and parallel arefurther analyzed in detail.

Example 1: Before analyzing approximately implemented fractors in series and parallel, an approximate implementation of fractor should be achieved. To facilitate the analyses, we only approximately implement a capacitive fractor in this example. Likewise, the result of inductive fractor is similar to that of capacitive fractor.

From (5.131) and (5.133), according to the binomial theorem, assume that $\alpha=1$ and $\beta=10^{-4}$, we can derive an approximate implementation of a certain band-pass filtering capacitive fractor $s^{-1/2}$. It follows that in vector form:

$$\begin{bmatrix} \begin{pmatrix} 2.2\cdot10^{-7}\cdot s^2+7.92\cdot10^{-3}\cdot s \\ +1.2 \end{pmatrix} & \begin{pmatrix} -6.6\cdot10^{-9}\cdot s^2-9.24\cdot10^{-4}\cdot s \\ -6.6\cdot10^{-1} \end{pmatrix} & \begin{pmatrix} 1.2\cdot10^{-10}\cdot s^2+7.92\cdot10^{-5}\cdot s \\ +2.2\cdot10^{-1} \end{pmatrix} \\ 0 & \begin{pmatrix} 1.2\cdot10^{-21}\cdot s^5+2.2\cdot10^{-16}\cdot s^4 \\ +7.92\cdot10^{-12}\cdot s^3+7.92\cdot10^{-8}\cdot s^2 \\ +2.2\cdot10^{-4}\cdot s+1.2\cdot10^{-1} \end{pmatrix} & 0 \\ \begin{pmatrix} 1.2\cdot10^{-32}\cdot s^8+7.92\cdot10^{-19}\cdot s^5 \\ +2.2\cdot10^{-7}\cdot s^2 \end{pmatrix} & \begin{pmatrix} -6.6\cdot10^{-29}\cdot s^7-9.24\cdot10^{-16}\cdot s^4 \\ -6.6\cdot10^{-5}\cdot s \end{pmatrix} & \begin{pmatrix} 2.2\cdot10^{-25}\cdot s^6+7.92\cdot10^{-13}\cdot s^3 \\ +1.2\cdot10^{-2} \end{pmatrix} \end{bmatrix}$$

$$\begin{bmatrix} s^{1/4} \\ s^{1/2} \\ s^{3/4} \end{bmatrix} = \begin{bmatrix} \begin{pmatrix} 10^{-12}\cdot s^3+4.95\cdot10^{-6}\cdot s^2 \\ +4.95\cdot10^{-2}\cdot s+1 \end{pmatrix} \\ \begin{pmatrix} 10^{-24}\cdot s^6+6.6\cdot10^{-19}\cdot s^5 \\ +4.95\cdot10^{-14}\cdot s^4+9.24\cdot10^{-10}\cdot s^3 \\ +4.95\cdot10^{-6}\cdot s^2+6.6\cdot10^{-3}\cdot s+1 \end{pmatrix} \\ \begin{pmatrix} 10^{-36}\cdot s^9+4.95\cdot10^{-22}\cdot s^6 \\ +4.95\cdot10^{-10}\cdot s^3+1 \end{pmatrix} \end{bmatrix}. \tag{5.138}$$

On the basis of the solution identification theorem of nonhomogeneous linear equations, we can derive the nonzero solutions $[s^{1/4}, s^{1/2}, s^{3/4}]^{\mathrm{T}}$ to (5.69). Therefore, from the Cramer's rule, it follows that:

$$\begin{bmatrix} s^{1/4} \\ s^{1/2} \\ s^{3/4} \end{bmatrix} = \begin{bmatrix} \dfrac{\begin{pmatrix} -1.44 \cdot 10^{-67} \cdot s^{16} - 1.2936 \cdot 10^{-61} \cdot s^{15} - 2.73768 \cdot 10^{-56} \cdot s^{14} - 2.0962 \cdot 10^{-51} \cdot s^{13} + 1.16662 \cdot 10^{-46} \cdot s^{12} + 7.54928 \cdot 10^{-42} \cdot s^{11} - 4.87615 \cdot 10^{-38} \cdot s^{10} \\ -4.709422 \cdot 10^{-33} \cdot s^9 - 1.90813 \cdot 10^{-29} \cdot s^8 + 4.294345 \cdot 10^{-25} \cdot s^7 + 2.3805359 \cdot 10^{-21} \cdot s^6 - 6.468197 \cdot 10^{-18} \cdot s^5 - 3.939 \cdot 10^{-14} \cdot s^4 \\ -3.93744 \cdot 10^{-12} \cdot s^3 + 1.15301 \cdot 10^{-7} \cdot s^2 + 4.4856 \cdot 10^{-5} \cdot s - 1.704 \cdot 10^{-2} \end{pmatrix}}{\begin{pmatrix} -1.728 \cdot 10^{-63} \cdot s^{15} - 4.45728 \cdot 10^{-57} \cdot s^{14} - 1.65581 \cdot 10^{-52} \cdot s^{13} + 4.40282 \cdot 10^{-48} \cdot s^{12} + 5.74295 \cdot 10^{-43} \cdot s^{11} + 2.72029 \cdot 10^{-39} \cdot s^{10} - 3.468 \cdot 10^{-38} \cdot s^9 \\ -3.224071 \cdot 10^{-30} \cdot s^8 + 3.22407 \cdot 10^{-26} \cdot s^7 + 3.467827 \cdot 10^{-22} \cdot s^6 - 2.72029 \cdot 10^{-19} \cdot s^5 - 5.74295 \cdot 10^{-15} \cdot s^4 - 4.402816 \cdot 10^{-12} \cdot s^3 \\ +1.65581 \cdot 10^{-8} \cdot s^2 + 1.45728 \cdot 10^{-5} \cdot s + 1.728 \cdot 10^{-3} \end{pmatrix}} \\ \dfrac{s^6 + 6.6 \cdot 10^5 \cdot s^5 + 4.95 \cdot 10^5 \cdot s^4 + 9.24 \cdot 10^{14} \cdot s^3 + 4.95 \cdot 10^{18} \cdot s^2 + 6.6 \cdot 10^{21} \cdot s + 10^{24}}{1.2 \cdot 10^3 \cdot s^5 + 2.2 \cdot 10^8 \cdot s^4 + 7.92 \cdot 10^{12} \cdot s^3 + 7.92 \cdot 10^{16} \cdot s^2 + 2.2 \cdot 10^{20} \cdot s + 1.2 \cdot 10^{23}} \\ \dfrac{\begin{pmatrix} 1.704 \cdot 10^{-64} \cdot s^{16} - 6.4856 \cdot 10^{-59} \cdot s^{15} - 1.15301 \cdot 10^{-53} \cdot s^{14} + 3.93744 \cdot 10^{-50} \cdot s^{13} + 3.939 \cdot 10^{-44} \cdot s^{12} + 6.468197 \cdot 10^{-40} \cdot s^{11} - 2.38036 \cdot 10^{-35} \cdot s^{10} \\ -4.2943449 \cdot 10^{-31} \cdot s^9 + 1.90813 \cdot 10^{-27} \cdot s^8 + 4.09422 \cdot 10^{-23} \cdot s^7 + 4.87615 \cdot 10^{-20} \cdot s^6 - 7.549277 \cdot 10^{-16} \cdot s^5 - 1.166616 \cdot 10^{-12} \cdot s^4 \\ +2.09616 \cdot 10^{-9} \cdot s^3 + 2.73768 \cdot 10^{-6} \cdot s^2 + 1.2936 \cdot 10^{-3} \cdot s + 1.44 \cdot 10^{-1} \end{pmatrix}}{\begin{pmatrix} -1.728 \cdot 10^{-63} \cdot s^{15} - 1.45728 \cdot 10^{-57} \cdot s^{14} - 1.65581 \cdot 10^{-52} \cdot s^{13} + 4.40282 \cdot 10^{-48} \cdot s^{12} + 5.74295 \cdot 10^{-43} \cdot s^{11} + 2.72029 \cdot 10^{-39} \cdot s^{10} - 3.468 \cdot 10^{-34} \cdot s^9 \\ -3.224071 \cdot 10^{-30} \cdot s^8 + 3.22407 \cdot 10^{-26} \cdot s^7 + 3.467827 \cdot 10^{-22} \cdot s^6 - 2.72029 \cdot 10^{-19} \cdot s^5 - 5.74295 \cdot 10^{-15} \cdot s^4 - 4.402816 \cdot 10^{-12} \cdot s^3 \\ +1.65581 \cdot 10^{-8} \cdot s^2 + 1.45728 \cdot 10^{-5} \cdot s + 1.728 \cdot 10^{-3} \end{pmatrix}} \end{bmatrix}$$

(5.139)

$[s^{1/4}, s^{1/2}, s^{3/4}]^T$ are the nonzero solutions of (69). For $s^{-1/2}$ is the reciprocal of $s^{1/2}$, thus from (5.70) it follows that:

$$\begin{aligned} s^{-1/2} &= \frac{1.2 \cdot 10^3 \cdot s^5 + 2.2 \cdot 10^8 \cdot s^4 + 7.92 \cdot 10^{12} \cdot s^3 + 7.92 \cdot 10^{16} \cdot s^2 + 2.2 \cdot 10^{20} \cdot s + 1.2 \cdot 10^{23}}{s^6 + 6.6 \cdot 10^5 \cdot s^5 + 4.95 \cdot 10^{10} \cdot s^4 + 9.24 \cdot 10^{10} \cdot s^3 + 4.95 \cdot 10^{18} \cdot s^2 + 6.6 \cdot 10^{21} \cdot s + 10^{24}} \\ &= \frac{-4.14656596 \cdot 10^2}{s + 5.76954805 \cdot 10^5} + \frac{1.13807119 \cdot 10^2}{s + 5.82842712 \cdot 10^4} + \frac{4.49732729 \cdot 10}{s + 1.69839637 \cdot 10^4} + \frac{2.64798451 \cdot 10}{s + 5.88790706 \cdot 10^3} \\ &\quad + \frac{1.95262146 \cdot 10}{s + 1.71572875 \cdot 10^3} + \frac{1.695554397 \cdot 10}{s + 1.73323801 \cdot 10^2}. \end{aligned}$$

(5.140)

Equation (5.140) is an approximate implementation of a band-pass filtering capacitive fractor $s^{-1/2}$. Equation (5.71) shows that all poles of the band-pass filtering capacitive fractor $s^{-1/2}$ are on the left half plane. Therefore, the band-pass filtering capacitive fractor $s^{-1/2}$ is a stable system. From (5.140), we can derive the corresponding analog cascade circuit of the band-pass filtering capacitive fractor $s^{-1/2}$, as shown in Figure 5.24.

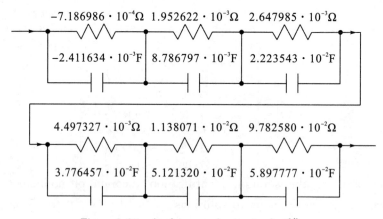

Figure 5.24 Analog cascade circuit of $s^{-1/2}$.

In Figure 5.24, from (5.132) and (5.133), we can implement a negative resistance or a negative capacitance by the active elements. Furthermore, substituting $j\omega$ for s in (5.140), where j denotes an imaginary unit, and ω denotes angular frequency, we can derive the Bode diagram of the aforementioned approximate implementation of capacitive fractor $s^{-1/2}$ and the ideal filter of capacitive fractor $s^{-1/2}$, as shown in Figure 5.25.

Figure 5.25 shows that the frequency characteristics of the approximate implementation of band-pass filtering capacitive fractor $s^{-1/2}$ in Figure 5.24 can approach those of the ideal filter of capacitive fractor $s^{-1/2}$ with a high degree of accuracy in the pass-band of $[10^2 \text{rad/sec}, 1.2 \cdot 10^4 \text{rad/sec}]$. Both the magnitude and the phase of the fractional-order reactance of the aforementioned approximate implementation of capacitive fractor $s^{-1/2}$ can approach those of the ideal filter of capacitive fractor $s^{-1/2}$ in the pass-band of $[10^2 \text{rad/sec}, 1.2 \cdot 10^4 \text{rad/sec}]$.

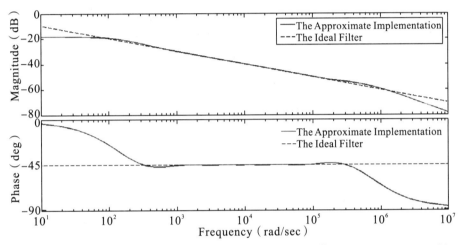

Figure 5.25 Bode diagram of approximate implementation of $s^{-1/2}$ capable of band-pass filtering and ideal filter of $s^{-1/2}$.

Example 2: For the convenience of illustration, approximately implemented v-order capacitive fractors in series are only discussed in this example. Likewise, the results of inductive fractors in series are similar to those of capacitive fractors in series.

Let me use the same two approximately implemented 1/2-order capacitive fractors $s^{-1/2}$ of Example 1 to the analysis of approximately implemented 1/2-order capacitive fractors in series. With regard to the same two approximately implemented 1/2-order capacitive fractors $s^{-1/2}$ in Figure 5.24, from (5.71), we can derive $v=1/2$, $q=0$, $p=1/2$, $c_1=c_2=c=1$, and $r_1=r_2=r=1$, where c_1 and c_2 are the capacitances of the first and second series $s^{-1/2}$, and r_1 and r_2 are the resistances of the first and second series $s^{-1/2}$, respectively. Thus, one can obtain the analog series circuit of the same two approximately implemented 1/2-order capacitive fractors $s^{-1/2}$ of Example 1, as shown in Figure 5.26.

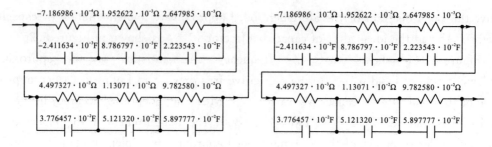

Figure 5.26 Analog series circuit of same two $s^{-1/2}$.

From (5.71), (5.140), and Figure 5.26, we can derive the 1/2-order capacitive reactance of the same two 1/2-order capacitive fractors $s^{-1/2}$ in their approximate implementations of Example 1, respectively. Thus, it follows that:

$$\begin{aligned}
{}_1F^c_{-1/2} &= V_i(s)/I_i(s) \\
&\approx c_1^{-1/2} r_1^{1/2} s^{-1/2} \\
&\stackrel{c_1=c=1, r_1=r=1}{=} s^{-1/2},
\end{aligned} \quad (5.141)$$

$$\begin{aligned}
{}_1F^c_{-1/2} &= V_i(s)/I_i(s) \\
&\approx c_2^{-1/2} r_2^{1/2} s^{-1/2} \\
&\stackrel{c_2=c=1, r_2=r=1}{=} s^{-1/2},
\end{aligned} \quad (5.142)$$

where ${}_1F^c_{-1/2}$ and ${}_2F^c_{-1/2}$ denote the first and second 1/2-order capacitive reactances of the first and second series of the same two 1/2-order approximately implemented capacitive fractors $s^{-1/2}$ connected in series in Figure 5.26, respectively. Thus, from (5.92), the following can be obtained:

$$\begin{aligned}
{}_1F_{-1/2} &= {}_2F_{-1/2} \\
&\approx c^{1/2} r^{1/2} P_{-1/2} \\
&\stackrel{c=1, r=1}{=} P_{-1/2},
\end{aligned} \quad (5.143)$$

where ${}_1F_{-1/2}$ and ${}_2F_{-1/2}$ denote the first and second 1/2-order capacitive fractances of the first and second series of the same two 1/2-order approximately implemented capacitive fractors $s^{-1/2}$ connected in series in Figure 5.26, respectively. For $c_1=c_2=c=1$ and $r_1=r_2=r=1$, from (5.103), it follows that:

$$\begin{aligned}
{}_{1+2}F^c_{-1/2} &= V_i(s)/I_i(s) \\
&\approx \sum_{j=1}^{2} c_j^{-1/2} r_j^{1/2} s^{-1/2} \\
&\stackrel{\substack{c_1=c_2=c,\\ r_1=r_2=r}}{=} c^{-1/2} r^{1/2} \Big(\sum_{j=1}^{2} 1\Big) s^{-1/2} \\
&= 2c^{-1/2} r^{1/2} s^{-1/2} \\
&\stackrel{c=1, r=1}{=} 2s^{-1/2},
\end{aligned} \quad (5.144)$$

where ${}_{1+2}F^c_{-1/2}$ denotes the 1/2-order capacitive reactance of the same two 1/2-order approximately implemented capacitive fractors $s^{-1/2}$ connected in series in Figure 5.26.

Thus, from (5.103), (5.104), and (5.144), we can derive the total fractance of the same two 1/2-order approximately implemented capacitive fractors $s^{-1/2}$ connected in series in Figure 5.9, given as:

$$F_{-1/2}^s \stackrel{\substack{c_1=c_2=c,\\r_1=r_2=r}}{\approx} c^{1/2}r^{1/2}\left(\sum_{j=1}^{2}1\right)\mathrm{P}_{-1/2}$$
$$= 2c^{1/2}r^{1/2}\mathrm{P}_{-1/2}$$
$$\stackrel{c=1,r=1}{=} 2\mathrm{P}_{-1/2}, \tag{5.145}$$

where F_{-v}^s denotes the total fractance of the same two 1/2-order approximately implemented capacitive fractors $s^{-1/2}$ connected in series in Figure 5.26. In addition, from (5.140) and (5.144), we can further derive the 1/2-order capacitive reactance of the same two 1/2-order approximately implemented capacitive fractors $s^{-1/2}$ connected in series in Figure 5.9, given as:

$$_{1+2}F_{-1/2}^c \approx 2s^{-1/2}$$
$$= 2 \cdot \frac{1.2\cdot 10^3 \cdot s^5 + 2.2\cdot 10^8 \cdot s^4 + 7.92\cdot 10^{12} \cdot s^3 + 7.92\cdot 10^{16} \cdot s^2 + 2.2\cdot 10^{20} \cdot s + 1.2\cdot 10^{23}}{s^6 + 6.6\cdot 10^5 \cdot s^5 + 4.95\cdot 10^{10} \cdot s^4 + 9.24\cdot 10^{14} \cdot s^3 + 4.95\cdot 10^{18} \cdot s^2 + 6.6\cdot 10^{21} \cdot s + 10^{24}}$$
$$= \frac{-8.29313192\cdot 10^2}{s+5.76954805\cdot 10^5} + \frac{2.27614238\cdot 10^2}{s+5.82842712\cdot 10^4} + \frac{8.99465458\cdot 10}{s+1.69839637\cdot 10^4} + \frac{5.29596902\cdot 10}{s+5.88790706\cdot 10^3}$$
$$+ \frac{3.90524292\cdot 10}{s+1.71572875\cdot 10^3} + \frac{3.39110794\cdot 10}{s+1.73323801\cdot 10^2}. \tag{5.146}$$

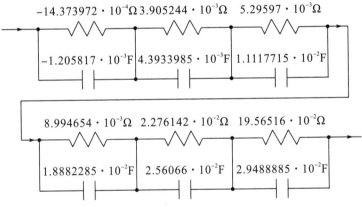

Figure 5.27 Equivalent circuit of analog series circuit of same two $s^{-1/2}$.

From (5.140) and (5.146), we can see that all poles of the same two 1/2-order approximately implemented capacitive fractors $s^{-1/2}$ connected in series in (5.146) are the same as all corresponding poles of each series approximately implemented capacitive fractor $s^{-1/2}$ in (5.140). Note that all poles of the arbitrary-order fractors connected in series are be able to keep the same as all corresponding poles of every series fractors. Furthermore, all poles of the same two 1/2-order approximately implemented capacitive fractors $s^{-1/2}$ connected in series are on the left half plane. Thus, the same two 1/2-order approximately implemented capacitive fractors $s^{-1/2}$ connected in series is a stable system. From (5.146), the corresponding equivalent circuit of the analog series circuit of the same two 1/2-order approximately implemented capacitive fractors $s^{-1/2}$ can be derived, as shown in Figure 5.27.

Comparing Figure 5.24 with Figure 5.27, we can further see that all resistances in Figure 5.27 are the double size of all corresponding resistances in Figure 5.24, while all capacitances in Figure 5.27 are the half size of all corresponding capacitances in Figure 5.24, respectively. Substituting $j\omega$ for s in (5.146), where j denotes an imaginary unit, and ω denotes angular frequency, we can derive the Bode diagram of the same two approximately implemented capacitive fractors $s^{-1/2}$ connected in series, as shown in Figure 5.28.

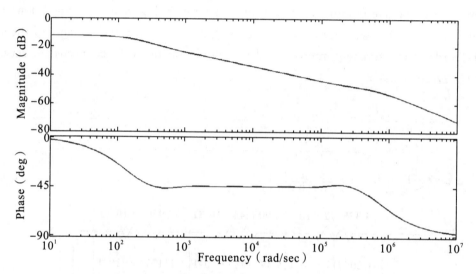

Figure 5.28 Bode diagram of analog series circuit of same two $s^{-1/2}$.

Comparing Figure 5.25 with Figure 5.28, we can see that the frequency characteristics of the same two approximately implemented capacitive fractors $s^{-1/2}$ connected in series are similar to those of each series approximately implemented capacitive fractor $s^{-1/2}$ with a high degree. The magnitude of the 1/2-order reactance of the same two approximately implemented capacitive fractors $s^{-1/2}$ connected in series increases by 6dB more than that of each series approximately implemented capacitive fractor $s^{-1/2}$, whereas the phase of the same two approximately implemented capacitive fractors $s^{-1/2}$ connected in series can quite approach that of each series approximately implemented capacitive fractor $s^{-1/2}$.

Example 3: For the convenience of illustration, approximately implemented v-order capacitive fractors in parallel are only discussed in this example. The results of inductive fractors in parallel are similar to those of capacitive fractors in parallel.

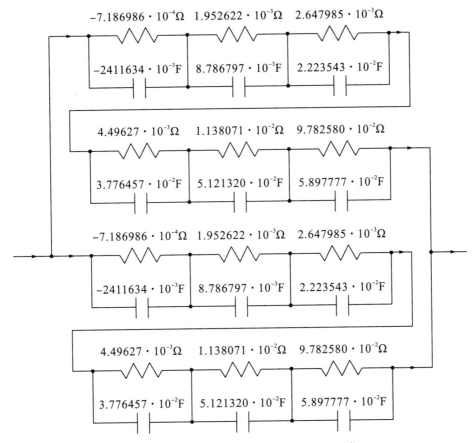

Figure 5.29 Parallel circuit of same two $s^{-1/2}$.

Let me use the same two approximately implemented 1/2-order capacitive fractors $s^{-1/2}$ of Example 1 to the analysis of approximately implemented 1/2-order capacitive fractors in parallel. With regard to the same two approximately implemented 1/2-order capacitive fractors $s^{-1/2}$ in Figure 5.24, from (5.71), we can derive $v=1/2$, $q=0$, $p=1/2$, $c_1=c_2=c=1$, and $r_1=r_2=r=1$, where c_1 and c_2 are the capacitances of the first and second parallel $s^{-1/2}$, and r_1 and r_2 are the resistances of the first and second parallel $s^{-1/2}$, respectively. Thus, one can obtain the analog parallel circuit of the same two approximately implemented 1/2-order capacitive fractors $s^{-1/2}$ of Example 1, as shown in Figure 5.29.

Because $c_1=c_2=c=1$ and $r_1=r_2=r=1$, from (5.71), (5.110), (5.140) — (5.143), and Figure 5.29, the following can be obtained:

$$_{1/\!/2}F^c_{-1/2} = V_i(s)/I_i(s)$$
$$\approx \Big(\sum_{j=1}^{2} c_j^{1/2} r_j^{-1/2}\Big)^{-1} s^{-1/2}$$
$$\stackrel{\substack{c_1=c_2=c,\\ r_1=r_2=r}}{=} c^{-1/2} r^{1/2} \Big(\sum_{j=1}^{2} 1\Big)^{-1} s^{-1/2}$$

$$= \frac{1}{2}c^{-1/2}r^{1/2}s^{-1/2}$$

$$\stackrel{c=1,r=1}{=} \frac{1}{2}s^{-1/2}, \qquad (5.147)$$

where $_{1/2}F^c_{-1/2}$ denotes the 1/2-order capacitive reactance of the same two 1/2-order approximately implemented capacitive fractors $s^{-1/2}$ connected in parallel in Figure 5.29. Thus, from (5.110), (5.111), and (5.147), we can derive the total fractance of the same two 1/2-order approximately implemented capacitive fractors $s^{-1/2}$ connected in parallel in Figure 5.29, given as:

$$F^s_{-1/2} \stackrel{\substack{c_1=c_2=c,\\r_1=r_2=r}}{=} c^{1/2}r^{1/2}\Big(\sum_{j=1}^{2}1\Big)^{-1}\mathrm{P}_{-1/2}$$

$$= \frac{1}{2}c^{1/2}r^{1/2}\mathrm{P}_{-1/2}$$

$$\stackrel{c=1,r=1}{=} \frac{1}{2}\mathrm{P}_{-1/2}, \qquad (5.148)$$

where F^s_{-v} denotes the total fractance of the same two 1/2-order approximately implemented capacitive fractors $s^{-1/2}$ connected in parallel in Figure 5.29. In addition, from (5.140) and (5.147), we can further derive the 1/2-order capacitive reactance of the same two 1/2-order approximately implemented capacitive fractors $s^{-1/2}$ connected in parallel in Figure 5.29, given as:

$$_{1/2}F^c_{-1/2} \approx \frac{1}{2}s^{-1/2}$$

$$= \frac{1}{2} \cdot \frac{1.2 \cdot 10^3 \cdot s^5 + 2.2 \cdot 10^8 \cdot s^4 + 7.92 \cdot 10^{12} \cdot s^3 + 7.92 \cdot 10^{16} \cdot s^2 + 2.2 \cdot 10^{20} \cdot s + 1.2 \cdot 10^{23}}{s^6 + 6.6 \cdot 10^5 \cdot s^5 + 4.95 \cdot 10^{10} \cdot s^4 + 9.24 \cdot 10^{14} \cdot s^3 + 4.95 \cdot 10^{18} \cdot s^2 + 6.6 \cdot 10^{21} \cdot s + 10^{24}}$$

$$= \frac{-2.07328298 \cdot 10^2}{s + 5.76954805 \cdot 10^5} + \frac{0.569035595 \cdot 10^2}{s + 5.82842712 \cdot 10^4} + \frac{2.248663645 \cdot 10}{s + 1.69839637 \cdot 10^4} + \frac{1.323992255 \cdot 10}{s + 5.88790706 \cdot 10^3}$$

$$+ \frac{0.97631073 \cdot 10}{s + 1.71572875 \cdot 10^3} + \frac{0.847776985 \cdot 10}{s + 1.73323801 \cdot 10^2}. \qquad (5.149)$$

From (5.140) and (5.149), we can see that all poles of the same two 1/2-order approximately implemented capacitive fractors $s^{-1/2}$ connected in parallel in (5.149) are the same as all corresponding poles of each parallel approximately implemented capacitive fractor $s^{-1/2}$ in (5.140). Note that all poles of the arbitrary-order fractors connected in parallel can keep the same as all corresponding poles of every parallel fractors. Furthermore, all poles of the same two 1/2-order approximately implemented capacitive fractors $s^{-1/2}$ connected in parallel are on the left half plane. Thus, the same two 1/2-order approximately implemented capacitive fractors $s^{-1/2}$ connected in parallel is a stable system. From (5.146), we can derive the corresponding equivalent circuit of the analog parallel circuit of the same two 1/2-order approximately implemented capacitive fractors $s^{-1/2}$, as shown in Figure 5.30.

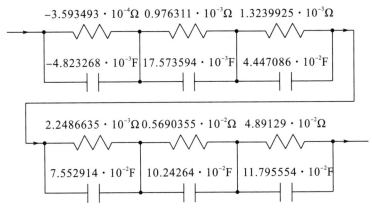

Figure 5.30 Equivalent circuit of analog parallel circuit of same two $s^{-1/2}$.

Comparing Figure 5.24 with Figure 5.30, we can further see that all resistances in Figure 5.30 are the half size of all corresponding resistances in Figure 5.24, and all capacitances in Figure 5.30 are the double size of all corresponding capacitances in Figure 5.24, respectively. Substituting jω for s in (5.146), where j denotes an imaginary unit, and ω denotes angular frequency, we can derive the Bode diagram of the same two approximately implemented capacitive fractors $s^{-1/2}$ connected in parallel, as shown in Figure 5.31.

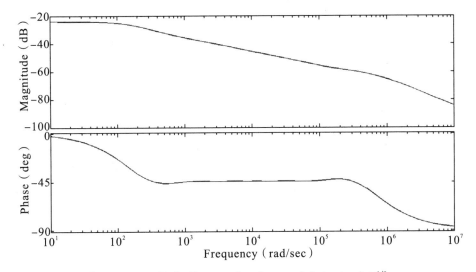

Figure 5.31 Bode diagram of analog parallel circuit of $s^{-1/2}$.

Comparing Figure 5.25 with Figure 5.31, we can see that the frequency characteristics of the same two approximately implemented capacitive fractors $s^{-1/2}$ connected in parallel are similar to those of each parallel approximately implemented capacitive fractor $s^{-1/2}$ with a high degree. The magnitude of the 1/2-order reactance of the same two approximately implemented capacitive fractors $s^{-1/2}$ connected in parallel decreases by 6dB less than that of each parallel approximately implemented capacitive fractor $s^{-1/2}$, whereas the phase of the same two approximately implemented capacitive fractors $s^{-1/2}$ connected in parallel can quite approach that of each parallel approximately implemented capacitive fractor $s^{-1/2}$.

5.2.5 CONCLUSIONS

Fractional calculus has evolved as an important, contemporary branch of mathematical analyses. The origin of fractional calculus was concurrent with that of integer-order calculus; however, until recently, the applications of fractional calculus were limited to mathematics. Now, fractional calculus appears to be a novel promising mathematical method for physical scientists and engineering technicians. The application of fractional calculus to circuits and systems, especially to a promising fractional-order circuit element called fractor, is an emerging and interesting discipline, in which insufficient studies have been implemented. Fractional calculus has been applied to circuits and systems mainly because of its inherent strength of long-term memory, non-locality, and weak singularity. The term "fractor" arose following the successful synthesis of a fractional-order capacitor or a fractional-order inductor in an analog circuit. The term "fractance", as a portmanteau of "the fractional-order impedance", refers to the fractional-order impedance of a fractor. Thus, it is natural to ponder what the measurement units and physical dimensions of fractance are. Until now, however, no efficient measurement units and physical dimensions of fractance have been proposed, which is a challenging theoretical problem. Motivated by this need, in this part, based on the aforementioned studies, here and in the companion paper (Part II), the measurement units and physical dimensions of fractance and rules for fractors in series and parallel are discussed in detail. We use a state-of-the-art mathematical method, fractional calculus, to analyze the proposed conceptual framework. In particular, part I proposed to introduce the position of purely ideal fractor in Chua's circuit axiomatic element system and the fractional-order reactance of fractor in its natural implementation. this part discusses fundamental issues, whereas Part II is devoted to the measurement units and physical dimensions of purely ideal fractance and the rules for fractors in series and parallel.

第6章 分抗元的阶—频特性[①]

本章主要讨论了一种新颖的电路元件——分抗元的阶次频率特性。分抗值是分抗的分数阶阻抗，是在模拟电路中分数阶微分或分数阶积分概念提出后的又一新概念。在本章中，我们研究了一些分抗元的电气特征，特别介绍了分抗元的阶次频率特性。首先，提出了分抗元的阶敏特性，研究了分抗元的阶次频率特性。其次，分析了分抗的时间常数，讨论了温度对分抗的影响。最后，通过数学分析和仿真结果，详细讨论了分抗电气特性的几个问题，特别是分抗的温度影响和时间常数。

6.1 Introduction

Fractional calculus has become an important novel branch of mathematical analyses [3], [176], [178], [230] − [231]. Fractional calculus is as old as the integer-order calculus, although until recently, its applications were exclusively in mathematics. Fractional calculus now seems to be a novel promising mathematical method for physical scientists and engineering technicians. Fractional calculus extends the concepts of difference and Riemann sums [3], [178] − [179]. The fractional differential of a Heaviside function is non-zero, whereas its integer-order differential must be zero. Fractional calculus of various functions possesses one obvious feature: the fractional calculus of most functions is equal to a power series, and the others are equal to the superposition or product of a certain function and a power function [3], [176], [178] − [179], [230] − [231]. Scientific study has shown that a fractional-order or fractional-dimensional approach is now the best description for many natural phenomena. It has obtained promising results and ideas demonstrating fractional calculus can be an interesting and useful tool in many scientific fields such as physical problems [11], [180], biomedical engineering [94], diffusion processes [181] − [183], viscoelasticity theory [126], fractal dynamics [184], fractional control [89] − [90], fractor and fractance [130], [171] − [173], [185], [196], [198] − [205] [227], [230], [232] − [234], fractional image processing [186] − [187], [189] − [190], and fractional neural networks [192].

The application of fractional calculus to signal analysis, signal processing, and circuits

---

[①] PU Yifei, ZHANG Ni, WANG Huai, et al. Order-Frequency Characteristics of a Promising Circuit Element: Fractor [J]. Journal of Circuits Systems and Computers, 2016, 25 (12): 17 pages, Article ID 1650156.

and systems, especially to fractor, is a promising discipline of study. Fractional calculus has been hybridized with signal analysis, signal processing, and circuits and systems mainly because of its inherent strength of long-term memory and non-locality. Following the success in the synthesis of a fractional differentiator and integrator in an analog circuit, the emergence of a promising circuit element was called fractor [130], [171] − [173], [185], [196], [198] − [205] [227], [230], [232] − [234]. Fractance means the fractional-order impedance of a fractor. The driving-point impedance function of a fractor is equal to its fractional-order reactance [204] − [205]. Consequently, let us denote fractor as the symbol ─┤W├─, in which F is the abbreviation of fractor [130], [171] − [173], [185], [196], [198] − [205] [227], [230], [232] − [234]. There are two types of fractor: capacitive fractor and inductive fractor. A capacitive fractor is a fractional-order capacitor, which involves a negative-order fractional differential operation. The fractional-order impedance of a capacitive fractor is its capacitive fractance. The electrical properties of a capacitive fractor fall in between those of a capacitor and those of a resistor [235] − [236]. Pu has derived the generic nonlinear relation between capacitance and resistance of the arbitrary order capacitive fractance [204] − [205]. Similarly, an inductive fractor is a fractional-order inductor, which involves a positive-order fractional differential operation. The fractional-order impedance of an inductive fractor is its inductive fractance. The electrical properties of an inductive fractor fall in between those of an inductor and those of a resistor [204] − [205]. Pu has derived the generic nonlinear relation between inductance and resistance of the arbitrary order inductive fractance [235] − [236]. With regard to capacitive fractor, we can see that it consists of ordinary capacitors and resistors in an analog circuit of the tree type [198], [230], two-circuits type [173], [185], H type [172], [185], net-grid type [171], [185], or some other infinite recursive types with extreme self-similar fractal structures. Similarly, an inductive fractor consists of a series of ordinary inductors and resistors with the same type of fractal structures as a capacitive fractor. In particular, the tree type [198], [230], two-circuits type [173], [185], H type [172], [185], net-grid type [171], [185] should be four natural implementations of fractor. What distinguishes these four aforementioned types from the other approximate implementations of fractor [130], [196], [227], [232], [199] − [203], [233] − [234] is that they consist of series of ordinary circuit elements in the form of natural fractal structures exhibiting extreme self-similarity. In contrast to other approximate implementations of fractor [130], [196], [227], [232], [199] − [203], [233] − [234], the floating point values of the capacitance, inductance, and resistance of these four aforementioned types of fractor [198], [230], [171] − [173], [185] are not required. In fact, there are zero errors between these four types of fractance [198], [230], [171] − [173], [185] with infinite recursive extreme self-similar structures and a purely ideal fractance. Devices manufactured using the other approximate implementations of fractor [130], [196], [227], [232], [199] — [203], [233] — [234] never represent a purely ideal fractor. In this chapter, we studied some electrical properties of a fractor. In particular, the order-frequency characteristics of a fractor are introduced.

The remainder of this chapter is organized as follows: Section 6.2 recalls the necessary mathematical background of fractional calculus. Section 6.3 discusses some electrical properties of a fractor. In Section 6.3, first, the order-sensitivity characteristics of a fractor are proposed. Second, the order-frequency characteristics of a fractor are studied. Third, the time constant of a fractor is analyzed. Fourth is the proposal for the temperature effect of a fractor. Section 6.4 reports the experiment's results and analysis. Through mathematical analysis and simulation results, we discuss in detail some issues of the electrical properties of a fractor, especially its temperature effect and time constant.

6.2 Related Work

This section includes a brief introduction to the basic definitions of fractional calculus.

Euclidean measure-based fractional calculus is more developed than the Hausdorff measure-based one. Because of its more advanced development, the Euclidean measure-based type of fractional calculus is widely used. The commonly used fractional calculus definitions are Grünwald-Letnikov, Riemann-Liouville, and Caputo [3], [176], [178], [230] — [231]. The Laplace transform of the v-order Riemann-Liouville differential operator is as $L[{}^{R-L}_{\ 0}D_x^v f(x)] = s^v L[f(x)] - \sum_{k=0}^{n-1} s^k [{}^{R-L}_{\ 0}D_x^{v-1-k} f(x)]_{x=0}$, where s denotes the Laplace operator. When $f(x)$ is causal signal, and its fractional primitives must also be zero, we can simplify the Laplace transform for ${}^{R-L}_{\ 0}D_x^v f(x)$ as $L[{}^{R-L}_{\ 0}D_x^v f(x)] = s^v L[f(x)]$. The Laplace transform of the v-order Caputo differential operator is as $L[{}^C_0 D_x^v f(x)] = s^v L[f(x)] - \sum_{k=0}^{n-1} s^k f^{(k)}(x)|_{x=0}$. When $f(x)$ is causal signal, and its fractional primitives must also be zero, we can simplify the Laplace transform for ${}^C_0 D_x^v f(x)$ as $L[{}^C_0 D_x^v f(x)] = s^v L[f(x)]$. In this case the three cited definitions of fractional derivatives are equivalent. In this chapter, we use indifferently the equivalent notations $D_x^v = {}^{G-L}_{\ 0}D_x^v = {}^{R-L}_{\ 0}D_x^v = {}^C_0 D_x^v$. Therefore, when $f(x)$ is causal signal, and its fractional primitives must also be zero, we can simplify the Laplace transform for $D^v f(x)$ as follows:

$$L[D^v f(x)] = s^v L[f(x)], \tag{6.1}$$

where s denotes the Laplace operator.

6.3 Order-Frequency Characteristics of Fractor

6.3.1 Order-Sensitivity Characteristics of Fractor

In this subsection, we propose the order-sensitivity characteristics of a fractor.

Let the input voltage and input current of a circuit element be equal to $V_i(s)$ and $I_i(s)$, respectively. The capacitive reactance of a capacitor [204] – [205] is as follows:

$$Z_{-1}^c(s) = V_i(s)/I_i(s) = c^{-1}s^{-1}, \qquad (6.2)$$

where c is equal to the capacitance of a capacitor. The fractional-order capacitive reactance of an arbitrary-order capacitive fractor in its natural implementation [204] – [205], F_{-v}^c, is as follows:

$$F_{-v}^c = F_{-(q+p)}^c = V_i(s)/I_i(s) = c^{-(q+p)}r^{1-p}s^{-(q+p)}, \qquad (6.3)$$

where $v = q + p$ is a positive real number, q is a positive integer, and $0 \leqslant p \leqslant 1$. The inductive reactance of an inductor [204] – [205] is as follows:

$$Z_1^l(s) = V_i(s)/I_i(s) = l^1 s^1, \qquad (6.4)$$

where l is equal to the inductance of an inductor. The fractional-order inductive reactance of an arbitrary-order inductive fractor in its natural implementation [204] – [205], F_v^l, is as follows:

$$F_v^l = F_{q+p}^l = V_i(s)/I_i(s) = l^{q+p} r^{1-p} s^{q+p}, \qquad (6.5)$$

where $v = q + p$ is a positive real number, q is a positive integer, and $0 \leqslant p \leqslant 1$.

From (6.1), (6.3), and (6.5), we can also see that the crux of the implementation of an arbitrary-order fractor is to perform a fractional integral operator s^{-v} or a fractional differential operator s^v in the form of an analog circuit. Suppose that v is a positive rational number. Thus, we can derive that v is equal to the summation of a positive integer and a positive rational fraction. For the fractional differential operator s^v, it follows that:

$$s^v = s^m s^{i/n}, \qquad (6.6)$$

where m, n, and i are positive integers, $n \neq 1$, $i = 1, 2, \cdots, n - 1$, and i/n is a positive rational fraction. v is an arbitrary-order. We can implement s^m by means of inductors or capacitors. Therefore, the key to implementing the fractional differential operator s^v is to implement the fractional differential operator $s^{i/n}$ in the form of an analog circuit.

According to circuits and systems theory, $s^{i/n}$ can be regarded as the transfer function of a i/n-order system that is equal to dividing its input by its output. Thus, we can rewrite the fractional differential operator $s^{i/n}$ as follows:

$$\begin{aligned} s^{i/n} &= B(s)/A(s) \\ &= \frac{b_\alpha s^\alpha + b_{\alpha-1} s^{\alpha-1} + \cdots + b_1 s + b_0}{s^\beta + a_{\beta-1} s^{\beta-1} + \cdots + a_1 s + a_0} \\ &= \frac{\sum_{k=0}^{\alpha} b_k s^{k-\beta}}{1 + \sum_{l=0}^{\beta-1} a_l s^{l-\beta}}, \end{aligned} \qquad (6.7)$$

where $B(s)$ and $A(s)$ denote the input and output of $s^{i/n}$, respectively, α, β, k, and l are positive integers, and a_l and b_k are constants. According to circuits and systems theory, all physically realizable systems must meet the condition of $\alpha \leqslant \beta$ in (6.7). Moreover, as we know, if $|\alpha - (\beta s)^{i/n}| < 1$, then from the binomial theorem, we have $\lim\limits_{j \to \infty} [\alpha - (\beta s)^{i/n}]^j$
$= \lim\limits_{j \to \infty} \sum\limits_{k=0}^{j} (-1)^k \binom{j}{k} \alpha^{j-k} \beta^{ik/n} s^{quo \langle ik/n \rangle} s^{rem \langle ik/n \rangle} = 0$, where α and β are positive numbers, $\binom{j}{k} = \dfrac{\Gamma(1+j)}{\Gamma(j-k+1)\Gamma(k+1)}$, Γ denotes the Gamma function, $quo()$ denotes quotient calculation, and $rem()$ denotes the remainder calculation. Therefore, if $j \to \infty$, we can derive the computational completeness of $s^{rem(ik/n)}$ as $\lim\limits_{j \to \infty} s^{rem(ik/n)} = s^{i/n} = s^{1/n}, s^{2/n}, \cdots, s^{n-1/n}$ corresponding to $i = 1, 2, \cdots, n-1$. Thus, according to the binomial theorem, we can derive that when $i = 1, 2, \cdots, n-1$, every denominator of each $s^{i/n}$ in (6.7) is uniformly equal, while every numerator of each $s^{i/n}$ in (6.7) is not equal, respectively. For example, if $n = 5$, the denominators of $s^{1/5}$, $s^{2/5}$, $s^{3/5}$, and $s^{4/5}$ in (6.7) are equal, while the numerators of $s^{1/5}$, $s^{2/5}$, $s^{3/5}$, and $s^{4/5}$ in (6.7) are not equal, respectively. Thus, we can see that the numerator of (6.7) determines the numerator i of the fractional-order i/n of the fractional differential operator $s^{i/n}$, while the denominator of (6.7) determines the denominator n of its fractional-order i/n. The variation of its numerator and denominator are both order-sensitive to the fractional differential operator $s^{i/n}$, but the variation of its denominator is relatively more order-sensitive than the variation of its numerator. From (6.7), according to Mason rules, the cascaded signal flow graph of the fractional differential operator $s^{i/n}$ is shown in Fig. 6.1.

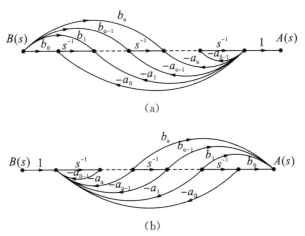

Fig. 6.1 Cascaded signal flow graph of fractional differential operator $s^{i/n}$: (a) Set rightmost s^{-1} as common part of $s^{i/n}$; (b) Set leftmost s^{-1} as common part of $s^{i/n}$.

From Fig. 6.1, according to the Mason rules, we can see that, if all forward paths and feedback paths of the fractional differential operator $s^{i/n}$ have common parts, each numerator term of (6.7) constitutes an individual forward path, while each denominator term of (6.7) except for 1 constitutes an individual feedback path. Figs. 6.1 (a) and 6.1 (b) are two equivalent cascaded signal flow graphs of the fractional differential operator $s^{i/n}$. We set the rightmost s^{-1} in Fig. 6.1 (a) and the leftmost s^{-1} in Fig. 6.1 (b) as a common part of the fractional differential operator $s^{i/n}$, respectively. From (6.7) and Fig. 6.1, we can also see that the fractional differential operator $s^{i/n}$ only owns three fundamental operations, namely, s^{-1}, scalar product, and addition. Hence, we only need the first-order integrator, multiplier, and summator to implement s^{-1}, scalar product, and addition, respectively. From Fig. 6.1, we can further see that the forward path of the cascaded signal flow graphs of the fractional differential operator $s^{i/n}$ determines the numerator i of its fractional-order i/n, while its feedback path determines the denominator n of its fractional-order i/n.

6.3.2 Order-Frequency Characteristics of Fractor

In this subsection, to preferably represent the characteristics of a fractor, we propose the order-frequency characteristics of a fractor.

From (6.3), and (6.5), the fractional-order reactance of the v-order capacitive fractor and the v-order inductive fractor can be collectively represented as follows:

$$H_p(s) = V_i(s)/I_i(s) = F_p s^p, \tag{6.8}$$

where p is a non-integer, which denotes the fractional-order of a fractor, and F_p denotes the fractance of a p-order fractor, which is defined as (6.3) and (6.5). When $p < 0$, $H_p(s)$ represents the $|p|$-order capacitive reactance; when $p > 0$, $H_p(s)$ represents the $|p|$-order inductive reactance. Substituting $j\omega$ for s in (6.8), we can derive the frequency characteristics of a fractor, where j denotes an imaginary unit, and ω denotes the angular frequency. It follows that:

$$H_p(j\omega) = V_i(j\omega)/I_i(j\omega) = F_p (j\omega)^p. \tag{6.9}$$

It follows that:

$$\begin{cases} A_p(\omega) = |H_p(j\omega)| = F_p |\omega|^p \\ \Theta_p(\omega) = \dfrac{p\pi}{2} \operatorname{sgn}(\omega) \end{cases}, \tag{6.10}$$

where $A_p(\omega)$ and $\Theta_p(\omega)$ denote the magnitude-frequency characteristic function and the phase-frequency characteristic function of a p-order fractor, respectively, and sgn( ) denotes the sign function. $A_p(\omega)$ can be further expressed in logarithmic form. From (6.10), it follows that:

$$\Lambda_p(\omega) = \ln A_p(\omega) = \ln F_p + p \ln |\omega|, \tag{6.11}$$

where ln( ) denotes the natural logarithm. From (6.10), and (6.11), we can see that $A_p(\omega)$ is collectively determined by three factors, namely, the fractance F_p, the fractional-order p, and the angular frequency ω. $\Theta_p(\omega)$ is only determined by a single factor, namely, the fractional-order p. Thus, we can further see that fractor is essentially a constant phase circuit element. The fractional-order p is essentially a constant phase exponent. From (6.10) and (6.11), it follows that:

$$p = \frac{d\Lambda_p(\omega)}{d\ln|\omega|} = \frac{d\ln A_p(\omega)}{d\ln|\omega|} = \frac{2\Theta_p(\omega)}{\pi \cdot \mathrm{sgn}(\omega)}. \tag{6.12}$$

Equation (6.12) shows that the fractional-order p is nothing to do with the angular frequency ω.

In fact, with regard to the fractor of any approximate implementation [130], [196], [227], [232], [199] − [203], [233] − [234], for the approximate error is inevitable, $\Theta_p(\omega)$ is also determined by the angular frequency ω. Thus, the fractor of any approximate implementation is no longer a constant phase circuit element. Thus, from (6.12), p is also determined by the angular frequency ω, namely, $p = p(\omega)$. Consequently, we suggest defining the order-frequency characteristics as the operational nature representation of a fractor, given as:

$$p(\omega) = \frac{d\Lambda_p(\omega)}{d\ln|\omega|} = \frac{d\Lambda_p(\omega)}{d|\omega|}\frac{d|\omega|}{d\ln|\omega|} = \frac{d\ln A(\omega)}{d\ln|\omega|} = \frac{|\omega|}{A(\omega)}\frac{dA(\omega)}{d|\omega|}. \tag{6.13}$$

Equation (6.13) shows that the order-frequency characteristics of a fractor, $p(\omega)$, can simultaneously represent two interrelated factors, the fractional-order and the frequency characteristics, of a fractor.

6.3.3 Time Constant of Fractor

In this subsection, the time constant of a fractor is studied.

From (6.5) and (6.7), we can regard the fractional differential operator $s^{i/n}$ as a driving-point impedance function of a i/n-order inductive fractor. Thus, $s^{i/n} = V_i(s)/I_i(s)$ denotes the fractional-order reactance, where $V_i(s) = B(s)$ is the input voltage and $I_i(s) = A(s)$ is the input current of a i/n-order inductive fractor in (6.7). If the fractional differential operator $s^{i/n}$ does not have multiple poles, it follows that:

$$I_i(s) = (s - \vartheta_1)\cdots(s - \vartheta_k)\cdots(s - \vartheta_n) \tag{6.14}$$

where $\vartheta_1, \cdots, \vartheta_k, \cdots, \vartheta_n$ are different simple roots of $I_i(s) = 0$ for $k = 1, 2, \cdots, n$, respectively. According to circuits and systems theory, all stable systems must meet the condition of $\vartheta_k < 0$. Thus, from (6.3) and (6.7), it follows that:

$$s^{i/n} = V_i(s)/I_i(s) = C_0 + \frac{C_1}{s - \vartheta_1} + \cdots + \frac{C_k}{s - \vartheta_k} + \cdots + \frac{C_n}{s - \vartheta_n}, \tag{6.15}$$

where $C_0 = \lim_{s \to \infty} s^{i/n}$ and $C_k = \lim_{s \to \vartheta_k}(s - \vartheta_k)s^{i/n}$. Thus, it follows that:

$$V_i(s) = C_0 I_i(s) + \sum_{k=1}^{n} \frac{C_k}{s - \vartheta_k} I_i(s). \tag{6.16}$$

Then, the input voltage of the kth series $C_k/(s - \vartheta_k)$ is as follows:

$$V_i^k(s) = \frac{C_k}{s - \vartheta_k} I_i(s). \tag{6.17}$$

The inverse Laplace transform of (6.17) is as follows:

$$\frac{d}{dt}V_i^k(t) = \vartheta_k V_i^k(t) + C_k I_i(t). \tag{6.18}$$

Thus, we can derive the state equation and the output equation of the fractional differential operator $s^{i/n}$ in vector form, respectively, as follows:

$$\begin{bmatrix} \dfrac{d}{dt}V_i^1(t) \\ \dfrac{d}{dt}V_i^2(t) \\ \vdots \\ \dfrac{d}{dt}V_i^n(t) \end{bmatrix} = \begin{bmatrix} \vartheta_1 & 0 & \cdots & 0 \\ 0 & \vartheta_2 & \cdots & 0 \\ \vdots & \vdots & & \vdots \\ 0 & 0 & \cdots & \vartheta_n \end{bmatrix} \begin{bmatrix} V_i^1(t) \\ V_i^2(t) \\ \vdots \\ V_i^n(t) \end{bmatrix} + \begin{bmatrix} C_1 \\ C_2 \\ \vdots \\ C_n \end{bmatrix} I_i(t), \qquad (6.19)$$

$$V_i(t) = \begin{bmatrix} 1 & 1 & \cdots & 1 \end{bmatrix} \begin{bmatrix} V_i^1(t) \\ V_i^2(t) \\ \vdots \\ V_i^n(t) \end{bmatrix} + C_0 I_i(t). \qquad (6.20)$$

From (6.20), we can see that the coefficient matrix of the state equation of the fractional differential operator $s^{i/n}$ is a diagonal matrix whose coefficients on the principal diagonal are equal to the poles of the driving-point impedance function of a i/n-order inductive fractor, respectively.

If the fractional differential operator $s^{i/n}$ has the r-order multiple poles, from (6.7), it follows that:

$$s^{i/n} = C_0 + \frac{C_1^1}{(s-\vartheta_1)} + \cdots + \frac{C_1^r}{(s-\vartheta_1)} + \frac{C_2}{(s-\vartheta_2)} + \cdots + \frac{C_n}{(s-\vartheta_n)} \qquad (6.21)$$

where $C_1^m = \lim\limits_{s \to \vartheta_1}(s-\vartheta_1)^m s^{i/n}$, $m = 1, 2, \cdots, r$. Thus, the state equation and the output equation of the fractional differential operator $s^{i/n}$ are the same as (6.19) and (6.20), respectively.

From (6.7), we can see that $s^{i/n}$ is equal to a nonlinear expression of the first-order differentiator s. From (6.15), (6.19), and (6.21), we can see further that the fractional differential operator $s^{i/n}$ is equivalent to a series circuit of the first-order integral analog circuit. The fractional differential operator $s^{i/n}$ is the result of the nonlinear interaction of the series circuit of the first-order integral analog circuit. Figure 6.2 shows the kth series first-order integral analog circuit of the fractional differential operator $s^{i/n}$.

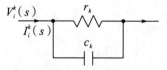

Fig. 6.2 The kth series first-order integral analog circuit.

In Fig. 6.2, we set r_k and c_k, respectively, as follows:

$$r_k = -C_k/\vartheta_k \, \Omega, \qquad (6.22)$$

$$c_k = 1/C_k \, \text{F}, \qquad (6.23)$$

where the measurement units of resistance and capacitance are Ohm (Ω) and Farad (F), respectively. Then, we can derive the input impedance of the kth series first-order integral analog circuit. It follows that:

$$Z^k(s) = V_i^k(s)/I_i^k(s) = \frac{C_k}{s - \vartheta_k}. \tag{6.24}$$

Thus, from (6.15), (6.19), and (6.21), the corresponding analog series circuit of the fractional differential operator $s^{i/n}$ is shown in Fig. 6.3.

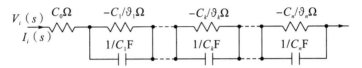

Fig. 6.3 Analog series circuit of fractional differential operator $s^{i/n}$.

From Fig. 6.3, we can regard the circuits of the fractional differential operator $s^{i/n}$ as a two-terminal analog network. Then, by short-circuiting the voltage source and open-circuiting its current source of the fractional differential operator $s^{i/n}$, we can convert $s^{i/n}$ into a series circuit of equivalent resistance r_e and equivalent capacitance c_e. Figure 6.4 shows the conversion circuit of the fractional differential operator $s^{i/n}$.

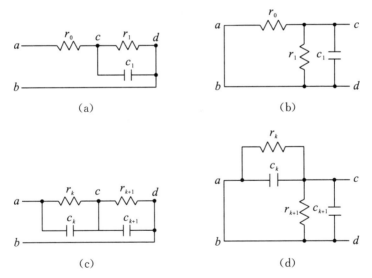

Fig. 6.4 Conversion circuit of fractional differential operator $s^{i/n}$: (a) Series circuit of r_0 with first series, (b) Conversion circuit of r_0 with first series, (c) Series circuit of kth with $(k+1)th$ series, (d) Conversion circuit of kth with $(k+1)th$ series.

From Fig. 6.4 (b), we can see that r_0 and r_1 connect in parallel in the conversion circuit of r_0 with the first series. From Fig. 6.4 (d), we can see that r_k and r_{k+1} connect in parallel, and c_k and c_{k+1} also connect in parallel in the conversion circuit of the kth with the $(k+1)th$ series. Thus, the equivalent resistance of the fractional differential operator $s^{i/n}$ is equal to:

$$r_e = r_0 \parallel r_1 \parallel \cdots \parallel r_n = \left(\sum_{k=1}^{n} r_k^{-1}\right)^{-1}. \tag{6.25}$$

The equivalent capacitance of the fractional differential operator $s^{i/n}$ is equal to:

$$c_e = c_1 \parallel c_2 \parallel \cdots \parallel c_n = \sum_{k=1}^{n} c_k. \tag{6.26}$$

Therefore, time constant T_F of the fractional differential operator $s^{i/n}$ is equal to:

$$T_F = |r_e c_e| s, \tag{6.27}$$

where is the measurement units of time is second (s).

6.3.4 Temperature Effect of Fractor

In this subsection, to implement a robust fractor, we need study the temperature effect of a fractor.

As we know, the resistance and capacitance of most fractors are subject to environmental variations such as temperature, supply voltage, and electromagnetic interference, which can affect their performance. Equations (6.12) and (6.13) show that the crux of the implementation of a robust fractor is to keep its fractional-order being the same under different environmental conditions, especially across varying temperatures. In order to eliminate the change of resistance or capacitance corresponding to temperature variation, on the one hand, the commonly used method is to connect in series a pair of resistors with opposite temperature coefficient, and to connect in parallel a pair of capacitors with opposite temperature coefficient. On the other hand, we can also defense against the temperature effect of a fractor by employing a set of specific resistor and capacitor, whose temperature coefficient are the reciprocal of each other. Let us set the temperature coefficient of resistance (TCR) and the temperature coefficient of capacitance (TCC) to be equal to constants ζ_r and ζ_c, respectively. We suggest setting $\zeta_c = 1/\zeta_r$. Thus, when temperature varies, we get $r_k = -\zeta_r C_k/\vartheta_k \Omega$, and $c_k = \zeta_c/C_k = 1/\zeta_r C_k \mathrm{F}$ in Fig. 6.2. From (6.24), it follows that:

$$Z_T^k(s) = V_i^k(s)/I_i^k(s) = \frac{\zeta_r C_k}{s - \vartheta_k}. \tag{6.28}$$

Substituting (6.28) into (6.15) or (6.21) and it follows that:

$$s_T^{i/n} = \zeta_r C_0 + \frac{\zeta_r C_1}{s - \vartheta_1} + \cdots + \frac{\zeta_r C_k}{s - \vartheta_k} + \cdots + \frac{\zeta_r C_n}{s - \vartheta_n} = \zeta_r s^{i/n}. \tag{6.29}$$

From (6.29), we can see that, if $\zeta_c = 1/\zeta_r$ is satisfied, temperature effect produces an effect on $s_T^{i/n} = \zeta_r s^{i/n}$. If $\zeta_c = 1/\zeta_r$ is satisfied, temperature effect can only change the fractance of a fractor, but cannot change the fractional-order i/n of the fractional differential operator $s^{i/n}$. In fact, $\zeta_c = 1/\zeta_r$ is a very strictly restricted condition for a set of resistor and capacitor. But with progress made in modern material science, some innovative materials maybe make it possible for us to find out the set of resistor and capacitor, whose temperature coefficients are the reciprocal of each other.

6.4 Experiment and Analysis

In this section, through mathematical analysis and simulation results, we discuss some issues of the order-frequency characteristics of a fractor.

Example 1: with regard to the temperature effect of a fractor, according to the binomial theorem, we can derive an approximate implementation algorithm of a certain band-pass filtering fractional integral operator $s^{-1/2}$, whose approximation pass-band is [0.1Hz, 12Hz]. We can specify the approximation pass-band according to the literature [19]. It follows that in vector form:

$$s^{-1/2} = \frac{1.2 \cdot 10^{3/2} \cdot s^5 + 2.2 \cdot 10^{7/2} \cdot s^4 + 7.92 \cdot 10^{9/2} \cdot s^3 + 7.92 \cdot 10^{11/2} \cdot s^2 + 2.2 \cdot 10^{13/2} \cdot s + 1.2 \cdot 10^{13/2}}{s^6 + 6.6 \cdot 10^2 \cdot s^5 + 4.95 \cdot 10^4 \cdot s^4 + 9.24 \cdot 10^5 \cdot s^3 + 4.95 \cdot 10^6 \cdot s^2 + 6.6 \cdot 10^6 \cdot s + 10^6}$$

$$= \frac{3.09352345 \cdot 10}{s + 5.76954805 \cdot 10^2} + \frac{3.5988971}{s + 5.82842712 \cdot 10} + \frac{1.42217976}{s + 1.69839637 \cdot 10} + \frac{8.373662263 \cdot 10^{-1}}{s + 5.88790706} + \frac{6.17473122 \cdot 10^{-1}}{s + 1.71572875}$$

$$+ \frac{5.3618124 \cdot 10^{-1}}{s + 1.73323801 \cdot 10^{-1}} \tag{6.30}$$

Equation (6.30) shows that all of the poles of the fractional integral operator $s^{-1/2}$ are on the left half plane. Thus, the fractional integral operator $s^{-1/2}$ is a stable system. From (6.22), (6.23) and Fig. 6.2, we can derive the corresponding analog series circuit of the fractional integral operator $s^{-1/2}$ as shown in Fig. 6.5.

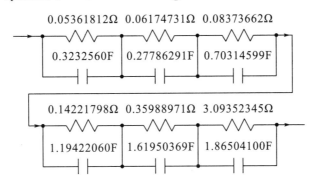

Fig. 6.5 Analog series circuit of fractional integral operator $s^{-1/2}$.

Substituting $j\omega$ for s in (6.30), we can derive the Bode diagram of the fractional integral operator $s^{-1/2}$ of the aforementioned approximate implementation and that of the ideal filter, respectively, as shown in Fig. 6.6.

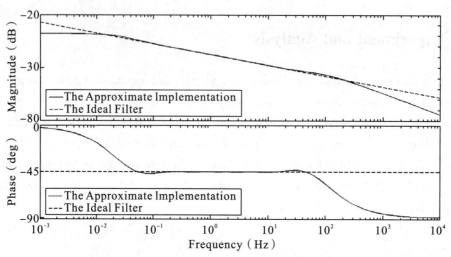

Fig. 6.6 Bode diagram of fractional integral operator $s^{-1/2}$ of approximate implementation and that of ideal filter.

Figure 6.6 shows that the frequency characteristics of the fractional integral operator $s^{-1/2}$ of the aforementioned approximate implementation can approach those of the ideal filter with a high degree of accuracy in the pass-band of [0.1Hz, 12Hz]. Both the magnitude and the phase of the fractional integral operator $s^{-1/2}$ of the aforementioned approximate implementation can approach those of the ideal filter in the pass-band of [0.1Hz, 12Hz].

As in the aforementioned discussion, we set TCR and TCC be equal to constant ζ_r and ζ_c, respectively, and set $\zeta_c = 1/\zeta_r$. Without loss of generality, we suppose that $\zeta_r = 2$. Thus, from (6.29) and Fig. 6.6, we can get an analog series circuit of the fractional integral operator $s_T^{-1/2}$ of the temperature variation as shown in Fig. 6.7.

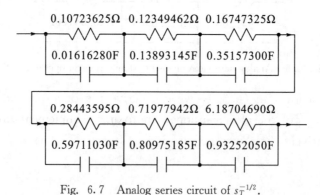

Fig. 6.7 Analog series circuit of $s_T^{-1/2}$.

Furthermore, from (6.29), and (6.30), we can derive the fractional integral operator $s_T^{-1/2}$ of the temperature variation as follows:

$$s_T^{-1/2} = \frac{2.4 \cdot 10^{3/2} \cdot s^5 + 4.4 \cdot 10^{7/2} \cdot s^4 + 1.584 \cdot 10^{11/2} \cdot s^3 + 1.584 \cdot 10^{13/2} \cdot s^2 + 4.4 \cdot 10^{13/2} \cdot s + 2.4 \cdot 10^{13/2}}{s^6 + 6.6 \cdot 10^2 \cdot s^5 + 4.95 \cdot 10^4 \cdot s^4 + 9.24 \cdot 10^5 \cdot s^3 + 4.95 \cdot 10^6 \cdot s^2 + 6.6 \cdot 10^6 \cdot s + 10^6}$$

$$= \frac{6.18704690 \cdot 10}{s + 5.76954805 \cdot 10^2} + \frac{7.19779418}{s + 5.82842712 \cdot 10} + \frac{2.84435952}{s + 1.69839637 \cdot 10} + \frac{1.6747325}{s + 5.88790706} + \frac{1.23494624}{s + 1.71572875}$$

$$+\frac{1.07236249}{s+1.73323801 \cdot 10^{-1}}. \tag{6.31}$$

Substituting $j\omega$ for s in (6.31), we can derive the Bode diagram of the fractional integral operator $s^{-1/2}$ of the approximate implementation in (6.30) and the fractional integral operator $s_T^{-1/2}$ of the temperature variation in (6.31) as shown in Fig. 6.8.

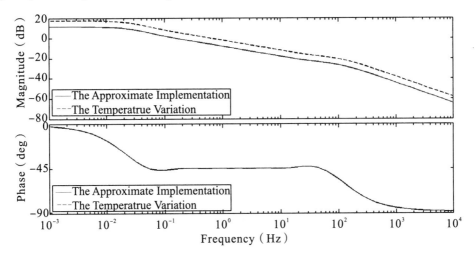

Fig. 6.8 Bode diagram of $s^{-1/2}$ and $s_T^{-1/2}$.

Figure 6.8 shows that the magnitude of the fractional integral operator $s_T^{-1/2}$ of the temperature variation in (6.31) obviously deviates from that of the fractional integral operator $s^{-1/2}$ of the approximate implementation in (6.30). Meanwhile, the phase of the former coincides exactly with that of the latter. The temperature effect can only change the magnitude of the fractional integral operator $s^{-1/2}$ of the approximate implementation in (6.30), but cannot change its phase. As shown in (6.29), we can see that if we set $\zeta_c = 1/\zeta_r$, the temperature effect cannot change the fractional-order of a fractor.

Example 2: with regard to the time constant of a fractor, according to the binomial theorem, we can derive an approximate implementation algorithm of a certain band-pass filtering fractional integral operator $s^{-1/2}$, whose approximation pass-band is $[10^2 \text{Hz}, 1.2 \cdot 10^4 \text{Hz}]$. We can specify the approximation pass-band according to the literature [232]. It follows that in vector form:

$$\begin{bmatrix} \begin{pmatrix} 2.2 \cdot 10^{-7} \cdot s^2 + 7.92 \cdot 10^{-3} \cdot s \\ +1.2 \end{pmatrix} & \begin{pmatrix} -6.6 \cdot 10^{-9} \cdot s^2 - 9.24 \cdot 10^{-4} \cdot s \\ -6.6 \cdot 10^{-1} \end{pmatrix} & \begin{pmatrix} 1.2 \cdot 10^{-10} \cdot s^2 + 7.92 \cdot 10^{-5} \cdot s \\ +2.2 \cdot 10^{-1} \end{pmatrix} \\ 0 & \begin{pmatrix} 1.2 \cdot 10^{-21} \cdot s^5 + 2.2 \cdot 10^{-16} \cdot s^4 \\ +7.92 \cdot 10^{-12} \cdot s^3 + 7.92 \cdot 10^{-8} \cdot s^2 \\ +2.2 \cdot 10^{-4} \cdot s + 1.2 \cdot 10^{-1} \end{pmatrix} & 0 \\ \begin{pmatrix} 1.2 \cdot 10^{-32} \cdot s^8 + 7.92 \cdot 10^{-19} \cdot s^5 \\ +2.2 \cdot 10^{-7} \cdot s^2 \end{pmatrix} & \begin{pmatrix} -6.6 \cdot 10^{-29} \cdot s^7 - 9.24 \cdot 10^{-16} \cdot s^4 \\ -6.6 \cdot 10^{-5} \cdot s \end{pmatrix} & \begin{pmatrix} 2.2 \cdot 10^{-25} \cdot s^6 + 7.92 \cdot 10^{-13} \cdot s^3 \\ +1.2 \cdot 10^{-2} \end{pmatrix} \end{bmatrix}$$

$$\begin{bmatrix} s^{1/4} \\ s^{1/2} \\ s^{3/4} \end{bmatrix} = \begin{bmatrix} \begin{pmatrix} 10^{-12} \cdot s^3 + 4.95 \cdot 10^{-6} \cdot s^2 \\ +4.95 \cdot 10^{-2} \cdot s + 1 \end{pmatrix} \\ \begin{pmatrix} 10^{-24} \cdot s^6 + 6.6 \cdot 10^{-19} \cdot s^5 \\ +4.95 \cdot 10^{-14} \cdot s^4 + 9.24 \cdot 10^{-10} \cdot s^3 \\ +4.95 \cdot 10^{-6} \cdot s^2 + 6.6 \cdot 10^{-3} \cdot s + 1 \end{pmatrix} \\ \begin{pmatrix} 10^{-36} \cdot s^9 + 4.95 \cdot 10^{-22} \cdot s^6 \\ +4.95 \cdot 10^{-10} \cdot s^3 + 1 \end{pmatrix} \end{bmatrix}. \quad (6.32)$$

On the basis of the solution identification theorem of nonhomogeneous linear equations, we can derive the nonzero solutions $[s^{1/4}, s^{1/2}, s^{3/4}]^T$ to (6.32). From the Cramer's rule, we can see that $[s^{1/4}, s^{1/2}, s^{3/4}]^T$ are the nonzero solutions of (6.32). For $s^{-1/2}$ is the reciprocal of $s^{1/2}$, thus from (6.32) it follows that:

$$\begin{aligned} s^{-1/2} &= \frac{1.2 \cdot 10^3 \cdot s^5 + 2.2 \cdot 10^8 \cdot s^4 + 7.92 \cdot 10^{12} \cdot s^3 + 7.92 \cdot 10^{16} \cdot s^2 + 2.2 \cdot 10^{20} \cdot s + 1.2 \cdot 10^{23}}{s^6 + 6.6 \cdot 10^5 \cdot s^5 + 4.95 \cdot 10^{10} \cdot s^4 + 9.24 \cdot 10^{14} \cdot s^3 + 4.95 \cdot 10^{18} \cdot s^2 + 6.6 \cdot 10^{21} \cdot s + 10^{24}} \\ &= \frac{-4.14656596 \cdot 10^2}{s + 5.76954805 \cdot 10^5} + \frac{1.13807119 \cdot 10^2}{s + 5.82842712 \cdot 10^4} + \frac{4.49732729 \cdot 10}{s + 1.69839637 \cdot 10^4} + \frac{2.64798451 \cdot 10}{s + 5.88790706 \cdot 10^3} \\ &\quad + \frac{1.95262146 \cdot 10}{s + 1.71572875 \cdot 10^3} + \frac{1.69555397 \cdot 10}{s + 1.73323801 \cdot 10^2}. \end{aligned} \quad (6.33)$$

Equation (6.33) is an approximate implementation of a high-pass filtering fractional integral operator $s^{-1/2}$. From (6.33), we can see that the all poles of the fractional integral operator $s^{-1/2}$ are on the right half plane. Thus, the fractional integral operator $s^{-1/2}$ is a stable system. From (6.22), (6.23), and Fig. 6.2, we can derive the corresponding analog series circuit of the high-pass filtering the fractional integral operator $s^{-1/2}$. Figure 6.9 shows this circuit.

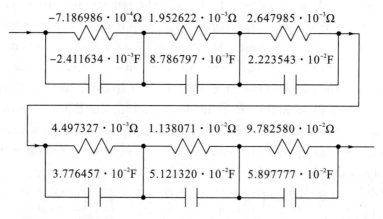

Fig. 6.9 Analog series circuit of fractional integral operator $s^{-1/2}$.

In Fig. 6.9, we can implement a negative resistance or negative capacitance by the active elements. From Fig. 6.4, (6.25), and (6.26), we can derive the equivalent resistance and capacitance of the high-pass filtering fractional integral operator $s^{-1/2}$. Thus, it follows that:

$$r_e = r_1 \parallel r_2 \parallel \cdots \parallel r_6 = -5.519323 \cdot 10^{-3} \Omega, \quad (6.34)$$

$$c_e = c_1 \parallel c_2 \parallel \cdots \parallel c_6 = 0.1765661 \text{F}. \quad (6.35)$$

Thus, from (6.27), we can derive the time constant of the high-pass filtering fractional integral operator $s^{-1/2}$. Thus, it follows that:
$$T_F = |r_e c_e| = 9.745255 \cdot 10^{-4} \text{s}. \tag{6.36}$$
It indicates that the high-pass filtering fractional integral operator $s^{-1/2}$ can be in operative condition within $9.745255 \cdot 10^{-4}$s of receiving power.

Furthermore, if $j\omega$ substitutes for s in (6.33), we can derive the Bode diagram of the fractional integral operator $s^{-1/2}$ of the approximate implementation in (6.33) and that of the ideal filter as shown in Fig. 6.10.

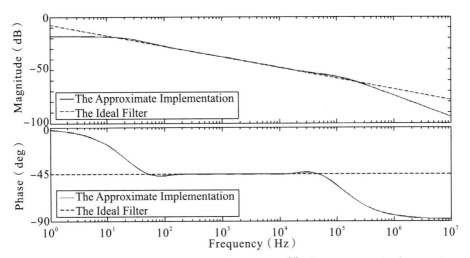

Fig. 6.10 Bode diagram of fractional integral operator $s^{-1/2}$ of approximate implementation and that of ideal filter.

Fig. 6.10 shows that the frequency characteristics of the previously mentioned approximate implementation of high-pass filtering fractional integral operator $s^{-1/2}$ can approach those of the ideal filter with a high degree of accuracy in the pass-band of $[10^2 \text{Hz}, 1.2 \cdot 10^4 \text{Hz}]$. Both the magnitude and phase of the aforementioned approximate implementation of high-pass filtering fractional integral operator $s^{-1/2}$ can approach those of the ideal filter in the pass-band of $[10^2 \text{Hz}, 1.2 \cdot 10^4 \text{Hz}]$.

6.5 Conclusions

Recently, fractional calculus has been considered an important novel branch of mathematical analyses. Fractional calculus is as old as integer-order calculus, although until recently, its application was exclusively in mathematics. Fractional calculus appears to be a novel promising mathematical method for physical scientists and engineering technicians. The application of fractional calculus to signal analysis, signal processing, and circuits and systems, especially to fractor, is a promising discipline of study. Fractional calculus has been hybridized with signal analysis, signal processing, and circuits and systems mainly because of its inherent strength of long-term memory and non-locality.

In this chapter, we mainly discussed the order-frequency characteristics of a promising circuit element: fractor. In particular, we introduced the order-sensitivity characteristics, order-frequency characteristics, time constant, and temperature effect of a fractor.

There are still many other interesting issues, such as defense against chip cloning attacks based on fractor, of the electrical performance of a fractor. It is intended that in-depth studies in these directions will be undertaken as future course of work.

第 7 章　分忆抗元：分数阶忆阻元[①]

本章主要讨论一种新颖的基本概念——分忆抗元（一种分数阶忆阻元）。这是一个关于介于忆阻元和电容或电导之间的分忆抗元的内插特性是什么，以及关于分抗元在蔡氏电路周期表中的位置在哪里的有趣理论问题。受此需求的激励，我们提出一种结合了分数阶电路元件和忆阻元的关于分忆抗元的新的基本概念。我们运用一种先进的数学方法——分数阶微积分，来分析我们所提出的基本概念。分忆抗元（the fracmemristor）是分数阶忆阻元（the fractional-order memristor）的一个合成词。分忆抗值（fracmemristance）是分忆抗元的分数阻抗。第一，我们讨论了分忆抗值和分抗值之间的区别和联系。第二，我们分别研究了1/2阶理想的容性分忆抗元和感性分忆抗元的分忆抗值。第三，我们分别研究了任意阶理想的容性分忆抗元和感性分忆抗元的分忆抗值。第四，我们分别用数值实现和模拟电路实现两种方式实现了分忆抗元，通过实验详细分析了其不挥发的记忆特性和非线性预测能力。与经典的一阶忆阻元相比，非线性预测特性是分忆抗元最重要的优点。

7.1　Introduction

The memristor was originally envisioned in 1971 by circuit theorist Chua as the missing nonlinear passive two-terminal electrical component relating electric charge and magnetic flux linkage [216]. The memristor was generalized to memristive systems in Chua's 1976 paper [217]. The memristor has non-volatility property [219] − [223], [237]. In 2008, a team at HP Labs claimed to have found Chua's missing memristor based on an analysis of a thin film of titanium dioxide [237], [238]. Williams argued that magnetic random access memory, phase change memory and resistive random access memory were memristor technologies [237], [238]. Chua argued for a broader definition that included all 2-terminal non-volatile memory devices based on resistance switching [220]. Chua has more recently argued that the definition could be generalized to cover all forms of two-terminal non-volatile memory devices based on resistance switching effects [219] − [223] although some experimental evidence contradicts this claim, since a non-passive nanobattery effect is observable in resistance switching memory [239]. Meuffels and Schroeder noted that one of the early memristor papers included a mistaken assumption regarding ionic conduction [240]. Meuffels and Soni furthermore noted that

---

[①]　PU Yifei, YUAN Xiao. Fracmemristor: Fractional-Order Memristor [J]. IEEE Access, 2016 (4): 1872−1888.

the dynamic state equations set up for a solely current-controlled memristor with the so-called non-volatility property [220] — [223] would allow the violation of Landauer's principle of the minimum amount of energy required to change "information" states in a system [241], [242]. The concept of a solely current-controlled memristor provides no physical mechanism enabling such a memristor system to erratically change its state just under the influence of white current noise [241], [243]. Nonlinear ionic drift models of the memristor have been proposed by other researchers [244]. As of 2014, the search continues for a model that balances the issues that Strukov's initial memristor modeling equations do not reflect the actual device physics well [245]. One of the resulting properties of the memristors and memristive systems is the existence of a pinched hysteresis effect [246]. It has been proven that some types of non-crossing pinched hysteresis curves cannot be described by the memristors [247]. There are titanium dioxide memristor [237], [238], [248], [249], polymeric memristor [250], [251], layered memristor [252], ferroelectric memristor [253], and spin memristive systems [254] — [259], respectively. Williams' solid-state memristors can be combined into devices called crossbar latches, which could replace transistors in future computers [260]. A simple electronic circuit [261] consisting of an LC network and a memristor was used to model experiments on adaptive behavior of unicellular organisms [262]. Versace and Chandler described the modular neural exploring traveling agent model [263], [264]. Application of the memristor crossbar structure in the construction of an analog soft computing system was demonstrated by Merrikh-Bayat and Shouraki [265]. Chua published a tutorial underlining the broad span of complex phenomena and applications [266]. Di Ventra extended the notion of memristive systems to capacitive and inductive elements in the form of memcapacitors and meminductors [242], [267]. In 2011, memristor-based content addressable memory was introduced [268]. Tse demonstrated printed memristive counters based on solution processing, with potential applications as low-cost packaging components [269]. Politecnico showed that a purely passive circuit, employing already-existing components [270]. Some Chaos circuits consisting of a fractional-order Chua's circuit and a memristor were introduced [271], [272]. Tenreiro Machado studied the generalization of the memristor in the perspective of the fractional-order systems [273]. In 2014, Abdelhouahad proposed the memfractor, which interpolates characteristics between the memristor and the memcapacitor, the meminductor or the second-order memristor [236].

    Fractional calculus has become an important novel branch of mathematical analyses [175], [178], [180]. Fractional calculus is as old as the integer-order calculus, although until recently, its applications were exclusively in mathematics. Scientific study has shown that a fractional-order or fractional dimensional approach is now the best description for many natural phenomena. Many scientific fields such as physics, bioengineering, diffusion processes, viscoelasticity theory, fractal dynamics, fractional control, signal processing, and image processing presently use fractional calculus [130], [171] − [173], [175], [178], [180], [187], [196] − [205] [227], [274] − [276], which has obtained promising results and ideas demonstrating these fractional mathematical operators can be interesting and useful tools.

The application of fractional calculus to analyzing the memristor is an emerging discipline of study in which few studies have been performed [236], [271] − [273]. In the scientific fields of latest signal analysis, signal processing, and circuits and systems, there are many issues on non-linear, non-causal, non-Gaussian, non-stationary, non-minimum phase, non-white additive noise, non-integer-dimensional, and non-integer-order characteristics needed to be analyzed and processed. The classical integer-order signal processing filters and circuit models cannot deal with the aforementioned non-problems efficiently. Fractional calculus has been hybridized with signal processing, circuits and systems, and material science mainly because of its inherent strength of long-term memory, non-locality, and weak singularity. A fractional-order or fractional dimensional system is now a powerful model for dealing with these non-problems. Following the success in the synthesis of the fractional-order differentiator and integrator in an analog circuit, the emergence of a promising fractional-order circuit element was called the fractor [130], [171] − [173], [187], [196] − [205], [227], [274] − [276]. The fractor is essentially a signal processing filter implementing the computation of fractional calculus. The term "fractance" refers to the fractional-order impedance of a fractor. The driving-point impedance function of a fractor is equal to its fractional-order reactance. The fractor has two types, namely, the capacitive fractor and the inductive fractor. The capacitive fractor is the fractional-order capacitor, which is of fractional integral operation. The fractional-order impedance of the capacitive fractor is the capacitive fractance. In a similar way, the inductive fractor is the fractional-order inductor, which is of fractional differential operation. The fractional-order impedance of the inductive fractor is the inductive fractance. As we know, in Chua's periodic table of all two-terminal circuit elements [216] − [221], the capacitive fractor is lying on the line segment between the capacitor and resistor. The electrical properties of the capacitive fractor fall in between the electrical properties of the capacitor and those of the resistor [130], [171] − [173], [187], [196] − [205], [227], [274] − [276]. In a similar way, in Chua's periodic table of all two-terminal circuit elements [216] − [221], the inductive fractor is lying on the line segment between the inductor and resistor. The electrical properties of the inductive fractor fall in between the electrical properties of the inductor and those of the resistor [130], [171] − [173], [187], [196] − [205], [227], [274] − [276]. From the Chua's axiomatic element system [216] − [221] and the constitutive relation, according to logical consistency, axiomatic completeness, and formal symmetry, we can assume that there should be a novel corresponding capacitive circuit element and inductive circuit element to the capacitive fractor and inductive fractor, respectively. Therefore, it is natural to ponder a challenging theoretical problem to determine what the fractional-order memristor interpolating characteristics between the memristor and capacitor or inductor are, and where the positions of the fractional-order memristor in the Chua's axiomatic element system [216] − [221] are. Motivated by this need, in this chapter, we proposed to introduce an interesting conceptual framework of the fracmemristor, which joins the concepts underlying the fractional-order circuit element and the meristor. We use a state-of-the-art mathematical method, fractional calculus, to analyze the proposed conceptual framework. In particular, in Chua's periodic table of all two-terminal circuit elements, the electrical properties of the capacitive fracmemristor should fall in between the electrical properties of the capacitor and those of the memristor. The electrical properties of the inductive fracmemristor should fall in between the electrical properties of the inductor and those of the memristor. The fracmemristor is the fractional-order memristor with predictable function. The predictable characteristics of the fracmemristor are a major advantage when comparing with the classical first-order memristor.

The remainder of this chapter is organized as follows: Section 7.2 recalls some preliminary concepts of the memristor and fractance. Section 3 discusses an interesting conceptual framework of the fracmemristor. In Section 7.3, first, the relationship between the fracmemristance and fractance is discussed. Second, the fracmemristance of the purely ideal 1/2-order capacitive fracmemristor and inductive fracmemristor is studied, respectively. The third step is the proposal for the fracmemristance of the purely ideal arbitrary-order capacitive fracmemristor and inductive fracmemristor, respectively. In Section 7.4, the fracmemristor is achieved by numerical implementation, and its non-volatility property of memory and nonlinear predictive ability are analyzed in detail experimentally. In Section 7.5, the conclusions of this manuscript are presented.

7.2 Related Work

This section includes a brief introduction to the necessary mathematical background of the memristor and fractance.

Chua proposed that there should be a fourth basic element M, which he called the memristor, for memory resistor, completing the set of relations with [216] − [223]:
$$\varphi[q(t)] = M[q(t)]q(t), \tag{7.1}$$
where φ and q denotes magnetic flux and quantity of electric charge, respectively. Equation (7.1) shows that a memristor is as any passive two-terminal circuit element that maintains a functional relation between the time integral of current and the time integral of voltage. The slope of this function is called the memristance, $R[q(t)]$, and is similar to variable resistance [216] − [223]. It follows that:
$$V_i(t) = \mathrm{d}\varphi(q)/\mathrm{d}q I_i(t) = [M(q) + q\mathrm{d}M(q)/\mathrm{d}q]I_i(t) = R[q(t)]I_i(t)$$
$$= H[q(t)] * I_i(t), \tag{7.2}$$

where $V_i(t)$ and $I_i(t)$ denote the instantaneous value of the input voltage and input current of a memristor, respectively. The symbol $*$ denotes convolution, $R[q(t)] = [M(q) + q\mathrm{d}M(q)/\mathrm{d}q]$ and $H[q(t)]$ denote the memristance and transmission function of a memristor, respectively. Note that in the time-invariant case, the constitutive relation of a memristor can be simply expressed as a set of points in the φ-q plane. (7.2) shows that in the case of small-signal behavior about Q, the Laplace transform of $V_i(t) = R[q(t)]I_i(t)$ is equal to $V_i(s) = 1/2\pi L\{R[q(t)]\} * I_i(s)$, which implements a convolution in Laplace transform domain and is sometimes too difficult to be calculated. For the convenience of analysis, let's assume that the transmission function of a memristor, $H[q(t)]$, satisfies $V_i(t) = R[q(t)]I_i(t) = H[q(t)] * I_i(t)$ (i.e. $R[q(t)] = \{H[q(t)] * I_i(t)\}/I_i(t)$). Thus, the Laplace transform of $V_i(t) = H[q(t)] * I_i(t)$, $V_i(s) = r[q(s)]I_i(s)$, implements a multiplication in Laplace transform domain, where $r[q(s)] = L\{H[q(t)]\}$ is the reactance of this memristor. Recently, Chua argued for a broader definition that included all 2-terminal non-volatile memory devices based on resistance switching [220] — [223]. Figure 7.1 shows the periodic table of all two-terminal circuit elements [216] — [223].

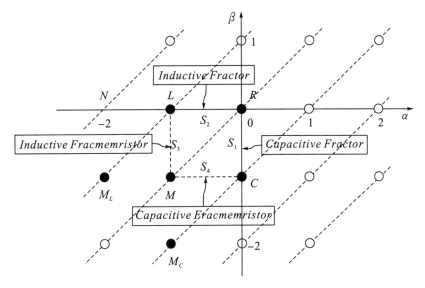

Fig. 7.1 Periodic table of all two-terminal circuit elements.

In Fig. 7.1, let us denote the Chua's axiomatic element and its corresponding electrical characteristics as the symbol $C^{(\alpha,\beta)}$, in which α denotes the voltage exponent, β denotes the current exponent. α and β are equal to the order of the time derivative of $v(t)$ and $i(t)$, respectively. $D_t^\alpha v(t)$ and $D_t^\beta i(t)$ are collectively referred to as the Chua's constitutive variables, where D denotes the differential operator. (α,β) is referred to the Chua's plane of the Chua's axiomatic element system. C, R, L, M, M_L, and M_C denote the capacitor, resistor, inductor, memristor, meminductor, and memcapacitor, respectively. The symbol O denotes the other postulated eletments of the Chua's axiomatic element system. The all Chua's axiomatic elements have the element independence property. Thus, $(D_t^\alpha v(t), D_t^\beta i(t))$ establishes a corresponding constitutive relation. It follows that [218] − [221]:

$$D_t^\alpha v(t) - C^{\langle\alpha,\beta\rangle} D_t^\beta i(t) = 0, \qquad (7.3)$$

where $\alpha \in \mathbf{R}$ and $\beta \in \mathbf{R}$. It follows that [216] − [223], [236], [273]:

$$C^{\langle\alpha,\beta\rangle} = \begin{cases} C, & \text{if } \alpha = 0, \beta = -1 \\ R, & \text{if } \alpha = 0, \beta = 0 \\ L, & \text{if } \alpha = -1, \beta = 0 \\ M, & \text{if } \alpha = -1, \beta = -1 \end{cases}, \qquad (7.4)$$

where C, R, L, and M denote the electrical characteristics of the capacitor, resistor, inductor, and memristor, respectively.

With respect to a capacitive fractor, in Fig. 7.1, the capacitive fractor is lying on the line segment, S_1, between C and R. The order of a capacitive fractance can be extended to the whole field of negative real numbers. Pu has derived the generic nonlinear relation between the capacitance and resistance of the arbitrary-order capacitive fractance [204], [205]. It follows that:

$$\begin{aligned} F_{-v}^c &= F_{-\langle m+p\rangle}^c \\ &= V_i(s)/I_i(s) \\ &= c^{-\langle m+p\rangle} r^{1-p} s^{-\langle m+p\rangle}, \end{aligned} \qquad (7.5)$$

where $v = m + p$ is a positive real number, m is a positive integer, and $0 \leqslant p \leqslant 1$. F_{-v}^c, c, r, and $c^{(m+p)} r^{1-p}$ denote the driving-point impedance function, capacitance, resistance, and capacitive fractance of a purely ideal v-order capacitive fractor, respectively. The driving-point impedance function of a capacitive fractor is equal to its fractional-order capacitive reactance. Equation (7.5) represents the fractional-order capacitive reactance of the purely ideal arbitrary-order capacitive fractor. In addition, with respect to an inductive fractor, in Fig. 7.1, the inductive fractor is lying on the line segment, S_2, between L and R. The order of an inductive fractance can be extended to the whole field of positive real numbers. Pu has derived the generic nonlinear relation between inductance and resistance of the arbitrary-order inductive fractance [204], [205]. It follows that:

$$\begin{aligned} F_v^l &= F_{m+p}^l \\ &= V_i(s)/I_i(s) \\ &= l^{m+p} r^{1-p} s^{m+p}, \end{aligned} \qquad (7.6)$$

where $v = m + p$ is a positive real number, m is a positive integer, and $0 \leqslant p \leqslant 1$. F_v^l, l, r, and $l^{m+p} r^{1-p}$ denote the driving-point impedance function, inductance, resistance, and inductive fractance of a purely ideal v-order inductive fractor, respectively. The driving-point impedance function of an inductive fractor is equal to its fractional-order inductive reactance. Equation (7.6) represents the fractional-order inductive reactance of the purely ideal arbitrary-order inductive fractor.

7.3 Fracmemristor

7.3.1 Relationship between Fracmemristance and Fractance

In this subsection, we discuss the relationship between the fracmemristance and fractance.

From Fig. 7.1 and the constitutive relation, according to logical consistency, axiomatic completeness, and formal symmetry, we can see that there should be the corresponding capacitive fracmemristor and inductive fracmemristor to the capacitive fractor and inductive fractor, respectively. The term "fracmemristor" is a portmanteau of "the fractional-order memristor." In Fig. 7.1, the capacitive fracmemristor should be lying on the line segment, S_4, between C and M. The inductive fracmemristor should be lying on the line segment, S_3, between L and M. The term "fractance" refers to the fractional-order impedance of a fractor. The fractional-order impedance of the capacitive fractor and inductive fractor are the capacitive fractance and inductive fractance, respectively. In a similar way, the term "fracmemristance" refers to the fractional-order impedance of a fracmemristor. The fractional-order impedance of the capacitive fracmemristor and inductive fracmemristor are the capacitive fracmemristance and inductive fracmemristance, respectively. In addition, from Fig. 7.1, we can further see that the electrical properties of the capacitive fracmemristor should fall in between the electrical properties of the capacitor and those of the memristor. The electrical properties of the inductive fracmemristor should fall in between the electrical properties of the inductor and those of the memristor. The difference between the electrical properties of the memristor and those of the resistor is a major difference between the electrical properties of the fracmemristor and those of the fractor. In addition, equations (7.1) and (7.2) show that the memristor definition is based solely on the fundamental circuit variables of current and voltage and their time-integrals, just like the resistor, capacitor, and inductor. The ideal memristor, for memory resistor, is a special case of generic memristor when $R[q(t)]$ depends only on quantity of electric charge. $R[q(t)]$ (measurement unit of the memristance in Ohm similarly to resistance) is the incremental resistance. Therefore, in a similar way, we can achieve the fracmemristor refer to the aforementioned implementations of the fractor [130], [171] − [173], [187], [196] − [205], [227], [274] − [276].

As we know, the tree-type [175], [198], two-circuits-type [173], H-type [172], net-grid-type [171] should be four natural implementations of the fractor. What distinguishes these four aforementioned types [171] − [173], [175], [198] from the other approximate implementations [171] − [173], [175], [198] of the fractor is that the formers consist of series of ordinary circuit elements in the form of natural fractal structures exhibiting extreme self-similarity. In contrast to other approximate implementations [130], [196] − [203], [227], [274] − [276] of the fractor, the floating point values of the capacitance, inductance, and resistance of these four aforementioned types [171] − [173], [175], [198] of the fractor are not required. In fact, there are zero errors between these four aforementioned types [171] − [173], [175], [198] of fractance with infinite recursive extreme self-similar structures and an ideal fractance. Devices manufactured using the other approximate implementations [130], [196] − [203], [227], [274] − [276] of the fractor never represent purely ideal fractance. In a similar way, the aforementioned tree-type, two-circuits-type, H-type, and net-grid-type should also be the natural implementations of the fracmemristor. Because natural implementations usually indicate some fundamental rules of various circuit elements, we study the fracmemristor mainly from the perspective of the fracmemristor of natural implementations. With respect to a purely ideal fractor, the generic electrical characteristics of a purely ideal fractor can be derived from the aforementioned four natural implementations of the fractor [204], [205]. The fractances of the aforementioned four natural implementations of the purely ideal fractor [171] − [173], [175], [198] have the same expression of the electrical properties as (7.5) and (7.6). In a similar way, without loss of generality, we can only employ the net-grid-type fracmemristor in this chapter to discuss the generic electrical characteristics of a purely ideal fracmemristor.

7.3.2 Fracmemristance of Purely Ideal 1/2-Order Fracmemristor

In this subsection, we discuss the fracmemristance of the purely ideal 1/2-order fracmemristor.

The term "fracmemristance" refers to the fractional-order impedance of a fracmemristor. Let us denote the v-order driving-point impedance function (the v-order reactance) of the v-order fracmemristor as the symbol FM_v, which is the abbreviation of fracmemristor. For convenience of illustration, in this chapter, FM_v is also directly referred to as the v-order fracmemristor. Figure 7.2 shows the structural representation of a purely ideal 1/2-order net-grid-type fracmemristor, $FM_{1/2}$.

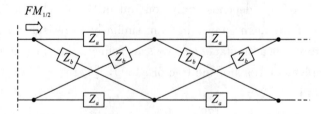

Fig. 7.2 A purely ideal 1/2-order net-grid-type fracmemristor.

From Fig. 7.2, we can see that the purely ideal 1/2-order net-grid-type fracmemristor is of extreme self-similar fractal structure with the series connection of infinitely repeated net-grid-type structures, where Z_a denotes the memristor and Z_b denotes the classical passive capacitor or inductor. The number of Z_a and Z_b is equal to two-fold the number of layers. Figure 7.3 shows its equivalent circuit.

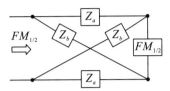

Fig. 7.3 Equivalent circuit of purely ideal 1/2-order net-grid-type fracmemristor.

For convenience, in this chapter, Z_a and Z_b are also directly referred to as the driving-point impedance functions (the reactances) of a memristor and a capacitor or inductor, respectively. Suppose the currents of Z_a and Z_b are equal to $i_a(s)$ and $i_b(s)$, respectively. Let the input voltage and input current of $FM_{1/2}$ be equal to $V_i(s)$ and $I_i(s)$, respectively. From Fig. 7.3, according to Kirchoff's current law and Kirchoff's voltage law, we can obtain the following:

$$\begin{cases} Z_a i_a + Z_b i_b = V_i \\ (Z_a + FM_{1/2}) i_a - (FM_{1/2} + Z_b) i_b = 0 \end{cases}. \tag{7.7}$$

According to the Cramer's rule of linear algebra, it follows that:

$$\begin{cases} i_a(s) = \dfrac{(Z_b + FM_{1/2}) V_i}{\begin{vmatrix} Z_a + FM_{1/2} & -(FM_{1/2} + Z_b) \\ Z_a & Z_b \end{vmatrix}} \\ i_b(s) = \dfrac{(Z_a + FM_{1/2}) V_i}{\begin{vmatrix} Z_a + FM_{1/2} & -(FM_{1/2} + Z_b) \\ Z_a & Z_b \end{vmatrix}} \end{cases}. \tag{7.8}$$

Hence, $FM_{1/2}$ is equal to:

$$\begin{aligned} FM_{1/2} &= \frac{V_i(s)}{I_i(s)} \\ &= \frac{V_i(s)}{i_a(s) + i_b(s)} \\ &= \frac{2 Z_a Z_b + FM_{1/2}(Z_a + Z_b)}{2 FM_{1/2} + Z_a + Z_b}. \end{aligned} \tag{7.9}$$

From (7.9), it follows that:

$$FM_{1/2} = V_i(s)/I_i(s) = (Z_a Z_b)^{1/2}. \tag{7.10}$$

First, with regard to a purely ideal 1/2-order capacitive fracmemristor, $FM^c_{-1/2}$, let the memristance and capacitance of its fracmemristance be equal to $R[g(t)]$ and c, respectively, where q denotes quantity of electric charge. Assuming $r[q(s)] = L\{H[q(t)]\}$ is the reactance of the memristor. Suppose the initial energy of the electric element of the purely ideal 1/2-order capacitive fracmemristor is equal to zero. Then, in the Laplace transform domain, $Z_a = r(q)$ and $Z_b = 1/cs$. From (7.10), it follows that:

$$FM^c_{-1/2} = \xi^{-1/2} s^{-1/2}, \tag{7.11}$$

where $\xi = c/r(q)$. Equation (7.11) shows that $c^{1/2} r(q)^{1/2}$ denotes the 1/2-order capacitive fracmemristance of the purely ideal 1/2-order capacitive fracmemristor. The 1/2-order capacitive fracmemristor is essentially the $-1/2$-order fracmemristor. Equation (7.11) represents the 1/2-order capacitive driving-point impedance function (the 1/2-order capacitive reactance) of the purely ideal 1/2-order capacitive fracmemristor. From (7.11), we can derive the relation between input voltage $V_i(s)$ and input current $I_i(s)$ of the purely ideal 1/2-order capacitive fracmemristor. It follows that:

$$V_i(s) = \xi^{-1/2} s^{-1/2} I_i(s). \tag{7.12}$$

The inverse Laplace transform of (7.12) is as follows:

$$V_i(t) = c^{-1/2} L^{-1}\{[r(q)]^{1/2}\} * [D_t^{-1/2} I_i(t)], \tag{7.13}$$

where symbol $*$ denotes convolution. From (7.13), we can see that there is a positive nonlinear correlation between $V_i(t)$ and the 1/2-order fractional integral of $I_i(t)$. From (7.12), however, it follows that:

$$I_i(s) = \xi^{1/2} s^{1/2} V_i(s). \tag{7.14}$$

The inverse Laplace transform of (7.14) is as follows:

$$I_i(t) = c^{1/2} L^{-1}\{[r(q)]^{-1/2}\} * [D_t^{1/2} V_i(t)]. \tag{7.15}$$

From (7.15), we can see that there is a positive nonlinear correlation between $I_i(t)$ and the 1/2-order fractional derivative of $V_i(t)$.

Second, with regard to a purely ideal 1/2-order inductive fracmemristor, $FM_{1/2}^l$, let the memristance and inductance of its fracmemristance be equal to $r(q)$ and l, respectively. Suppose the initial energy of the electric element of the purely ideal 1/2-order inductive fracmemristor is equal to zero. Then, in the Laplace transform domain, $Z_a = r(q)$ and $Z_b = ls$. From (7.10), it follows that:

$$FM_{1/2}^l = \xi^{1/2} s^{1/2}, \tag{7.16}$$

where $\xi = l \cdot r(q)$. Equation (7.16) shows that $l^{1/2} r(q)^{1/2}$ denotes the 1/2-order inductive fracmemristance of the purely ideal 1/2-order inductive fracmemristor. The 1/2-order inductive fracmemristor is essentially the 1/2-order fracmemristor. Equation (7.16) represents the 1/2-order inductive driving-point impedance function (the 1/2-order inductive reactance) of the purely ideal 1/2-order inductive fracmemristor. In a similar way, from (7.16), we can derive the relation between input voltage $V_i(s)$ and input current $I_i(s)$ of the purely ideal 1/2-order inductive fracmemristor. It follows that, respectively:

$$V_i(s) = \zeta^{1/2} s^{1/2} I_i(s), \tag{7.17}$$

$$I_i(s) = \zeta^{-1/2} s^{-1/2} V_i(s). \tag{7.18}$$

From (7.17) and (7.18), we can see that $V_i(t)$ is positively correlated to the 1/2-order fractional derivative of $I_i(t)$ and $I_i(t)$ is positively correlated to the 1/2-order fractional integral of $V_i(t)$.

7.3.3 Fracmemristance of Purely Ideal Arbitrary-Order Fracmemristor

In this subsection, we further discuss the fracmemristance of the purely ideal arbitrary-order fracmemristor. The purely ideal arbitrary-order net-grid-type fracmemristor is a recursively nested structure of the aforementioned purely ideal 1/2-order net-grid-type fracmemristor.

First, by an extension of aforementioned logic, we can derive the fracmemristance of the purely ideal arbitrary-order capacitive fracmemristor. Suppose $Z_a = r(q)$ and $Z_b = FM^c_{-1/2}$ in Fig. 7.2. From (7.10) and (7.11), it follows that:

$$F^c_{-1/4} = [r(q)]^{1/2} \xi^{-1/4} s^{-1/4}. \tag{7.19}$$

And then, suppose $Z_a = r(q)$ and $Z_b = FM^c_{-1/2^{n-1}}$ in Fig. 7.2, where n is a positive integer. From (7.10), it follows that:

$$FM^c_{-1/2^n} = [r(q)]^{\frac{2^{n-1}-1}{2^{n-1}}} \xi^{\frac{-1}{2^n}} s^{\frac{-1}{2^n}}. \tag{7.20}$$

In a similar way, suppose $Z_a = 1/cs$ and $Z_b = FM^c_{-1/2}$ in Fig. 7.2. From (7.10) and (7.11), it follows that:

$$FM^c_{-3/4} = c^{-1/2} \xi^{-1/4} s^{-3/4}. \tag{7.21}$$

And then, suppose $Z_a = 1/cs$ and $Z_b = FM^c_{-(2^{n-1}-1)/2^{n-1}}$ in Fig. 7.2. From (7.10), it follows that:

$$FM^c_{-(2^n-1)/2^n} = c^{\frac{-(2^{n-1}-1)}{2^{n-1}}} \xi^{\frac{-1}{2^n}} s^{\frac{-(2^n-1)}{2^n}}. \tag{7.22}$$

In a similar way, suppose $Z_a = 1/cs$ and $Z_b = FM^c_{-1/2^n}$ in Fig. 7.2. From (7.10) and (7.20), it follows that:

$$FM^c_{-(2^n+1)/2^{n+1}} = c^{\frac{-1}{2}} [r(q)]^{\frac{2^{n-1}-1}{2^n}} \xi^{\frac{-1}{2^{n+1}}} s^{\frac{-(2^n+1)}{2^{n+1}}}. \tag{7.23}$$

Suppose $Z_a = 1/cs$ and $Z_b = FM^c_{-(2^n+1)/2^{n+1}}$ in Fig. 7.2. From (7.10) and (7.23), it follows that:

$$FM^c_{-(2^{n+1}+2^n+1)/2^{n+2}} = c^{\frac{-3}{4}} [r(q)]^{\frac{2^{n-1}-1}{2^{n+1}}} \xi^{\frac{-1}{2^{n+2}}} s^{\frac{-(2^{n+1}+2^n+1)}{2^{n+2}}}. \tag{7.24}$$

And then, suppose $Z_a = 1/cs$ and $Z_b = FM^c_{-(2^{n+k-1}-2^n+1)/2^{n+k-1}}$ in Fig. 7.2, where k is a positive integer. From (10), it follows that:

$$FM^c_{-(2^{n+k}-2^n+1)/2^{n+k}} = c^{\frac{-(2^k-1)}{2^k}} [r(q)]^{\frac{2^{n-1}-1}{2^{n+k-1}}} \xi^{\frac{-1}{2^{n+k}}} s^{\frac{-(2^{n+k}-2^n+1)}{2^{n+k}}}. \tag{7.25}$$

In a similar way, suppose $Z_a = r(q)$ and $Z_b = FM^c_{-(2^n-1)/2^n}$ in Fig. 7.2. From (7.10) and (7.22), it follows that:

$$FM^c_{-(2^n-1)/2^{n+1}} = [r(q)]^{\frac{1}{2}} c^{\frac{-(2^{n-1}-1)}{2^n}} \xi^{\frac{-1}{2^{n+1}}} s^{\frac{-(2^n-1)}{2^{n+1}}}. \tag{7.26}$$

Suppose $Z_a = r(q)$ and $Z_b = FM^c_{-(2^n-1)/2^{n+1}}$ in Fig. 7.2. From (7.10) and (7.26), it follows that:

$$FM^c_{-(2^n-1)/2^{n+2}} = [r(q)]^{\frac{3}{4}} c^{\frac{-(2^{n-1}-1)}{2^{n+1}}} \xi^{\frac{-1}{2^{n+2}}} s^{\frac{-(2^n-1)}{2^{n+2}}}. \tag{7.27}$$

And then, suppose $Z_a = r(q)$ and $Z_b = FM^c_{-(2^n-1)/2^{n+k-1}}$ in Fig. 7.2, where k is a positive integer. From (7.10), it follows that:

$$FM^c_{-(2^n-1)/2^{n+k}} = [r(q)]^{\frac{2^k-1}{2^k}} c^{\frac{-(2^{n-1}-1)}{2^{n+k-1}}} \xi^{\frac{-1}{2^{n+k}}} s^{\frac{-(2^n-1)}{2^{n+k}}}. \tag{7.28}$$

Furthermore, if n and k are non-negative integers, respectively, it follows that:

$$\frac{1}{2} \leqslant \frac{2^{n+k}-2^n+1}{2^{n+k}} = p \leqslant 1, \tag{7.29}$$

$$0 \leqslant \frac{2^n - 1}{2^{n+k}} = p \leqslant \frac{1}{2}. \tag{7.30}$$

From (7.25) and (7.29), if $1/2 \leqslant p \leqslant 1$, it follows that:

$$FM_{-p}^c = c^{\frac{(2^n-1)(1-p)-(2^n-1)}{2^n-1}} [r(q)]^{\frac{(2^n-1)(1-p)}{2^n-1}} s^{-p}$$
$$= c^{-p} [r(q)]^{1-p} s^{-p}. \tag{7.31}$$

From (7.28) and (7.30), if $0 \leqslant p \leqslant 1/2$, it follows that:

$$FM_{-p}^c = c^{\frac{-(2^n-1)p}{2^n-1}} [r(q)]^{\frac{(2^n-1)(1-p)}{2^n-1}} s^{-p}$$
$$= c^{-p} [r(q)]^{1-p} s^{-p}. \tag{7.32}$$

From (7.31) and (7.32), we can see that no matter $1/2 \leqslant p \leqslant 1$ or $0 \leqslant p \leqslant 1/2$, FM_{-p}^c has the same analytical expression, $c^{-p} [r(q)]^{1-p} s^{-p}$. Equations (7.31) and (7.32) do collectively analytically represent the fractional-order driving-point impedance function of the purely ideal p-order capacitive fracmemristor, where $0 \leqslant p \leqslant 1$. Thus, from (7.31) and (7.32), if $0 \leqslant p \leqslant 1$, it follows that:

$$FM_{-p}^c = c^{-p} [r(q)]^{1-p} s^{-p}. \tag{7.33}$$

In addition, as we know, the driving-point impedance function (the capacitive reactance) of a capacitor is as follows:

$$Z_{-1}^c(s) = V_i(s)/I_i(s)$$
$$= c^{-1} s^{-1}, \tag{7.34}$$

where c is equal to the capacitance of a capacitor. From (7.25), we can see that with regard to the capacitor, $V_i(t)$ is in direct ratio to the first-order integral of $I_i(t)$ and $I_i(t)$ is in direct ratio to the first-order derivative of $V_i(t)$. A capacitor implements the first-order integral operator. Thus, from (7.25), the capacitive driving-point impedance function (the capacitive reactance) of a cascade system of the m-stage first-order integrators is as follows:

$$Z_{-m}^c(s) = V_i(s)/I_i(s)$$
$$= c^{-m} s^{-m}, \tag{7.35}$$

where m is a positive integer.

Therefore, the order of a purely ideal capacitive fracmemristor can be extended to the whole field of negative real numbers. We can naturally implement a purely ideal arbitrary-order capacitive fracmemristor in a cascade manner. From (7.33) and (7.35), the v-order capacitive driving-point impedance function (the v-order reactance) of the purely ideal v-order capacitive fracmemristor is as follows:

$$FM_{-v}^c = FM_{-(m+p)}^c$$
$$= V_i(s)/I_i(s)$$
$$= c^{-(m+p)} [r(q)]^{1-p} s^{-(m+p)}, \tag{7.36}$$

where $v = m + p$ is a positive real number, m is a positive integer, and $0 \leqslant p \leqslant 1$. Equation (7.36) shows that $c^{(m+p)} [r(q)]^{1-p}$ denotes the v-order capacitive fracmemristance of the purely ideal v-order capacitive fracmemristor. The v-order capacitive fracmemristor is essentially the $-v$-order fracmemristor. FM^c_{-v} is the v-order capacitive driving-point impedance function (the v-order capacitive reactance) of the purely ideal v-order capacitive fracmemristor. From (7.9), (7.12), and (7.36), we can see that the purely ideal v-order capacitive fracmemristor implements a cascade system of a v-order fractional integral operator and a fractional power function of memristance. Comparing with (7.5) and (7.36), we can see that the measurement unit and physical dimension of the purely ideal capacitive fracmemristance are the same as those of the purely ideal capacitive fractance [204], [205], because the memristance and resistance have the same measurement unit and physical dimension [216] − [223]. In Fig. 7.1, the capacitive fracmemristor should be lying on the line segment, S_4, between C and M. In particular, in Chua's periodic table of all two-terminal circuit elements, the electrical properties of the capacitive fracmemristor should fall in between the electrical properties of the capacitor and those of the memristor. The electrical properties of the inductive fracmemristor should fall in between the electrical properties of the inductor and those of the memristor. The capacitive fracmemristor can be considered in a certain way as an interpolation of the memristor and capacitor. The fractional-order capacitive driving-point impedance function of a capacitive fracmemristor is equal to its fractional-order capacitive reactance. Equation (7.36) represents the fractional-order capacitive reactance of a purely ideal arbitrary-order capacitive fracmemristor. From (7.36), we can also see that if $r[q(s)] = L\{H[q(t)]\}$ is equal to an arbitrary power function of s, the v-order capacitive fracmemristor will be turned into a traditional fractor. If $v = 0$, the v-order capacitive fracmemristor will be turned into a traditional first-order memristor. The capacitive fracmemristor can be converted to a fractor or memristor in some given conditions.

Second, in a similar way, we can derive the fracmemristance of the purely ideal arbitrary-order inductive fracmemristor. Suppose $Z_a = r(q)$ and $Z_b = FM^l_{1/2}$ in Fig. 7.2. From (7.10) and (7.16), it follows that:

$$FM^l_{1/4} = [r(q)]^{1/2} \zeta^{1/4} s^{1/4}. \tag{7.37}$$

And then, suppose $Z_a = r(q)$ and $Z_b = FM^l_{1/2^{n-1}}$ in Fig. 7.2, where n is a positive integer. From (7.10), it follows that:

$$FM^l_{1/2^n} = [r(q)]^{\frac{2^{n-1}-1}{2^{n-1}}} \zeta^{\frac{1}{2^n}} s^{\frac{1}{2^n}}. \tag{7.38}$$

In a similar way, suppose $Z_a = ls$ and $Z_b = FM^l_{1/2}$ in Fig. 7.2. From (7.10) and (7.16), it follows that:

$$FM^l_{3/4} = l^{1/2} \zeta^{1/4} s^{3/4}. \tag{7.39}$$

And then, suppose $Z_a = ls$ and $Z_b = FM^l_{(2^{n-1}-1)/2^{n-1}}$ in Fig. 7.2. From (7.10), it follows that:

$$FM^l_{(2^n-1)/2^n} = l^{\frac{2^{n-1}-1}{2^{n-1}}} \zeta^{\frac{1}{2^n}} s^{\frac{2^n-1}{2^n}}. \tag{7.40}$$

In a similar way, suppose $Z_a = ls$ and $Z_b = FM^l_{1/2^n}$ in Fig. 7.2. From (7.10) and (7.38), it follows that:

$$FM^l_{(2^n+1)/2^{n+1}} = l^{\frac{1}{2}} [r(q)]^{\frac{2^{n-1}-1}{2^n}} \zeta^{\frac{1}{2^{n+1}}} s^{\frac{2^n+1}{2^{n+1}}}. \tag{7.41}$$

And then, suppose $Z_a = ls$ and $Z_b = FM^l_{(2^{n+k-1}-2^n+1)/2^{n+k-1}}$ in Fig. 7.2, where k is a positive integer. From (7.10), it follows that:

$$FM^l_{(2^{n+k}-2^n+1)/2^{n+k}} = l^{\frac{2^k-1}{2^k}} [r(q)]^{\frac{2^{n-1}-1}{2^{n+k-1}}} \zeta^{\frac{1}{2^{n+k}}} s^{\frac{2^{n+k}-2^n+1}{2^{n+k}}}. \tag{7.42}$$

In a similar way, suppose $Z_a = r(q)$ and $Z_b = FM^l_{(2^n-1)/2^n}$ in Fig. 7.2. From (7.10) and (7.40), it follows that:

$$FM^l_{(2^n-1)/2^{n+1}} = [r(q)]^{\frac{1}{2}} l^{\frac{2^{n-1}-1}{2^n}} \zeta^{\frac{1}{2^{n+1}}} s^{\frac{2^n-1}{2^{n+1}}}. \tag{7.43}$$

And then, suppose $Z_a = r(q)$ and $Z_b = FM^l_{(2^n-1)/2^{n+k-1}}$ in Fig. 7.2, where k is a positive integer. From (7.10), it follows that:

$$FM^l_{(2^n-1)/2^{n+k}} = [r(q)]^{\frac{2^k-1}{2^k}} l^{\frac{2^{n-1}-1}{2^{n+k-1}}} \zeta^{\frac{1}{2^{n+k}}} s^{\frac{2^n-1}{2^{n+k}}}. \tag{7.44}$$

Thus, in a similar way, we can derive the fractional-order driving-point impedance function of the purely ideal p-order inductive fracmemristor, where $0 \leqslant p \leqslant 1$. From (7.42) and (7.44), if $0 \leqslant p \leqslant 1$, it follows that:

$$FM^l_p = l^p [r(q)]^{1-p} s^p. \tag{7.45}$$

In addition, as we know, the driving-point impedance function (the inductive reactance) of an inductor is as follows:

$$\begin{aligned} Z^l_1(s) &= V_i(s)/I_i(s) \\ &= l^1 s^1, \end{aligned} \tag{7.46}$$

where l is equal to the inductance of an inductor. An inductor implements the first-order differential operator. Thus, from (7.46), the inductive driving-point impedance function (the inductive reactance) of a cascade system of the m-stage first-order differentiators is as follows:

$$\begin{aligned} Z^l_m(s) &= V_i(s)/I_i(s) \\ &= l^m s^m, \end{aligned} \tag{7.47}$$

where m is a positive integer.

Therefore, the order of a purely ideal inductive fracmemristance can be extended to the whole field of positive real numbers. We can naturally implement a purely ideal arbitrary-order inductive fracmemristance in a cascade manner. From (7.45) and (7.47), the v-order inductive driving-point impedance function (the v-order inductive reactance) of the purely ideal v-order inductive fracmemristor is as follows:

$$\begin{aligned} FM^l_v &= FM^l_{m+p} \\ &= V_i(s)/I_i(s) \\ &= l^{m+p} [r(q)]^{1-p} s^{m+p}, \end{aligned} \tag{7.48}$$

where $v = m + p$ is a positive real number, m is a positive integer, and $0 \leqslant p \leqslant 1$. Equation (7.48) shows that $l^{m+p} [r(q)]^{1-p}$ denotes the v-order inductive fracmemristance of the purely ideal v-order inductive fracmemristor. The v-order inductive fracmemristor is essentially the v-order fracmemristor. FM^l_v is the v-order inductive driving-point impedance function (the v-order inductive reactance) of the purely ideal v-order inductive fracmemristor. From (7.9), (7.12), and (7.48), we can see that the purely ideal v-order inductive fracmemristor implements a cascade system of a v-order fractional differential operator and a fractional power function of memristance. Comparing with (7.6) and (7.48), we can see that the measurement unit and physical dimension of the purely ideal inductive fracmemristance are the same as those of the purely ideal inductive fractance [204], [205], because the memristance and resistance have the same measurement unit and physical dimension [216] − [223]. In Fig. 7.1, the inductive fracmemristor should be lying on the line segment, S_3, between L and M. In particular, in Chua's periodic table of all two-terminal circuit elements, the electrical properties of the inductive fracmemristor should fall in between the electrical properties of the inductor and those of the memristor. The inductive fracmemristor can be considered in a certain way as an interpolation of the memristor and inductor. The fractional-order inductive driving-point impedance function of an inductive fracmemristor is equal to its fractional-order inductive reactance. Equation (7.48) represents the fractional-order inductive reactance of a purely ideal arbitrary-order inductive fracmemristor. From (7.48), we can also see that if $r[q(s)] = L\{H[q(t)]\}$ is equal to an arbitrary power function of s, the v-order inductive fracmemristor will be turned into a traditional fractor. If $v=0$, the v-order inductive fracmemristor will be turned into a traditional first-order memristor. The inductive fracmemristor can be converted to a fractor or memristor in some given conditions.

Third, as we know, the commonly used fractional calculus definitions are Grünwald-Letnikov, Riemann-Liouville, and Caputo [47] − [49], respectively. Suppose that ${}^{G-L}_0D^v_x$, ${}^{R-L}_0D^v_x$, and ${}^C_0D^v_x$ denote the Grünwald-Letnikov defined, the Riemann-Liouville defined, and the Caputo defined fractional differential operator, respectively. The Laplace transform of the v-order Riemann-Liouville differential operator is $L[{}^{R-L}_0D^v_x f(x)] = s^v L[f(x)] - \sum_{k=0}^{n-1} s^k [{}^{R-L}_0D^{v-1-k}_x f(x)]_{x=0}$, where $n-1 \leqslant v < n$ is a non-integer, $f(x)$ is a differintegrable function [175], [178], [180], $[0, x]$ is the duration of $f(x)$, and s denotes the Laplace operator. When $f(x)$ is causal signal and its fractional primitives must also be zero, we can simplify the Laplace transform for ${}^{R-L}_0D^v_x f(x)$ as $L[{}^{R-L}_0D^v_x f(x)] = s^v L[f(x)]$. Further, the Laplace transform of the v-order Caputo differential operator is $L[{}^C_0D^v_x f(x)] = s^v L[f(x)] - \sum_{k=0}^{n-1} s^k f^{(k)}(x)|_{x=0}$. When $f(x)$ is causal signal and its fractional primitives must also be zero, we can simplify the Laplace transform for ${}^C_0D^v_x f(x)$ as $L[{}^C_0D^v_x f(x)] = s^v L[f(x)]$. Thus, in this case the three cited definitions of fractional derivatives are equivalent. From (7.36) and (7.48), we can see that when the memristance, $R[q(t)]$, is causal signal and its fractional primitives must also be zero, we can use indifferently three equivalent cited definitions of fractional calculus to the fracmemristor.

7.3.4 Branch-Current Analysis of Purely Ideal Arbitrary-Order Fracmemristor

In this subsection, from Kirchhoff's current law and Kirchhoff's voltage law, the branch-current of the purely ideal arbitrary-order fracmemristor is analyzed.

First, with respect to the purely ideal 1/2-order fracmemristor, from (7.8), it follows that:

$$\frac{i_a(s)}{i_b(s)} = \frac{Z_b + FM_{1/2}}{Z_a + FM_{1/2}}, \tag{7.49}$$

where $i_a(s)$ and $Z_a = r[q(s)]$ denote the branch-current and the reactance of the memristor in a purely ideal 1/2-order fracmemristor in the Laplace transform domain, respectively. $i_b(s)$ and Z_b denote the branch-current and the capacitive or inductive reactance of the corresponding capacitor or inductor in a purely ideal 1/2-order fracmemristor in the Laplace transform domain, respectively. As aforementioned discussion, if $i_b(s)$ and Z_b represent the branch-current of the capacitor and the capacitive reactance, respectively, $FM_{1/2}$ represents a purely ideal 1/2-order capacitive fracmemristor. If $i_b(s)$ and Z_b represent the branch-current of the inductor and the inductive reactance, respectively, $FM_{1/2}$ represents a purely ideal 1/2-order inductive fracmemristor. Thus, from (7.9), it follows that:

$$I_i(s) = i_a(s) + i_b(s), \tag{7.50}$$

where $I_i(s)$ denotes the input current of the purely ideal 1/2-order fracmemristor. Thus, from (7.49) and (7.50), it follows that:

$$i_a(s) = \frac{FM_{1/2} + Z_b}{2FM_{1/2} + Z_a + Z_b} I_i(s). \tag{7.51}$$

From (7.10) and (7.51), it follows that:

$$i_a(s) = \left[1 - \frac{Z_a + (Z_a Z_b)^{1/2}}{(Z_a^{1/2} + Z_b^{1/2})^2}\right] I_i(s). \tag{7.52}$$

Second, with respect to the purely ideal 1/4-order fracmemristor, from (7.19) and (7.37), its equivalent circuit can be shown as given in Fig. 7.4.

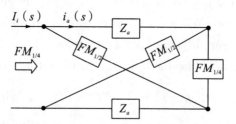

Fig. 7.4 Equivalent circuit of purely ideal 1/4-order net-grid-type fracmemristor.

In Fig. 7.4, $I_i(s)$ denotes the input current of a purely ideal 1/4-order fracmemristor. $i_a(s)$ and $Z_a = r[q(s)]$ denote the branch-current and the reactance of the memristor in a purely ideal 1/4-order fracmemristor in the Laplace transform domain, respectively. As aforementioned discussion, if $FM_{1/2}$ denotes a purely ideal 1/2-order capacitive fracmemristor, $FM_{1/4}$ is a corresponding purely ideal 1/4-order capacitive fracmemristor. If $FM_{1/2}$ denotes a purely ideal 1/2-order inductive fracmemristor, $FM_{1/4}$ is a corresponding purely ideal 1/4-order inductive fracmemristor. In a similar way, from (7.10) and Fig. 7.4, the following can be obtained:

$$i_a(s) = \left[1 - \frac{Z_a + Z_a^{3/4}Z_b^{1/4}}{(Z_a^{1/2} + Z_a^{1/2}Z_b^{1/2})^2}\right]I_i(s), \tag{7.53}$$

where Z_b denotes the capacitive or inductive reactance of the corresponding capacitor or inductor in $FM_{1/2}$.

Third, with respect to the purely ideal p-order fracmemristor, from (7.33) and (7.45), the following can be obtained:

$$FM_p = Z_a^{1-p}Z_b^p, \tag{7.54}$$

where $0 \leqslant p \leqslant 1$. $Z_a = r[q(s)]$ and Z_b denote the reactance of the memristor and the capacitor or inductor in a purely ideal p-order fracmemristor in the Laplace transform domain, respectively. From (7.54), the equivalent circuit of a purely ideal $p/2$-order fracmemristor can be shown as given in Fig. 7.5.

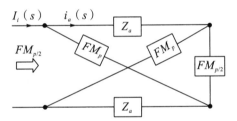

Fig. 7.5 Equivalent circuit of purely ideal $p/2$-order net-grid-type fracmemristor.

In Fig. 7.5, $I_i(s)$ denotes the input current of a purely ideal $p/2$-order fracmemristor. $i_a(s)$ and $Z_a = r[q(s)]$ denote the branch-current and the reactance of the memristor in a purely ideal $p/2$-order fracmemristor in the Laplace transform domain, respectively. As aforementioned discussion, if FM_p denotes a purely ideal p-order capacitive fracmemristor, $FM_{p/2}$ is a corresponding purely ideal $p/2$-order capacitive fracmemristor. If FM_p denotes a purely ideal p-order inductive fracmemristor, $FM_{p/2}$ is a corresponding purely ideal $p/2$-order inductive fracmemristor. In a similar way, from (7.54) and Fig. 7.5, the following can be obtained:

$$i_a(s) = \left[1 - \frac{Z_a + Z_a^{(2-p)/2}Z_b^{p/2}}{(Z_a^{1/2} + Z_a^{1-p}Z_b^p)^2}\right]I_i(s). \tag{7.55}$$

Moreover, from (7.35) and (7.47), we can naturally implement a purely ideal arbitrary-order capacitive or inductive fracmemristor by connecting a purely ideal $p/2$-order fracmemristor with a m-stage first-order integrators or differentiators in a cascade manner. With regard to a series circuit, the input current for each stage first-order integrator or differentiator is the same as the input current of a purely ideal $p/2$-order fracmemristor, $I_i(s)$. Thus, in the Laplace transform domain, (7.55) represents the branch-current of the memristor of the purely ideal $(m + p/2)$-order, arbitrary-order, fracmemristor.

7.3.5 Analog Circuit Realization of Arbitrary-Order Fracmemristor in Its Natural Implementation

In this subsection, we discuss the analog circuit realization of the arbitrary-order fracmemristor in its natural implementation.

As aforementioned discussion, similar to the fractor, the tree-type [175], [198], two-circuits-type [173], H-type [172], net-grid-type [171] should be four natural implementations of the fracmemristor. Therefore, from Fig. 7.2, we can use the memristor and capacitor or inductor to realize the corresponding arbitrary-order capacitive or inductive fracmemristor in its natural implementation. In order to simplify the analysis, we take the 1/2-order capacitive fracmemristor in the net-grid-type analog circuit as an example for discussion. We can use the memristor and capacitor to achieve the analog circuit realization of a purely ideal fractional-order capacitive fracmemristor. In Fig. 7.2, we let Z_a and Z_b to be a memristor and a capacitor, respectively. As shown in Fig. 7.6, from (7.11) and (7.36), we construct a 1/2-order capacitive fracmemristor in a series connection of k-layer repeated net-grid-type structures.

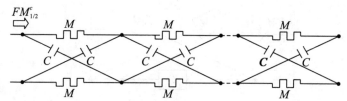

Fig. 7.6 Analog circuit of 1/2-order capacitive fracmemristor in series connection of k-layer repeated net-grid-type structures.

In Fig. 7.6, the symbol of ⌐⌐ denotes the memristor [216] - [223] with the same memristance, the symbol of C denotes the capacitor with the same capacitance. The memristor can be obtained by the independent electronic component. Moreover, when the number of layers of the repeated net-grid-type structures increases, the approximation precision of the 1/2-order capacitive fracmemristor also increases. The purely ideal arbitrary-order net-grid-type fracmemristor is a recursively nested structure of the aforementioned purely ideal 1/2-order net-grid-type fracmemristor.

Furthermore, for the memristance $R[q(t)]$ can be theoretically an arbitrary variable resistance [216] − [223], thus $r[q(s)]$ is also arbitrary. In (7.36) and (7.48), the approximate implementation of $[r(q)]^{1-p}$ and $L^{-1}\{[r(q)]^{1-p}\}$ for arbitrary $r[q(s)]$ are difficult and complex.

7.4 Analog Circuit Implementation of Fractional-Order Memristor: Arbitrary-Order Lattice Scaling Fracmemristor[①]

In this part, based on fractional calculus, the fractional-order memristor, an arbitrary-order fracmemristor, is proposed to be implemented in the form of a lattice scaling analog circuit. Since the concept of the memristor is generalized from the classic integer-order memristor to that of the fractional-order memristor, fracmemristor, it is natural to ponder a challenging theoretical problem to propose a circuit theoretic methodology to achieve an arbitrary-order memristor by using the ordinary memristor and capacitor or inductor in the form of an analog circuit. Motivated by this need, we propose an interesting analog circuit implementation method of an arbitrary-order memristor. The first step is the proposal for a novel feasible analog circuit implementation of an arbitrary-order lattice scaling fracmemristor. In particular, the hardware achievement of this arbitrary-order lattice scaling fracmemristor is mathematically derived and analyzed in detail. Secondly, the approximation performance, electrical characteristics, especially fingerprint, and analog circuit achievement of an arbitrary-order fracmemristor are analyzed in detail experimentally, respectively. The main contribution of this part is the proposal for the first preliminary attempt of a feasible hardware achievement of an arbitrary-order fracmemristor and for the recognition of the fingerprint of fracmemristor.

7.4.1 INTRODUCTION

The memristor was originally envisioned [216] and was generalized to memristive systems [217] by circuit theorist Chua as a missing nonlinear passive two-terminal electrical component [218], which has non-volatility property [219] — [223], [237] — [238]. A broader definition of the memristor was argued that it could cover all forms of two-terminal non-volatile memory devices based on resistance switching effects [219] — [223], [239]. A solely current-controlled memristor with the so-called non-volatility property [220] — [223] cannot enable such a memristor system to erratically change its state just under the influence of white current noise [241], [243], whose dynamic state equations allow the violation of Landauer's principle of the minimum amount of energy [241], [242]. The pinched hysteresis effect of the memristors and memristive systems [246], [247], the nonlinear ionic drift models of the memristors [244] and Strukov's initial memristor modeling equations [245] were further studied, respectively. Moreover, the memristors can be generally classified into five categories: the titanium dioxide

---

① PU Yifei, YUAN Xiao, YU Bo. Analog Circuit Implementation of Fractional-Order Memristor: Arbitrary-Order Lattice Scaling Fracmemristor [J]. IEEE Transactions on Circuits and Systems I: Regular Papers, accepted to be published, 2017. (SCI IF: 2.407) 论文链接: http://ieeexplore.ieee.org/document/8263247/.

memristor [237] — [238], [248] — [249], the polymeric memristor [250], [251], the layered memristor [252], the ferroelectric memristor [253] and the spin memristive systems [254] — [259], respectively. The electrical properties of the aforementioned memristors and memristive systems are of the classic integer-order.

Fractional calculus has become an important novel branch of mathematical analyses [175], [178]. For physical scientists and engineering technicians, fractional calculus is now a novel useful mathematical method mainly because of its inherent strength of long-term memory, non-locality and weak singularity. Many scientific fields such as the fractional diffusion processes [182], [215], fractional viscoelasticity [126], fractal dynamics [184], fractional control [89], fractional image processing [187], fractional signal processing [192], fractional neural networks [399], [499], fractional circuits and systems [225], [276], [405], [638], etc., presently use fractional calculus and has obtained some promising results. However, the application of fractional calculus to analyzing the memristor is an emerging discipline of study in which few studies have been performed [224], [236], [271] — [273], [639] — [643]. From the Chua's axiomatic element system [216] — [221] and according to constitutive relation, logical consistency, axiomatic completeness and formal symmetry, we can assume that there should be a capacitive fractional-order memristor and an inductive fractional-order memristor corresponding to the capacitive fractor and inductive fractor, respectively. The concept of the memristor was generalized preliminarily from the classic integer-order memristor to that of the fractional-order one [224], [236], [271] — [273], [639] — [643]. In particular, the literature [224] derived the generic fractional-order driving-point impedance functions of an arbitrary-order capacitive fracmemristor and inductive one in their natural implementations, respectively. There are two types of the fractional-order memristor: the capacitive fractional-order memristor and the inductive fractional-order memristor [224]. The terms "fracmemristor" and "fracmemristance" are two portmanteaus of "the fractional-order memristor" and "the fractional-order memristance", respectively [224]. The dynamic characteristics of the frammeristor based chaotic systems depend on not only its circuit parameters but also its initial state as well as its fractional-order. The frammeristor based chaotic behavior is a promising scope of study. Therefore, based on the aforementioned previous researches of the fractional-order memristor [224], [236], [271] — [273], [639] — [643], it is natural to ponder a challenging theoretical problem to propose a circuit theoretic methodology to implement a fractional-order memristor by using the ordinary memristor and capacitor or inductor in the form of an analog circuit. Motivated by this need, in this part, we used fractional calculus to propose an interesting analog circuit implementation method of an arbitrary-order memristor. The literature [224] only derived the generic fractional-order driving-point impedance functions of an arbitrary-order capacitive fracmemristor and inductive one in their natural implementations, respectively. Comparing to the previous work of fracmemristor [224],

[236], [271] — [273], [639] — [643], the main contribution of this part is the proposal for the first preliminary attempt of a feasible hardware achievement of an arbitrary-order fracmemristor and for the recognition of the fingerprint of fracmemristor. The major advantage of the proposed lattice scaling approach is that we can use ordinary memristor and capacitor or inductor to easily achieve an arbitrary-order fracmemristor.

The remainder of this part is organized as follows: Section 2 recalls some preliminary concepts of the fractor and the fracmemristor. In Section 3, at first, the circuit configurations of an arbitrary-order capacitive lattice scaling fracmemristor is proposed and mathematically derived. The second step is the proposal for an arbitrary-order compensatory capacitive lattice scaling fracmemristor. In Section 4, the approximation performance, electrical characteristics and analog circuit achievement of an arbitrary-order fracmemristor are experimentally analyzed in detail, respectively. In Section 5, the conclusions of this manuscript are presented.

7.4.2 RELATED WORK

This section presents a brief introduction to the necessary electrical background of the fractor and the fracmemristor.

The memristor, M, completes the set of relations with [216]:

$$\varphi[q(t)] = M[q(t)]q(t), \tag{7.56}$$

where φ, q and t denote the magnetic flux, quantity of electric charge and time variable, respectively. The slope of this function is called the memristance being similar to variable resistance [216], given as:

$$V_i(t) = d\varphi(q)/dq I_i(t) = [M(q) + q dM(q)/dq]I_i(t) = R[q(t)]I_i(t)$$
$$= H[q(t)] * I_i(t), \tag{7.57}$$

where $V_i(t)$ and $I_i(t)$ denote the instantaneous value of the input voltage and input current of a memristor, respectively. The symbol $*$ denotes convolution, $R[q(t)] = [M(q) + q dM(q)/dq]$ and $H[q(t)]$ denote the memristance and transmission function of a memristor, respectively. Note that in the time-invariant case, the constitutive relation of a memristor can be simply expressed as a set of points in the φ-q plane. We can regard this relationship as a generalized Ohm's law [218]. Moreover, Chua's periodic table of all two-terminal circuit elements [219] — [223] can be shown as given in Fig. 7.7.

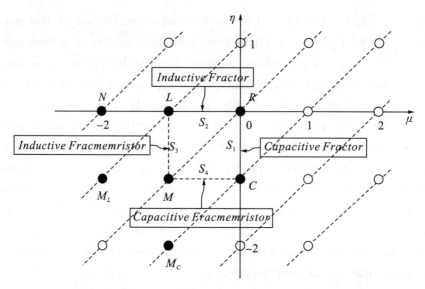

Fig. 7.7 Chua's periodic table of all two-terminal circuit elements.

In Fig. 7.7, μ and η denote the voltage exponent and current exponent, which are equal to the order of the time derivative of $V_i(t)$ and $I_i(t)$, respectively. (μ, η) is referred to the Chua's plane of the Chua's axiomatic element system. C, R, L, M, M_L and M_C denote the capacitor, resistor, inductor, memristor, meminductor and memcapacitor, respectively. O denotes the other postulated elements of the Chua's axiomatic element system. Note that at first, small-signal behavior method is an efficient approach to study a resistive nonlinear network [644] − [646]. To give some physical meaning to each Chua's axiomatic element in Fig. 7.1 (including memristor), it is convenient to examine its small-signal behavior about an operating point Q on its associated $V_i^{\langle\mu\rangle} - I_i^{\langle\eta\rangle}$ curve. Assuming that a Chua's axiomatic element is characterized by $V_i^{\langle\mu\rangle} = f(I_i^{\langle\eta\rangle})$ and its small-signal behavior about Q by described $V_i^{\langle\mu\rangle}(t) = m_Q I_i^{\langle\eta\rangle}(t)$, where m_Q denotes the slope $f'[I_i^{\langle\eta\rangle}]$ at Q. Therefore, the small-signal reactance of this Chua's axiomatic element can be defined by taking Laplace transform of $V_i^{\langle\mu\rangle}(t) = m_Q I_i^{\langle\eta\rangle}(t)$. Thus, we obtain $L\{V_i(t)\} = Z(s)L\{I_i(t)\}$, where $Z(s) = s^{\eta-\mu} m_Q$, $L\{\}$ denotes Laplace transform and s denotes a complex variable of the Laplace transform. We can interpret the small-signal reactance $Z(s) = s^{\eta-\mu} m_Q$ as the impedance of an associated linearized element about Q [217] − [219]. For instance, based on certain algebraic properties of the Laplace operator, nonlinear n-port decomposition is achieved [647]. Finite-time synchronization of fractional-order memristor-based neural networks with time delays was considered by using Laplace transform, such as the generalized Gronwall inequality and Mittag-Leffler functions [648]. Secondly, (7.57) shows that in the case of small-signal behavior about Q, the Laplace transform of $V_i(t) = R[q(t)]I_i(t)$ is equal to $V_i(s) = 1/2\pi L\{R[q(t)]\} * I_i(s)$, which implements a convolution in Laplace transform domain and is sometimes too difficult to be calculated. For the convenience of analysis, let's assume that the transmission function of a memristor, $H[q(t)]$, satisfies $V_i(t) = R[q(t)]I_i(t) = H[q(t)] * I_i(t)$ (i.e. $R[q(t)] = \{H[q(t)] * I_i(t)\}/I_i(t)$). Thus, the Laplace transform of $V_i(t) = H[q(t)] * I_i(t)$, $V_i(s) = r[q(s)]I_i(s)$, implements a multiplication in Laplace transform domain, where $r[q(s)] = L\{H[q(t)]\}$ is the reactance of this memristor. In addition, Fig. 7.1 shows that at first, the capacitive fracmemristor is lying on the line segment, S_4, between C and M. The electrical properties of the capacitive fracmemristor fall in between those of the capacitor and those of the memristor [224]. Secondly, the inductive fracmemristor is lying on the line segment, S_3, between L and M. The electrical properties of the inductive fracmemristor fall in between those of the inductor and those of the memristor [224]. Furthermore, Pu derived the generic fractional-order driving-point impedance functions of an arbitrary-order capacitive fractor [225], [405], inductive fractor [225], [405], capacitive fracmemristor [224] and inductive fracmemristor [224] in their natural implementations, respectively, given as:

$$F^c_{-v} = F^c_{-(\eta+p)} = V_i(s)/I_i(s) = c^{-v} r^{1-p} s^{-v}, \quad (7.58)$$

$$F^l_v = F^l_{\eta+p} = V_i(s)/I_i(s) = l^v r^{1-p} s^v, \quad (7.59)$$

$$FM^c_{-v} = FM^c_{-(\eta+p)} = V_i(s)/I_i(s) = c^{-v} \{r[q(s)]\}^{1-p} s^{-v}, \quad (7.60)$$

$$FM^l_v = FM^l_{\eta+p} = V_i(s)/I_i(s) = l^v \{r[q(s)]\}^{1-p} s^v, \quad (7.61)$$

where $v = \eta + p$ is a non-negative real number, η is a non-negative integer and $0 \leqslant p \leqslant 1$. Note that at first, if $v=0$, then $\eta=0$ and $p=0$. Equations (7.58) — (7.61) identically degenerate to the driving-point impedance function of R in Fig. 7.1. Secondly, if v is a positive integer, then $\eta = v-1$ and $p=1$. Equations (7.58) — (7.61) degenerate to the driving-point impedance functions of C, L, C and L in Fig. 7.7, respectively. Thirdly, if $0 < v < 1$, then $\eta = 0$ and $0 < p < 1$. Equations (7.58) — (7.61) denote the $0 < v < 1$ order driving-point impedance functions of the capacitive fractor, inductive fractor, capacitive fracmemristor and inductive fracmemristor in Fig. 7.7, respectively. Fourthly, if $v > 1$ is a positive fraction, then $\eta = [v]$ and $0 < p < 1$, where $[\ ]$ denotes rounding operation. Equations (7.58) — (7.61) denote the corresponding $v > 1$ order driving-point impedance functions. F^c_{-v}, F^l_v, FM^c_{-v} and FM^l_v denote the fractional-order driving-point impedance function of a purely ideal v-order capacitive fractor, inductive fractor, capacitive fracmemristor and inductive fracmemristor, respectively. c, l and r denote the capacitance, inductance and resistance, respectively. $c^v r^{1-p}$, $l^v r^{1-p}$, $c^v L^{-1}\{[r(q)]^{1-p}\}$ and $l^v L^{-1}\{[r(q)]^{1-p}\}$ denote the capacitive fractance, inductive fractance, capacitive fracmemristance, inductive fracmemristance, respectively, where $L^{-1}\{\ \}$ denotes inverse Laplace transform. For instance, with regard to a nonlinear time dependent current-controlled memristor, let's $I_i(t) = \sin(at)u(t)$ in (7.57), where a is frequency and $u(t)$ is a Heaviside function. Then, $I_i(s) = a/(s^2 + a^2)$ and $q(t) = D_t^{-1} I_i(t) = -(1/a)\cos(at)u(t)$. And supposing that $M[q(t)] = q(t)/2 + 1/a$. Thus, $R[q(t)] = [-(1/a)\cos(at) + 1/a]u(t)$ and $V_i(t) = [-\sin(2at)/(2a) + \sin(at)/a]u(t)$. And assuming the initial state of this memristor is zero, thus $V_i(s) = r[q(s)]I_i(s) = 3a^2/[(s^2+4a^2)(s^2+a^2)]$. Then, we have $r[q(s)] = V_i(s)/I_i(s) = 3a/(s^2+4a^2)$. Therefore, $H[q(t)] = L^{-1}\{r[q(s)]\} = (3/2)\sin(2at)u(t)$, where $L^{-1}\{\}$ denotes inverse Laplace transform. Let's assume that $v = \eta + 2/3$, thus η is a positive integer and $p = 2/3$. Then, in (7.60) and (7.61), $L^{-1}\{[r(q)]^{1-p}\} = L^{-1}\{[r(q)]^{1/3}\} = (3a)^{1/3}\sqrt{\pi}/\Gamma(1/3)[t/(4a)]^{-1/6}J_{-1/6}(2at)u(t)$, where $J_{-1/6}(t)$ is a Bessel function of the first kind extended to non-integer orders by one of Schläfli's integrals. From aforementioned discussion, in the case of small-signal behavior about Q, we can approximately apply Laplace transform to analyze a circuit consisted of the memristor [217] − [219]. The generic fractional-order driving-point impedance functions of an arbitrary-order capacitive fracmemristor and inductive one in their natural implementations can be described as (7.60) and (7.61), respectively [224]. The measurement units and physical dimensions of the fracmemristance are the same as those of the fractance [225], [405], [224].

This section is the proposal for a novel feasible analog circuit implementation of an arbitrary-order lattice scaling fracmemristor.

According to the characteristics of fractional calculus and Laplace transform, provided the transmission function of the memristor $H[q(t)]$ is a causal function and its fractional primitives are zero, we can simplify the inverse Laplace transforms of (7.60) and (7.61), given as:

$$V_i(t) = c^{-v}L^{-1}\{[r(q)]^{1-p}\} * [D_t^{-v}I_i(t)], \qquad (7.62)$$
$$V_i(t) = l^v L^{-1}\{[r(q)]^{1-p}\} * [D_t^v I_i(t)], \qquad (7.63)$$

where D_t^{-v} denotes the v-order integral with respect to t, and D_t^v denotes the v-order differential with respect to t. Equations (7.62) and (7.63) show that since the transmission function of a memristor $H[q(t)]$ can be arbitrary [216] − [223], $r[q(s)]$ is correspondingly arbitrary. Thus, in (7.60) − (7.63), the approximate implementation of $\{r[q(s)]\}^{1-p}$ and $L^{-1}\{[r(q)]^{1-p}\}$ are very difficult. The hardware achievement method of an arbitrary-order fracmemristor cannot directly employ that of an arbitrary-order fractor [225], [276], [405], [638]. For conciseness, in this part, we only discuss a novel feasible analog circuit implementation of an arbitrary-order capacitive lattice scaling fracmemristor in detail. An arbitrary-order inductive lattice scaling fracmemristor can be achieved in a similar way.

The circuit configurations of a v-order capacitive lattice scaling fracmemristor can be shown in Fig. 7.8.

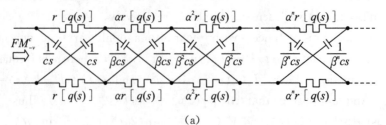

(a)

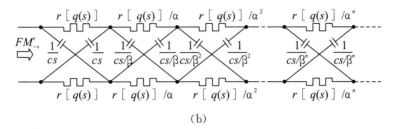

(b)

Fig. 7.8　Circuit configurations of v-order capacitive lattice scaling fracmemristor: (a) v-order low-pass filtering capacitive lattice scaling fracmemristor; (b) v-order high-pass filtering capacitive lattice scaling fracmemristor.

In Fig. 7.8, $0<v<1$ is an arbitrary positive rational number, the symbol of ⎍⎍ denotes the memristor, and α and β represent two positive scaling factors, respectively. The amount of memristors or capacitors the v-order capacitive lattice scaling fracmemristor is equal to the double amount of layers. Figure 7.8 shows that the v-order capacitive lattice scaling fracmemristor has a cascade circuit with a series connection of infinitely repeated lattice structures.

To simplify discussion, we useiterative solution method to analyze the fractional-order reactance of the v-order low-pass filtering capacitive lattice scaling fracmemristor. The circuit configuration of a 1/2-order low-pass filtering capacitive lattice scaling fracmemristor can be shown in Fig. 7.9.

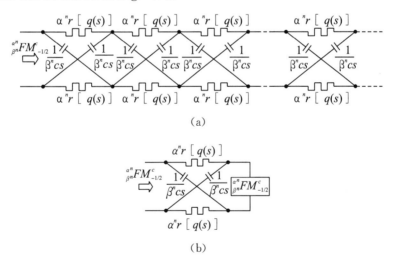

Fig. 7.9　Circuit configuration of a 1/2-order low-pass filtering capacitive lattice scaling fracmemristor: (a) Circuit configuration; (b) Equivalent circuit.

Figure 7.9 shows that the positive scaling factor of the reactance of a memristor and that of the capacitance of its every series circuit, α^n and β^n, are identical, respectively. The every series circuit of the 1/2-order low-pass filtering capacitive lattice scaling fracmemristor in Fig. 7.9 (a) has the same circuit configuration as that of the $(n+1)th$ series circuit of the v-order one in Fig. 7.8 (a). Provided the number of the series circuit is infinite, $\frac{\alpha^n}{\beta^n}FM^c_{-1/2}$ is the 1/2-order capacitive reactance of the 1/2-order low-pass filtering capacitive lattice scaling fracmemristor in Fig. 7.9. As we know, Kirchhoff's Current Law (KCL) and Kirchhoff's Voltage Law (KVL) describe the topological constraints, which construct the basic theory of circuits and systems [649], [650]. KCL (the algebraic sum of currents in a circuit meeting at a point is zero) and KVL (the algebraic sum of the voltage around any closed network is zero) can be described as two constraint algebraic equations, which merely depend on the topological structure and have nothing to do with the electrical characteristics of the circuit devices in a circuit. Since the memristor is a nonlinear circuit element, a circuit consisted of the memristor is also a nonlinear circuit. In classic theory of circuits and systems, the model of a nonlinear circuit can be described as a constraint equation $f(q, \varphi, I, V, t) = 0$. Since a nonlinear circuit consisted of memristor still satisfies the law of conservation of charge and the law of conservation of energy, $f(q, \varphi, I, V, t)$ conforms to KCL and KVL [217] − [219], [649], [650]. With regard to the four circuit variables in $f(q, \varphi, I, V, t)$, there are two differential mapping relationships, $dq/dt = I$ and $d\varphi/dt = V$, between (q and φ) and (I and V), respectively. From (7.56), (7.57) and Fig. 7.7, we can observe that in the Chua's axiomatic element system, the resistor, capacitor, inductor and memristor satisfy ($\mu=0$, $\eta=0$), ($\mu=0$, $\eta=-1$), ($\mu=-1$, $\eta=0$) and ($\mu=-1$, $\eta=-1$), respectively [217] − [219]. A memristor in the small-signal behavior about Q can be described as $V_i^{(-1)}(t) = m_Q I_i^{(-1)}(t)$, thus its small-signal reactance $Z(s) = m_Q$ can be interpreted as the impedance of an associated linearized element about Q [217] − [219]. Therefore, in the case of small-signal behavior about Q, the electrical characteristics of a circuit consisted of the memristor can be analyzed according to KCL and KVL [217] − [219], [649], [650].

In Fig. 7.9 (b), let's assume that $i_r(s)$ and $i_c(s)$ are the currents of $\alpha^n r[q(s)]$ and $1/(\beta^n cs)$, respectively. $V_i(s)$ and $I_i(s) = i_r(s) + i_c(s)$ denote the input voltage and input current of ${}_{\beta^n}^{\alpha^n} FM_{-1/2}^c$, respectively. Therefore, in the case of small-signal behavior about Q, according to KCL and KVL, we obtain from Fig. 7.9 the following relationship:

$$\begin{cases} \alpha^n r[q(s)] i_r(s) + [1/(\beta^n cs)] i_c(s) - V_i(s) = 0 \\ \{\alpha^n r[q(s)] + {}_{\beta^n}^{\alpha^n} FM_{-1/2}^c\} i_r(s) - [{}_{\beta^n}^{\alpha^n} FM_{-1/2}^c + 1/(\beta^n cs)] i_c(s) = 0. \end{cases} \quad (7.64)$$

Thus, according to the Cramer's rule of linear algebra, from (7.64), we get:

$$\begin{cases} i_r(s) = \dfrac{[1/(\beta^n cs) + {}_{\beta^n}^{\alpha^n} FM_{-1/2}^c] V_i(s)}{\begin{vmatrix} \alpha^n r[q(s)] + {}_{\beta^n}^{\alpha^n} FM_{-1/2}^c & -[{}_{\beta^n}^{\alpha^n} FM_{-1/2}^c + 1/(\beta^n cs)] \\ \alpha^n r[q(s)] & 1/(\beta^n cs) \end{vmatrix}} \\ i_c(s) = \dfrac{[\alpha^n r[q(s)] + {}_{\beta^n}^{\alpha^n} FM_{-1/2}^c] V_i(s)}{\begin{vmatrix} \alpha^n r[q(s)] + {}_{\beta^n}^{\alpha^n} FM_{-1/2}^c & -[{}_{\beta^n}^{\alpha^n} FM_{-1/2}^c + 1/(\beta^n cs)] \\ \alpha^n r[q(s)] & 1/(\beta^n cs) \end{vmatrix}} \end{cases} \quad (7.65)$$

Then, from (7.65), ${}_{\beta^n}^{\alpha^n} FM_{-1/2}^c$ can be derived as:

$$_{\beta^n}^{\alpha^n}FM_{-1/2}^c = \frac{V_i(s)}{I_i(s)} = \frac{2\alpha^n r[q(s)][1/(\beta^n cs)] + _{\beta^n}^{\alpha^n}FM_{-1/2}^c\{\alpha^n r[q(s)] + 1/(\beta^n cs)\}}{2_{\beta^n}^{\alpha^n}FM_{-1/2}^c + \alpha^n r[q(s)] + 1/(\beta^n cs)}.$$
(7.66)

Thus, from (7.66), the following can be obtained:
$$_{\beta^n}^{\alpha^n}FM_{-1/2}^c = V_i(s)/I_i(s) = (\beta^n c)^{-1/2}\{\alpha^n r[q(s)]\}^{1/2} s^{-1/2}.$$
(7.67)

From (7.67), we can observed that at first, there is a negative 1/2-order complex variable of the Laplace transform $s^{-1/2}$ (a 1/2-order integral capacitive operator), however, there is also a $\{r[q(s)]\}^{-1/2}$ in the formula of $_{\beta^n}^{\alpha^n}FM_{-1/2}^c$. Thus, being more complex than a pure fractional-element (ideal capacitive fractor), the cascade circuit with a series connection of infinitely repeated lattice scaling structures (when $m \to \infty$) in Fig. 7.9 achieves a 1/2-order low-pass filtering nonlinear capacitive operation. Secondly, as subsequent mathematically derivation, provided the number of the series circuit is infinite, the fractional-order $v = 1/2$ of a 1/2-order low-pass filtering capacitive lattice scaling fracmemristor can be also solely derived from (7.80). Thirdly, as subsequent further demonstration in following Example 1, in a certain pass-band, both the magnitude and the phase of the 1/2-order capacitive reactance of the 1/2-order capacitive lattice scaling fracmemristor with m series circuits, $_{\beta^n}^{\alpha^n}FM_{-1/2}^c$, can approach those of the ideal 1/2-order capacitive fracmemristor with a high degree of accuracy. The larger m is, the wider the pass-band of the 1/2-order capacitive lattice scaling fracmemristor is. From aforementioned discussion, in the case of small-signal behavior about Q, according to KCL and KVL, the electrical characteristics of a 1/2-order low-pass filtering capacitive lattice scaling fracmemristor can be derived as (7.67).

Furthermore, the infinitely iterative circuit configurations of the v-order low-pass filtering capacitive lattice scaling fracmemristor can be shown in Fig. 7.10.

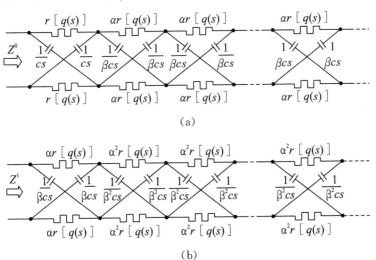

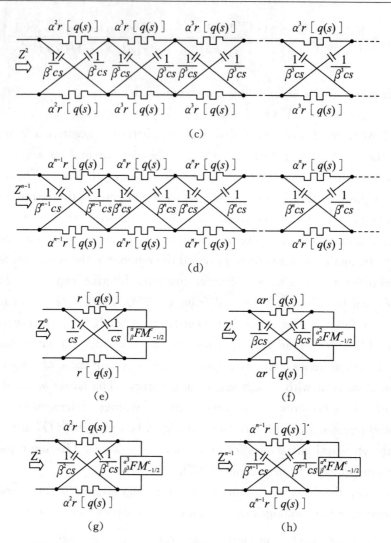

Fig. 7.10 Infinitely iterative circuit configurations of v-order low-pass filtering capacitive lattice scaling fracmemristor: (a) First iterative circuit; (b) Second iterative circuit; (c) Third iterative circuit; (d) nth iterative circuit; (e) Equivalent circuit of first iterative circuit; (f) Equivalent circuit of second iterative circuit; (g) Equivalent circuit of third iterative circuit; (h) Equivalent circuit of nth iterative circuit.

In Fig. 7.10, Z^0, Z^1, Z^2 and Z^{n-1} denote the driving-point capacitive impedance function of the first, the second, the third, and the nth iterative circuit of the v-order low-pass filtering capacitive lattice scaling fracmemristor, respectively. In Fig. 7.10 (e), let's assume that $i_r(s)$ and $i_c(s)$ are the currents of $r[q(s)]$ and c, respectively. $V_i(s)$ and $I_i(s)$ are the input voltage and the input current of Z^0, respectively. Thus, from Fig. 7.10 (e), according to KCL and KVL, we obtain:

$$\begin{cases} r[q(s)]i_r(s) + [1/(cs)]i_c(s) - V_i(s) = 0 \\ (r[q(s)] + {}_\beta^\alpha FM^c_{-1/2})i_r(s) - \{{}_\beta^\alpha FM^c_{-1/2} + [1/(cs)]\}i_c(s) = 0 \end{cases} \quad (7.68)$$

Then, according to the Cramer's rule of linear algebra, from (7.68), we have:

$$\begin{cases} i_r(s) = \dfrac{\{[1/(cs)] +{}_\beta^a FM_{-1/2}^c\} V_i(s)}{\begin{vmatrix} r[q(s)] +{}_\beta^a FM_{-1/2}^c & -\{[1/(cs)] +{}_\beta^a FM_{-1/2}^c\} \\ r[q(s)] & [1/(cs)] \end{vmatrix}} \\ i_c(s) = \dfrac{\{r[q(s)] +{}_\beta^a FM_{-1/2}^c\} V_i(s)}{\begin{vmatrix} r[q(s)] +{}_\beta^a FM_{-1/2}^c & -\{[1/(cs)] +{}_\beta^a FM_{-1/2}^c\} \\ r[q(s)] & [1/(cs)] \end{vmatrix}} \end{cases}. \quad (7.69)$$

Then, (7.69) can be rewritten as:

$$Z^0(s) = \frac{V_i(s)}{I_i(s)} = \frac{V_i(s)}{i_r(s)+i_c(s)} = \frac{2r[q(s)](1/cs) +{}_\beta^a FM_{-1/2}^c \{r[q(s)] + 1/cs\}}{2{}_\beta^a FM_{-1/2}^c + r[q(s)] + 1/cs}. \quad (7.70)$$

Substituting (7.67) into (7.70), we obtain:

$$Z^0(s) = \frac{2r[q(s)]/(cs) + \sqrt{\alpha r[q(s)]/(\beta cs)}\{r[q(s)] + 1/(cs)\}}{2\sqrt{\alpha r[q(s)]/(\beta cs)} + r[q(s)] + 1/(cs)}. \quad (7.71)$$

Further, from Figs. 7.10 (b), 7.10 (c), 7.10 (b) and 7.10 (c), we get $Z^1(s) = \alpha Z^0(\alpha\beta s)$ and $Z^2(s) = \alpha Z^1(\alpha\beta s)$, respectively. In a similar way, from Figs. 7.10 (d) and 7.10 (h), we obtain:

$$\begin{aligned} Z^{n-1}(s) &= \frac{2\alpha^{n-1} r[q(s)]/(\beta^{n-1} cs) + \sqrt{\alpha^n r[q(s)]/(\beta^n cs)}\{\alpha^{n-1} r[q(s)] + 1/(\beta^{n-1} cs)\}}{2\sqrt{\alpha^n r[q(s)]/(\beta^n cs)} + \alpha^{n-1} r[q(s)] + 1/(\beta^{n-1} cs)} \\ &= \alpha \frac{2\alpha^{n-2} r[q(s)]/[\beta^{n-2} c(\alpha\beta s)] + \sqrt{\alpha^{n-1} r[q(s)]/[\beta^{n-1} c(\alpha\beta s)]}\{\alpha^{n-2} r[q(s)] + 1/[\beta^{n-2} c(\alpha\beta s)]\}}{2\sqrt{\alpha^{n-1} r[q(s)]/\beta^{n-1} c(\alpha\beta s)} + \alpha^{n-2} r[q(s)] + 1/\beta^{n-2} c(\alpha\beta s)} \\ &= \alpha Z^{n-2}(\alpha\beta s). \end{aligned} \quad (7.72)$$

Equations (7.71) and (7.72) show that the v-order low-pass filtering capacitive lattice scaling fracmemristor shown in Fig. 7.8 (a) can be treated as an infinitely successively nested structure of the first, the second, the third, ⋯ and the nth iterative circuit ($n \to \infty$) shown in Fig. 7.10. Thus, provided $n \to \infty$, the v-order capacitive reactance of the v-order low-pass filtering capacitive lattice scaling fracmemristor is equal to the limiting value of the recursion equation of successively nested $Z^0(s)$, $Z^1(s)$, $Z^2(s)$, ⋯ and $Z^{n-1}(s)$. Hence, from (7.71) and (7.72), if $n \to \infty$, the v-order iterative capacitive reactance of the v-order low-pass filtering capacitive lattice scaling fracmemristor can be derived as:

$$FM_{-v}^{nc}(s) = \frac{2r[q(s)]/(cs) + \alpha FM_{-v}^{(n-1)c}(\alpha\beta s)\{r[q(s)] + 1/(cs)\}}{2\alpha FM_{-v}^{(n-1)c}(\alpha\beta s) + r[q(s)] + 1/(cs)}, \quad (7.73)$$

where $FM_{-v}^{nc}(s)$ and $FM_{-v}^{(n-1)c}(s)$ are the driving-point capacitive impedance functions of the v-order low-pass filtering capacitive lattice scaling fracmemristor with n series circuits and that with $(n-1)$ series circuits, respectively. Equation (7.73) is essentially a specific continued fraction expansion. Therefore, from Figs. 7.8 and 7.10, provided $n \to \infty$, it follows that:

$$FM_{-v}^c(s) = \lim_{n\to\infty} FM_{-v}^{nc}(s) = \lim_{n\to\infty} FM_{-v}^{(n-1)c}(s). \quad (7.74)$$

Then, from (7.74), (7.73) can be rewritten as:

$$FM^c_{-v}(s) = \frac{2r[q(s)]/(cs) + \alpha FM^2_{-v}(\alpha\beta s)\{r[q(s)] + 1/(cs)\}}{2\alpha FM^c_{-v}(\alpha\beta s) + r[q(s)] + 1/(cs)}. \tag{7.75}$$

In (7.75), $\alpha\beta$ is actually the fractal scaling factor of the v-order low-pass filtering capacitive lattice scaling fracmemristor. Equation (7.75) is the irregular iterative scaling equation of the v-order low-pass filtering capacitive lattice scaling fracmemristor, which accords with standard dynamical scaling law [407] — [409], [417], [651]. Hence, the solution of (7.75) can be given as:

$$FM^c_{-v}(s) = \kappa(s)s^{-v}, \tag{7.76}$$

where $\kappa(s)$ is a scalar factor. Although the irregular iterative scaling equation maybe has multiple solutions in pure mathematics, the true solution of (7.75) must accord with the actual circuit of the v-order low-pass filtering capacitive lattice scaling fracmemristor shown in Fig. 7.8 (a). All other multiple solutions of (7.75) are spurious solutions. From Fig. 7.8 (a), by means of nested iterations, the generic fractional-order driving-point impedance functions of an arbitrary-order capacitive fracmemristor in their natural implementations can be derived as (7.60) [224]. Thus, with regard to the v-order low-pass filtering capacitive lattice scaling fracmemristor, substituting (7.71) into (7.72) recursively, by means of mathematical induction, we can obtain the limiting value of $\kappa(s)$ when $n \to \infty$:

$$\kappa(s) \stackrel{n \to \infty}{\approx} c^{-v}\{r[q(s)]\}^{1-p}, \tag{7.77}$$

where $v = \eta + p$ that is the same as that of (7.60) and (7.61). Equations (7.76) and (7.77) are consistent with (7.60). Further, substituting (7.76) into (7.75) results in:

$$\kappa s^{-v} = \frac{2r[q(s)]/(cs) + \alpha\kappa(\alpha\beta s)^{-v}\{r[q(s)] + 1/(cs)\}}{2\alpha\kappa(\alpha\beta s)^{-v} + r[q(s)] + 1/(cs)} = \frac{2r[q(s)] + \alpha\kappa(\alpha\beta s)^{-v}(r[q(s)]cs + 1)}{2\alpha\kappa(\alpha\beta s)^{-v}cs + r[q(s)]cs + 1}. \tag{7.78}$$

Since $0 < v < 1$, when $s \to 0$, we have $s^{-v}s \to 0$ and $\kappa(s) \stackrel{s \to 0}{\approx} \kappa(\alpha\beta s)$. Then, when $s \to 0$ (low-pass filtering), (7.78) can be simplified as:

$$\kappa s^{-v} = \frac{2r[q(s)] + \alpha\kappa(\alpha\beta s)^{-v}(r[q(s)]cs + 1)}{2\alpha\kappa(\alpha\beta s)^{-v}cs + r[q(s)]cs + 1} \stackrel{s \to 0}{\approx} \frac{\alpha\kappa(\alpha\beta s)^{-v}(r[q(s)]cs + 1)}{r[q(s)]cs + 1} = \alpha\kappa(\alpha\beta s)^{-v}. \tag{7.79}$$

Therefore, when $s \to 0$ (low-pass filtering), the solution of (7.79) can be derived as:

$$v = \log(\alpha)/[\log(\alpha) + \log(\beta)], \tag{7.80}$$

where $\log()$ is a logarithm. Equation (7.80) shows that the fractional-order of the v-order low-pass filtering capacitive lattice scaling fracmemristor shown in Fig. 7.8 merely essentially depends on its two positive scaling factors (α and β), but has nothing to do with its reactance of a memristor and capacitance ($r[q(s)]$ and c). In particular, with regard to Fig. 4, the positive scaling factor of the reactance of a memristor and that of the capacitance of its every series circuit, α^n and β^n, are identical, respectively. Then, comparing Fig. 7.8 (a) with Fig. 7.9, we can observe that if we substitute $\alpha^n r[q(s)]$ and $1/(\beta^n cs)$ for $r[q(s)]$ and $1/(cs)$ in Fig. 7.8 (a), respectively, Fig. 7.9 is equivalent to Fig. 7.8 (a) with two positive scaling factors $\alpha = 1$ and $\beta = 1$. Thus, from (7.80), provided $\alpha \to 1$ and $\beta \to 1$, we can derive that $\lim\limits_{\alpha \to 1, \beta \to 1} v = 1/2$, which is consistent with the derivation of (7.67). Substituting $\alpha = 1$ and $\beta = 1$ and $v = 1/2$ into (7.76), (7.77) and (7.78) gives:

$$FM^c_{-v}(s) \stackrel{\alpha=1,\beta=1}{\approx} FM^c_{-1/2}(s) = \kappa s^{-1/2} = c^{-1/2}\{r[q(s)]\}^{1/2} s^{-1/2}. \quad (7.81)$$

Equation (7.81) shows that if $\alpha = 1$ and $\beta = 1$, (7.79) can be accurately simplified as $\kappa s^{-v} \stackrel{s\to 0}{=} \alpha \kappa (\alpha \beta s)^{-v}|_{\alpha=1,\beta=1,v=1/2}$. The 1/2-order low-pass filtering capacitive lattice scaling fracmemristor is a special case of the purely ideal v-order capacitive fracmemristor [224]. From aforementioned discussion, in the case of small-signal behavior about Q, according to KCL and KVL, the fractional-order v of an arbitrary-order low-pass filtering capacitive lattice scaling fracmemristor can be determined by means of altering its two positive scaling factors (α and β) in (7.80).

In addition, (7.79) further shows that to improve its approximation accuracy, an impedance $-2r[q(s)]/\{r[q(s)]cs+1]\}$, should be compensatory in the first series circuit of the v-order low-pass filtering capacitive lattice scaling fracmemristor shown in Fig. 7.8 (a), as illustrated in Fig. 7.11.

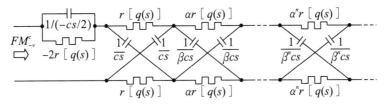

Fig. 7.11 Circuit configuration of v-order compensatory low-pass filtering capacitive lattice scaling fracmemristor.

In Fig. 7.11, negative reactance of a memristor and negative capacitance can be achieved by negative impedance converter. Thus, from Fig. 7.11, the fractional-order driving-point impedance function of the v-order compensatory low-pass filtering capacitive lattice scaling fracmemristor can be derived as $\kappa s^{-v} = \dfrac{2r[q(s)]+\alpha\kappa(\alpha\beta s)^{-v}(r[q(s)]cs+1)}{2\alpha\kappa(\alpha\beta s)^{-v}cs+r[q(s)]cs+1} - \dfrac{2r[q(s)]}{r[q(s)]cs+1} \stackrel{s\to 0}{=} \alpha\kappa(\alpha\beta s)^{-v}$. Thus, by adding the first compensatory series circuit of the v-order low-pass filtering capacitive lattice scaling fracmemristor, (7.80) can be accurately derived.

In a similar way, with respect to the v-order high-pass filtering capacitive lattice scaling fracmemristor shown in Fig. 7.8 (b),

$$FM^c_{-v}(s) = \dfrac{2r[q(s)]/(cs)+(1/\alpha)FM^c_{-v}[s/(\alpha\beta)]\{r[a(s)]+1/(cs)\}}{2(1/\alpha)FM^c_{-v}[s/(\alpha\beta)]+r[q(s)]+1/(cs)} \text{ and } FM^c_{-v}(s) =$$

$\kappa(s)s^{-v}$ can be derived. Since $0 < v < 1$, when $s \to \infty$, we have $s^{-v} \to 0$, $\kappa(s) \stackrel{s\to\infty}{\approx} \kappa(\alpha\beta s)$. Thus, when $s \to \infty$ (high-pass filtering), we have:

$$\kappa s^{-v} = \frac{2r[q(s)]/(cs) + (1/\alpha)\kappa[s/(\alpha\beta)]^{-v}\{r[q(s)]+1/(cs)\}}{2(1/\alpha)\kappa[s/(\alpha\beta)]^{-v}+r[q(s)]+1/(cs)}$$

$$\overset{s\to\infty}{\approx} \frac{(1/\alpha)\kappa[s/(\alpha\beta)]^{-v}(r[q(s)]cs+1)}{r[q(s)]cs+1} = (1/\alpha)\kappa(s/(\alpha\beta))^{-v}. \quad (7.82)$$

Then, when $s\to\infty$ (high-pass filtering), the solution of (7.82) can be derived as $v=\lg(\alpha)/[\lg(\alpha)+\lg(\beta)]$. In particular, if $\alpha=1$ and $\beta=1$, $\kappa s^{-v} \overset{s\to\infty}{=} (1/\alpha)\kappa[s/(\alpha\beta)]^{-v}|_{\alpha=1,\beta=1,v=1/2}$. The classical 1/2-order high-pass filtering capacitive lattice scaling fracmemristor is a special case of the purely ideal v-order capacitive fracmemristor [224]. In addition, (7.82) further shows that to improve its approximation accuracy, an impedance, $-2r[q(s)]/\{r[q(s)]cs+1\}$, should be compensatory in the first series circuit of the v-order high-pass filtering capacitive lattice scaling fracmemristor shown in Fig. 7.8 (b), as illustrated in Fig. 7.12.

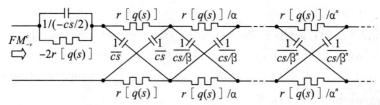

Fig. 7.12 Circuit configuration of v-order compensatory high-pass filtering capacitive lattice scaling fracmemristor.

Comparing Fig. 7.12 with Fig. 7.11, we can see that the first compensatory series circuit of the v-order compensatory high-pass filtering capacitive lattice scaling fracmemristor is the same as that of the v-order compensatory low-pass filtering one.

7.5 Experiment and Analysis

7.5.1 Approximation Performance of Arbitrary-Order Capacitive Lattice Scaling Fracmemristor

In this subsection, the effect of the number of series circuits on the approximation performance of an arbitrary-order low-pass filtering capacitive lattice scaling fracmemristor is analyzed, which is similar to that of an arbitrary-order high-pass filtering one.

Example 1: Without loss of generality, the approximate implementation of a 2/3-order low-pass filtering capacitive lattice scaling fracmemristor is illustrated. In particular, from (7.80), let's set $\alpha=4$ and $\beta=2$, we obtain $v=2/3$. Suppose the capacitance $c=1$. Equation (7.57) shows that the memristance $R[q(t)]$ and transmission function $H[q(t)]$ change with the input current $I_i(t)$ of a memristor. For the convenience of illustration, let's set the transmission function of a memristor $H[q(t)]=\delta(t)$, where $\delta(t)$ is an impulse signal. Thus, the reactance of this memristor $r[q(s)]=L\{H[q(t)]\}$. Thus, from Fig. 7.8 (a), the Bode diagrams of the fractional-order capacitive reactances of this 2/3-order low-pass filtering capacitive lattice scaling fracmemristor with m series circuits and the corresponding 2/3-order compensatory one can be shown as given in Fig. 7.13, respectively.

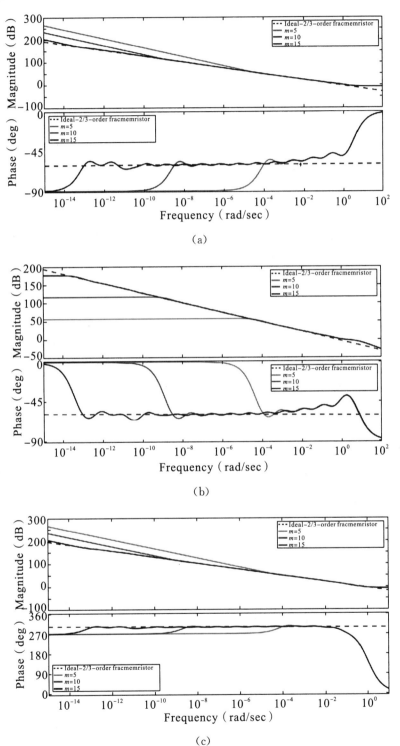

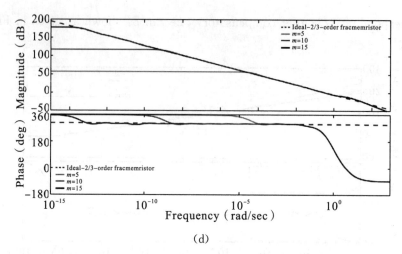

(d)

Fig. 7.13 Bode diagrams of fractional-order capacitive reactances of a 2/3-order low-pass filtering capacitive lattice scaling fracmemristor with m series circuits and corresponding 2/3-order compensatory one: (a) Open-circuit capacitive reactance $Z_{-2/3}^{om}$; (b) Short-circuit capacitive reactance $Z_{-2/3}^{sm}$; (c) Compensatory open-circuit capacitive reactance $Z_{-2/3}^{om}$; (d) Compensatory short-circuit capacitive reactance $Z_{-2/3}^{sm}$.

Figure 7.13 shows that at first, the optimum value of m is closely related to the pass-band of actually design. Both the magnitude and the phase of $Z_{-2/3}^{om}$ and $Z_{-2/3}^{sm}$ of the 2/3-order low-pass filtering capacitive lattice scaling fracmemristor with m series circuits and those of the corresponding 2/3-order compensatory one can approach those of the ideal 2/3-order capacitive fracmemristor with a high degree of accuracy in a certain pass-band. The larger m is, the wider the pass-band of the 2/3-order low-pass filtering capacitive lattice scaling fracmemristor is. When $m=15$, the pass-band is $[10^{-13} \text{rad/sec}, 10^{0} \text{rad/sec}]$. Secondly, the approximation performance of the 2/3-order compensatory low-pass filtering capacitive lattice scaling fracmemristor is more precise and smoother than that of the corresponding 2/3-order low-pass filtering one. As shown in Fig. 7.11 and Fig. 7.13, (7.79) converts to the regularized iterative scaling equation of a v-order compensatory low-pass filtering capacitive lattice scaling fracmemristor, which can be more accurately simplified as $\kappa s^{-v} = \alpha \kappa (\alpha \beta s)^{-v}$ when $s \to 0$. The intrinsic effect of the compensatory circuit in the first series circuit is to change the pole-zero location of the v-order low-pass filtering capacitive lattice scaling fracmemristor by means of zero pole cancellation. From aforementioned discussion, the larger m is, the wider the pass-band of the fractional-order capacitive lattice scaling fracmemristor is.

7.5.2 Electrical Characteristics of Fracmemristor

In this subsection, to recognize whether a circuit element is fracmemristor or not, the electrical characteristics of the fracmemristor, especially the fingerprint of a fracmemristor, are studied.

Example 2: Without loss of generality, let's take a current-controlled capacitive fracmemristor as an example to analyze the electrical characteristics of the fracmemristor. Let's assume that the input causal current sources $I_i(t) = \sin(at)u(t)$ applied across a memristor and a fracmemristor are identical, where a is frequency and $u(t)$ is a Heaviside function. Thus, $I_i(s) = a/(s^2 + a^2)$ and the corresponding quantity of electric charge $q(t) = D_t^{-1} I_i(t) = -(1/a)\cos(at)u(t)$. And assuming that $M[q(t)] = q(t)/2 + 1/a$. Thus, from (7.57), we have:

$$R[q(t)] = [M(q) + q\mathrm{d}M(q)/\mathrm{d}q] = [-(1/a)\cos(at) + 1/a]u(t). \quad (7.83)$$

Then, from (7.57) and (7.83), the instantaneous value of the input voltage of a memristor can be derived as:

$$V_i(t) = R[q(t)]I_i(t) = H[q(t)] * I_i(t)$$
$$= [-[1/(2a)]\sin(2at) + (1/a)\sin(at)]u(t), \quad (7.84)$$

where $R[q(t)] = \{H[q(t)] * I_i(t)\}/I_i(t)$. And assuming the initial state of this memristor is zero, thus the Laplace transform of (7.84) is as follows:

$$V_i(s) = r[q(s)]I_i(s) = 3a^2/[(s^2 + 4a^2)(s^2 + a^2)], \quad (7.85)$$

where the reactance of this memristor $r[q(s)] = L\{H[q(t)]\}$. For $I_i(s) = a/(s^2 + a^2)$, form (7.85), we have:

$$r[q(s)] = 3a/(s^2 + 4a^2). \quad (7.86)$$

The inverse Laplace transform of (7.86) is the transmission function $L^{-1}\{r[q(s)]\} = H[q(t)] = (3/2)\sin(2at)u(t)$ of a memristor. Further, in (7.60) and (7.62), for $v = \eta + \rho$, from (7.86), we obtain:

$$L^{-1}\{[r(q)]^{1-p}\} = (3a)^{1-p}\sqrt{\pi}/\Gamma(1-p)[t/(4a)]^{(1/2-p)}J_{(1/2-p)}(2at)u(t), \quad (7.88)$$

where $J_{(1/2-p)}(t)$ is a Bessel function of the first kind extended to non-integer orders by one of Schläfli's integrals. Then, substituting (7.88) into (7.62), the voltage-current relation equation of a v-order capacitive fracmemristor can be derived.

Therefore, from (7.57), (7.60), (7.62), (7.84) and (7.88), the constitutive relation of a fracmemristor can be analyzed in the $V_i - I_i$ plane. At first, let's illustrate the effect of the fractional-order v on the electrical characteristics of a fracmemristor. Let's set the frequency $a = 1$ rad/s and the time duration of $I_i(t)$ be equal to 6π. To illustrate the $V_i(t)$ curves of the memristor and the fracmemristor in the same plot, the experimental values of the fracmemristor are divided by 1000. To avoid completely overlapping the $V_i - I_i$ curve of a fracmemristor with that of a memristor and inconvenience of demonstration, let's set $v = 0.001$ when $v = 0$. Thus, the $V_i - I_i$ curve of the ideal v-order capacitive fracmemristor can be shown in Fig. 7.14.

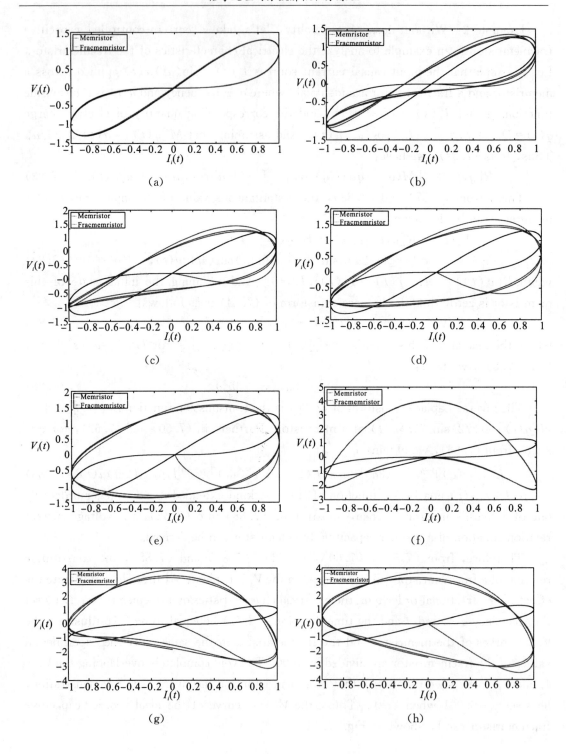

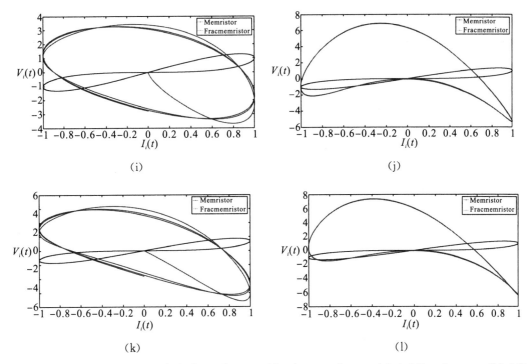

Fig. 7.14 $V_i - I_i$ curve of ideal v-order capacitive fracmemristors: (a) 0.001-order one; (b) 1/6-order one; (c) 1/3-order one; (d) 1/2-order one; (e) 2/3-order one; (f) 1-order one; (g) 4/3-order one; (h) 3/2-order one; (i) 5/3-order one; (j) 2-order one; (k) 5/2-order one; (l) 3-order one.

From Fig. 7.14, (7.58), (7.60), (7.62) and (7.88), we can observe that if the initial state of an ideal capacitive fracmemristor is zero and it is stimulated by a bipolar periodic signal with zero starting value, at first, when the fractional-order $v \to 0$, from (7.62) and (7.88), an ideal capacitive fracmemristor degenerates to an ideal memristor, which is lying on the point of M in Fig. 7.1. Thus, in Fig. 7.14 (a), the $V_i - I_i$ curve of an ideal 0.001-order capacitive fracmemristor almost overlaps that of an ideal memristor, which has multiple-valued Lissajous curves for all $I_i(t)$ except when it passes through the pinched point (0, 0). Secondly, when $0 < v$ is not a positive integer, from (7.62) and (7.88), the electrical properties of a capacitive fracmemristor fall in between those of the capacitor and those of the memristor [224]. The capacitive fracmemristor is lying on the line segment, S_4, between C and M in Fig. 7.1. In particular, on the one hand, when $0 < v < v_{th}$, the effect of memristor on the electrical properties of a capacitive fracmemristor is larger than that of capacitor, where v_{th} is a critical order that nonlinearly depends on the fractional-order v, the system parameter and original state of a capacitive fracmemristor as well as the amplitude and frequency of $I_i(t)$. Then, in Fig. 7.14 (b), the $V_i - I_i$ curve of an ideal 1/6-order capacitive fracmemristor has also a pinched hysteresis loop start from the original point of (0, 0). However, the pinched point of the multiple-valued Lissajous curves is no longer fixed at the point of (0, 0), but continuously drifts from the original point (0, 0) to the top of one side lobe of the $V_i - I_i$

curve along with the increasing of v. In Fig. 7.14 (c), when $v \to v_{th} \approx 1/3$, the pinched point disappears. On the other hand, when $v_{th} \leqslant v$, the effect of capacitor on the electrical properties of a capacitive fracmemristor is larger than that of memristor. Therefore, in Figs. 7.14 (d), 7.14 (e), 7.14 (g), 7.14 (h), 7.14 (i) and 7.14 (k), the $V_i - I_i$ curve of an ideal v-order capacitive fracmemristor is a nonlinear elliptic hysteresis loop, which has multiple-valued twisted ellipse curves for all $I_i(t)$ start from the original point of (0, 0). Thirdly, when v is a positive integer, from (7.60), (7.62) and (7.88), an ideal v-order capacitive fracmemristor degenerates to an integer-order integrator, whose electrical properties merely possess those of the integer-order integral [224]. The capacitive fracmemristor is lying on the point of C in Fig. 7.1. Then, we certainly obtain $V_i(t+T) = 0$ and $V_i(t+T/2) \neq 0$, where T is the period of the bipolar periodic signal. Thus, in Figures. 7.14 (f), 7.14 (j) and 7.14 (l), the $V_i - I_i$ curve of an ideal v-order capacitive fracmemristor is a nonlinear multiple-valued twisted elliptic hysteresis loop, which certainly passes through the point of (0, 0).

Secondly, let's illustrate the effect of the frequency a of the input causal current sources $I_i(t)$ on the electrical characteristics of a fracmemristor. Thus, keeping the aforementioned parameter settings unchanged, the electrical characteristics of an ideal 2/3-order capacitive fracmemristor with $a = 2$ rad/s, $a = 3$ rad/s, $a = 4$ rad/s and $a = 100$ rad/s, can be shown in Fig. 7.15.

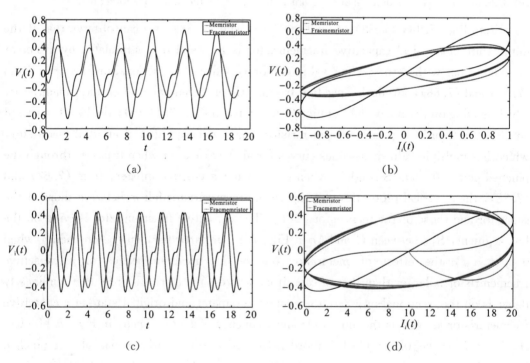

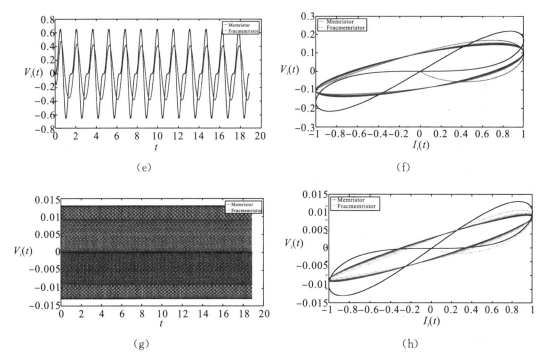

Fig. 7.15 Electrical characteristics of an ideal 2/3-order capacitive fracmemristors: (a) $V_i - t$ curve with $a = 2\text{rad/s}$; (b) $V_i - I_i$ curve with $a = 2\text{rad/s}$; (c) $V_i - t$ curve with $a = 3\text{rad/s}$; (d) $V_i - I_i$ curve with $a = 3\text{rad/s}$; (e) $V_i - t$ curve with $a = 4\text{rad/s}$; (f) $V_i - I_i$ curve with $a = 4\text{rad/s}$; (c) $V_i - t$ curve with $a = 100\text{rad/s}$; (d) $V_i - I_i$ curve with $a = 100\text{rad/s}$.

From Fig. 7.14 (e), Fig. 7.15, (7.58), (7.60), (7.62) and (7.88), we can observe that at first, the input voltage of acapacitive fracmemristor nonlinearly consists of both the fundamental component and other harmonic component of $I_i(t)$, whose weight nonlinearly depends on the fractional-order v, the system parameter and original state of a capacitive fracmemristor as well as the amplitude and frequency of $I_i(t)$. Therefore, the responses of the fracmemristor are nonlinearly periodic. In pure mathematical computations, the time duration of $s_1(t) * s_2(t)$ should be equal to the summation of the time duration of $s_1(t)$ and that of $s_2(t)$, where $s_1(t)$ and $s_2(t)$ are two analytic functions. However, for real causal circuits and systems, if we ignore the discharge process of the capacitor in a fracmemristor, when $I_i(t)$ has been powered off, a fracmemristor should not continue to run and its fracmemristance should remain unchanged. Therefore, although (7.62) has a mathematical computation of convolution, the time duration of $V_i(t)$ across a fracmemristor should be equal to the time duration of $I_i(t)$. Thus, in Fig. 7.15, the curve of $V_i(t)$ across a fracmemristor persists for the time of $t = 6\pi$. Secondly, in (7.62) and (7.88), the weight of each harmonic component of $I_i(t)$ nonlinearly is in inverse proportion to the frequency a of $I_i(t)$. The corresponding lobe area of the nonlinear elliptic hysteresis loop in the $V_i - I_i$ plane can be calculated by $\int V_i \mathrm{d}I_i$. Thus, in Figs. 7.15 (a), 7.15 (c), 7.15 (e) and 7.15 (g), the amplitude of the $V_i - t$ curve of a fracmemristor decreases with the increasing of the frequency a of $I_i(t)$. Meanwhile, in Figs. 7.15 (b), 7.15 (d), 7.15 (f) and 7.15 (h), the corresponding lobe area of the nonlinear elliptic hysteresis loop in the $V_i - I_i$ plane decreases with the increasing of the frequency a of $I_i(t)$. In particular, when $a \to \infty$, the nonlinear elliptic hysteresis loop of a fracmemristor in the $V_i - I_i$ plane shrinks to a straight line segment.

Therefore, from aforementioned discussion, the fingerprint of a fracmemristor can be summarized as four points: if the initial state of an ideal capacitive fracmemristor is zero and it is stimulated by a bipolar periodic signal with zero starting value, at first, if the fractional-order v is equal to zero, it degenerates to a memristor. Its $V_i - I_i$ curve is a pinched hysteresis loop, which is the same as that of the corresponding memristor. Secondly, if v is a positive fraction, when $0 < v < v_{th}$, its $V_i - I_i$ curve is also a pinched hysteresis loop, however, whose pinched point continuously drifts from the original point (0, 0) to the top of one side lobe with the increasing of v; Moreover, when $v_{th} \leq v$, the pinched point disappears, its $V_i - I_i$ curve is a nonlinear elliptic hysteresis loop started from the original point of (0, 0). Thirdly, if v is a positive integer, it degenerates to an integer-order integrator. Its $V_i - I_i$ curve is a nonlinear multiple-valued twisted elliptic hysteresis loop, which certainly passes through the point of (0, 0). Fourthly, the corresponding lobe area of its nonlinear elliptic hysteresis loop in the $V_i - I_i$ plane decreases with the increasing of the frequency a of $I_i(t)$. When $a \to \infty$, the nonlinear elliptic hysteresis loop of a fracmemristor in the $V_i - I_i$ plane shrinks to a straight line segment.

7.5.3 Analog circuit achievement of an arbitrary-order fracmemristor

In this subsection, without loss of generality, a 2/3-order high-pass filtering capacitive lattice scaling fracmemristor is implemented in the form of analog circuit.

Example 3: For the verification of the actualexistence of a fracmemristor, we implement a fracmemristor in the form of analog circuit, which satisfied its corresponding electrical characteristics in Example 2, respectively. From (7.80), when $\alpha = 4$ and $\beta = 2$, we obtain $v = 2/3$. Let's suppose that the memristance and capacitance in Fig. 7.12 are $R[q(t)]$ and $c = 16\mu F$, respectively. Since $R[q(t)] = \{H[q(t)] * I_i(t)\}/I_i(t)$ and $r[q(s)] = L\{H[q(t)]\}$ in (7.57), the memristance $R[q(t)]$ keeps up the same multiple growths as the reactance $r[q(s)]$ of a memristor. Thus, from Fig. 7.8 (b) and Fig. 7.12, we can implement a 2/3-order high-pass filtering capacitive lattice scaling fracmemristor with 5 open-circuit series circuits and a 2/3-order compensatory one, as shown in Fig. 7.16.

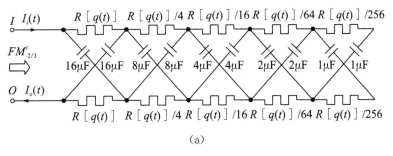

(a)

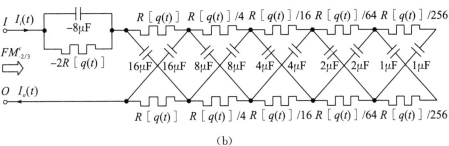

(b)

Fig. 7.16 Analog series circuit of a 2/3-order high-pass filtering capacitive lattice scaling fracmemristor: (a) 2/3-order one; (b) 2/3-order compensatory one.

In Fig. 7.16, I and O are the input port and the output port of the analog series circuit of a 2/3-order high-pass filtering capacitive lattice scaling fracmemristor, respectively. Figure 7.16 shows that at first, in Fig. 7.16 (b), the negative memristance and the negative capacitance can be achieved by negative impedance converter. Secondly, a key technical problem of the analog circuit implementation of a 2/3-order high-pass filtering capacitive lattice scaling fracmemristor is to synchronously achieve the scaling memristance $R[q(t)]/\alpha^n$ without being affected by the branch-current or branch-voltage of its $(n+1)th$ series circuit. Thus, we should implement a floating voltage-controlled memristor that implements $R[q(t)] = [-(1/a)\cos(at) + 1/a]u(t)$ in (7.83), as shown in Fig. 7.17.

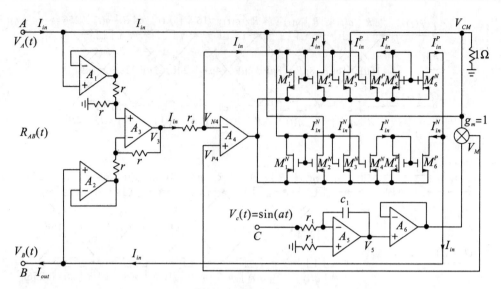

Fig. 7.17 A floating voltage-controlled memristor.

In Fig. 7.17, A, B and C are the input port, the output port and the control voltage source input port of a floating voltage-controlled memristor, respectively. A_1, A_2 and A_6 are three voltage followers, A_3 is a subtracter, A_4 is a voltage-current converter and A_5 is a phase-reversing integrator. $V_A(t)$, $V_B(t)$, $V_C(t)$, I_{in} and I_{out} are the input voltage source, output voltage source, control voltage source, input current and output current of a floating voltage-controlled memristor, respectively. M_1^P, M_2^P, M_3^P, M_4^P, M_5^N and M_6^N are a multichannel positive Current Mirror (CM) copies the positive part of I_{in}. M_1^N, M_2^N, M_3^N, M_4^N, M_5^P, and M_6^P are a multichannel negative CM copies the negative part of I_{in}. Thus, we obtain $I_{in} = I_{out}$. Let's set $r_s = 1/a$, $r_1 c_1 = 1$ and multiplier gain $g_M = 1$. Thus, according to the virtual short and virtual off electrical characteristics of an operational amplifier, we can derive that $V_3 = V_A - V_B$, $V_5 = -1/(r_1 c_1) \int \sin(at) \, dt = 1/a \cos(at) u(t)$, $V_{CM} = -I_{in}$, $V_{N4} = V_{P4} = V_M = g_M V_{CM} V_5 = -1/a \cos(at) I_{in} u(t)$ and $I_{in} = (V_3 - V_{N4})/r_s$. Thus, we have:

$$V_3 = V_A - V_B = [-1/a \cos(at) + 1/a] I_{in} u(t). \tag{7.89}$$

Thus, from (7.89), we obtain $R_{AB}(t) = R[q(t)] = [-(1/a) \cos(at) + 1/a] u(t)$. Note that for the convenience of actual implementation, we can also use a programmable voltage-controlled resistance to achieve $R_{AB}(t) = R[q(t)] = [-(1/a) \cos(at) + 1/a] u(t)$ directly.

Thus, in Fig. 7.17, let's set $a = 100 \text{Hz}$, $r = 1\Omega$, $r_s = 0.01\Omega$, $r_1 = 1\text{M}\Omega$, $c = c_1 = 1\mu\text{F}$ and $I_i(t) = I_o(t) = \sin(at) u(t)$. Let's select the type of the operational amplifiers $A_1 - A_6$, N-type MOSFET $M_1^P - M_6^P$, P-type MOSFET $M_1^N - M_6^N$ and multiplier are OP37G, ALD1116PAL, ALD1117PAL and AD633ANZ, respectively. Therefore, from Fig. 7.16 and Fig. 7.17, using commonly used PCB design software of Multisim13, we can simulate the actual analog series circuits of a 2/3-order high-pass filtering capacitive lattice scaling fracmemristor and a 2/3-order compensatory one, whose electrical characteristics can be shown in Fig. 7.18.

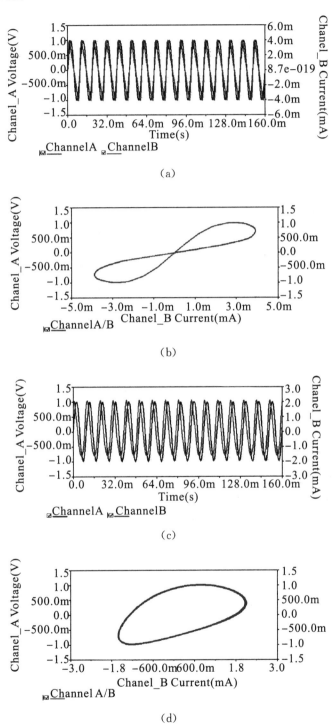

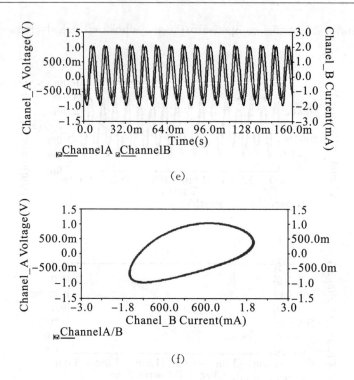

Fig. 7.18. Electrical characteristics of actual analog series circuits of a 2/3-order high-pass filtering capacitive lattice scaling fracmemristors: (a) $V_i - t$ curve of memristor in 2/3-order capacitive fracmemristor; (b) $V_i - I_t$ curve of memristor in 2/3-order capacitive fracmemristor; (c) $V_i - t$ curve of 2/3-order capacitive fracmemristor; (d) $V_i - I_i$ curve of 2/3-order capacitive fracmemristor; (e) $V_i - t$ curve of 2/3-order compensatory capacitive fracmemristor; (f) $V_i - I_i$ curve of 2/3-order compensatory capacitive fracmemristor.

In Fig. 7.18 (a), Channel_A is the input voltage $[V_A(T) - V_B(T)]$ and Channel_B is the input current $I_{in}(t)$ in Fig. 7.17. In Fig. 7.18 (b), Channel_B is the input current $I_{in}(t)$ and Channel_A is the input voltage $[V_A(t) - V_B(t)]$ in Fig. 7.17. In Figs. 7.18 (c) and 7.18 (e), Channel_A is the input voltage $[V_I(t) - V_O(t)]$ and Channel_B is the input current $I_{in}(t)$ in Fig. 7.16. In Figs. 7.18 (d) and 7.18 (f), Channel_B is the input current $I_{in}(t)$ and Channel_A is the input voltage $[V_I(t) - V_O(t)]$ in Fig. 7.16. Comparing Fig. 7.18 with Fig. 7.14 (e) and Fig. 7.15, we can observe that at first, the electrical characteristics of a floating voltage-controlled memristor in Fig. 7.17 is approximately identical to those of the memristor in (7.83). The $V_i - t$ curves and $V_i - I_i$ curves of an actual 2/3-order capacitive fracmemristor accord with those of a theoretical ideal one. Secondly, the electrical characteristics of actual analog series circuits of a 2/3-order high-pass filtering capacitive lattice scaling fracmemristor in Fig. 7.16 are approximately identical to those of a corresponding ideal 2/3-order capacitive fracmemristor in (7.62) and (7.88). For the memristor in a 2/3-order capacitive fracmemristor, there is a pinched hysteresis loop in the $V_i - I_i$ plane, which passes through the pinched point (0, 0). For the corresponding 2/3-order capacitive fracmemristor, there is a nonlinear twisted ellipse curve in the $V_i - I_i$ plane. Note that since we let Multisim13 doesn't begin to illustrate until circuits perform stably, the nonlinear twisted ellipse curves in Figs. 7.18 (d) and 7.18 (f) don't start from the point of (0, 0). Thirdly, in comparison with an ideal 2/3-order capacitive fracmemristor, the approximation performance of the 2/3-order compensatory high-pass filtering capacitive lattice scaling fracmemristor is more precise and smoother than that of the corresponding 2/3-order high-pass filtering one. Therefore, the simulation results of Example 3 effectively verify the actual existence of a fracmemristor.

Example 4: For further analysis the electrical characteristics of an arbitrary-order capacitive fracmemristor, we should firstly achieve a versatile floating equivalent circuit of memristor achieved by a voltage-controlled linear resistor. A floating voltage-controlled linear resistor can be achieved by Junction Field Effect Transistor (JFET), as shown in Fig. 7.19.

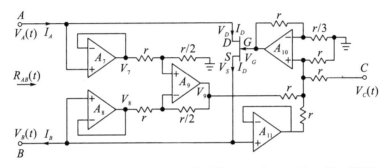

Fig. 7.19 A floating voltage-controlled linear resistor achieved by JFET.

In Fig. 7.19, A, B and C are the input port, the output port and the control voltage source input port of a floating voltage-controlled mirror linear resistor achieved by JFET, respectively. A_7, A_8 and A_{11} are three voltage followers, A_9 is a subtracter, A_{10} is a phase-identical adder, $V_C(t)$ is a control voltage source and I_D is the drain current of a JFET. Thus, according to the virtual short and virtual off electrical characteristics of an operational amplifier, we can derive that $V_7 = V_A$, $V_8 = V_B$, $V_D = V_A$, $V_S = V_B$, $V_9 = (V_A - V_B)/2$, $V_G = (V_A + V_B)/2 + V_B + V_C$, and $I_A = I_D = I_B$. Thus, $V_{GS} = (V_A - V_B)/2 + V_C = V_{DS}/2 + V_C$. Further, for a N-channel JFET, it should satisfy $V_{DS} > 0$. When $0 < V_{GS} < V_P$ and $|V_{DS} - V_{GS}| < V_P$, it works in variable resistance region (triode region), in which its drain-source current can be expressed as $I_D = \{2I_{DSS}[(V_{GS} - V_p)V_{DS} - V_{DS}^2/2]\}/V_P^2$, where I_{DSS} is the saturation current at zero gate-source voltage and V_P is the pinch-off voltage of a JFET. Then, we obtain:

$$R_{AB}(T) = R_{DS}(T) = V_{DS}/I_{DS} = V_P^2/[2I_{DSS}(V_{GS} - V_P - V_{DS}/2)] = V_P^2/[2I_{DSS}(V_C - V_P)].$$
(7.90)

Equation (7.90) shows that a floating voltage-controlled linear resistor achieved by JFET in Fig. 7.19 is linearly controlled by $V_C(t)$. In theory, the control voltage source of a floating voltage-controlled linear resistor, $V_C(t)$, can be arbitrary in the variable resistance region of a JFET except for $V_C = V_P$. Note that with regard to a N-channel JFET, since $V_{DS} > 0$, $0 < V_{GS} < V_P$ and $|V_{GD}| = |V_{GS} - V_{DS}|$, from (7.90), V_A, V_B and V_C should satisfy $V_A > V_B$, $-(V_A - V_B)/2 < V_C < V_P - (V_A - V_B)/2$ and $|V_C - (V_A - V_B)/2| < V_P$. Note that since the control voltage source $V_C(T)$ is generated by an equivalent circuit of the memristor, the memristance $R[q(t)]$ of a floating voltage-controlled linear resistor achieved by JFET cannot be controlled by its input potential difference $(V_A - V_B)$. No matter what $(V_A - V_B)$, the memristance $R[q(t)]$ of a floating voltage-controlled linear resistor achieved by JFET is always controlled by $V_C(t)$. Thus, in this case, the floating voltage-controlled linear resistor achieved by JFET is actually a three-port mirror memristor.

Further, using the aforementionedfloating voltage-controlled linear resistor, we can achieve a versatile floating equivalent circuit of memristor. Then, $R_{AB}(t) = R[q(t)]$. Without loss of generality, let's choose a control voltage source $V_C(t)$ achieved by operational amplifier and multiplier, as shown in Fig. 7.20.

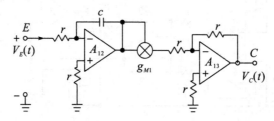

Fig. 7.20 A control voltage source $V_C(t)$.

In Fig. 7.20, E are the input port of a control voltage source. The port C in Fig. 7.20 is identical with the port C in Fig. 7.18. A_{12} is a phase-reversing integrator and A_{13} is a phase-reversing proportioner. g_M is the multiplier gain of a multiplier. Thus, we have:

$$V_C(t) = -g_M[-1/(rc)\int V_E(t)dt]^2 u(t). \tag{7.91}$$

To further analyze the electrical characteristics of the fractional-order capacitive fracmemristor, let's construct a 1/2-order compensatory high-pass filtering capacitive lattice scaling fracmemristor. In (7.80), let's set $\alpha = 2$ and $\beta = 2$. Then, we obtain $v = 1/2$. In (7.90), let's set $V_E(t) = \sin(at) u(t)$, $g_M = 1$, $r = 1\Omega$ and $c = 100\mu F$. Thus, from Fig. 7.12, Fig. 7.19 and Fig. 15, using Multisim13, we can simulate the actual analog series circuits of a 1/2-order compensatory high-pass filtering capacitive lattice scaling fracmemristor, whose electrical characteristics can be shown in Fig. 7.21.

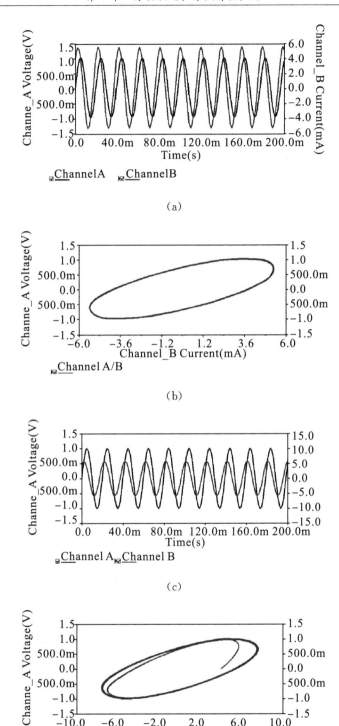

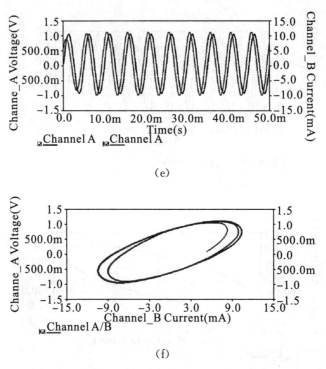

Fig. 7.21 Electrical characteristics of actual analog series circuits of a 1/2-order compensatory high-pass filtering capacitive lattice scaling fracmemristor: (a) $V_i - t$ curve when $a = 50\text{Hz}$; (b) $V_i - I_i$ curve when $a = 50\text{Hz}$; (c) $V_i - I_i$ curve when $a = 100\text{Hz}$; (d) $V_i - t$ curve when $a = 100\text{Hz}$; (e) $V_i - I_i$ curve when $a = 200\text{Hz}$; (f) $V_i - I_i$ curve when $a = 200\text{Hz}$.

In Figs. 7.21 (a), 7.21 (c) and 7.21 (e), Channel _ A is the input voltage $[V_I(t) - V_O(t)]$ and Channel _ B is the input current $I_{in}(t)$ in Fig. 7.12. In Figs. 7.21 (b), 7.21 (d) and 7.21 (f), Channel _ B is the input current $I_{in}(t)$ and Channel _ A is the input voltage $[V_I(t) - V_O(t)]$ in Fig. 7.12. Example 4 shows that at first, the responses of a 1/2-order capacitive fracmemristor are nonlinearly periodic which vary with different driving frequency. From Fig. 7.7 and Fig. 7.12, we can see that a 1/2-order compensatory high-pass filtering capacitive lattice scaling fracmemristor can be considered in a certain way as a nonlinear interpolation of the memristor and capacitor. Equation (7.62) shows that the v-order capacitive fracmemristor implements the convolution of $L^{-1}\{[r(q)]^{1-p}\}$ and $D_t^{-v}I_i(t)$. Thus, in Figs. 7.21 (a), 7.21 (c) and 7.21 (e), the time variations of the voltage across a 1/2-order capacitive fracmemristor depend on the convolution of its input current history and its fractional calculus. Further, according to the theory of fractional calculus, the fractional integral of the input current across a 1/2-order compensatory capacitive fracmemristor, $D_t^{-1/2}I_i(t)$, suppresses its high frequency singular noise. From (7.60) and (7.62), we can see that the electrical characteristics of a capacitive fractor provide a physical mechanism of the fractional-order smoothness, which enables a capacitive fracmemristor to protect its memory states under the influence of noise. Secondly, the $V_i - t$ curves and $V_i - I_i$ curves of an actual 1/2-order capacitive fracmemristor accord with those of a theoretical ideal one. Comparing with Fig. 7.14 (d) and Figs. 7.21 (d) and 7.21 (f), we can see that with regard to a 1/2-order compensatory capacitive fracmemristor, there should be a nonlinear twisted ellipse curve in the $V_i - I_i$ plane, which start from the point of (0, 0). Note that since we let Multisim13 doesn't begin to illustrate until circuits perform stably, the nonlinear twisted ellipse curves in Figs. 7.21 (b), 7.21 (d) and 7.21 (f) don't start from the point of (0, 0). Therefore, the simulation results of Example 4 effectively verify the actual existence of a fracmemristor.

Example 5: For further discussion, from Fig. 7.12, Fig. 7.19 and Fig. 15, keeping the parameter settings of Example 4 unchanged, let's use Tektronix oscilloscope TDS 1012C − EDU to demonstrate, the physical realization of a 1/2-order high-pass filtering capacitive lattice scaling fracmemristor can be achieved, as shown in Fig. 7.22.

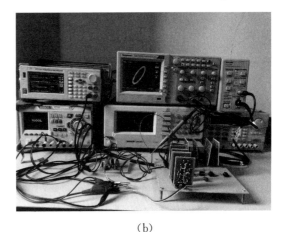

(a)　　　　　　　　　　　　　　　(b)

Fig. 7.22　Physical realization of a 1/2-order high-pass filtering capacitive lattice scaling fracmemristor: (a) $V_i - t$ curve; (b) $V_i - I_i$ curve.

Therefore, from Fig. 7.22, the actual characteristics of the physical realization of a 1/2-order high-pass filtering capacitive lattice scaling fracmemristor can be tested, as shown in Fig. 7.23.

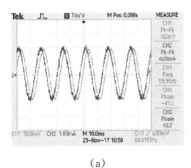
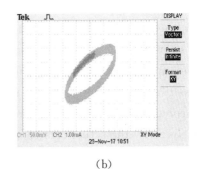

(a)　　　　　　　　　　　　　　　(b)

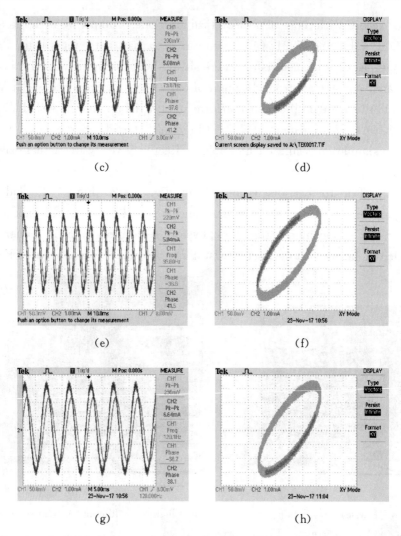

Fig. 7.23 Actual electrical characteristics of physical realization of a 1/2-order high-pass filtering capacitive lattice scaling fracmemristor: (a) $V_i - t$ curve when $a=60$Hz; (b) $V_i - I_i$ curve when $a=60$Hz; (c) $V_i - I_i$ curve when $a=80$Hz; (d) $V_i - t$ curve when $a=80$Hz; (e) $V_i - I_i$ curve when $a=100$Hz; (f) $V_i - t$ curve when $a=100$Hz; (g) $V_i - I_i$ curve when $a=80$Hz; (h) $V_i - t$ curve when $a=80$Hz.

Note that to obtain the enough number of points of the experimental results in Fig. 7.23, let's set the time persist of display to be infinite. Comparing Fig. 7.23 with Fig. 7.21, we can see that the actual characteristics of the physical realization of a 1/2-order high-pass filtering capacitive lattice scaling fracmemristor perfectly consistent with those of its theoretical results. These physical realization experimental results of Example 5 also effectively verify the actual existence of a fracmemristor.

7.6 Conclusions and Discussion

From aforementioned mathematical derivation, experimental and analysis, we can observe that at first, the proposed arbitrary-order lattice scaling fracmemristor is a feasible hardware achievement of an arbitrary-order memristor. In addition, there should be some other type arbitrary-order scaling fracmemristors in their natural implementations, such as an arbitrary-order tree type scaling fracmemristor, arbitrary-order two-circuit type scaling fracmemristor, arbitrary-order H type scaling fracmemristor [225], [405], and so on. Secondly, the electrical characteristics of an arbitrary-order fracmemristor depend on the convolution of its input current history and its fractional calculus. Thus, a capacitive fracmemristor and inductive fracmemristor can be considered in a certain way as a nonlinear interpolation of the memristor and capacitor and that of the memristor and inductor, respectively. Thus, the fingerprint of an arbitrary-order fracmemristor is different from that of a memristor. Thirdly, from the mathematical analysis of the relationship between the memristance and transmission function of a memristor, $R[q(t)]$ and $H[q(t)]$, in (7.57), we can further see that since authors' clerical mistake, the reactance of a memristor $r[q(s)]$ is the Laplace transform of $H[q(t)]$ rather than that of $R[q(t)]$ in literature [224]. However, when $r[q(s)]$ is considered as the Laplace transform of $H[q(t)]$ in (7.60) and (7.61), the generic fractional-order driving-point impedance functions of an arbitrary-order capacitive fracmemristor and inductive fracmemristor in their natural implementations derived in literature [224] are still correct. Fourthly, from the physical analysis of real causal circuits and systems in Example 2, although there is a convolution in (7.62) in pure mathematics, the time duration of $V_i(t)$ across a fracmemristor should be equal to the time duration of $I_i(t)$. Thus, since the insufficient understanding of authors' preliminary research, the prediction characteristics of a fracmemristor concluded in literature [224] are non-existent.

Fracmemristor extends the concepts of the classical memristor. Until recently, the applications of fracmemristor were mainly in the domain of physics and mathematics [224], [236], [271] − [273], [639] − [643]. With respect to the state-of-the-art analog circuit implementation and application of the fracmemristor, there are many other fascinating issues else need to be further studied. For instance, at first, there should be some other circuit configurations of an arbitrary-order fracmemristors, as well as the fracmemristor based adaptive intelligence and learning systems, neural networks and weighted feedback systems, spiking-timing-dependent plasticity experimentation, chaotic systems, etc. Secondly, rules for series and parallel connections of two fracmemristors are more profound than those of two memristors [652], [653]. From (7.60) and (7.61), we can see that the fracmemristance nonlinearly changes with the instantaneous value of the reactance of the memristor contained in each fracmemristor. Further, the fractional-order of each fracmemristor could be also different. Thus, when two fracmemristors are connected in series or parallel, the electrical characteristics of the composite circuit of fracmemristor are quite complicated, which extremely depend on the fractional-order, circuit parameters, the frequency and amplitude of the input voltage or input current of a fracmemristor. Thirdly, the circuit design of the fracmemristor is also more complex than that of the memristor [654], [650]. From Fig. 7.8, we can observe that a v-order capacitive lattice scaling fracmemristor is a floating two-port circuit element, which can be feasibly embedded in a fracmemristive circuit. In a similar way to the memristive circuit [650], we can deal with frequency-domain model of the fracmemristor, and achieve its steady-state analysis by means of the harmonic-balance method [650]. Since the fractional-order of a fracmemristive filter can be chosen arbitrarily, the fractional adaptive multi-scale filtering capability of a fracmemristor based filter is a major advantage that is superior to the conventional memristive filters and LTI filters. It will be further discussed in detail in our future work.

第 8 章　任意分数阶神经型脉冲振荡器

首先，本章在分析 1/2 阶网格型模拟分抗电路自相似特点的基础上提出了主值分抗电路的概念。其次，论述了 1/2 阶网格型主值分抗电路与 T 型、Π 型、桥 T 型主值分抗电路之间的等效转换，用晶体谐振体和差接变量器实现 1/2 阶网格型主值分抗电路的方法，以及 1/2 阶网格型模拟分抗电路模型的频率特性。最后，提出并论述了用模拟分抗电路构造任意分数阶仿生神经型脉冲振荡器的概念与模型。实验中，以 1/2 阶网格型模拟分抗电路为例构造了 1/2 阶仿生神经型脉冲振荡器，其输出的仿生神经脉冲信号比较正确地模仿了实际生物神经脉冲振荡信号的波形特征。

8.1　问题提出

神经科学的研究表明，神经元之间的信息传递不是以静电感应方式进行的，而是一种电脉冲刺激。处于兴奋状态的神经元所发放的电脉冲频率不仅取决于刺激强度，而且取决于神经元的本征特性。不论哪种类型的生物神经元，所发放的电脉冲都是等幅、恒宽的离散脉冲信号。细胞膜的电位是连续变化的电信号。

构造各种仿生神经型脉冲振荡器的研究成果可以为后天由于创伤或疾病导致失明的盲人重建部分视觉而设计助视器，用仿生神经脉冲信号对大脑视皮层进行电脉冲刺激提供人工视觉。为了能够在空间和时间上组成正确的图案，必须正确仿效视神经冲动对视皮层的刺激。同理，还可以设计出针对聋人的助听器，针对下肢瘫痪病人的助行器。上述仿生器的实现依赖于对生物神经元冲动产生的电脉冲信号的正确仿效。R. W. Newcomb 等人在神经元的电特性方面做了大量的研究工作[278]。Kiruthi 在 1983 年提出了滞迟神经型脉冲振荡器[279]后，相继出现了一些神经型振荡器[280-281]。但是，上述这些神经型脉冲振荡器的输出波形或者是方波，或者是矩形波，其电脉冲宽度和脉冲频率受到输入刺激强度—膜电位的控制。这与生物神经元发放恒定脉宽、非矩形波非方波电脉冲信号的实际情况是不相符的。

用普通的电路元件如电阻、电容或电感按照某种高度自相似的分形结构可以构造出模拟分抗电路来实现信号的分数阶演算（分数阶微积分）。在第 4 章中笔者已经提出并论述了网格型、两回路型、H 型三种任意分数阶模拟分抗电路[171-174]，那么，如何运用这些模拟分抗电路来构造任意分数阶仿生神经型脉冲振荡器，使其输出的仿生神经脉冲信号能够比较正确地模仿实际生物神经脉冲振荡信号的波形特征[179]，是本章研究的主要目的。

8.2 1/2阶网格型模拟分抗的等效实现及其电路特性

由第4章的分析可知,网格型1/2阶递归模拟分抗电路在电路结构上是本章所研究的四种1/2阶分抗电路中最优的一种。不失一般性,本章以网格型模拟分抗电路为例来研究如何构造分数阶神经型脉冲振荡器。由于网格型1/2阶递归模拟分抗电路是由平衡电桥构成的,这种平衡的条件的要求是十分苛刻的,实际加工时难以实现。因此,为了解决这个问题,就必须对其进行电路等效。

8.2.1 1/2阶网格型模拟分抗的晶体谐振体实现

由图4.11、图4.12可知,1/2阶网格型模拟分抗电路模型是由网格型平衡电路单元无限重复构成的,于是可以仿照DFT变换中类似的概念,把图4.11、图4.12中无限重复的网格型对称平衡电路单元称为网格型模拟主值分抗电路[193]。与此同理,把其他类型模拟分抗电路模型中无限重复的电路单元称为其他类型的模拟主值分抗电路。如果令 $Z_a = Z_{1M}$,$Z_b = Z_{2M}$,则1/2阶网格型分抗的网格型模拟主值分抗电路如图8.1所示。

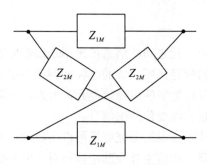

图8.1 网格型模拟主值分抗电路

因此,对1/2阶网格型分抗某些电路特性的研究可以转化为对其网格型模拟主值分抗电路的研究。于是,对网格型分抗与其他类型模拟分抗之间的转换研究就可以转化为对网格型主值分抗与其他类型主值分抗之间的转换研究,这样就避免了电路的高度自相似性给研究带来的不必要的麻烦。

T型、Π型、桥T型主值分抗电路分别如图8.2、图8.3、图8.4所示。

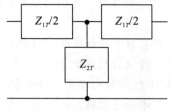

图8.2 T型主值分抗电路

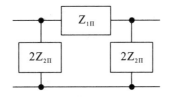

图 8.3 Ⅱ 型主值分抗电路

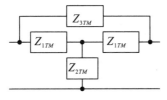

图 8.4 桥 T 型主值分抗电路

显然，如果两个四端网络的特性阻抗和传输常数相等，则这两个四端网络彼此等效[193,282−289]。于是可以推得 1/2 阶网格型模拟主值分抗电路和 T 型主值分抗电路之间的转换公式为

$$Z_{1T} = 2Z_{1M} \tag{8.1}$$

$$Z_{2T} = \frac{Z_{2M} - Z_{1M}}{2} \tag{8.2}$$

1/2 阶网格型模拟主值分抗电路和 Ⅱ 型主值分抗电路之间的转换公式为

$$Z_{1\Pi} = \frac{2Z_{1M}Z_{2M}}{Z_{2M} - Z_{1M}} \tag{8.3}$$

$$Z_{2\Pi} = \frac{Z_{2M}}{2} \tag{8.4}$$

1/2 阶网格型模拟主值分抗电路和桥 T 型主值分抗电路之间的转换公式为

$$Z_{1TM} = \sqrt{Z_{1M}Z_{2M}} \tag{8.5}$$

$$Z_{2TM} = \frac{Z_{2M} - \sqrt{Z_{1M}Z_{2M}}}{2} \tag{8.6}$$

$$Z_{3TM} = \frac{2Z_{1M}\sqrt{Z_{1M}Z_{2M}}}{\sqrt{Z_{1M}Z_{2M}} - Z_{1M}} \tag{8.7}$$

显然，把 T 型、Ⅱ 型、桥 T 型转换成等效的网格型模拟分抗电路都是可能的。但是，相反的变换并不是永远可能的。只有当网格型主值分抗能用一臂阻抗 Z_{2M} 与另一臂阻抗 Z_{1M} 串联得到可实现的阻抗时，网格型主值分抗电路才可以转换为等效的 T 型主值分抗。只有当网格型主值分抗的两臂阻抗 Z_{2M} 和 Z_{1M} 并联后得到可实现的阻抗时，网格型主值分抗电路才可以转换为等效的 Ⅱ 型主值分抗。

众所周知，晶体谐振体的等效电路如图 8.5 所示[193,282−289]。

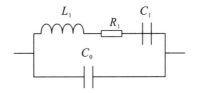

图 8.5 晶体谐振体的等效电路

图 8.5 中，等效电感 L_1 和等效电容 C_1 的数值取决于晶体片的几何尺寸以及晶体的某些物理常数。损耗电阻 R_1 与晶体薄片的加工过程、夹紧方法有关。C_0 是电极支架以及组合体的静电容。令 $Z_{1M}=r$ 或 $Z_{1M}=c$，$Z_{2M}=c$ 或 $Z_{2M}=r$，其等效的 T 型、Ⅱ型、桥 T 型主值分抗电路中必定出现复阻抗的情况。适当地调整晶体薄片的切割方式以及加工工艺，可以得到与之相应的复阻抗。由于晶体谐振体具有高度的稳定度和品质因数，因此制成的晶体分抗具有很好的电气特性。

8.2.2　1/2 阶网格型模拟分抗的差接变量器实现

众所周知，在通信系统中，如果适当调节变量器的变换系数可以使四端网络与负载相匹配，从而消除两个四端网络之间的电流耦合，达到消除一个纵向不对称的非平衡四端网络对其他电路的影响的目的。信号处理中的变量器与电力变压器相比，主要的差别是前者不像电力变压器那样工作在一个频率上，而是工作在很宽的频带内。变量器的结构如图 8.6 所示[193,282-289]。

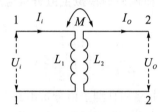

图 8.6　变量器的结构

忽略变量器线圈和铁心内的损失与漏磁以及线圈的匝间电容，假设变量器的线圈电阻与其通过的电流大小无关，根据 KVL 可得

$$\begin{cases} U_i = I_i j\omega L_1 - I_O j\omega M \\ U_O = I_i j\omega M - I_O j\omega L_2 \end{cases} \quad (8.8)$$

假设没有漏磁，即漏磁系数 $\sigma=0$，那么绕组之间的耦合系数 $k=\sqrt{1-\sigma}=1=M/\sqrt{L_1 L_2}$，$M=\sqrt{L_1 L_2}$。将它们代入式（8.8），便可求得其电压变换系数为 $n=U_i/U_O=\sqrt{L_1/L_2}$。令次级绕组的负载阻抗为 Z_O，则可求得次级绕组的输出电流为 $I_O=U_O/Z_O=U_i/nZ_O$。于是求得初级绕组的输入电流为 $I_i=U_i/j\omega L_1+U_i/n^2 Z_O$，这样便可求得其电流的比值为 $I_O/I_i=j\omega L_1 n/(n^2 Z_O+j\omega L_1)$。当次级绕组短路时，$Z_O=0$，$I_O/I_i=n$。于是得 1↔1 端的短路输入阻抗为 $Z_{i1SC}=U_i/I_i=n^2 Z_O$。

网格型主值分抗需要平衡的条件是很苛刻的。当采用变量器来构造差接格型分抗电路时，不仅可以避免苛刻的电路平衡条件，而且可使其包含的元件数量减半。本章用差接格型电路对 1/2 阶网格型模拟主值分抗进行电路等效，如图 8.7 所示[193,282-289]。

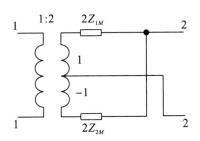

图 8.7 1/2 阶差接格型主值分抗电路

可以求得 1/2 阶网格型模拟主值分抗电路 1↔1 端的开路输入阻抗 Z_{i1CC} 以及短路输入阻抗 Z_{i1SC} 分别为

$$\begin{cases} Z_{i1CC} = (Z_{1M} + Z_{2M})/2 \\ Z_{i1SC} = 2Z_{1M}Z_{2M}/(Z_{1M} + Z_{2M}) \end{cases} \quad (8.9)$$

相应地，对于差接格型主值分抗电路而言，可以推得

$$\begin{cases} Z'_{i1CC} = 2(Z_{1M} + Z_{2M})n^2 = (Z_{1M} + Z_{2M})/2 \\ Z'_{i1SC} = 2Z_{1M}2Z_{2M}/(2Z_{1M} + 2Z_{2M}) = 2Z_{1M}Z_{2M}/(Z_{1M} + Z_{2M}) \end{cases} \quad (8.10)$$

对比式（8.9）和（8.10）可知，$Z_{i1CC} = Z'_{i1CC}$，$Z_{i1SC} = Z'_{i1SC}$，所以两电路等效。另外，本章还提出了两种用变量器实现的 1/2 阶网格型主值分抗等效电路，如图 8.8 所示[193,282−289]。

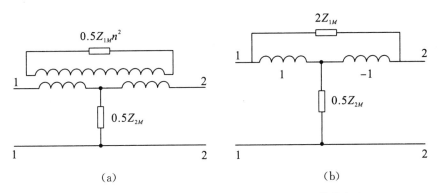

图 8.8 用变量器实现的 1/2 阶网格型主值分抗等效电路

同理，其推导从略。可见，在图 8.8（b）中，没有差接变量器，所以这种等效电路在实际生产时更容易加工。

8.2.3 1/2 阶网格型模拟主值分抗的频率特性

令四端网络的固有传输常数为 $g = b + ja$，其中，b 为电路功率、电压或电流的固有衰耗，a 为电路电压或电流的固有相移。对于 1/2 阶网格型模拟主值分抗，可得 $\operatorname{th}\dfrac{g}{2} = \sqrt{Z_{1M}/Z_{2M}}$，于是得 $\operatorname{th}\dfrac{g}{2} = \dfrac{\operatorname{sh}b}{\operatorname{ch}b + \cos a} + j\dfrac{\sin a}{\operatorname{ch}b + \cos a} = \sqrt{\dfrac{Z_{1M}}{Z_{2M}}}$ [193,282−289]。

令 $Z_{1M} = r$，$Z_{2M} = 1/j\omega C$，可得

$$\begin{cases} \dfrac{\mathrm{sh}^2 b - \sin^2 a}{(\mathrm{ch}\,b + \cos a)^2} = 0 \\ \dfrac{2\mathrm{sh}\,b\,\sin a}{(\mathrm{ch}\,b + \cos a)^2} = \omega r c \end{cases} \quad (8.11)$$

显然，$\mathrm{ch}\,b = \dfrac{\mathrm{e}^b + \mathrm{e}^{-b}}{2} \geqslant 1$。当 $b=0$ 时，$\mathrm{ch}\,b = 1$，若相移 $a = \pi$，则 $\mathrm{ch}\,b + \cos a = 0$，式 (8.11) 无意义。当 $\mathrm{sh}^2 b - \sin^2 a = 0$ 时，式 (8.11) 成立。所以 $\mathrm{ch}^2 b = 1 + \mathrm{sh}^2 b = 1 + \sin^2 a$，代入式 (8.11) 得

$$\frac{\mathrm{sh}\,b\,\sin a}{a + \mathrm{ch}\,b\,\cos a} = \omega r c \quad (8.12)$$

在实际无源电路中固有衰耗 $b \geqslant 0$，所以 $\mathrm{sh}\,b = \dfrac{\mathrm{e}^b - \mathrm{e}^{-b}}{2} \geqslant 0$，$\mathrm{sh}\,b = \sin a$（$\sin a \geqslant 0$，$0 \leqslant a \leqslant \pi$），代入式 (8.12)，得 $\dfrac{\sin^2 a}{1 + \cos a \sqrt{1 + \sin^2 a}} = \omega r c$，即

$$\frac{\sin^2 a}{1 \pm \sqrt{1 - \sin^4 a}} = \omega r c \quad (8.13)$$

等式中出现 ± 号的根本原因在于 $\sqrt{j} = \mathrm{e}^{\pi/2 + 2k\pi}$，$k=0,1$ 存在两个复对称的根。当 $\omega = 0$ 时，可得 $\mathrm{sh}\,b = 0$，$\sin a = 0$，即 $b = 0$，$a = 0$。此时 1/2 阶网格型模拟主值分抗电路没有固有损耗和固有相移。如前所述，1/2 阶网格型模拟分抗电路的输入阻抗（特性阻抗）为

$$Z_{1/2} = (Z_{1M} Z_{2M})^{1/2} = \sqrt{\frac{r}{\omega c}} \mathrm{e}^{j\frac{-\pi/2 + 2k\pi}{2}} = \begin{cases} \sqrt{\dfrac{r}{\omega c}} \mathrm{e}^{-j\frac{\pi}{4}}, & k=0 \\ \sqrt{\dfrac{r}{\omega c}} \mathrm{e}^{j\frac{3\pi}{4}}, & k=1 \end{cases} \quad (8.14)$$

可见，当 $k=0$ 时，电路呈现电容性，在 $\omega=0$ 点网格型分抗由电感性转变为电容性，$\omega=0$ 为电流谐振；当 $k=1$ 时，电路呈现电感性，在 $\omega=0$ 点网格型分抗由电容性转变为电感性，$\omega=0$ 为电压谐振。

令 $Z_{1M} = 1/j\omega c$，$Z_{2M} = r$，可得

$$\begin{cases} \dfrac{\mathrm{sh}^2 b - \sin^2 a}{(\mathrm{ch}\,b + \cos a)^2} = 0 \\ \dfrac{2\mathrm{sh}\,b\,\sin a}{(\mathrm{ch}\,a + \cos a)^2} = -1/\omega r c \end{cases} \quad (8.15)$$

于是得 $\dfrac{\mathrm{sh}\,b\,\sin a}{1 + \mathrm{ch}\,b\,\cos a} = -1/\omega r c \xrightarrow{\omega \to \infty} 0$。可见，当 $k=0$ 时，$\omega = \infty$ 为电流谐振；当 $k=1$ 时，$\omega = \infty$ 为电压谐振。

8.3 基于分数阶演算的分数阶神经型振荡器

不失一般性，本章以声带外展肌——环杓后肌的神经电脉冲信号为例，如图 8.9 所示。

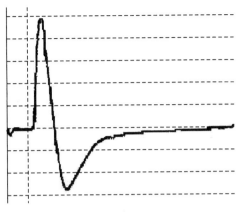

图 8.9　环杓后肌的神经电脉冲信号

为了真实地模拟生物神经元的行为特征，本章提出了用模拟分抗构造仿生分数阶神经型脉冲振荡器模型[193]，如图 8.10 所示。

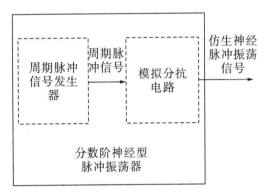

图 8.10　分数阶神经型脉冲振荡器模型

周期脉冲信号发生器产生脉冲宽度和脉冲频率可调的电脉冲。模拟分抗电路对信号进行分数阶微分，其分数阶数和内部电路元件参数根据待模拟的神经脉冲波形特征作相应调整。

8.4　实验仿真及结果分析

不失一般性，本章以 1/2 阶网格型模拟分抗电路为例构造出 1/2 阶分数阶神经型脉冲振荡器进行实验。同理，用其他分数阶次的模拟分抗电路可以构造出其他分数阶次的分数阶神经型脉冲振荡器。本章采用的 1/2 阶网格型模拟分抗电路的级数仍是 5 级，其结构如图 4.20 所示。

本章利用周期脉冲信号发生器分别产生了近似的 δ、三角尖波两种信号。适当调整网格型 1/2 阶模拟分抗电路中的元件参数，产生了相应的仿生神经脉冲信号。用 Orcad 电路软件进行仿真实验，在示波器上的实验结果如图 8.11 所示。

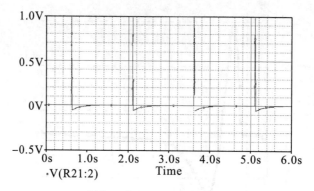

(a) 周期δ信号的 1/2 阶微分结果 1

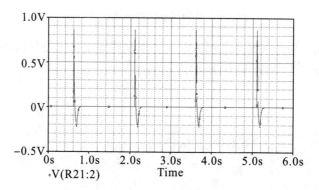

(b) 周期δ信号的 1/2 阶微分结果 2

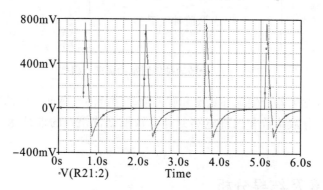

(c) 周期三角尖波的 1/2 阶微分结果 1

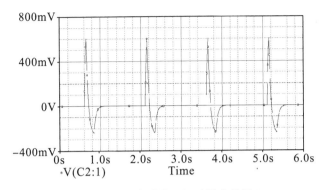

(d) 周期三角尖波的 1/2 阶微分结果 2

图 8.11 分数阶神经脉冲信号实验结果

对比图 8.11 和图 8.9 可以发现，本章提出的分数阶神经型脉冲振荡器能够比较正确地模仿真实生物神经冲动。

自然地，通过实验，另外一个新的问题摆在面前。不同生物的不同神经元、不同生物的相同神经元乃至同一生物相同的神经元，由于这样和那样的差别，其产生的神经脉冲振荡信号在波形上存在差别。那么，用什么分数阶神经型脉冲振荡器及什么样的内部参数可以各自正确地描述它们呢？它们之间存在什么样的物理关系呢？这是还需要进一步深入研究的课题。

8.5 小结

笔者在本章中提出了分数阶仿生神经型脉冲振荡器的模型与概念，并论述了怎样用模拟分抗电路构造任意分数阶仿生神经型脉冲振荡器模型。实验中，1/2 阶仿生神经型脉冲振荡器输出的仿生神经脉冲信号比较正确地模仿了实际生物神经脉冲振荡信号的波形特征。但是，这方面还有许多工作需要进一步研究，比如：

(1) 诸如混沌理论、分形理论、分数阶傅立叶变换等许多新兴学科的研究结果越来越促使人们猜想自然界的许多自然现象是不是本来就是分数维的呢？用传统的整数阶的方法解决不好的许多自然现象，用分数阶的方法是不是可以得到比较好的解决呢？用模拟分抗电路构造仿生分数阶神经型脉冲振荡器，其实验结果能够比较正确地模仿真实生物神经冲动。于是我们猜想，实际的生物神经冲动的内在本质规律是不是本来就是分数维的呢？因此用分数阶的方法才能比较好地逼近其真实面目呢？

(2) 如上所述，$1/n$ 阶模拟分抗的输入阻抗（分抗）存在 n 个关于复平面中单位圆对称的两个复根，它们的幅值相同，相位不同。那么，这 n 个复根各自在什么情况下分别出现呢？怎么利用这个性质来构造基于相位记忆的分数阶联想记忆器呢？笔者还在对上述这些问题作进一步的研究。

(3) 用什么分数阶神经型脉冲振荡器及什么样的内部参数可以各自正确地描述各种神经元产生的神经脉冲振荡信号？

第 9 章　分数阶 Hopfield 神经网络：分数阶动态联想递归神经网络

多层动态联想神经网络与双层联想记忆神经网络之间既有相似，又有区别。多层动态联想神经网络是不仅有反馈，更有跨连接的任意结构，在结构上比双层联想记忆神经网络复杂得多。有跨连接任意结构的多层动态联想神经网络的客观存在决定了对其突触权值训练算法研究的必要性。本章从物理结构模型上分析了多层动态联想神经网络与经典的 Hopfield 神经网络存在的相似之处，提出了将其转换为经典的 Hopfield 神经网络进行训练的方法，并从数学上证明了运用基于广义 Hebb 规则的学习算法对它进行训练的稳定性，其仿真实验的收敛效果良好。

如上所述，可以将多层动态联想神经网络转换为经典的 Hopfield 神经网络进行训练。于是，可以很自然地想到，是否可以将经典的 Hopfield 神经网络从传统的整数阶推广到分数阶领域来进行研究呢？本章的另一个研究目的就是将传统的基于整数阶微积分的经典 Hopfield 神经网络拓广到基于分数阶微积分的分数阶 Hopfield 神经网络，即用模拟分抗元件来构造任意分数阶的多层动态联想神经网络。换言之，本章将只在传统的整数维空间中研究的多层动态联想神经网络模型推广到分数维空间中去。本章提出并论证了基于分数阶微积分的任意分数阶多层动态联想神经网络模型，提出并论证了其 Lyapunov 能量函数的收敛性，从而证明了它的稳定性以及联想记忆性。通过实验仿真与结果分析，很好地证实了上述理论论述。本章的研究表明，当分数阶的多层动态联想神经网络的阶数取为整数时，传统的基于整数阶微积分的经典 Hopfield 神经网络其实是分数阶的多层动态联想神经网络的一个特例。

9.1　问题提出

神经动力学把一种让时间通过反馈以隐含方式嵌入的神经网络视为非线性动力系统。非线性动力系统的稳定性是整个系统的特性，稳定性意味着在系统的各个独立部分之间存在某种形式的协调[290]。有界输入和有界输出（BIBO）稳定性准则非常适合于线性系统，但是，由于嵌入神经元结构之中的饱和非线性使得所有的这种非线性动态系统都是 BIBO 稳定的，所以 BIBO 稳定性准则不适用于神经网络。非线性动态系统的稳定性都是在 Lyapunov 意义下的稳定。

众所周知，所有图灵机（Turing）都可建立在用 sigmoid 激活函数的神经元上的完全连接递归网络模型[291]。对于有外部输入的非线性自回归模型（NARX），若具有一个隐藏层单元，其激活函数为有界和单侧饱和（BOSS）且有一个线性输出神经元，那么不计线性延迟，它可以模拟用完全连接的具有 BOSS 激活函数的递归网络[292]。另外，有一个隐藏层神经元组，激活函数为 BOSS 函数及一个线性输出神经元的 NARX 网络是和 Turing

等价的[293]。但是，当网络的体系结构受到限制时，递归网络的计算能力就不再成立。由此可见，神经网络中神经元的构造方式是和训练网络的学习算法紧密联系的。因此，用于神经网络设计的学习规则是被构造出来的[294]。那么，如何对多层动态联想神经网络触突权值的学习算法进行构造，并证明其稳定性呢？

众所周知，经典的 Hopfield 神经网络以及其他多层动态联想神经网络都是建立在整数阶微积分运算的基础之上的[277,295−296]。由第 5 章的分析可知，对脉冲信号分数阶微分的结果与实际的生物神经脉冲非常相似，由模拟分抗电路元件可以构造出分数阶的神经型脉冲振荡器。进而，可以很自然地想到，能否用模拟分抗电路来构造分数阶的多层动态联想神经网络呢？能否在以上基础之上构造分数阶混沌神经网络呢[298]？本章的主要工作集中在研究和构造任意分数阶的多层动态联想神经网络。

9.2 多层动态联想神经网络

一般有三种神经网络的图形表示法，即方框图[299]、信号流图[299]和结构图[300]。方框图提供神经网络的功能描述，信号流图提供网络中完全的信号流描述，结构图描述网络布局。

多层动态联想神经网络是既有反馈又有跨连接的任意结构，这种结构存在其客观必要性[301]。该网络中的反馈可以从多层感知器的输出层到输入层，也可以由网络的隐层神经元到输入层。当感知器有多个隐层时，全局反馈的可能形式甚至可以进一步扩大[294]。使用电解存储器和 VLSI 技术在多层神经网络中实现可变权值是困难的[302]，但随着集成度的增加，现有的技术水平可完全实现图 9.1 中的多层神经网络[301]。

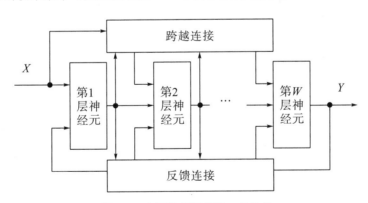

图 9.1　多层神经网络的一般结构

作为其特例，递归多层感知器（RMLP）的每一个计算层对它的邻近层有一个反馈，它有一个或多个隐藏层，这种静态多层感知器比那些使用单个隐藏层的感知器更有效和节约[303]。

显然，多层动态联想神经网络存在的必要性决定了对其突触权值训练算法研究的必要性。不失一般性，本章以三层动态联想神经网络（即只有一个隐藏层）为例进行研究，其一般结构的方框图如图 9.2 所示。

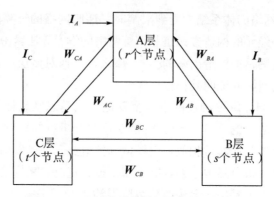

图 9.2 三层动态联想神经网络的一般结构

虽然神经网络的输入没有累加器、偏置向量和传输函数，但是它和隐藏层之间却包括待学习训练的突触权值矩阵，所以一些学者也把神经网络的输入看作一层，本章采用此观点。在图 9.2 中，可以任取 A、B、C 层之一为神经网络的输入层、隐藏层、输出层。图 9.2 中各层的外界输入向量用 I 表示，任意两层之间的权值矩阵为 W。通常，每层的输入个数并不一定等于该层中神经元的数目。在多层动态联想神经网络中，若第 i 层有 R 个输出，第 $i+1$ 层有 S 个神经元，则第 i 层神经元的输出向量通过权值矩阵 $_{ji}W$ 反馈给第 $i+1$ 层神经元：

$$_{ji}W = \begin{bmatrix} _iW_{1,1} & _iW_{1,2} & \cdots & _iW_{1,R} \\ _iW_{2,1} & _iW_{2,2} & \cdots & _iW_{2,R} \\ \vdots & \vdots & & \vdots \\ _iW_{S,1} & _iW_{S,2} & \cdots & _iW_{S,R} \end{bmatrix} \tag{9.1}$$

矩阵 $_{ji}W$ 中元素的行下标代表该权值相应连接输出的目的神经元，列下标代表该权值相应连接的输入源神经元。

9.3 基于广义 Hebb 规则的多层动态联想神经网络学习算法

9.3.1 理论分析

9.3.1.1 从物理模型上分析

众所周知，任何算法和理论都有其深刻的物理解释，所以本节首先从物理模型上分析三层动态联想神经网络的学习算法。令图 9.2 中 A 层为输入层，B 层为隐藏层，C 层为输出层，可将其转化为结构图，如图 9.3 所示。

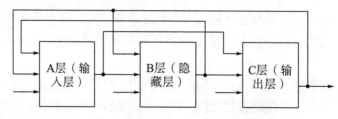

图 9.3 三层动态联想神经网络的结构图

如何对图 9.3 所示的三层动态联想神经网络进行训练呢？显然，三层动态联想神经网络的网络结构和通常的双向联想存储器有一定的相似性，其任何两层间的相互关系类似于双向联想神经网络两层之间的相互关系。一种很自然的想法是：是否可以将多层动态联想网络的任意一层看作双向联想存储器的任意一层，把多层动态联想网络的其余各层看作双向联想存储器的另外一层，采用基于外积法的双向联想存储器的学习规则来训练三层乃至多层动态联想网络的突触权值矩阵呢？多层动态联想神经网络中任意两层间的等效信号流图如图 9.4 所示。

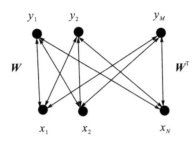

图 9.4　多层动态联想神经网络中任意两层间的等效信号流图

基于上面这种想法，多层动态联想网络突触权值的训练算法可采用基于外积法的双向联想存储器的学习规则来训练。令其等效的双层联想记忆神经网络（BAM）存的是一组向量对 $(\boldsymbol{x}_i, \boldsymbol{y}_i)$，$i=1, 2, \cdots, p$，可正向或逆向联想，正向通过 $\boldsymbol{W}$，逆向通过 $\boldsymbol{W}^{\mathrm{T}}$，如图 9.4 所示。设 $\boldsymbol{x} \in \{-1, +1\}^N$，$\boldsymbol{y} \in \{-1, +1\}^M$，令其阈值为 0 的情况[302]，一般其学习规则为

$$\begin{cases} \boldsymbol{W} = \sum_{K=1}^{p} \boldsymbol{x}^{\mathrm{T}} \boldsymbol{y} \\ \boldsymbol{W} = \sum_{K=1}^{p} \boldsymbol{y}^{\mathrm{T}} \boldsymbol{x} \end{cases} \quad (9.2)$$

其系统的 Lyapunov 函数为

$$E(\boldsymbol{x}, \boldsymbol{y}) = -\frac{1}{2} \boldsymbol{x} \boldsymbol{W} \boldsymbol{y}^{\mathrm{T}} - \frac{1}{2} \boldsymbol{y} \boldsymbol{W}^{\mathrm{T}} \boldsymbol{x}^{\mathrm{T}} = -\boldsymbol{x} \boldsymbol{W} \boldsymbol{y}^{\mathrm{T}} \quad (9.3)$$

以三层动态联想记忆神经网络为例，若输入层 A 共有 r 个神经元，隐藏层 B 共有 s 个神经元，输出层 C 共有 t 个神经元，其相应各层的输出向量 $\boldsymbol{a}$，$\boldsymbol{b}$，$\boldsymbol{c}$ 分别为

$$\boldsymbol{a} = f(\boldsymbol{W}_{\mathrm{BA}}^{\mathrm{T}} \boldsymbol{b} + \boldsymbol{W}_{\mathrm{CA}} \boldsymbol{c}) \quad (9.4)$$

$$\boldsymbol{b} = f(\boldsymbol{W}_{\mathrm{AB}}^{\mathrm{T}} \boldsymbol{a} + \boldsymbol{W}_{\mathrm{CB}} \boldsymbol{c}) \quad (9.5)$$

$$\boldsymbol{c} = f(\boldsymbol{W}_{\mathrm{AC}}^{\mathrm{T}} \boldsymbol{a} + \boldsymbol{W}_{\mathrm{BC}} \boldsymbol{b}) \quad (9.6)$$

具体运用式（9.3）训练该网络突触权值时，为了方便使用，可以规定其相应各层的输入向量矩阵和输出向量矩阵分别为式（9.7）、（9.8）、（9.9），使其构成增广矩阵。

$$\boldsymbol{X} = (\boldsymbol{a}, \boldsymbol{b}, 0), \quad \boldsymbol{Y} = (0, 0, \boldsymbol{c}) \quad (9.7)$$

$$\boldsymbol{X} = (0, \boldsymbol{b}, \boldsymbol{c}), \quad \boldsymbol{Y} = (\boldsymbol{a}, 0, 0) \quad (9.8)$$

$$\boldsymbol{X} = (\boldsymbol{a}, 0, \boldsymbol{c}), \quad \boldsymbol{Y} = (0, \boldsymbol{b}, 0) \quad (9.9)$$

但是，图 9.3 中不仅存在前馈和反馈连接，而且存在跨越连接。显然，其网络结构比通常的双向联想存储器复杂得多，因此，当多层动态联想网络的层数很多时，简单套用双

向联想存储器的学习规则来训练多层动态联想网络不一定会收敛。本章从"简单性原则"出发,先画出多层动态联想神经网络的等效信号流图,如图 9.5 所示[297]。

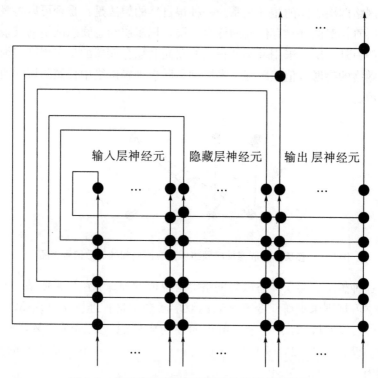

图 9.5 多层动态联想神经网络的等效信号流图

分析图 9.5 可知,多层动态联想神经网络和 Hopfield 神经网络有一定的相似性。于是,本章提出[297]:将多层动态联想神经网络等效为经典的 Hopfield 神经网络,只取经典的 Hopfield 神经网络输出层与多层动态联想神经网络对应神经元的输出作为多层动态联想神经网络的输出。这样就可以把对多层动态联想神经网络训练这样一个复杂的问题巧妙地转化为对其等效 Hopfield 网络的训练,从而使问题得到简化。显然,当其等效 Hopfield 神经网络的训练收敛时,等效 Hopfield 神经网络对应神经元的输出恰好就是对多层动态联想神经网络训练收敛时的输出;等效 Hopfield 神经网络对应各神经元间的突触权值矩阵恰好就是对多层动态联想神经网络训练收敛时各神经元间的突触权值矩阵。很自然地,本章采用改进的基于广义 Hebb 规则的训练算法对其突触权值进行训练。

9.3.1.2 从数学推导上分析

Hebb 假设一条突触两侧的两个神经元同时被激活,那么突触的强度将会增大。换言之,如果一个正的输入 p_j 产生一个正的输出 a_i,那么应该增加 w_{ij} 的值[298]。令学习速度为 α,如果只利用包含被更新权值的神经网络层的信号,于是可得

$$w_{ij}(n) = w_{ij}(n-1) + \alpha a_i(n) p_j(n) \tag{9.10}$$

式 (9.10) 没有考虑连续提交输入使权值趋于无限大与实际生物系统的悖论,所以需要加入权值的衰减项,即

$$w_{ij}(n) = w_{ij}(n-1) + \alpha a_i(n) p_j(n) - \gamma w_{ij}(n-1) \tag{9.11}$$

如果刺激没有不断重复,联想就会丢失。因此,规定只有当神经元是活跃时才允许权

值衰减，这样权值仍被限制，但遗忘被减到最小[305]。令迭代次数为 n，衰减速度为 γ，可得

$$w_{ij}(n) = w_{ij}(n-1) + \alpha a_i(n) p_j(n) - \gamma a_i(n) w_{ij}(n-1) \tag{9.12}$$

令新的权值的学习速度与旧的权值的衰减速度相同，故 $\gamma = \alpha$。于是可得

$$w_{ij}(n) = w_{ij}(n-1) + \alpha a_i(n)[p_j(n) - w_{ij}(n-1)] \tag{9.13}$$

写成向量形式为

$$_i\boldsymbol{W}(n) = {_i\boldsymbol{W}}(n-1) + \alpha a_i(n)[p(n) - {_i\boldsymbol{W}}(n-1)] \tag{9.14}$$

如图 9.5 所示，若多层动态联想神经网络共有 m 个隐藏层，各隐藏层有 n_i（$i=1$，$2,\cdots,m$）个神经元，若干隐藏层共有 s 个神经元，输入层共有 r 个神经元，输出层共有 t 个神经元。具体用式（9.14）训练该网络时，须令其输入向量矩阵为式（9.15），输出向量矩阵为式（9.16），权值矩阵为式（9.17），即

$$\boldsymbol{X} = [x_1, \cdots, x_r, x_{r+1}, \cdots, x_{r+s}, x_{r+s+1}, \cdots, x_{r+s+t}] \tag{9.15}$$

$$\boldsymbol{Y} = [0, \cdots, 0, 0, \cdots, 0, y_{r+s+1}, \cdots, y_{r+s+t}] \tag{9.16}$$

$$\boldsymbol{W} = [\boldsymbol{W}_{r\times(r+s+t)}, \boldsymbol{W}_{n_1\times(r+t+s)}, \cdots, \boldsymbol{W}_{n_m\times(r+t+s)}, \boldsymbol{W}_{t\times(r+t+s)}] \tag{9.17}$$

令在图 9.5 中总共有 N 个神经元，各神经元均采用 $M-P$ 模型，且无自反馈 $w_{i,i} = 0$；令整个网络的状态为 $\boldsymbol{S} = [s_1, s_2, \cdots, s_N]^{\mathrm{T}}$，$n$ 时刻各神经元的电压累积为 $v_j(n)$，神经元 j 的状态输出为 $s_j(n)$，门限为 θ_j；令神经元的传输函数是符号函数 $\mathrm{sgn}[\cdot]$，各神经元节点的输出值归一化为 $+1$ 或 -1，于是有

$$\begin{cases} v_j = \sum_{i=1}^{N} w_{j,i} s_i - \theta_j \\ s_j = \mathrm{sgn}[v_j] \end{cases} \tag{9.18}$$

其状态变化方程为

$$s_j(n+1) = \mathrm{sgn}\Big[\sum_{i=1}^{N} w_{j,i} s_i(n) - \theta_j\Big] \tag{9.19}$$

令权值矩阵对称，故 $w_{i,j} = w_{j,i}$，且其对角线上的元素非负。显然，当 $\boldsymbol{S}(n+1) = \boldsymbol{S}(n)$ 时，网络的状态不再改变，神经网络趋于稳定，故有

$$s_j = \mathrm{sgn}\Big[\sum_{i=1}^{N} w_{j,i} s_i - \theta_j\Big], \quad j = 1, 2, \cdots, N \tag{9.20}$$

令系统的 Lyapunov 函数为

$$E(n) = -\frac{1}{2} \boldsymbol{S}^{\mathrm{T}}(n) \boldsymbol{W} \boldsymbol{S}(n) + \boldsymbol{S}^{\mathrm{T}}(n) \cdot \boldsymbol{\theta} \tag{9.21}$$

故有 $\Delta \boldsymbol{E}(n) = \boldsymbol{E}(n+1) - \boldsymbol{E}(n)$ 和 $\Delta \boldsymbol{S}(n) = \boldsymbol{S}(n+1) - \boldsymbol{S}(n)$ 成立。令任一 n 时刻只有一个神经元 i 改变其状态 s_i，故 $\Delta \boldsymbol{S}(n)$ 中唯一不为零的元素是 Δs_i，故有

$$\boldsymbol{E}(n+1) = -\frac{1}{2}\boldsymbol{S}(n)^{\mathrm{T}}\boldsymbol{W}\boldsymbol{S}(n) + \boldsymbol{S}(n)^{\mathrm{T}} \cdot \boldsymbol{\theta} - \frac{1}{2}\boldsymbol{S}(n)^{\mathrm{T}}\boldsymbol{W}\Delta\boldsymbol{S}(n) - \frac{1}{2}\Delta\boldsymbol{S}(n)^{\mathrm{T}}\boldsymbol{W}\boldsymbol{S}(n)$$

$$-\frac{1}{2}\Delta\boldsymbol{S}(n)^{\mathrm{T}}\boldsymbol{W}\Delta\boldsymbol{S}(n) + \Delta\boldsymbol{S}(n)^{\mathrm{T}} \cdot \boldsymbol{\theta} \tag{9.22}$$

忽略 $\Delta \boldsymbol{E}(n)$ 的二次项，可推得

$$\Delta \boldsymbol{E}(n) = -\Delta \boldsymbol{S}_i(n)^{\mathrm{T}} \Big[\sum_{i=1}^{N} w_{i,j} \boldsymbol{S}_j(n) - \boldsymbol{\theta}\Big] \tag{9.23}$$

$$\Delta s_i(n) = \begin{cases} 0, & s_i = \mathrm{sgn}\Big[\sum_{i=1}^{N} w_{i,j} s_j(n) - \theta_i \Big] \\ -2, s_i(n) = 1, \mathrm{sgn}(\cdot) = -1 \\ 2, & s_i(n) = -1, \mathrm{sgn}(\cdot) = 1 \end{cases} \quad (9.24)$$

因此，$\Delta E(n) \leqslant 0$，即 $E(n)$ 随状态变化且单调递减，它会收敛到能量的极小点，从而证明了本算法的收敛性。

9.3.2 实验仿真及结果分析

不失一般性，本章以每层只有一个神经元的三层动态联想神经网络进行仿真实验，其网络结构如图 9.6 所示。

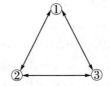

图 9.6 有三个神经元的动态联想神经网络

图中①②③分别为输入神经元、隐藏神经元、输出神经元。对于该三层动态联想神经网络，设状态向量为 $S = (s_1, s_2, s_3)$，给定的待存向量为 $S^{(1)} = (1, -1, 1)$ 和 $S^{(2)} = (-1, 1, -1)$，并设域值 θ_j 为 0。令其初始网络权值为零，用本算法对突触权值进行学习训练，其权值收敛过程的图形表示如图 9.7 所示。

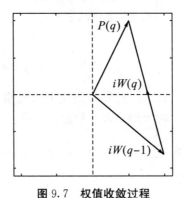

图 9.7 权值收敛过程

网络收敛后，其权值矩阵为

$$W = \begin{bmatrix} 0 & -\frac{2}{3} & +\frac{2}{3} \\ -\frac{2}{3} & 0 & -\frac{2}{3} \\ +\frac{2}{3} & -\frac{2}{3} & 0 \end{bmatrix} \quad (9.25)$$

为了便于在三维空间中观测网络的稳定性规律，本实验令状态向量中任意分量为常量，对其 Lyapunov 函数进行研究。任取其中 $E(s_1, s_2)$ 和 $E(s_1, s_3)$ 的等势曲线和二维曲面图以演示本算法的稳定性，其他类似，如图 9.8、图 9.9 所示。

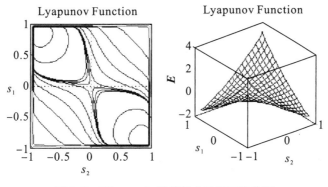

图 9.8　$E(s_1, s_2)$ 的等势曲线和二维曲面

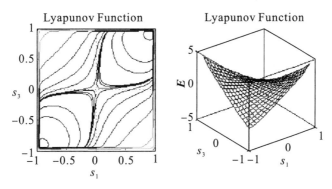

图 9.9　$E(s_1, s_3)$ 的等势曲线和二维曲面

可见，W 满足对称且对角线元素非负，所以有稳定点。网络有 3 个节点，所以共有 $2^3=8$ 个状态（域值取为 0）。其中只有（1，-1，1）和（-1，1，-1）是稳定状态，因为它满足式（9.20），其余状态都会收敛到与 Hamming 距离最小的稳定状态。该网络具有一定的纠错能力，当测试（-1，-1，1），（1，1，1），（1，-1，-1）时，它们会收敛到（1，-1，1）；当测试（1，1，-1），（-1，-1，-1），（-1，1，1）时，它们会收敛到（-1，1，-1）。

9.4　分数阶 Hopfield 神经网络：分数阶多层动态联想神经网络[①]

本节主要讨论一种新颖的基本概念——分数阶 Hopfield 神经网络（fractional Hopfield neural network，FHNN）。众所周知，分数阶微积分被引入人工神经网络主要是因为其具有的长时记忆与非定域性。一些学者为此做了有益的尝试并取得了显著的成效。因此，一种很自然的想法是：如何将一阶 Hopfield 神经网络推广到分数阶，以及如何运用分数阶微积分来实现 FHNN。我们引入一种新颖的数学方法——分数阶微积分，用以实现 FHNN。第一，我们用模拟电路形式实现分抗；第二，我们应用分抗和分数阶

---

① PU Yifei, ZHANG Yi, ZHOU Jiliu. Fractional Hopfield Neural Networks：Fractional DynamicAssociative Recurrent Neural Networks [J]. IEEE Transactions on Neural Networks and Learning Systems，2017，28（10）：2319－2333.

最速下降法实现 FHNN，构建其 Lyapunov 函数，并进一步分析其吸引子；第三，我们通过实验分析了其稳定性和收敛性，并进一步讨论了其在锁相回路和关于防伪的抗芯片克隆中的应用。我们研究的主要贡献在于：用模拟电路的形式和分数阶最速下降法提出了 FHNN，构造了其 Lyapunov 函数，证明了 Lyapunov 的稳定性，分析了其吸引子，并同时发现了 FHNN 在关于防伪的抗芯片克隆中的显著优势。

9.4.1 INTRODUCTION

It is well known that the classical first-order Hopfield neural networks, HNN, is one of the most influential neural networks [277], [295] — [296], [350]. The circuit configuration of HNN's first-order neuron is based on a first-order integral circuit. Each first-order neuron of HNN consists of one operational amplifier and its related capacitor and resistors. Each first-order neuron has the same circuit configuration. There are many classical applications of HNN in content addressable memory [295], analogue-to-digital converters [323], linear programming [296], and so on. Meanwhile, there are also many dynamic associative memories being closely related to HNN, such as the Li-Michel neural networks [324], [325], bidirectional associative memories [326], [327], and more. Furthermore, with the widespread application of HNN, some model modifications of HNN, such as the high-order Hopfield neural networks [328] — [331], fuzzy Hopfield neural networks [332] — [333], and stochastic Hopfield neural networks [334] — [335], are proposed, respectively. In addition, fractional calculus has been incorporated into artificial neural networks, mainly because of its long-term memory and nonlocality. Some researchers have made interesting attempts at fractional neural networks and gained competitive advantages over the integer-order neural networks. For instance, N. Özdemir et al. proposed a new type of activation function for a complex valued neural network [336]. Abdulaziz Alofi et al. studied the finite-time stability of Caputo fractional neural networks with distributed delay [337]. E. Kaslik discussed the stability analysis of the fractional-order neural networks of Hopfield type [338]. Ran Zhang et al. discussed a fractional-order financial system based on a fractional-order three-dimensional Hopfield type neural network [339]. M. A. Z. Raja et al. proposed stochastic techniques [340] — [341] as well as evolutionary techniques [342] — [343] for the solution of the fractional-order systems represented by fractional differential equations, respectively. In these approaches, feed-forward artificial neural networks are employed for accurate mathematical modeling. The advantage of these approaches is that the solution of fractional differential equations is available on the domain of continuous inputs unlike the other integer-order calculation based numerical techniques. Therefore, it is one naturally to ponder how to generalize HNN to the fractional-order ones, and how to implement the fractional Hopfield neural networks, FHNN, by means of fractional calculus. This manuscript discusses a novel conceptual framework: FHNN.

In over the past 300 years, fractional calculus has been an important novel branch of mathematical analyses [3], [176], [178] — [179], [230] — [231]. Fractional calculus is as old as the integer one, although up to recently, its application was exclusively in the field of mathematics. It looks to as if fractional calculus is a promising mathematical method for physical scientists and engineering technicians. Scientific study has shown that a fractional-order or fractional dimensional approach is now the best description for many natural phenomena. Fractional calculus has been used presently in many fields such as specific physical problems [11], [180], biomedical engineering [94], diffusion processes [181] — [183], viscoelasticity theory [126], fractal dynamics [184], and fractional control [89]. Unfortunately, its major application still focuses on describing the transient state of physical change, but seldom involves systemic evolution processes.

How to apply fractional calculus to signal analysis and processing, especially to neural networks, is an emerging discipline branch field of study and few studies have been seldom performed in this area. The properties of the fractional calculus of a signal are quite different from those of its integer-order calculus [343] — [345]. Therefore, the fractional differential can nonlinearly enhance the complex texture details of an image [186] — [188] and implement texture image denoising approaches [189] — [191], [346]. Following the success in the synthesis of a fractional differentiator in the form of an analog circuit, the emergence of a novel electrical circuit element has been named "fractor" [230], [345], [171] — [173], [193] — [197], [204] — [205]. As in our previous studies [204] — [205], an ideal fractor consists of an ordinary resistor and an ordinary capacitor or inductor in the form of an analog circuit on the tree-type [230], , two-circuit-type [173], [345], H-type [172], [345], net-grid-type [171], [193] — [194], [345], and other infinite recursive structures, which are of extreme self-similar fractal structure. On this basis, the first preliminary attempt on implementation of a fractional-order neural network of the Hopfield type by means of fractional calculus was reported [194]. Another prior study [192] had shown that, in fractional adaptive signal processing and fractional adaptive control, the fractional extreme point is quite different than a traditional integer-order extreme one such as the first-order stationary point. In order to seek the fractional extreme points of the energy norm, we have generalized the integer-order steepest descent approach to a fractional approach [192]. Based on the prior studies mentioned above [171] — [173], [187], [192] — [194], [204] — [205], [230], [345], we propose to introduce a novel mathematical method: fractional calculus to implement FHNN. A significant advantage of FHNN is that its attractors essentially relate to neuron's fractional-order. FHNN possesses the fractional-order-stability and fractional-order-sensitivity characteristics.

The rest of the manuscript is organized as follows: Section 2 recalls the necessary theoretical background of fractional calculus and fractional neural works. Section 3 implements FHNN and studies its stability and convergence. Firstly, we implement fractor in the form of an analog circuit. Secondly, we implement FHNN by utilizing fractor and the fractional steepest descent approach. Thirdly, we construct the Lyapunov function of FHNN. Fourthly, we analyze the attractors of FHNN. Section 4 reports the experiment results and analysis. First, we deduce numerical implementation of FHNN. Second, we analyze the stability and convergence of FHNN. Third, we study the applications of FHNN to the defense against chip cloning attacks for anti-counterfeiting. In Section 5, the conclusions of this manuscript are presented.

9.4.2 MATHEMATICAL BACKGROUND

This section presents a brief introduction to the necessary mathematical background of fractional calculus. The commonly used fractional calculus definitions are those of Grünwald-Letnikov, Riemann-Liouville, and Caputo [3], [176], [178], [230] − [231].

The Grünwald-Letnikov definition of fractional calculus, in a convenient form, for causal signal $s(x)$, is as follows:

$$^{G-L}_{a}D_x^v s(x) = \lim_{N \to \infty} \left\{ \frac{\left(\frac{x-a}{N}\right)^{-v}}{\Gamma(-v)} \sum_{k=0}^{N-1} \frac{\Gamma(k-v)}{\Gamma(k+1)} s\left(x - k\left(\frac{x-a}{N}\right)\right) \right\}, \qquad (9.26)$$

where $s(x)$ is a differintegrable function [3], [176], [178], [230] − [231], $[a, x]$ is the duration of $s(x)$, v is a non-integer, $\Gamma(\alpha) = \int_0^\infty e^{-x} x^{\alpha-1} dx$ is the Gamma function, and $^{G-L}_{a}D_x^v$ denotes the Grünwald-Letnikov defined fractional differential operator.

The Riemann-Liouville definition of the v-order integral, for causal signal $s(x)$, is as follows:

$$^{R-L}_{a}I_x^v s(x) = \frac{1}{\Gamma(v)} \int_a^x \frac{s(\tau)}{(x-\tau)^{1-v}} d\tau, \qquad (9.27)$$

where $v > 0$ and $^{R-L}_{a}I_x^v$ denotes the Riemann-Liouville left-sided fractional integral operator. The Riemann-Liouville definition of the v-order derivative is as follows:

$$^{R-L}_{a}D_x^v s(x) = \frac{1}{\Gamma(n-v)} \frac{d^n}{dx^n} \int_a^x \frac{s(\tau)}{(x-\tau)^{v-n+1}} d\tau, \qquad (9.28)$$

where $n - 1 \leqslant v < n$, and $^{R-L}_{a}D_x^v$ denotes the Riemann-Liouville left-handed fractional differential operator. The Laplace transform of the v-order Riemann-Liouville differential operator is $L[^{R-L}_{0}D_x^v s(x)] = S^v L[s(x)] - \sum_{k=0}^{n-1} S^k [^{R-L}_{0}D_x^{v-1-k} s(x)]_{x=0}$, where S denotes the Laplace operator. When $s(x)$ is a causal signal, and its fractional primitives are also required to be zero, we can simplify the Laplace transform for $^{R-L}_{0}D_x^v s(x)$ as $L[^{R-L}_{0}D_x^v s(x)] = S^v L[s(x)]$.

The Caputo definition of the v-order derivative for causal signal $s(x)$ is as follows:

$$^{C}_{a}D_x^v s(x) = \frac{1}{\Gamma(n-v)} \int_a^x (x-\tau)^{n-v-1} s^{(n)}(\tau) d\tau. \qquad (9.29)$$

where $0 \leqslant n-1 < v < n$, $n \in \mathbf{R}$, and ${}_a^C D_x^v$ denotes a Caputo defined fractional differential operator. From (4), we can see that ${}_a^C D_x^v$ is equivalent to the successive performance of an n-order differential and an $(n-v)$-order integral of signal $s(x)$. The Laplace transform of the v-order Caputo differential operator is $L[{}_0^C D_x^v s(x)] = S^v L[s(x)] - \sum_{k=0}^{n-1} S^k s^{(k)}(x)|_{x=0}$. When $s(x)$ is a causal signal, and its fractional primitives are also required to be zero, we can simplify the Laplace transform for ${}_0^C D_x^v s(x)$ as $L[{}_0^C D_x^v s(x)] = S^v L[s(x)]$. In this case the three cited definitions of fractional derivatives are equivalent. In this work, we use the equivalent notations $D_x^v = {}^G d D_x^v = {}^R d D_x^v = {}^C D_x^v$ in an arbitrary, interchangeable manner.

Nowadays, fractional calculus has been incorporated into artificial neural networks, mainly because of its long-term memory and nonlocality. Some remarkable progress in the studies on fractional neural networks not only validates them as fractional dynamic systems, but also gives interesting and practical suggestions for future research. For instance, firstly, the exploration of theoretical properties for fractional neural networks is needed. Fractional neural networks employed by fractional activation functions have modeling capability to achieve desired parametric learning, but structural mutation requires specific modification of the algorithm to represent additional complexity. Stability and multi-stability, bifurcations, and chaos of fractional neural networks should be investigated. Secondly, the verification of intrinsic differences in behavior between the fractional-order neural networks and the integer-order neural networks is desired. Behavioral differences are observed in experiments, but whether or not they could be attributed to the inherent differences of the fractional-order networks remains to be shown. Experiments and analyses might also be implemented on different topologies to detect further interesting relationships between these two types of neural networks. Thirdly, any biologically-inspired computational intelligence algorithms could be utilized to solve fractional differential equations virtually. With the exception of the stochastic techniques and evolutionary computations, the combination of an artificial neural network aided with other biological-inspired methods, such as an artificial bee colony, might also solve fractional differential equations efficiently.

9.4.3 FHNN AND ITS STABILITY AND CONVERGENCE

9.4.3.1 Implementation of Fractor in Form of an Analog Circuit

In this subsection, in order to implement FHNN in the form of an analog circuit, we first need to perform fractional calculus of the signal in an analog circuit processing. Following the success in the synthesis of a fractional differentiator in the form of an analog circuit, the emergence of a novel electrical circuit element was named "fractor" [171] — [173], [187], [193] — [194], [204] — [205], [230], [345]. As in our previous studies [204], [205], an ideal fractor consists of an ordinary resistor and an ordinary capacitor or inductor in the form of an analog circuit on the tree-type, two-circuits-type,

H-type, net-grid-type, and other infinite recursive structures, which are of extreme self-similar fractal structure. In this sense, fractance means the fractional-order impedance of a fractor. Consequently, let's sign identify fractor as a symbol of ─┤WW├─ , in which F is the abbreviation of fractor [171] — [173], [193] — [194], [204] — [205].

In above mentioned ideal fractor structures, the net-grid-type fractor has an optimal performance [171] — [173], [193] — [194], [204] — [205]. The structural representation of the 1/2-order net-grid-type fractor is shown in Fig. 9.10.

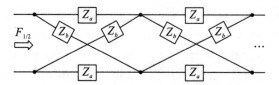

Fig. 9.10 1/2-order net-grid-type fractor.

In Fig. 9.10, $F_{1/2}$ denotes the driving-point impedance function of the 1/2-order net-grid-type fractor. From Fig. 9.10, we can see that the 1/2-order net-grid-type fractor is of extreme self-similar fractal structure with the series connection of infinitely repeated net-grid-type structure where Z_a and Z_b are impedance. The number of Z_a and Z_b is equal to two-fold the number of layers. Then, its equivalent circuit is shown in Fig. 9.11.

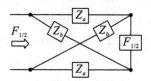

Fig. 9.11 Equivalent circuit of 1/2-order net-grid-type fractor.

Suppose the current of Z_a and Z_b is equal to i_a and i_b. Let the input voltage and input current of $F_{1/2}$ be equal to $V_i(S)$ and $I_i(S)$, respectively. From Fig. 9.11, according to Kirchoff's current law and Kirchoff's voltage law:

$$\begin{cases} Z_a i_a + Z_b i_b = V_i \\ (Z_a + F_{1/2}) i_a - (F_{1/2} + Z_b) i_b = 0 \end{cases}. \quad (9.30)$$

According to the Cramer's rule in linear algebra, it follows that:

$$\begin{cases} i_a = \dfrac{(Z_b + F_{1/2})V_i}{\begin{vmatrix} Z_a + F_{1/2} & -(F_{1/2} + Z_b) \\ Z_a & Z_b \end{vmatrix}} \\ i_b = \dfrac{(Z_a + F_{1/2})V_i}{\begin{vmatrix} Z_a + F_{1/2} & -(F_{1/2} + Z_b) \\ Z_a & Z_b \end{vmatrix}} \end{cases}. \quad (9.31)$$

Hence, $F_{1/2}$ is equal to:

$$F_{1/2} = \frac{V_i}{i_a + i_b} = \frac{2Z_a Z_b + F_{1/2}(Z_a + Z_b)}{2F_{1/2} + Z_a + Z_b}. \quad (9.32)$$

From (9.32), it follows that:

$$F_{1/2} = (Z_a Z_b)^{1/2}. \qquad (9.33)$$

For the convenience of discussion, we only discuss the issue of capacitive fractor in the following. Inductive fractor behaves in a similar way. Suppose the initial energy of the electric element of capacitive fractor is equal to zero. Let the resistance and capacitance of capacitive fractor be equal to r and c, respectively. Then, in the Laplace transform domain, $Z_a = r$, and $Z_b = 1/(cS)$ where S is the Laplace operator. From (9.33), it follows that:

$$F_{1/2} = \zeta^{1/2} S^{-1/2}, \qquad (9.34)$$

where $\zeta = r/c$. From (9.34), we can derive the relationship between input voltage $V_i(S)$ and input current $I_i(S)$ of the 1/2-order capacitive fractor. It follows that:

$$V_i(S) = \zeta^{1/2} S^{-1/2} I_i(s). \qquad (9.35)$$

The inverse Laplace transform of (9.35) is as follows:

$$V_i(t) = \frac{\zeta^{1/2}}{\Gamma(1/2)} \int_0^t \frac{I_i(\tau)}{(t-\tau)^{1/2}} d\tau, \qquad (9.36)$$

where Γ is the Gamma function. From (9.29) and (9.36), we can see that $V_i(t)$ is in direct ratio to the 1/2-order fractional integral of $I_i(t)$. On the other hand, from (9.35), it follows that:

$$I_i(S) = \zeta^{-1/2} S^{1/2} V_i(S). \qquad (9.37)$$

The inverse Laplace transform of (9.37) is as follows:

$$I_i(t) = \frac{1}{\zeta^{1/2} \Gamma(1/2)} \int_0^t \frac{V_i^{(1)}(\tau)}{(t-\tau)^{1/2}} d\tau. \qquad (9.38)$$

From (9.29) and (9.38), we can see that $I_i(t)$ is in direct ratio to the 1/2-order fractional differential of $V_i(t)$.

Similarly, with regard to the v-order capacitive fractor, let us use F_v to denote its driving-point impedance function. We can further derive that $V_i(t)$ is generally in direct ratio to the v-order fractional integral of $I_i(t)$, whereas $I_i(t)$ is generally in direct ratio to the v-order fractional differential of $V_i(t)$ [204], [205]. Thus:

$$I_i(t) = \frac{1}{\zeta^v \Gamma(n-v)} \int_0^t (t-\tau)^{n-v-1} V_i^{(n)}(\tau) d\tau, \qquad (9.39)$$

where $\zeta = r^{(1-p)/v}/c$, $v = q + p$ is a positive real number, q is a positive integer, $0 \leqslant p \leqslant 1$, $V_i(t)$ is input voltage, and $I_i(t)$ is input current of the v-order capacitive fractor [204], [205].

9.4.3.2 Implementation of FHNN

In this subsection, we implement FHNN. On the basis of the aforementioned fractor, we can implement FHNN in the form of an analog circuit by utilizing fractor and the fractional steepest descent approach [192]. FHNN model is shown in Fig. 9.12.

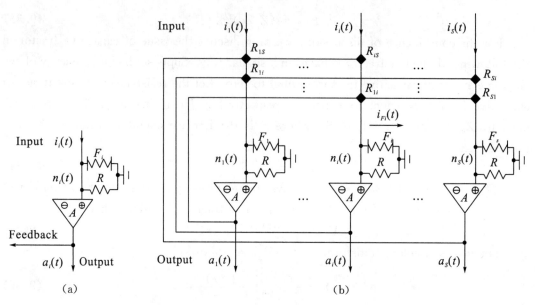

Fig. 9.12 FHNN model: (a) Circuit configuration for the fractional neuron of FHNN, (b) Circuit configuration of FHNN.

Looking at Fig. 9.12 and comparing the FNN with HNN model [277], [295] − [296], [350], it is apparent that the circuit configuration of FHNN's fractional neuron is striking dissimilar to the circuit configuration of HNN's first-order neuron. In Fig. 9.12, each fractional neuron of FHNN consists of one operational amplifier A and its related fractor ─⫼F⫼─, and resistors. Each fractional neuron has the same circuit configuration. From (9.39), we know that fractor ─⫼F⫼─ implements the v-order fractional calculus. We set $F_1 = F_{v_1}$, $F_i = F_{v_i}$ and $F_s = F_{v_s}$, i.e. the fractional-order of FHNN's neuron is equal to v_1, v_i, and v_s, respectively. Thus, input current $i_{F_i}(t)$ is generally in direct ratio to the v_i-order fractional differential of the input voltage $n_i(t)$ of the v_i-order F_{v_i}. In Laplace transform domain, set $F_{v_i}(S)$ denotes the v_i-order $F_{v_i}(t)$. Therefore, we can derive the operation rule of FHNN from Kirchhoff's current law. It follows that:

$$I_{F_i}(S) = \frac{N_i(S)}{F_{v_i}(S)} = \sum_{j=1}^{S} T_{i,j} \cdot A_j(S) - \frac{N_i(S)}{Z_i} + I_i(S), \qquad (9.40)$$

where $I_{F_i}(S)$, $N_i(S)$, $F_{v_i}(S)$, $A_j(S)$, and $I_i(S)$ are the Laplace transforms of $i_{F_i}(t)$, $n_i(t)$, $F_{v_i}(t)$, $a_j(t)$, and $i_i(t)$, respectively. The function $n_i(t)$ is the input voltage of the ith fractional neuron's A amplifier, $a_i(t)$ is the output voltage of the ith fractional neuron, and $i_i(t)$ is the input electric current of the ith fractional neuron. $R_{i,j}$ is the feedback resistor being connected with the output of the jth fractional neuron and the input of the ith one. Thus, conductance has $T_{i,j} = 1/R_{i,j}$. It is assumed that the circuit of FHNN is symmetric, so that $T_{i,j} = T_{j,i}$. We set $1/Z_i = 1/R + \sum_{j=1}^{S}(1/R_{i,j})$.

Furthermore, using (14) the inverse Laplace transform of (9.40) is as follows:

$$i_{F_i}(t) = K_i \frac{d^{v_i} n_i(t)}{dt^{v_i}}$$

$$= \frac{K_i}{\Gamma(n-v_i)} \int_0^\tau (t-\tau)^{n-v_i-1} n_i^{(n)}(\tau) d\tau$$

$$= \sum_{j=1}^S T_{i,j} \cdot a_j(t) - \frac{n_i(t)}{Z_i} + i_i(t), \tag{9.41}$$

where $K_i = \zeta_i^{-v_i}$ is positive constant. With respect to capacitive fractor, $\zeta_i = r^{(1-p_i)/v_i}/c$, $v_i = q_i + p_i$ is a positive real number, q_i is a positive integer, and $0 \leqslant p_i \leqslant 1$. Multiplying both sides of (9.41) by Z_i, it follows that:

$$Z_i K_i \frac{d^{v_i} n_i(t)}{dt^{v_i}} = \sum_{j=1}^S Z_i T_{i,j} \cdot a_j(t) - n_i(t) + Z_i i_i(t). \tag{9.42}$$

Define $\chi_i = Z_i K_i$ as a positive constant, and set $w_{i,j} = Z_i T_{i,j}$, $b_i = Z_i i_i$. Thus, it follows that:

$$\chi_i \frac{d^{v_i} n_i(t)}{dt^{v_i}} = -n_i(t) + \sum_{j=1}^S w_{i,j} a_j(t) + b_i(t). \tag{9.43}$$

Thus, it follows that in vector form:

$$\boldsymbol{\chi} \frac{d^v n(t)}{dt^v} = -n(t) + \boldsymbol{W} a(t) + \boldsymbol{b}, \tag{9.44}$$

where $\boldsymbol{\chi} = [\chi_1 \ \cdots \ \chi_i \ \cdots \ \chi_s]^T$, $\boldsymbol{v} = [v_1 \ \cdots \ v_i \ \cdots \ v_s]^T$, $\boldsymbol{W} = [w_{i,j}]_{s \times s}$ is the weighting matrix of FHNN, and $\boldsymbol{b} = [b_1 \ b_2 \ \cdots \ b_s]^T$. Both (9.43) and (19) are the state equation of FHNN.

As we know, the state equation of HNN [277], [295] − [296], [350] is given as $\varepsilon_i \frac{dn_i(t)}{dt} = -n_i(t) + \sum_{j=1}^S w_{i,j} a_j(t) + b_i(t)$, where $\varepsilon_i = R_i c$, and c is capacitance. Comparing the state equation of FHNN with that of HNN, we can see that $d^{v_i} n_i(t)/dt^{v_i}$ appears in the former, while $dn_i(t)/dt$ exists in the latter. The former is an explicitly FPDE, but the latter is the first-order one. Furthermore, from (9.26) − (9.29), it can be seen that since fractional differential is nonlocal and has weakly singular kernel, it provides an excellent method for the description of long-term memory and the nonlocality of nonlinear dynamic processes. Therefore, an FPDE is used to describe nonlinear dynamic systems, such as FHNN, which can be characterized by power-law nonlocality, power-law long-term memory, fractal property and chaotic behavior, because the arbitrary-order of the fractional differential represents an additional degree of freedom to fit specific behavior. Another important characteristic is that the fractional differential depends not only on the local conditions of evaluated time, but also on the all history of signal. This factor is often useful since FHNN has long-term memory and any evaluation point depends on the past values of signal. $d^{v_i} n_i(t)/dt^{v_i}$ is a nonlocal expression with respect to the nonlocality of the fractional differential. Some of the derived nonlocal voltage yields stationary charges that in turn can be converted into nonlocal conserved charges. Just as in the aforementioned discussion, A is an operational amplifier. In terms of electrical circuits, f represents the transfer function of the nonlinear amplifier A with negligible response time. Set $f(0)=0$. It is also convenient to define the inverse output-input relation, f^{-1}. Thus it has:

$$a_i(t) = f[n_i(t)]. \tag{9.45}$$

Thus, it follows that in vector form:

$$\boldsymbol{a}(t) = f[\boldsymbol{n}(t)]. \tag{9.46}$$

It is assumed that the analytic increasing function f has the inverse function f^{-1}. Thus, f^{-1} is also an analytic increasing function. Then, it follows that:

$$n_i(t) = f^{-1}[a_i(t)]. \tag{9.47}$$

Thus, it follows that in vector form:

$$\boldsymbol{n}(t) = f^{-1}[\boldsymbol{a}(t)]. \tag{9.48}$$

9.4.3.3 Construction of Lyapunov Function of FHNN

In this subsection, we construct the Lyapunov function of FHNN making use of the variable gradient method. Practical applications of FHNN heavily depend on its dynamical behaviors, such as Lyapunov stability and asymptotic stability. For the existence and uniqueness of the solution of the state equation of FHNN should be proved before studying its asymptotical stability, we prove the existence and uniqueness of the fractional-order equilibrium point of (18) before constructing the Lyapunov function of FHNN, respectively.

Firstly, we prove the existence of the fractional-order equilibrium point of (9.43). Let's assume that the transfer function f of FHNN satisfies:

$$\begin{cases} |f(n)| \leqslant M, & n \in \mathbf{R} \\ |f(m) - f(n)| \leqslant K|m-n|, & m \in \mathbf{R} \end{cases} \tag{9.49}$$

where M and K are two nonnegative constants, R denotes the field of real numbers. Equation (9.49) shows that the transfer function f of FHNN satisfies boundedness and Lipschitz continuity. Assume further that $\boldsymbol{n}^* = [n_1^* \quad n_2^* \quad \cdots \quad n_S^*]^\mathrm{T}$ is a fractional-order equilibrium point of (9.43), so that from (9.43) and (9.45), it follows that:

$$n_i^*(t) = \sum_{j=1}^{S} w_{i,j} f[n_j^*(t)] + b_i(t). \tag{9.50}$$

Thus, it follows that in vector form:

$$\boldsymbol{n}^* = \boldsymbol{W}\boldsymbol{F}(\boldsymbol{n}^*) + \boldsymbol{b}, \tag{9.51}$$

where $\boldsymbol{F}(\boldsymbol{n}^*) = [f(n_1^*) \quad f(n_2^*) \quad \cdots \quad f(n_S^*)]^\mathrm{T}$. Let's assume that a mapping $\boldsymbol{G}$ satisfies:

$$\boldsymbol{G}(\boldsymbol{n}) = \boldsymbol{W}\boldsymbol{F}(\boldsymbol{n}) + \boldsymbol{b}, \tag{9.52}$$

where $\boldsymbol{n} = [n_1 \quad n_2 \quad \cdots \quad n_S]^\mathrm{T}$. From (9.49), we can see that $\boldsymbol{F}(\boldsymbol{n})$ is a uniform continuous mapping from $\boldsymbol{R}^n$ to $\boldsymbol{R}^n$. Thus, $\boldsymbol{G}(\boldsymbol{n})$ is also a uniform continuous mapping from $\boldsymbol{R}^n$ to $\boldsymbol{R}^n$. From (9.49) and (9.52), it follows that:

$$\|\boldsymbol{G}(\boldsymbol{n})\|^2 = \sum_{i=1}^{S} \Big[\sum_{j=1}^{S} w_{i,j} f(n_j) + b_j\Big]^2$$

$$\leqslant \sum_{i=1}^{S} \Big[\sum_{j=1}^{S} |w_{i,j}| M + |b_i|\Big]^2 \tag{9.53}$$

$$= \rho^2,$$

where $\| \ \|$ denotes Euclid norm. From (9.53) we can see that $\Phi = \{n \mid \|n\| \leqslant \rho\}$ is a bounded convex set, and $G(n)$ is a uniform continuous mapping from Φ to Φ. Thus, from Brouwer's fixed-point theorem, we can derive that $\exists n^* \in \Phi$ to enable $G(n^*) = n^*$ to be set up. Thus, (9.50) and (9.51) are set up. n^* is a fractional-order equilibrium point of (9.43).

Secondly, we construct the Lyapunov function of FHNN making use of the variable gradient method. From (9.43) — (9.46), we can see that FHNN is a fractional nonlinear system. The Lyapunov function of FHNN is to solve a kind of fractional asymptotic stability. From (9.43), define an analytic function $\rho_i(t)$:

$$\rho_i(t) = \chi_i \frac{d^{v_i} n_i(t)}{dt^{v_i}} = -n_i(t) + \sum_{j=1}^{S} w_{i,j} a_j(t) + b_i(t). \tag{9.54}$$

$\rho_i(t)$ is a differintegrable function [3], [176], [178], [230] — [231]. According to characteristics of fractional calculus:

$$D_t^v \psi(t) = \sum_{n=0}^{\infty} \frac{D_t^n \psi(t)}{\Gamma(1+n-v)} t^{n-v}, \tag{9.55}$$

where D_t^v is the v-order fractional differential, $\psi(t)$ is analytic function, and D_t^n is the n-order differential operator. Then, from (9.55) and Faà di Bruno formula [179], it follows that:

$$D_t^v \psi[\varphi(t)] = \frac{t^{-v}}{\Gamma(1-v)} \psi[\varphi(t)] + \sum_{n=1}^{\infty} \binom{v}{n} \frac{t^{n-v} n!}{\Gamma(1+n-v)} \sum_{m=1}^{n} D_\varphi^m \psi \sum \prod_{k=1}^{n} \frac{1}{P_k!} \left(\frac{D_t^k \varphi(t)}{k!}\right)^{P_k}, \tag{9.56}$$

where $\psi = \psi[\varphi(t)]$ is the composite function, $\binom{v}{n} = \frac{(-1)^{-n} \Gamma(n-v)}{\Gamma(-v) \Gamma(1+n)}$, D_φ^m and D_t^k are the integer-order differential operators, and P_k satisfies:

$$\begin{cases} \sum_{k=1}^{n} k P_k = n \\ \sum_{k=1}^{n} P_k = m \end{cases} . \tag{9.57}$$

The third summation notation $\sum$ in (9.56) denotes the summation of the corresponding $\left\{ \prod_{k=1}^{n} \frac{1}{P_k!} \left[\frac{D_t^k \varphi(t)}{k!}\right]^{P_k} \right\} \bigg|_{m=1 \to n}$ of all of the combinations of $P_k |_{m=1 \to n}$ that satisfy the requirement of (9.57). From (9.28) and (9.56), we can see that the fractional differential of the composite function is equal to an infinite sum. Therefore, we suggest constructing the Lyapunov function of FHNN as follows:

$$E[a(t)] = -\sum_{i=1}^{S} D_t^{-v_i} \{(D_{n_i}^1 f)[D_t^{1-v_i}(\rho_i(t)/\chi_i)]\}^2 = -\sum_{i=1}^{S} D_t^{-v_i} [D_t^1 a_i(t)]^2$$

$$= \frac{-t^{v_i}}{\Gamma(1+v_i)} \sum_{i=1}^{S} (D_t^1 a_i)^2 - \sum_{i=1}^{S} \sum_{n=1}^{\infty} \binom{-v_i}{n} \frac{t^{n+v_i} n!}{\Gamma(1+n+v_i)}$$

$$\sum_{m=1}^{n} D_{D_t^1 a_i}^{m_1} (D_t^1 a_i)^2 \sum_{k=1}^{n} \prod_{k=1}^{n} \frac{1}{P_k!} \Big(\frac{D_t^{k+1} a_i}{k!} \Big)^{P_k}, \tag{9.58}$$

where $D_t^1 a_i(t) = (D_{n_i}^1 f)[D_t^{1-v_i}(\rho_i(t)/\chi_i)]$, $D_t^{-v_i}$ is the v_i-order fractional integral operator. From (9.54) and (9.58), we can see that to enable (9.58) to be set up, $D_t^{v_i} n_i(t)$ must to be an inverse operation of $D_t^{-v_i} n_i(t)$. Therefore, from the combination rules of fractional calculus [3], [176], [178], [230] − [231], the additional condition to guarantee the existence of Lyapunov function is $w_{i,j} = w_{j,i}$. FHNN's weighting matrix W is an symmetric matrix. For $D_{D_t^1 a_i}^1 (D_t^1 a_i)^2 = 2 D_t^2 a_i$, $D_{D_t^1 a_i}^2 (D_t^1 a_i)^2 = 2$, and $D_{D_t^1 a_i}^{m_1} (D_t^1 a_i)^2 \overset{m \geq 3}{\equiv} 0$. Thus, (9.58) can be simplified as follows:

$$E[\boldsymbol{a}(t)] = \frac{-t^{v_i}}{\Gamma(1+v_i)} \sum_{i=1}^{S} (D_t^1 a_i)^2 - \binom{-v_i}{1} \frac{2 t^{1+v_i}}{\Gamma(2+v_i)} \sum_{i=1}^{S} (D_t^1 a_i)(D_t^2 a_i)$$
$$- \sum_{i=1}^{S} \sum_{n=2}^{\infty} \binom{-v_i}{n} \frac{2 t^{n+v_i} n!}{\Gamma(1+n+v_i)}$$
$$\Big\{ D_t^1 a_i \Big[\prod_{k=1}^{n} \frac{1}{P_k!} \Big(\frac{D_t^{k+1} a_i}{k!} \Big)^{P_k} \Big]_{m=1} + \Big[\prod_{k=1}^{n} \frac{1}{P_k!} \Big(\frac{D_t^{k+1} a_i}{k!} \Big)^{P_k} \Big]_{m=2} \Big\}. \tag{9.59}$$

From (9.58) or (9.59), implement the v_i-order fractional differential of $E[a_i(t)]$ on the both sides of (9.58) or (9.59), thus it follows that:

$$D_t^v E = \frac{d^v E}{d t^v} = - \sum_{i=1}^{S} [D_t^1 a_i(t)]^2 \leqslant 0. \tag{9.60}$$

Hence we can see that $\frac{d^v E}{dt^v}$ is a negative semidefinite function. $E(t)$ is a valid Lyapunov function of FHNN. When $D_t^{v_i} E = \frac{d^{v_i} E}{dt^{v_i}} = 0$, the system "energy" of FHNN is unchanging and its system has reached steady state. From (9.60), we can obtain equivalent form of $D_t^{v_i} E = \frac{d^{v_i} E}{dt^{v_i}} = 0$ as follows:

$$D_t^1 a_i(t) = \frac{d a_i(t)}{dt} = 0. \tag{9.61}$$

Thus, it follows that in vector form:

$$D_t^1 \boldsymbol{a}(t) = \frac{d \boldsymbol{a}(t)}{dt} = 0. \tag{9.62}$$

From (9.60), (9.61), and (9.62), we can determine the equilibrium points of FHNN's Lyapunov function according to LaSalle's invariance theorem and the fractional steepest descent approach [192]. For the fractional steepest descent approach is different from the first-order steepest descent approach. Its every optimal searching adjustment step is on the negative direction of Lyapunov function's fractional gradient but not of its first-order one. The equilibrium points of FHNN's Lyapunov function are the potential attractors of FHNN.

9.4.3.4 Analysis of Attractors of FHNN

In this subsection, we analyze the attractors of FHNN.

As in the aforementioned discussion, we can see that the equilibrium points of FHNN's Lyapunov function are the potential attractors of FHNN. From (9.55) and (9.58):

$$D_{a_i}^{v_i} E = \frac{d^{v_i} E}{da_i^{v_i}} = -\sum_{n=0}^{\infty}\left[\frac{a_i^{n-v_i} D_t^{n-v_i}[D_t^1 a_i(t)]^2}{\Gamma(1+n-v_i)}\right]. \tag{9.63}$$

Therefore, it follows that in vector form:

$$D_a^v E = \frac{d^v E}{da^v} = -\sum_{i=1}^{S}\sum_{n=0}^{\infty}\left[\frac{a_i^{n-v_i} D_t^{n-v_i}[D_t^1 a_i(t)]^2}{\Gamma(1+n-v_i)}\right], \tag{9.64}$$

where $v = [v_1 \cdots v_i \cdots v_S]^T$. From (9.61) and (9.63), we have $D_{v_i}^{v_i} E = \frac{d^{v_i} E}{da_i^{v_i}} = 0$, only when $D_t^1 a_i(t) = \frac{da_i(t)}{dt} = 0$. In vector form, we have $\frac{d^v E}{da^v} = 0$, when $D_t^1 a(t) = \frac{da(t)}{dt} = 0$. Thus, when $a(t)$ satisfies (9.62), the equilibrium points of FHNN's Lyapunov function are the attractors of FHNN. Note that a significant advantage of FHNN is that its attractors essentially relate to neuron's fractional-order. FHNN possesses the fractional-order-stability and fractional-order-sensitivity characteristics.

9.4.3.5 Implementation of training algorithm for FHNN

In this subsection, we implement a training algorithm based on the supervised Hebb rule for FHNN.

We can implement a training algorithm based on the Hebb rule for FHNN. As we know, the Hebb rule and Storkey rule are two efficient training algorithms for HNN [304], [347] − [348]. It is desirable for a learning rule to be both local and incremental [349]. Similar to HNN, in fact, a design procedure based on FHNN's Lyapunov function is used to determine FHNN's weighting matrix. Suppose that we want to store a set of prototype patterns in a FHNN. When an input pattern is presented to a FHNN, its output should then converge to the prototype pattern closest to the input pattern. Let's assume that the prototype patterns of FHNN are $\{p_1, p_2, \cdots, p_q, \cdots, p_Q\}$, where the elements of the vectors $p_q = [p_{q_1} \cdots p_{q_i} \cdots p_{q_j} \cdots p_{q_s}]^T$ are restricted to be ± 1. Assume further that $Q \ll S$, so that the state space is large enough, and the prototype patterns are well distributed in this space. In order for a FHNN to be able to recall the prototype patterns, the prototype patterns must be minima of FHNN's Lyapunov function. We propose an appropriate quadratic performance index as follows:

$$J(a) = -\sum_{i=1}^{S} D_t^{-v_i}\left\{\frac{(D_{n_i}^1 f)^2}{\chi_i^2}\left[D_t^{1-v_i}\left(\sum_{j=1}^{S}\sum_{q=1}^{Q} p_{q_i} p_{q_j} a_j\right)\right]^2\right\}, \tag{9.65}$$

where the elements of the vectors $a = [a_1 \cdots a_j \cdots a_S]^T$ are restricted to be ± 1, and $(D_{n_i}^1 f)^2/\chi_i^2$ is a positive constant. From (9.65), we evaluate the performance index at a random input pattern a, which is presumably not close to any prototype pattern.

$\sum_{j=1}^{S} \sum_{q=1}^{Q} p_{q_i} p_{q_j} a_j$ in (9.65) is an inner product between a prototype pattern and the input pattern. The inner product will increase as the input pattern moves closer to a prototype pattern. However, if the input pattern is not close to any prototype pattern, all terms of $\sum_{j=1}^{S} \sum_{q=1}^{Q} p_{q_1} p_{q_j} a_j$ in (9.65) will be small. Thus, $J(\boldsymbol{a})$ will be largest (least negative) when $\boldsymbol{a}$ is not close to any prototype pattern, and will be smallest (most negative) when $\boldsymbol{a}$ is equal to any one of the prototype patterns. Assume that the prototype patterns are orthogonal. We further evaluate the performance index at one of the prototype patterns as follows:

$$J(\boldsymbol{p}_k) = -\sum_{i=1}^{S} D_t^{-v_i} \left\{ \frac{(D_{n_i}^1 f)^2}{\chi_i^2} \left[D_t^{1-v_i} \left(\sum_{j=1}^{S} \sum_{q=1}^{Q} p_{q_i} p_{q_j} p_{k_j} \right) \right]^2 \right\}$$
$$= -\sum_{i=1}^{S} D_t^{-v_i} \left[\frac{(D_{n_i}^1 f)^2}{\chi_i^2} (D_t^{1-v_i} S)^2 \right]. \quad (9.66)$$

From the properties of the fractional calculus and (9.66), we can derive that $J(\boldsymbol{a})$ is minimized at the prototype patterns. We use the supervised Hebb rule to compute FHNN's weighting matrix (with target patterns being the same as input patterns) as follows:

$$\boldsymbol{W} = \sum_{q=1}^{Q} \boldsymbol{p}_q (\boldsymbol{p}_q)^{\mathrm{T}}, \quad (9.67)$$

where $\boldsymbol{W}$ is the weighting matrix of FHNN, and $w_{i,j} = \sum_{q=1}^{Q} p_{q_i} p_{q_j}$. From (9.45), we can see that if f is an high-gain arc-tangent function, $D_t^{1-v_i}(n_i) = D_t^{1-v_i}[f^{-1}(a_i)] = 0$. From (9.54), (9.58), and (9.67), and set the bias $b_i(t)$ to zero, FHNN's high-gain Lyapunov function is as follows:

$$E(\boldsymbol{a}) = -\sum_{i=1}^{S} D_t^{-v_i} (D_t^1 a_i)^2$$
$$= -\sum_{i=1}^{S} D_t^{-v_i} \left\{ \frac{(D_{n_i}^1 f)^2}{\chi_i^2} \left[D_t^{1-v_i} \left(\sum_{j=1}^{S} \sum_{q=1}^{Q} p_{q_i} p_{q_j} a_j \right) \right]^2 \right\}$$
$$= J(\boldsymbol{a}). \quad (9.68)$$

Thus, FHNN's high-gain Lyapunov function is indeed equal to the quadratic performance index for the content-addressable memory problem. FHNN output will tend to converge to the stored prototype patterns. In particular, from Fig. 9.12, (9.43), and (9.44), we can see that the diagonal elements of FHNN's weighting matrix are set to zero. From (9.67), since the elements of each $\boldsymbol{p}_q$ are restricted to be ± 1, all of the diagonal elements of $\boldsymbol{W}$ will be equal to Q that is the number of prototype patterns. Thus, we can zero the diagonal by subtracting Q times the identity matrix as follows:

$$\boldsymbol{W} = \sum_{q=1}^{Q} \boldsymbol{p}_q (\boldsymbol{p}_q)^{\mathrm{T}} - Q\boldsymbol{I}, \quad (9.69)$$

where $\boldsymbol{I}$ is the identity matrix. Note that there will be at least two minima of the performance index for each prototype pattern. If $\boldsymbol{p}_q$ is a prototype pattern, then $-\boldsymbol{p}_q$ will be also in the space spanned by the prototype patterns. Therefore, each prototype patterns will be one of the corners of the hypercube $\{\boldsymbol{a}: -1 < a_j < 1\}$. These corners will include the prototype patterns, but they will also include some linear combinations of the prototype patterns. There will also be a number of other minima (spurious patterns) of the FHNN's Lyapunov function that do not correspond to prototype patterns. We can use an improved design method [324] that is guaranteed to minimize the number of spurious patterns.

9.4.4 EXPERIMENT AND ANALYSIS

9.4.4.1 Numerical Implementation of FHNN

In this subsection, we achieve the numerical implementation of FHNN before analyze its stability and convergence.

Firstly, suppose there are only two fractional neurons of FHNN. Thus, $S=2$. From Fig. 9.12, we can see that the output of either FHNN's fractional neuron feeds back to the input of the other through a feedback resistor. Thus, we set $R_{1,2} = R_{2,1} = 1\ \Omega$ (ohm), $R_{1,1} = R_{2,2} = \infty\ \Omega$, and also set $R = 1\ \Omega$ and $i_1 = i_2 = 0$ on FHNN's fractional neuron. Thus, it follows that $T_{1,2} = T_{2,1} = 1$ s (siemens), $T_{1,1} = T_{2,2} = 0$ s, $Z_1 = Z_2 = 1/2$, and $b_1 = b_2 = 0$. Furthermore, we set $r = 1\ \Omega$, $c = 1$ nF (nanofarad) of fractor, $F_i = F_{v_i}$ in (9.39) and (9.41). Thus, it follows that $K_i = \xi_i^{-v_i} = (r^{1-p/v_i}/c)^{v_i} = 1$, $\chi_1 = \chi_2 = Z_1 K_1 = Z_2 K_2 = 1/2$, $w_{1,2} = w_{2,1} = 1/2$, and $w_{1,1} = w_{2,2} = 0$. Therefore, we have the weighting matrix in (9.44) as follows:

$$\boldsymbol{W} = \begin{bmatrix} 0 & 1/2 \\ 1/2 & 0 \end{bmatrix}. \tag{9.70}$$

Suppose that the transfer function of operational amplifier $\boldsymbol{A}$ is $f(\tau) = 2/\pi \arctan(\gamma\pi\tau/2)$. Thus, from (9.45) and (9.47), it follows that, respectively:

$$a_1(t) = \frac{2}{\pi} \arctan\left[\frac{\gamma\pi n_1(t)}{2}\right], \tag{9.71}$$

$$a_2(t) = \frac{2}{\pi} \arctan\left[\frac{\gamma\pi n_2(t)}{2}\right], \tag{9.72}$$

$$n_1(t) = \frac{2}{\gamma\pi} \tan\frac{\pi a_1(t)}{2}, \tag{9.73}$$

$$n_2(t) = \frac{2}{\gamma\pi} \tan\frac{\pi a_2(t)}{2}, \tag{9.74}$$

where γ is the gain coefficient of the transfer function of the operational amplifier $\boldsymbol{A}$.

From (9.26), when $\Delta t \to 0$, we have:

$$D_t^v s(t) \cong \sum_{k=0}^{n} \frac{\Gamma(k-v)}{\Delta t^v \Gamma(-v)\Gamma(k+1)} s(t - k\Delta t)$$

$$\cong \sum_{k=0}^{n} \frac{(-1)^k \Gamma(1+v)}{\Delta t^v \Gamma(k+1)\Gamma(v-k+1)} s(t - k\Delta t), \tag{9.75}$$

where n is a large positive integer. Equation (9.75) is the approximate calculation of the v-order fractional calculus when $\Delta t \to 0$ [187], i. e., fractional backward difference. Therefore, from (9.75), when $v = n$, it follows that:

$$D_t^n s(t) \cong \sum_{k=0}^{n} \frac{\Gamma(k-n)}{\Delta t^n \Gamma(-n) \Gamma(k+1)} s(t - k\Delta t). \tag{9.76}$$

Equation (9.76) is the approximate calculation of the n-order integer-order calculus when $\Delta t \to 0$. Thus, from (9.76), when $v = 1$, it becomes:

$$D_t^1 s(t) \cong \frac{1}{\Delta t} [s(t) - s(t - \Delta t)]. \tag{9.77}$$

Equation (9.77) is the approximate calculation of the first-order differential when $\Delta t \to 0$, i. e., the first-order backward difference. From (9.75) and (9.77), we can see that fractional difference has nonlocal characteristics. $D_t^v s(t)$ is not only correlated with $s(t)$ and $s(t - \Delta t)$, but also with $s(t - k\Delta t)$. It is quite different from the first-order calculus.

Then, from (9.43), (9.44), and (9.70):

$$\frac{d^{v_1} n_1(t)}{dt^{v_1}} = -2 n_1(t) + a_2(t), \tag{9.78}$$

$$\frac{d^{v_2} n_2(t)}{dt^{v_2}} = -2 n_2(t) + a_1(t). \tag{9.79}$$

Thus, from (9.75), (9.78) and (9.79), using the fractional forward difference, when $\Delta t \to 0$, we have, respectively:

$$n_1(t + \Delta t) \cong -2 n_1(t) \Delta t^{v_1} + a_2(t) \Delta t^{v_1}$$
$$- \sum_{k=0}^{n} \frac{\Gamma(k - v_1 + 1)}{\Gamma(-v_1) \Gamma(k+2)} n_1(t - k\Delta t), \tag{9.80}$$

$$n_2(t + \Delta t) \cong -2 n_2(t) \Delta t^{v_2} + a_1(t) \Delta t^{v_2}$$
$$- \sum_{k=0}^{n} \frac{\Gamma(k - v_2 + 1)}{\Gamma(-v_2) \Gamma(k+2)} n_2(t - k\Delta t). \tag{9.81}$$

Then (9.80) and (9.81) are taken into (9.71) and (9.72), resulting in the numerical computation of $a_1(t + \Delta t)$ and $a_2(t + \Delta t)$, respectively.

Furthermore, from (9.58) and (9.75), thus we have the numerical computation of FHNN's Lyapunov function as follows:

$$E[\boldsymbol{a}(t)] = -\sum_{i=1}^{2} D_t^{-v_i} [D_t^1 a_i(t)]^2$$
$$\cong -\sum_{i=1}^{2} \sum_{k=0}^{n} \frac{(-1)^k \Gamma(1 - v_i) \Delta t^{v_i}}{\Gamma(k+1) \Gamma(1 - v_i - k)} [D_t^1 a_i(t - k\Delta t)]^2. \tag{9.82}$$

Thus, take (9.71), (9.72), (9.77), (9.80), and (9.81) into (9.82), we can numerically compute the Lyapunov function of FHNN. In order to keep stability and convenience of its numerical computation, we set $n = 60$ and $\Delta t = 0.0001$ in the following examples. Because it is impossible to obtain the first sixty system initial values before running FHNN, we set $n_i(0 + k\Delta t) \stackrel{k=1 \sim 59}{=} n_i(0)$. Hence, there are 59 arbitrary initial values for the system initial states of FHNN. In order to display its actual operation law, we start showing the experimental results of FHNN's Lyapunov function from the 61th calculation in the following examples.

Secondly, we suppose there are three fractional neurons of FHNN, thus, $S=3$. In order to avoid any of the eight corners of the hypercube being the saddle point of FHNN, we set the weighting matrix **W** of FHNN to be an asymmetric matrix. To keep the circuit of FHNN symmetric, we set $R_{1,2}=R_{2,1}=1$ Ω, $R_{1,3}=R_{3,1}=2$ Ω, $R_{2,3}=R_{3,2}=3$ Ω, and $R_{1,1}=R_{2,2}=R_{3,3}=\infty$ Ω. We also set $R=1$ Ω and $i_1=i_2=i_3=0$ on FHNN's fractional neuron in Fig. 9.12. Thus, it results that $T_{1,2}=T_{2,1}=1$ s, $T_{1,3}=T_{3,1}=1/2$ s, $T_{2,3}T_{3,2}=1/3$ s, $T_{1,1}=T_{2,2}=T_{3,3}=0$ s, and $b_1=b_2=b_3=0$. Furthermore, we also set $r=1$ Ω, $c=1$ nF of fractor, $F_i=F_{v_i}$ in (9.39) and (9.41). Thus, it has $Z_1=2/5$, $Z_2=3/7$, $Z_3=6/11$, $K_i=\zeta_i^{-v_i}=(r^{1-p/v_i}/c)^{v_i}=1$, $\chi_1=Z_1K_1=2/5$, $X_2=Z_2K_2=3/7$, $X_3=Z_3K_3=6/11$, $w_{1,2}=2/5$, $w_{1,3}=1/5$, $w_{2,1}=3/7$, $w_{2,3}=1/7$, $w_{3,1}=3/11$, $w_{3,2}=2/11$ and $w_{1,1}=w_{2,2}=w_{3,3}=0$. Therefore, we have the weighting matrix in (9.44) as follows:

$$\mathbf{W}=\begin{vmatrix} 0 & 2/5 & 1/5 \\ 3/7 & 0 & 1/7 \\ 3/11 & 2/11 & 0 \end{vmatrix}. \quad (9.83)$$

Thus, from (9.43), (9.44), and (9.83):

$$\frac{d^{v_1}n_1(t)}{dt^{v_1}}=-\frac{5}{2}n_1(t)+a_2(t)+\frac{1}{2}a_3(t), \quad (9.84)$$

$$\frac{d^{v_2}n_2(t)}{dt^{v_2}}=-\frac{7}{3}n_2(t)+a_1(t)+\frac{1}{3}a_3(t), \quad (9.85)$$

$$\frac{d^{v_3}n_3(t)}{dt^{v_3}}=-\frac{11}{6}n_3(t)+\frac{1}{2}a_1(t)+\frac{1}{3}a_2(t). \quad (9.86)$$

Then, from (9.45), (9.58), and (8.75), we have the numerical computation of $a_1(t+\Delta t)$, $a_2(t+\Delta t)$, $a_3(t+\Delta t)$, and $E(t+\Delta t)$, respectively. Similarly, we can achieve the numerical implementation of FHNN when S is equal to any positive integer.

9.4.4.2 Analysis of Stability and Convergence of FHNN

In this subsection, we analyze the stability and convergence of FHNN. We evaluate the convergence trajectory performance of FHNN's output and its Lyapunov function, and further study its equilibrium points and attractors.

Example 1: suppose there are only two fractional neurons of FHNN, and each neuron has the same fractional-order. Thus, $S=2$ and $v_1=v_2=v$. We set $\gamma=1.40$. From (9.82), the Lyapunov function of FHNN can be shown in Fig. 9.4 when $v=1.50$, $v=3.50$, $v=5.50$, and $v=7.50$.

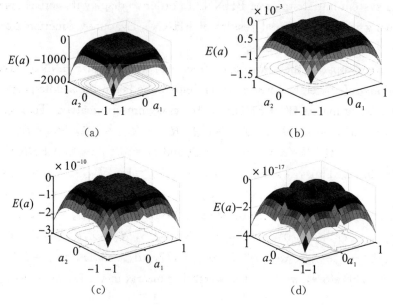

Fig. 9.13 Lyapunov function of FHNN ($\gamma=1.40$): (a) $v=1.50$, (b) $v=3.50$, (c) $v=5.50$, (d) $v=7.50$.

From (9.71) and (9.72), it can be seen that the output voltage of FHNN is limited to $\{\boldsymbol{a}: -1 \leqslant a_i \leqslant 1\}$ by the transfer function of the operational amplifier A. Thus, from Fig. 9.13, we can see that the Lyapunov function of FHNN is limited to a minimum at any of the four corners of hypercube. Therefore, we have the minimum of FHNN's Lyapunov function when (a_1, a_2) is equal to $(-1, -1)$, $(-1, 1)$, $(1, -1)$ and $(1, 1)$, respectively. From Fig. 9.13, we can also see that $(0, 0)$ is the saddle point of FHNN's Lyapunov function, and $a_1=0$ or $a_2=0$ is the ridge of its Lyapunov function. The ridge of FHNN's Lyapunov function is more pronounced when the fractional-order v of FHNN's neuron is greater.

In the above example, we select $v=7.50$ and $\gamma=1.40$. The contour line of FHNN's Lyapunov function can be shown in Fig. 9.14.

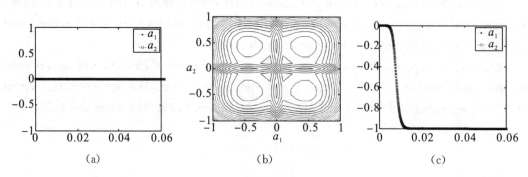

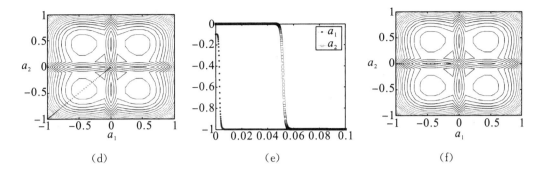

Fig. 9.14 Contour line of FHNN's Lyapunov function ($v=7.50$, $\gamma=1.40$): (a) Convergence trajectory of FHNN's output (saddle point: $n_1(0)=0.00$, $n_2(0)=0.00$), (b) Contour line of FHNN's Lyapunov function ($n_1(0)=0.00$, $n_2(0)=0.00$), (c) Convergence trajectory of FHNN's output ($n_1(0)=-0.0001$, $n_2(0)=-0.0001$), (d) Contour line of FHNN's Lyapunov function ($n_1(0)=-0.0001$, $n_2(0)=-0.0001$), (e) Convergence trajectory of FHNN's output ($n_1(0)=-0.10$, $n_2(0)=0.00$), (f) Contour line of FHNN's Lyapunov function ($n_1(0)=-0.10$, $n_2(0)=0.00$).

From Fig. 9.14 (a) and 9.14 (b), we can see that the output of FHNN (a_1, a_2) is identically equal to (0, 0) when the input of FHNN (n_1, n_2) is equal to (0, 0). Hence, (0, 0) is the saddle point of the Lyapunov function of FHNN. Also, from Fig. 9.14 (c) and 9.14 (d), we can also see that when (n_1, n_2) is equal to (-0.0001, -0.0001), which is very close to the saddle point (0, 0), (a_1, a_2) converges to one of the equilibrium points (-1, -1) of FHNN. From the discussion on (9.60) — (9.61) mentioned above, we determine the equilibrium points of FHNN's Lyapunov function according to LaSalle's invariance theorem and fractional steepest descent approach [192]. Each of its optimal searching adjustment step is in the negative direction of Lyapunov function's fractional gradient but not of its first-order one. Thus, its convergence trajectory easy passes through the first-order local minimum and maximum points of FHNN's Lyapunov function. Furthermore, from Fig. 9.14 (e) and 9.14 (f), we can further see that when (n_1, n_2) is equal to (-0.10, 0.00), which is on the ridge of FHNN's Lyapunov function, the convergence trajectory of (a_1, a_2) passes along the ridge of FHNN's Lyapunov function $a_2=0$ firstly, and then passes along the boundary of the hypercube $a_1=0$. It finally converges to one of the equilibrium points of FHNN, i.e. (-1, -1).

In order to analyze the effect of FHNN's fractional-order v on its stability and convergence, we set the fractional-order of the neuron to $v=0.50$, $v=1.50$ and $v=2.50$, respectively. Its convergence trajectory performance can be shown in Fig. 9.15.

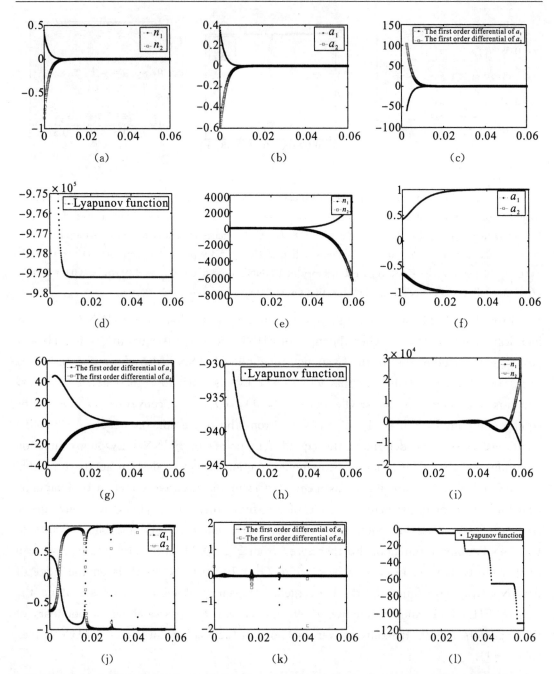

Fig. 9.15 Effect of fractional-order v of neuron on stability and convergence of FHNN ($\gamma=1.40$, $n_1(0)=0.50$, $n_2(0)=-1.00$): (a) $n_1(t)$ and $n_2(t)$ ($v=0.50$), (b) $a_1(t)$ and $a_2(t)$ ($v=0.50$), (c) First-order differential of $a_1(t)$ and $a_2(t)$ ($v=0.50$), (d) Lyapunov function of FHNN ($v=0.50$), (e) $n_1(t)$ and $n_2(t)$ ($v=1.50$), (f) $a_1(t)$ and $a_2(t)$ ($v=1.50$), (g) First-order differential of $a_1(t)$ and $a_2(t)$ ($v=1.50$), (h) Lyapunov function of FHNN ($v=1.50$), (i) $n_1(t)$ and $n_2(t)$ ($v=2.50$), (j) $a_1(t)$ and $a_2(t)$ ($v=2.50$), (k) First-order differential of $a_1(t)$ and $a_2(t)$ ($v=2.50$), (l) Lyapunov function of FHNN ($v=2.50$).

From Fig. 9.15 (a) — (d), it can be seen that when $v=0.50$, $n_1(t)$ and $n_2(t)$

converge to zero gradually. Meanwhile, $a_1(t)$, $a_2(t)$, and their first-order differential converge to zero correspondingly. From Fig. 9.15 (e) — (h), it can be also seen that when $v=1.50$, $n_1(t)$ and $n_2(t)$ nonlinearly amplify step-by-step. Consequently, $a_1(t)$ and $a_2(t)$ are limited to $\{a: -1 \leqslant a_i \leqslant 1\}$ by the transfer function of the operational amplifier A. Therefore, (a_1, a_2) converges to $(1, -1)$, and their first-order differential converges to zero accordingly. From Fig. 9.15 (i) — (l), it can be further seen that when $v=2.50$, $n_1(t)$ and $n_2(t)$ nonlinearly amplify gradually, while at the same time, alternate between positive and negative. Affected by the crossfade alternation of $n_1(t)$ and $n_2(t)$, $a_1(t)$ and $a_2(t)$ periodically alternate between 1 and -1, respectively. Similarly, we can summarize the corresponding relationship between FHNN's fractional-order and its convergence using mathematical induction. It can be shown in Table 9.1.

Table 9.1 Corresponding relationship between FHNN's fractional-order and its convergence.

Fractional Order v	Convergent yes/no	Fractional Order v	Convergent yes/no
$0<v<1$	yes	$4<v<5$	no
$v=1$	yes	$v=5$	no
$1<v<2$	yes	$5<v<6$	yes
$v=2$	no	$v=6$	no
$2<v<3$	no	$2k-2<v<2k-1$	no
$v=3$	no	$v=2k-1$	no
$3<v<4$	yes	$2k-1<v<2k$	yes
$v=4$	no	$v=2k$	no

From Table 9.1, we can see that FHNN is convergent when $0.00<v<1.00$, $v=1.00$ and $2k-1<v<2k$, where k is a positive integer. Particularly, FHNN is the classical first-order HNN when $v=1.00$. HNN is a special case of FHNN.

Per the aforementioned discussion, the output of FHNN is limited to $\{a: -1 \leqslant a_i \leqslant 1\}$ by the transfer function of the operational amplifier A. Hence, we focus on discussing the case in which the input voltage of FHNN is limited to $\{n: -1 \leqslant n_i \leqslant 1\}$. For the case of $\{n: n_i < -1\}$ or $\{n: 1 < n_i\}$ the convergence trajectory performance of FHNN is similar. Then, from (46), (47), (55), (56), and (57), we have the convergence trajectory performance of FHNN. Then, we have corresponding relationship between FHNN's input and its output. It can be shown in Table 9.2.

Table 9.2 Corresponding relationship between FHNN's input and its output: (a) $0.00<v<1.00$, $\gamma=1.40$, (b) $1<v<2$, $\gamma=1.40$, (c) $3<v<4$, $\gamma=1.40$, (d) $2k-1<v<2k$, $\gamma=1.40$.

(a)

Input n_1, n_2	Output a_1, a_2	Input n_1, n_2	Output a_1, a_2
−1.0, −1.0	0.0, 0.0	0.5, 0.0	0.0, 0.0
−0.5, −1.0	0.0, 0.0	1.0, 0.0	0.0, 0.0
0.0, −1.0	0.0, 0.0	−1.0, 0.5	0.0, 0.0
0.5, −1.0	0.0, 0.0	−0.5, 0.5	0.0, 0.0
1.0, −1.0	0.0, 0.0	0.0, 0.5	0.0, 0.0
−1.0, −0.5	0.0, 0.0	0.5, 0.5	0.0, 0.0
−0.5, −0.5	0.0, 0.0	1.0, 0.5	0.0, 0.0
0.0, −0.5	0.0, 0.0	−1.0, 1.0	0.0, 0.0
0.5, −0.5	0.0, 0.0	−0.5, 1.0	0.0, 0.0
1.0, −0.5	0.0, 0.0	0.0, 1.0	0.0, 0.0
−1.0, 0.0	0.0, 0.0	0.5, 1.0	0.0, 0.0
−0.5, 0.0	0.0, 0.0	1.0, 1.0	0.0, 0.0

(b)

Input n_1, n_2	Output a_1, a_2	Input n_1, n_2	Output a_1, a_2
−1.0, −1.0	−1.0, −1.0	0.5, 0.0	1.0, 1.0
−0.5, −1.0	−1.0, −1.0	1.0, 0.0	1.0, 1.0
0.0, −1.0	−1.0, −1.0	−1.0, 0.5	−1.0, 1.0
0.5, −1.0	1.0, −1.0	−0.5, 0.5	−1.0, 1.0
1.0, −1.0	1.0, −1.0	0.0, 0.5	1.0, 1.0
−1.0, −0.5	−1.0, −1.0	0.5, 0.5	1.0, 1.0
−0.5, −0.5	−1.0, −1.0	1.0, 0.5	1.0, 1.0
0.0, −0.5	−1.0, −1.0	−1.0, 1.0	−1.0, 1.0
0.5, −0.5	1.0, −1.0	−0.5, 1.0	−1.0, 1.0
1.0, −0.5	1.0, −1.0	0.0, 1.0	1.0, 1.0
−1.0, 0.0	−1.0, −1.0	0.5, 1.0	1.0, 1.0
−0.5, 0.0	−1.0, −1.0	0.5, 1.0	1.0, 1.0

(c)

Input n_1, n_2	Output a_1, a_2	Input n_1, n_2	Output a_1, a_2
−1.0, −1.0	−1.0, −1.0	0.5, 0.0	1.0, 1.0
−0.5, −1.0	−1.0, −1.0	1.0, 0.0	1.0, 1.0

(Continued)

Input n_1, n_2	Output a_1, a_2	Input n_1, n_2	Output a_1, a_2
0.0, −1.0	−1.0, −1.0	−1.0, 0.5	−1.0, 1.0
0.5, −1.0	1.0, −1.0	−0.5, 0.5	−1.0, 1.0
1.0, −1.0	1.0, −1.0	0.0, 0.5	1.0, 1.0
−1.0, −0.5	−1.0, −1.0	0.5, 0.5	1.0, 1.0
−0.5, −0.5	−1.0, −1.0	1.0, 0.5	1.0, 1.0
0.0, −0.5	−1.0, −1.0	−1.0, 1.0	−1.0, 1.0
0.5, −0.5	1.0, −1.0	−0.5, 1.0	−1.0, 1.0
1.0, −0.5	1.0, −1.0	0.0, 1.0	1.0, 1.0
−1.0, 0.0	−1.0, −1.0	0.5, 1.0	1.0, 1.0
−0.5, 0.0	−1.0, −1.0	1.0, 1.0	1.0, 1.0

(d)

Input n_1, n_2	Output a_1, a_2	Input n_1, n_2	Output a_1, a_2
−1.0, −1.0	−1.0, −1.0	0.5, 0.0	1.0, 1.0
−0.5, −1.0	−1.0, −1.0	1.0, 0.0	1.0, 1.0
0.0, −1.0	−1.0, −1.0	−1.0, 0.5	−1.0, 1.0
0.5, −1.0	1.0, −1.0	−0.5, 0.5	−1.0, 1.0
1.0, −1.0	1.0, −1.0	0.0, 0.5	1.0, 1.0
−1.0, −0.5	−1.0, −1.0	0.5, 0.5	1.0, 1.0
−0.5, −0.5	−1.0, −1.0	1.0, 0.5	1.0, 1.0
0.0, −0.5	−1.0, −1.0	−1.0, 1.0	−1.0, 1.0
0.5, −0.5	1.0, −1.0	−0.5, 1.0	−1.0, 1.0
1.0, −0.5	1.0, −1.0	0.0, 1.0	1.0, 1.0
−1.0, 0.0	−1.0, −1.0	0.5, 1.0	1.0, 1.0
−0.5, 0.0	−1.0, −1.0	1.0, 1.0	1.0, 1.0

From Table 9.2, we can see that corresponding to various inputs (n_1, n_2), output (a_1, a_2) converges to (0, 0) when $0.00 < v < 1.00$. Furthermore, output (a_1, a_2) converges to (−1, −1), (−1, 1), (1, −1) or, (1, 1) when $1.00 < v < 2.00$, $3.00 < v < 4.00$ and $2k-1 < v < 2k$, where k is a positive integer. Hence, (−1, −1), (−1, 1), (1, −1), and (1, 1) are four attractors of FHNN when $1.00 < v < 2.00$, $3.00 < v < 4.00$ and $2k-1 < v < 2k$. Furthermore, it should be noted that FHNN is the classical first-order HNN when $v = 1.00$. HNN is a special case of FHNN. As we all known, the convergence rule of HNN is related to operational amplifier's gain coefficient γ and weighting matrix $\boldsymbol{W}$

of HNN [277], [295] − [296], [350]. To avoid repetition, its convergence rule is not described in this article. Then, the convergence trajectory of FHNN in two-dimensional space can be shown in Fig. 9.16.

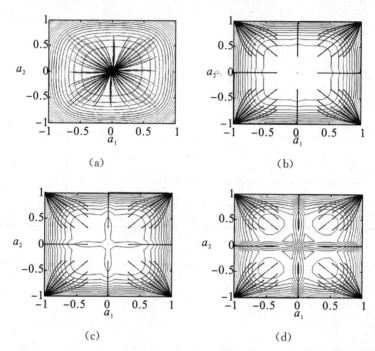

Fig. 9.16 Convergence trajectory of FHNN in two-dimensional space. (a) $0.00<v<1.00$, $\gamma=1.40$, (b) $1.00<v<2.00$, $\gamma=1.40$, (c) $3.00<v<4.00$, $\gamma=1.40$, (d) $2k-1<v<2k$, $\gamma=1.40$.

From Fig. 9.16, we can see that, on the one hand, $(0, 0)$ is the single attractor of FHNN when $0.00<v<1.00$. On the other hand, $(-1, -1)$, $(-1, 1)$, $(1, -1)$, and $(1, 1)$ are four different equilibrium points or attractors, and $(0, 0)$ is the saddle point of FHNN's Lyapunov function when $1.00<v<2.00$, $3.00<v<4.00$, and $2k-1<v<2k$. Meanwhile, FHNN's fractional-order v can also affect the rate of FHNN's convergence. Furthermore, the convergence trajectory of FHNN easily passes through the first-order local minimum and maximum points of FHNN's Lyapunov function. When the input is on the ridge of FHNN's Lyapunov function, the convergence trajectory of the output passes along the ridge of FHNN's Lyapunov function firstly, and then passes along the boundary of the hypercube. It finally converges to one of FHNN's equilibrium points or attractors. Compared to the classical first-order HNN, it is known that, in general, a double-neuron HNN usually associatively memorizes two characteristics [277], [295] − [296], [350]. Hence, we can further see that FHNN has a stronger associative memory than HNN. As mentioned earlier, the double-neuron FHNN can associatively memorize four characteristics at most.

Example 2: suppose there are three fractional neurons of FHNN, and each of its neuron has the same fractional-order. Thus, $S=3$ and $v_1=v_2=v_3=v$. We also set $\gamma=$

1.40. Firstly, suppose that the weighting matrix **W** of FHNN is a symmetric matrix. From (9.83) — (9.86), corresponding relationship between FHNN's input and its output can be shown in Table 9.3.

Table 9.3 Corresponding relationship between FHNN's input and its output ($S=3$, $2k-1<v<2k$, $\gamma=1.40$).

Input n_1, n_2, n_3	Output a_1, a_2, a_3	Input n_1, n_2, n_3	Output a_1, a_2, a_3
−1.0, −1.0, −1.0	−1.0, −1.0, −1.0	1.0, 0.0, 0.0	1.0, 1.0, 1.0
0.0, −1.0, −1.0	−1.0, −1.0, −1.0	−1.0, 1.0, 0.0	−1.0, 1.0, −1.0
1.0, −1.0, −1.0	1.0, −1.0, −1.0	0.0, 1.0, 0.0	1.0, 1.0, 1.0
−1.0, 0.0, −1.0	−1.0, −1.0, −1.0	1.0, 1.0, 0.0	1.0, 1.0, 1.0
0.0, 0.0, −1.0	−1.0, −1.0, −1.0	−1.0, −1.0, 1.0	−1.0, −1.0, 1.0
1.0, 0.0, −1.0	1.0, 1.0, −1.0	0.0, −1.0, 1.0	−1.0, −1.0, 1.0
−1.0, 1.0, −1.0	−1.0, 1.0, −1.0	1.0, −1.0, 1.0	1.0, −1.0, 1.0
0.0, 1.0, −1.0	1.0, 1.0, −1.0	−1.0, 0.0, 1.0	−1.0, −1.0, 1.0
1.0, 1.0, −1.0	1.0, 1.0, −1.0	0.0, 0.0, 1.0	1.0, 1.0, 1.0
−1.0, −1.0, 0.0	−1.0, −1.0, −1.0	1.0, 0.0, 1.0	1.0, 1.0, 1.0
0.0, −1.0, 0.0	−1.0, −1.0, −1.0	−1.0, 1.0, 1.0	−1.0, 1.0, 1.0
1.0, −1.0, 0.0	1.0, −1.0, 1.0	0.0, 1.0, 1.0	1.0, 1.0, 1.0
−1.0, 0.0, 0.0	−1.0, −1.0, −1.0	1.0, 1.0, 1.0	1.0, 1.0, 1.0

From Table 9.3, we can see that corresponding to various inputs (n_1, n_2, n_3), the output (a_1, a_2, a_3) converges to (−1, −1, −1), (1, −1, −1), (−1, 1, −1), (1, 1, −1), (−1, −1, 1), (1, −1, 1), (−1, 1, 1), or (1, 1, 1) when $2k-1<v<2k$, where k is a positive integer. Then, the convergence trajectory of FHNN in three-dimensional space can be shown in Fig. 9.17.

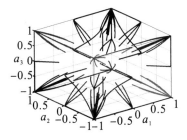

Fig. 9.17 Convergence trajectory of FHNN in three-dimensional space ($S=3$, $2k-1<v<2k$, $\gamma=1.40$).

From Fig. 9.17, we can see that (−1, −1, −1), (1, −1, −1), (−1, 1, −1), (1, 1, −1), (−1, −1, 1), (1, −1, 1), (−1, 1, 1), and (1, 1, 1) are eight different equilibrium points or attractors, and (0, 0, 0) is the saddle point of FHNN's Lyapunov function when $2k-1<v<2k$, where k is positive integer. In Fig. 9.17, we use different colored (red, blue, green, and purple) convergence trajectories to demonstrate eight domains of attraction corresponding to aforementioned eight attractors of FHNN, respectively. Compared to the classical first-order HNN, it is known that in general, the trio-neuron HNN usually associatively memorizes two characteristics [277], [295] – [296], [350]. Hence, we can further see that FHNN has a stronger associative memory than HNN. Per the aforementioned discussion, the trio-neuron FHNN can associatively memorize eight characteristics at most.

Example 3: suppose there are two fractional neurons of FHNN, but each neuron has a different fractional-order. Thus, $S=2$ and $v_1 \neq v_2$. We set $v_1=0.50$, $v_2=1.50$ and $v_1=1.50$, $v_2=0.50$, respectively. We also set $\gamma=1.40$. Then, from (9.71), (9.72), (9.80), (9.81), and (9.82), the corresponding relationship between the input and output of FHNN can be shown in Table 9.4.

Table 9.4 Corresponding relationship between FHNN's input and its output ($\gamma=1.40$): (a) $v_1=0.50$, $v_2=1.50$, (b) $v_1=1.50$, $v_2=0.50$.

(a)

Input n_1, n_2	Output a_1, a_2	Input n_1, n_2	Output a_1, a_2
−1.0, −1.0	−0.0685, −1.0	0.5, 0.0	0.0685, 1.0
−0.5, −1.0	−0.0685, −1.0	1.0, 0.0	0.0685, 1.0
0.0, −1.0	−0.0685, −1.0	−1.0, 0.5	0.0685, 1.0
0.5, −1.0	−0.0685, −1.0	−0.5, 0.5	0.0685, 1.0
1.0, −1.0	−0.0685, −1.0	0.0, 0.5	0.0685, 1.0
−1.0, −0.5	−0.0685, −1.0	0.5, 0.5	0.0685, 1.0
−0.5, −0.5	−0.0685, −1.0	1.0, 0.5	0.0685, 1.0
0.0, −0.5	−0.0685, −1.0	−1.0, 1.0	0.0685, 1.0
0.5, −0.5	−0.0685, −1.0	−0.5, 1.0	0.0685, 1.0
1.0, −0.5	−0.0685, −1.0	0.0, 1.0	0.0685, 1.0
−1.0, 0.0	−0.0685, −1.0	0.5, 1.0	0.0685, 1.0
−0.5, 0.0	−0.0685, −1.0	1.0, 1.0	0.0685, 1.0

(b)

Input n_1, n_2	Output a_1, a_2	Input n_1, n_2	Output a_1, a_2
−1.0, −1.0	−1.0, −0.0685	0.5, 0.0	1.0, 0.0685
−0.5, −1.0	−1.0, −0.0685	1.0, 0.0	1.0, 0.0685
0.0, −1.0	−1.0, −0.0685	−1.0, 0.5	−1.0, −0.0685
0.5, −1.0	1.0, 0.0685	−0.5, 0.5	−1.0, −0.0685

(**Continued**)

Input n_1, n_2	Output a_1, a_2	Input n_1, n_2	Output a_1, a_2
1.0, −1.0	1.0, 0.0685	0.0, 0.5	1.0, 0.0685
−1.0, −0.5	−1.0, −0.0685	0.5, 0.5	1.0, 0.0685
−0.5, −0.5	−1.0, −0.0685	1.0, 0.5	1.0, 0.0685
0.0, −0.5	−1.0, −0.0685	−1.0, 1.0	−1.0, −0.0685
0.5, −0.5	1.0, 0.0685	−0.5, 1.0	−1.0, −0.0685
1.0, −0.5	1.0, 0.0685	0.0, 1.0	1.0, 0.0685
−1.0, 0.0	−1.0, −0.0685	0.5, 1.0	1.0, 0.0685
−0.5, 0.0	−1.0, −0.0685	1.0, 1.0	1.0, 0.0685

From Table 9.4, we can see that on the one hand, corresponding to various inputs (n_1, n_2), the output (a_1, a_2) converges to $(-0.0685, -1.0)$ or $(0.0685, 1.0)$ when $v_1 = 0.50$ and $v_2 = 1.50$. On the other hand, corresponding to various inputs (n_1, n_2), the output (a_1, a_2) converges to $(-1.0, -0.0685)$ or $(1.0, 0.0685)$, when $v_1 = 1.50$ and $v_2 = 0.50$, respectively. Thus, it can be seen that the coordinates of FHNN's equilibrium points or attractors has a $\pi/2$ counterclockwise rotation while the values of v_1 and v_2 have been exchanged. Then, FHNN's Lyapunov function and convergence trajectory of FHNN's output and are shown in Fig. 9.18.

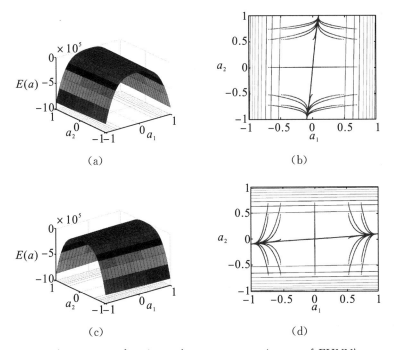

Fig. 9.18 FHNN's Lyapunov function and convergence trajectory of FHNN's output ($\gamma = 1.40$): (a) Lyapunov function ($v_1 = 0.50$, $v_2 = 1.50$), (b) Convergence trajectory ($v_1 = 0.50$, $v_2 = 1.50$), (c) Lyapunov function ($v_1 = 1.50$, $v_2 = 0.50$), (d) Convergence trajectory ($v_1 = 1.50$, $v_2 = 0.50$).

From Fig. 9.18, it can be seen that, on the one hand, $(-0.0685, -1.0)$ and $(0.0685, 1.0)$ are two different equilibrium points of FHNN's Lyapunov function or attractors of FHNN when $v_1=0.50$ and $v_2=1.50$. We have global maximum of FHNN's Lyapunov function at the points of $(-0.0685, -1.0)$ and $(0.0685, 1.0)$. On the other hand, $(-1.0, -0.0685)$ and $(1.0, 0.0685)$ are two different attractors of FHNN when $v_1=1.50$ and $v_2=0.50$. We have a global maximum of FHNN's Lyapunov function at the points of $(-1.0, -0.0685)$ and $(1.0, 0.0685)$. In either case, $(0, 0)$ is the saddle point of FHNN. Compared with Fig. 9.16 and Fig. 9.18, we can further see that there are only two, but not four, attractors of FHNN whenever its two neurons satisfy $0<v_i<1$ and $2k-1<v_j<2k$, where k is the positive integer, and $i \neq j$. Therefore, we can change the number of the attractors of FHNN by means of altering the fractional-order of the neuron.

Furthermore, we set $v_1=0.25$, $v_2=1.50$ and $v_1=0.75$, $v_2=1.50$. Then, the corresponding relationship between the input and output of FHNN can be shown in Table 9.5.

Table 9.5 Corresponding relationship between input and output of FHNN ($\gamma=1.40$): (a) $v_1=0.25$, $v_2=1.50$, (b) $v_1=0.75$, $v_2=1.50$.

(a)

Input n_1, n_2	Output a_1, a_2	Input n_1, n_2	Output a_1, a_2
$-1.0, -1.0$	$-0.1672, -1.0$	$0.5, 0.0$	$0.1672, 1.0$
$-0.5, -1.0$	$-0.1672, -1.0$	$1.0, 0.0$	$0.1672, 1.0$
$0.0, -1.0$	$-0.1672, -1.0$	$-1.0, 0.5$	$0.1672, 1.0$
$0.5, -1.0$	$-0.1672, -1.0$	$-0.5, 0.5$	$0.1672, 1.0$
$1.0, -1.0$	$-0.1672, -1.0$	$0.0, 0.5$	$0.1672, 1.0$
$-1.0, -0.5$	$-0.1672, -1.0$	$0.5, 0.5$	$0.1672, 1.0$
$-0.5, -0.5$	$-0.1672, -1.0$	$1.0, 0.5$	$0.1672, 1.0$
$0.0, -0.5$	$-0.1672, -1.0$	$-1.0, 1.0$	$0.1672, 1.0$
$0.5, -0.5$	$-0.1672, -1.0$	$-0.5, 1.0$	$0.1672, 1.0$
$1.0, -0.5$	$-0.1672, -1.0$	$0.0, 1.0$	$0.1672, 1.0$
$-1.0, 0.0$	$-0.1672, -1.0$	$0.5, 1.0$	$0.1672, 1.0$
$-0.5, 0.0$	$-0.1672, -1.0$	$1.0, 1.0$	$0.1672, 1.0$

(b)

Input n_1, n_2	Output a_1, a_2	Input n_1, n_2	Output a_1, a_2
$-1.0, -1.0$	$-0.0322, -1.0$	$0.5, 0.0$	$0.0322, 1.0$

(Continued)

Input n_1, n_2	Output a_1, a_2	Input n_1, n_2	Output a_1, a_2
−0.5, −1.0	−0.0322, −1.0	1.0, 0.0	0.0322, 1.0
0.0, −1.0	−0.0322, −1.0	−1.0, 0.5	0.0322, 1.0
0.5, −1.0	−0.0322, −1.0	−0.5, 0.5	0.0322, 1.0
1.0, −1.0	−0.0322, −1.0	0.0, 0.5	0.0322, 1.0
−1.0, −0.5	−0.0322, −1.0	0.5, 0.5	0.0322, 1.0
−0.5, −0.5	−0.0322, −1.0	1.0, 0.5	0.0322, 1.0
0.0, −0.5	−0.0322, −1.0	−1.0, 1.0	0.0322, 1.0
0.5, −0.5	−0.0322, −1.0	−0.5, 1.0	0.0322, 1.0
1.0, −0.5	−0.0322, −1.0	0.0, 1.0	0.0322, 1.0
−1.0, 0.0	−0.0322, −1.0	0.5, 1.0	0.0322, 1.0
−0.5, 0.0	−0.0322, −1.0	1.0, 1.0	0.0322, 1.0

From Table 9.5, it can be seen that, on the one hand, (−0.1672, −1.0) and (0.1672, 1.0) are two different equilibrium points of FHNN's Lyapunov function or attractors of FHNN when $v_1 = 0.25$ and $v_2 = 1.50$. On the other hand, (−0.0322, −1.0) and (0.0322, 1.0) are two different attractors of FHNN when $v_1 = 0.75$ and $v_2 = 1.50$. Therefore, we can change the value of FHNN's attractors by means of altering the fractional-order of the neuron.

Example 4: suppose there are three fractional neurons of FHNN, but each neuron has different fractional-order. Thus, $S = 3$ and $v_1 \neq v_2 \neq v_3$. We also set $\gamma = 1.40$. Then, from (9.83) − (9.86), the corresponding relationship between the input and the output of FHNN according to various permutation of v_1, v_2 and v_3 can be shown in Table 9.6.

Table 9.6 Corresponding relationship between input and output of FHNN ($S = 3$, $\gamma = 1.40$): (a) $v_1 = 0.25$, $v_2 = 0.75$, $v_3 = 1.50$, (b) $v_1 = 0.25$, $v_2 = 1.50$, $v_3 = 0.75$, (c) $v_1 = 0.75$, $v_2 = 0.25$, $v_3 = 1.50$, (d) $v_1 = 0.75$, $v_2 = 1.50$, $v_3 = 0.25$, (e) $v_1 = 1.50$, $v_2 = 0.25$, $v_3 = 0.75$, (f) $v_1 = 1.50$, $v_2 = 0.75$, $v_3 = 0.25$.

(a)

Input n_1, n_2, n_3	Output a_1, a_2, a_3	Input n_1, n_2, n_3	Output a_1, a_2, a_3
−1.0, −1.0, −1.0	−0.0805, −0.0132, −1.0	1.0, 0.0, 0.0	0.0805, 0.0132, 1.0
0.0, −1.0, −1.0	−0.0805, −0.0132, −1.0	−1.0, 1.0, 0.0	0.0805, 0.0132, 1.0
1.0, −1.0, −1.0	−0.0805, −0.0132, −1.0	0.0, 1.0, 0.0	0.0805, 0.0132, 1.0
−1.0, 0.0, −1.0	−0.0805, −0.0132, −1.0	1.0, 1.0, 0.0	0.0805, 0.0132, 1.0
0.0, 0.0, −1.0	−0.0805, −0.0132, −1.0	−1.0, −1.0, 1.0	0.0805, 0.0132, 1.0

(**Continued**)

Input n_1, n_2, n_3	Output a_1, a_2, a_3	Input n_1, n_2, n_3	Output a_1, a_2, a_3
1.0, 0.0, −1.0	−0.0805, −0.0132, −1.0	0.0, −1.0, 1.0	0.0805, 0.0132, 1.0
−1.0, 1.0, −1.0	−0.0805, −0.0132, −1.0	1.0, −1.0, 1.0	0.0805, 0.0132, 1.0
0.0, 1.0, −1.0	−0.0805, −0.0132, −1.0	−1.0, 0.0, 1.0	0.0805, 0.0132, 1.0
1.0, 1.0, −1.0	−0.0805, −0.0132, −1.0	0.0, 0.0, 1.0	0.0805, 0.0132, 1.0
−1.0, −1.0, 0.0	−0.0805, −0.0132, −1.0	1.0, 0.0, 1.0	0.0805, 0.0132, 1.0
0.0, −1.0, 0.0	−0.0805, −0.0132, −1.0	−1.0, 1.0, 1.0	0.0805, 0.0132, 1.0
1.0, −1.0, 0.0	−0.0805, −0.0132, −1.0	0.0, 1.0, 1.0	0.0805, 0.0132, 1.0
−1.0, 0.0, 0.0	−0.0805, −0.0132, −1.0	1.0, 1.0, 1.0	0.0805, 0.0132, 1.0

(b)

Input n_1, n_2, n_3	Output a_1, a_2, a_3	Input n_1, n_2, n_3	Output a_1, a_2, a_3
−1.0, −1.0, −1.0	−0.1556, −1.0, −0.0133	1.0, 0.0, 0.0	0.1556, 1.0, 0.0133
0.0, −1.0, −1.0	−0.1556, −1.0, −0.0133	−1.0, 1.0, 0.0	0.1556, 1.0, 0.0133
1.0, −1.0, −1.0	−0.1556, −1.0, −0.0133	0.0, 1.0, 0.0	0.1556, 1.0, 0.0133
−1.0, 0.0, −1.0	−0.1556, −1.0, −0.0133	1.0, 1.0, 0.0	0.1556, 1.0, 0.0133
0.0, 0.0, −1.0	−0.1556, −1.0, −0.0133	−1.0, −1.0, 1.0	−0.1556, −1.0, −0.0133
1.0, 0.0, −1.0	−0.1556, −1.0, −0.0133	0.0, −1.0, 1.0	−0.1556, −1.0, −0.0133
−1.0, 1.0, −1.0	0.1556, 1.0, 0.0133	1.0, −1.0, 1.0	−0.1556, −1.0, −0.0133
0.0, 1.0, −1.0	0.1556, 1.0, 0.0133	−1.0, 0.0, 1.0	0.1556, 1.0, 0.0133
1.0, 1.0, −1.0	0.1556, 1.0, 0.0133	0.0, 0.0, 1.0	0.1556, 1.0, 0.0133
−1.0, −1.0, 0.0	−0.1556, −1.0, −0.0133	1.0, 0.0, 1.0	0.1556, 1.0, 0.0133
0.0, −1.0, 0.0	−0.1556, −1.0, −0.0133	−1.0, 1.0, 1.0	0.1556, 1.0, 0.0133
1.0, −1.0, 0.0	−0.1556, −1.0, −0.0133	0.0, 1.0, 1.0	0.1556, 1.0, 0.0133
−1.0, 0.0, 0.0	−0.1556, −1.0, −0.0133	1.0, 1.0, 1.0	0.1556, 1.0, 0.0133

(c)

Input n_1, n_2, n_3	Output a_1, a_2, a_3	Input n_1, n_2, n_3	Output a_1, a_2, a_3
−1.0, −1.0, −1.0	−0.0177, −0.0567, −1.0	1.0, 0.0, 0.0	0.0177, 0.0567, 1.0
0.0, −1.0, −1.0	−0.0177, −0.0567, −1.0	−1.0, 1.0, 0.0	−0.0177, −0.0567, −1.0
1.0, −1.0, −1.0	−0.0177, −0.0567, −1.0	0.0, 1.0, 0.0	0.0177, 0.0567, 1.0
−1.0, 0.0, −1.0	−0.0177, −0.0567, −1.0	1.0, 1.0, 0.0	0.0177, 0.0567, 1.0
0.0, 0.0, −1.0	−0.0177, −0.0567, −1.0	−1.0, −1.0, 1.0	0.0177, 0.0567, 1.0

(Continued)

Input n_1, n_2, n_3	Output a_1, a_2, a_3	Input n_1, n_2, n_3	Output a_1, a_2, a_3
1.0, 0.0, −1.0	−0.0177, −0.0567, −1.0	0.0, −1.0, 1.0	0.0177, 0.0567, 1.0
−1.0, 1.0, −1.0	−0.0177, −0.0567, −1.0	1.0, −1.0, 1.0	0.0177, 0.0567, 1.0
0.0, 1.0, −1.0	−0.0177, −0.0567, −1.0	−1.0, 0.0, 1.0	0.0177, 0.0567, 1.0
1.0, 1.0, −1.0	−0.0177, −0.0567, −1.0	0.0, 0.0, 1.0	0.0177, 0.0567, 1.0
−1.0, −1.0, 0.0	−0.0177, −0.0567, −1.0	1.0, 0.0, 1.0	0.0177, 0.0567, 1.0
0.0, −1.0, 0.0	−0.0177, −0.0567, −1.0	−1.0, 1.0, 1.0	0.0177, 0.0567, 1.0
1.0, −1.0, 0.0	0.0177, 0.0567, 1.0	0.0, 1.0, 1.0	0.0177, 0.0567, 1.0
−1.0, 0.0, 0.0	−0.0177, −0.0567, −1.0	1.0, 1.0, 1.0	0.0177, 0.0567, 1.0

(d)

Input n_1, n_2, n_3	Output a_1, a_2, a_3	Input n_1, n_2, n_3	Output a_1, a_2, a_3
−1.0, −1.0, −1.0	−0.0327, −1.0, −0.0614	1.0, 0.0, 0.0	0.0327, 1.0, 0.0614
0.0, −1.0, −1.0	−0.0327, −1.0, −0.0614	−1.0, 1.0, 0.0	0.0327, 1.0, 0.0614
1.0, −1.0, −1.0	−0.0327, −1.0, −0.0614	0.0, 1.0, 0.0	0.0327, 1.0, 0.0614
−1.0, 0.0, −1.0	−0.0327, −1.0, −0.0614	1.0, 1.0, 0.0	0.0327, 1.0, 0.0614
0.0, 0.0, −1.0	−0.0327, −1.0, −0.0614	−1.0, −1.0, 1.0	−0.0327, −1.0, −0.0614
1.0, 0.0, −1.0	0.0327, 1.0, 0.0614	0.0, −1.0, 1.0	−0.0327, −1.0, −0.0614
−1.0, 1.0, −1.0	0.0327, 1.0, 0.0614	1.0, −1.0, 1.0	−0.0327, −1.0, −0.0614
0.0, 1.0, −1.0	0.0327, 1.0, 0.0614	−1.0, 0.0, 1.0	−0.0327, −1.0, −0.0614
1.0, 1.0, −1.0	0.0327, 1.0, 0.0614	0.0, 0.0, 1.0	0.0327, 1.0, 0.0614
−1.0, −1.0, 0.0	−0.0327, −1.0, −0.0614	1.0, 0.0, 1.0	0.0327, 1.0, 0.0614
0.0, −1.0, 0.0	−0.0327, −1.0, −0.0614	−1.0, 1.0, 1.0	0.0327, 1.0, 0.0614
1.0, −1.0, 0.0	−0.0327, −1.0, −0.0614	0.0, 1.0, 1.0	0.0327, 1.0, 0.0614
−1.0, 0.0, 0.0	−0.0327, −1.0, −0.0614	1.0, 1.0, 1.0	0.0327, 1.0, 0.0614

(e)

Input n_1, n_2, n_3	Output a_1, a_2, a_3	Input n_1, n_2, n_3	Output a_1, a_2, a_3
−1.0, −1.0, −1.0	−1.0, −0.1595, −0.0179	1.0, 0.0, 0.0	1.0, 0.1595, 0.0179
0.0, −1.0, −1.0	−1.0, −0.1595, −0.0179	−1.0, 1.0, 0.0	−1.0, −0.1595, −0.0179
1.0, −1.0, −1.0	1.0, 0.1595, 0.0179	0.0, 1.0, 0.0	1.0, 0.1595, 0.0179
−1.0, 0.0, −1.0	−1.0, −0.1595, −0.0179	1.0, 1.0, 0.0	1.0, 0.1595, 0.0179
0.0, 0.0, −1.0	−1.0, −0.1595, −0.0179	−1.0, −1.0, 1.0	−1.0, −0.1595, −0.0179

(Continued)

Input n_1, n_2, n_3	Output a_1, a_2, a_3	Input n_1, n_2, n_3	Output a_1, a_2, a_3
1.0, 0.0, −1.0	1.0, 0.1595, 0.0179	0.0, −1.0, 1.0	1.0, 0.1595, 0.0179
−1.0, 1.0, −1.0	−1.0, −0.1595, −0.0179	1.0, −1.0, 1.0	1.0, 0.1595, 0.0179
0.0, 1.0, −1.0	−1.0, −0.1595, −0.0179	−1.0, 0.0, 1.0	−1.0, −0.1595, −0.0179
1.0, 1.0, −1.0	1.0, 0.1595, 0.0179	0.0, 0.0, 1.0	1.0, 0.1595, 0.0179
−1.0, −1.0, 0.0	−1.0, −0.1595, −0.0179	1.0, 0.0, 1.0	1.0, 0.1595, 0.0179
0.0, −1.0, 0.0	−1.0, −0.1595, −0.0179	−1.0, 1.0, 1.0	−1.0, −0.1595, −0.0179
1.0, −1.0, 0.0	1.0, 0.1595, 0.0179	0.0, 1.0, 1.0	1.0, 0.1595, 0.0179
−1.0, 0.0, 0.0	−1.0, −0.1595, −0.0179	1.0, 1.0, 1.0	1.0, 0.1595, 0.0179

(f)

Input n_1, n_2, n_3	Output a_1, a_2, a_3	Input n_1, n_2, n_3	Output a_1, a_2, a_3
−1.0, −1.0, −1.0	−1.0, −0.0328, −0.0895	1.0, 0.0, 0.0	1.0, 0.0328, 0.0895
0.0, −1.0, −1.0	−1.0, −0.0328, −0.0895	−1.0, 1.0, 0.0	−1.0, −0.0328, −0.0895
1.0, −1.0, −1.0	1.0, 0.0328, 0.0895	0.0, 1.0, 0.0	1.0, 0.0328, 0.0895
−1.0, 0.0, −1.0	−1.0, −0.0328, −0.0895	1.0, 1.0, 0.0	1.0, 0.0328, 0.0895
0.0, 0.0, −1.0	−1.0, −0.0328, −0.0895	−1.0, −1.0, 1.0	−1.0, −0.0328, −0.0895
1.0, 0.0, −1.0	1.0, 0.0328, 0.0895	0.0, −1.0, 1.0	−1.0, −0.0328, −0.0895
−1.0, 1.0, −1.0	−1.0, −0.0328, −0.0895	1.0, −1.0, 1.0	1.0, 0.0328, 0.0895
0.0, 1.0, −1.0	1.0, 0.0328, 0.0895	−1.0, 0.0, 1.0	−1.0, −0.0328, −0.0895
1.0, 1.0, −1.0	1.0, 0.0328, 0.0895	0.0, 0.0, 1.0	1.0, 0.0328, 0.0895
−1.0, −1.0, 0.0	−1.0, −0.0328, −0.0895	1.0, 0.0, 1.0	1.0, 0.0328, 0.0895
0.0, −1.0, 0.0	−1.0, −0.0328, −0.0895	−1.0, 1.0, 1.0	−1.0, −0.0328, −0.0895
1.0, −1.0, 0.0	1.0, 0.0328, 0.0895	0.0, 1.0, 1.0	1.0, 0.0328, 0.0895
−1.0, 0.0, 0.0	−1.0, −0.0328, −0.0895	1.0, 1.0, 1.0	1.0, 0.0328, 0.0895

From Table 9.6, we can see that, on the one hand, with regard to the multi-neuron FHNN, the equilibrium points of FHNN's Lyapunov function or attractors of FHNN varies with the fractional-order of the neuron. On the other hand, compared to Table 9.3, Fig. 9.17 and Table 9.6, we can further see that there are only two, but not eight, attractors of FHNN whenever any two neurons of the trio-neuron FHNN satisfy $0 < v_i < 1$ and $2k - 1 < v_j < 2k$, where k is a positive integer, and $i \neq j$. Therefore, we can simultaneously change both the value and number of the attractors of FHNN by means of altering the fractional-order of the neuron. Furthermore, in any above mentioned case, (0, 0, 0) is the saddle point of FHNN.

9.4.4.3 Application of FHNN to Defense against Chip Cloning Attacks for Anti-Counterfeiting

In this subsection, we analyze the application of FHNN to the defense against chip cloning attacks for anti-counterfeiting. From the aforementioned discussion, it can be seen that FHNN has a stronger associative memory than HNN. We can obviously apply FHNN to pattern recognition, similar to HNN. To avoid repetition, the application of FHNN to pattern recognition is not described in this report. Therefore, we propose a novel promising application case of FHNN in brief.

We apply FHNN to defense against chip cloning attacks for anti-counterfeiting. Copyright is in crisis nowadays. Levies or legal penalties only patch the holes in an already leaky system. The flaw lies not only in the attitude toward copyright in our society, but also in anti-counterfeiting technology. In many cases, identification and qualification are important methods for anti-counterfeiting. Encryption and digital watermarking technology have become mature, but there is not an effective method to identify the pirate of electronic copies. Defense against chip cloning attacks technology for anti-counterfeiting is an emerging discipline that has not been studied yet in depth. Based on aforementioned features of FHNN, we can analyze the defense against chip cloning attacks properties of FHNN.

In the first case, suppose there are only two fractional neurons of FHNN, and each neuron has the same fractional-order. Thus, $v_1 = v_2 = v$. From (9.70) — (9.72) and (9.80) — (9.82), FHNN's stability and convergence in the neighborhood of even-order $v = 2k$ can be shown in Fig. 9.19.

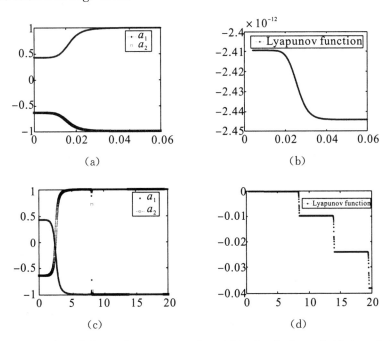

Fig. 9.19 FHNN's stability and convergence in neighborhood of $v = 6.00$ ($\gamma = 1.40$, $n_1(0) =$

-0.50, $n_2(0)=-1.00$). (a) Time response curve of FHNN's output when $v=5.999999$, (b) Time response curve of FHNN's Lyapunov function when $v=5.999999$, (c) Time response curve of FHNN's output when $v=6.000000$, (d) Time response curve of FHNN's Lyapunov function when $v=6.000000$.

From Fig. 9.19, we can see that in the neighborhood of even-order $v=2k$, FHNN is convergent when v is only less than $2k$ part per million, but it is not convergent when $v=2k$, where k is a positive integer. Furthermore, FHNN's stability and convergence in the neighborhood of $v=1.00$ can be shown in Fig. 9.20.

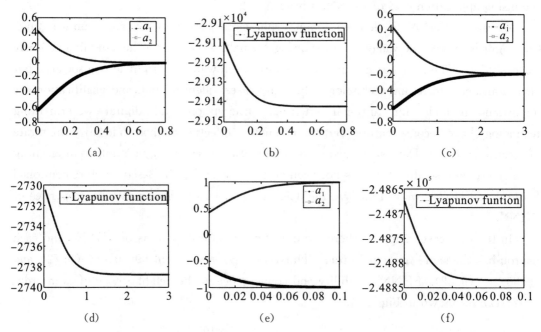

Fig. 9.20 FHNN's stability and convergence in neighborhood of $v=1.00$ ($\gamma=1.40$, $n_1(0)=0.50$, $n_2(0)=-1.00$). (a) Time response curve of FHNN's output when $v=0.99$, (b) The time response curve of FHNN's Lyapunov function when $v=0.99$, (c) The time response curve of FHNN's output when $v=1.00$, (d) The time response curve of FHNN's Lyapunov function when $v=1.00$, (e) The time response curve of FHNN's output when $v=1.01$, (f) The time response curve of FHNN's Lyapunov function when $v=1.01$.

In Fig. 9.20, in order to get higher distinguish, we set the weighting matrix in (9.44) as $\boldsymbol{W}=\begin{bmatrix} 0 & 1 \\ 1 & 0 \end{bmatrix}$. From Fig. 9.20, we can see that, the attractors of FHNN are (0, 0), $(-0.1823, -0.1823)$ and $(1.00, -1.00)$ when $v=0.99$, $v=1.00$ and $v=1.01$, respectively.

In the second case, suppose there are also two fractional neurons of FHNN, but each of its neurons has a different fractional-order. Thus, $v_1 \neq v_2$. The corresponding relationship between FHNN's input and its output according to the weak variation of the fractional-order of the neuron can be shown in Table 9.7.

Table 9.7 Corresponding relationship between input and output of FHNN according to weak variation of fractional-order of the neuron ($\gamma=1.40$): (a) $v_1=0.498$, $v_2=1.50$, (b) $v_1=0.499$, $v_2=1.50$, (c) $v_1=0.501$, $v_2=1.50$, (d) $v_1=0.502$, $v_2=1.50$.

(a)

Input n_1, n_2	Output a_1, a_2	Input n_1, n_2	Output a_1, a_2
−1.0, −1.0	−0.0690, −1.0	0.5, 0.0	0.0690, 1.0
−0.5, −1.0	−0.0690, −1.0	1.0, 0.0	0.0690, 1.0
0.0, −1.0	−0.0690, −1.0	−1.0, 0.5	0.0690, 1.0
0.5, −1.0	−0.0690, −1.0	−0.5, 0.5	0.0690, 1.0
1.0, −1.0	−0.0690, −1.0	0.0, 0.5	0.0690, 1.0
−1.0, −0.5	−0.0690, −1.0	0.5, 0.5	0.0690, 1.0
−0.5, −0.5	−0.0690, −1.0	1.0, 0.5	0.0690, 1.0
0.0, −0.5	−0.0690, −1.0	−1.0, 1.0	0.0690, 1.0
0.5, −0.5	−0.0690, −1.0	−0.5, 1.0	0.0690, 1.0
1.0, −0.5	−0.0690, −1.0	0.0, 1.0	0.0690, 1.0
−1.0, 0.0	−0.0690, −1.0	0.5, 1.0	0.0690, 1.0
−0.5, 0.0	−0.0690, −1.0	1.0, 1.0	0.0690, 1.0

(b)

Input n_1, n_2	Output a_1, a_2	Input n_1, n_2	Output a_1, a_2
−1.0, −1.0	−0.0680, −1.0	0.5, 0.0	0.0680, 1.0
−0.5, −1.0	−0.0680, −1.0	1.0, 0.0	0.0680, 1.0
0.0, −1.0	−0.0680, −1.0	−1.0, 0.5	0.0680, 1.0
0.5, −1.0	−0.0680, −1.0	−0.5, 0.5	0.0680, 1.0
1.0, −1.0	−0.0680, −1.0	0.0, 0.5	0.0680, 1.0
−1.0, −0.5	−0.0680, −1.0	0.5, 0.5	0.0680, 1.0
−0.5, −0.5	−0.0680, −1.0	1.0, 0.5	0.0680, 1.0
0.0, −0.5	−0.0680, −1.0	−1.0, 1.0	0.0680, 1.0
0.5, −0.5	−0.0680, −1.0	−0.5, 1.0	0.0680, 1.0
1.0, −0.5	−0.0680, −1.0	0.0, 1.0	0.0680, 1.0
−1.0, 0.0	−0.0680, −1.0	0.5, 1.0	0.0680, 1.0
−0.5, 0.0	−0.0680, −1.0	1.0, 1.0	0.0680, 1.0

(c)

Input n_1, n_2	Output a_1, a_2	Input n_1, n_2	Output a_1, a_2
−1.0, −1.0	−0.0688, −1.0	0.5, 0.0	0.0688, 1.0
−0.5, −1.0	−0.0688, −1.0	1.0, 0.0	0.0688, 1.0
0.0, −1.0	−0.0688, −1.0	−1.0, 0.5	0.0688, 1.0
0.5, −1.0	−0.0688, −1.0	−0.5, 0.5	0.0688, 1.0
1.0, −1.0	−0.0688, −1.0	0.0, 0.5	0.0688, 1.0
−1.0, −0.5	−0.0688, −1.0	0.5, 0.5	0.0688, 1.0
−0.5, −0.5	−0.0688, −1.0	1.0, 0.5	0.0688, 1.0
0.0, −0.5	−0.0688, −1.0	−1.0, 1.0	0.0688, 1.0
0.5, −0.5	−0.0688, −1.0	−0.5, 1.0	0.0688, 1.0
1.0, −0.5	−0.0688, −1.0	0.0, 1.0	0.0688, 1.0
−1.0, 0.0	−0.0688, −1.0	0.5, 1.0	0.0688, 1.0
−0.5, 0.0	−0.0688, −1.0	1.0, 1.0	0.0688, 1.0

(d)

Input n_1, n_2	Output a_1, a_2	Input n_1, n_2	Output a_1, a_2
−1.0, −1.0	−0.0683, −1.0	0.5, 0.0	0.0683, 1.0
−0.5, −1.0	−0.0683, −1.0	1.0, 0.0	0.0683, 1.0
0.0, −1.0	−0.0683, −1.0	−1.0, 0.5	0.0683, 1.0
0.5, −1.0	−0.0683, −1.0	−0.5, 0.5	0.0683, 1.0
1.0, −1.0	−0.0683, −1.0	0.0, 0.5	0.0683, 1.0
−1.0, −0.5	−0.0683, −1.0	0.5, 0.5	0.0683, 1.0
−0.5, −0.5	−0.0683, −1.0	1.0, 0.5	0.0683, 1.0
0.0, −0.5	−0.0683, −1.0	−1.0, 1.0	0.0683, 1.0
0.5, −0.5	−0.0683, −1.0	−0.5, 1.0	0.0683, 1.0
1.0, −0.5	−0.0683, −1.0	0.0, 1.0	0.0683, 1.0
−1.0, 0.0	−0.0683, −1.0	0.5, 1.0	0.0683, 1.0
−0.5, 0.0	−0.0683, −1.0	1.0, 1.0	0.0683, 1.0

From Table 9.4 and Table 9.7, we can further see that the value of the attractors of FHNN change very little while the fractional-order of the neuron v_1 varies weakly.

From the aforementioned two cases, it can be seen that the value of the attractors of FHNN essentially relate to the fractional-order of the neuron v_i. Furthermore, from (9.43), it can be also seen that the fractional-order of the neuron v_i relates to its v_i-order fractor in essence. With regard to the neuron's v_i-order fractor, from (9.39), its driving-point impedance function F_{v_i} is generally in direct ratio to $\zeta_i = r^{(1-p_i)/v_i}/c$, where $v_i = q_i$

$+p_i$ is a positive real number, q_i is a positive integer, and $0 \leqslant p_i \leqslant 1$. Thus, in a word, the values of the attractors of FHNN are changing while the values of the resistors or capacitors of the fractor vary.

Actually, according to the electronic manufacturing technology at present, we have not been able to manufacture two resistor or capacitor with identical value. It's luck in the midst of sadness. To date, no one has been able to manufacture two FHNNs with identical values of the attractors. Therefore, we can apply FHNN to defense against chip cloning attacks for anti-counterfeiting, and this will be discussed in our future work.

9.4.5 CONCLUSIONS

How to apply fractional calculus to signal analysis and processing, especially to neural networks, is an emerging discipline branch field of study and few studies have been seldom performed in this area. Fractional calculus has been incorporated into artificial neural networks, mainly because of its long-term memory and nonlocality. Therefore, it is natural to think about how to generalize the first-order Hopfield neural networks to the fractional-order ones, and how to implement FHNN by means of fractional calculus. The paper is mainly to discuss a novel conceptual framework: fractional Hopfield neural networks. Therefore, it is one naturally to ponder how to generalize HNN to the fractional-order ones, and how to implement FHNN by means of fractional calculus. This manuscript presents a novel conceptual framework: FHNN. We propose to introduce a novel mathematical method: fractional calculus to implement FHNN. We implement FHNN by utilizing fractor and the fractional steepest descent approach, construct its Lyapunov function, and further analyze its attractors. We apply fractional calculus to implement FHNN, mainly because of its long-term memory and nonlocality. The main contribution of our work is to propose FHNN in the form of an analog circuit by utilizing fractor and the fractional steepest descent approach, construct its Lyapunov function, prove its Lyapunov stability, analyze its attractors, and apply FHNN to the defense against chip cloning attacks for anti-counterfeiting. The arbitrary-order of FHNN represents an additional degree-of-freedom to fit a specific behavior such as power-law long-term memory, or power-law nonlocality. A significant advantage of FHNN is that its attractors essentially relate to neuron's fractional-order. FHNN possesses the fractional-order-stability and fractional-order-sensitivity characteristics. We can apply FHNN to the defense against chip cloning attacks for anti-counterfeiting.

From the aforementioned discussion, we can also see that there are many other problems that need to be further studied. For example, how to construct an FHNN that is convergent when $2k-2 < v < 2k-1$, how to construct the arbitrary-order fractor, how to construct fractional chaotic neural networks, how to implement the analog circuit realization of defense against chip cloning attacks based on FHNN, and so on. These will be discussed in our future work.

第10章　分数阶硬件安全：基于分数阶 Hopfield 神经网络的抗芯片克隆

本章讨论了分数阶 Hopfield 神经网络（FHNNs）在抗芯片克隆方面的最新应用，并深入分析了该方法优于目前效果最好的物理抗克隆技术（PUF）的原因。在过去的十年中，物理抗克隆功能已经发展成为最好的硬件安全类型之一。然而，物理抗克隆功能的发展受到其实施成本和温度变化的影响明显，并受到电磁干扰效应、熵量等因素的限制。我们拟引入新颖的数学方法和物理模型来研发一些新的机制，以克服物理抗克隆技术的弱点。基于此，我们提出将分数阶 Hopfield 神经网络用于抗芯片克隆。第一，实现了任意阶次分数阶 Hopfield 神经网络。第二，对分数阶 Hopfield 神经网络的实施成本进行说明。第三，在周边环境温度变化时，分数阶 Hopfield 神经网络性能不受影响。第四，分析了在电磁干扰的条件下，分数阶 Hopfield 神经网络的电磁性能的稳定性。第五，研究了分数阶 Hopfield 神经网络的熵量。第六，用实验验证了任意阶次分数阶 Hopfield 神经网络的分抗的带宽以及抗芯片克隆能力。实验详细说明了在抗芯片克隆中，分数阶 Hopfield 神经网络抗电磁干扰和抵抗温度变化的能力。本章指出并论证了分数阶 Hopfield 神经网络在抗芯片克隆应用上的可行性，其性能优于物理抗克隆技术。分数阶 Hopfield 神经网络的显著优点在于它的实施成本远低于 PUF，在不同温度条件下分数阶 Hopfield 神经网络的电气性能比物理抗克隆技术稳定得多，在电磁干扰条件下的电气性能稳定性也比物理抗克隆技术更高，并且其熵量明显高于具有相同等级电路规模的物理抗克隆技术的熵量。

10.1 Defense Against Chip Cloning Attacks Based on Fractional Hopfield Neural Networks[①]

10.1.1 Introduction

It is a well-known fact that due to the big increase in the demands of applications and advanced silicon technology, design productivity is failing to keep pace. The globalization of design process has established itself as an inevitable solution for faster and efficient design, in which the design of modern Systems-on-Chip (SoC) relies heavily on reusable Intellectual Property (IP) cores as a practical solution. Security issues in IP's can be either in the form of IP piracy/IP counterfeit or embedded malicious logic. The security

---

① PU Yifei, ZHANG Yi, ZHOU Jiliu. Defense against Chip Cloning Attacks Based on Fractional Hopfield Neural Networks [J]. International Journal of Neural Systems, 2016, 27 (4): 28 pages, Article ID 1750003.

countermeasures of IP protection can be divided into four main categories: data fingerprinting [351], data watermarking [352], [353], Ending Piracy of Integrated Circuits (EPIC) [354], and data tagging [355] − [359]. With respect to data tagging, the defend against chip cloning attacks presents robust security countermeasures designed to safeguard the authenticity of chips and more specifically to protect them from chip cloning attempts. Modern resilient security systems require a difficult-to-clone physical module integrated in core system units as basic security anchors. One of the best types of hardware security is the Physically Unclonable Function (PUF). Physical One-Way Function (POWF) [360] and PUF [361], [362] are a unique physical entity that is embodied in a chip of silicon during the manufacturing process of Integrated Circuit (IC). The pre-processing or post-processing error correction PUFs can be utilized as a key for cryptographic purposes [363], [364]. Nowadays, the PUF entropy is used to construct cryptographic keys, chip identifiers, or the Challenge-Response Pairs (CRP) in a chip authentication mechanism [365] − [370]. To prevent a rogue designer or vendor from studying challenge to get the response of a PUF, there should be an adequate number of CRP [371]. Different measuring results can be yielded by using some especially controllable techniques during the manufacturing process of a PUF [372], [373]. Delay based PUFs such as Ring Oscillator PUFs (ROPUF) [372], [373] and Arbiter PUFs (APUF) [362], [374] are the most popular PUFs. There are some other widely used PUFs, such as memory based PUFs [364], [365], [375] and coating PUFs [376], [377]. However, the efficiency of a PUF depends on its implementation cost [372], [373], its reliability [362], [363], [374], its resiliency to attacks [378] − [381], and the amount of entropy in it [382] that is limited by the circuit resources available. As a result, generating longer keys or larger sets of CRPs may increase a PUF circuit cost [382]. It is also important to ensure that the environment is suitable for the requisite security level [363]. The development of the PUFs has been somewhat limited by its relatively high implementation cost and its working environment effects. Therefore, it is imperative to discover, through promising mathematical methods and physical modules, some novel mechanisms to overcome the aforementioned weaknesses of the PUFs.

Some remarkable progress in the studies on the implementations of neural networks not only validates them as dynamic systems and powerful computational tools, but also gives us some interesting and practical suggestions for future research [383] − [398]. Liao et al. investigate the problem of global and robust stability of a class of interval Hopfield neural networks that have time-varying delays [383]. Al-Alawi presented the hardware architecture of a probabilistic logic neuron [384]. Bénédic et al. presented a new approach for the implementation of threshold logic functions with binary-output cellular neural networks [385]. Strack et al. discussed a multi-layer multi-column model of the cortex that uses four different neuron types and short-term plasticity dynamics [386]. To reproduce accurately the membrane voltage dynamics of a biological neuron, Wang et al.

presented a generalized leaky integrate-and-fire neuron model [387]. In addition, until now, several researchers have concentrated upon the domain of Spiking Neural Networks (SNN) [388] – [398]. In particular, Ghosh-Dastidar and Adeli presented a state-of-the-art review of the spiking neurons and provides insight into their evolution as the third generation neural networks [395]. Ghosh-Dastidar and Adeli studied a novel multi-SNN model, in which information from one neuron is transmitted to the next in the form of multiple spikes via multiple [396]. Ghosh-Dastidar and Adeli developed an efficient SNN model for epilepsy and epileptic seizure detection using ElectroEncephaloGrams (EEGs) [397]. Furthermore, based on the authors' groundbreaking research, Adeli and Ghosh-Dastidar introduced a novel automated EEG-based diagnosis of neurological disorders by means of neural networks, wavelets, and chaos theory [398]. Inspired by the aforementioned work on neural networks, to reduce the cost of implementation, to achieve temperature variation effect much more stable, to make electromagnetic interference effect much more robust, and to significantly improve the amount of entropy, it is natural to ponder how to hardware implement the arbitrary-order Fractional Hopfield Neural Networks (FHNNs) [194], [399] based on fractional calculus and the fractor, and how to apply the FHNNs to the defense against chip cloning attacks.

Fractionalcalculus has become an important branch of mathematical analysis [230], [400]. Fractional calculus is as old as integral calculus, but until recently, it is still seldom known by most of mathematicians and physical scientists in engineering fields. Nowadays, fractional calculus appears to be a promising mathematical method for physical scientists and engineering technicians too [230], [400]. The properties of the fractional calculus of a signal are quite different from those of its integer-order calculus [230], [400]. The fractional differential, except for Caputo defined fractional calculus, of a Heaviside function is non-zero, whereas its integer-order differential must be zero. In fact, most of the special functions of mathematical physics are differintegrable series. Fractional calculus extends and unifies the concepts of difference quotients and Riemann sums [230], [400].

The issue of how to efficiently apply fractional calculus for the purpose of signal processing, image processing, circuits and systems, artificial intelligence, and information security, especially for the purpose of IP protection such as the defense against chip cloning attacks, is an emerging area of research [126], [130], [181] – [182], [187], [192], [194], [200], [204] – [205], [226], [230], [399] – [402]. In scientific fields, there are many issues on non-linear, non-causal, non-Gaussian, non-stationary, non-minimum-phase, non-white-noise, non-integer-dimensional, and non-integer-order natural phenomena needed to be analyzed and processed. Scientific study has shown that a fractional-order or a fractionaldimensional approach is the best description for the aforementioned "non-" problems. Some interesting results have demonstrated that fractional calculus can be a promising and useful tool in many scientific fields [126],

[130], [181] - [182], [187], [192], [194], [200], [204] - [205], [226], [230], [399] - [402]. Fractional calculus has been used in scientific study primarily owing to its inherent advantages of long-term memory, non-locality, and weak singularity. In the fields of the fractional-order circuits and systems and the fractional-order artificial intelligence, the concept of the fractance, as the fractional-order impedance of a promising circuit element, the fractor, arose following the successful synthesis of a fractional differentiator or fractional integrator in the form of an analog circuit [130], [194], [200], [204] - [205], [226], [230], [399] - [400]. There are two types of the fractor: the capacitive fractor and the inductive fractor. In particular, the tree type, two-circuit type, H type, and net-grid type should be four natural implementations of the fractor [204], [205]. In contrast to other approximate implementations of the fractor [130], [200], [226], [402], the floating point values of the capacitance, inductance, and resistance of these four aforementioned types of the fractor [204], [205] are not required. Further, Pu et al. proposed to introduce fractional calculus and the fractor to achieve a novel conceptual framework: the FHNNs [194], [399]. In this chapter, based on fractional calculus and the fractor [130], [194], [200], [204] - [205], [226], [230], [399] - [400], we proposed a feasible and promising application of the FHNNs [194], [399] to the defense against chip cloning attacks. The FHNNs can have a much greater capability to distinguish circuit devices from one another and have the detectable characteristics of ultralow cloning deviation. This is due to that as a result of the values of the resistance and capacitance of the fractors in the neurons of a FHNN of the chip has been cloned are altered, the values of the attractors of a FHNN are changed when the fractional-orders of its neurons are altered [194], [399]. The FHNNs possess the fractional-order-stability characteristics. A significant advantage of the FHNNs is that its attractors are essentially related to the fractional-order of the corresponding neurons of a FHNN. The main contribution of our work is to propose a feasible and promising application of the FHNNs to the defense against chip cloning attacks, which is superior to PUFs.

This manuscript is organized as follows: Section 2 reviews the necessarymathematical preliminaries of fractional calculus, fractance, and the FHNNs. Section 3 proposes applying the FHNNs to the defend against chip cloning attacks. First, we implement the arbitrary-order fractor of a FHNN. Second, we describe the implementation cost of the FHNNs. Third, we propose the achievement of the constant-order performance of a FHNN when ambient temperature varies. Fourth, we analyze the electrical performance stability of the FHNNs under electromagnetic disturbance conditions. Fifth, we study the amount of entropy of the FHNNs. Section 4 reports the experimental results and analysis. In particular, the capabilities of defense against chip cloning attacks, anti-electromagnetic interference, and anti-temperature variation of a FHNN are illustrated experimentally in detail. In Section 5, the conclusions of this manuscript are presented.

10.1.2 Mathematical Preliminaries

In this section, we give a brief necessary introduction of the definitions of fractional calculus, fractance, and the FHNNs.

The most commonly used fractional calculus definitions are Grünwald-Letnikov, Riemann-Liouville, and Caputo [230], [400], respectively. We can perform the fractional calculus of a signal in the form of an analog circuit, the fractor. Let's represent a fractor with the symbol ─W─, in which F is the abbreviation of fractor. In this chapter, let's use the notation F_v as a v-order fractor. With respect to a capacitive fractor, Pu has derived the generic nonlinear relationship between the capacitance and resistance of the arbitrary-order capacitive fractance [204], [205], which can be given as:

$$F^c_{-v} = F^c_{-(m+p)} = V_i(s)/I_i(s) = c^{-(m+p)}r^{1-p}s^{-(m+p)}, \quad (10.1)$$

where $v = m + p$ is a positive real number, m is a positive integer, $0 \leqslant p \leqslant 1$, and s denotes the Laplace operator. F^c_{-v}, $V_i(s)$, $I_i(s)$, c, r, and $c^{m+p}r^{1-p}$ denote the driving-point impedance function, input voltage, input current, capacitance, resistance, and capacitive fractance of a purely ideal v-order capacitive fractor in the Laplace transform domain, respectively. Equation (10.1) represents the fractional-order capacitive reactance of the purely ideal arbitrary-order capacitive fractor. In addition, with respect to an inductive fractor, Pu has derived the generic nonlinear relationship between inductance and resistance of the arbitrary-order inductive fractance [204], [205], which can be given as:

$$F^l_v = F^l_{m+p} = V_i(s)/I_i(s) = l^{m+p}r^{1-p}s^{m+p}, \quad (10.2)$$

where $v = m + p$ is a positive real number, m is a positive integer, and $0 \leqslant p \leqslant 1$. F^l_v, $V_i(s)$, $I_i(s)$, l, r, and $l^{m+p}r^{1-p}$ denote the driving-point impedance function, input voltage, input current, inductance, resistance, and inductive fractance of a purely ideal v-order inductive fractor in the Laplace transform domain, respectively. Equation (10.2) represents the fractional-order inductive reactance of the purely ideal arbitrary-order inductive fractor. Thus, according to the Caputo definition of the v-order derivative, the inverse Laplace transform of (10.1) is as follows:

$$I_i(t) = \frac{1}{\zeta^v \Gamma(n-v)} \int_0^t (t-\tau)^{n-v-1} V_i^{(n)}(\tau) d\tau, \quad (10.3)$$

where $\zeta = r^{(1-p)/v}/C$ is a constant, Γ denotes the Gamma function, and $0 \leqslant n-1 < v < n$, $n \in \mathbf{R}$. $V_i(t)$ and $I_i(t)$ are the inverse Laplace transform of $V_i(s)$ and $I_i(s)$, respectively. Equation (10.3) shows that for a capacitive fractor, $V_i(t)$ is generally in direct ratio to the v-order fractional integral of $I_i(t)$, whereas $I_i(t)$ is generally in direct ratio to the v-order fractional differential of $V_i(t)$.

On the basis of fractional calculus and the fractor, the first preliminary attempt to implement the FHNNs was reported in Refs. [194] [399]. The FHNNs can be implemented by utilizing fractor in the form of an analog circuit, as shown in Fig. 10.1.

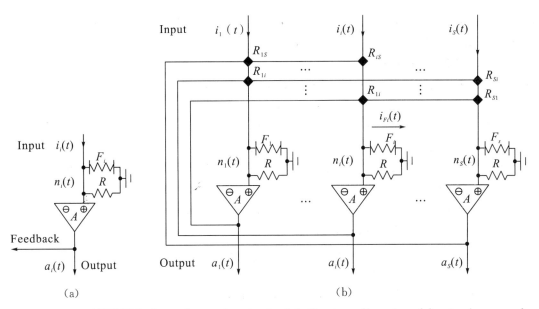

Fig. 10.1　AFHNN in form of an analog circuit: (a) Circuit configuration of fractional neuron of a FHNN; (b) Circuit configuration of a FHNN.

In Fig. 10.1, each fractional neuron of a FHNN consists of one operational amplifier A and its related fractor ─〰─, and resistors. Each fractional neuron has the same circuit configuration. S is the amount of the fractional neurons of a FHNN. The function $n_i(t)$ denotes the input voltage of the *ith* fractional neuron's A amplifier, $a_i(t)$ denotes the output voltage of the *ith* fractional neuron, and $i_i(t)$ denotes the input electric current of the *ith* fractional neuron. $R_{i,j}$ denotes the feedback resistor being connected with the output of the *jth* fractional neuron and the input of the *ith* one. $a_i(t) = f[n_i(t)]$ and $n_i(t) = f^{-1}[a_i(t)]$, where f denotes the input-output function of A, f^{-1} denotes the inverse output-input function of A, and $f(0) = 0$. Equation (10.1) shows that fractor ─〰─ implements the v-order fractional calculus. We set $F_1 = F_{v_1}$, $F_i = F_{v_i}$ and $F_S = F_{v_S}$, i.e. the fractional-orders of a FHNN's neurons are equal to v_1, v_i, and v_S, respectively. Thus, input current $i_{F_i}(t)$ is generally in direct ratio to the v_i-order fractional differential of the input voltage $n_i(t)$ of the v-order F_{v_i}.

From Fig. 10.1 and (10.3), we can derive the operation rule of the FHNNs according to Kirchhoff's current law and Kirchhoff's voltage law [194], [399], given as:

$$\chi_i \frac{d^{v_i} n_i(t)}{dt^{v_i}} = -n_i(t) + \sum_{j=1}^{S} \overline{w}_{i,j} a_j(t) + b_i(t), \tag{10.4}$$

where $\chi_i = Z_i K_i$, $1/Z_i = 1/R + \sum_{j=1}^{S}(1/R_{i,j})$, $K_i = \zeta_i^{-v_i}$, $\overline{w}_{i,j} = Z_i T_{i,j}$, $b_i = Z_i i_i$, $T_{i,j} = 1/R_{i,j}$, and $T_{i,j} = T_{j,i}$. With respect to capacitive fractor, $\zeta_i = r^{(1-p_i)/v_i}/c$, $v_i = q_i + p_i$ is a positive real number, q_i is a positive integer, and $0 \leqslant p_i \leqslant 1$. It is assumed that the circuit of the FHNNs is symmetric, so that $T_{i,j} = T_{j,i}$ and $T_{i,i} = 0$. Equation (10.4) shows that the operation rule of the FHNNs is a fractional-order partial differential equation. In this chapter, for the convenience of illustration, let's set $v_1 = v_i = v_S$.

The FHNNs can have a much greater capability to distinguish devices from one another and have the detectable characteristics of ultralow cloning deviation. This is due to that as a result of the values of the resistance and capacitance of the fractors in the neurons of a FHNN of the chip has been cloned are altered, the values of the attractors of a FHNN are changed when the fractional-orders of its neurons are altered [194], [399]. The FHNNs possess the fractional-order-stability characteristics. A significant advantage of the FHNNs is that its attractors are essentially related to the fractional-order of the corresponding neurons of a FHNN. Actually, because of the inherent nature of materials and the present state of electronic manufacturing technology, we have not been able to achieve two resistors or capacitors with identical values during the manufacturing process of IC. To date, no one has been able to create two FHNNs with identical attractor values. The FHNNs can be included in a chip design without modifications to the manufacturing process. An individual defense against chip cloning attacks based on the FHNNs is easy to make but practically impossible to duplicate, even given the exact manufacturing process that produced it. In this respect, the FHNNs are a hardware implementation of the physical unclonable genes of chips. Equations (10.1) and (10.2) show that the fractional-order, v, of a fractor in essence is the phase of the corresponding fractional integrator or fractional differentiator in frequency domain. Therefore, the clue of application of the FHNNs to the defense against chip cloning attacks for anti-piracy is in the use of phase (more invariant than amplitude) for physical system recognition.

Contrasting chip counterfeiting through hardware solutions is a very difficult problem, since the hardware solution must provide enough differences between different devices, but at the same time be stable enough when the environmental conditions change. In this chapter, we propose a novel efficient method to implement hardware security mechanisms into chips, to prevent cloning attacks. The method applies a framework (named FHNNs) introduced by our previous work [194], [399], which is superior to the PUFs. The framework is based on the proposed arbitrary-order fractor of a FHNN as electronic circuits, and uses the recalled notion of fractional calculus as well as that of the fractor.

10.1.3 Defense against Chip Cloning Attacks Based on FHNNs

10.1.3.1 Arbitrary-Order Fractor of FHNNs

This section is the proposal for the analog circuit realization of an arbitrary-order FHNN. In particular, the analog circuit realization of the arbitrary-order fractor of a FHNN is discussed in detail.

Figure 10.1 shows thatthe analog circuit realization of an arbitrary-order fractor is the crux of the analog circuit realization of an arbitrary-order FHNN. Further, from (10.1) and (10.2), we can also see that the key of the analog circuit realization of the arbitrary-order fractor is to perform the fractional integral operator s^{-v} and fractional differential operator s^v in the form of an analog circuit. Suppose that v is a positive rational number. Thus, we can derive that v is equal to a positive integer plus a positive rational fraction. It follows that $s^{-v}=s^{-m}s^{-i/n}$, where m, n and i are positive integers, $n \neq 1$, $i=1, 2, \cdots, n-1$, and $0<i/n<1$ is a positive rational fraction. Therefore, the key to implementation of the fractional integral operator s^{-v} is to implement the fractional integral operator $s^{-i/n}$ in an analog circuit. The fractional integral operator $s^{-i/n}$ is the reciprocal of the fractional differential operator $s^{i/n}$. Suppose $|\alpha - (\beta s)^{i/n}|<1$, then it follows from the binomial theorem that:

$$\lim_{j \to \infty}[\alpha - (\beta s)^{i/n}]^j = \lim_{j \to \infty}\sum_{k=0}^{j} \alpha^{j-k}\beta^{ik/n} s^{quo(ik/n)} s^{rem(ik/n)} = 0, \qquad (10.5)$$

where α and β are both positive numbers, $\binom{j}{k} = \dfrac{\Gamma(1+j)}{\Gamma(j-k+1)\Gamma(k+1)}$, $quo()$ denotes quotient calculation, and $rem()$ denotes the remainder calculation. Therefore, if $j \to \infty$, we can derive the computational completeness of $s^{rem(ik/n)}$ as $\lim_{j \to \infty} s^{rem(jk/n)} = s^{i/n} = s^{1/n}$, $s^{2/n}, \cdots, s^{n-1/n}$ corresponding to $i=1, 2, \cdots, n-1$. We then define $\lim_{j \to \infty} F_{i/n}^j(s) = \lim_{j \to \infty} s^{rem(ik/n)}$ and $F_{i/n}(s) = s^{i/n}$. If $j \to \infty$, it follows that:

$$\lim_{j \to \infty} F_{i/n}^j(s) = \lim_{j \to \infty} s^{rem(ik/n)} = F_{i/n}(s) = s^{i/n}. \qquad (10.6)$$

From (10.5) and (10.6), when $j \to \infty$, by ergodic calculations $i=1, 2, \cdots, n-1$ and merging similar terms, it follows that in vector form:

$$\begin{bmatrix} w_{1,1}(s) & \cdots & w_{1,i-1}(s) & w_{1,i}(s) & w_{1,i+1}(s) & \cdots & w_{1,n-1}(s) \\ \vdots & & \vdots & \vdots & \vdots & & \vdots \\ w_{i-1,1}(s) & \cdots & w_{i-1,i-1}(s) & w_{i-1,i}(s) & w_{i-1,i+1}(s) & \cdots & w_{i-1,n-1}(s) \\ w_{i,1}(s) & \cdots & w_{i,i-1}(s) & w_{i,i}(s) & w_{i,i+1}(s) & \cdots & w_{i,n-1}(s) \\ w_{i+1,1}(s) & \cdots & w_{i+1,i-1}(s) & w_{i+1,i}(s) & w_{i+1,i+1}(s) & \cdots & w_{i+1,n-1}(s) \\ \vdots & & \vdots & \vdots & \vdots & & \vdots \\ w_{n-1,1}(s) & \cdots & w_{n-1,i-1}(s) & w_{n-1,i}(s) & w_{n-1,i+1}(s) & \cdots & w_{n-1,n-1}(s) \end{bmatrix}$$

$$\begin{bmatrix} F_{1/n}^j(s) \\ \vdots \\ F_{i-1/n}^j(s) \\ F_{i/n}^j(s) \\ F_{i+1/n}^j(s) \\ \vdots \\ F_{n-1/n}^j(s) \end{bmatrix} = \begin{bmatrix} \psi_1(s) \\ \vdots \\ \psi_{i-1}(s) \\ \psi_i(s) \\ \psi_{i+1}(s) \\ \vdots \\ \psi_{n-1}(s) \end{bmatrix},$$

where $\boldsymbol{W}(s) = [w_{x,y}(s)]_{(n-1) \times (n-1)}$ is the coefficient matrix of (10.7), $\boldsymbol{W}_{x,y}(s)$ is a $(n-1) \times (n-1)$ matrix, $w_{x,y}(s)$ is correlated with i, n, α, β, and $s^{quo(ik/n)}$, $\boldsymbol{\psi}(s) = [\psi_1(s) \cdots \psi_i(s) \cdots \psi_{n-1}(s)]^T$ is the nonhomogeneous constant matrix of (10.7). Equation

(10.7) is a system of nonhomogeneous linear equations in vector form. Suppose that $s^{quo(ik/n)}$ is a set of constants and $F_{i/n}^{j}(s)$ is a set of independent variables in (10.7). On the basis of the solution identification theorem of nonhomogeneous linear equations, we can derive the nonzero solutions $F_{i/n}^{j}(s)$ for (10.7). Then, from (10.6), if $j \to \infty$, one can obtain:

$$\lim_{j \to \infty} \begin{bmatrix} F_{1/n}^{j}(s) \\ \vdots \\ F_{i-1/n}^{j}(s) \\ F_{i/n}^{j}(s) \\ F_{i+1/n}^{j}(s) \\ \vdots \\ F_{n-1/n}^{j}(s) \end{bmatrix} = \begin{bmatrix} s^{1/n} \\ \vdots \\ s^{i-1/n} \\ s^{i/n} \\ s^{i+1/n} \\ \vdots \\ s^{s-1/n} \end{bmatrix}. \tag{10.8}$$

From (10.6) and (10.7), we can see that $F_{i/n}^{j}(s)$ depends on $s^{quo(ik/n)}$. For $quo(ik/n)$ is a positive integer, we can implement $[s^{1/n}, s^{2/n}, \cdots, s^{i/n}, \cdots, s^{n-1/n}]^{T}$ by means of capacitors.

Furthermore, from (10.5), we can see that $s^{qou(ik/n)}$ obviously depends on i, k, and n. We can derive the ergodic computing of $s^{quo(ik/n)}$ as $s^{quo(ik/n)} \stackrel{k=0,1,\cdots,j}{=} s^{0}, s^{1}, \cdots, s^{quo(ij/n)-1}, s^{quo(ij/n)}$, where $s^{quo(ik/n)} \stackrel{k=0}{\equiv} s^{0}$ and $s^{rem(ik/n)} \stackrel{k=0}{\equiv} s^{0}$. For convenience of illustration, let's compute the case of $k=0$ in (10.5) separately. We set $j = \chi n$, where χ is a positive integer. Thus, it follows that:

$$s^{quo(ik/n)} \stackrel{k=1,2,\cdots,n}{=} s^{0}, s^{1}, \cdots, s^{n-2}, s^{n-1}, \tag{10.9}$$

$$s^{quo(ik/n)} \stackrel{k=(\chi-1)n+1,(\chi-1)n+2,\cdots,\chi n}{=} s^{i(\chi+1)}, s^{1+i(\chi-1)}, \cdots, s^{n-2+i(\chi-1)}, s^{n-1+i(\chi-1)}. \tag{10.10}$$

Then, if $n=2\lambda$, $k=1, 2, \cdots, n$, or $n=2\lambda-1$, $k=1, 2, \cdots, n$, from (9), the computation rules of $rem(ik/n)$ and $quo(ik/n)$ are shown in Figs. 10.2 and 10.3, respectively.

第10章 分数阶硬件安全：基于分数阶Hopfield神经网络的抗芯片克隆

Remainder i \ k	1	2	...	λ	...	$n-2$	$n-1$	$n=2\lambda$ $\lambda=2\zeta$
1	$1/n$	$2/n$	...	λ/n	...	$n-2/n$	$n-1/n$	0
2	$2/n$	$4/n$	...	0	...	$n-4/n$	$n-2/n$	0
3	$3/n$	$6/n$	...	λ/n	...	$n-6/n$	$n-3/n$	0
4	$4/n$	$8/n$	...	0	...	$n-8/n$	$n-4/n$	0
...	...	...	...	...	...	...	...	...
$\lambda-2$	$\lambda-2/n$	$2\lambda-4/n$	...	0	...	$4/n$	$\lambda+2/n$	0
$\lambda-1$	$\lambda-1/n$	$2\lambda-2/n$	...	λ/n	...	$2/n$	$\lambda+1/n$	0
λ	λ/n	0	...	0	...	0	λ/n	0
$\lambda+1$	$\lambda+1/n$	$2/n$	...	λ/n	...	$2\lambda-2/n$	$\lambda-1/n$	0
$\lambda+2$	$\lambda+2/n$	$4/n$	...	0	...	$2\lambda-4/n$	$\lambda-2/n$	0
...	...	...	...	...	...	...	...	...
$n-4$	$n-4/n$	$n-8/n$	...	0	...	$8/n$	$4/n$	0
$n-3$	$n-3/n$	$n-6/n$	...	λ/n	...	$6/n$	$3/n$	0
$n-2$	$n-2/n$	$n-4/n$	...	0	...	$4/n$	$2/n$	0
$n-1$	$n-1/n$	$n-2/n$	...	λ/n	...	$2/n$	$1/n$	0

(a)

Quotient k\i	n−1	n−2	n−3	n−4	n−5	⋮	λ+1	λ	λ−1	⋮	5	4	3	2	1	n=2λ, λ=2ζ
1	0	0	0	0	0	⋮	0	0	0	⋮	0	0	0	0	0	1
2	1	1	1	1	1	⋮	1	1	0	⋮	0	0	0	0	0	2
3	2	1	1	1	2	⋮	1	1	1	⋮	0	0	0	0	0	3
4	3	2	2	2	3	⋮	2	2	1	⋮	0	0	0	0	0	4
5	4	3	3	3	3	⋮	2	2	2	⋮	1	0	0	0	0	5
⋮	⋮	⋮	⋮	⋮	⋮		⋮	⋮	⋮		⋮	⋮	⋮	⋮	⋮	⋮
λ−1	λ−2	λ−2	λ−3	λ−3	λ−4	⋮	Δ−1	Δ−1	Δ−1	⋮	2	1	1	0	0	λ−1
λ	λ−1	λ−1	λ−2	λ−2	λ−3	⋮	Δ	Δ=λ/2	Δ−1	⋮	2	2	1	1	0	λ
λ+1	λ	λ−1	λ−1	λ−2	λ−2	⋮	Δ+1	Δ	Δ−1	⋮	2	2	1	1	0	λ+1
⋮	⋮	⋮	⋮	⋮	⋮		⋮	⋮	⋮		⋮	⋮	⋮	⋮	⋮	⋮
n−5	n−6	n−7	n−8	n−9	n−9	⋮	λ−2	λ−3	λ−4	⋮	3	3	2	1	0	n−5
n−4	n−5	n−6	n−7	n−8	n−9	⋮	λ−2	λ−2	λ−3	⋮	4	3	2	1	0	n−4
n−3	n−4	n−5	n−6	n−7	n−8	⋮	λ−1	λ−2	λ−3	⋮	4	3	2	1	0	n−3
n−2	n−3	n−4	n−5	n−6	n−7	⋮	λ−1	λ−1	λ−2	⋮	4	3	2	1	0	n−2
n−1	n−2	n−3	n−4	n−5	n−6	⋮	λ	λ−1	λ−2	⋮	4	3	2	1	0	n−1

(b)

Remainder k \ i	1	2	...	λ	...	$n-2$	$n-1$	$n=2\lambda$ $\lambda=2\zeta-1$
1	$1/n$	$2/n$	...	λ/n	...	$n-2/n$	$n-1/n$	0
2	$2/n$	$4/n$	...	0	...	$n-4/n$	$n-2/n$	0
3	$3/n$	$6/n$	...	λ/n	...	$n-6/n$	$n-3/n$	0
4	$4/n$	$8/n$	...	0	...	$n-8/n$	$n-4/n$	0
...	...	...	...	...	...	...	...	...
$\lambda-2$	$\lambda-2/n$	$2\lambda-4/n$	...	λ/n	...	$4/n$	$\lambda+2/n$	0
$\lambda-1$	$\lambda-1/n$	$2\lambda-2/n$	...	0	...	$2/n$	$\lambda+1/n$	0
λ	λ/n	0	...	λ/n	...	0	λ/n	0
$\lambda+1$	$\lambda+1/n$	$2/n$	...	λ/n	...	$2\lambda-2/n$	$\lambda-1/n$	0
$\lambda+2$	$\lambda+2/n$	$4/n$	...	λ/n	...	$2\lambda-4/n$	$\lambda-2/n$	0
...	...	...	...	...	...	...	...	...
$n-4$	$n-4/n$	$n-8/n$	...	0	...	$8/n$	$4/n$	0
$n-3$	$n-3/n$	$n-6/n$	...	λ/n	...	$6/n$	$3/n$	0
$n-2$	$n-2/n$	$n-4/n$	...	0	...	$4/n$	$2/n$	0
$n-1$	$n-1/n$	$n-2/n$	...	λ/n	...	$2/n$	$1/n$	0

(c)

Quotient / i \ k	1	2	3	4	5	...	λ−1	λ	λ+1	...	n−5	n−4	n−3	n−2	n−1	n=2λ, λ=2ζ−1
1	0	0	0	0	0	...	0	0	0	...	0	0	0	0	0	1
2	0	0	0	0	0	...	0	1	1	...	1	1	1	1	1	2
3	0	0	0	0	0	...	1	1	1	...	2	2	2	2	2	3
4	0	0	0	0	0	...	1	2	2	...	3	3	3	3	3	1
5	0	0	0	0	1	...	2	2	2	...	3	4	4	4	4	5
...	...	...	...	...	...	...	...	...	...	...	...	...	...	...	...	...
λ−1	0	0	1	1	2	...	Δ−1	Δ	Δ	...	λ−4	λ−3	λ−3	λ−2	λ−2	λ−1
λ	0	1	1	2	2	...	Δ	Δ=λ−1/2	Δ+1	...	λ−3	λ−3	λ−2	λ−1	λ−1	λ
λ+1	0	1	1	2	2	...	Δ	Δ+1	Δ+1	...	λ−3	λ−2	λ−2	λ−1	λ	λ+1
...	...	...	...	...	...	...	...	...	...	...	...	...	...	...	...	...
n−5	0	1	1	2	3	...	λ−4	λ−3	λ−2	...	n−9	n−9	n−8	n−7	n−6	n−5
n−4	0	1	2	3	4	...	λ−3	λ−2	λ−2	...	n−9	n−8	n−7	n−6	n−5	n−4
n−3	0	1	2	3	4	...	λ−3	λ−2	λ−1	...	n−8	n−7	n−6	n−5	n−4	n−3
n−2	0	1	2	3	4	...	λ−2	λ−1	λ−1	...	n−7	n−6	n−5	n−4	n−3	n−2
n−1	0	1	2	3	4	...	λ−2	λ−1	λ	...	n−6	n−5	n−4	n−3	n−2	n−1

(d)

Fig. 10.2 The computation rules of $rem(ik/n)$ and $quo(ik/n)$ ($n=2\lambda$, $k=1,2,\cdots,n$, where λ and ζ are positive integers): (a) The remainder of $\lambda=2\zeta$; (b) The quotient of $\lambda=2\zeta$; (c) The remainder of $\lambda=2\zeta-1$; (d) The quotient of $\lambda=2\zeta-1$.

第 10 章 分数阶硬件安全：基于分数阶 Hopfield 神经网络的抗芯片克隆

Remainder i \ k	1	2	...	$\lambda-1$	λ	...	$n-2$	$n-1$	$n=2\lambda-1$ $\lambda=2\zeta$
1	$1/n$	$2/n$	...	$\lambda-1/n$	λ/n	...	$n-2/n$	$n-1/n$	0
2	$2/n$	$4/n$	...	$2\lambda-2/n$	$2\lambda/n$	...	$n-4/n$	$n-2/n$	0
3	$3/n$	$6/n$	...	$\lambda-2/n$	$\lambda+1/n$	...	$n-6/n$	$n-3/n$	0
...	...	$2\lambda-4/n$	...	$\Delta_1=$ $rem(\lambda^2-3\lambda+2/n)$	$\Delta_2=$ $rem(\lambda^2-2\lambda/n)$	...	$3/n$	$\lambda+1/n$	0
$\lambda-2$	$\lambda-2/n$	$2\lambda-2/n$	...	$\Delta_3=$ $rem(\lambda^2-2\lambda+1/n)$	$\Delta_4=$ $rem(\lambda^2-\lambda/n)$	...	$1/n$	λ/n	0
$\lambda-1$	$\lambda-1/n$	$1/n$	...	$2\lambda-1-\Delta_3$	$2\lambda-1-\Delta_4$	...	$2\lambda-2/n$	$\lambda-1/n$	0
λ	λ/n	$3/n$	...	$2\lambda-1-\Delta_1$	$2\lambda-1-\Delta_2$	...	$2\lambda-4/n$	$\lambda-2/n$	0
$\lambda+1$	$\lambda+1/n$	...	...	...	...	...	...	...	...
...	...	$n-6/n$	...	$\lambda+1/n$	$\lambda-2/n$	...	$6/n$	$3/n$	0
$n-3$	$n-3/n$	$n-4/n$	...	$1/n$	$2\lambda-2/n$	...	$4/n$	$2/n$	0
$n-2$	$n-2/n$	$n-2/n$	...	λ/n	$\lambda-1/n$	...	$2/n$	$1/n$	0
$n-1$	$n-1/n$								

(a)

Quotient $k \backslash i$	1	2	3	4	5	...	$\lambda-1$	λ	$\lambda+1$	...	$n-5$	$n-4$	$n-3$	$n-2$	$n-1$	$n=2\lambda-1$, $\lambda=2\zeta$
1	0	0	0	0	0	...	0	0	0	...	0	0	0	0	0	1
2	0	0	0	0	0	⋱	0	1	1	...	1	1	1	1	1	2
3	0	0	0	0	0	⋱	1	1	1	...	2	2	2	2	2	3
4	0	0	0	0	1	⋱	1	2	2	...	2	3	3	3	3	4
5	0	0	0	1	1	⋱	2	2	3	...	3	3	4	4	4	5
...	...	⋱	⋱	⋱	⋱	...	⋮	⋮	⋮	...	⋮	⋮	⋮	⋮	⋮	⋮
$\lambda-1$	0	0	1	1	2	...	$\Delta-1$	$\Delta-1$	Δ	...	$\lambda-4$	$\lambda-3$	$\lambda-3$	$\lambda-2$	$\lambda-2$	$\lambda-1$
λ	0	1	1	2	2	...	$\Delta-1$	$\Delta=\lambda/2$	Δ	...	$\lambda-3$	$\lambda-3$	$\lambda-2$	$\lambda-2$	$\lambda-1$	λ
$\lambda+1$	0	1	1	2	3	...	Δ	Δ	$\Delta+1$	...	$\lambda-3$	$\lambda-2$	$\lambda-1$	$\lambda-1$	λ	$\lambda+1$
...	⋮	⋮	⋮	⋮	⋮	...	⋮	⋮	⋮	...	⋮	⋮	⋮	⋮	⋮	⋮
$n-5$	0	1	2	2	3	...	$\lambda-4$	$\lambda-3$	$\lambda-3$	...	$n-9$	$n-8$	$n-8$	$n-7$	$n-6$	$n-5$
$n-4$	0	1	2	3	3	...	$\lambda-3$	$\lambda-3$	$\lambda-2$	...	$n-8$	$n-8$	$n-7$	$n-6$	$n-5$	$n-4$
$n-3$	0	1	2	3	4	...	$\lambda-3$	$\lambda-2$	$\lambda-1$	...	$n-8$	$n-7$	$n-6$	$n-5$	$n-4$	$n-3$
$n-2$	0	1	2	3	4	...	$\lambda-2$	$\lambda-2$	$\lambda-1$	...	$n-7$	$n-6$	$n-5$	$n-4$	$n-3$	$n-2$
$n-1$	0	1	2	3	4	...	$\lambda-2$	$\lambda-1$	λ	...	$n-6$	$n-5$	$n-4$	$n-3$	$n-2$	$n-1$

(b)

Remainder i \ k	1	2	...	$\lambda-1$	λ	...	$n-2$	$n-1$	$n=2\lambda-1$ $\lambda=2\zeta-1$
1	$1/n$	$2/n$	...	$\lambda-1/n$	λ/n	...	$n-2/n$	$n-1/n$	0
2	$2/n$	$4/n$	...	$2\lambda-2/n$	$2\lambda/n$	...	$n-4/n$	$n-2/n$	0
3	$3/n$	$6/n$	...	$\lambda-2/n$	$\lambda+1/n$	...	$n-6/n$	$n-3/n$	0
...	...	...	...	...	...	...	...	...	...
$\lambda-2$	$\lambda-2/n$	$2\lambda-4/n$	...	$\Delta_1 = rem(\lambda^2-3\lambda+2/n)$	$\Delta_2 = rem(\lambda^2-2\lambda/n)$	...	$3/n$	$\lambda+1/n$	0
$\lambda-1$	$\lambda-1/n$	$2\lambda-2/n$	...	$\Delta_3 = rem(\lambda^2-2\lambda+1/n)$	$\Delta_4 = rem(\lambda^2-\lambda/n)$	...	$1/n$	λ/n	0
λ	λ/n	$1/n$	...	$2\lambda-1-\Delta_3$	$2\lambda-1-\Delta_4$	...	$2\lambda-2/n$	$\lambda-1/n$	0
$\lambda+1$	$\lambda+1/n$	$3/n$	...	$2\lambda-1-\Delta_1$	$2\lambda-1-\Delta_2$	...	$2\lambda-4/n$	$\lambda-2/n$	0
...	...	...	...	...	...	...	...	...	...
$n-3$	$n-3/n$	$n-6/n$	...	$\lambda+1/n$	$\lambda-2/n$	...	$6/n$	$3/n$	0
$n-2$	$n-2/n$	$n-4/n$	...	$1/n$	$2\lambda-2/n$	...	$4/n$	$2/n$	0
$n-1$	$n-1/n$	$n-2/n$	...	λ/n	$\lambda-1/n$	...	$2/n$	$1/n$	0

(c)

Quotient i \ k	1	2	3	4	5	...	$\lambda-1$	λ	$\lambda+1$	...	$n-5$	$n-4$	$n-3$	$n-2$	$n-1$	$n=2\lambda-1$ $\lambda=2\zeta-1$
1	0	0	0	0	0	⋮	0	0	0	⋮	0	0	0	0	0	1
2	0	0	0	0	0	⋮	0	1	1	⋮	1	1	1	1	1	2
3	0	0	0	0	0	⋱	1	1	1	⋮	2	2	2	2	2	3
4	0	0	0	0	1	⋱	1	2	2	⋮	3	3	3	3	3	4
5	0	0	0	1	1	⋱	2	2	3	⋮	3	3	4	4	4	5
...	⋮	⋮	⋮	⋮	⋮	⋱	⋮	⋮	⋮	⋮	⋮	⋮	⋮	⋮	⋮	⋮
$\lambda-1$	0	0	1	1	2	⋮	$\Delta-1$	Δ	Δ	⋮	$\lambda-4$	$\lambda-3$	$\lambda-3$	$\lambda-2$	$\lambda-2$	$\lambda-1$
λ	0	0	1	2	2	⋮	Δ	$\Delta=\lambda-1/2$	$\Delta+1$	⋮	$\lambda-3$	$\lambda-3$	$\lambda-2$	$\lambda-2$	$\lambda-1$	λ
$\lambda+1$	0	0	1	2	3	⋮	Δ	$\Delta+1$	$\Delta+2$	⋮	$\lambda-3$	$\lambda-2$	$\lambda-1$	$\lambda-1$	λ	$\lambda+1$
...	⋮	⋮	⋮	⋮	⋮	⋮	⋮	⋮	⋮	⋮	⋮	⋮	⋮	⋮	⋮	⋮
$n-5$	0	1	2	2	3	⋮	$\lambda-4$	$\lambda-3$	$\lambda-3$	⋮	$n-9$	$n-8$	$n-8$	$n-7$	$n-6$	$n-5$
$n-4$	0	1	2	3	3	⋮	$\lambda-3$	$\lambda-3$	$\lambda-2$	⋮	$n-8$	$n-8$	$n-7$	$n-6$	$n-5$	$n-4$
$n-3$	0	1	2	3	4	⋮	$\lambda-3$	$\lambda-2$	$\lambda-1$	⋮	$n-8$	$n-7$	$n-6$	$n-5$	$n-4$	$n-3$
$n-2$	0	1	2	3	4	⋮	$\lambda-2$	$\lambda-2$	$\lambda-1$	⋮	$n-7$	$n-6$	$n-5$	$n-4$	$n-3$	$n-2$
$n-1$	0	1	2	3	4	⋮	$\lambda-2$	$\lambda-1$	λ	⋮	$n-6$	$n-5$	$n-4$	$n-3$	$n-2$	$n-1$

(d)

Fig. 10.3　The computation rules of $rem(ik/n)$ and $quo(ik/n)$, ($n=2\lambda-1, k=1,2,\cdots,n$, where λ and ζ are positive integers): (a) The remainder of $\lambda=2\zeta$; (b) The quotient of $\lambda=2\zeta$; (c) The remainder of $\lambda=2\zeta-1$; (d) The quotient of $\lambda=2\zeta-1$.

For the convenience of comprehension, Figs. 10.2 and 10.3 only show the case of $k=1, 2\cdots, n$. From (10.10), we can see that if $k=(\chi-1)n+1, (\chi-1)n+2, \cdots, \chi n$, the computation rules of $rem(ik/n)$ are the same as in the case of $k=1, 2, \cdots, n$. However, $quo(ik/n)$ is a periodically increasing function. Its period is equal to n. The $quo(ik/n)$ of $k=(\chi-1)n+1, (\chi-1)n+2, \cdots, \chi n$ is equal to the corresponding $quo(ik/n)$ of $k=1, 2, \cdots, n$ plus $i(\chi-1)$. From Figs. 10.2 and 10.3, we can see that the computation rules of $rem(ik/n)$ and $quo(ik/n)$ are regular whenever n is either an even or odd number. From Figs. 10.2 (a) and 10.2 (b), we can see that if $n=2\lambda$, $\lambda=2\zeta$, where λ and ζ are positive integers, either $rem(ik/n)$ or $quo(ik/n)$ of $k=1, 2, \cdots, n-1$ is a $(n-1)\times(n-1)$ symmetric matrix. In addition, either $rem(ik/n)$ or $quo(ik/n)$ of $k=1, 2, \cdots, n-1$ is also an symmetric matrix around the axis of $k=\lambda$ or $i=\lambda$. The $rem(ik/n)$ on the axis of $k=\lambda$ or $i=\lambda$ permutates circularly between λ/n and 0. The sum of the two corresponding $rem(ik/n)$ of $k=1, 2, \cdots, n-1$ around the axis of $k=\lambda$ or $i=\lambda$ is equal to n/n. The $quo(ik/n)$ on the axis of $k=\lambda$ or $i=\lambda$ increasingly permutates circularly from $0, 1, 1, 2, 2$ to $\lambda-1, \lambda-1$. The sum of the two corresponding $quo(ik/n)$ of $k=1, 2, \cdots, n-1$ around the axis of $k=\lambda$ or $i=\lambda$ is equal to $i-1$ or $k-1$. Furthermore, the $rem(ik/n)$ of $k=n$ is equal to 0. The $quo(ik/n)$ of $k=n$ is equal to i. From Figs. 10.2 (c) and 10.2 (d), we can see that if $n=2\lambda$, $\lambda=2\zeta-1$, the computation rules of $rem(ik/n)$ and $quo(ik/n)$ are the same as the case of $n=2\lambda$, $\lambda=2\zeta$. For the parity of the conversion of λ, either the $rem(\lambda^2/n)$ or $quo(\lambda^2/n)$ of $\lambda=2\zeta$ is different from the case of $\lambda=2\zeta-1$. From Fig. 10.3, we can see that if $n=2\lambda-1$, the computation rules of $rem(ik/n)$ and $quo(ik/n)$ are the similar to the case of $n=2\lambda$. From Figs. 10.3 (a) and 10.3 (c), we can see that the sum of two corresponding $rem(ik/n)$ on the axes of $k=\lambda-1$ and $k=\lambda$ or $i=\lambda-1$ and $i=\lambda$ is equal to n/n. From Figs. 10.3 (b) and 10.3 (d), we can also see that the $quo(ik/n)$ on the axis of $k=\lambda-1$ or $i=\lambda-1$ is an increasing circular permutation from $0, 0, 1, 1, 2, 2$ to $\lambda-2, \lambda-2$. The $quo(ik/n)$ on the axis of $k=\lambda$ or $i=\lambda$ is also an increasing circular permutation from $0, 1, 1, 2, 2$ to $\lambda-2, \lambda-1$.

Therefore, from Figs. 10.2 and 10.3, we can derive the coefficient matrix $\boldsymbol{W}(s)$ and nonhomogeneous constant matrix $\boldsymbol{\Psi}(s)$ of (10.7). From the aforementioned discussion, we see that either $rem(ik/n)$ or $quo(ik/n)$ is a $(n-1)\times(n-1)$ symmetric matrix. Thus, the corresponding matrices of $rem(ik/n)$ and $quo(ik/n)$ can be diagonalizable. The eigenvectors of their corresponding matrices are orthometric. Further, because elementary row transformation or column transformation of a matrix cannot change its rank, the coefficient matrix $\boldsymbol{W}(s)$ of (10.7) is also orthometric. Thus, $|\boldsymbol{W}(s)|\neq 0$, where $|\ |$ denotes determinant calculation. In addition, the nonhomogeneous constant matrix $\boldsymbol{\Psi}(s)$ of (10.7) can be derived by merging similar terms of $s^{quo(ik/n)}$ corresponding to $rem(ik/n)=0$ in Figs. 10.2 and 10.3, and summation of the $s^{quo(ik/n)}\stackrel{k=0}{=}s^0$ terms. The weighted value of $s^{quo(ik/n)}$ depends on i, k, and n corresponding to $rem(ik/n)=0$, according to the computation rule of (10.5). It is worth noting that the symbols for nonhomogeneous constant terms require a sign change when they are transposed from left to right in (10.7). Thus, from Cramer's rule, we can derive the unique nonzero solution $F_{i/n}^j(s)$ to (10.7), given as:

$$s^{i/n} = \lim_{j \to \infty} F_{i/n}^j(s) = \lim_{j \to \infty} \frac{1}{|\boldsymbol{W}(s)|} \begin{bmatrix} w_{1,1}(s) & \cdots & w_{1,i-1}(s) & \psi_1(s) & w_{1,i+1}(s) & \cdots & w_{1,n-1}(s) \\ \vdots & & \vdots & \vdots & \vdots & & \vdots \\ w_{i-1,1}(s) & \cdots & w_{i-1,i-1}(s) & \psi_{i-1}(s) & w_{i-1,i+1}(s) & \cdots & w_{i-1,n-1}(s) \\ w_{i,1}(s) & \cdots & w_{i,i-1}(s) & \psi_i(s) & w_{i,i+1}(s) & \cdots & w_{i,n-1}(s) \\ w_{i+1,1}(s) & \cdots & w_{i+1,i-1}(s) & \psi_{i+1}(s) & w_{i+1,i+1}(s) & \cdots & w_{i+1,n-1}(s) \\ \vdots & & \vdots & \vdots & \vdots & & \vdots \\ w_{n-1,1}(s) & \cdots & w_{n-1,i-1}(s) & \psi_{n-1}(s) & w_{n-1,i+1}(s) & \cdots & w_{n-1,n-1}(s) \end{bmatrix}. \quad (10.11)$$

Equation (10.11) shows that it is a universal formula for the approximate implementation of the arbitrary-order fractional differential operator $s^{i/n}$. Every denominator of the fractional differential operator $s^{i/n}$ is uniformly equal to $|\boldsymbol{W}(s)|$, where $i = 1, 2, \cdots, n-1$. Further, the numerator of (10.11) determines the numerator i of the fractional-order n of the fractional differential operator $s^{i/n}$, whereas the denominator of (10.11) determines the denominator n of its fractional-order i/n. The variation of its numerator and denominator are both order-sensitive to the fractional differential operator $s^{i/n}$, but the variation of its denominator is relatively more order-sensitive than the variation of its numerator.

10.1.3.2 Implementation Cost of FHNNs

This subsection is the description of the implementation cost of the FHNNs, which is considerably lower than that of the PUFs.

Equation (10.11) shows that if $I_i(s) = \prod_{k=1}^{m}(s + \mu_k)$, the following is true:

$$s^{i/n} = V_i(s)/I_i(s) = C_0 + \sum_{k=1}^{m} \frac{C_k}{s + \mu_k}, \quad (10.12)$$

where $-\mu_k$ is the simple root of $I_i(s) = 0$, $C_0 = \lim_{s \to \infty} s^{i/n}$, and $C_k = \lim_{s \to \mu_k}(s + \mu_k)s^{i/n}$. In this case, $s^{i/n}$ has m simple poles. Equation (10.12) shows that $s^{i/n}$ is a stable system if $s^{i/n}$ is satisfies $\mu_k > 0$. Furthermore, if $I_i(s) = (s + \mu_1)^h \prod_{k=2}^{m}(s + \mu_k)$, the following is true:

$$s^{i/n} = V_i(s)/I_i(s) = C_0 + \sum_{r=1}^{h} \frac{C_1^r}{s + \mu_1} + \sum_{k=2}^{m} \frac{C_k}{s + \mu_k}, \quad (10.13)$$

where $-\mu_1$ is the a h-order multiple root of $I_i(s) = 0$, and $C_1^r = \lim_{s \to \mu_1}(s + \mu_1)^r s^{i/n}$. In this case, $s^{i/n}$ has h multiple poles. Equations (10.12) and (10.13) show that $s^{i/n}$ is equivalent to a series circuit of the first-order integral analog circuit. The fractor can be described as the result of a nonlinear interaction of the series circuit of the first-order integral analog circuit, which is consistent with fractional calculus theory: fractional calculus is the result of a continuous interpolation of the integer-order calculus [230], [400]. Equation (10.12) shows that the *kth* series first-order integral analog circuit can be shown in Fig. 10.4.

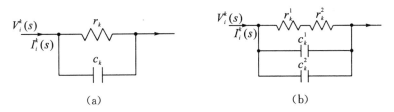

Fig. 10.4 *kth* series circuit: (a) Theoretical design, (b) Improved design for eliminating temperature variation effect.

Figures 10.4 (a) and (b) are the theoretical design and the improved design for eliminating temperature variation effect of the *kth* series circuit, respectively. Figure 10.4 (b) is discussed in next subsection. From Fig. 10.4 (a) and (10.12), r_k and c_k satisfy the conditions, given as, respectively:

$$r_k = C_k/\mu_k \Omega, \qquad (10.14)$$

$$c_k = 1/C_k \mathrm{F}, \qquad (10.15)$$

where the measurement units of resistance and capacitance are Ohm (Ω) and Farad (F), respectively. The input impedance of the *kth* series first-order integral analog circuit can be derived from (10.14) and (10.15), given as:

$$Z^k(s) = V_i^k(s)/I_i^k(s) = \frac{C_k}{s+\mu_k}. \qquad (10.16)$$

Thus, from (10.12) and (10.14) − (10.16), we can derive the correspondinganalog series circuit of $s^{i/n}$. It is shown in Fig. 10.5.

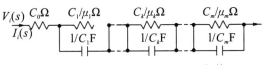

Fig. 10.5 Analog series circuit of $s^{i/n}$.

Equations (10.14) − (10.16) and Fig. 10.5 shows that theresistances of $s^{i/n}$ intrinsically subject to the numerator and poles of $s^{i/n}$ at the same time, whereas the capacitances of $s^{i/n}$ only intrinsically subject to the numerator of $s^{i/n}$. From (10.14) − (10.16) and Figs. 10.1 and 10.5, we can further see that since a FHNN is essentially a fractional-order intelligent analog circuit with the striking functions of nonlinear self-computing, reasoning, judging, and memorizing, the extraction of a physical measurement such as output voltage from the responses of a FHNN is straightforward. Thus, the hardware implementation of an arbitrary-order FHNN only requires some ordinary resistors, capacitors, and operational amplifiers (or triode amplifiers), which is easy and feasible and needs not to change the production process of IC. However, since a PUF evaluation requires an additional dedicated function module for physical measurement, the extraction of a key from the responses of a PUF is not straightforward. Different measuring results can be yielded by using some especially controllable techniques during the manufacturing process of a PUF [372], [373]. Thus, the hardware implementation of PUFs needs to change the production process of IC. The implementation cost of the FHNNs is considerably lower than that of the PUFs.

10.1.3.3 Temperature Variation Effect of FHNNs

This subsection is the proposal for the achievement of the constant-order performance of a FHNN when ambient temperature varies.

It is important that the output voltage of every neuron of a FHNN remains stable when it is subject to temperature variations. Equation (10.4) shows that the achievement of the constant-order performance of a FHNN is the crux of the maintenance of its output voltage stability when ambient temperature varies. Further, from Fig. 10.1 and (10.4), we can also see that the key of the achievement of the constant-order performance of a FHNN is to perform the constant-order performance of the fractor of its every neuron under different temperature conditions. In order to eliminate the change of the resistance or capacitance of a fractor caused by ambient temperature variations, at least three methods can be used to achieve near-perfect reliability. Let's use symbols ζ_r and ζ_c to denote the Temperature Coefficient of Resistance (TCR) and the Temperature Coefficient of Capacitance (TCC), respectively. At first, the commonly used method is to connect in series a pair of resistors with opposite TCR, and to connect in parallel a pair of capacitors with opposite TCC in the fractor of a FHNN. In Figs. 10.4 (a) and (b), $r_k^1 = r_k^2 = r_k/2$, $c_k^1 = c_k^2 = c_k/2$, the ζ_r of r_k^1 is opposite to that of r_k^2, and the ζ_c of c_k^1 is opposite to that of c_k^2. Secondly, the temperature effect of a fractor can be eliminated by employing a set of specific resistor and capacitor, whose temperature coefficients are the reciprocal of each other, $\zeta_c = 1/\zeta_r$. Hence, in Fig. 10.4 (b), when ambient temperature varies, from (10.14) and (10.15), one can obtain:

$$r_k = \zeta_r C_k/\mu_k \Omega, \qquad (10.17)$$

$$c_k = \zeta_c/C_k = 1/\zeta_r C_k \text{F}. \qquad (10.18)$$

Then, substitution of (10.17) and (10.18) into (10.16) results in:

$$Z_T^k(s) = V_i^k(s)/I_i^k(s) = \frac{\zeta_r C_k}{s + \mu_k}. \qquad (10.19)$$

Substituting (10.19) into (10.12), the following can be obtained:

$$s_T^{i/n} = \zeta_r C_0 + \zeta_r \sum_{k=1}^{m} \frac{C_k}{s + \mu_k} = \zeta_r s^{i/n}, \qquad (10.20)$$

where $s_T^{i/n}$ denotes a temperature influenced $s^{i/n}$. From (10.20), we can see that if $\zeta_c = 1/\zeta_r$, the temperature effect of a fractor can only change the magnitude of the fractional-order reactance of $s^{i/n}$, but cannot change its fractional-order i/n. If $\zeta_c = 1/\zeta_r$, the constant-order performance of the fractor can be achieved under different temperature conditions. Note that $\zeta_c = 1/\zeta_r$ is a strictly restricted theoretical condition for a set of resistor and capacitor. It is not a real condition until now. But with progress made in prospective material science, some innovative materials maybe make it possible for us to satisfy this condition. Thirdly, the temperature effect of the tree type, two-circuit type, H type, and net-grid type fractors [204], [205] can be eliminated themselves because of their specific recursive types with extreme self-similar fractal structures. For the convenience of illustration, we only employ the net-grid-type fractor in this chapter to analyze the constant-order performance of these four natural implementations of the fractor [204], [205] when ambient temperature varies. The structural representation of the 1/2-order net-grid type fractor can be shown in Fig. 10.6.

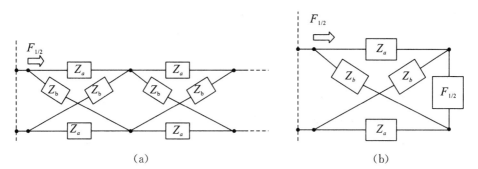

Fig. 10.6 Structural representation of the 1/2-order net-grid type fractor: (a) 1/2-order net-grid type fractor; (b) Equivalent circuit of the 1/2-order net-grid type fractor.

From Fig. 10.6, we can derive the 1/2-order reactance $F_{1/2}$ of a 1/2-order net-grid type fractor [204], [205], given as:

$$F_{1/2} = V_i(s)/I_i(s) = (Z_a Z_b)^{1/2}, \tag{10.21}$$

where Z_a and Z_b are the reactances of two ordinary passive circuit elements, resistor and capacitor/inductor. For the convenience of illustration, Z_a and Z_b are also referred to as two ordinary passive circuit elements themselves. In particular, with regard to the 1/2-order capacitive fractor, we set $Z_a = r$ and $Z_b = 1/(cs)$ as a resistor and a capacitor, respectively. In this case, from (10.21), $F_{1/2}^c = (r/c)^{1/2} s^{-1/2}$, where $F_{1/2}^c$ denotes the 1/2-order capacitive reactance of a 1/2-order net-grid type capacitive fractor. Hence, in Fig. 10.6, when ambient temperature varies, from (10.21), one can obtain:

$$_T F_{1/2}^c = \left(\frac{\zeta_r r}{\zeta_c c}\right)^{1/2} s^{-1/2} = \left(\frac{\zeta_r}{\zeta_c}\right)^{1/2} \left(\frac{r}{c}\right)^{1/2} s^{-1/2}, \tag{10.22}$$

where $_T F_{1/2}^c$ denotes temperature influenced $F_{1/2}^c$. Equation (10.22) shows that whatever the values of ζ_r and ζ_c are, the temperature effect can only change the magnitude of the 1/2-order reactance, but cannot change the fractional-order, 1/2, of the 1/2-order net-grid type fractor. Hence, the FHNNs employing the tree type, two-circuit type, H type, and net-grid type fractors have much more stable electrical performance than that of the PUFs under different temperature conditions.

10.1.3.4　Electromagnetic Interference Effect of FHNNs

This subsection is the analysis of the electrical performance stability of the FHNNs under electromagnetic disturbance conditions, which is much more robust than that of the PUFs.

As we know, all circuits are subject to electromagnetic disturbance, which could affect their electrical performance. The FHNNs are no exception, but their electromagnetic interference effect can be eliminated by themselves, which is attributed to the fractional-order-stability characteristics of the FHNNs. As our previous study [194], [399], a FHNN is essentially the fractional-order neural system with a significant advantage, whose attractors are intrinsically related to the fractional-order of its corresponding neurons. Only if the fractional-orders the corresponding neurons of a FHNN keep unchanged, its attractors and domains of attraction remain stable. Therefore, the electrical dynamic behavior of every neuron of a FHNN can converge automatically to its corresponding steady state (attractor), as long as its electrical performance deviation caused by electromagnetic interference is not beyond its corresponding domain of attraction. Hence, compared to the electromagnetic interference effect of the PUFs [362], [363], [374], one of real strengths of a FHNN is not only able to achieve the different performance from another FHNN, but also able to implement the same performance itself under electromagnetic disturbance conditions.

10.1.3.5 Amount of Entropy of FHNNs

This subsection is the description of theamount of entropy of the FHNNs, which is significantly higher than that of the PUFs with the same rank circuit scale.

With respect to a FHNN, from (10.4), the following can be obtained:

$$\chi_i \frac{d^{v_i} f^{-1}[a_i(t)]}{dt^{v_i}} = -f^{-1}[a_i(t)] + \sum_{j=1}^{S} \overline{w}_{i,j} a_j(t) + b_i(t). \qquad (10.23)$$

Thus, it follows that in vector form:

$$\chi \frac{d^v f^{-1}[\boldsymbol{a}(t)]}{dt^v} = -f^{-1}[\boldsymbol{a}(t)] + \overline{\boldsymbol{W}} \boldsymbol{a}(t) + \boldsymbol{b}, \qquad (10.24)$$

where $\chi = [\chi_1, \cdots, \chi_i, \cdots, \chi_S]^T$, $\boldsymbol{v} = [v_1, \cdots, v_i, \cdots, v_S]^T$, $\boldsymbol{a} = [a_1, \cdots, a_i, \cdots a_S]^T$ and $\boldsymbol{b} = [b_1, \cdots, b_i, \cdots, b_S]^T$, and $\overline{\boldsymbol{W}} = [\overline{w}_{i,j}]_{S \times S}$ is the weighting matrix of a FHNN. Figure 10.1 and (10.24) show that the FHNNs exploit the small mismatches across identically designed analog circuits, which are intrinsically dependent on the manufacture deviations of the resistors, capacitors, and amplifiers of the FHNNs. The output voltage deviation of $a_i(t)$ substantially subjects to the manufacture deviations of the resistors and capacitors of the fractor of the *ith* fractional neuron, and those of the feedback resistors being connected with the outputs of other all fractional neurons and the input of the *ith* one, and that of the amplifier of the *ith* fractional neuron, and as well as the output voltage deviations of other all fractional neurons feeding back to the input of the *ith* one. The substantive characteristics of the deviations are stochastic properties. The stochastic characteristics of the manufacture deviations of the *ith* fractional neuron of a FHNN can be given as:

$$D_i, p_i : \begin{cases} [r_1^i, r_2^i, \cdots, r_X^i, c_1^i, c_2^i, \cdots, c_Y^i, A^i]_{1\times(X+Y+1)} \\ [p_r(r_1^i), p_r(r_2^i), \cdots, p_r(r_X^i), p_c(c_1^i), p_c(c_2^i), \cdots, p_c(c_Y^i), p_A(A^i)]_{1\times(X+Y+1)} \end{cases},$$

(10.25)

where D_i and p_i denote the probability space of the manufacture deviations (random variables) and the corresponding joint probability density of the *ith* fractional neuron, respectively. r_x^i, c_y^i, and A^i denote the continuous random variable of the manufacture deviation of the *xth* resistor, that of the *yth* capacitor, and that of the amplifier of the *ith* fractional neuron, respectively. $p_r(r_x^i)$, $p_c(c_y^i)$, and $p_A(A^i)$ denote the probability density of the manufacture deviation of the *xth* resistor, that of the *yth* capacitor, and that of the amplifier of the *ith* fractional neuron, respectively. X and Y denote the amount of resistors and that of capacitors of the *ith* fractional neuron, respectively. Without loss of generality, it is assumed that the manufacture procedures of the resistors, capacitors, and amplifier of the *ith* fractional neuron are independent and identically distributed, respectively, so that:

$$p_i(r_1^i, r_2^i, \cdots, r_X^i, c_1^i, c_2^i, \cdots, c_Y^i, A^i) = \left[\prod_{x=1}^{X} p_r(r_x^i)\right]\left[\prod_{y=1}^{Y} p_c(c_y^i)\right] p_A(A^i)$$
$$\stackrel{p_r(r_x^i)=p_r(r), p_c(c_y^i)=p_c(c), p_A(A^i)=p_A(A)}{=} p_r^X p_c^Y p_A. \quad (10.26)$$

Further, since the FHNNs are fractional dynamic associative recurrent neural networks, a FHNN converts the manufacture deviations of the *ith* fractional neuron into its corresponding output voltage deviations, which will be feed back to the input of the *jth* fractional neuron. The output voltage of the *ith* fractional neuron does not be feed back to the input of itself, thus $i \neq j$. Hence, the stochastic characteristics of the manufacture deviations of a FHNN can be given as:

$$D_J, p_J : \begin{cases} \begin{bmatrix} D_1 & D_1 D_2 & \cdots & D_1 D_S \\ D_2 D_1 & D_2 & \cdots & D_2 D_S \\ \vdots & \vdots & & \vdots \\ D_S D_1 & D_S D_2 & \cdots & D_S \end{bmatrix}_{S\times S} \\ \begin{bmatrix} p_1 & p_1 p_2 & \cdots & p_1 p_S \\ p_2 p_1 & p_2 & \cdots & p_2 p_S \\ \vdots & \vdots & & \vdots \\ p_S p_1 & p_S p_2 & \cdots & p_S \end{bmatrix}_{S\times S} \end{cases} \quad (10.27)$$

where D_J and p_J denote the probability space of the joint feedback deviations and the corresponding joint probability density of a FHNN, respectively. $D_i D_j$ represents that D_i and D_j are concurrent events. Equation (10.27) shows that the ergodic pair-wise combination of the feedback deviations of all fractional neurons of a FHNN constitutes a $S \times S$ square matrix D_J, in which every matrix element is a joint probability event. In a similar way, the corresponding joint probability density of D_J is also a $S \times S$ square matrix p_J, in which every matrix element is a joint probability density of $D_i D_j$. Without loss of generality, it is further assumed that the manufacture procedure of every fractional neuron is independent and identically distributed. Thus, from (10.26) and (10.27), we can derive the joint probability density of a FHNN, given as:

$$p_J = \prod_{i=1}^{S} p_i^{2S-1} = \prod_{i=1}^{S} \{[p_r(r_x^i)]^X [p_c(c_y^i)]^Y p_A(A^i)\}^{2S-1}$$
$$\stackrel{p_{i=1 \to S} = p_i}{=} p_i^{(2S-1) \times S} = (p_r^X p_c^Y p_A)^{(2S-1) \times S}. \tag{10.28}$$

Therefore, from (10.29), we can derive the amount of entropy of a FHNN, given as:

$$E = -\iint_{R^3} p_J \log(p_J) \, dr \, dc \, dA$$
$$= -(2S^2 - S) \iint_{R^3} p_r^X p_c^Y p_A \log(p_r^X p_c^Y p_A) \, dr \, dc \, dA, \tag{10.29}$$

where log() denotes the logarithmic function. When the bases of log() are equal to 2, 3, natural base, and 10, the measurement units of the amount of entropy are Bit, Tet, Nat, and Det, respectively. Note that (10.29) shows that the amount of entropy of a FHNN increases quadratically with the increase of the amount of its fractional neurons S, whereas the amount of entropy of a PUF increases linearly with the increase of its circuit resources available [382]. Hence, the amount of entropy of the FHNNs is significantly higher than that of the PUFs with the same rank circuit scale.

10.1.4 Experiment and Analysis

10.1.4.1 Pass-Band Width of Fractor of FHNNs

In this subsection, in order to achieve the convergence of the fractor of an arbitrary-order FHNN, we discuss its pass-band width and ascertain the value of parameters α and β.

Example 1: Equation (10.5) shows that $|\alpha - (\beta s)^{i/n}| < 1$ is the conditional convergence condition of an arbitrary-order fractor. Let's set $s = \sigma + \eta\omega$, where σ denotes a real number, η denotes an imaginary unit, and ω denotes angular frequency. Set $\sigma = \rho\cos\theta$ and $\omega = \rho\sin\theta$, where ρ is a polar radius and θ is a polar angle. Thus, in polar coordinates, $s = \rho(\cos\theta + \eta\sin\theta)$. According to De Moivre's theorem, $s^\mu = \rho^\mu(\cos\mu\theta + \eta\sin\mu\theta)$. Thus, we can derive the boundary curve of the conditional convergence domain of an arbitrary-order fractor, given as:

$$|\alpha - \rho^\mu \beta^\mu \cos\mu\theta - \eta\rho^\mu \beta^\mu \sin\mu\theta| = 1, \tag{10.30}$$

where $\mu = i/n$, and $i = 1, 2, \cdots, n-1$. Then, from (10.30), the following can be obtained:

$$\rho = \frac{1}{\beta}(\alpha\cos\mu\theta + \sqrt{1 - \alpha^2 \sin^2\mu\theta})^{1/\mu}, \tag{10.31}$$

where $0 < \alpha \leq |\csc\mu\theta|$, β is a positive number, $\theta \in [2k\pi, 2\pi + 2k\pi]$, and $k = 0, 1, 2, \cdots$. Let's reject the extraneous root $\rho = \frac{1}{\beta}(\alpha\cos\mu\theta - \sqrt{1 - \alpha^2 \sin^2\mu\theta})^{1/\mu}$ of (10.30), because it cannot maintain the validity of (10.30) for the entire interval $\theta \in [2k\pi, 2\pi + 2k\pi]$. Further, if $0 < \alpha \leqslant 1 \leqslant |\csc\mu\theta|$, it can always maintain the validity of $1 - \alpha^2 \sin^2\mu\theta \geqslant \cos^2\mu\theta \geqslant 0$ for arbitrary μ and all of $\theta \in [2k\pi, 2\pi + 2k\pi]$. Thus, for the convenience of the approximate implementation of $s^\mu = s^{i/n}$, let's set $0 < \alpha \leqslant 1$.

First, from (10.31), if $\alpha = 1.0$, $\beta = 1.0$, the conditional convergence domain of the approximate implementation of $s^\mu = s^{i/n}$ in polar coordinates can be shown as given in Fig. 10.7.

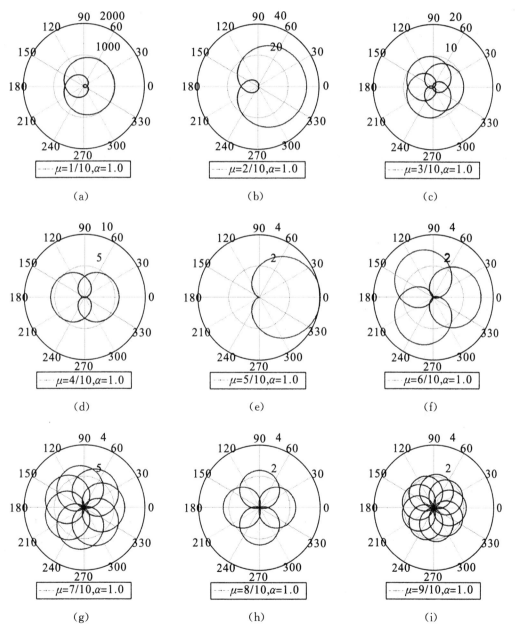

Fig. 10.7 The conditional convergence domain of the approximate implementation of $s^\mu = s^{i/n}$ in polar coordinates ($\alpha = 1.0$ and $\beta = 1.0$): (a) $\mu = 1/10$; (b) $\mu = 2/10$; (c) $\mu = 3/10$; (d) $\mu = 4/10$; (e) $\mu = 5/10$;

(f) $\mu=6/10$; (g) $\mu=7/10$; (h) $\mu=8/10$; (i) $\mu=9/10$.

Figure 10.7 shows that the conditional convergence domain of the approximate implementation of $s^\mu = s^{i/n}$ is the region being bounded by its closed curve ρ. The variation of boundary curve ρ depends on the fractional-order $\mu=i/n$. ρ is more complex when μ is larger. ρ represents a family of flower curves. If $\mu=2/10$, ρ is a Pascal spiral. If $\mu=5/10$, ρ is a cardioid. Furthermore, if $\theta=\pi/2+k\pi$, $S=\eta\omega$. Hence, if $\theta=\pi/2+k\pi$, the Laplace transform is equal to the spectral density of the Fourier transform. The pass-band width of a fractor depends primarily on the pass-band width of the approximate implementation of $s^\mu = s^{i/n}$. Thus, from (10.31), we can derive the pass-band width of a fractor F_μ, given as:

$$B_{F_\mu} = \frac{1}{\beta}\max\left(\alpha\cos\mu(\frac{\pi}{2}+k\pi)+\sqrt{1-\alpha^2\sin^2\mu\left(\frac{\pi}{2}+k\pi\right)}\right)^{1/\mu}, \quad (10.32)$$

where max() denotes the maximum value function.

Second, from (10.31), if $\alpha=0.5$ and $\beta=0.1$, the conditional convergence domain of the approximate implementation of $s^\mu = s^{i/n}$ in polar coordinates can be shown as given in Fig. 10.8.

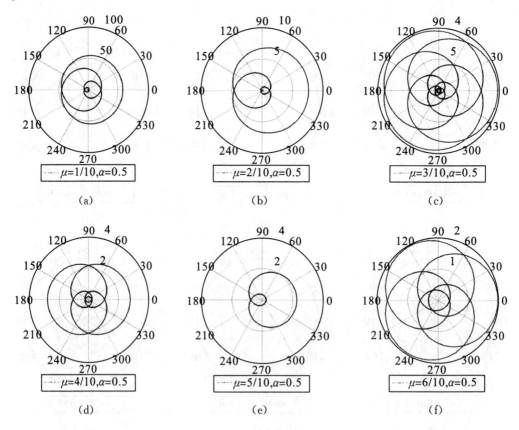

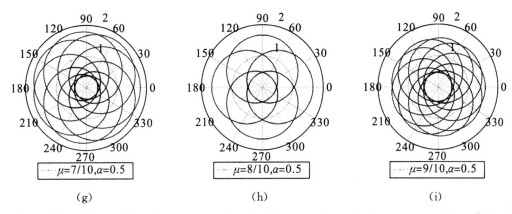

Fig. 10.8 The conditional convergence domain of the approximate implementation of $s^\mu = s^{i/n}$ in polar coordinates ($\alpha = 0.5$ and $\beta = 1.0$): (a) $\mu = 1/10$; (b) $\mu = 2/10$; (c) $\mu = 3/10$; (d) $\mu = 4/10$; (e) $\mu = 5/10$; (f) $\mu = 6/10$; (g) $\mu = 7/10$; (h) $\mu = 8/10$; (i) $\mu = 9/10$.

By comparing Figs. 10.7 and 10.8, we can further see that the petal of rose curve ρ is more complex when α is smaller. However, the variation of α cannot greatly change the pass-band width of a fractor. The order of magnitude of its pass-band width cannot change obviously with α.

Third, from (10.31), if $\alpha = 1.0$ and $\beta = 0.1$, the conditional convergence domain of the approximate implementation of $s^\mu = s^{i/n}$ in polar coordinates can be shown as given in Fig. 10.9.

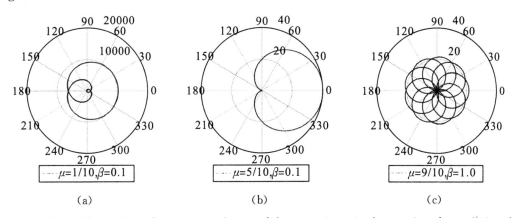

Fig. 10.9 The conditional convergence domain of the approximate implementation of $s^\mu = s^{i/n}$ in polar coordinates ($\alpha = 1.0$ and $\beta = 0.1$): (a) $\mu = 1/10$; (b) $\mu = 5/10$; (c) $\mu = 9/10$.

By comparing Figs. 10.7 and 10.9, we can see that the variation of β can sharply change the pass-band width of a fractor. Equation (10.31) shows that the pass-band width of a fractor linearly increases with β. Hence, in order to get the wider pass-band width of a fractor, let's set β as a positive number that is as small as possible according to actual conditions.

10.1.4.2 Defense against Chip Cloning Attacks Capability of FHNNs

In this subsection, the capabilities of defense against chip cloning attacks, anti-electromagnetic interference, and anti-temperature variation of a FHNN are illustrated experimentally in detail.

Example 2: We analyze the capability of defense against chip cloning attacks of the FHNNs. For the convenience of illustration, without loss of generality, let's set the fractional-order of a FHNN to be the 1/2-order. Equation (10.6) shows that for $\lim_{j\to\infty} F_{i/n}^j(s) = s^{i/n}$, the limit symbol may be removed when j is sufficiently large. Equation (10.7) shows that when $i=2$ and $n=4$, $s^{i/n} \overset{i=2,n=4}{=} s^{1/2}$. Further, Fig. 10.9 (b) shows that when $\alpha = 1.0$ and $\beta = 0.1$, we can get an enough pass-band width of a 1/2-order band-pass fractor $s^{1/2}$. Thus, let's set $j=12$, $n=4$, $\alpha=1.0$, and $\beta=0.1$ in (10.5). From (10.7), (10.8), (10.11), and Figs. 10.3 (a) and 10.3 (b), On the basis of the solution identification theorem of the nonhomogeneous linear equations, we can derive the nonzero solutions of $[s^{1/4}, s^{1/2}, s^{3/4}]^T$. Thus, the following can be obtained:

$$s^{-1/2} = \frac{1.2 \cdot 10^{3/2} \cdot s^5 + 2.2 \cdot 10^{7/2} \cdot s^4 + 7.92 \cdot 10^{9/2} \cdot s^3 + 7.92 \cdot 10^{11/2} \cdot s^2 + 2.2 \cdot 10^{13/2} \cdot s + 1.2 \cdot 10^{13/2}}{s^6 + 6.6 \cdot 10^2 \cdot s^5 + 4.95 \cdot 10^4 \cdot s^4 + 9.24 \cdot 10^5 \cdot s^3 + 4.95 \cdot 10^6 \cdot s^2 + 6.6 \cdot 10^6 \cdot s + 10^6}$$

$$= \frac{3.09352345 \cdot 10}{s + 5.76954805 \cdot 10^2} + \frac{3.5988971}{s + 5.82842712 \cdot 10} + \frac{1.42217976}{s + 1.69839637 \cdot 10} + \frac{8.373662263 \cdot 10^{-1}}{s + 5.88790706}$$

$$+ \frac{6.17473122 \cdot 10^{-1}}{s + 1.71572875} + \frac{5.3618124 \cdot 10^{-1}}{s + 1.73323801 \cdot 10^{-1}}, \quad (10.33)$$

where $s^{-1/2}$ is the reciprocal of $s^{1/2}$. Equation (10.33) shows that all of the poles of $s^{-1/2}$ are on the left half plane in the Laplace transform domain. $s^{-1/2}$ is a stable system. Equations (10.14) and (10.15) show that if C_k is multiplied by a constant, we can not only adjust the values of r_k and c_k according to actual conditions but also keep the order of $s^{i/n}$ being unchanged. Then, from (10.14) and (10.15), we can derive an analog series circuit of $10^3 \cdot s^{-1/2}$, as shown in Fig. 10.10.

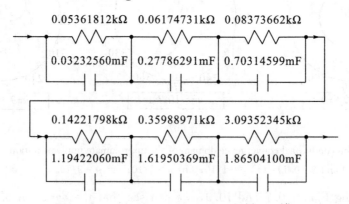

Fig. 10.10 Analog series circuit of $10^3 \cdot s^{-1/2}$.

Thus, from Figs. 10.1 and 10.10, we can implement a 1/2-order band-pass FHNN with a 1/2-order band-pass fractor [194], [399]. For the convenience of illustration, let's suppose a FHNN has two neurons with identical fractional-order. The fractional-order of either neuron of a 1/2-order FHNN is equal to $v=0.5$. Thus, we can implement either neuron of a 1/2-order FHNN with $10^3 \cdot s^{-1/2}$. To make it in such a way that the output convergence trajectories of two 1/2-order neurons do not overlap, in Fig. 10.1, let's set the feedback resistances $R_{12}=300\Omega$, $R_{21}=250\Omega$, and the shunt resistances $R=0.1\Omega$. The selected chip type of a Voltage mode Operational Amplifier (VOA) is an ultralow offset voltage operational amplifier OP37G. Note that we can use an Operational Transconductance Amplifier (OTA) to implement arbitrary resistance and capacitance, as shown in Fig. 10.11.

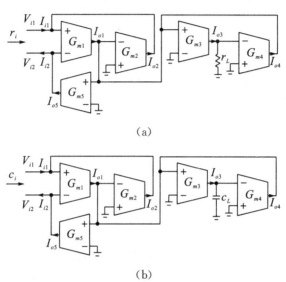

Fig. 10.11 Equivalent OTA circuits for resistor and capacitor: (a) Resistor; (b) Capacitor.

In Fig. 10.11, G_m is the amplifier gain of OTA. From Fig. 10.11, according to Kirchhoff's current law, we can derive the input resistance r_i and the input capacitance c_i in the time domain, given as:

$$r_i = \frac{G_{m1}G_{m2}}{G_{m3}G_{m4}} r_L, \tag{10.34}$$

$$c_i = \frac{G_{m1}G_{m2}}{G_{m3}G_{m4}} c_L. \tag{10.35}$$

Equations (10.34) and (10.35) show that we can adjust G_{m1}, G_{m2}, G_{m3}, and G_{m4} to achieve capacitance and resistance of arbitrary value. Further, temperature variation has effects on the G_m of OTA, but the effects of temperature variation on r_i and c_i are restrained because the power of G_m is equal in (10.34) and (10.35). Thus, from Figs. 10.1 and 10.10, let's utilize commonly used Printed Circuit Board (PCB) design software of Cadence, OrCAD v16.6 to simulate the analog circuit realization of a 1/2-order FHNN of original chip, as shown in Fig. 10.12.

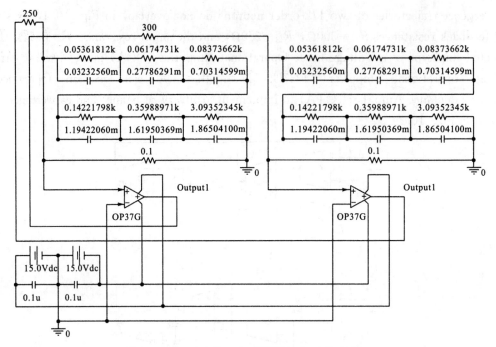

Fig. 10.12 A 1/2-order FHNN of original chip.

By comparing Figs. 10.1 and 10.12, we can see that the input currents of a 1/2-order FHNN of original chip are set to be $i_1 = i_2 = 0$ in Fig. 10.12. The responses of a PUF can be created to its corresponding inputs. Different from the PUFs, the responses of a FHNN can be generated not only to its various inputs but also to its initial state without any input. A FHNN is essentially the fractional-order neural system with the fractional-order-stability characteristics, whose attractors are intrinsically related to the fractional-order of its corresponding neurons [194], [399]. Only if the fractional-orders the corresponding neurons of a FHNN keep unchanged, its attractors and domains of attraction remain stable. Therefore, only if the initial state of a FHNN without any input, (0V, 0V), is not on its steady state position, the electrical dynamic behavior of every neuron of a FHNN can converge automatically to its corresponding attractor. Note that since the responses of a FHNN can be generated by its initial state without any input, which can prevent a rogue designer or vendor from studying the nonlinear relationship between the responses and inputs of a PUF by machine learning.

From Fig. 10.12, we can then get the convergence trajectory performance of the outputs of a 1/2-order FHNN of original chip, which can be shown in Fig. 10.13.

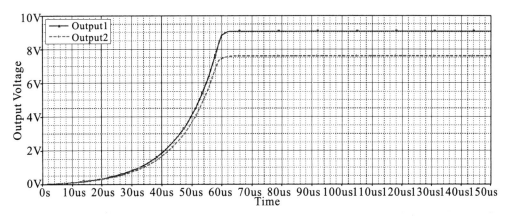

Fig. 10.13   Time response curves of outputs of a 1/2-order FHNN of original chip.

Figure 10.13 shows that the outputs of a 1/2-order FHNN of original chip are gradually convergent. Corresponding to the initial state (0V, 0V) and the input (0, 0) of a 1/2-order FHNN of original chip, its outputs converge to (9.050432V, 7.588827V). Hence, (9.050432V, 7.588827V) is one of the attractors of a 1/2-order FHNN of original chip.

According to the electronic manufacturing technology at present, no chip cloning attacker has been able to manufacture two resistors or capacitors with actually identical values. To date, no one has been able to manufacture two FHNNs with identical values of the attractors. Therefore, we can apply the FHNNs to the defense against chip cloning attacks for anti-piracy. Without loss of generality, let's suppose that the resistance or capacitance manufactured by a chip cloning attacker has a weak variation according to Gaussian distribution. Let's assume the Gaussian distribution of resistance variation of a FHNN of the chip has been cloned as $N(0, 10^{-4})$, and its Gaussian distribution of capacitance variation as $N(0, 2 \cdot 10^{-8})$. From Fig. 10.10, we can get two random chip cloned samples of $10^3 \cdot s_c^{-1/2}$, as shown in Fig. 10.14.

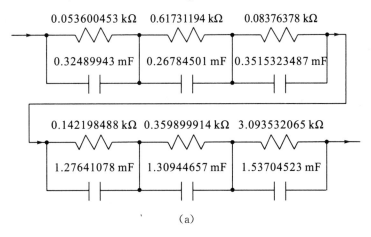

(a)

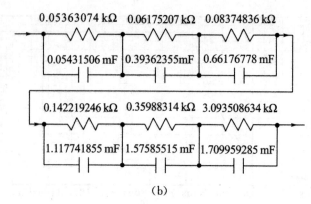

(b)

Fig. 10.14 Two random chip cloned samples of $10^3 \cdot s_c^{-1/2}$: (a) First chip cloned sample; (b) Second chip cloned sample.

By comparing Figs. 10.10 and 10.14, we can see that the variation of resistance or capacitance of a FHNN of the chip has been cloned is limited in the range of 10^{-4} to 10^{-2}. The deviation of a FHNN of the chip has been cloned is relatively small. Thus, from Fig. 10.14 (a), we can derive the driving-point impedance function of the first chip cloned sample of $10^3 \cdot s_c^{-1/2}$, given as:

$$10^3 \cdot s_c^{-1/2} = \frac{3.07787552155 \cdot 10^4}{s + 573.90880479} + \frac{3.7335024455 \cdot 10^3}{s + 60.479997407} + \frac{2.8446884155 \cdot 10^3}{s + 33.97195440182}$$
$$+ \frac{7.8344684528 \cdot 10^2}{s + 5.50953008205} + \frac{7.63681406934 \cdot 10^2}{s + 2.121927171682} + \frac{6.50598941649 \cdot 10^2}{s + 0.2103097223651}$$
$$= \frac{39554.6732704 \cdot s^5 + 8754255.6833498 \cdot s^4 + 421126745.34194 \cdot s^3 + 5159211007.40894 \cdot s^2 + 16795885375.585 \cdot s + 11001695199.6904}{s^6 + 676.202523273 \cdot s^5 + 61515.855384454 \cdot s^4 + 1627244.4931 \cdot s^3 + 9996433.061404 \cdot s^2 + 15816280.1975 \cdot s + 2899205.1488927}$$

(10.36)

Substituting $j\omega$ for s in (10.36), we can derive the Bode diagram of the approximately implemented $10^3 \cdot s^{-1/2}$ of original chip and that of the approximately implemented $10^3 \cdot s_c^{-1/2}$ of the chip has been cloned, as shown in Fig. 10.15.

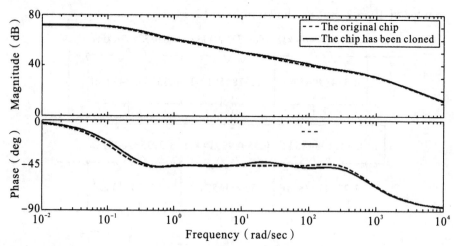

Fig. 10.15 Bode diagram of $10^3 \cdot s^{-1/2}$ of original chip and that of $10^3 \cdot s_c^{-1/2}$ of chip has been cloned.

Figure 10.15 shows that for the manufacturing error of capacitance and resistance, the frequency characteristics of the approximately implemented $10^3 \cdot s_c^{-1/2}$ of the chip has been cloned deviate from that of the approximately implemented $10^3 \cdot s^{-1/2}$ of original chip. In the pass-band of $[10^1 \text{Hz}, 2.4 \cdot 10^2 \text{Hz}]$, the magnitude of the approximately implemented $10^3 \cdot s_c^{-1/2}$ of the chip has been cloned slightly deviates from that of the approximately implemented $10^3 \cdot s^{-1/2}$ of original chip. Meanwhile, the phase of the former also slightly deviates from that of the latter. There are a slight deviation ΔA and a small included angle $\Delta \theta$ between the magnitude and the phase of the former and that of the latter, respectively. Substituting $\eta \omega$ for s, we can derive $10^3 \cdot s^{-1/2} \stackrel{s=\eta\omega}{=} 10^3 \cdot \omega^{-1/2} e^{\eta(-\pi/4)}$ and $10^3 \cdot s_c^{-1/2} \stackrel{s=\eta\omega}{=} 10^3 \cdot (\omega^{1/2} + \Delta A) e^{\eta(-\pi/4 + \Delta \theta)}$. Thus, we can see that the approximately implemented $10^3 \cdot s_c^{-1/2}$ of the chip has been cloned, in essence, slightly changes the fractional-order of the approximately implemented $10^3 \cdot s^{-1/2}$ of original chip. The fractional-order of the approximately implemented $10^3 \cdot s_c^{-1/2}$ of the chip has been cloned is no longer equal to $-1/2$. Therefore, we can implement two neurons of a 1/2-order FHNN of the chip has been cloned with two aforementioned chip cloned samples of $10^3 \cdot s_c^{-1/2}$ in Fig. 10.14, as shown in Fig. 10.16.

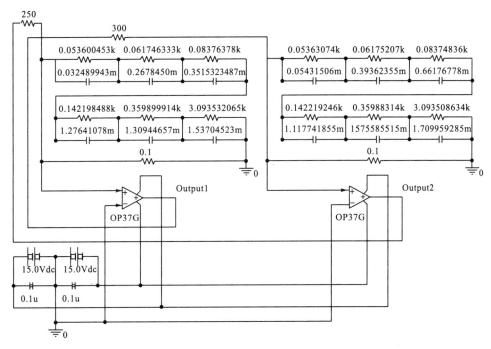

Fig. 10.16　A 1/2-order FHNN of chip has been cloned.

By comparing Figs. 10.1 and 10.16, we can see that since the responses of a FHNN can be generated to its initial state without any input, (0V, 0V), the input currents of a 1/2-order FHNN of the chip has been cloned are set to be $i_1 = i_2 = 0$ in Fig. 10.16. The time response curve of the outputs of a 1/2-order FHNN of the chip has been cloned is shown in Fig. 10.17.

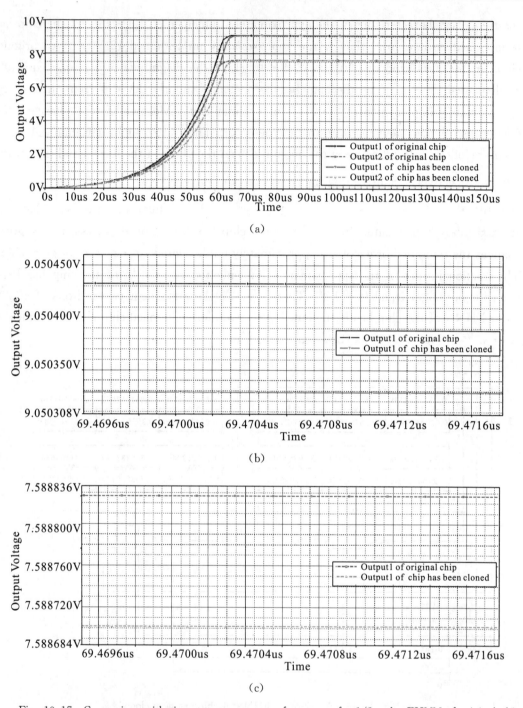

Fig. 10.17 Comparison with time response curves of outputs of a 1/2-order FHNN of original chip and that of a chip has been cloned: (a) Comparison with output voltages of original chip and that of chip has been cloned; (b) Comparison with output1 voltage of original chip and that of chip has been cloned; (c) Comparison with output2 voltage of original chip and that of chip has been cloned.

Figure 10.17 shows that the outputs of a 1/2-order FHNN of the chip has been cloned are gradually convergent. Corresponding to Corresponding to the initial state (0V, 0V) and the input (0, 0) of a 1/2-order FHNN of the chip has been cloned, its outputs converge to (9.050329V, 7.588701V). Hence, (9.050329V, 7.588701V) is one of the attractors of a 1/2-order FHNN of the chip has been cloned.

By comparing Figs. 10.13 and 10.17, we can see that the values of the attractors of a 1/2-order FHNN of the chip has been cloned deviate slightly from those of a 1/2-order FHNN of original chip. The deviation between the outputs of a 1/2-order FHNN of the chip has been cloned and those of an original chip is (−0.103mV, −0.126mV). Let's define a threshold that can allow discern between real physical and fake one as the seventy percent of the absolute value of the output deviation between each other. In this example, the threshold is (0.07mV, 0.08mV). The FHNNs can easy detect a negative two orders of magnitude, or even smaller, deviation of the chip has been cloned. The values of the attractors of the FHNNs essentially relate to the corresponding fractional-order of its neurons. Since the values of the resistance and capacitance of the fractors in the neurons of a FHNN of the chip has been cloned are changed, the values of the attractors of a FHNN are changed while the fractional-orders of its neurons are changed [194], [399]. Through measuring the output voltages of a FHNN, we can easy determine whether the chip of a FHNN has been chip cloned or not. Consequently, we can apply the FHNNs to the defense against chip cloning attacks for anti-piracy.

Example 3: We analyze the capability of anti-electromagnetic interference of the FHNNs. It is assumed that the amplitude peak, frequency, and duration of time of a suffered random electromagnetic interference of the neurons a 1/2-order FHNN are (±2V, ±4V), 1MHz, and 30us, respectively. To avoid changing the circuit configuration and circuit parameters of a 1/2-order FHNN, the suffered random electromagnetic interference is fed into the negative input of the operational amplifier OP37G. Thus, a 1/2-order FHNN of chip has been electromagnetically interfered can be shown in Fig. 10.18.

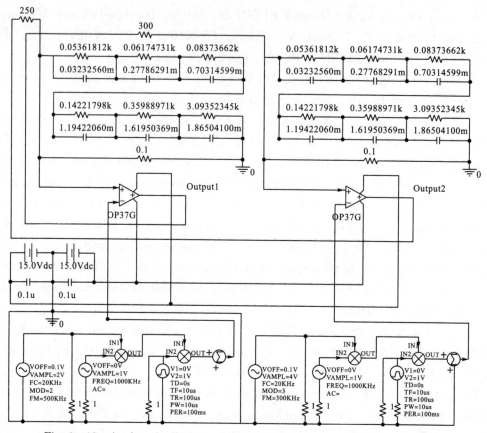

Fig. 10.18　A 1/2-order FHNN of chip has been electromagnetically interfered.

From Fig. 10.18, we can get the time response curve of the outputs of a 1/2-order FHNN of the chip has been electromagnetically interfered, as shown in Fig. 10.19.

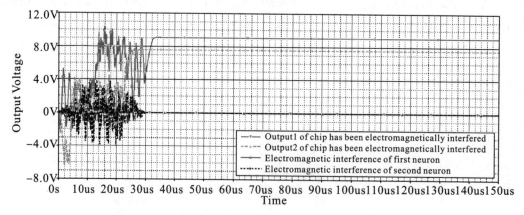

(a)

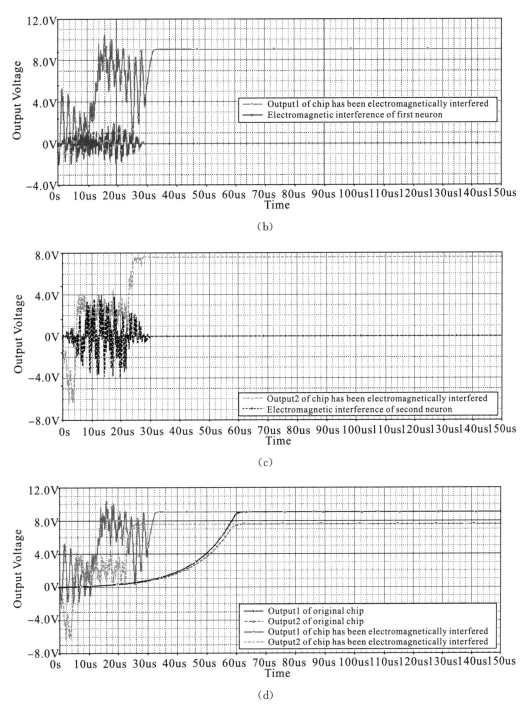

Fig. 10.19 Comparison with time response curves of outputs of a 1/2-order FHNN of original chip and that of a chip has been electromagnetically interfered: (a) Outputs of chip has been electromagnetically interfered; (b) Output of first neuron of chip has been electromagnetically interfered; (c) Output of second neuron of chip has been electromagnetically interfered; (d) Outputs of original chip and that of chip has been electromagnetically interfered.

From Figs. 10.19 (a), 10.19 (b), and 10.19 (c), we can see that provided the suffered deviation caused by random disturbance is still in the domain of attraction, the time response curve of the corresponding neuron of a 1/2-order FHNN of the chip has been electromagnetically interfered can converge gradually to its steady state. By comparing Figs. 10.13 and 10.19 (d), we can see that the steady state of a 1/2-order FHNN of chip has been electromagnetically interfered and that of original chip are identical, i.e. (9.050432V, 7.588827V). Because of the fractional-order-stability characteristics of the FHNNs [194], [399], the electromagnetic interference effect can be eliminated by themselves.

Example 4: The realistic simulation of a 1/2-order FHNN in the form of actual analog circuit is discussed. In particular, the anti-temperature variation capability of the FHNNs employing the net-grid type fractor [204], [205] is illustrated experimentally in detail. From Fig. 10.6, we can use the resistor and capacitor to achieve the analog circuit realization of the 1/2-order net-grid type capacitive fractor. In Fig. 10.6, let $Z_a = r$ and $Z_b = 1/(cs)$ to be a resistor and a capacitor, and set the resistance and capacitance as $r = 1\Omega$ and $c = 1\text{pF}$, respectively. Thus, from Fig. 10.6, we can construct a 1/2-order capacitive fractor in a series connection of 10-layer repeated net-grid-type structures [204], [205], as shown in Fig. 10.20.

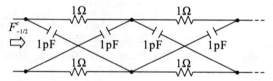

Fig. 10.20　1/2-order capacitive fractor of 10-layer repeated net-grid-type structures.

To make it in such a way that the output convergence trajectories of two 1/2-order neurons do not overlap, in Fig. 10.1, let's set the feedback resistances $R_{12} = 2\Omega$, $R_{21} = 1\Omega$, and the shunt resistances $R = 1\Omega$. With respect to the resistance and capacitance, their deviations of actual manufacture are within 5%. The actual TCR and TCC are $\zeta_r = 100\text{ppm}/℃$ and $\zeta_c = 85\text{ppm}℃$, respectively. The selected chip type of a VOA is a classic UA741CN KQQ520, whose operating voltage is ±15V. The operating temperature range of resistor and capacitor is −55℃~155℃, and that of operational amplifier is 0℃~70℃. From (10.4), (10.23), and (10.24), we can see that temperature variation has also effects on the input-output function f of operational amplifier, but the effects of temperature variation on the output voltages of the FHNNs, $a(t)$, are restricted as a linear transformation because of the homogeneous property of fractional calculus. Then, from Figs. 10.1 and 10.20, we can achieve the actual analog circuits of four independent 1/2-order FHNNs, as shown in Fig. 10.21.

(a)   (b)

Fig. 10.21 Actual analog circuits of four independent 1/2-order FHNNs: (a) Four independent 1/2-order FHNNs; (b) A 1/2-order FHNN.

At the test temperature of 27℃, the actual output voltages of the four independent 1/2-order FHNNs in Fig. 10.21 are (75.075000mV, 50.049930mV), (75.098451mV, 50.100260mV), (75.050730mV, 50.081644mV), and (75.120578mV, 50.074701mV), respectively. The corresponding actual output voltages of each FHNN are different from each other. Thus, through measuring the actual output voltages of the FHNNs, we can easily determine whether the chip of a FHNN has been cloned or not. In particular, by means of the actual analog circuit of a 1/2-order FHNN in Fig. 10.21(b), we can acquire the actual measurement data of the effects of temperature variation on its output voltages in the operating temperature range of 10℃~50℃, as shown in Table 10.1.

Table 10.1 Actual measurement data of effects of temperature variation on output voltages of a 1/2-order FHNN.

Temperature (℃)	Output Voltages (Output1, Output2) (mV)
10	(74.947325, 49.964846)
15	(74.984863, 49.989871)
20	(75.022403, 50.014896)
27	(75.075000, 50.049930)
30	(75.097471, 50.064945)
35	(75.135010, 50.089975)
40	(75.172550, 50.114998)
45	(75.210878, 50.140020)
50	(75.247625, 50.165050)

Thus, from Table 10.1, we can obtain the nonlinear fitting curve of the actual measurement data of the effects of temperature variation on the output voltages of the actual analog circuit of a 1/2-order FHNN, as shown in Fig. 10.22.

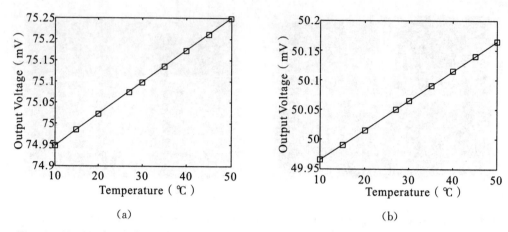

Fig. 10.22 Nonlinear fitting curve of actual measurement data of effects of temperature variation on output voltages of actual analog circuit of a 1/2-order FHNN: (a) Output1, (b) Output2.

Figure 10.22 and (10.22) show that whatever the TCR, TCC and temperature coefficient of operational amplifier vary, the effects of temperature variation on the output voltages of a 1/2-order FHNN employing the net-grid type fractor [204], [205] are sure to accord with linear unbiased rules, whose line slopes are linearly in direct proportion to $(\zeta_r/\zeta_c)^{1/2}$. In a similar way, we can derive that whatever the TCR, TCC and temperature coefficient of operational amplifier vary, the effects of temperature variation on the output voltages of a v-order FHNN employing the tree type, two-circuit type, H type, and net-grid type fractors [204], [205] are sure to accord with linear unbiased rules, whose line slopes are linearly in direct proportion to $(\zeta_r/\zeta_c)^v$. Therefore, by detecting whether the output voltages of the chip of a FHNN are on the linear lines of the effects of temperature variation on its output voltages or not, we can easily determine whether the chip of a FHNN has been cloned or not.

10.1.5 Conclusions

How to apply the fractional-order neural networks to hardware security is an emerging discipline branch field of study and few studies have been seldom performed in this area. In this chapter, a feasible and promising application of the FHNNs to the defense against chip cloning attacks, which is superior to the PUFs. This is attributed to the fractional-order-stability characteristics of the FHNNs. The clue of application of the FHNNs to the defense against chip cloning attacks for anti-piracy is in the use of phase (more invariant than amplitude) for physical system recognition. The analog circuit of a FHNN is presented as the physical embodiment of some fractors. The fundamental idea behind the FHNNs is that a fractors based neural networks can be described in terms of attractors that are in turn very sensitive to process variations in the resistance and capacitance values that comprise fractor. This sensitivity can then be leveraged to provide a high degree of entropy that can be applied for mitigating cloning. At first, since the hardware implementation of an arbitrary-order FHNN only requires some ordinary resistors, capacitors, and operational amplifiers (or triode amplifiers), it is easy and feasible and needs not to change the production process of IC. Thus, the implementation cost of the FHNNs, which is considerably lower than that of the PUFs. Secondly, since the v-order reactance of a v-order tree type, two-circuit type, H type, or net-grid type fractor is linearly in direct proportion to $(\zeta_r/\zeta_c)^v$ the FHNNs employing the tree type, two-circuit type, H type, and net-grid type fractors have much more stable electrical performance than that of the PUFs under different temperature conditions. Thirdly, since the electrical dynamic behavior of every neuron of a FHNN can converge automatically to its corresponding steady state (attractor), as long as its electrical performance deviation caused by electromagnetic interference is not beyond its corresponding domain of attraction. Thus, the electrical performance stability of the FHNNs under electromagnetic disturbance conditions is much more robust than that of the PUFs. Fourthly, since the amount of entropy of a FHNN increases quadratically with the increase of the amount of its fractional neurons, whereas the amount of entropy of a PUF increases linearly with the increase of its circuit resources available, the amount of entropy of the FHNNs is significantly higher than that of the PUFs with the same rank circuit scale.

From the aforementioned discussion, we can also see that there are many other problems that need to be further studied. For example, how to improve the stability of supply voltage, how to reduce thermal noise effect, how to prolong the service life of circuit components, how to further improve the capability of the defense against chip cloning attacks of the FHNNs by means of the fractional-order memristor, and so on. These will be discussed in our future work.

10.2 Analog Circuit Realization of Arbitrary-Order Fractional Hopfield Neural Networks: A Novel Application of Fractor to Defense against Chip Cloning Attacks[①]

基于分数阶微积分，我们提出了一种用于实现任意阶次分数阶 Hopfield 神经网络的模拟电路的任意阶分抗，并首次将其应用于抗芯片克隆攻击。分抗是实现分数阶 Hopfield 神经网络最重要的电路组件，任意阶分抗的硬件实现是将分数阶 Hopfield 神经网络应用于抗芯片克隆的关键。我们提出了一种基于任意阶分数阶 Hopfield 神经网络的模拟电路，详细分析了任意阶分数阶 Hopfield 神经网络的任意阶网格型电容缩放分抗的硬件实现，分析了分数阶 Hopfield 神经网络的任意阶分抗的逼近性能，用模拟电路实现了分数阶 Hopfield 神经网络，并通过实验详细说明了其抗芯片克隆的能力。本节的主要贡献在于首次在硬件上实现了用任意阶次分数阶 Hopfield 神经网络进行抗芯片克隆。

---

① PU Yifei. Analog Circuit Realization of Arbitrary-Order Fractional Hopfield Neural Networks: A Novel Application of Fractor to Defense against Chip Cloning Attacks [J]. IEEE Access, 2016 (4): 5417-5435.

10.2.1 INTRODUCTION

Fractional calculus is created as old as the integer-order calculus, but until recently, it is seldom known by most of mathematicians and physical scientists in engineering fields. The commonly used fractional calculus definitions are Grünwald-Letnikov, Riemann-Liouville, and Caputo, respectively [178], [230]. Nowadays, fractional calculus appears to be a promising mathematical approach for physical scientists and engineering technicians [178], [230]. Some promising results and ideas have demonstrated that a fractional-order or fractionaldimensional approach is the best description for many natural phenomena such as diffusion processes [181], [183], bioengineering [94], viscoelasticity theory [126], [184], nonlinear systems [403], [400], control [89], circuits and systems [130], [171] — [173], [193], [195] — [203], [208], [215], [225], [274], [404] — [409], image processing [187], [189], and neural networks [192], [194], [338], [399], [410] — [415].

The issue of how to applythe fractor to information forensics and security, especially to the defense against chip cloning attacks, is an emerging area of research [192], [194], [399], [410]. In the field of the fractional-order circuits and systems, the concept of the fractor arose following the successful implementation of the fractional-order differentiator or integrator in the form of an analog circuit. The fractance is the fractional-order impedance of a fractor [130], [171] — [173], [193], [195] — [203], [208], [215], [225], [274], [404] — [409]. Let's denote a fractor with the symbol ⊣𝑀𝑀⊢^{F}, in which F is the abbreviation of fractor. There are two types of the fractor: capacitive fractor and inductive fractor. In particular, the tree type, two-circuit type, H type, net-grid type should be four discovered natural implementations of the fractor [171] — [173], [193], [225], [405]. Compared to other approximate implementations of fractor [130], [195] — [203], [208], [215], [274], [404], [406] — [409], the floating point values of the capacitance, inductance, and resistance of these four natural implementation types of the fractor are not required [171] - [173], [193], [225], [405]. Pu derived the generic fractional-order capacitive reactance and inductive reactance of the purely ideal arbitrary-order capacitive fractor and inductive fractor, respectively [225], [405]. On the basis of the fractor and fractional calculus, the first preliminary attempt on implementation of a fractional-order neural network of the Hopfield type was originally envisioned in 2005 by Pu [194]. Pu et al. proposed to introduce fractional calculus to achieve a novel conceptual framework: Fractional Hopfield Neural Networks (FHNNs) [192], [194], [399], [410]. The FHNNs have a fascinating capability to distinguish hardware circuits from one another, which attributes to its property of the fractional-order-stability. The values of the attractors of a FHNN change with the fractional-order of its corresponding neurons [192], [194], [399], [410]. Therefore, to apply the FHNNs to the defense against chip cloning attacks, we should achieve the analog circuit realization of the arbitrary-order FHNNs in the first place. this chapter is the proposal for the analog circuit realization of the arbitrary-order FHNNs that can be applied to defense against chip cloning attacks. In particular, the analog circuit achievement of the arbitrary-order net-grid-type scaling fractor of a FHNN is discussed in detail. The main contribution of this chapter is the proposal for the first preliminary attempt of a feasible hardware implementation of the arbitrary-order FHNNs for defense against chip cloning attacks.

10.2.2 RELATED WORK

This section presents a brief necessary introduction of the FHNNs.

Pu et al. proposed to introduce fractional calculus to implement a novel conceptual framework: the FHNNs, which can be implemented by the fractor in the form of an analog circuit [192], [194], [399], [410]. A FHNN model can be shown in Fig. 10.23.

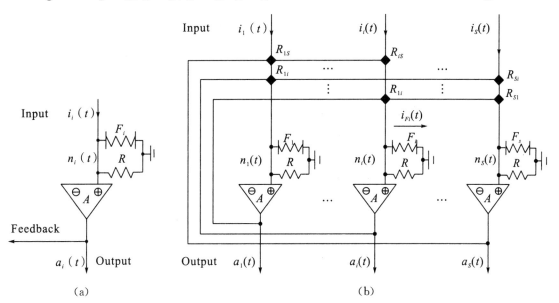

Fig. 10.23 AFHNN model: (a) Fractional-order neuron of a FHNN; (b) A FHNN.

In Fig. 10.23, S represents the number of the fractional-order neurons, $n_i(t)$ epresents the input voltage of the amplifier A of the ith fractional-order neuron, $a_j(t)$ represents the output voltage of the jth fractional-order neuron, and $i_i(t)$ represents the input electric current of the ith fractional-order neuron. $R_{i,j}$ represents the feedback resistor that connects with the output of the jth fractional-order neuron and the input of the ith one. Figure 10.1 shows that every fractional-order neuron of a FHNN consists of one operational amplifier A and its related fractor ─\|F\|─ and resistors. Each fractional-order neuron has the same circuit configuration. If the fractional-order of FHNN's neuron is equal to v_1, v_i, and v_s, respectively, $F_1 = F_{v_1}$, $F_i = F_{v_i}$, and $F_S = F_{v_s}$. Therefore, from Kirchhoff's current law, we can derive the operation rule of a FHNN [192], [194],

[399], [410], which can be given as:

$$\chi_i \frac{d^{v_i} n_i(t)}{dt^{v_i}} = -n_i(t) + \sum_{j=1}^{S} w_{i,j} a_j(t) + b_i(t), \tag{10.37}$$

where $\chi_i = Z_i K_i$, $1/Z_i = 1/R + \sum_{j=1}^{S}(1/R_{i,j})$, $K_i = \xi_i^{-v_i}$, $w_{i,j} = Z_i T_{i,j}$, $b_i = Z_i i_i$, $T_{i,j} = 1/R_{i,j}$, and $T_{i,j} = T_{j,i}$. For capacitive fractor, $\xi_i = r^{(1-p_i)/v_i}/c$, $v_i = q_i + p_i$ is a positive real number, q_i is a positive integer, and $0 \leqslant p_i \leqslant 1$ [225], [405]. Equation (10.37) shows that the operation rule of a FHNN is intrinsically a fractional-order partial differential equation.

10.2.3 ANALOG CIRCUIT REALIZATION OF ARBITRARY-ORDER FHNNs

This section is the proposal for the analog circuit realization of the arbitrary-order FHNNs that can be applied to defense against chip cloning attacks. Since the fractor is the most important circuit component needed to implement the FHNNs [192], [194], [399], [410], the hardware achievement of the arbitrary-order fractor is the primary task and crux for the state-of-the-art application of the FHNNs to defense against chip cloning attacks. Due to the length limitation of this chapter, in this chapter, the hardware achievement of the arbitrary-order net-grid-type capacitive scaling fractor of the arbitrary-order FHNNs is chiefly analyzed in detail. The net-grid-type scaling fractor has a considerably simple circuit structure that can extremely easily implement the arbitrary-order fractor.

Two equivalent circuit configurations of the arbitrary-order low-pass filtering net-grid-type scaling capacitive fractor can be shown in Fig. 10.24.

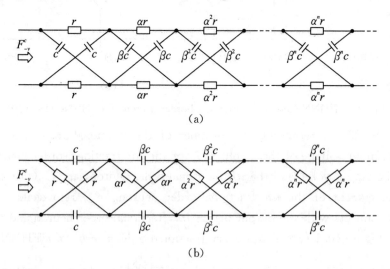

Fig. 10.24 Two equivalent circuit configurations of arbitrary-order low-pass filtering net-grid-type capacitive scaling fractor: (a) First equivalent circuit configuration of v-order low-pass filtering net-grid-type capacitive scaling fractor; (b) Second equivalent circuit configuration of v-order low-pass filtering net-grid-type capacitive scaling fractor.

Further, two equivalent circuit configurations of the arbitrary-order high-pass filtering net-grid-type scaling capacitive fractor can be shown in Fig. 10.25.

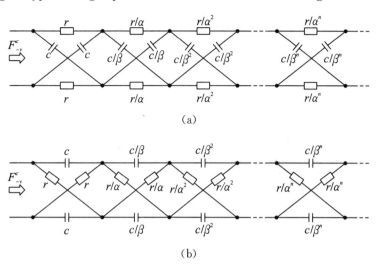

Fig. 10.25 Two equivalent circuit configurations of arbitrary-order high-pass filtering net-grid-type capacitive scaling fractor: (a) First equivalent circuit configuration of v-order high-pass filtering net-grid-type capacitive scaling fractor; (b) Second equivalent circuit configuration of v-order high-pass filtering net-grid-type capacitive scaling fractor.

Further, two equivalent circuit configurations of the arbitrary-order low-pass filtering net-grid-type scaling inductive fractor can be shown in Fig. 10.26.

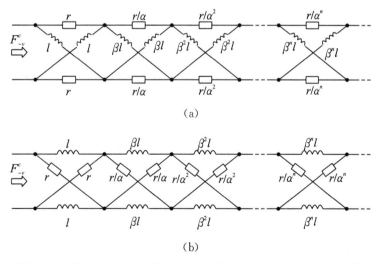

Fig. 10.26 Two equivalent circuit configurations of arbitrary-order low-pass filtering net-grid-type inductive scaling fractor: (a) First equivalent circuit configuration of v-order low-pass filtering net-grid-type inductive scaling fractor; (b) Second equivalent circuit configuration of v-order low-pass filtering net-grid-type inductive scaling fractor.

Further, two equivalent circuit configurations of the arbitrary-order high-pass filtering net-grid-type scaling inductive fractor can be shown in Fig. 10.27.

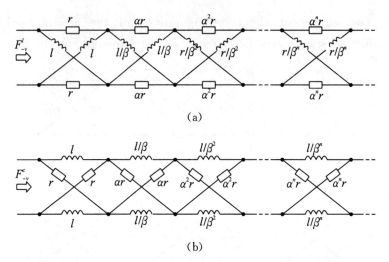

Fig. 10.27 Two equivalent circuit configurations of arbitrary-order high-pass filtering net-grid-type inductive scaling fractor: (a) First equivalent circuit configuration of v-order high-pass filtering net-grid-type inductive scaling fractor; (b) Second equivalent circuit configuration of v-order high-pass filtering net-grid-type inductive scaling fractor.

In Figs. 10.24—27, $0 < v < 1$ is an arbitrary positive rational number. F^c_{-v} is the v-order capacitive reactance of the v-order net-grid-type capacitive scaling fractor. Two arbitrary-order low-pass filtering net-grid-type scaling capacitive fractors in Figs 10.24 (a) and 10.24 (b) have the identical v-order capacitive reactance. Two arbitrary-order high-pass filtering net-grid-type scaling capacitive fractors in Figs 10.25 (a) and 10.25 (b) have the identical v-order capacitive reactance. F^l_v is the v-order inductive reactance of a v-order net-grid-type inductive scaling fractor. Two arbitrary-order low-pass filtering net-grid-type scaling inductive fractors in Figs 10.26 (a) and 10.26 (b) have the identical v-order inductive reactance. Two arbitrary-order high-pass filtering net-grid-type scaling inductive fractors in Figs 10.27 (a) and 10.27 (b) have the identical v-order inductive reactance. r, c, and l represent the resistance, capacitance, and inductance. α and β represent two positive scaling constants, respectively. Figures 10.24—27 show that the v-order net-grid-type scaling fractor has an extreme self-similar fractal structure with a series connection of infinitely repeated net-grid type structures. With regard to the v-order net-grid-type capacitive scaling fractor, the amount of resistors or capacitors is equal to the double amount of layers. In a similar way, with regard to the v-order net-grid-type inductive scaling fractor, the amount of resistors or inductors is also equal to the double amount of layers.

For the convenience of illustration, in this chapter, the v-order capacitive reactance of the v-order low-pass filtering net-grid-type capacitive scaling fractor is chiefly derived in detail. In a similar way, the v-order capacitive reactance of the v-order high-pass filtering net-grid-type capacitive scaling fractor, the v-order inductive reactance of the v-order low-pass filtering net-grid-type inductive scaling fractor, and that of the v-order high-pass filtering net-grid-type inductive scaling fractor can be easily derived. Let's choose transmission parameters (or *ABCD* parameters) [416] as a set of parameters to restrict the voltages and currents at the input port and those at the output port of the $(n+1)th$ series circuit of the v-order low-pass filtering net-grid-type capacitive scaling fractor shown in Fig. 10.24 (a), as shown in Fig. 10.28.

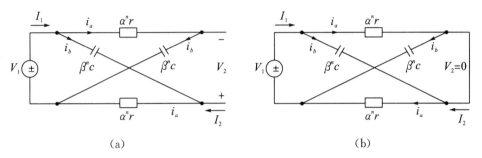

(a)   (b)

Fig. 10.28 Circuit analysis of $(n+1)th$ series circuit of v-order low-pass filtering net-grid-type capacitive scaling fractor: (a) Finding A and C of $(n+1)th$ series circuit; (b) Finding B and D of $(n+1)th$ series circuit.

Figure 10.28 shows that the transmission parameters of the $(n+1)th$ series circuit of the v-order low-pass filtering net-grid-type capacitive scaling fractor can be given as:

$$\begin{bmatrix} V_1 \\ I_1 \end{bmatrix} = \begin{bmatrix} A_{n+1} & B_{n+1} \\ C_{n+1} & D_{n+1} \end{bmatrix} \begin{bmatrix} V_2 \\ -I_2 \end{bmatrix}$$

$$= [T_{n+1}] \begin{bmatrix} V_2 \\ -I_2 \end{bmatrix}. \tag{10.38}$$

Equation (10.38) provides a direct relationship between the sending-end variables, V_1 and I_1, and the receiving-end variables, V_2 and $-I_2$. The two-port parameters in (10.38) provide a measure of how the $(n+1)th$ series circuit of the v-order low-pass filtering net-grid-type capacitive scaling fractor transmits voltage and current from a source to a load. From Fig. 10.28 and (10.38), the transmission parameters of the $(n+1)th$ series circuit of the v-order low-pass filtering net-grid-type capacitive scaling fractor are determined as, respectively:

$$A_{n+1} = \left. \frac{V_1}{V_2} \right|_{I_2=0}, \tag{10.39}$$

$$B_{n+1} = \left. -\frac{V_1}{I_2} \right|_{V_2=0}, \tag{10.40}$$

$$C_{n+1} = \left. \frac{I_1}{V_2} \right|_{I_2=0}, \tag{10.41}$$

$$D_{n+1} = \left. -\frac{I_1}{I_2} \right|_{V_2=0}, \tag{10.42}$$

where A_{n+1}, B_{n+1}, C_{n+1}, and D_{n+1} are the open-circuit voltage ratio, negative short-circuit transfer impedance, open-circuit transfer admittance, and negative short-circuit current ratio of the $(n+1)th$ series circuit of the v-order low-pass filtering net-grid-type capacitive scaling fractor, respectively.

To determine A_{n+1} and C_{n+1}, from (10.39) and (10.31), we let the output port open as shown in Fig. 10.28 (a) so that $I_2 = 0$ and place a voltage source V_1 at the input port. Thus, according to Kirchoff's current law and Kirchoff's voltage law, from Fig. 10.28 (a), the following can be obtained:

$$\begin{cases} \alpha^n r i_a + i_b/(\beta^n cs) - V_1 = 0 \\ \alpha^n r i_a - i_b/(\beta^n cs) - V_2 = 0 \\ i_a + i_b - I_1 = 0 \\ -i_a + i_b = 0 \end{cases}, \quad (10.43)$$

where s denotes the Laplace operator. Thus, from (10.43), we obtain:

$$A_{n+1} = \frac{V_1}{V_2}\bigg|_{I_2=0}$$
$$= \frac{\alpha^n r + 1/(\beta^n cs)}{\alpha^n r - 1/(\beta^n cs)}, \quad (10.44)$$

$$C_{n+1} = \frac{I_1}{V_2}\bigg|_{I_2=0}$$
$$= \frac{2}{\alpha^n r - 1/(\beta^n cs)}. \quad (10.45)$$

To determine B_{n+1} and D_{n+1}, from (10.41) and (10.42), we let the output port short-circuit as shown in Fig. 10.28 (b) so that $V_2 = 0$ and place a voltage source V_1 at the input port. Thus, according to Kirchoff's current law and Kirchoff's voltage law, from Fig. 10.28 (b), the following can be obtained:

$$\begin{cases} \alpha^n r i_a + i_b/(\beta^n cs) - V_1 = 0 \\ \alpha^n r i_a - i_b/(\beta^n cs) = 0 \\ i_a + i_b - I_1 = 0 \\ i_a - I_2 - i_b = 0 \end{cases}. \quad (10.46)$$

Thus, from (10.46), we obtain:

$$D_{n+1} = -\frac{I_1}{I_2}\bigg|_{V_2=0}$$
$$= \frac{\alpha^n r + 1/(\beta^n cs)}{\alpha^n r - 1/(\beta^n cs)}, \quad (10.47)$$

$$B_{n+1} = -\frac{V_1}{I_2}\bigg|_{V_2=0}$$
$$= \frac{2\alpha^n r/(\beta^n cs)}{\alpha^n r - 1/(\beta^n cs)}. \quad (10.48)$$

Thus, according to the characteristics of hyperbolic function, from (10.38), (10.44), (10.45), (10.47), and (10.48), we have:

$$[T_{n+1}] = \begin{bmatrix} A_{n+1} & B_{n+1} \\ C_{n+1} & D_{n+1} \end{bmatrix}$$

$$= \begin{bmatrix} \dfrac{\alpha^n r + 1/(\beta^n cs)}{\alpha^n r - 1/(\beta^n cs)} & \dfrac{2\alpha^n r/(\beta^n cs)}{\alpha^n r - 1/(\beta^n cs)} \\ \dfrac{2}{\alpha^n r - 1/(\beta^n cs)} & \dfrac{\alpha^n r + 1/(\beta^n cs)}{\alpha^n r - 1/(\beta^n cs)} \end{bmatrix}$$

$$= \begin{bmatrix} \dfrac{1 + 1/(\alpha^n \beta^n rcs)}{1 - 1/(\alpha^n \beta^n rcs)} & \dfrac{2/(\beta^n cs)}{1 - 1/(\alpha^n \beta^n rcs)} \\ \dfrac{2/(\alpha^n r)}{1 - 1/(\alpha^n \beta^n rcs)} & \dfrac{1 + 1/(\alpha^n \beta^n rcs)}{1 - 1/(\alpha^n \beta^n rcs)} \end{bmatrix}$$

$$= \begin{bmatrix} \cosh(\xi_n) & \Psi_n \sinh(\xi_n) \\ \sinh(\xi_n)/\Psi_n & \cosh(\xi_n) \end{bmatrix}, \tag{10.49}$$

where $\xi_n = \operatorname{arccosh}\left[\dfrac{1+1/(\alpha^n \beta^n rcs)}{1-1/(\alpha^n \beta^n rcs)}\right]$ and $\Psi_n = \sqrt{\alpha^n r/(\beta^n cs)}$.

Since the transmission parameters for the overall cascaded network are equal to the product of the transmission parameters for the individual transmission parameters of each cascaded network [416], from Fig. 10.24 (a), we can derive the transmission parameters for the v-order low-pass filtering net-grid-type capacitive scaling fractor, given as:

$$\begin{aligned} [T_{-v}^m] &= \begin{bmatrix} A_{-v}^m & B_{-v}^m \\ C_{-v}^m & D_{-v}^m \end{bmatrix} \\ &= \prod_{n=0}^{m} [T_{n+1}] \\ &= \prod_{n=0}^{m} \begin{bmatrix} \cosh(\xi_n) & \Psi_n \sinh(\xi_n) \\ \sinh(\xi_n)/\Psi_n & \cosh(\xi_n) \end{bmatrix}, \end{aligned} \tag{10.50}$$

where m is the amount of the series circuits of the v-order low-pass filtering net-grid-type capacitive scaling fractor. Since the driving-point capacitive impedance function of the v-order low-pass filtering net-grid-type capacitive scaling fractor is its v-order capacitive reactance, the v-order open-circuit capacitive reactance (Z_{-v}^{om}) and the v-order short-circuit capacitive reactance (Z_{-v}^{sm}) of the v-order low-pass filtering net-grid-type capacitive scaling fractor with m series circuits can be determined as follows, respectively:

$$Z_{-v}^{om} = \dfrac{A_{-v}^m}{C_{-v}^m}, \tag{10.51}$$

$$Z_{-v}^{sm} = \dfrac{B_{-v}^m}{D_{-v}^m}. \tag{10.52}$$

Unfortunately, (10.49) and (10.50) show that the general analytical solutions of (10.50), (10.51), and (10.52) shall be very complex and difficult when m is large. Therefore, to simplify the related solutions, it is natural to ponder what the fractional-order iterative reactance of the v-order low-pass filtering net-grid-type capacitive scaling fractor is.

The circuit configuration of a 1/2-order low-pass filtering net-grid-type capacitive scaling fractor [171] - [173], [193], [225], [405] can be shown in Fig. 10.29.

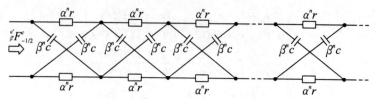

Fig. 10.29 Circuit configuration of a 1/2-order low-pass filtering net-grid-type capacitive scaling fractor.

In Fig. 10.29, $_{\beta^n}^{\alpha^n}F^c_{-1/2}$ is the 1/2-order capacitive reactance of the 1/2-order low-pass filtering net-grid-type capacitive scaling fractor whose positive scaling constant of the resistance (α^n) and that of the capacitance (β^n) of its every series circuit are identical, respectively. Comparing Fig. 10.24 (a) with Fig. 10.29, we can see that the every series circuit of the 1/2-order low-pass filtering net-grid-type capacitive scaling fractor has the same circuit configuration as that of the $(n+1)th$ series circuit of the v-order low-pass filtering net-grid-type capacitive scaling fractor. Thus, the transmission parameters of the arbitrary series circuit of the 1/2-order low-pass filtering net-grid-type capacitive scaling fractor are identical with those of the $(n+1)$ th series circuit of the v-order low-pass filtering net-grid-type capacitive scaling fractor. Thus, from (10.49), the characteristic equation of $[T_{n+1}]$ can be derived as:

$$|\lambda E - T_{n+1}| = \begin{vmatrix} \lambda - \cosh(\xi_n) & -\Psi_n \sinh(\xi_n) \\ -\sinh(\xi_n)/\Psi_n & \lambda - \cosh(\xi_n) \end{vmatrix} = 0, \qquad (10.53)$$

where λ is the eigenvalue of $[T_{n+1}]$. E is an identity matrix. Thus, from (10.53), the following is true:

$$\begin{cases} \lambda_1 = \cosh(\xi_n) - \sinh(\xi_n) = e^{-\xi_n} \\ \lambda_2 = \cosh(\xi_n) + \sinh(\xi_n) = e^{\xi_n} \end{cases}, \qquad (10.54)$$

where λ_1 and λ_2 are two different eigenvalues of $[T_{n+1}]$. Substituting λ_1 and λ_2 into $(\lambda E - T_{n+1})X = 0$, one of the eigenvector of T_{n+1} corresponding to λ_1 and λ_2 can be derived as:

$$X = \begin{bmatrix} -\Psi_n & \Psi_n \\ 1 & 1 \end{bmatrix}, \qquad (10.55)$$

where X is one of the eigenvector of T_{n+1}. Thus, form (10.49) and (10.55), we obtain:

$$[T_{n+1}] = \begin{bmatrix} -\Psi_n & \Psi_n \\ 1 & 1 \end{bmatrix} \begin{bmatrix} e^{-\xi_n} & 0 \\ 0 & e^{\xi_n} \end{bmatrix} \begin{bmatrix} -\Psi_n & \Psi_n \\ 1 & 1 \end{bmatrix}^{-1}$$

$$= \begin{bmatrix} -\Psi_n & \Psi_n \\ 1 & 1 \end{bmatrix} \begin{bmatrix} e^{-\xi_n} & 0 \\ 0 & e^{\xi_n} \end{bmatrix} \begin{bmatrix} -1/(2\Psi_n) & 1/2 \\ 1/(2\Psi_n) & 1/2 \end{bmatrix}. \qquad (10.56)$$

With regard to the 1/2-order low-pass filtering net-grid-type capacitive scaling fractor shown in Fig. 10.29, $[T_{n+1}]=[T_{n+2}]$ can be easily derived. Thus, from (10.56), we get:

$$[T_{n+1}][T_{n+2}] = \begin{bmatrix} -\Psi_n & \Psi_n \\ 1 & 1 \end{bmatrix} \begin{bmatrix} e^{-\xi_n} & 0 \\ 0 & e^{\xi_n} \end{bmatrix} \begin{bmatrix} e^{-\xi_n} & 0 \\ 0 & e^{\xi_n} \end{bmatrix} \begin{bmatrix} -\Psi_n & \Psi_n \\ 1 & 1 \end{bmatrix}^{-1}$$

$$= \begin{bmatrix} -\Psi_n & \Psi_n \\ 1 & 1 \end{bmatrix} \begin{bmatrix} e^{-\xi_n} & 0 \\ 0 & e^{\xi_n} \end{bmatrix} \begin{bmatrix} e^{-\xi_n} & 0 \\ 0 & e^{\xi_n} \end{bmatrix} \begin{bmatrix} -1/(2\Psi_n) & 1/2 \\ 1/(2\Psi_n) & 1/2 \end{bmatrix}.$$

$$(10.57)$$

Thus, from (10.50) and (10.57), we can see that the recurrence relations of $[T_{-1/2}^m]$ = $\prod_{n=0}^{m}[T_{n+1}]$ is the periodical matrix chain-multiplication of $[T_{n+1}][T_{n+2}]$. Substituting (10.57) into (10.50), the following can be obtained:

$$[T_{-1/2}^m] = \begin{bmatrix} A_{-1/2}^m & B_{-1/2}^m \\ C_{-1/2}^m & D_{-1/2}^m \end{bmatrix}$$

$$= \prod_{n=0}^{m}[T_{n+1}]$$

$$= \begin{bmatrix} -\Psi_n & \Psi_n \\ 1 & 1 \end{bmatrix} \left\{ \prod_{n=0}^{m} \begin{bmatrix} e^{-\zeta_n} & 0 \\ 0 & e^{\zeta_n} \end{bmatrix} \right\} \begin{bmatrix} -\Psi_n & \Psi_n \\ 1 & 1 \end{bmatrix}^{-1}$$

$$= \begin{bmatrix} -\Psi_n & \Psi_n \\ 1 & 1 \end{bmatrix} \begin{bmatrix} (e^{-\zeta_n})^m & 0 \\ 0 & (e^{\zeta_n})^m \end{bmatrix} \begin{bmatrix} -\Psi_n & \Psi_n \\ 1 & 1 \end{bmatrix}^{-1}$$

$$= \begin{bmatrix} -\Psi_n & \Psi_n \\ 1 & 1 \end{bmatrix} \begin{bmatrix} e^{-m\zeta_n} & 0 \\ 0 & e^{m\zeta_n} \end{bmatrix} \begin{bmatrix} -1/(2\Psi_n) & 1/2 \\ 1/(2\Psi_n) & 1/2 \end{bmatrix}. \quad (10.58)$$

Substituting (10.49) into (10.58), the following can be obtained:

$$[T_{-1/2}^m] = \begin{bmatrix} A_{-1/2}^m & B_{-1/2}^m \\ C_{-1/2}^m & D_{-1/2}^m \end{bmatrix}$$

$$= \begin{bmatrix} \cosh(m\zeta_n) & \Psi_n \sinh(m\zeta_n) \\ \sinh(m\zeta_n)/\Psi_n & \cosh(m\zeta_n) \end{bmatrix}. \quad (10.59)$$

Thus, form (10.59), the 1/2-order open-circuit capacitive reactance ($Z_{-1/2}^{om}$) and the 1/2-order short-circuit capacitive reactance ($Z_{-1/2}^{sm}$) of the 1/2-order low-pass filtering net-grid-type capacitive scaling fractor with m series circuits can be determined as follows, respectively:

$$Z_{-1/2}^{om} = \frac{A_{-1/2}^m}{C_{-1/2}^m}$$

$$= \sqrt{\frac{\alpha^n r}{\beta^n cs}} \coth(m\zeta_n), \quad (10.60)$$

$$Z_{-1/2}^{sm} = \frac{B_{-1/2}^m}{D_{-1/2}^m}$$

$$= \sqrt{\frac{\alpha^n r}{\beta^n cs}} \tanh(m\zeta_n). \quad (10.61)$$

Thus, from (10.60) and (10.61), the following is true:

$$_{\beta^n}^{\alpha^n} F_{-1/2}^c = \lim_{m \to \infty} Z_{-1/2}^{om}$$

$$= \lim_{m \to \infty} \sqrt{\frac{\alpha^n r}{\beta^n cs}} \coth(m\zeta_n)$$

$$= \sqrt{\frac{\alpha^n r}{\beta^n cs}}, \quad (10.62)$$

$$_{\beta^n}^{\alpha^n} F_{-1/2}^c = \lim_{m \to \infty} Z_{-1/2}^{sm}$$

$$= \lim_{m \to \infty} \sqrt{\frac{\alpha^n r}{\beta^n cs}} \tanh(m\zeta_n)$$

$$= \sqrt{\frac{\alpha^n r}{\beta^n cs}}. \tag{10.63}$$

Equations (10.62) and (10.63) show that the cascade circuit with a series connection of infinitely repeated net-grid type structures shown in Fig. 10.29 achieves the 1/2-order low-pass filtering net-grid-type capacitive scaling fractor whose 1/2-order open-circuit capacitive reactance is the same as its 1/2-order short-circuit capacitive reactance.

Further, the infinitely iterative circuit configurations of the v-order low-pass filtering net-grid-type capacitive scaling fractor can be shown in Fig. 10.30.

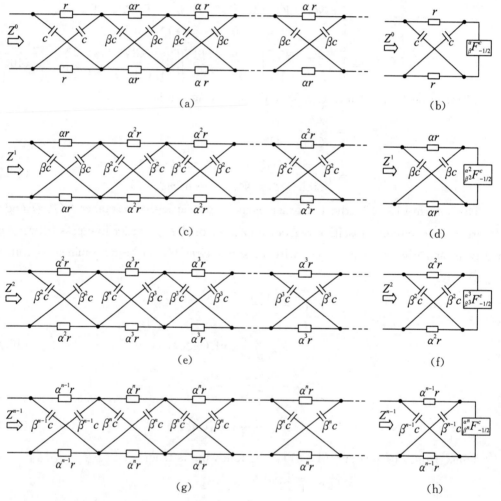

Fig. 10.30 Infinitely iterative circuit configurations of v-order net-grid-type capacitive scaling fractor: (a) First iterative circuit; (b) Equivalent circuit of first iterative circuit; (c) Second iterative circuit; (d) Equivalent circuit of second iterative circuit; (e) Third iterative circuit; (f) Equivalent circuit of third iterative circuit; (g) nth iterative circuit; (h) Equivalent circuit of nth iterative circuit.

In Fig. 10.30, Z^0, Z^1, Z^2, and Z^{n-1} are the driving-point capacitive impedance function of the first iterative circuit, the second iterative circuit, the third iterative circuit, and the nth iterative circuit of the v-order low-pass filtering net-grid-type capacitive scaling fractor, respectively. In Fig. 10.30 (b), let's assume that the currents of r and c are $i_r(s)$ and $i_c(s)$, respectively. The input voltage and input current of Z^0 are $V_i(s)$ and $I_i(s)$, respectively. Hence, from Fig. 10.30 (b), according to Kirchoff's current law and Kirchoff's voltage law, we get:

$$\begin{cases} ri_r + (1/cs)i_c = V_i \\ (r + {}_{\beta}^{\alpha}F^c_{-1/2})i_r - ({}_{\beta}^{\alpha}F^c_{-1/2} + 1/cs)i_c = 0 \end{cases}. \tag{10.64}$$

According to the Cramer's rule of linear algebra, from (10.64), we obtain:

$$\begin{cases} i_r = \dfrac{(1/cs + {}_{\beta}^{\alpha}F^c_{-1/2})V_i}{\begin{vmatrix} r + {}_{\beta}^{\alpha}F^c_{-1/2} & -({}_{\beta}^{\alpha}F^c_{-1/2} + 1/cs) \\ r & 1/cs \end{vmatrix}} \\ i_c = \dfrac{(r + {}_{\beta}^{\alpha}F^c_{-1/2})V_i}{\begin{vmatrix} r + {}_{\beta}^{\alpha}F^c_{-1/2} & -({}_{\beta}^{\alpha}F^c_{-1/2} + 1/cs) \\ r & 1/cs \end{vmatrix}} \end{cases}. \tag{10.65}$$

Thus, (10.65) can be rewritten as:

$$\begin{aligned} Z^0(s) &= \frac{V_i(s)}{I_i(s)} \\ &= \frac{V_i(s)}{i_r(s) + i_c(s)} \\ &= \frac{2r(1/cs) + {}_{\beta}^{\alpha}F^c_{-1/2}(r + 1/cs)}{2{}_{\beta}^{\alpha}F^c_{-1/2} + r + 1/cs}. \end{aligned} \tag{10.66}$$

Thus, from (10.63) and (10.66), the following is true:

$$Z^0(s) = \frac{\dfrac{2r}{cs} + \sqrt{\dfrac{\alpha r}{\beta cs}}\left(r + \dfrac{1}{cs}\right)}{2\sqrt{\dfrac{\alpha r}{\beta cs}} + r + \dfrac{1}{cs}}. \tag{10.67}$$

In a similar way, from Figs. 10.30 (d), 10.30 (f), and 10.30 (h), the following can be obtained, respectively:

$$\begin{aligned} Z^1(s) &= \frac{\dfrac{2\alpha r}{\beta cs} + \sqrt{\dfrac{\alpha^2 r}{\beta^2 cs}}\left(\alpha r + \dfrac{1}{\beta cs}\right)}{2\sqrt{\dfrac{\alpha^2 r}{\beta^2 cs}} + \alpha r + \dfrac{1}{\beta cs}} \\ &= \frac{\alpha^2 \dfrac{2r}{c(\alpha\beta s)} + \alpha^2 \sqrt{\dfrac{\alpha r}{\beta c(\alpha\beta s)}}\left[r + \dfrac{1}{c(\alpha\beta s)}\right]}{\alpha\left[2\sqrt{\dfrac{\alpha r}{\beta c(\alpha\beta s)}} + r + \dfrac{1}{c(\alpha\beta s)}\right]} \\ &= \alpha \frac{\dfrac{2r}{c(\alpha\beta s)} + \sqrt{\dfrac{\alpha r}{\beta c(\alpha\beta s)}}\left[r + \dfrac{1}{c(\alpha\beta s)}\right]}{2\sqrt{\dfrac{\alpha r}{\beta c(\alpha\beta s)}} + r + \dfrac{1}{c(\alpha\beta s)}} \end{aligned}$$

$$= \alpha Z^0(\alpha\beta s), \tag{10.68}$$

$$Z^2(s) = \frac{\dfrac{2\alpha^2 r}{\beta^2 cs} + \sqrt{\dfrac{\alpha^3 r}{\beta^3 cs}}\left(\alpha^2 r + \dfrac{1}{\beta^2 cs}\right)}{2\sqrt{\dfrac{\alpha^3 r}{\beta^3 cs}} + \alpha^2 r + \dfrac{1}{\beta^2 cs}}$$

$$= \frac{\alpha^2 \dfrac{2\alpha r}{\beta c(\alpha\beta s)} + \alpha^2 \sqrt{\dfrac{\alpha^2 r}{\beta^2 c(\alpha\beta s)}}\left[\alpha r + \dfrac{1}{\beta c(\alpha\beta s)}\right]}{\alpha\left[2\sqrt{\dfrac{\alpha^2 r}{\beta^2 c(\alpha\beta s)}} + \alpha r + \dfrac{1}{\beta c(\alpha\beta s)}\right]}$$

$$= \alpha \frac{\dfrac{2\alpha r}{\beta c(\alpha\beta s)} + \sqrt{\dfrac{\alpha^2 r}{\beta^2 c(\alpha\beta s)}}\left[\alpha r + \dfrac{1}{\beta c(\alpha\beta s)}\right]}{2\sqrt{\dfrac{\alpha^2 r}{\beta^2 c(\alpha\beta s)}} + \alpha r + \dfrac{1}{\beta c(\alpha\beta s)}}$$

$$= \alpha Z^1(\alpha\beta s), \tag{10.69}$$

$$Z^{n-1}(s) = \frac{\dfrac{2\alpha^{n-1} r}{\beta^{n-1} cs} + \sqrt{\dfrac{\alpha^n r}{\beta^n cs}}\left(\alpha^{n-1} r + \dfrac{1}{\beta^{n-1} cs}\right)}{2\sqrt{\dfrac{\alpha^n r}{\beta^n cs}} + \alpha^{n-1} r + \dfrac{1}{\beta^{n-1} cs}}$$

$$= \frac{\alpha^2 \dfrac{2\alpha^{n-2} r}{\beta^{n-2} c(\alpha\beta s)} + \alpha^2 \sqrt{\dfrac{\alpha^{n-1} r}{\beta^{n-1} c(\alpha\beta s)}}\left[\alpha^{n-2} r + \dfrac{1}{\beta^{n-2} c(\alpha\beta s)}\right]}{\alpha\left[2\sqrt{\dfrac{\alpha^{n-2} r}{\beta^{n-1} c(\alpha\beta s)}} + \alpha^{n-2} r + \dfrac{1}{\beta^{n-2} c(\alpha\beta s)}\right]}$$

$$= \alpha \frac{\dfrac{2\alpha^{n-2} r}{\beta^{n-2} c(\alpha\beta s)} + \sqrt{\dfrac{\alpha^{n-1} r}{\beta^{n-1} c(\alpha\beta s)}}\left[\alpha^{n-2} r + \dfrac{1}{\beta^{n-2} c(\alpha\beta s)}\right]}{2\sqrt{\dfrac{\alpha^{n-1} r}{\beta^{n-1} c(\alpha\beta s)}} + \alpha^{n-2} r + \dfrac{1}{\beta^{n-2} c(\alpha\beta s)}}$$

$$= \alpha Z^{n-2}(\alpha\beta s). \tag{10.70}$$

Comparing Fig. 10.24 (a) with Fig. 10.30, we can further see that the v-order low-pass filtering net-grid-type capacitive scaling fractor can be treated as an infinitely successively nested structure of the first iterative circuit, the second iterative circuit, the third iterative circuit, $\cdots$, the nth iterative circuit ($n \to \infty$) shown in Fig. 10.30. Therefore, the v-order capacitive reactance of the v-order low-pass filtering net-grid-type capacitive scaling fractor is equal to the limiting value of the recursion equation of successively nested $Z^0(s)$, $Z^1(s)$, $Z^2(s)$, $\cdots$, and $Z^{n-1}(s)$ when $n \to \infty$. Hence, from (10.67) — (10.70), the v-order capacitive reactance of the v-order low-pass filtering net-grid-type capacitive scaling fractor can be derived as:

$$F_{-v}^{c}(s) = \lim_{n\to\infty} \cfrac{\dfrac{2r}{cs} + \alpha\cfrac{\dfrac{2r}{c(\alpha\beta s)} + \left\{\cdots\alpha\cfrac{\dfrac{2\alpha^{n-2}r}{\beta^{n-2}c(\alpha\beta s)} + \sqrt{\dfrac{\alpha^{n-1}r}{\beta^{n-1}c(\alpha\beta s)}}\left[\alpha^{n-2}r + \dfrac{1}{\beta^{n-2}c(\alpha\beta s)}\right]}{2\sqrt{\dfrac{\alpha^{n-1}r}{\beta^{n-1}c(\alpha\beta s)}} + \alpha^{n-2}r + \dfrac{1}{\beta^{n-2}c(\alpha\beta s)}}\right\}\left[r+\dfrac{1}{c(\alpha\beta s)}\right]}{2\left\{\cdots\alpha\cfrac{\dfrac{2\alpha^{n-2}r}{\beta^{n-2}c(\alpha\beta s)} + \sqrt{\dfrac{\alpha^{n-1}r}{\beta^{n-1}c(\alpha\beta s)}}\left[\alpha^{n-2}r + \dfrac{1}{\beta^{n-2}c(\alpha\beta s)}\right]}{2\sqrt{\dfrac{\alpha^{n-1}r}{\beta^{n-1}c(\alpha\beta s)}} + \alpha^{n-2}r + \dfrac{1}{\beta^{n-2}c(\alpha\beta s)}}\right\} + r + \dfrac{1}{c(\alpha\beta s)}} \left(r+\dfrac{1}{cs}\right)}{2\alpha\cfrac{\dfrac{2r}{c(\alpha\beta s)} + \left\{\cdots\alpha\cfrac{\dfrac{2\alpha^{n-2}r}{\beta^{n-2}c(\alpha\beta s)} + \sqrt{\dfrac{\alpha^{n-1}r}{\beta^{n-1}c(\alpha\beta s)}}\left[\alpha^{n-2}r + \dfrac{1}{\beta^{n-2}c(\alpha\beta s)}\right]}{2\sqrt{\dfrac{\alpha^{n-1}r}{\beta^{n-1}c(\alpha\beta s)}} + \alpha^{n-2}r + \dfrac{1}{\beta^{n-2}c(\alpha\beta s)}}\right\}\left[r+\dfrac{1}{c(\alpha\beta s)}\right]}{2\left\{\cdots\alpha\cfrac{\dfrac{2\alpha^{n-2}r}{\beta^{n-2}c(\alpha\beta s)} + \sqrt{\dfrac{\alpha^{n-1}r}{\beta^{n-1}c(\alpha\beta s)}}\left[\alpha^{n-2}r + \dfrac{1}{\beta^{n-2}c(\alpha\beta s)}\right]}{2\sqrt{\dfrac{\alpha^{n-1}r}{\beta^{n-1}c(\alpha\beta s)}} + \alpha^{n-2}r + \dfrac{1}{\beta^{n-2}c(\alpha\beta s)}}\right\} + r + \dfrac{1}{c(\alpha\beta s)}} + r + \dfrac{1}{cs}}$$

(10.71)

Equation (10.71) isessentially a specific continued fraction expansion. Thus, from (35), the fractional-order iterative reactance of the v-order low-pass filtering net-grid-type capacitive scaling fractor can be derived as:

$$F_{-v}^{nc}(s) = \frac{\dfrac{2r}{cs} + \alpha F_{-v}^{(n-1)c}(\alpha\beta s)\left(r+\dfrac{1}{cs}\right)}{2\alpha F_{-v}^{(n-1)c}(\alpha\beta s) + r + \dfrac{1}{cs}}, \quad (10.72)$$

where $F_{-v}^{nc}(s)$ and $F_{-v}^{(n-1)c}(s)$ are the driving-point capacitive impedance functions of the v-order low-pass filtering net-grid-type capacitive scaling fractor with n series circuits and that with $(n-1)$ series circuits, respectively. From Fig. 10.24, when $n \to \infty$, the following is true:

$$\begin{aligned} F_{-v}^{c}(s) &= \lim_{n\to\infty} F_{-v}^{nc}(s) \\ &= \lim_{n\to\infty} F_{-v}^{(n-1)c}(s). \end{aligned} \quad (10.73)$$

Thus, from (10.73), (10.72) can be rewritten as:

$$F_{-v}^{c}(s) = \frac{\dfrac{2r}{cs} + \alpha F_{-v}^{c}(\alpha\beta s)\left(r+\dfrac{1}{cs}\right)}{2aF_{-v}^{c}(\alpha\beta S) + r + \dfrac{1}{cs}}. \quad (10.74)$$

In (10.74), $\alpha\beta$ is actually the fractal scaling factor of the v-order low-pass filtering net-grid-type capacitive scaling fractor. Equation (10.74) is the non-regularized iterative scaling equation of the v-order low-pass filtering net-grid-type capacitive scaling fractor. Equation (10.74) shows that it accords with standard dynamical scaling law [406] — [409], [417]. Thus, the solution of (10.74) can be given as:

$$F_{-v}^{c}(s) = \kappa s^{-v}, \quad (10.75)$$

where κ is a scalar factor. Substituting (10.75) into (10.74) gives:

$$\kappa s^{-v} = \frac{\dfrac{2r}{cs} + \alpha\kappa(\alpha\beta s)^{-v}\left(r+\dfrac{1}{cs}\right)}{2\alpha\kappa(\alpha\beta s)^{-v} + r + \dfrac{1}{cs}}$$

$$= \frac{2r + \alpha\kappa(\alpha\beta s)^{-v}(rcs + 1)}{2\alpha k(\alpha\beta s)^{-v}cs + rcs + 1}, \tag{10.76}$$

where $1/rc$ is the eigen frequency of the v-order low-pass filtering net-grid-type scaling capacitive fractor. Since $0 < v < 1$, when $s \to 0$, $s^{-v}s \to 0$. Thus, when $s \to 0$ (low-pass filtering), (10.76) can be simplified as:

$$\begin{aligned} \kappa s^{-v} &= \frac{2r + \alpha\kappa(\alpha\beta s)^{-v}(rcs + 1)}{2\alpha\kappa(\alpha\beta s)^{-v}cs + rcs + 1} \\ &\overset{s \to 0}{\approx} \frac{\alpha\kappa(\alpha\beta s)^{-v}(rcs + 1)}{rcs + 1} \\ &= \alpha\kappa(\alpha\beta s)^{-v}. \end{aligned} \tag{10.77}$$

Hence, when $s \to 0$ (low-pass filtering), the solution of (10.77) can be derived as:

$$v = \frac{\lg(\alpha)}{\lg(\alpha) + \lg(\beta)}, \tag{10.78}$$

where $\lg()$ is a logarithm. Equation (10.78) shows that with regard to the v-order low-pass filtering net-grid-type scaling capacitive fractor shown in Fig. 10.24, its fractional-order merely essentially depends on its two positive scaling constants (α and β), but has nothing to do with its resistance and capacitance (r and c). Therefore, the fractional-order of the v-order low-pass filtering net-grid-type scaling capacitive fractor can be arbitrarily changed by means of altering its two positive scaling constants (α and β).

In addition, from (10.78), we can further see that when $\alpha = 1$ and $\beta = 1$, $v = 1/2$. Substituting $\alpha = 1$, $\beta = 1$, and $v = 1/2$ into (10.75) and (10.76) results in:

$$\begin{aligned} F_{-v}^{c}(s) &\overset{\alpha=1,\beta=1}{=} F_{-1/2}^{c}(s) \\ &= \kappa s^{-1/2} \\ &= (c/r)^{-1/2} s^{-1/2}. \end{aligned} \tag{10.79}$$

Equation (10.79) shows that when $\alpha = 1$, $\beta = 1$, (40) can be accurately simplified as $\kappa s^{-v} = \alpha\kappa(\alpha\beta s)^{-v}|_{\alpha=1,\beta=1,v=1/2}$. The classical 1/2-order net-grid-type capacitive fractor is a purely ideal capacitive fractor [171] − [173], [193], [225], [405].

In a similar way, at first, with regard to the v-order high-pass filtering net-grid-type scaling capacitive fractor shown in Fig. 10.25, the fractional-order iterative reactance of the v-order high-pass filtering net-grid-type scaling capacitive fractor can be derived as:

$$F_{-v}^{nc}(s) = \frac{\dfrac{2r}{cs} + \dfrac{1}{\alpha}F_{-v}^{(n-1)c}\left(\dfrac{s}{\alpha\beta}\right)\left(r + \dfrac{1}{cs}\right)}{2\dfrac{1}{\alpha}F_{-v}^{(n-1)c}\left(\dfrac{s}{\alpha\beta}\right) + r + \dfrac{1}{cs}}, \tag{10.80}$$

where $F_{(-v)}^{nc}(s)$ and $F_{-v}^{(n-1)c}(s)$ are the driving-point capacitive impedance functions of the v-order high-pass filtering net-grid-type scaling capacitive fractor with n series circuits and that with $(n-1)$ series circuits, respectively. From Fig. 10.25, when $n \to \infty$, the following is true:

$$\begin{aligned} F_{-v}^{c}(s) &= \lim_{n \to \infty} F_{-v}^{nc}(s) \\ &= \lim_{n \to \infty} F_{-v}^{(n-1)c}(s). \end{aligned} \tag{10.81}$$

Thus, from (10.81), (10.80) can be rewritten as:

$$F_{-v}^{c}(s) = \frac{\frac{2r}{cs} + \frac{1}{\alpha}F_{-v}^{c}\left(\frac{s}{\alpha\beta}\right)\left(r + \frac{1}{cs}\right)}{2\frac{1}{\alpha}F_{-v}^{(n-1)c}\left(\frac{s}{\alpha\beta}\right) + r + \frac{1}{cs}}, \tag{10.82}$$

where $1/rc$ is the eigen frequency of the v-order high-pass filtering net-grid-type scaling capacitive fractor. Equation (10.82) is the non-regularized iterative scaling equation of the v-order high-pass filtering net-grid-type scaling capacitive fractor. Thus, the solution of (10.82) can be given as:

$$F_{-v}^{c}(s) = \kappa s^{-v}. \tag{10.83}$$

Since $0 < v < 1$, when $s \to \infty$, $s^{-v} \to 0$. Thus, when $s \to \infty$ (high-pass filtering), from (10.83), (10.82) can be simplified as:

$$\kappa s^{-v} = \frac{\frac{2r}{cs} + \frac{1}{\alpha}\kappa\left(\frac{s}{\alpha\beta}\right)^{-v}\left(r + \frac{1}{cs}\right)}{2\frac{1}{\alpha}\kappa\left(\frac{s}{\alpha\beta}\right)^{-v} + r + \frac{1}{cs}}$$

$$\stackrel{s \to \infty}{\approx} \frac{(1/\alpha)\kappa[s/(\alpha\beta)]^{-v}(rcs + 1)}{rcs + 1}$$

$$= (1/\alpha)\kappa[s/(\alpha\beta)]^{-v}. \tag{10.84}$$

Hence, when $s \to \infty$ (high-pass filtering), the solution of (10.84) can be derived as:

$$v = \frac{\lg(\alpha)}{\lg(\alpha) + \lg(\beta)}. \tag{10.85}$$

From (10.83) and (10.84), we can see that when $\alpha = 1$, $\beta = 1$, (10.84) can be accurately simplified as $\kappa s^{-v} = (1/\alpha)\kappa[s/(\alpha\beta)]^{-v}|_{\alpha=1,\beta=1,v=1/2}$. The classical 1/2-order net-grid-type capacitive fractor is a purely ideal capacitive fractor [171] − [173], [193], [225], [405].

Secondly, with regard to the v-order low-pass filtering net-grid-type scaling inductive fractor shown in Fig. 10.26, the fractional-order iterative reactance of the v-order low-pass filtering net-grid-type inductive scaling fractor can be derived as:

$$F_{v}^{nl}(s) = \frac{2rls + (1/\alpha)F_{v}^{(n-1)l}(\alpha\beta s)(r + ls)}{2(1/\alpha)F_{v}^{(n-1)l}(\alpha\beta s) + r + ls}, \tag{10.86}$$

where $F_{v}^{nl}(s)$ and $F_{v}^{(n-1)l}(s)$ are the driving-point inductive impedance functions of the v-order low-pass filtering net-grid-type inductive scaling fractor with n series circuits and that with $(n-1)$ series circuits, respectively. From Fig. 10.26, when $n \to \infty$, the following is true:

$$F_{v}^{l}(s) = \lim_{n \to \infty} F_{v}^{nl}(s)$$

$$= \lim_{n \to \infty} F_{v}^{(n-1)l}(s). \tag{10.87}$$

Thus, from (10.87), (10.86) can be rewritten as:

$$F_{s}^{l}(s) = \frac{2rls + (1/\alpha)F_{v}^{l}(\alpha\beta s)(r + ls)}{2(1/\alpha)F_{v}^{l}(\alpha\beta s) + r + ls}, \tag{10.88}$$

where r/l is the eigen frequency of the v-order low-pass filtering net-grid-type scaling inductive fractor. Equation (10.88) is the non-regularized iterative scaling equation of the v-order low-pass filtering net-grid-type inductive scaling fractor. Thus, the solution of (10.88) can be given as:

$$F_v^l(s) = \kappa s^v. \tag{10.89}$$

Since $0 < v < 1$, when $s \to \infty$, $s^v \to 0$. Thus, when $s \to 0$ (low-pass filtering), from (10.89), (10.88) can be simplified as:

$$\kappa s^v = \frac{2rls + (1/\alpha)\kappa(\alpha\beta s)^v(r + ls)}{2(1/\alpha)\kappa(\alpha\beta s)^v + r + ls}$$

$$\stackrel{s \to \infty}{\approx} \frac{(1/\alpha)\kappa(\alpha\beta s)^v(r + ls)}{r + ls}$$

$$= (1/\alpha)\kappa(\alpha\beta s)^v. \tag{10.88}$$

Hence, when $s \to 0$ (low-pass filtering), the solution of (10.90) can be derived as:

$$v = \frac{\lg(\alpha)}{\lg(\alpha) + \lg(\beta)}. \tag{10.91}$$

From (10.90) and (10.91), we can see that when $\alpha = 1$, $\beta = 1$, (10.90) can be accurately simplified $\kappa s^v = (1/\alpha)\kappa(\alpha\beta s)^v|_{\alpha=1,\beta=1,v=1/2}$. The classical 1/2-order net-grid-type inductive fractor is a purely ideal inductive fractor [171] − [173], [193], [225], [405].

Thirdly, with regard to the v-order high-pass filtering net-grid-type scaling inductive fractor shown in Fig. 10.27, the fractional-order iterative reactance of the v-order high-pass filtering net-grid-type inductive scaling fractor can be derived as:

$$F_v^{nl}(s) = \frac{2rls + \alpha F_v^{(n-1)l}\left(\frac{s}{\alpha\beta}\right)(r + ls)}{2\alpha F_v^{(n-1)l}\left(\frac{s}{\alpha\beta}\right) + r + ls}, \tag{10.92}$$

where $F_v^{nl}(s)$ and $F_v^{(n-1)l}(s)$ are the driving-point inductive impedance functions of the v-order high-pass filtering net-grid-type inductive scaling fractor with n series circuits and that with $(n-1)$ series circuits, respectively. From Fig. 10.27, when $n \to \infty$, the following is true:

$$F_v^l(s) = \lim_{n \to \infty} F_v^{nl}(s)$$

$$= \lim_{n \to \infty} F_v^{(n-1)l}(s). \tag{10.93}$$

Thus, from (10.93), (10.92) can be rewritten as:

$$F_v^l(s) = \frac{2rls + \alpha F_v^l\left(\frac{s}{\alpha\beta}\right)(r + ls)}{2\alpha F_v^l\left(\frac{s}{\alpha\beta}\right) + r + ls}, \tag{10.94}$$

where r/l is the eigen frequency of the v-order high-pass filtering net-grid-type scaling inductive fractor. Equation (10.94) is the non-regularized iterative scaling equation of the v-order high-pass filtering net-grid-type inductive scaling fractor. Thus, the solution of (10.94) can be given as:

$$F_v^l(s) = \kappa s^v. \tag{10.95}$$

Since $0 < v < 1$, when $s \to \infty$, $s^v/s \to 0$. Thus, when $s \to \infty$ (high-pass filtering), from (10.95), (10.94) can be simplified as:

$$\kappa s^v = \frac{2rls + \alpha\kappa[s/(\alpha\beta)]^v(r+ls)}{2\alpha k[s/(\alpha\beta)]^v + r + ls}$$
$$\stackrel{s\to\infty}{\approx} \frac{\alpha\kappa[s/(\alpha\beta)]^v(r/s+l)}{r/s+l}$$
$$= \alpha\kappa[s/(\alpha\beta)]^v. \tag{10.96}$$

Hence, when $s \to \infty$ (high-pass filtering), the solution of (10.96) can be derived as:

$$v = \frac{\lg(\alpha)}{\lg(\alpha) + \lg(\beta)}. \tag{10.97}$$

From (10.95) and (10.96), we can see that when $\alpha = 1$, $\beta = 1$, (10.96) can be accurately simplified as $\kappa s^v = \alpha\kappa[s/(\alpha\beta)]^v|_{\alpha=1,\beta=1,v=1/2}$. The classical 1/2-order net-grid-type inductive fractor is a purely ideal inductive fractor [171] − [173], [193], [225], [405].

10.2.4 EXPERIMENT AND ANALYSIS

10.2.4.1 Approximation Performance of arbitrary-order Fractor of FHNNs

In this subsection, in order to realize the arbitrary-order FHNNs in the form of an analog circuit that can be applied to defense against chip cloning attacks, the approximation performance of the arbitrary-order fractor of the FHNNs needs to be analyzed firstly.

Example 1: Let's analyze the approximation performance of an arbitrary-order net-grid-type scaling fractor. At first, with regard to the arbitrary-order low-pass filtering net-grid-type capacitive scaling fractor shown in Fig. 10.24, for the convenience of the numerical iterative analysis of the approximation performance of the arbitrary-order fractor, the method of diagonalization of matrix is used to illustrate the matrix chain-multiplication of (10.50), (10.51), and (10.52). Thus, from (10.49) and (10.56), the following can be obtained:

$$\begin{bmatrix} -\Psi_n & \Psi_n \\ 1 & 1 \end{bmatrix}^{-1} \begin{bmatrix} -\Psi_{n+1} & \Psi_{n+1} \\ 1 & 1 \end{bmatrix} = \begin{bmatrix} -\Psi_n & \Psi_n \\ 1 & 1 \end{bmatrix} \begin{bmatrix} -1/(2\Psi_{n+1}) & 1/2 \\ 1/(2\Psi_{n+1}) & 1/2 \end{bmatrix}$$
$$= \begin{bmatrix} \Psi_{n+1}/(2\Psi_n) + 1/2 & -\Psi_{n+1}/(2\Psi_n) + 1/2 \\ -\Psi_{n+1}/(2\Psi_n) + 1/2 & \Psi_{n+1}/(2\Psi_n) + 1/2 \end{bmatrix}$$
$$= \begin{bmatrix} \sqrt{\alpha/\beta}/2 + 1/2 & -\sqrt{\alpha/\beta}/2 + 1/2 \\ -\sqrt{\alpha/\beta}/2 + 1/2 & \sqrt{\alpha/\beta}/2 + 1/2 \end{bmatrix}. \tag{10.98}$$

Equation (10.98) shows that $\begin{bmatrix} -\Psi_n & \Psi_n \\ 1 & 1 \end{bmatrix}^{-1} \begin{bmatrix} -\Psi_{n+1} & \Psi_{n+1} \\ 1 & 1 \end{bmatrix}$ is a constant matrix, which is only related to two positive scaling constants (α and β) of the v-order low-pass filtering net-grid-type capacitive scaling fractor. Thus, from (10.56) and (10.98), we get:

$$[T_{n+1}][T_{n+2}] = \begin{bmatrix} -\Psi_n & \Psi_n \\ 1 & 1 \end{bmatrix} \begin{bmatrix} e^{-\xi_n} & 0 \\ 0 & e^{\xi_n} \end{bmatrix} \begin{bmatrix} \sqrt{\alpha/\beta}/2+1/2 & -\sqrt{\alpha/\beta}/2+1/2 \\ -\sqrt{\alpha/\beta}/2+1/2 & \sqrt{\alpha/\beta}/2+1/2 \end{bmatrix}$$
$$\begin{bmatrix} e^{-\xi_{n+1}} & 0 \\ 0 & e^{\xi_{n+1}} \end{bmatrix} \begin{bmatrix} -\Psi_{n+1} & \Psi_{n+1} \\ 1 & 1 \end{bmatrix}^{-1}. \quad (10.99)$$

Then, from (10.50) and (10.99), we can see that therecurrence relations of $[T^m] = \prod_{n=0}^{m}[T_{n+1}]$ is the periodical matrix chain-multiplication of $[T_{n+1}][T_{n+2}]$. Substituting (10.99) into (10.50), the following can be obtained:

$$[T^m] = \begin{bmatrix} A^m & B^m \\ C^m & D^m \end{bmatrix}$$
$$= \prod_{n=0}^{m}[T_{n+1}]$$
$$= \begin{bmatrix} -\Psi_0 & \Psi_0 \\ 1 & 1 \end{bmatrix} \begin{bmatrix} e^{-\xi_0} & 0 \\ 0 & e^{\xi_0} \end{bmatrix}[\Delta]\begin{bmatrix} e^{-\xi_1} & 0 \\ 0 & e^{\xi_1} \end{bmatrix}[\Delta]\begin{bmatrix} e^{-\xi_2} & 0 \\ 0 & e^{\xi_2} \end{bmatrix}[\Delta]\begin{bmatrix} e^{-\xi_3} & 0 \\ 0 & e^{\xi_3} \end{bmatrix}\cdots$$
$$[\Delta]\begin{bmatrix} e^{-\xi_{m-1}} & 0 \\ 0 & e^{\xi_{m-1}} \end{bmatrix}\begin{bmatrix} -\Psi_{m-1} & \Psi_{m-1} \\ 1 & 1 \end{bmatrix}^{-1}$$
$$= \begin{bmatrix} -\Psi_0 & \Psi_0 \\ 1 & 1 \end{bmatrix}\begin{bmatrix} e^{-\xi_0} & 0 \\ 0 & e^{\xi_0} \end{bmatrix}[\Delta]\begin{bmatrix} e^{-\xi_1} & 0 \\ 0 & e^{\xi_1} \end{bmatrix}[\Delta]\begin{bmatrix} e^{-\xi_2} & 0 \\ 0 & e^{\xi_2} \end{bmatrix}[\Delta]\begin{bmatrix} e^{-\xi_3} & 0 \\ 0 & e^{\xi_3} \end{bmatrix}\cdots$$
$$[\Delta]\begin{bmatrix} e^{-\xi_{m-1}} & 0 \\ 0 & e^{\xi_{m-1}} \end{bmatrix}\begin{bmatrix} -1/(2\Psi_{m-1}) & 1/2 \\ 1/(2\Psi_{m-1}) & 1/2 \end{bmatrix}, \quad (10.100)$$

where $[\Delta] = \begin{bmatrix} \sqrt{\alpha/\beta}/2+1/2 & -\sqrt{\alpha/\beta}/2+1/2 \\ -\sqrt{\alpha/\beta}/2+1/2 & \sqrt{\alpha/\beta}/2+1/2 \end{bmatrix}$. Without loss of generality, the 1/3-order low-pass filtering net-grid-type capacitive scaling fractor is illustrated. Let's set $\alpha = 2$ and $\beta = 4$. Then, from (10.78), $v = 1/3$. From (10.49) − (10.52) and (10.100), the Bode diagram of the 1/3-order open-circuit capacitive reactance ($Z_{-1/3}^{om}$) and the 1/3-order short-circuit capacitive reactance ($Z_{-1/3}^{sm}$) of the 1/3-order low-pass filtering net-grid-type capacitive scaling fractor with m series circuits can be shown as given in Fig. 10.31.

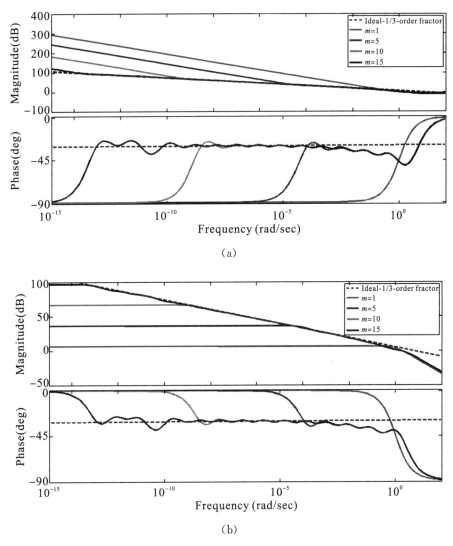

Fig. 10.31 Bode diagram of $Z^{om}_{-1/3}$ and $Z^{sm}_{-1/3}$ of 1/3-order low-pass filtering net-grid-type capacitive scaling fractor with m series circuits: (a) $Z^{om}_{-1/3}$; (b) $Z^{sm}_{-1/3}$.

Fig. 10.31 shows that both the magnitude and the phase of $Z^{om}_{-1/3}$ and $Z^{sm}_{-1/3}$ of the 1/3-order low-pass filtering net-grid-type capacitive scaling fractor with m series circuits can approach that of the 1/3-order capacitive fractor with a high degree of accuracy in a certain pass-band (for instance, when $m=15$, the pass-band is $[10^{-13}\,\text{rad/sec},\ 10^0\,\text{rad/sec}]$). The larger m is, the wider the pass-band of the 1/3-order low-pass filtering net-grid-type capacitive scaling fractor is.

Secondly, with regard to the 1/3-order high-pass filtering net-grid-type capacitive scaling fractor shown in Fig. 10.25, let's also set $\alpha=2$ and $\beta=4$. Then, from (10.85), $v=1/3$. In a similar way, the Bode diagram of the 1/3-order open-circuit capacitive reactance ($Z^{om}_{-1/3}$) and the 1/3-order short-circuit capacitive reactance ($Z^{sm}_{-1/3}$) of the 1/3-order high-pass filtering net-grid-type capacitive scaling fractor with m series circuits can be shown as given in Fig. 10.32.

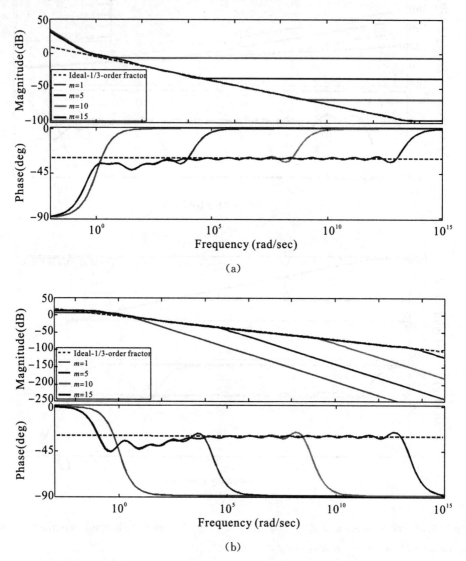

Fig. 10.32 Bode diagram of $Z^{om}_{-1/3}$ and $Z^{sm}_{-1/3}$ of 1/3-order high-pass filtering net-grid-type capacitive scaling fractor with m series circuits: (a) $Z^{om}_{-1/3}$; (b) $Z^{sm}_{-1/3}$.

Fig. 10.32 shows that both the magnitude and the phase of $Z^{om}_{-1/3}$ and $Z^{sm}_{-1/3}$ of the 1/3-order high-pass filtering net-grid-type capacitive scaling fractor with m series circuits can approach that of the 1/3-order capacitive fractor with a high degree of accuracy in a certain pass-band (for instance, when $m=15$, the pass-band is [10^0 rad/sec, 10^{13} rad/sec]). The larger m is, the wider the pass-band of the 1/3-order high-pass filtering net-grid-type capacitive scaling fractor is. Further, comparing Fig. 10.31 with Fig. 10.32, we can see that when $\alpha=2$, $\beta=4$, and m of the 1/3-order low-pass filtering net-grid-type capacitive scaling fractor and the 1/3-order high-pass filtering one have the same values, the pass-band of the former is the reciprocal of the latter.

Thirdly, with regard to the 1/3-order low-pass filtering net-grid-type inductive scaling fractor shown in Fig. 10.26, let's also set $\alpha=2$ and $\beta=4$. Then, from (10.92), $v=1/3$. In a similar way, the Bode diagram of the 1/3-order open-circuit inductive reactance ($Z^{om}_{-1/3}$) and the 1/3-order short-circuit inductive reactance ($Z^{sm}_{-1/3}$) of the 1/3-order low-pass filtering net-grid-type inductive scaling fractor with m series circuits can be shown as given in Fig. 10.33.

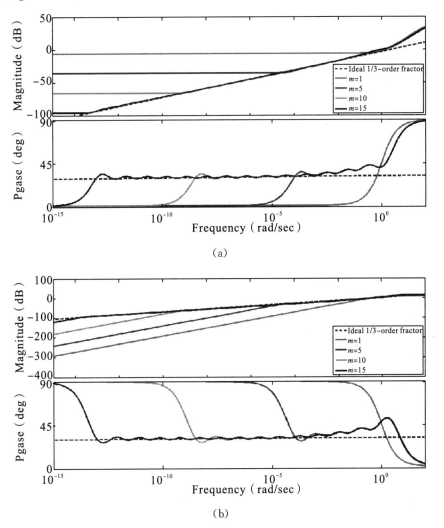

Fig. 10.33. Bode diagram of $Z^{om}_{-1/3}$ and $Z^{sm}_{-1/3}$ of 1/3-order low-pass filtering net-grid-type inductive scaling fractor with m series circuits: (a) $Z^{om}_{-1/3}$; (b) $Z^{sm}_{-1/3}$.

Fig. 10.33 shows that both the magnitude and the phase of $Z^{om}_{-1/3}$ and $Z^{sm}_{-1/3}$ of the 1/3-order low-pass filtering net-grid-type inductive scaling fractor with m series circuits can approach that of the 1/3-order inductive fractor with a high degree of accuracy in a certain pass-band (for instance, when $m=15$, the pass-band is [10^{-13} rad/sec, 10^0 rad/sec]). The larger m is, the wider the pass-band of the 1/3-order low-pass filtering net-grid-type inductive scaling fractor is.

Fourthly, with regard to the 1/3-order high-pass filtering net-grid-type inductive scaling fractor shown in Fig. 10.27, let's also set $\alpha = 2$ and $\beta = 4$. Then, from (61), $v = 1/3$. In a similar way, the Bode diagram of the 1/3-order open-circuit inductive reactance ($Z^{om}_{-1/3}$) and the 1/3-order short-circuit inductive reactance ($Z^{sm}_{-1/3}$) of the 1/3-order high-pass filtering net-grid-type inductive scaling fractor with m series circuits can be shown as given in Fig. 10.34.

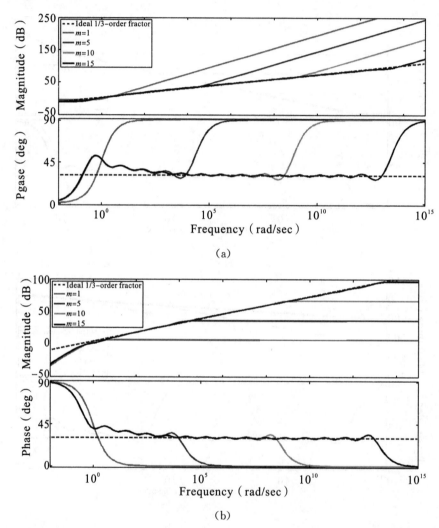

Fig. 10.34 Bode diagram of $Z^{om}_{-1/3}$ and $Z^{sm}_{-1/3}$ of 1/3-order high-pass filtering net-grid-type inductive scaling fractor with m series circuits: (a) $Z^{om}_{-1/3}$; (b) $Z^{sm}_{-1/3}$.

Fig. 10.34 shows that both the magnitude and the phase of $Z^{om}_{-1/3}$ and $Z^{sm}_{-1/3}$ of the 1/3-order high-pass filtering net-grid-type inductive scaling fractor with m series circuits can approach that of the 1/3-order inductive fractor with a high degree of accuracy in a certain pass-band (for instance, when $m = 15$, the pass-band is $[10^0 \text{ rad/sec}, 10^{13} \text{ rad/sec}]$). The larger m is, the wider the pass-band of the 1/3-order high-pass filtering net-grid-type inductive scaling fractor is. Further, comparing Fig. 10.33 with Fig. 10.34, we can see that when $\alpha=2$, $\beta=4$, and m of the 1/3-order low-pass filtering net-grid-type inductive scaling fractor and the 1/3-order high-pass filtering one have the same values, the pass-band of the former is the reciprocal of the latter.

10.2.4.2 Analog circuit realization of 1/3-Order FHNN

In thissubsection, the arbitrary-order FHNNs are achieved by analog circuit realization and its ability of defense against chip cloning attacks is illustrated in detail experimentally. Without loss of generality, a 1/3-order high-pass FHNN is achieved by analog circuit realization.

Example 2: Let's implement a FHNN of original chip. From Figs. 10.1 and 10.3, we can implement a 1/3-order high-pass FHNN with a 1/3-order high-pass filtering net-grid-type capacitive scaling fractor. Let's set $\alpha=2$ and $\beta=4$. Then, from Fig. 10.25 and (85), $v=1/3$. Suppose the resistance and capacitance of Fig. 10.25 are $r=16\Omega$ and $c=256\text{pF}$, respectively. Thus, from Fig. 10.25, the analog series circuit of a 1/3-order high-pass filtering net-grid-type scaling capacitive fractor can be shown in Fig. 10.35.

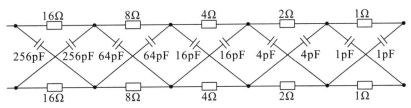

Fig. 10.35 Analog series circuit of a 1/3-order high-pass filtering net-grid-type capacitive scaling fractor.

Without loss of generality, we suppose the FHNN to be achieved has two neurons with identical fractional-order. The fractional-order of either neuron of the FHNN is equal to $v_1=v_2=1/3$. Thus, we can implement either neuron of a 1/3-order FHNN with the 1/3-order high-pass filtering net-grid-type capacitive scaling fractor shown in Fig. 10.35. For the convenience of illustration, we make the convergence trajectories of the two outputs of the FHNN do not overlap. With respect to circuit configuration of the FHNN shown in Fig. 10.1, let's set the input currents $i_1=i_2=0$, the feedback resistances $R_{12}=300\Omega$ and $R_{21}=250\Omega$, and the shunting resistances $R=0.01\Omega$. Let's choose the integrated operational amplifier of OP37G to implement either neuron of the FHNN. Using commonly used PCB design software of Cadence OrCAD v16.6, the analog circuit realization of a 1/3-order FHNN of original chip can be simulated as shown in Fig. 10.36.

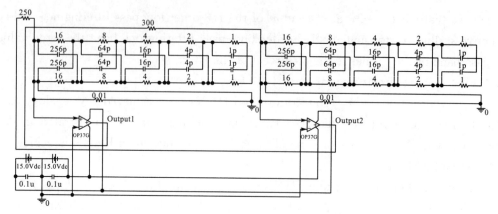

Fig. 10.36 Analog circuit realization of a 1/3-order FHNN of original chip.

From Fig. 10.36, the convergence trajectory performance of either neuron of the 1/3-order FHNN of original chip can be obtained. Thus, the time response curves of two outputs of the 1/3-order FHNN of original chip can be shown in Fig. 10.37.

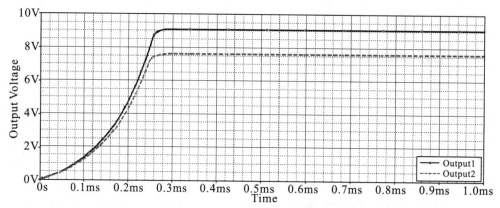

Fig. 10.37 Time response curves of two outputs of a 1/3-order FHNN of original chip.

Fig. 10.37 shows that two outputs of the 1/3-order FHNN of original chip are all gradually convergent. Corresponding to the input (0, 0) of the 1/3-order FHNN of original chip, its two outputs converge to (9.0450 V, 7.5850 V). Therefore, (9.0450 V, 7.5850 V) is one of the attractors of the 1/3-order FHNN of original chip.

Example 3: Let's demonstrate the core competence of defense against chip cloning attacks of the FHNNs. The FHNNs can very easily detect a negative three orders of magnitude, or even smaller, deviation of the chip cloning attacks. With respect to the current electronic manufacturing technology, no chip cloning attacker can implement two resistors or capacitors of the FHNNs with actually identical reactance value. In other words, no attacker can achieve two FHNNs with the same values of their attractors. The FHNN possesses the fractional-order-stability characteristics. The values of the attractors of a FHNN substantially have bearing on the fractional-orders of its neurons. Thus, comparing with the FHNN of original chip, the values of the attractors of a FHNN of chip cloning attacks shall be slightly changed as the result of its manufacture errors of resistors and capacitors. By means of detecting the output voltages of a FHNN, we can very easily determine whether it has been chip cloned or not. Then, we can apply the FHNN to defense against chip cloning attacks [192], [194], [399], [410].

Without loss of generality, it is assumed that the manufacture errors of the resistance and capacitance of chip cloning attacks are all subject to Gaussian distribution. The Gaussian distributions of the resistance error and the capacitance error of chip cloning attacks obey $N(0, 0.01)$ and $N(0, 0.81)$, respectively. Thus, from Fig. 10.35, two 1/3-order high-pass filtering net-grid-type scaling capacitive fractors of chip cloning attacks can be shown in Fig. 10.38.

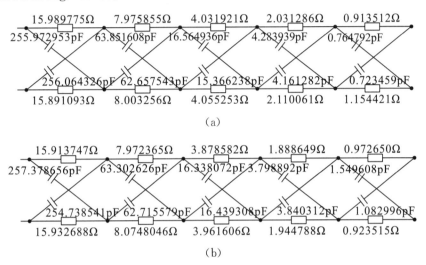

Fig. 10.38 Two 1/3-order high-pass filtering net-grid-type capacitive scaling fractors of chip cloning attacks: (a) First sample of chip cloning attacks; (b) Second sample of chip cloning attacks.

Comparing Fig. 10.35 with Fig. 10.38, we can see that the errors of resistance and capacitance of two 1/3-order high-pass filtering net-grid-type scaling capacitive fractors of chip cloning attacks are relatively small. As a matter of fact, chip cloning attacks slightly change the fractional-order of the fractional-order fractor. The fractional-orders of two 1/3-order high-pass filtering net-grid-type scaling capacitive fractors of chip cloning attacks are no longer equal to 1/3.

Using two 1/3-order high-pass filtering net-grid-type scaling capacitive fractors of chip cloning attacks, a 1/3-order FHNN with two neurons of chip cloning attacks can be implemented as shown in Fig. 10.39.

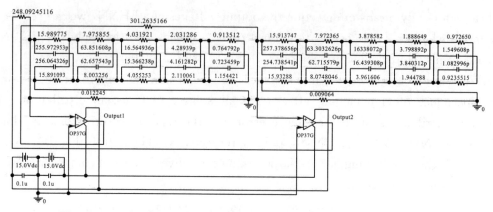

Fig. 10.39 A 1/3-order FHNN with two neurons of chip cloning attacks.

From Fig. 10.39, the convergence trajectory performance of either neuron of the 1/3-order FHNN of chip cloning attacks can be gotten. Thus, the time response curves of two outputs of the 1/3-order FHNN of chip cloning attacks can be shown in Fig. 10.40.

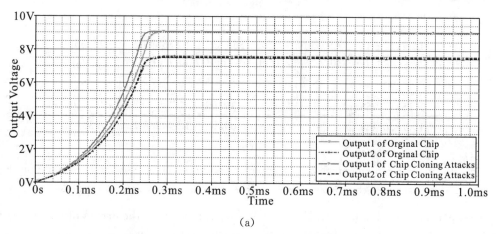

(a)

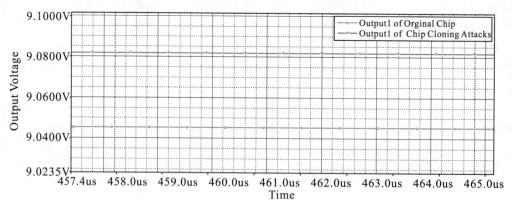

(b)

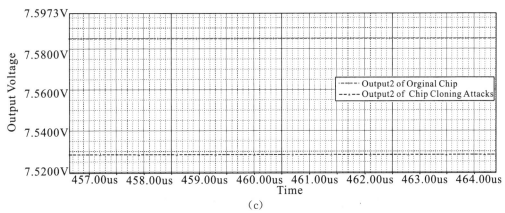

Fig. 10.40 Comparison with time response curves of outputs of a 1/3-order FHNN with two neurons of original chip and that of chip cloning attacks: (a) Comparison with output voltages of original chip and that of chip cloning attacks; (b) Comparison with output1 voltage of original chip and that of chip cloning attacks; (c) Comparison with output2 voltage of original chip and that of chip cloning attacks.

Fig. 10.40 shows that corresponding to the input (0, 0) of the 1/3-order FHNN of chip cloning attacks, its two outputs converge to (9.0820 V, 7.5280 V). Therefore, (9.0820 V, 7.5280 V) is one of the attractors of the 1/3-order FHNN of chip cloning attacks. Comparing Fig. 10.15 with Fig. 10.18, we can see that the values of two output voltages of the 1/3-order FHNN of chip cloning attacks deviate slightly from those of the 1/3-order FHNN of original chip. Thus, by means of detecting the output voltages of a FHNN, we can very easily determine whether it has been chip cloned or not.

10.2.5 CONCLUSIONS

In this chapter, based on fractional calculus, the arbitrary-order fractor is proposed to implement the analog circuit realization of the arbitrary-order FHNNs and the fractor based FHNNs are proposed to apply to defense against chip cloning attacks. From aforementioned illustration, we can see that the FHNNs have a significant and promising intrinsic capability to distinguish chip devices from one another. The analog circuit of the FHNNs is presented as the physical embodiment of the fractor. The fundamental idea behind the FHNN is that a fractor based neural networks can be described in terms of attractors that are in turn very sensitive to process variations in the reactance values of resistors and capacitors that comprise fractor. This sensitivity can then be leveraged to provide a high degree of entropy that can be applied for mitigating cloning. The FHNNs can be applied to defense against chip cloning attacks.

With respect to thestate-of-the-art application of the FHNNs to defense against chip cloning attacks, there are many fascinating issues else need to be further discussed, such as the implementation cost, the constant-order performance, the electrical performance stability under electromagnetic disturbance, and the amount of entropy of the FHNNs. It will be further discussed in detail in my future work.

第 11 章 图像处理的分数阶微分算子：
分数阶微分掩模

11.1 二维数字图像信号分数阶微分的数值实现

本章分别从生物视觉神经模型和信号处理两个角度系统论述了图像分数阶微分的感受野模型及其侧抑制原理；提出并论述了二维数字图像处理中各种分数阶微分掩模及其运算规则。首先，推导并分析分数阶微积分与其他著名的信号时—频分析之间的关系，从而分析在一定条件下二维分数阶微分的可分离性。在此基础上分析分数阶微积分在信号调制解调方面应用的理论依据。其次，提出分数阶微积分在动力学系统中的物理意义，定义分数阶稳定系数，提出并论述了图像分数阶微积分的感受野模型，分别从生物视觉神经模型和信号处理两个角度分析了它对图像纹理细节产生的特殊马赫现象。最后，提出并论述数字图像处理中各种分数阶微分掩模及其运算规则。实验结果表明，对提取和加强图像灰度变化不大的平滑区域中的纹理细节而言，图像分数阶微分运算的效果明显优于图像整数阶微分运算。当处理纹理细节信息丰富的图像时，这一点尤为突出。根据对实验结果的分析和对比，尝试了一些图像分数阶微分的应用。

11.1.1 问题提出

一般来说，图像信号具有高度自相似性。在数字图像中，邻域内的像素与像素之间的灰度值具有很大的相关性。对一维信号进行分数阶微分处理既可加强信号的高频信息，又可在一定程度上保留信号的低频和直流信息。另外，分数阶微积分是分形学说的数学基础之一，而图像中高度自相似的分形信息通常是以复杂的纹理细节信息表现的。那么，对于二维图像信号进行分数阶微积分处理，能否既加强其灰度变化相对较大的高频边缘信息，又尽量保持其低频轮廓信息和非线性加强其平滑区域中灰度相对变化不大的纹理细节信息呢？围绕这个问题，本章展开了深入的研究和系统的论述。

本章的研究主要集中在四个方面：①推导并分析分数阶微积分与其他著名的信号时—频分析之间的关系，分析在一定条件下二维分数阶微分的可分离性；②分析分数阶微积分在信号调制解调方面应用的理论依据；③提出分数阶微积分在动力学系统中的物理意义，并定义分数阶稳定系数；④提出并论述图像分数阶微积分的感受野模型，分别从生物视觉神经模型和信号处理两个角度分析它对图像纹理细节信息产生的特殊马赫现象；⑤提出并论述数字图像处理中各种分数阶微分掩模及其运算规则。

11.1.2　分数阶微积分与其他时—频分析之间的关系推导

为了深刻理解将分数阶微积分运用于二维图像处理的原因、机理及其算法规则，应首先对信号进行分数阶微分运算的作用、性质及其与其他经典时—频分析之间的关系做更深入的研究。

p 阶分数阶 Fourier 变换是定义在 u' 域的信号 $s(u')$ 的一个线性积分运算[78-79,322]，即

$$s_F^p(u) = \int_{-\infty}^{+\infty} K_p(u,u')s(u')\mathrm{d}u' \tag{11.1}$$

式中，分数阶 Fourier 变换的核函数（基函数）为 $K_p(u,u') \equiv A_\alpha \exp[j\pi(u^2\cot\alpha - 2uu'\csc\alpha + u'^2\cot\alpha)]$。$A_\alpha \equiv \sqrt{1-j\cot\alpha}$，旋转角 $\alpha \equiv p\pi/2$，$p \neq 2n$，n 是整数。显然，当 $p=4n$，$\alpha=2n\pi$ 时，$K_p(u,u') \equiv \delta(u-u')$；当 $p=4n\pm 2$，$\alpha=(2n\pm 1)\pi$ 时，$K_p(u,u') \equiv \delta(u+u')$。

对于任意平方可积函数或信号 $s(t) \in L^2(R)$，其连续子波变换定义为

$$Ws(\tau,a) = <s(t), \Psi_{\tau,a}(t)> = \frac{1}{2\pi}\int \hat{s}(\omega)\hat{\Psi}_{\tau,a}(\omega)\mathrm{d}\omega$$

$$= \frac{1}{\sqrt{|a|}}\int_{-\infty}^{+\infty} s(t)\Psi^*(\frac{t-\tau}{a})\mathrm{d}t \tag{11.2}$$

式中，τ，a，$\Psi(t)$，$\Psi_{\tau,a}(t)$ 分别为扫描时间、尺度因子、母波、子波。本章用母波构造能量恒等化子波，即

$$\Psi_{\tau,a}(t) = 1/\sqrt{a}\,\Psi(\frac{t-\tau}{a}) \stackrel{FT}{\Longleftrightarrow} \hat{\Psi}_{\tau,a}(\omega) = \sqrt{a}\,\Psi(\hat{a}\omega)\mathrm{e}^{-j\omega\tau} \tag{11.3}$$

在式（11.1）中，取 $u'=t$，即在时间域内考察信号 $s(t)$。令 $\tau = t\sec(\alpha)$，代入式（11.1），可得信号 $s(t)$ 的 p 阶分数阶 Fourier 变换为[78]

$$s_F^p(u) = \frac{C(\alpha)}{\exp[j\pi\tau^2\sin^2(\alpha)]}\int_{-\infty}^{+\infty}\exp\left[j\pi(\frac{t-\tau}{\tan^2(\alpha)})^2\right]s(t)\mathrm{d}t \tag{11.4}$$

式中，$C(\alpha)$ 是以 α 为参数的常数。对比式（11.2），（11.4）知，信号 $s(t)$ 的 p 阶分数阶 Fourier 变换 $s_F^p(u)$ 是以尺度因子 $a=\tan^2(\alpha)$ 的子波变换。令其母波函数 $\Psi(t) = \exp(j\pi t^2)$，故其子波函数为

$$\Psi_{\tau,a}(t) = \frac{1}{\sqrt{\tan^2(\alpha)}}\Psi(\frac{t-\tau}{\tan^2(\alpha)})$$

$$= \frac{1}{|\tan(\alpha)|}\exp\left[j\pi(\frac{t-\tau}{\tan^2(\alpha)})^2\right] \tag{11.5}$$

所以，信号 $s(t)$ 的子波变换 $Ws(\tau,a)$ 与其 p 阶分数阶 Fourier 变换 $s_F^p(u)$ 的关系为[78-79,322]

$$Ws(\tau,a) = \frac{\exp[j\pi\tau^2\sin^2(\alpha)]s_F^p(u)}{C(\alpha)|\tan(\alpha)|}$$

$$= \frac{\exp[j\pi\tau^2\sin^2(p\pi/2)]s_F^p(u)}{C(p\pi/2)|\tan(p\pi/2)|} \tag{11.6}$$

由式（3.19）知，信号 $s(t)$ 的分数阶微积分与其子波变换的转换公式为

$$D_v[s(\tau)] = \lim_{a \to 0}\frac{E(v)Ws(\tau,a)}{a^v} \tag{11.7}$$

由式（11.7）和（11.6）可以推知，信号 $s(t)$ 的 v 阶分数阶微积分与其 p 阶分数阶 Fourier 变换的转换关系为[313]

$$s^{(v)}(\tau) = \lim_{p \to 0} \Delta(v,p,\tau) s_F^p(u) \tag{11.8}$$

式中，$\Delta(v,p,\tau) = \dfrac{E(v)\exp[j\pi\tau^2 \sin^2(p\pi/2)]}{\tan^{2v}(p\pi/2)C(p\pi/2)|\tan(p\pi/2)|}$ 是以 (v,p) 为参数，时间 s 为变量的正态函数。可见，信号 $s(t)$ 的 v 阶分数阶微积分 $s^{(v)}(t)$ 可由 $s(t)$ 的 $p\to 0$ 阶分数阶 Fourier 变换 $s_F^p(u)$ 在时间 $\tau = t\sec(\alpha)$ 处的正态补偿。

那么，信号的分数阶微积分与其分数阶 Fourier 变换的转换公式在信号与信息系统理论中的物理意义是什么呢？作者认为，信号 $s(t)$ 的一阶 Fourier 变换是将 $s(t)$ 的 Wigner 分布逆时针旋转 $\alpha = \pi/2$ 得到由时间 t 轴变到频率 ω 轴的表示形式，即一函数在与时间 t 轴夹角为 $\pi/2$ 的频率 ω 轴上得出。与此相对应，信号 $s(t)$ 的 p 阶 Fourier 变换是将 $s(t)$ 的 Wigner 分布逆时针旋转 $\alpha = p\pi/2$ 得到由时间 t 轴变到频率 ω 轴的表示形式，即一函数在与时间 t 轴夹角为 $p\pi/2$ 的频率 ω 轴上得出，如图 11.1 所示。

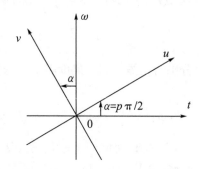

图 11.1　信号分数阶 Fourier 变换的 Wigner 分布旋转图

当信号分数阶 Fourier 变换 $s_F^p(u)$ 的阶次 $p\to 0$ 时，将 $s(t)$ 的 Wigner 分布逆时针旋转 $\alpha = p\pi/2 \xrightarrow{p\to 0} 0$ 得到分数阶 Fourier 变换。可见，当 $p\to 0$ 时，信号 $s(t)$ 的 Wigner 分布几乎没有旋转，此时信号的分数阶 Fourier 变换失去了对信号频率的分析能力，而只具有对信号时间细节变化的分析能力，因而不是完整意义上的时—频分析。这恰好和信号的分数阶微积分只是对信号做时间细节分析的物理含义相一致。这就是式（11.8）的定性解释，这个物理解释类似于对式（3.19）的定性解释。换言之，一方面，当分数阶 Fourier 变换的阶数 $p\to 0$ 时，信号的分数阶 Fourier 变换从对信号的时—频分析退化为只对信号进行时间分析的变换，这恰好与信号的分数阶微积分只是对信号进行时间分析的物理意义相一致；另一方面，如式（11.8）所示，信号 $s(t)$ 的 v 阶分数阶微积分 $s^{(v)}(t)$ 可由 $s(t)$ 的 $p\to 0$ 阶分数阶 Fourier 变换 $s_F^p(u)$ 在时间 $\tau = t\sec(\alpha)$ 处的正态补偿，所以 $s^{(v)}(t)$ 对信号具有极强的分析时间细节的能力，这恰好与信号的分数阶微分是对信号进行高通滤波，加强高频信息的物理意义相吻合。这就是式（11.8）的物理含义。

式（11.7）和（11.8）是一座桥梁，它找到了分数阶微积分与典型的线性时频分析（子波变换、短时 Fourier 变换等）以及典型的二次型时频分析（分数阶 Fourier 变换、Wigner-Ville 分布、Radon-Wigner 分布、模糊函数等）之间相互转换的关系。

11.1.3 分数阶微积分在信号调制解调方面应用的理论分析

适合用作时频聚集性评价的典型非平稳信号是线性调频信号（LFM）。幅度为 1 的 LFM 信号 $s(t) = \exp(j(\omega_0 t + at^2/2))$ 的 Wigner–Ville 分布为[78-79,322]

$$WV_{LFM}(t,\omega) = 2\pi\delta[\omega - (\omega_0 + at)] \tag{11.9}$$

可见，单位分量的 Wigner–Ville 分布为沿直线 $\omega = (\omega_0 + at)$ 分布的冲激线谱，其 Wigner–Ville 分布的幅值均出现在表示信号的瞬时频率变化率的直线上。

由此可知，信号 $s(t)$ 的 p 阶分数阶 Fourier 变换是 $s(t)$ 的 Wigner–Ville 分布旋转 $\alpha = p\pi/2$ 后得到的。由式（11.8）可知，信号 $s(t)$ 的 v 阶分数阶微积分逼近于 $s(t)$ 的 Wigner–Ville 分布的正态加权。这就为分数阶微积分应用于信号的调制解调找到了理论依据。

11.1.4 分数阶微积分在动力学系统中的物理意义探究

不失一般性，采用第 3 章中推导的基于子波变换的数值算法[157-159]实现正态高斯信号的 $\phi(t) = \frac{1}{\sqrt{2\pi}} e^{\frac{t^2}{2}}$，$-\infty < t < \infty$ 的各分数阶微分，实验结果如图 11.2 所示。

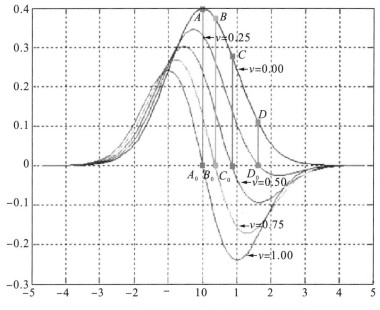

图 11.2 正态高斯信号各分数阶微分的零值点

由图 11.2 可见，点 A_0 处 $\phi^{(1)}(t)=0$，对应于原信号 $\phi(t)$ 的 A 点。点 B_0 处 $\phi^{(0.75)}(t)=0$，对应于 $\phi(t)$ 的 B 点。点 C_0 处 $\phi^{(0.5)}(t)=0$，对应于 $\phi(t)$ 的 C 点。点 D_0 处 $\phi^{(0.25)}(t)=0$，对应于 $\phi(t)$ 的 D 点。显然，$\phi(t)$ 是向上凸的，在点 A 处物体处于不稳定平衡，当有极小的力作用于物体时，物体就会从点 A 向下非匀加速滑落。当物体分别置于 $\phi(t)$ 的 B，C，D 点时，物体会以不同的初始瞬时加速度向下滑落。物体从 $\phi(t)$ 的 A，B，C，D 点向下滑，由于势能的不同，到达 $\phi(t)$ 曲线最低点时，物体将具有不同的动能 W。显然，$W_A > W_B > W_C > W_D$。也就是说，函数 $\phi(t)$ 的分数阶微分的阶次 v 是对 $\phi(t)$ 曲线上

各点稳定性的一种度量。由此，作者提出并定义分数阶稳定系数的理论与概念[313]。如果函数的凸凹性用 η 表示为

$$\eta = \begin{cases} 1, & \text{向上凸} \\ -1, & \text{向下凸} \end{cases} \quad (11.10)$$

定义该函数曲线上各点的分数阶稳定系数为

$$\rho = \begin{cases} 1-v, & \text{when } \eta = 1 \\ v, & \text{when } \eta = -1 \end{cases} \quad (11.11)$$

式中，v 是分数阶微分的阶次。曲线上某点的分数阶稳定性系数 ρ 越高，该点越稳定。在图 11.2 中，A，B，C，D 点的稳定系数分别为 $\rho_A=0$，$\rho_B=0.25$，$\rho_C=0.5$，$\rho_D=0.75$。判定点 A，B，C，D 的稳定性依次升高，这与物理实际情况相一致。

另外，由图 11.2 还可知，对于信号的一阶微分等于零的地方，信号的分数阶微分一般不为零。于是可以自然地想到将关于分数阶稳定性系数的定义和结论推广到二维曲面上去。不难理解，由于一阶微分在包括局部极值点和全局最值点的所有驻点处为零，所以那些采用基于一阶梯度算子的各种学习训练算法就容易陷入局部极值点。与此相反，这些驻点处的分数阶微分却一般不为零。那么，采用基于分数阶梯度的学习训练算法，就可以很容易地跳出这些局部极值点。作者正试图将分数阶梯度算子和整数阶梯度算子相结合提出新的全局最优的学习训练算法。换言之，首先，运用分数阶微积分的阶次 v 来作为判断诸如自适应信号处理、神经网络、遗传算法训练收敛到极小点附近的指标。其次，构造分数阶梯度，提出基于分数阶微分的分数阶自适应信号处理理论。关于这个问题还有待进一步研究。

11.1.5　图像分数阶微积分的侧抑制原理分析

符合侧抑制原理是作者提出将分数阶微积分应用于二维图像处理的根本原因。理论和实验表明，对二维图像进行分数阶微积分可以很好地提取其纹理细节信息。这一点对于纹理丰富的二维图像，效果尤为突出[313]。要深刻理解二维图像分数阶微积分的原理及其算法规则，就必须首先从信号处理角度深刻理解它背后所蕴涵的神经视觉原理。

11.1.5.1　马赫带

1868 年，马赫（Ernst Mach）用一个黑白颜色组成的马赫圆盘做了关于马赫带（Mach band）现象的实验[213-219]，如图 11.3 所示。

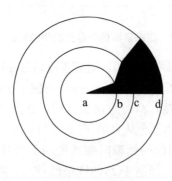

图 11.3　马赫圆盘

当圆盘快速旋转时,可以在圆盘三个不同的灰度区的交界处看到亮度比较分明的两条窄环。亮环在 b 点上,暗环在 c 点上,亮环和暗环称为马赫带或马赫环。在物理上,虽然 ab 和 cd 区的亮度都是均匀的,然而看起来 b 点却比 ab 区亮,而 c 点又比 cd 区暗,如图 11.4 所示。马赫带效应的存在说明生物的感觉变化并不完全对应于相应的光强变化。

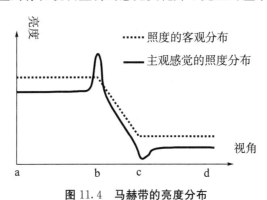

图 11.4 马赫带的亮度分布

11.1.5.2 侧抑制原理

1932 年,哈特兰和格雷厄姆在鲨眼上用微电极记录了单根神经纤维的电脉冲[314-320],其实验如图 11.5 所示。

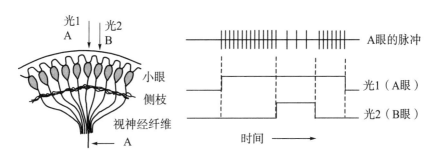

图 11.5 鲨眼的侧抑制作用示意图

当光照射鲨眼上的一个小眼 A 而引起兴奋时,再用光照射邻近的小眼 B,小眼 A 的脉冲发放频率就下降。这是由于小眼 B 的兴奋抑制了邻近的小眼 B 的兴奋,刺激小眼 A 也会抑制小眼 B 的兴奋。这种作用称为侧抑制现象。另外,哈特兰还发现,如果小眼 C 远离 A 而邻近 B,当小眼 B 对小眼 A 产生抑制作用时,再光照另一个小眼 C,则小眼 B 对小眼 A 的抑制便减弱了,这种作用称为去抑制现象。

令第 j 个细胞对第 p 个细胞的抑制作用大小用抑制系数 K_{pj} 表示,抑制阈值为 r_{pj}^0,e_p 是第 p 个感受细胞单独光照下的反应(以脉冲频率计量),r_p 是其他细胞受光照情况下第 p 个细胞的反应,其相互抑制作用描述如下式:

$$r_p = e_p - \sum_{j=1}^{n} K_{pj}(r_j - r_{pj}^0), p = 1,2,3,\cdots,n \tag{11.12}$$

只有当第 j 个细胞的兴奋值 r_j 超过 r_{pj}^0 时,才能对第 p 个细胞产生抑制作用。输入的亮度空间分布经过侧抑制网络处理之后,神经网络输出端相当于亮度大的一边有一个高峰,而亮度低的一边出现一个凹谷,从而使亮度的空间变化在神经网络中形成马赫效应。

11.1.5.3 视网膜神经节细胞感受野及其数学模型

1938年，哈特兰在蛙眼单根视神经纤维（即神经节细胞的轴突）上记录到电反应，并首次将一个神经元所反应（支配）的刺激区域定义为神经元的感受野（receptive field，又译为受纳野）[213-219]。

如图11.6所示，一个视神经节细胞和许多视觉感受器（光感受体）相连接，一条视神经纤维在视网膜上所支配的感受野一般是由中心的兴奋区和周边的抑制区所组成的同心圆结构，它们在功能上是互相拮抗的。

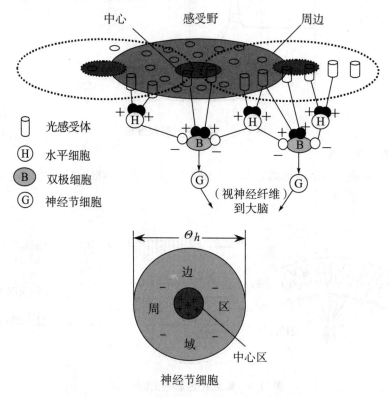

图11.6 视网膜神经节细胞感受野模式图

Rodieck在1965年提出了关于同心圆拮抗式感受野的数学模型。Rodieck模型又被称为高斯差模型（difference of two Gaussians，简记为DOG式）。它还被称为中心—周边拮抗式的Rodieck DOG经典感受野模型，简称经典DOG感受野模型[314-320]，如图11.7所示。它由一个兴奋作用强的中心机制和一个作用较弱但面积更大的抑制性周边机制所构成。这两个相互拮抗的兴奋—抑制机制在生理学中又称颉颃作用，即神经网络的侧抑制作用。它们都具有高斯分布的性质，但中心机制具有更高的峰值敏感度，而二者方向相反，故是相减的关系。

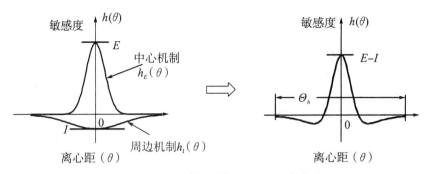

图 11.7 视网膜神经节细胞 Rodieck 感受野模型

令视神经节细胞感受野距其中心点的视角大小（离心距）为 θ，用 $h_E(\theta)$ 描述中心的兴奋机制（由参量 E 和 σ_E 来控制），用 $h_I(\theta)$ 描述周边的抑制性机制（由参量 I 和 σ_I 来控制），用 $h(\theta)$ 描述一个视神经节细胞的同心圆感受野的拮抗总体特性，它们之间的数学关系为

$$h(\theta) = h_E(\theta) - h_I(\theta) = E\exp\left(-\frac{1}{2}\left|\frac{\theta}{\sigma_E}\right|^2\right) - I\exp\left(-\frac{1}{2}\left|\frac{\theta}{\sigma_I}\right|^2\right) \quad (11.13)$$

如图 11.7 所示，经典 DOG 感受野模型（11.13）是偶对称的，即

$$h(\theta) = h(-\theta) \quad (11.14)$$

在数学上，虽然式（11.13）所表示的模型是非紧支撑的，但它却是一个快速衰减的函数。因此，截断其拖尾可得到紧支撑函数来刻画细胞的紧支撑感受野的敏感性分布特征。一般用感受野视角宽度 Θ_h 来表征同心圆式感受野大小。其实，感受野视角宽度 Θ_h 就是图 11.6 中所画同心圆式拮抗感受野的最外圆周的视角直径，即

$$\Theta_h : 6 \sim 8\max\{\sigma_E, \sigma_I\} \quad (11.15)$$

11.1.5.4　侧抑制原理与边缘提取的数学模型

20 世纪 70 年代末，Levine 和 Shefner 做了一个实验，用亮暗对比边缘刺激图形落在给光—中心神经节细胞感受野不同部位时，检测其神经纤维上的放电频率的变化情况来研究神经节细胞的反应[314−321]，实验结果如图 11.8 所示。

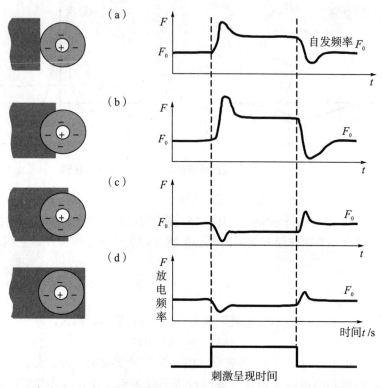

图 11.8 亮暗对比边缘刺激图形落在给光—中心神经节细胞感受野不同部位时的反应

图 11.8 中,左边是亮暗对比边缘刺激图形落在感受野不同部位的情形,右边是对应神经节细胞反应放电频率变化。图 11.8 (a) 为对比边刺激物位于感受野之外时,该感受野全部受到光的照射的情况。图 11.8 (b) 为对比边刺激物位置移动到使感受野中心全部受光照射,而感受野周边仅有一部分受到光照射的情况。此时对比边刺激能引起神经节细胞比图 11.8 (a) 更为强烈的反应。这就说明神经节细胞对某个特定位置的对比边缘刺激的反应要比同样光强度的覆盖整个感受野的弥散光所引起的反应更强。图 11.8 (c) 为感受野只有周边一部分受到光刺激的情况,其反应就完全不同于图 11.8 (a) 和 (b) 两种情况。它在刺激开始的瞬间反应为抑制性,在刺激撤去的瞬间反应为兴奋性,而且整个反应幅度都比较小。图 11.8 (d) 则是整个感受野全部没入阴影之中的情况。可见,在时间上,感受野对刺激的瞬间变化很敏感,这表现在给光与撤光瞬间,发放频率有很大的起伏;在空间上,感受野对明暗对比边的刺激特别敏感,这表现在刺激呈现的平稳时期(即除去给光和撤光的两头时期)。神经节细胞的这种对于空间明暗对比边特别敏感的性质,就可以用来解释马赫带现象[314-321]。

在图 11.9 的 a 位置上,$\theta = \theta_a$,令其对应的感受野敏感性为 $h(\theta - \theta_a)$,其附近区域的物理光强是 $s(\theta_a)$,故该处感受野所对应的神经节细胞的主观知觉光强为

$$\begin{aligned}
y(\theta_a) &= \int_\theta s(\theta) h(\theta - \theta_a) \mathrm{d}\theta \\
&= s(\theta_a) \int_\theta h(\theta) \mathrm{d}\theta \\
&= s(\theta_a) S_h
\end{aligned} \quad (11.16)$$

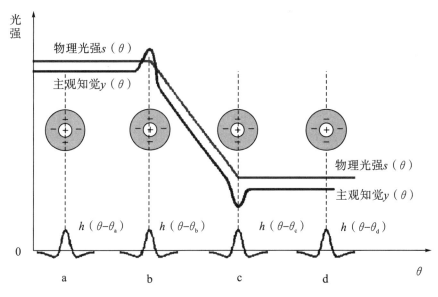

图 11.9 马赫带现象的生理学基础

人类的感受野是给光—中心兴奋型：$0<S_h<1$，故有
$$y(\theta_a) < s(\theta_a) \tag{11.17}$$

同理，在 d 位置上，$\theta=\theta_d$，令其感受野敏感性为 $h(\theta-\theta_d)$，其对应的主观知觉光强为
$$y(\theta_d) = \int_\theta s(\theta)h(\theta-\theta_d)\mathrm{d}\theta = s(\theta_d)S_h < s(\theta_d) \tag{11.18}$$

在 b 位置上，$\theta=\theta_b$，当感受野区域宽度小于物理光强线性变化的区域时，有

$$\begin{aligned}
y(\theta_b) &= \int_\theta s(\theta)h(\theta-\theta_b)\mathrm{d}\theta \\
&= \int_{-\infty}^{\theta_b} s(\theta_b)h(\theta-\theta_b)\mathrm{d}\theta + \int_{\theta_b}^{+\infty}[-K(\theta-\theta_b)+s(\theta_b)]h(\theta-\theta_b)\mathrm{d}\theta \\
&= \int_{-\infty}^{0} s(\theta_b)h(\theta)\mathrm{d}\theta + \int_{0}^{+\infty}[-K\theta+s(\theta_b)]h(\theta)\mathrm{d}\theta \\
&= s(\theta_b)S_h - K\int_{0}^{+\infty}\theta h(\theta)\mathrm{d}\theta
\end{aligned} \tag{11.19}$$

式中，K 是图 11.9 中物理光强线性变化的（绝对）斜率大小，即
$$K = \frac{s(\theta_b)-s(\theta_c)}{\theta_c-\theta_b} > 0 \tag{11.20}$$

而 $\gamma_h = \int_0^{\theta_h/2}\theta h(\theta)\mathrm{d}\theta = \int_0^\infty \theta h(\theta)\mathrm{d}\theta = E\sigma_E^2 - I\sigma_I^2$，于是可推得：

$$\begin{aligned}
y(\theta_b) &= s(\theta_b)S_h + K\frac{S_h}{\sqrt{\pi}} = s(\theta_a)S_h + K\gamma_h \\
&= y(\theta_a) + \delta_p > y(\theta_a)
\end{aligned} \tag{11.21}$$

式中，

$$\delta_p = K\gamma_h = \frac{s(\theta_b) - s(\theta_c)}{\theta_c - \theta_b}(I\sigma_I^2 - E\sigma_E^2) \tag{11.22}$$

将 δ_p 称为上冲峰高（overshoot peak）。

同理，在 c 位置上，$\theta = \theta_c$，我们有

$$\begin{aligned} y(\theta_c) &= s(\theta_c)S_h - K\gamma_h \\ &= y(\theta_d) - \delta_v < y(\theta_d) \end{aligned} \tag{11.23}$$

式中，

$$\delta_v = K\gamma_h = \frac{s(\theta_b) - s(\theta_c)}{\theta_c - \theta_b}(I\sigma_I^2 - E\sigma_E^2) \tag{11.24}$$

将 δ_v 称为下凹谷深（downdrop valley）。

对比式（11.22）和式（11.24）知，在理论上马赫带的上冲峰高 δ_p 同下凹谷深 δ_v 的幅度大小是一样的，他们都正比于物理光强线性变化斜率 K 和感受野的拮抗特性，即

$$\gamma_h = E\sigma_E^2 - I\sigma_I^2 \tag{11.25}$$

式（11.16）、（11.18）、（11.19）、（11.23）的统一表达式为

$$y(\vartheta) = \int_\theta s(\theta)h(\theta - \vartheta)d\theta \tag{11.26}$$

对于经典 DOG 感受野模型而言，由式（11.15），式（11.26）可写成 $y(\vartheta) = \int_\theta s(\theta)h(\vartheta - \theta)d\theta$，再作变量代换得

$$y(\theta) = \int_\vartheta s(\vartheta)h(\theta - \vartheta)d\vartheta = s(\theta) * h(\theta) \tag{11.27}$$

这是一个卷积运算。

对于一维信号的卷积，可以将其自变量视为时间变量，不失一般性，式（11.27）改写为

$$y(t) = \int_\tau s(\tau)h(t - \tau)d\tau = s(t) * h(t) \tag{11.28}$$

这样就能够用信号处理的观点来分析和理解马赫带问题了。

11.1.5.5 图像信号分数阶微积分的拮抗特性与纹理细节提取

由第 3 章论述可知，对于任意平方可积的能量型信号 $s(t) \in L^2(\mathbf{R})$，其 Fourier 变换为 $\hat{s}(\omega) = \int_R s(t)\exp(-i\omega t)dt$。根据 Fourier 变换的性质[58,60]，其一阶导数 $s'(t)$ 在频率域的等价形式为 $\boldsymbol{D}s(t) \overset{FT}{\Leftrightarrow} (\hat{\boldsymbol{D}}s)(\omega) = (i\omega) \cdot \hat{s}(\omega) = \hat{d}(\omega)\hat{s}(\omega)$。推而广之可得 v 阶分数阶微分在 Fourier 变换域内的等价形式为

$$\begin{cases} \boldsymbol{D}_v s(t) = \dfrac{d^v s(t)}{dt^v}, & v \in \mathbf{R}^+ \\ \qquad \Updownarrow \\ (\boldsymbol{D}_v \hat{s})(\omega) = (i\omega)^v \cdot \hat{s}(\omega) = \hat{d}_v(\omega) \cdot \hat{s}(\omega), & v \in \mathbf{R}^+ \end{cases} \tag{11.29}$$

v 阶微分算子 $\boldsymbol{D}_v = \boldsymbol{D}^v$ 是 v 阶微分乘子函数 $\hat{d}(\omega) = (i\omega)^v$ 的乘性算子，在复数域中其指数形式为

第11章 图像处理的分数阶微分算子：分数阶微分掩模

$$\begin{cases} \hat{d}(\omega) = (i\omega)^v = \hat{a}_v(\omega) \cdot \exp(i\theta_v(\omega)) = \hat{a}(\omega) \cdot \hat{p}_v(\omega) \\ \hat{a}_v(\omega) = |\omega|^v, \hat{\theta}_v(\omega) = \dfrac{v\pi}{2}\mathrm{sgn}(\omega) \end{cases} \quad (11.30)$$

由式（11.30）可知，从通信调制角度看，作者认为信号分数阶微分的物理意义可以理解为广义的调幅调相，其振幅随频率呈分数阶幂指数变化，其相位是频率的广义 Hilbert 变换。

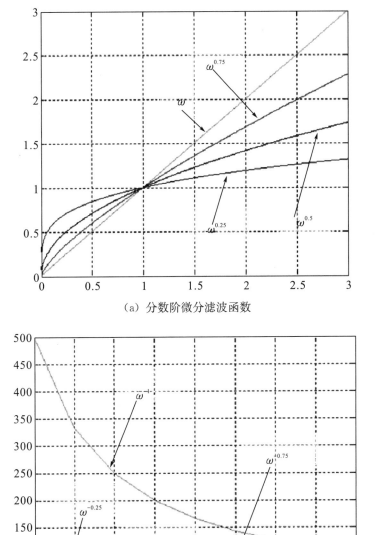

(a) 分数阶微分滤波函数

(b) 分数阶积分滤波函数

图 11.10 分数阶微积分滤波函数为 $|\hat{d}_v(\omega)| = \omega^v$

由图 11.10 可知，从信号处理角度看，v 阶微积分运算其实是对信号进行线性时不变滤波，其滤波函数为 $\hat{d}_v(\omega)=(\mathrm{i}\omega)^v=|\omega|^v \cdot \exp(\mathrm{i}\theta_v(\omega))$。于是可以得到如下结论：

(1) 当 $v=0$ 时，它是全通滤波器，$\hat{d}_v(\omega)\equiv 1 \Leftrightarrow d_v(t)=\delta(t)$。

(2) 当 $v<0$ 时，它是积分器，$\lim\limits_{|\omega|\to 0}|\hat{d}_v(\omega)|\to\infty$，$\hat{d}_v(\omega)$ 是奇异低通滤波。$|v|$ 越大，低通特性越明显，突出了信号的低频成分，压制了高频成分，所以积分后信号变得更加平滑，这有利于观测信号的总体变化趋势，但抹去了信号的变化细节。相对而言，微分滤波器振幅特性的通频带较宽，而积分滤波器振幅特性的通频带较窄。

(3) 当 $v>0$ 时，是对信号作微分运算，$\lim\limits_{|\omega|\to 0}|\hat{d}_v(\omega)|\to\infty$，$\hat{d}_v(\omega)$ 是奇异高通滤波器。v 越大，通频带越窄，高通特性越明显，相应加强了信号 $s(t)$ 的高频成分，相对压制了其低频成分，这有利于突出信号的细节，但抗高频干扰成分性能较差。

(4) 当 $0<v<1$ 时，在 $\omega>1$ 段，分数阶微分对于信号高频成分的幅度加强没有一阶微分大。这就意味着，在 $\omega>1$ 段，图像信号的分数阶微分没有一阶微分或二阶微分对于图像边缘的幅度加强大。但在接近于零频 $0<\omega<1$ 这样的甚低频段，信号的分数阶微分不像一阶微分对信号的甚低频段成分进行大幅的线性衰减，而是进行一种非线性衰减，其幅度衰减没有一阶微分那么大。在 $0<\omega<1$ 甚低频段，随着微分阶数 v 的减小，分数阶微分对信号的甚低频段成分的幅度衰减越小，当 $v\to 0$ 时，不进行任何衰减。所以，对于图像灰度变化不大的平滑区域而言，若采用诸如基于一阶微分的 Sobel 算子去处理，那么经过一阶微分后图像平滑区域中的灰度变化不大的纹理细节信息必然会遭到大幅的线性衰减，其结果约等于零，于是就不能很好地检测出图像平滑区域的纹理细节信息。与此相反，经过分数阶微分处理后图像平滑区域中灰度变化不大的纹理细节信息没有遭到大幅的线性衰减，反而在一定程度上进行了非线性的保留。这便是分数阶微分比整数阶微分更有利于提取图像平滑区域中的纹理细节信息在信号处理方面的本质解释。

式 (11.30) 对应的时域形式为

$$d_v(t)=a_v(t)*p_v(t)=\frac{1}{2\pi}\int_{-\infty}^{+\infty}(\mathrm{i}\omega)^v \cdot \mathrm{e}^{\mathrm{i}\omega t}\mathrm{d}\omega \tag{11.31}$$

$$a_v(t)=\int_{-\infty}^{+\infty}\hat{a}(\omega)\cdot \mathrm{e}^{\mathrm{i}\omega t}\mathrm{d}\omega=\frac{1}{\pi}\int_{0}^{+\infty}|\omega|^v \cdot \cos(\mathrm{i}\omega t)\mathrm{d}\omega \tag{11.32}$$

$$p_v(t)=\frac{1}{2\pi}\int_{-\infty}^{+\infty}\hat{p}_v(\omega)\cdot \mathrm{e}^{\mathrm{i}\omega t}\mathrm{d}\omega=\cos\frac{v\pi}{2}\cdot\delta(t)-\sin\frac{v\pi}{2}\cdot\frac{1}{\pi t} \tag{11.33}$$

若把 $d_v(t)$ 视为滤波器的冲击响应，由式 (11.32)、(11.33) 知，微分滤波器的振幅特性是偶函数，相位特性是奇函数。因此，我们只需讨论 $\omega>0$ 时微积分滤波器的特性即可。显然，信号 $s(t)$ 的分数阶微积分在时域中的卷积形式为

$$\boldsymbol{D}_v s(t)=d_v(t)*s(t)=a_v(t)*p_v(t)*s(t)$$

$$=\left[\cos\frac{v\pi}{2}\cdot\delta(t)-\sin\frac{v\pi}{2}\cdot\frac{1}{\pi t}\right]*\int_{-\infty}^{+\infty}a_v(t-\tau)s(\tau)\mathrm{d}\tau \tag{11.34}$$

由上述可知，Rodieck 高斯差经典感受野模型 (difference of two Gaussians) 是偶对称中心—周边拮抗式。数字图像处理中经典的基于一阶微分的 Sobel 算子、Prewitt 算子和 Roberts 算子，以及基于二阶微分的 Laplace 算子，乃至基于子波变换的墨西哥帽算子

和 Bubble 算子，实际上就是对 Rodieck 感受野模型在数学上进行仿生实现[306-312]。受此启发，作者很自然地想到能否找到一种更有利于分析图像纹理细节的 Rodieck 感受野模型，从而构造一个基于分数阶微分的二维图像算子来更好地加强富于纹理细节图像中的纹理细节信息，找出其中高度自相似的分形信息[313]。

令同心圆感受野的拮抗总体特性为 $h(t)$，由式（11.28）知，t 时刻的主观知觉光强 $y(t)$ 是 t 时刻的物理光强 $s(t)$ 与拮抗总体特性 $h(t)$ 的卷积，即

$$y(t) = \int_\tau s(\tau)h(t-\tau)\mathrm{d}\tau = s(t)*h(t) \tag{11.35}$$

对比式（11.34）和（11.35）知，如果对图像信号进行分数阶微积分，其滤波器的冲击响应 $d_v(t)$ 等于它所对应的仿生感受野的拮抗总体特性 $h(t)$，故有

$$h(t) = d_v(t) \tag{11.36}$$

如上所述，当 $0<v<1$ 时，在 $\omega>1$ 段，分数阶微分对于信号高频成分的幅度加强没有一阶微分大。但在接近于零频 $0<\omega<1$ 这样的甚低频段，信号的分数阶微分不像一阶微分那样对信号的甚低频段成分进行大幅的线性衰减，而是进行一种非线性衰减，其衰减的幅度没有一阶微分那么大。所以，对于 $0<v<1$ 阶分数阶微分而言，图像分数阶微分对应的仿生 Rodieck 感受野模型的中心机制 $h_E^{fra}(t)$ 应该比整数阶微分所对应的中心机制 $h_E^{int}(t)$ 在中心位置幅度更小，同时，图像分数阶微分所对应的周边机制 $h_I^{fra}(t)$ 比整数阶微分所对应的周边机制 $h_I^{int}(t)$ 中心幅度更小，拖的尾巴更长，紧支撑性相对变弱。所以，图像分数阶微分所对应的拮抗总体特性 $h^{fra}(t)$ 比整数阶微分所对应的拮抗总体特性 $h^{int}(t)$ 中心幅度更小，拖的尾巴更长，紧支撑性相对变弱。图像分数阶微分所对应的感受野视角宽度 Θ_h^{fra} 大于整数阶微分所对应的感受野视角宽度 Θ_h^{int}[313]，于是有

$$\Theta_h^{fra} > \Theta_h^{int} \tag{11.37}$$

分数阶微分与整数阶微分的仿生 Rodieck 感受野模型对比如图 11.11 所示。

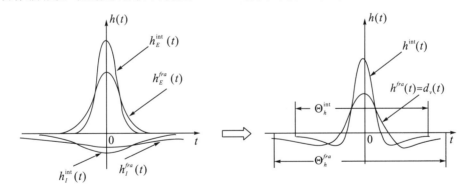

图 11.11 分数阶微分与整数阶微分的仿生 Rodieck 感受野模型对比

如式（11.34）、（11.35）、（11.36）和（11.37），用分数阶微分所对应的仿生 Rodieck 感受野模型对空间明暗对比边的处理结果与用整数阶微分对应的仿生 Rodieck 感受野模型的处理结果相对比，其产生不同的马赫带现象如图 11.12 所示。

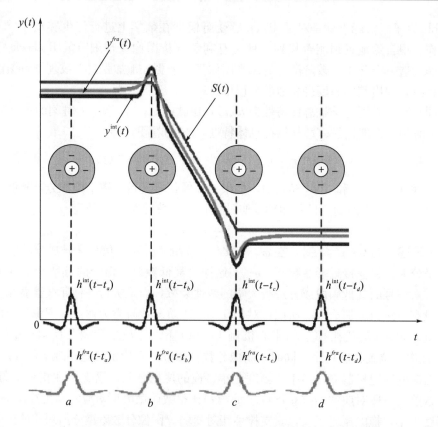

图 11.12 分数阶微分和整数阶微分对应仿生 Rodieck 感受野模型所产生马赫带现象对比

结合图 11.12，对于图像分数阶微分对应的仿生 Rodieck 感受野模型可以得到以下结论：

(1) 由于图像分数阶微分所对应的拮抗总体特性 $h^{fra}(t)$ 比整数阶微分所对应的拮抗总体特性 $h^{int}(t)$ 的中心幅度更小，所以按照式（11.35）卷积后，在空间上，分数阶微分所对应的仿生感受野对明暗对比边所产生的上冲峰高（overshoot peak）δ_p^{fra} 小于整数阶微分所对应的仿生感受野所产生的上冲峰高 δ_p^{int}，即 $\delta_p^{fra} < \delta_p^{int}$；同理，下凹谷深（downdrop valley）$\delta_v^{fra} < \delta_v^{int}$。换言之，图像分数阶微分没有整数阶微分对图像中灰度变化较大的高频边缘成分敏感，即图像分数阶微分对图像中灰度变化较大的高频边缘成分产生的马赫带现象没有整数阶微分明显。

(2) 除上冲峰和下凹谷处外，分数阶微分仿生感受野所产生的主观知觉光强 $y^{fra}(t)$ 一般比整数阶微分产生的 $y^{int}(t)$ 大，它们都比物理光强 $s(t)$ 小。这说明不论是对图像作分数阶微分还是作整数阶微分处理，图像的整体平均灰度都会下降，图像整体上都会变暗，但分数阶微分使图像变暗的程度没有整数阶微分强。

(3) 如果进行线性卷积的两个连续信号持续期有限，其线性卷积结果所得连续信号的持续期为相卷积的两连续信号的持续期之和（离散信号的点数则为相卷积的两信号的点数之和减1）。由式（11.37）知 $\Theta_h^{fra} > \Theta_h^{int}$，所以当分数阶微分仿生感受野产生上冲峰高时，会经历一个缓慢爬坡上升的过程。当整数阶微分仿生感受野产生上冲峰高时，是一个迅速陡然爬坡上升的过程；当分数阶微分仿生感受野释放下凹谷深时，也会经历一个缓慢的释放过程。当整数阶微分仿生感受野释放下凹谷深时,却会来得陡然而迅捷;同样,当分数阶微分仿生感受野释放上冲峰高和产生下凹谷深时,都比整数阶微分仿生感受野平缓得多。换言之,分数阶微分对于图像灰度值变化不大的平滑区域中的纹理细节信息的加强幅度比整数阶微分更大。图像分数阶微分对图像平滑区域的处理是在极大保留其低频轮廓信息的基础上,非线性加强其中灰度值跃变幅度和频率相对不大的高频纹理细节信息。所以,图像分数阶微分既保留了图像低频轮廓信息,加强了灰度值跃变幅度较大的高频边缘信息,同时还加强了灰度值跃变幅度较小的高频纹理细节信息;而图像整数阶微分在加强灰度值跃变幅度较大的高频边缘信息的同时,极大压制了低频轮廓信息和灰度值跃变幅度较小的高频纹理细节信息。这便是基于分数阶微分的图像算子比基于整数阶微分的图像算子更有利于分析和强化图像纹理细节信息在生物视觉侧抑制原理方面的本质解释。

如前所述,人类的感受野是给光—中心兴奋型,即人类的感受野是偶对称的。但是,并不是所有动物的感受野都是偶对称的。自然进化,适者生存,老鹰为了能够在高空中快速定位地上快速奔跑的猎物,猫为了迅速跟踪高速逃窜的老鼠,它们的视觉感受器为了有利于去除摄取图像的运动模糊,长期自然进化使它们的感受野是不对称的,如图 11.13 所示。

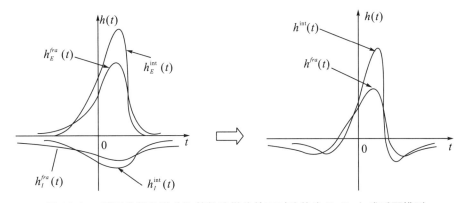

图 11.13 基于分数阶微分和整数阶微分的不对称仿生 Rodieck 感受野模型

分析图 11.13 可知,基于分数阶微分的不对称仿生 Rodieck 感受野模型不仅继承了对图像纹理细节信息的良好分析能力,而且继承了基于整数阶微分的不对称仿生 Rodieck 感受野模型消除图像运动模糊的能力。因此,对于连续二维图像而言,构造不对称的二维分数阶微分滤波器;对于二维数字图像而言,构造不对称的基于分数阶微分的掩模(模板),对消除图像中的运动模糊具有特殊的理论和现实意义。怎样构造出基于不对称 Rodieck 感受野模型的分数阶微分图像运算掩模(模板)算子?如何将这种不对称的图像分数阶微分应用到汽车的车牌的去运动模糊之中?如何用光学器件实现连续图像的对称和不对称分数阶微分?作者还在做进一步的研究。

11.1.6 二维数字图像分数阶微分的数值实现

11.1.6.1 理论分析

要构造出二维图像分数阶微分的数值实现算法,必须首先深刻理解信号分数阶微分对信号作用的规律。不失一般性,本章以锯齿波和方波信号的 $v=0.5$ 阶微分为例进行研究，其实验结果如图 11.14、图 11.15 所示。

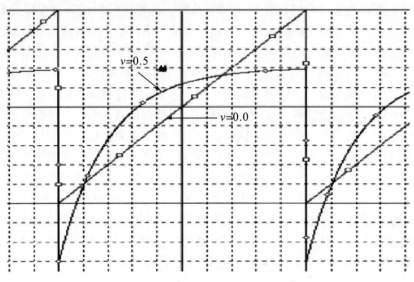

图 11.14　锯齿波信号的 $v=0.5$ 阶微分

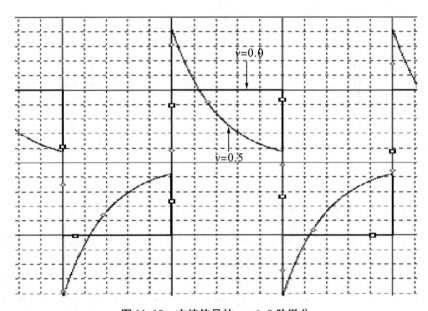

图 11.15　方波信号的 $v=0.5$ 阶微分

分析图 11.14、图 11.15 可知，尽管数学上函数的分数阶微分可以有不同的定义，但是各种分数阶微分定义应该满足以下几点性质[313]：

(1) 在平坦段（图像灰度值不变的区域）的分数阶微分值是由对应奇异跳变处的极大值渐趋于零（Riemann-Liouville 定义除外），而平坦段的任意整数阶微分必为零。这是分数阶微分相对于整数阶微分的显著区别之一。

(2) 在灰度阶梯或斜坡的起始点处的分数阶微分值非零，起到了加强高频信息的作用。

(3) 沿斜坡的分数阶微分值非零，亦非常数，而沿斜坡的整数阶微分值为常数。

第 11 章 图像处理的分数阶微分算子:分数阶微分掩模

可见,分数阶微分在加强图像信号高频边缘和纹理细节信息的同时,也保留了部分低频轮廓信息。因此,与整数阶微分相比,用分数阶微分处理图像信号,既有利于提取边缘以及纹理细节,又有利于保持轮廓。对于纹理丰富的图像信号,包含高度自相似的分形信息,在进行图像纹理细节加强处理时,用分数阶微分在理论上优于用整数阶微分。另外,由于分数阶微分在图像灰度平坦的区域不一定为零,所以图像进行分数阶微分后,会得到比整数阶微分相对较粗的边缘过渡。

由第 3 章的推导可知,Grümwald-Letnikov 定义是从研究连续函数整数阶导数的经典定义出发,将微积分的阶数与因次由整数扩大到分数推衍而来的。$\forall v \in \mathbf{R}$(包括分数),令其整数部分为 $[v]$,若函数或信号 $s(t) \in [a, t]$ ($a < t$, $a \in \mathbf{R}$, $t \in \mathbf{R}$) 存在 $m+1$ ($m \in \mathbf{Z}$) 阶连续导数。当 $v > 0$ 时,m 至少取 $[v]$,则定义 v 阶导数为

$${}_a^G\mathbf{D}_t^v s(t) \triangleq \lim_{h \to 0} s_h^{(v)}(t) \triangleq \lim_{\substack{h \to 0 \\ nh = t-a}} h^{-v} \sum_{r=0}^{n} \begin{bmatrix} -v \\ r \end{bmatrix} s(t-rh) \tag{11.38}$$

式中,$\begin{bmatrix} -v \\ r \end{bmatrix} = \dfrac{(-v)(-v+1)\cdots(-v+r-1)}{r!}$。若将组合数 $\begin{pmatrix} g \\ r \end{pmatrix} = \dfrac{(g)(g-1)\cdots(g+r-1)}{r!}$ 中 g 扩展为任意实数(包括分数),则 $\begin{pmatrix} -g \\ r \end{pmatrix} = (-1)^r \begin{bmatrix} g \\ r \end{bmatrix}$。

为使 $s_h^{(-v)}(t)$ 达到非零极限,须当 $h \to 0$ 时,$n \to \infty$,故令 $h = \dfrac{t-a}{n}$,于是 $n = \left[\dfrac{t-a}{h}\right]$。

对式 (11.38) 先运用数学归纳法,再作分部积分,得

$${}_a^G\mathbf{D}_t^v s(t) = \sum_{k=0}^{m} \frac{s^{(k)}(a)(t-a)^{-v+k}}{\Gamma(-v+k+1)} + \frac{1}{\Gamma(-v+m+1)} \int_a^t (t-\tau)^{-v+m} s^{(m+1)}(\tau) d\tau \tag{11.39}$$

式中,Gamma 函数 $\Gamma(\alpha) = \int_0^\infty e^{-x} x^{\alpha-1} dx = (\alpha-1)!$。

根据式 (11.38),若一元信号 $s(t)$ 的持续期为 $t \in [a, t]$,将信号持续期 $[a, t]$ 按单位等分间隔 $h = 1$ 进行等分,则 $n = \left[\dfrac{t-a}{h}\right]^{h=1} = [t-a]$。可以推导出一元信号 $s(t)$ 分数阶微分的差值表达为[313]

$$\frac{d^v s(t)}{dt^v} \approx s(t) + (-v)s(t-1) + \frac{(-v)(-v+1)}{2}s(t-2)$$
$$+ \frac{(-v)(-v+1)(-v+2)}{6}s(t-3) + \cdots + \frac{\Gamma(-v+1)}{n!\Gamma(-v+n+1)}s(t-n) \tag{11.40}$$

高维分数阶 Fourier 变换是可以分离的(其变换核是可以分离的)[322],即

$$F^{p_x, p_y} s(x, y) = F^{p_y} F^{p_x} s(x, y) = F^{p_x} F^{p_y} s(x, y) \tag{11.41}$$

由式 (11.8) 知,信号 $s(t)$ 的 v 阶分数阶微积分 $s^{(v)}(t)$ 可由 $s(t)$ 的 $p \to 0$ 阶分数阶 Fourier 变换 $s_F^p(u)$ 在时间 $\tau = t \sec(\alpha)$ 处的正态补偿,所以二维图像信号 $s(x, y)$ 对 (x, y) 的分数阶微分在一定条件下也应该是可以分离的。具体需要什么样严格的数学限制条件二维分数阶微分才可分离,还需要严格的数学推导,作者还在做进一步的深入研究。但

是，可以假设常见的灰度取值有限的二维实图像信号的二维 $0<v<1$ 阶分数阶微分是可以分离的。所以，本章姑且将二维数字图像信号 $s(x,y)$ 的分数阶微分视为沿两个空间轴分别进行偏分数阶微分。

由于计算机处理的是数字量，其值有限，图像信号灰度的最大变化量是有限的，二维数字图像信号灰度变化发生的最短距离是在两相邻像素之间，因此二维数字图像在 x 和 y 轴方向上的持续时间（图像矩阵的规模）只可能以像素为单位进行度量，$s(x,y)$ 的最小等分间隔只可能是 $h=1$。若二维数字图像信号中 x 和 y 的持续期分别为 $x\in[x_1, x_2]$ 和 $y\in[y_1, y_2]$，则最大等分数分别为 $n_x=\left[\dfrac{x_2-x_1}{h}\right]\stackrel{h=1}{=}[x_2-x_1]$ 和 $n_y=\left[\dfrac{y_2-y_1}{h}\right]\stackrel{h=1}{=}[y_2-y_1]$；若二维数字图像是一个正方阵，则 $n=n_x=n_y$。可见，对于数字图像而言，即使分数阶微分掩模的尺度大到等于数字图像本身的尺度，也只可能是对其分数阶微分解析值的最大逼近，而永远不可能等于其分数阶微分的解析值。本章以构造 3×3 图像分数阶微分掩模为例进行推导说明。针对数字图像处理，本章提出并定义 $s(x,y)$ 的偏分数阶微分为[313]

$$\frac{\partial^v s(x,y)}{\partial x^v}\approx s(x,y)+(-v)s(x-1,y)+\frac{(-v)(-v+1)}{2}s(x-2,y) \quad (11.42)$$

$$\frac{\partial^v s(x,y)}{\partial y^v}\approx s(x,y)+(-v)s(x,y-1)+\frac{(-v)(-v+1)}{2}s(x,y-2) \quad (11.43)$$

在 x 轴方向上的相对逼近误差为

$$\varepsilon_x^v s(x,y)=\frac{(-v)(-v+1)(-v+2)}{6}s(x-3,y)+\cdots+\frac{\Gamma(-v+1)}{n!\Gamma(-v+n+1)}s(x-n,y)$$

$$(11.44)$$

在 y 轴方向上的相对逼近误差为

$$\varepsilon_y^v s(x,y)=\frac{(-v)(-v+1)(-v+2)}{6}s(x,y-3)+\cdots+\frac{\Gamma(-v+1)}{n!\Gamma(-v+n+1)}s(x,y-n)$$

$$(11.45)$$

需要说明的是，式（11.44）和（11.45）中的相对逼近误差不是相对于图像分数阶微分的解析解，而是相对于分数阶微分掩模的尺度大小等于二维数字图像的尺度时的数值解（对于其解析解的最佳逼近）。

在图像像素点 (x,y) 上的分数阶梯度是通过一个二维列向量定义的，即

$$\nabla^v s=\begin{bmatrix}G_x^v\\G_y^v\end{bmatrix}=\begin{bmatrix}\dfrac{\partial^v s}{\partial x^v}\\[2mm]\dfrac{\partial^v s}{\partial y^v}\end{bmatrix} \quad (11.46)$$

定义分数阶梯度向量的模值为

$$mag(\nabla^v s)=\left[G_x^{v^2}+G_y^{v^2}\right]^{\frac{1}{2}} \quad (11.47)$$

尽管分数阶梯度本身是个线性算子，由于用到了平方和开方，所以分数阶梯度向量的模值显然不是线性的。另外，式（11.46）中定义的分数阶梯度向量不是旋转不变的（各向同性），但是分数阶梯度向量的模值却是各向同性的。本章定义在实际操作中，用绝对

值来代替平方根运算,近似求分数阶梯度的模值,即

$$mag(\nabla^v s) = |G_x^v| + |G_y^v| \tag{11.48a}$$

这样就简单地保持了灰度的相对变化,并且各向同性的特性一般就不存在了。

由式(11.42)、(11.43),作者分别定义在 x,y 方向上的分数阶微分掩模[313]如图 11.16 所示。

0	$\frac{v^2-v}{2}$	0
0	$-v$	0
0	1	0

0	0	0
$\frac{v^2-v}{2}$	$-v$	1
0	0	0

(a) x 方向上 (b) y 方向上

图 11.16 x,y 方向上的分数阶微分掩模

显然,用图 11.16 中定义的分数阶微分掩模处理数字图像具有 π 角旋转的各向同性结果。计算机数字图像处理是以对图像的离散像素直接处理为基础的,这种空间滤波的机理就是在待处理的数字图像中逐点移动掩模(模板)。针对这种定义的分数阶微分掩模的运算规则如图 11.17 所示。

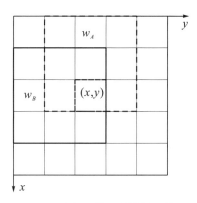

图 11.17 分数阶微分掩模的运算规则

图 11.17 中,掩模 w_A,w_B 分别是 x 方向上和 y 方向上的分数阶微分掩模。在 x 方向上和 y 方向上的运算规则分别为

$$g_x^v(x,y) = \sum_{z=-a}^{a}\sum_{k=-2b}^{0} w_A(z,k)s(x+z,y+k) \tag{11.48b}$$

$$g_y^v(x,y) = \sum_{z=-2a}^{0}\sum_{k=-b}^{b} w_B(z,k)s(x+z,y+k) \tag{11.48c}$$

显然,依照这种分数阶微分掩模的定义,其运算规则处理较为复杂。为了简化计算和便于处理,作者重新定义图像信号 $s(x,y)$ 的偏分数阶微分为[313]

$$\frac{\partial^v s(x,y)}{\partial x^v} \approx (-v)s(x+1,y) + s(x,y) + \frac{(-v)(-v+1)}{2}s(x-2,y) \tag{11.49}$$

$$\frac{\partial^v s(x,y)}{\partial y^v} \approx (-v)s(x,y+1) + s(x,y) + \frac{(-v)(-v+1)}{2}s(x,y-2) \quad (11.50)$$

相应的 x，y 方向上新的分数阶微分掩模如图 11.18 所示。

0	$\frac{v^2-v}{2}$	0
0	1	0
0	$-v$	0

(a) x 方向上

0	0	0
$\frac{v^2-v}{2}$	1	$-v$
0	0	0

(b) y 方向上

图 11.18 x，y 方向上新的分数阶微分掩模

相应的，针对新的分数阶微分掩模，其新的运算规则如图 11.19 所示。

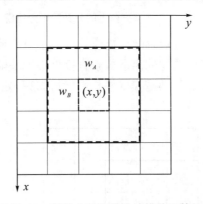

图 11.19 新的分数阶微分掩模的运算规则

在 $M \times N$ 的数字图像 $s(x,y)$ 中，用滤波器掩模 $w_{m \times n}$ 进行线性滤波，在 x 方向上和 y 方向上运算规则均为

$$g_{x\,or\,y}^v(x,y) = \sum_{z=-a}^{a} \sum_{k=-b}^{b} w_{A\,or\,B}(z,k) s(x+z, y+k) \quad (11.51)$$

为了 $w_{m \times n}$ 有明确的中心点，令 $m = 2a+1$，$n = 2b+1$ 为奇数。故 $a = (m-1)/2$，$b = (n-1)/2$。为了得到一幅完整的经过滤波处理的图像，必须对 $x = 0, 1, \cdots, M-1$，$y = 0, 1, \cdots, N-1$ 依次应用上面的公式。对于 $w_{m \times n}$ 的方形掩模，当掩模中心点距图像边缘小于 $(n-1)/2$ 个像素时，掩模的行或列就会处于图像平面之外。为了获得最佳的滤波效果，本章在运算时，使滤波掩模中心距原图像边缘的距离不小于 $(n-1)/2$ 个像素，这样就未对距原图像边缘 $(n-1)/2$ 行或列的像素进行分数阶微分。

为了使本章定义的分数阶微分掩模处理数字图像尽量具有 $\pi/4$ 角旋转的各向同性结果，并且为了对处理后得到的图像进行适当平滑，作者提出了 x，y 方向上改进的分数阶微分掩模[313]，如图 11.20 所示。

第11章 图像处理的分数阶微分算子：分数阶微分掩模

$\frac{v^2-v}{8}$	$\frac{v^2-v}{2}$	$\frac{v^2-v}{8}$
0	$\frac{3}{2}$	0
$-\frac{v}{4}$	$-v$	$-\frac{v}{4}$

(a) x 方向上

$\frac{v^2-v}{8}$	0	$-\frac{v}{4}$
$\frac{v^2-v}{2}$	$\frac{3}{2}$	$-v$
$\frac{v^2-v}{8}$	0	$-\frac{v}{4}$

(b) y 方向上

图 11.20 x，y 方向上分数阶微分的 $\pi/4$ 旋转掩模

使用 x，y 方向上改进的分数阶微分掩模的目的是通过突出 x，y 点的作用，达到平滑的目的。其中，权重 $4=2^2$，突出了掩模中心行或列的作用。

同时为了使分数阶微分后像素点 (x,y) 的灰度值的范围尽量不要超过 $[0,255]$，在总体上维持原图 $s(x,y)$ 的灰度直方图分布的包络，将模板的每一项都除以 $3/2$，如图 11.21 所示。

$\frac{v^2-v}{12}$	$\frac{v^2-v}{3}$	$\frac{v^2-v}{12}$
0	1	0
$-\frac{v}{6}$	$-\frac{2v}{3}$	$-\frac{v}{6}$

(a) x 方向上

$\frac{v^2-v}{12}$	0	$-\frac{v}{6}$
$\frac{v^2-v}{3}$	1	$-\frac{2v}{3}$
$\frac{v^2-v}{12}$	0	$-\frac{v}{6}$

(b) y 方向上

图 11.21 x，y 方向上分数阶微分的 $\pi/4$ 改进旋转掩模

另外，由于图像的灰度一般都是平滑过渡的，图像信号具有高度的自相似性，图像的相邻像素点之间具有很强的相关性，所以，一种更极端的做法，将不对称的 3×3 邻域内的每个像素点对于像素点 (x,y) 的影响都考虑进去。这样更有利于利用图像相邻像素点之间的自相关性，借助分数阶微分来加强纹理细节信息，从而提取图像信号的分形信息。作者提出了 x，y 方向上进一步改进的分数阶微分掩模[313]，如图 11.22 所示。

$\frac{v^2-v}{6}$	$\frac{v^2-v}{6}$	$\frac{v^2-v}{6}$
$-\frac{v}{5}$	$-\frac{v}{5}$	$-\frac{v}{5}$
$-\frac{v}{5}$	1	$-\frac{v}{5}$

(a) x 方向上

$\frac{v^2-v}{6}$	$-\frac{v}{5}$	$-\frac{v}{5}$
$\frac{v^2-v}{6}$	$-\frac{v}{5}$	1
$\frac{v^2-v}{6}$	$-\frac{v}{5}$	$-\frac{v}{5}$

(b) y 方向上

图 11.22 x，y 方向邻域内的分数阶微分掩模

在图 11.22 中，由于与像素点 (x, y) 相隔一个像素点的像素有 5 个，相隔两个像素点的像素点只有 3 个，为了使该模板中常数 1，$-v$，v^2-v 三项的系数（权重）的比例关系仍满足式 (11.42) 和 (11.43)，同时为了使分数阶微分后像素点 (x, y) 的灰度值的范围尽量不要超过 $[0, 255]$，在总体上维持原图 $s(x, y)$ 的灰度直方图分布的包络，所以将 $-v$ 的系数改为 1/5，将 v^2-v 的系数改为 1/6。

最后，为了完成在图像 $s(x, y)$ 右对角方向以及左对角方向上的分数阶微分运算，达到增强算法的抗旋转性能，作者分别定义在图像右对角方向以及左对角方向上的分数阶微分掩模[313]如图 11.23 所示。

0	0	1
0	$-v$	0
$\frac{v^2-v}{2}$	0	0

(a) 右对角方向上

$\frac{v^2-v}{2}$	0	0
0	$-v$	0
0	0	1

(b) 左对角方向上

图 11.23 右对角方向以及左对角方向上新的分数阶微分掩模

为了简化计算和便于处理，可以重新定义图像右对角方向以及左对角方向上偏分数阶微分[212]，如图 11.24 所示。

0	0	$-v$
0	1	0
$\frac{v^2-v}{2}$	0	0

(a) 右对角方向上

$\frac{v^2-v}{2}$	0	0
0	1	0
0	0	$-v$

(b) 左对角方向上

图 11.24 右对角方向以及左对角方向上的分数阶微分掩模

综上所述，本章提出的所有分数阶微分掩模中的系数总和不为零，这表明在图像灰度值恒定或变化不大的区域的响应不为零。这正好是分数阶微分算子为提取图像纹理细节信息所期望得到的性质。

同时，本章提出的分数阶微分掩模不是完全对称的。由图 11.2 知，高斯信号的分数阶微分结果拖着长长的负尾巴的不对称波，类似一个"孤波"。由于分数阶微分的仿生 Rodieck 感受野也是一个高斯差模型，所以分数阶微分掩模在理论上就应该是不完全对称的。图像的分数阶微分其实本来就有一定程度的去运动模糊的功能。

11.1.6.2 实验仿真及结果分析

在具体操作时，为了使图像分数阶微分处理具有更好的旋转不变性，本章同时选取在

x、y、右对角、左对角方向上的四种分数阶微分掩模算子分别对图像 $s(x,y)$ 中像素点 (x,y) 及其邻域进行运算,然后比较四种运算的结果,将其中最大值作为像素点 (x,y) 分数阶微分的灰度值。为了得到一幅更加清晰的加强了纹理细节的图像,需要将原图像与其分数阶微分结果图对应像素点的灰度值进行点对点的叠加。对于 RGB 彩色图像的分数阶微分处理,需要对其 R,G,B 分量分别进行分数阶微分,然后再合成为 RGB 彩色图像。本章以下的实验选取的在 x、y、右对角、左对角方向上的四种分数掩模算子分别如图 11.22 (a)、图 11.22 (b)、图 11.23 (a) 和图 11.23 (b) 所示。

有一定旋转角的 BRIDGE5 是常用检验图像算子旋转不变性的测试图像,如图 11.25 所示。本章分别用基于整数阶微分和基于分数阶微分的数字图像增强算子对 BRIDGE5 进行处理,实验结果对比如图 11.26 所示。

图 11.25　BRIDGE5

(a) sobel 算子

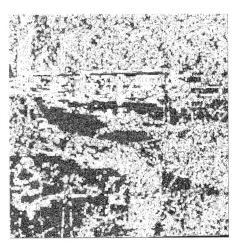

(b) Gauss_Laplace 算子

(c) Kirsch 算子　　　　　　　　　　　(d) $v=0.5$

图 11.26　BRIDGE5 的整数阶微分和分数阶微分结果图实验对比

由实验结果对比可知，首先，本章构造的二维图像分数阶微分算子具有较强的旋转不变性。其次，与基于整数阶微分的传统数字图像增强算子相比，分数阶微分算子在加强图像高频边缘信息以及尽量保留图像低频轮廓信息的同时，加强了图像平滑区域中的纹理细节信息。对于诸如 BRIDGE5 这样富于纹理细节信息的图像而言，其纹理细节加强的效果尤为突出。

为了验证和分析分数阶微分算子对图像纹理细节信息的侧抑制作用，本章分别对 32 位 RGB 虎步彩图进行分数阶微分和整数阶微分对比实验，为了显示一幅更加清晰的加强了纹理细节的图像，这里已将原图像与其分数阶微分结果图对应像素点的灰度值进行了点对点的叠加。其实验结果如图 11.27 所示。

(a) 原图　　　　　　　　　　　(b) $v=0.5$ 阶分数阶微分

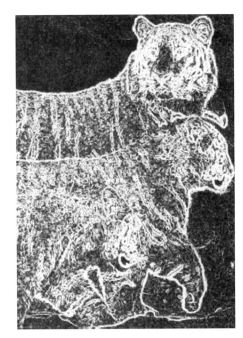

(c) sobel 算子　　　　　　　　　(d) Kirsch 算子

图 11.27　虎步 32 位 RGB 彩图整数阶微分与分数阶微分实验前后对比

可见，图像分数阶微分后具有复杂纹理细节信息的老虎皮毛的毛丝更加清晰了，雪地上的雪斑更加明显了，这是从人眼的角度定性分析的结果。为了从定量的角度进行分析，本章分别对虎步图分数阶和整数阶微分前后的投影特征进行了对比分析，如图 11.28、图 11.29 所示。

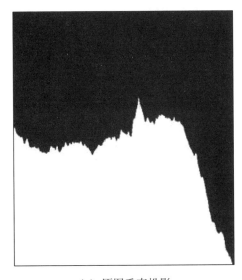
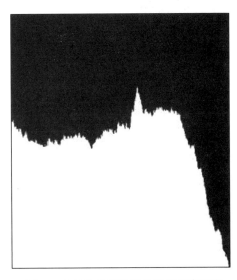

(a) 原图垂直投影　　　　　　　(b) $v=0.5$ 阶分数阶微分后的垂直投影

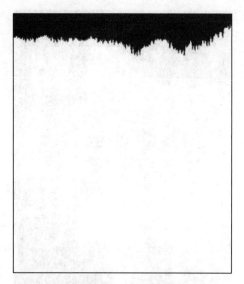

(c) sobel 算子后的垂直投影

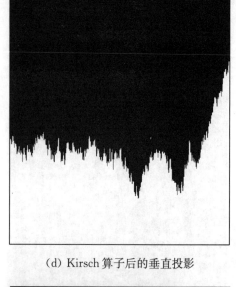

(d) Kirsch 算子后的垂直投影

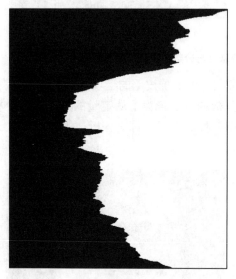

(e) 原图水平投影

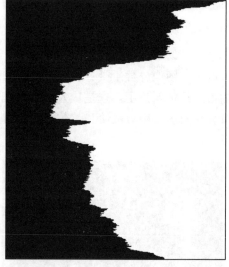

(f) $v=0.5$ 阶分数阶微分后的水平投影

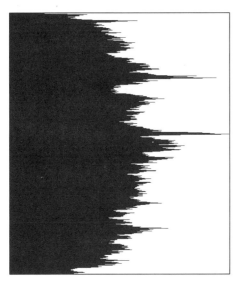
(g) sobel算子后水平投影

(h) Kirsch算子后的水平投影

图 11.28　虎步 32 位 RGB 彩图分数阶微分与整数阶微分实验前后灰度投影对比

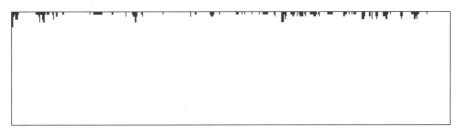
(a) 原图垂直投影

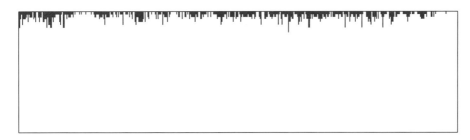
(b) $v=0.5$ 阶分数阶微分后的垂直投影

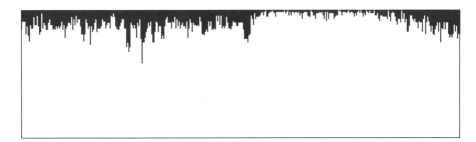
(c) sobel算子后的垂直投影

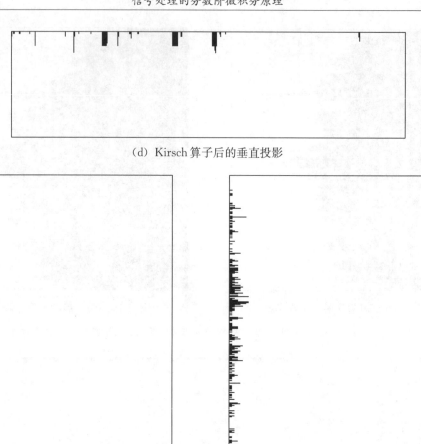

(d) Kirsch 算子后的垂直投影

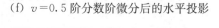

(e) 原图水平投影　　　　　　　(f) $v=0.5$ 阶分数阶微分后的水平投影

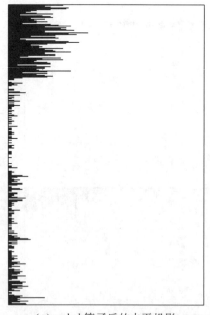

(g) sobel 算子后的水平投影　　　　　　　(h) Kirsch 算子后的水平投影

图 11.29　虎步 32 位 RGB 彩图分数阶微分与整数阶微分实验前后二值图投影对比

由图 11.28 和图 11.29 可知，首先，无论是对于灰度投影还是二值图投影，与图像整数阶微分算子相同，图像分数阶微分算子也加强了图像中的高频边缘信息，使原图中灰度变化较大的图像边缘得到了强化。但是，对于灰度变化较为剧烈的图像边缘信息，图像分数阶微分对它们的加强强度没有整数阶微分大，所以不会出现黑白分明的边缘。其次，图像分数阶微分在一定程度上保持了原图的灰度直方图分布的包络，图像整数阶微分却没有保持原图的灰度直方图分布的包络。所以，图像分数阶微分在一定程度上保持了原图的低频轮廓信息，而图像整数阶微分却不能保持原图的低频轮廓信息。最后，不论图像分数阶微分还是整数阶微分，都使原图的灰度投影或二值投影增加了许多"毛刺"，但是，只有图像分数阶微分才是在原图的灰度投影或二值投影的包络上增加了"毛刺"，图像的整数阶微分改变了原图对应的灰度投影或二值投影的包络。所以，图像分数阶微分能使原图平滑区域中灰度相对变化不太剧烈的纹理细节信息得到非线性的加强，而图像整数阶微分对于它们微分的结果趋近于零。图像整数阶微分改变原图对应灰度投影或二值投影的包络，从而改变了原图灰度直方图分布，使得处理后的结果要么灰度值很高，要么灰度值很低（即要么是黑，要么是白），从而丢掉了大量的纹理细节信息。

为了从人眼的角度定性地、更加清晰地验证分数阶微分算子对图像纹理细节信息的侧抑制作用，本章对 BABOO 8 位灰度图和人像一 32 位 RGB 彩图进行分数阶微分。实验结果如图 11.30、图 11.31 所示。

 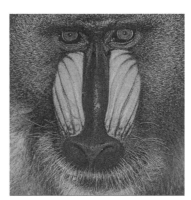

(a) BABOO 原图　　　　　　　(b) BABOO 的分数阶微分

图 11.30　BABOO 8 位灰度图分数阶微分实验对比

（a）人像一原图　　　　　　　（b）人像一的分数阶微分

图 11.31　人像一 32 位 RGB 彩图分数阶微分实验对比

由图 11.30 和图 11.31 可见，图像分数阶微分后，BABOO 和人像一的毛发都变得更加清晰了。特别值得一提的是，在人像一原图中，甚至连人的肉眼都不容易发现的散落在空中和衣服上的若干根头发，在分数阶微分后都得到了适当的加强，使得人的肉眼易于发现它们。这不禁使人想到，图像分数阶微分是否可以应用于对诸如指纹识别等某些需要提取纹理细节信息的项目之中呢？

本章还对其他几幅图作了图像分数阶微分前后的主观视觉效果对比，如图 11.32～图 11.38 所示。

（a）老虎行猎原图　　　　　　　（b）老虎行猎的分数阶微分

图 11.32　老虎行猎 32 位 RGB 彩图分数阶微分实验对比

(a)　人像二原图　　　　　　　　　　(b)　人像二的分数阶微分

图 11.33　人像二 32 位 RGB 彩图分数阶微分实验对比

(a)　人像三原图　　　　　　　　　　(b)　人像三的分数阶微分

图 11.34　人像三 32 位 RGB 彩图分数阶微分实验对比

(a) 三峡原图　　　　　　　　　(b) 三峡的分数阶微分

图 11.35　三峡 32 位 RGB 彩图分数阶微分实验对比

(a) 山地自行车原图　　　　　　(b) 山地自行车的分数阶微分

图 11.36　山地自行车 32 位 RGB 彩图分数阶微分实验对比

(a) 山间小屋原图　　　　　　　　(b) 山间小屋的分数阶微分

图 11.37　山间小屋 32 位 RGB 彩图分数阶微分实验对比

(a) 破卵原图　　　　　　　　　　(b) 破卵的分数阶微分

图 11.38　破卵 32 位 RGB 彩图分数阶微分实验对比

由于图像分数阶微分具有较强的纹理细节信息增强的作用，于是很自然地可以想到将分数阶微分应用于生物医学图像、医学 CT 图像、卫星遥感图像的纹理细节信息增强处理之中，如图 11.39 所示。

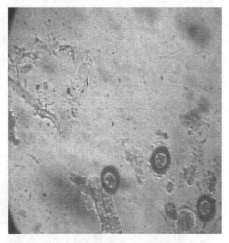

(a) 组织切片一原图

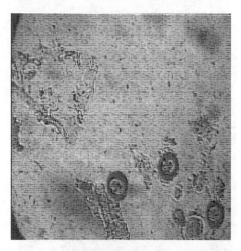

(b) 组织切片一的分数阶微分

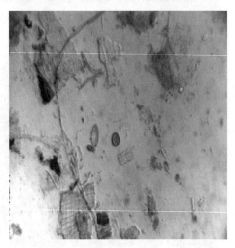

(c) 组织切片二原图

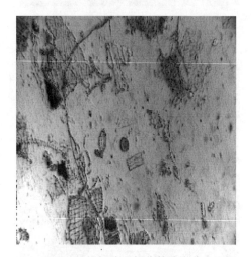

(d) 组织切片二的分数阶微分

第11章 图像处理的分数阶微分算子：分数阶微分掩模

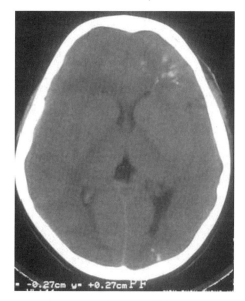

(e) 脑CT一原图

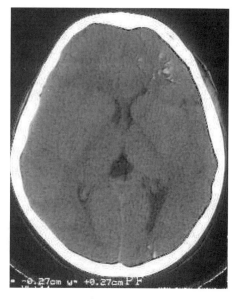

(f) 脑CT一的分数阶微分

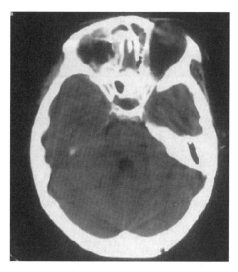

(g) 脑CT二原图

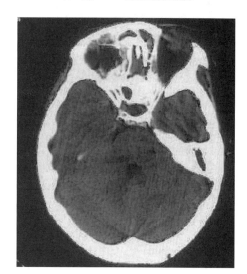

(h) 脑CT二的分数阶微分

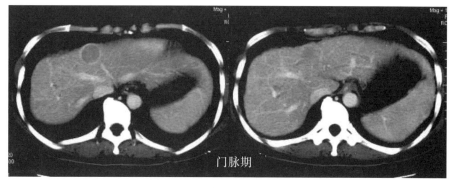

(i) 肝癌CT一原图

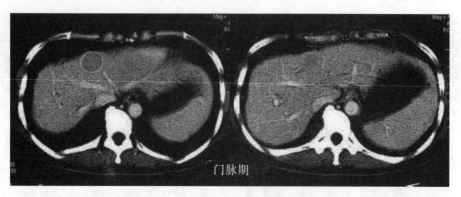

(j) 肝癌 CT 一的分数阶微分

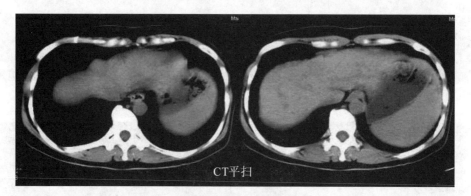

(k) 肝癌 CT 二原图

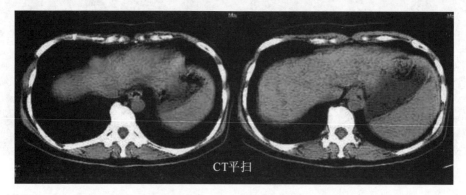

(l) 肝癌 CT 二的分数阶微分

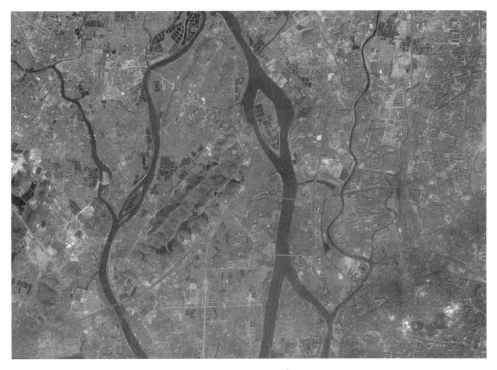

(m) 广州市卫星遥感原图

(n) 广州市卫星遥感的分数阶微分

(o) 上海市卫星遥感 RGB 彩图原图

(p) 上海市卫星遥感 RGB 彩图的分数阶微分

图 11.39 生物医学图像、医学 CT 图像、卫星遥感图像的分数阶微分实验对比

可见，卫星遥感图像、医学图像的分数阶微分在保持整体轮廓信息的同时，很好地非

线性加强了其高频纹理细节信息，从而更有利于对这些图像纹理细节信息的判读。

11.1.7 小结

将分数阶微积分应用于现代信号处理，特别是数字图像处理之中，在国内外还是一个刚刚起步的研究课题。本章从信号处理和生物视觉模型两个角度系统论述了分数阶微分提取和强化图像纹理细节信息的本质原因和独特优势，从计算机数字处理数字图像的方式和分数阶微积分数值算法两方面详细论述了如何构造图像分数阶微分掩模及其数值实现。本章的工作只是在这个领域进行了一些初步尝试和探究，还有许多问题有待进一步研究，诸如：

（1）将分数阶微积分应用于信号的调制解调技术，与传统的基于子波变换的过零点检测技术有什么联系与区别？

（2）如何运用分数阶微积分的阶次 v 作为判断诸如自适应信号处理、神经网络、遗传算法训练收敛到极小点附近的指标？同时，如何构造分数阶梯度来跨过局部最小点，以实现基于分数阶微分的分数阶自适应信号处理和分数阶自适应控制？

（3）如何构造出更好的二维分数阶微分掩模，从而在一定程度上消除图像的运动模糊？

（4）由 RGB 各分量之间存在相关性，同时由于其灰度值限制在 0~255 之间，因此对其各分量进行非线性加强后，叠加得到的 RGB 彩色图像有可能失真。如何在 YCrCb 和 HSV 色彩空间进行分数阶微分？当灰度值超出 0~255 范围时，如何进行多尺度分数阶微分以及如何进行与分数阶阶次相关的非线性补偿？

（5）如何将分数阶微分应用于三维图像的处理？

（6）如何用光学器件实现连续图像的分数阶微分？

（7）如何在群上进行图像信号的分数阶微分？

（8）进一步研究高维分数阶微分的可分离性。

（9）如何将数字图像的分数阶微分技术应用于生物医学图像处理、人脸检测、遥感图像处理、指纹识别、文物字画的真伪鉴别中？

11.2 Fractional Differential Mask：A Fractional Differential-Based Approach for Multiscale Texture Enhancement[①]

在本节中，我们拟构造一组高精度的分数阶微分掩模。基于 Grümwald-Letnikov 和 Riemann-Liouville 这两个常用的分数阶微分定义式，我们提出了 6 个分数阶微分掩模，并分别给出了 8 个方向上每个分数阶微分掩模的结构和参数，即负 x 坐标轴方向、正 x 轴坐标方向、负 y 轴坐标方向、正 y 轴坐标方向、左下对角线方向、左上对角线方向、右下对角线方向和右上对角线方向。此外，通过理论推导和实验分析，我们证明了在所提出的 6 个分数阶微分掩模中，第 2 个掩模具备最佳的性质，并进一步讨论了多尺度分数阶

---

① PU Yifei, ZHOU Jiliu, YUAN Xiao. Fractional Differential Mask：A Fractional Differential Based Approach for Multi-scale Texture Enhancement [J]. IEEE Transactions on Image Processing，2010，19（2）：491-511.

微分掩模用于纹理增强的能力。实验表明，对于富含纹理的数字图像而言，基于分数阶微分的方法在平滑区域中非线性增强复杂纹理细节的能力明显优于传统基于整数阶的算法。

11.2.1 INTRODUCTION

Texture enhancement is an important issue in many areas like pattern recognition, image restoration, medical imaging processing, robotics, interpretation of image data and remote sensing. Despite a family of techniques has been greatly developed over the last couple of decades, there are only a few reliable methods for texture-enhancing are presented. Multi-resolution techniques [418] − [428] seem to be attractive approaches for many applications. As far as texture-enhancing technique concerned, it has gained some improvement by integrating multi traditional approaches in recent years; however, it in fact is just the information synthesis for all the processed results. When the resolution is low, texture enhancement can be reliably dealt with in real-time. As the resolution increases, it becomes infeasible due to excessive storage and computational requirements. Note that the above-mentioned traditional texture-enhancing algorithms are in essence the integral differential based approaches. We propose to introduce a novel mathematic method—fractional differential to image processing, and to implement fractional differential based multiscale masks for texture-enhancing that hold the features of convenient, fast and efficient.

In the last 300 years, fractional differential has experienced the increasing interest and has been an important branch of mathematical analyses [176] − [177], [230], [429]. It, however, is still seldom known by mathematicians and physical scientists in engineering fields. In general, fractional differential in Euclid space has extended the step from integral to fractional [3], [430]. Moreover, the random variable in physical process can be viewed as the displacement of particle random movement. So, fractional differential can be applied to analyze the physical states and processes of objects in Euclidean space [11], [19], [23], [83], [94], [142], [180], [431] − [432]. The fractional differential functions have two obvious features. One is that for major functions, it is power function. And the other is that for the rest functions, it is the iterative adding or the multiplying product of certain function and power function [3], [176] − [177], [230], [429] − [430]. Do these features pre-show some natural evolution rule? It has been proven by scientific findings that many fractional order or dimensional mathematic approaches are the best description for natural phenomena [89], [126], [184], [433]. Fractional differential in Euclidean space has been used in many fields like dispersion processing, viscoelastic theory, and random fractal dynamics. Most of the researches on fractional differential application focus on the temporal state of physical change, however, the systemic evolvement processes are seldom involved [11], [19], [23], [83], [89], [94], [126], [142], [180], [184], [431] − [433].

How to apply fractional differential to the latest signal analyzing and processing [89],

[106], [130] − [131], [157], [172], [185], [193], [433], in particular to digital image processing [110], [186], [434] − [435], is a burgeoning subject branch under discussion. It's known that integral differential based approaches are effective for describing Euclid integral dimension space. In the fields of traditional signal analyzing and processing, especially in singularities inspecting and extracting, integral differential based algorithms have been wildly applied. However, the essence of modern signal analyzing and processing is to study the signals that feature nonlinear, non-causal, non-minimum phase systems, non-Gaussian, non-even, non integral differential and non-white additive noise [436]. Many findings show that fractional based algorithms [78], [437] − [441] are the powerful approaches for dealing with non-problems. In addition, fractional differential based algorithms can also be used for analyzing and processing the above-mentioned non-problems in signal processing [11], [19], [23], [83], [89], [94], [106], [110], [126], [130] − [131], [142], [157], [172], [180], [184] − [186], [193], [431] − [435].

To better deal with such non-problems inspires our thirst for such novel mathematical method and forms our compelling reason and original aspirations, and then, we try to introduce fractional differential to image processing. In image processing, the grey-level value between neighboring pixels is greatlycorrelated and highly self-similar. Most of the fractal structures are the evolutionary process (for example: fractal growth) and evolutionary final product (for example: fracture). Such fractal-like structure is often expressed by the complex texture detail features. Fractional differential is an effective mathematical method for dealing with fractal problems [80], [442]. So, it is natural to consider whether we can apply fractional differential to enhance complex fractal-like texture details. Most typical operators such as first-order Sobel, Prewitt and Roberts operators, second-order Laplace and wavelet-based Mexico cap operator and Bubble operator, are in fact the integral differential based bionic implementing of Rodieck receptive field model in mathematics. At present, a well texture-analyzing method is 2−D Hermite functions, whose most important property is that they are the eigenfunctions of Fourier transform. Thus when signal is changed into a Hermite functions series, it will give the information of Fourier transform of the signal at the same time [443] − [445]. In contrast with the property of integral differential that the direct current or low-frequency signal is zero, that of fractional differential is usually nonzero. Fractional differential of image has special Mach effect, and its antagonism characteristics lead it to have a special bionic vision receptive field model. In addition, fractional differential masks can be used for non-linearly enhancing complex fractal-like texture details [186], [434] − [435]. Therefore, we could implement the Grümwald-Letnikov based fractional differential mask and the circuits of digital filter [186], [434] − [435]. Fractional differential mask can furthest preserve the low-frequency contour feature in those smooth areas, and non-linearly keep high-frequency marginal feature in those areas that grey-level changes greatly, and also enhance texture details in those areas that grey-level does not change evidently [157], [185]. Furthermore, by taking different approximate approaches, we propose another 5 novel fractional differential masks which have different precision and convergence speed. Finally, by comparing the proposed six masks, we demonstrated the best performed mask.

This chapter organized as follows: Section 2 recalls on the necessary theoretical background of fractional differential, gives the main definitions of Grümwald-Letnikov and Riemann-Liouville in Euclidean space, and presents the geometric meaning and physical meaning. Section 3 deduces and gives 6 fractional differential masks. Section 4 reports the nonlinearly enhancing capability of fractional differential based algorithm from the viewpoints of telecommunication and signal processing, and the relative errors of the 6 fractional differential masks by theoretic analyzing and experiments. We illustrate that the operator $YiFeiPU$-2 has the best performance in its precision and convergence. In contrast with traditional approaches for texture-enhancement, we further study the capability of texture-enhancing and segmenting of multiscale fractional differential masks.

11.2.2 RELATED WORK

The purpose of this section is to recall on the necessary theoretical background about the application of fractional differential in signal processing, and to make this chapter self-contained for the reader not familiar with such approach.

It is known that fractal theory has greatly changed the traditional viewpoint of measurement, because fractal geometric denies the existence of Newton-Leibniz derivative. The Hausdorff based fractal theory is still an incomplete mathematic theory after more than 90 years development, but until now its theoretic system is not established and far from perfect. Among the whole family of Hausdorff measurement, the theory of fractional differential in Euclidean space is the comparatively completed. Thus, it is required Euclidean measurement in mathematics. The commonly used definitions under Euclidean measurement are Grümwald-Letnikov definition and Riemann-Liouville definition [3], [176] − [177], [230], [430].

v-order Grümwald-Letnikov based fractional differential can be expressed by

$$D_{G-L}^v s(x) = \frac{d^v}{[d(x-a)]^v} s(x) \bigg|_{G-L} = \lim_{N \to \infty} \left\{ \frac{\left(\frac{x-a}{N}\right)^{-v}}{\Gamma(-v)} \sum_{k=0}^{N-1} \frac{\Gamma(k-v)}{\Gamma(k+1)} s\left(x - k\left(\frac{x-a}{N}\right)\right) \right\} \tag{11.52}$$

where the duration of signal $s(x)$ is $[a, x]$, and v is any real number (including fraction). D_{G-L}^v denotes Grümwald-Letnikov based fractional differential operator. Grümwald-Letnikov definition in Euclidean space shown in (11.52) has extended the step from integral to fractional, and the order of differential is from integral to fractional. Grümwald-Letnikov based fractional differential is easy for calculating, because the only need coefficient is the discrete sampling of $s\left(x-k\left(\frac{x-a}{N}\right)\right)$ that is related to signal itself, and the derivative or integral value of signal $s(x)$ is not required.

Indeed, according to Grümwald-Letnikov definition, it has

$$\frac{d^n}{dx^n}\frac{d^v}{[d(x-a)]^v}s(x) = \frac{d^{n+v}}{[d(x-a)]^{n+v}}s(x) \tag{11.53}$$

where both v and n are any real numbers.

For $v<0$, Riemann-Liouville based v-order integral reads as

$$D_{R-L}^v s(x) = \frac{d^v}{[d(x-a)]^v}s(x)\bigg|_{R-L} = \frac{1}{\Gamma(-v)}\int_a^x (x-\xi)^{-v-1}s(\xi)d\xi, v<0 \tag{11.54}$$

Here, D_{R-L}^v denotes Riemann-Liouville based fractional differential operator. By choosing the proper integer n that satisfies $v-n<0$, one can study v-order differential for $v\geqslant 0$. Then, for $v\geqslant 0$, Riemann-Liouville based v-order fractional differential can be expressed as

$$D_{R-L}^v s(x) = \frac{d^v}{[d(x-a)]^v}s(x)\bigg|_{R-L} = \frac{d^n}{dx^n}\frac{d^{v-n}}{[d(x-a)]^{v-n}}s(x)\bigg|_{R-L}$$

$$= \sum_{k=0}^{n-1}\frac{(x-a)^{k-v}s^{(k)}(a)}{\Gamma(k-v+1)} + \frac{1}{\Gamma(n-v)}\int_a^x \frac{s^{(n)}(\xi)}{(x-\xi)^{v-n+1}}d\xi, 0\leqslant v<n \tag{11.55}$$

In our previous works [186], we already proposed that the geometric meaning of fractional derivative is the generalized slope of its function curve for what is called fractional slope, and fractional integral is generalized Euclidean measurement for shaping image that is called fractional Euclidean measurement. The physical meaning of fractional derivative is fractional flow or fractional speed, which is the continuous fractional measurement for the speed and the directions of distance changing, while the fractional order v is in fact the fractional description for temporal balance state that is called fractional equilibrium coefficient. In the paper [157], [185], we also know that the physical meaning of fractional differential is the generalized amplitude-and-phase modulation that is called fractional amplitude-and-phase modulation. Here, the amplitude of original signal is changing with its frequency as fractional power exponent, while the phase is generalized Hilbert transform of frequency. In addition, from (11.52) the fractional difference of signal $s(x)$ can be expressed as

$$\Delta^v s(x) = \frac{1}{\Gamma(-v)}\sum_{k=0}^{N-1}\frac{\Gamma(k-v)}{\Gamma(k+1)}s(x-k(\frac{x-a}{N})) \tag{11.56}$$

Fractional difference is the generalized form of difference, which is called generalized difference. After all, integral difference is the special case when the order takes integer number [186].

11.2.3 THEORETIC ANALYZING FOR IMPLEMENTING FRACTIONAL DIFFERENTIAL MASKS

Based on the above definitions and conclusions, in this section, we willfatherly discuss 6 fractional masks and algorithms.

11.2.3.1 SixFractional Differential Masks and Its Algorithms

Without losing generality, suppose $a=0$, one divides the duration of $s(x)$ into N equal shares, here, the duration belongs to $[0, x]$. Thus, there are $N+1$ nodes. The $N+1$ causal pixels can be given by

$$\begin{cases} s_N \equiv s(0) \\ s_{N-1} \equiv s(x/N) \\ \vdots \\ s_k \equiv s(x - kx/N) \\ \vdots \\ s_0 \equiv s(x) \end{cases} \tag{11.57}$$

Since the digital image processed is in the medium and the verge can be expanded by the periodical way, we also can depose non-causal pixels. It is given by

$$\begin{cases} s_0 \equiv s(x) \\ s_{-1} \equiv s(x + x/N) \\ \vdots \\ s_{-k} \equiv s(x + kx/N) \\ \vdots \\ s_{-N} \equiv s(2x) \end{cases} \tag{11.58}$$

11.2.3.2 The First Fractional Differential Mask

When N is big enough, one can get rid of the limits symbol and rewrite (11.52) as [186], [434] − [435].

$$\begin{aligned} \left.\frac{d^v}{dx^v}s(x)\right|_{AL-1} &\cong \frac{x^{-v}N^v}{\Gamma(-v)} \sum_{k=0}^{N-1} \frac{\Gamma(k-v)}{\Gamma(k+1)} s\left(x - \frac{kx}{N}\right) \\ &= \frac{x^{-v}N^v}{\Gamma(-v)} \sum_{k=0}^{N-1} \frac{\Gamma(k-v)}{\Gamma(k+1)} s_k \end{aligned} \tag{11.59}$$

It is the proximate expression, which simplifies fractional differential tomultiplication and add. Here, the numerical algorithm of fractional differential is noted as $AL-1$. For two-dimensional image signal, it has the followings two expressions. [186], [434] − [435].

$$\begin{aligned} \frac{\partial^v s(x,y)}{\partial x^v} &\cong s(x,y) + (-v)s(x-1,y) + \frac{(-v)(-v+1)}{2}s(x-2,y) \\ &+ \frac{(-v)(-v+1)(-v+2)}{6}s(x-3,y) + \cdots \\ &+ \frac{\Gamma(n-v-1)}{(n-1)!\Gamma(-v)}s(x-n+1,y) \end{aligned} \tag{11.60}$$

$$\begin{aligned} \frac{\partial^v s(x,y)}{\partial y^v} &\cong s(x,y) + (-v)s(x,y-1) + \frac{(-v)(-v+1)}{2}s(x,y-2) \\ &+ \frac{(-v)(-v+1)(-v+2)}{6}s(x,y-3) + \cdots \\ &+ \frac{\Gamma(n-v-1)}{(n-1)!\Gamma(-v)}s(x,y-n+1) \end{aligned} \tag{11.61}$$

To obtain the fractional differential on the eight symmetric directions and make the fractional differential masks have anti-rotation capability, 8 fractional differential masks which are respectively on the directions of negative x-coordinate, negative y-coordinate, positive x-coordinate, positive y-coordinate, left downward diagonal, right upward diagonal, left upward diagonal, and right downward diagonal are implemented. They are correspondingly denoted by W_x^-, W_y^-, W_x^+, W_y^+, W_{LDD}, W_{RUD}, W_{LUD} and W_{RDD} [186], [434] – [435]. It is noted as *YiFeiPU*-1 operator. (see Fig. 11.40)

	⋯	0	C_{s_n}	0	⋯	
	⋯	0	$C_{s_{n-1}}$	0	⋯	
		⋮	⋮	⋮		
	⋯	0	C_{s_k}	0	⋯	
		⋮	⋮	⋮		
	⋯	0	C_{s_1}	0	⋯	
	⋯	0	C_{s_0}	0	⋯	

(a)

⋮	⋮		⋮		⋮	⋮
0	0	⋯	0	⋯	0	0
C_{s_n}	$C_{s_{n-1}}$	⋯	C_{s_k}	⋯	C_{s_1}	C_{s_0}
0	0	⋯	0	⋯	0	0
⋮	⋮		⋮		⋮	⋮

(b)

	⋯	0	C_{s_0}	0	⋯	
	⋯	0	C_{s_1}	0	⋯	
		⋮	⋮	⋮		
	⋯	0	C_{s_k}	0	⋯	
		⋮	⋮	⋮		
	⋯	0	$C_{s_{n-1}}$	0	⋯	
	⋯	0	C_{s_n}	0	⋯	

(c)

⋮	⋮		⋮		⋮	⋮
0	0	⋯	0	⋯	0	0
C_{s_0}	C_{s_1}	⋯	C_{s_k}	⋯	$C_{s_{n-1}}$	C_{s_n}
0	0	⋯	0	⋯	0	0
⋮	⋮		⋮		⋮	⋮

(d)

				⋯	0	C_{s_0}
			⋯	0	C_{s_1}	0
			⋰	⋰	0	⋮
	⋯	0	C_{s_k}	0	⋯	
⋮	0	⋰	⋰			
0	$C_{s_{n-1}}$	0	⋯			
C_{s_n}	0	⋯				

(e)

				⋯	0	C_{s_n}
			⋯	0	$C_{s_{n-1}}$	0
			⋰	⋰	0	⋮
	⋯	0	C_{s_k}	0	⋯	
⋮	0	⋰	⋰			
0	C_{s_1}	0	⋯			
C_{s_0}	0	⋯				

(f)

C_{s_n}	0	...			
0	$C_{s_{n-1}}$	0	...		
⋮	0	⋱	⋱		
	⋮	0	C_{s_k}	0	⋮
		⋮	⋱	⋱	0	⋮
			...	0	C_{s_1}	0
				...	0	C_{s_0}

(g)

C_{s_0}	0	...			
0	C_{s_1}	0	...		
⋮	0	⋱	⋱		
⋮		0	C_{s_k}	0	⋮
⋮		⋱	⋱	0	⋮
		...	0	$C_{s_{n-1}}$	0
			...	0	C_{s_n}

(h)

Fig. 11.40 Fractional differential mask on the eight directions. (a) W_x^- (b) W_y^- (c) W_x^+ (d) W_y^+ (e) W_{LDD} (f) W_{RUD} (g) W_{LUD} (h) W_{RDD}.

C_{s_0} is the mask coefficient on interest pixel and it has $s_0 = s(x)$. When $k \to n = 2m - 2$, one can implement $(2m-1) \times (2m-1)$ fractional differential mask. To make sure fractional differential mask has the certain centre, in general, n is taken even number. From (11.60) and (11.61), we can see that the mask coefficients of fractional differential operator $YiFeiPU$-1 are given by

$$\begin{cases} C_{s_0} = 1 \\ C_{s_1} = -v \\ \vdots \\ C_{s_k} = \dfrac{\Gamma(k-v)}{k!\Gamma(-v)} \\ \vdots \\ C_{s_{n-1}} = \dfrac{\Gamma(n-v-1)}{(n-1)!\Gamma(-v)} \\ C_{s_n} = \dfrac{\Gamma(n-v)}{n!\Gamma(-v)} \end{cases} \quad (11.62)$$

Since digital image processing is based on the direct processing for discrete pixels, the algorithm of fractional differential mask is also taking airspace filtering of mask convolution [185] − [186], [434] − [435].

11.2.3.3 The Second Fractional Differential Mask

To capture faster speed and better precision of convergence, (11.59) can be rewritten as

$$\left.\frac{d^v}{dx^v}s(x)\right|_{G-L} \simeq \frac{x^{-v}N^v}{\Gamma(-v)} \sum_{k=0}^{N-1} \frac{\Gamma(k-v)}{\Gamma(k+1)} s\left(x + \frac{vx}{2N} - \frac{kx}{N}\right) \quad (11.63)$$

Let us compare (11.62) to (11.63), (11.63) has introduced the signal value of $s(x)$ on non-node besides supposing $v=0, \pm 2, \pm 4, \cdots$. Taking three neighboring nodes as $s(x + \frac{x}{N} - \frac{kx}{N})$, $s(x - \frac{kx}{N})$, $s(x - \frac{x}{N} - \frac{kx}{N})$, and using Lagrange 3-point interpolation expression, we get

$$s(\xi) \cong \frac{(\xi - x + \frac{kx}{N})(\xi - x + \frac{x}{N} + \frac{kx}{N})}{2x^2/N^2} s(x + \frac{x}{N} - \frac{kx}{N})$$

$$- \frac{(\xi - x - \frac{x}{N} + \frac{kx}{N})(\xi - x + \frac{x}{N} + \frac{kx}{N})}{x^2/N^2} s(x - \frac{kx}{N})$$

$$+ \frac{(\xi - x - \frac{x}{N} + \frac{kx}{N})(\xi - x + \frac{kx}{N})}{2x^2/N^2} s(x - \frac{x}{N} - \frac{kx}{N}) \quad (11.64)$$

Assume $\xi = x + \frac{vx}{2N} - \frac{kx}{N}$ and do fractional interpolation, it has

$$s(x + \frac{vx}{2N} - \frac{kx}{N}) \cong (\frac{v}{4} + \frac{v^2}{8})s(x + \frac{x}{N} - \frac{kx}{N}) + (1 - \frac{v^2}{4})s(x - \frac{kx}{N})$$

$$+ (\frac{v^2}{8} - \frac{v}{4})s(x - \frac{x}{N} - \frac{kx}{N})$$

$$= (\frac{v}{4} + \frac{v^2}{8})s_{k-1} + (1 - \frac{v^2}{4})s_k + (\frac{v^2}{8} - \frac{v}{4})s_{k+1} \quad (11.65)$$

Then, taking (11.65) into (11.63), it has

$$\left.\frac{d^v}{dx^v}s(x)\right|_{AL-2} \cong \frac{x^{-v}N^v}{\Gamma(-v)} \sum_{k=0}^{N-1} \frac{\Gamma(k-v)}{\Gamma(k+1)} \left[s_k + \frac{v}{4}(s_{k-1} - s_{k+1}) + \frac{v^2}{8}(s_{k-1} - 2s_k + s_{k+1}) \right]$$

(11.66)

Indeed, the expression can only get the approximated value due it simplifies fractional differential to multiplication and add. Here, the numerical algorithm of fractional differential is noted as $AL-2$. In contrast with $AL-1$, $AL-2$ have no requirements for $N \to \infty$.

As we known, the processed object of computer or digital filter is the limit number, the biggest variable of grey-level of image signal is also limited, and the shortest changing distance of grey-level is one pixel. Thus, the smallest distance of two-dimensional digital image $s(x, y)$ on x-coordinate and y-coordinate is also one pixel. Here, the duration is the dimension of image matrix. Assume the duration of x and y respectively are $[0, x]$ and $[0, y]$, the uniformly distance on x-coordinate and y-coordinate are $h_x = \frac{x}{N} = 1$ and $h_y = \frac{y}{N} = 1$, and its biggest divided number are $N_x = \left[\frac{x}{h_x}\right]^{h_x=1} = [x]$ and $N_y = \left[\frac{y}{h_y}\right]^{h_y=1} = [y]$. As a consequence, even the fractional differential mask is as big as the original digital image and it has $N = \min(N_x, N_y)$, the computing result will be extremely approximate to the analytic value and cannot equal to it.

When $k = n \leqslant N - 1$, from (11.66), the anterior $n+2$ approximate backward difference of fractional partial differential respectively on negative x-and y-coordinate, are expressed as

$$\frac{\partial^v s(x,y)}{\partial x^v} \cong (\frac{v}{4}+\frac{v^2}{8})s(x+1,y)+(1-\frac{v^2}{2}-\frac{v^3}{8})s(x,y)$$

$$+\frac{1}{\Gamma(-v)}\sum_{k=1}^{n-2}\left[\frac{\Gamma(k-v+1)}{(k+1)!}\cdot\left(\frac{v}{4}+\frac{v^2}{8}\right)+\frac{\Gamma(k-v)}{k!}\cdot(1-\frac{v^2}{4})\right.$$

$$\left.+\frac{\Gamma(k-v-1)}{(k-1)!}\cdot\left(-\frac{v}{4}+\frac{v^2}{8}\right)\right]s(x-k,y)$$

$$+\left[\frac{\Gamma(n-v-1)}{(n-1)!\Gamma(-v)}\cdot\left(1-\frac{v^2}{4}\right)+\frac{\Gamma(n-v-2)}{(n-2)!\Gamma(-v)}\cdot\left(-\frac{v}{4}+\frac{v^2}{8}\right)\right]$$

$$s(x-n+1,y)+\frac{\Gamma(n-v-1)}{(n-1)!\Gamma(-v)}\cdot\left(-\frac{v}{4}+\frac{v^2}{8}\right)s(x-n,y)$$

(11.67)

$$\frac{\partial^v s(x,y)}{\partial y^v} \cong (\frac{v}{4}+\frac{v^2}{8})s(x,y+1)+(1-\frac{v^2}{2}-\frac{v^3}{8})s(x,y)$$

$$+\frac{1}{\Gamma(-v)}\sum_{k=1}^{n-2}\left[\frac{\Gamma(k-v+1)}{(k+1)!}\cdot\left(\frac{v}{4}+\frac{v^2}{8}\right)+\frac{\Gamma(k-v)}{k!}\cdot(1-\frac{v^2}{4})\right.$$

$$\left.+\frac{\Gamma(k-v-1)}{(k-1)!}\cdot\left(-\frac{v}{4}+\frac{v^2}{8}\right)\right]s(x,y-k)$$

$$+\left[\frac{\Gamma(n-v-1)}{(n-1)!\Gamma(-v)}\cdot\left(1-\frac{v^2}{4}\right)+\frac{\Gamma(n-v-2)}{(n-2)!\Gamma(-v)}\cdot\left(-\frac{v}{4}+\frac{v^2}{8}\right)\right]$$

$$s(x,y-n+1)+\frac{\Gamma(n-v-1)}{(n-1)!\Gamma(-v)}\cdot\left(-\frac{v}{4}+\frac{v^2}{8}\right)s(x,y-n) \quad (11.68)$$

Similarly, the fractional differential mask respectively on the eight symmetric directions can be implemented, which is noted as *YiFeiPU*-2. (see Fig. 11.41)

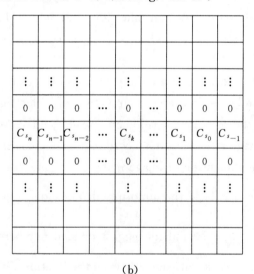

第 11 章　图像处理的分数阶微分算子：分数阶微分掩模

Fig. 11.41 Fractional differential operator $YiFeiPU$-2 on eight directions. (a) W_x^- (b) W_y^- (c) W_x^+ (d) W_y^+ (e) W_{LDD} (f) W_{RUD} (g) W_{LUD} (h) W_{RDD}.

Here $C_{s_{-1}}$ is the mask coefficient on non-causal pixel $s_{-1} = s(x + x/N)$, and C_{s_0} is the mask coefficient of on intrest pixel $s_0 = s(x)$. (11.67) and (11.68) show that, when $k \to n = 1$, it has 3×3 mask; when $k \to n = 3$, it has 5×5 mask; and when $k \to n = 2m - 1$, it has $(2m+1) \times (2m+1)$ mask. n is usually taken odd number, the coefficient of the fractional differential mask $YiFeiPU$-2 is given by

$$\begin{cases} C_{s_{-1}} = \dfrac{v}{4} + \dfrac{v^2}{8} \\ C_{s_0} = 1 - \dfrac{v^2}{2} - \dfrac{v^3}{8} \\ C_{s_1} = -\dfrac{5v}{4} + \dfrac{5v^3}{16} + \dfrac{v^4}{16} \\ \vdots \\ C_{s_k} = \dfrac{1}{\Gamma(-v)} \left[\dfrac{\Gamma(k-v+1)}{(k+1)!} \cdot \left(\dfrac{v}{4} + \dfrac{v^2}{8} \right) + \dfrac{\Gamma(k-v)}{k!} \cdot \left(1 - \dfrac{v^2}{4} \right) \right. \\ \qquad \left. + \dfrac{\Gamma(k-v-1)}{(k-1)!} \cdot \left(-\dfrac{v}{4} + \dfrac{v^2}{8} \right) \right] \\ \vdots \\ C_{s_{n-2}} = \dfrac{1}{\Gamma(-v)} \left[\dfrac{\Gamma(n-v+1)}{(n-1)!} \cdot \left(\dfrac{v}{4} + \dfrac{v^2}{8} \right) + \dfrac{\Gamma(n-v-2)}{(n-2)!} \cdot \left(1 - \dfrac{v^2}{4} \right) \right. \\ \qquad \left. + \dfrac{\Gamma(n-v-3)}{(n-3)!} \cdot \left(-\dfrac{v}{4} + \dfrac{v^2}{8} \right) \right] \\ C_{s_{n-1}} = \dfrac{\Gamma(n-v-1)}{(n-1)!\Gamma(-v)} \cdot \left(1 - \dfrac{v^2}{4} \right) + \dfrac{\Gamma(n-v-2)}{(n-2)!\Gamma(-v)} \cdot \left(-\dfrac{v}{4} + \dfrac{v^2}{8} \right) \\ C_{s_n} = \dfrac{\Gamma(n-v-1)}{(n-1)!\Gamma(-v)} \cdot \left(-\dfrac{v}{4} + \dfrac{v^2}{8} \right) \end{cases}$$

(11.69)

(11.69) shows that fractional differential operator $YiFeiPU$-2 is a sparse matrix, whose coefficient is $n+2$ nonzero numbers. Moreover, all the coefficients are the function of fractional differential order v. For instance, if $v = 0.5$ and $n = 7$, they have $C_{s_{-1}} \cong 0.15625$, $C_{s_0} \cong 0.85938$, $C_{s_1} \cong -0.58203$, $C_{s_2} \cong -0.080078$, $C_{s_3} \cong -0.052979$, $C_{s_4} \cong -0.030762$, $C_{s_5} \cong 0.0036621$. It also can prove that the sum of the coefficients is not zero, which has remarkable difference between fractional differential mask and integral one.

It is known that digital image processing is based on direct processing for discrete pixel, and the algorithm is also taking airspace filtering of mask convolution. The principle of airspace filter is moving the mask pixel by pixel. As far as the properties of grey image and color image are concerned, they have great differences. Thus, the algorithm of fractional differential mask also has two kinds. One is for grey image and the other is for color image. Let us introduce the algorithm for grey image in the first. As for

$n_x \times n_y$ digital grey image $s(x, y)$, we do convoluting filter respectively on the above 8 directions by using *YiFeiPU*-2 in $(2m+1) \times (2m+1)$ mask. The algorithms of W_x^-, W_y^-, W_x^+, W_y^+, W_{LDD}, W_{RUD}, W_{LUD} and W_{RDD} are respectively given by

$$s_{x^-}^{(v)}(x,y) = \sum_{i=-2m}^{1} \sum_{j=-m}^{m} W_x^-(i,j) s(x+i, y+j) \tag{11.70}$$

$$s_{x^+}^{(v)}(x,y) = \sum_{i=-1}^{2m} \sum_{j=-m}^{m} W_x^+(i,j) s(x+i, y+j) \tag{11.71}$$

$$s_{y^-}^{(v)}(x,y) = \sum_{i=-m}^{m} \sum_{j=-2m}^{1} W_y^-(i,j) s(x+i, y+j) \tag{11.72}$$

$$s_{y^+}^{(v)}(x,y) = \sum_{i=-m}^{m} \sum_{j=-1}^{2m} W_y^+(i,j) s(x+i, y+j) \tag{11.73}$$

$$s_{LDD}^{(v)}(x,y) = \sum_{i=-1}^{2m} \sum_{j=-2m}^{1} W_{LDD}(i,j) s(x+i, y+j) \tag{11.74}$$

$$s_{RUD}^{(v)}(x,y) = \sum_{i=-2m}^{1} \sum_{j=-1}^{2m} W_{RUD}(i,j) s(x+i, y+j) \tag{11.75}$$

$$s_{LUD}^{(v)}(x,y) = \sum_{i=-2m}^{1} \sum_{j=-2m}^{1} W_{LUD}(i,j) s(x+i, y+j) \tag{11.76}$$

$$s_{RDD}^{(v)}(x,y) = \sum_{i=-1}^{2m} \sum_{j=-1}^{2m} W_{RDD}(i,j) s(x+i, y+j) \tag{11.77}$$

As for digital color image, the algorithm is similar to that for grey image, but the R, G, B components should respectively do fractional differential. Due there are correlations among RGB elements and grey-level value is usually limited in [0, 255], the elements of RGB will be nonlinearly enhanced when order v is big. The correlation of weight R, G and B may be destroyed, which leads to the color distortion. Therefore, we often do fractional differential of digital color image in color space of HSI, or only do fractional differential of I weight (intensity or brightness) of HSI.

11.2.3.4 The Third Fractional Differential Mask

Base on Riemann-Liouville definition, we set the duration of $s(x)$ as $[0, x]$ and get

$$\left. \frac{d^v}{dx^v} s(x) \right|_{R-L} = \frac{1}{\Gamma(-v)} \int_0^x \frac{s(\xi)}{(x-\xi)^{v+1}} d\xi = \frac{1}{\Gamma(-v)} \int_0^x \frac{s(x-\xi)}{\xi^{v+1}} d\xi, v<0 \tag{11.78}$$

To simplify the calculating, one can change the continuous integral to discrete sum of products. Divide integral duration $[0, x]$ into N equal parts, when N is big enough, it has

$$\left. \frac{d^v}{dx^v} s(x) \right|_{R-L} = \frac{1}{\Gamma(-v)} \sum_{k=0}^{N-1} \int_{kx/N}^{(kx+x)/N} \frac{s(x-\xi)}{\xi^{v+1}} d\xi, v<0 \tag{11.79}$$

Then the integral in $[0, x]$ is changed as the sum of integral in N subsections. When N big enough, it has

$$\int_{kx/N}^{(kx+x)/N} s(\xi) d\xi \cong \frac{s\left(\frac{kx}{N}\right) + s\left(\frac{kx+x}{N}\right)}{2} \int_{kx/N}^{(kx+x)/N} d\xi = \frac{s\left(\frac{kx}{N}\right) + s\left(\frac{kx+x}{N}\right)}{2} \cdot \frac{x}{N}$$

$$\tag{11.80}$$

Then, (11.80) is rewritten as

$$\int_{kx/N}^{(kx+x)/N} \frac{s(x-\xi)}{\xi^{v+1}} d\xi \cong \frac{s\left(x-\frac{kx}{N}\right)+s\left(x-\frac{kx+x}{N}\right)}{2} \int_{kx/N}^{(kx+x)/N} \frac{1}{\xi^{v+1}} d\xi$$

$$= \frac{s_k + s_{k+1}}{-2v} \cdot \left[\left(\frac{kx+x}{N}\right)^{-v} - \left(\frac{kx}{N}\right)^{-v}\right] \quad (11.81)$$

Taking (11.81) into (11.79), we get

$$\frac{d^v}{dx^v} s(x)\bigg|_{AL-3} \cong \frac{x^{-v} N^v}{\Gamma(-v)} \sum_{k=0}^{N-1} \frac{s_k + s_{k+1}}{-2v}[(k+1)^{-v} - k^{-v}], v < 0 \quad (11.82)$$

(11.82) simplifies fractional differential as multiplication and add, and the algorithm is noted as $AL-3$.

When $k = n \leq N-1$, the anterior $n+1$ approximate backward difference of fractional partial differential on negative x- and y-coordinate are expressed as

$$\frac{\partial^v s(x,y)}{\partial x^v} \cong \frac{1}{\Gamma(-v)(-2v)} s(x,y) + \frac{1}{\Gamma(-v)(-2v)} \sum_{k=1}^{n-1}((k+1)^{-v} - (k-1)^{-v})s(x-k,y)$$

$$+ \frac{1}{\Gamma(-v)(-2v)}(n^{-v} - (n-1)^{-v})s(x-n,y), v < 0 \quad (11.83)$$

$$\frac{\partial^v s(x,y)}{\partial y^v} \cong \frac{1}{\Gamma(-v)(-2v)} s(x,y) + \frac{1}{\Gamma(-v)(-2v)} \sum_{k=1}^{n-1}((k+1)^{-v} - (k-1)^{-v})s(x,y-k)$$

$$+ \frac{1}{\Gamma(-v)(-2v)}(n^{-v} - (n-1)^{-v})s(x,y-n), v < 0 \quad (11.84)$$

In the same way, one may implement the fractional mask respectively on the eight symmetric directions, which are noted as $YiFeiPU$-3. In contrast with operator $YiFeiPU$-1, it has the same structure but different coefficient. The mask coefficient of $YiFeiPU$-3 operator is given by

$$\begin{cases} C_{s_0} = \dfrac{1}{\Gamma(-v)(-2v)} \\ C_{s_1} = \dfrac{2^{-v}}{\Gamma(-v)(-2v)} \\ \vdots \\ C_{s_k} = \dfrac{(k+1)^{-v} - (k-1)^v}{\Gamma(-v)(-2v)}, v < 0 \\ \vdots \\ C_{s_{n-1}} = \dfrac{n^{-v} - (n-2)^{-v}}{\Gamma(-v)(-2v)} \\ C_{s_n} = \dfrac{n^{-v} - (n-1)^{-v}}{\Gamma(-v)(-2v)} \end{cases} \quad (11.85)$$

(11.83) and (11.84) show that, when $k \to n = 2$, one can get 3×3 mask; when $k \to n = 4$, one can get 5×5 mask; and when $k \to n = 2m$, one can get $(2m+1) \times (2m+1)$ mask. Here n is often taken even number.

Similar to the algorithm of $YiFeiPU$-2, by $(2m+1) \times (2m+1)$ mask, $YiFeiPU$-3 respectively did convolution filter on the eight directions. The algorithm of W_x^-, W_y^-,

W_x^+, W_y^+, W_{LDD}, W_{RUD}, W_{LUD} and W_{RDD} can be given by

$$s_{x^-}^{(v)}(x,y) = \sum_{i=-2m}^{0}\sum_{j=-m}^{m} W_x^-(i,j)s(x+i,y+j) \tag{11.86}$$

$$s_{x^+}^{(v)}(x,y) = \sum_{i=0}^{2m}\sum_{j=-m}^{m} W_x^+(i,j)s(x+i,y+j) \tag{11.87}$$

$$s_{y^-}^{(v)}(x,y) = \sum_{i=-m}^{m}\sum_{j=-2m}^{0} W_y^-(i,j)s(x+i,y+j) \tag{11.88}$$

$$s_{y^+}^{(v)}(x,y) = \sum_{i=-m}^{m}\sum_{j=0}^{2m} W_y^+(i,j)s(x+i,y+j) \tag{11.89}$$

$$s_{LDD}^{(v)}(x,y) = \sum_{i=0}^{2m}\sum_{j=-2m}^{0} W_{LDD}(i,j)s(x+i,y+j) \tag{11.90}$$

$$s_{RUD}^{(v)}(x,y) = \sum_{i=-2m}^{0}\sum_{j=0}^{2m} W_{RUD}(i,j)s(x+i,y+j) \tag{11.91}$$

$$s_{LUD}^{(v)}(x,y) = \sum_{i=-2m}^{0}\sum_{j=-2m}^{0} W_{LUD}(i,k)s(x+i,y+j) \tag{11.92}$$

$$s_{RDD}^{(v)}(x,y) = \sum_{i=0}^{2m}\sum_{j=0}^{2m} W_{RDD}(i,j)s(x+i,y+j) \tag{11.93}$$

For color image, the algorithm of $YiFeiPU$-3 is the same as $YiFeiPU$-2.

11.2.3.5 The Fourth Fractional Differential Mask

To make the operator moreprecise, we can improve $YiFeiPU$-3. By taking two neighboring pixel $s(x-\frac{kx}{N})$ and $s(x-\frac{x}{N}-\frac{kx}{N})$ and using two-point linear interpolation equation, one has

$$s(\gamma) \cong \frac{\gamma-x+\frac{x}{N}+\frac{kx}{N}}{x/N}s(x-\frac{kx}{N}) + \frac{\gamma-x+\frac{kx}{N}}{-x/N}s(x-\frac{x}{N}-\frac{kx}{N}) \tag{11.94}$$

Let $\gamma = x-\xi$ and take into (11.94), it has

$$s(x-\xi) \cong (1+k-\frac{N\xi}{x})s(x-\frac{kx}{N}) + (\frac{N\xi}{x}-k)s(x-\frac{x}{N}-\frac{kx}{N})$$

$$= (1+k-\frac{N\xi}{x})s_k + (\frac{N\xi}{x}-k)s_{k+1} \tag{11.95}$$

Then, we get

$$\int_{kx/N}^{(kx+x)/N}\frac{s(x-\xi)}{\xi^{v+1}}d\xi \cong \int_{kx/N}^{(kx+x)/N}\frac{[(1+k-\frac{N\xi}{x})s_k + (\frac{N\xi}{x}-k)s_{k+1}]}{\xi^{v+1}}d\xi \tag{11.96}$$

Taking (11.95) into (11.79), it has

$$\frac{d^v}{dx^v}s(x)\bigg|_{AL-4} \cong \frac{1}{\Gamma(-v)}\sum_{k=0}^{N-1}\int_{kx/N}^{(kx+x)/N}\frac{[(1+k-N\xi/x)s_k + (N\xi/x-k)s_{k+1}]}{\xi^{v+1}}d\xi$$

$$= \frac{x^{-v}N^v}{\Gamma(-v)}\sum_{k=0}^{N-1}\bigg[\frac{(1+k)s_k - ks_{k+1}}{-v}((k+1)^{-v}-k^{-v})$$

$$+ \frac{s_{k+1}-s_k}{1-v}((k+1)^{1-v}-k^{1-v})\bigg], v<0 \tag{11.97}$$

By taking linear interpolated method, the algorithm is though more complex but more accurate. The fractional differential is also simplified as multiplying and adding. It is noted as $AL-4$. When $k=n \leqslant N-1$, the anterior $n+1$ approximate backward difference of fractional partial differential on negative x- and y-coordinate are expressed as

$$\frac{\partial^v s(x,y)}{\partial x^v} \cong \frac{1}{\Gamma(-v)(v^2-v)} s(x,y) + \frac{1}{\Gamma(-v)(v^2-v)}$$
$$\sum_{k=1}^{n-1} ((k+1)^{1-v} - 2k^{(1-v)} + (k-1)^{1-v}) s(x-k,y)$$
$$+ \frac{1}{\Gamma(-v)(v^2-v)} ((1-v)n^{-v} - n^{1-v} + (n-1)^{1-v}) s(x-n,y), v<0$$

(11.98)

$$\frac{\partial^v s(x,y)}{\partial y^v} \cong \frac{1}{\Gamma(-v)(v^2-v)} s(x,y) + \frac{1}{\Gamma(-v)(v^2-v)}$$
$$\sum_{k=1}^{n-1} ((k+1)^{1-v} - 2k^{(1-v)} + (k-1)^{1-v}) s(x,y-k)$$
$$+ \frac{1}{\Gamma(-v)(v^2-v)} ((1-v)n^{-v} - n^{1-v} + (n-1)^{1-v}) s(x,y-n), v<0$$

(11.99)

Similarly, one can implement the fractional masks respectively on the eight central symmetric directions, which are noted as *YiFeiPU*-4. It has the same structure and different coefficient with *YiFeiPU*-3. The mask coefficient of operator *YiFeiPU*-4 is given by

$$\begin{cases} C_{s_0} = \dfrac{1}{\Gamma(-v)(v^2-v)} \\ C_{s_1} = \dfrac{2^{1-v}-2}{\Gamma(-v)(v^2-v)} \\ \vdots \\ C_{s_k} = \dfrac{(k+1)^{1-v} - 2k^{1-v} + (k-1)^{1-v}}{\Gamma(-v)(v^2-v)}, v<0 \\ \vdots \\ C_{s_{n-1}} = \dfrac{n^{1-v} - 2(n-1)^{1-v} + (n-2)^{1-v}}{\Gamma(-v)(v^2-v)} \\ C_{s_n} = \dfrac{(1-v)n^{-v} - n^{1-v} + (n-1)^{1-v}}{\Gamma(-v)(v^2-v)} \end{cases}$$

(11.100)

(11.98) and (11.99) show that, when $k \to n = 2$, it has 3×3 mask; when $k \to n = 4$, it has 5×5 mask; and when $k \to n = 2m$, it has $(2m+1) \times (2m+1)$ mask. n is often taken even number. In addition, the algorithm of *YiFeiPU*-4 is the same as that of *YiFeiPU*-3.

11.2.3.6 The Fifth Fractional Differential Mask

We know that Riemann-Liouville based fractional integral algorithm requires $v<0$. However, we often adopt fractional differential when $0 \leqslant v < 1$ to enhance the texture details [185] - [186], [434] - [435]. So, we should implement Riemann-Liouville based fractional differential algorithm with $0 \leqslant v < 1$. Let $a = 0$ and $n = 1$, from (11.55) one get

$$\left.\frac{\mathrm{d}^v}{\mathrm{d}x^v}s(x)\right|_{R-L} = \left.\frac{\mathrm{d}^1}{\mathrm{d}x^1}\frac{\mathrm{d}^{v-1}}{\mathrm{d}x^{v-1}}s(x)\right|_{R-L}$$

$$= \frac{x^{-v}s(0)}{\Gamma(1-v)} + \frac{1}{\Gamma(1-v)}\int_0^x \frac{s^{(1)}(\xi)}{(x-\xi)^v}\mathrm{d}\xi$$

$$\cong \frac{1}{\Gamma(1-v)}\left[\frac{s(0)}{x^v} + \sum_{k=0}^{N-1}\int_{kx/N}^{(kx+x)/N}\left(\frac{\mathrm{d}}{\mathrm{d}\xi}s(x-\xi)\right)\frac{\mathrm{d}\xi}{\xi^v}\right], 0 \leqslant v < 1 \quad (11.101)$$

Refer to the difference equation of first-order derivative, we can get

$$\int_{kx/N}^{(kx+x)/N}\left(\frac{\mathrm{d}}{\mathrm{d}\xi}s(x-\xi)\right)\frac{\mathrm{d}\xi}{\xi^v} \cong \frac{s(x-\frac{kx}{N}) - s(x-\frac{x}{N}-\frac{kx}{N})}{x/N}\int_{kx/N}^{(kx+x)/N}\frac{\mathrm{d}\xi}{\xi^v}$$

$$= \frac{x^{-v}N^v}{1-v}(s_k - s_{k+1})[(k+1)^{1-v} - k^{1-v}] \quad (11.102)$$

Taking (11.102) into (11.101), it has

$$\left.\frac{\mathrm{d}^v}{\mathrm{d}x^v}s(x)\right|_{AL-5} \cong \frac{x^{-v}N^v}{\Gamma(2-v)}\left[\frac{(1-v)s_N}{N^v} + \sum_{k=0}^{N-1}(s_k - s_{k+1})((k+1)^{1-v} - k^{1-v})\right], 0 \leqslant v < 1$$

(11.103)

This expression is noted as $AL - 5$. When $k = n \leqslant N - 1$, the anterior $n + 1$ approximate backward differences of fractional partial differential on negative x- and y-coordinate are given by

$$\frac{\partial^v s(x,y)}{\partial x^v} \cong \frac{1}{\Gamma(2-v)}s(x,y) + \frac{1}{\Gamma(2-v)}\sum_{k=1}^{n-1}((k+1)^{1-v} - 2k^{(1-v)} + (k-1)^{1-v})s(x-k,y)$$

$$+ \frac{1}{\Gamma(2-v)}((1-v)n^{-v} - n^{1-v} + (n-1)^{1-v})s(x-n,y), 0 \leqslant v < 1$$

(11.104)

$$\frac{\partial^v s(x,y)}{\partial y^v} \cong \frac{1}{\Gamma(2-v)}s(x,y) + \frac{1}{\Gamma(2-v)}\sum_{k=1}^{n-1}((k+1)^{1-v} - 2k^{(1-v)} + (k-1)^{1-v})s(x,y-k)$$

$$+ \frac{1}{\Gamma(2-v)}((1-v)n^{-v} - n^{1-v} + (n-1)^{1-v})s(x,y-n), 0 \leqslant v < 1$$

(11.105)

We can implement the fractional differential masks respectively on the eight central symmetric directions, which are noted as *YiFeiPU*-5. Comparing with operator *YiFeiPU*-3, it has same structure and different mask coefficient. The mask coefficient of operator *YiFeiPU*-5 is expressed by

$$\begin{cases} C_{s_0} = \dfrac{1}{\Gamma(2-v)} \\ C_{s_1} = \dfrac{2^{1-v} - 2}{\Gamma(2-v)} \\ \vdots \\ C_{s_k} = \dfrac{(k+1)^{1-v} - 2k^{1-v} + (k-1)^{1-v}}{\Gamma(2-v)}, 0 \leqslant v < 1 \\ \vdots \\ C_{s_{n-1}} = \dfrac{n^{1-v} - 2(n-1)^{1-v} + (n-2)^{1-v}}{\Gamma(2-v)} \\ C_{s_n} = \dfrac{(1-v)n^{-v} - n^{1-v} + (n-1)^{1-v}}{\Gamma(2-v)} \end{cases} \quad (11.106)$$

(11.104) and (11.105) show that, when $k \to n = 2$, it has 3×3 mask; when $k \to n = 4$, it has 5×5 mask; and when $k \to n = 2m$, it has $(2m+1) \times (2m+1)$ mask. n is often taken even number. Moreover, the algorithm of *YiFeiPU*-5 is the same with *YiFeiPU*-3.

11.2.3.7 The Sixth Fractional Differential Mask

The fractional differential in $1 \leqslant v < 2$ is the commonly used to enhance verge and texture details in digital image processing. Thus, we should implement Riemann-Liouville based fractional differential algorithm when $1 \leqslant v < 2$. We set $a = 0$ and $n = 1$, and it has

$$\begin{aligned}\dfrac{d^v}{dx^v}s(x)\big|_{R-L} &= \dfrac{d^2}{dx^2}\dfrac{d^{v-2}}{[d(x-a)]^{v-2}}s(x)\big|_{R-L} \\ &= \dfrac{x^{-v}s(0)}{\Gamma(1-v)} + \dfrac{s^{1-v}s^{(1)}(0)}{\Gamma(2-v)} + \dfrac{1}{\Gamma(2-v)}\int_0^x \dfrac{s^{(2)}(\xi)}{(x-\xi)^{v-1}}d\xi \\ &\cong \dfrac{1}{\Gamma(2-v)}\bigg[\dfrac{(1-v)s(0)}{x^v} + \dfrac{s^{(1)}}{x^{v-1}} + \sum_{k=1}^{N-1}\int_{kx/N}^{(kx+x)/N}\dfrac{s^{(2)}(x-\xi)}{\xi^{v-1}}d\xi\bigg], 1 \leqslant v < 2 \end{aligned}$$
(11.107)

Refer to the difference expression of first-order and second-order derivative, one has

$$s^{(1)}(0) = \dfrac{s(x/N) - s(0)}{x/N} = \dfrac{N}{x}(s_{N-1} - s_N) \quad (11.108)$$

$$\int_{kx/N}^{(kx+x)/N}\dfrac{s^{(2)}(x-\xi)}{\xi^{v-1}}d\xi = \dfrac{s\left(x + \dfrac{x}{N} - \dfrac{kx}{N}\right) - 2s\left(x - \dfrac{kx}{N}\right) + s\left(x - \dfrac{x}{N} - \dfrac{kx}{N}\right)}{(x/N)^2}\int_{kx/N}^{(kx+x)/N}\dfrac{d\xi}{\xi^{v-1}}$$
$$= \dfrac{x^{-v}N^v}{2-v}(s_{k-1} - 2s_k + s_{k+1})[(k+1)^{2-v} - k^{2-v}] \quad (11.109)$$

Taking (11.108) and (11.109) into (11.107), we get

$$\dfrac{d^v s(x)}{dx^v}\bigg|_{AL-6} = \dfrac{x^{-v}N^v}{\Gamma(3-v)}\bigg[\dfrac{(1-v)(2-v)s_N}{N^v} + \dfrac{(2-v)(s_{N-1} - s_N)}{N^{v-1}}$$
$$+ \sum_{k=0}^{N-1}(s_{k-1} - 2s_k + s_{k+1})((k+1)^{2-v} - k^{2-v})\bigg], 1 \leqslant v < 2 \quad (11.110)$$

Simplify fractional differential as multiplication and add, and note the algorithm as $AL-6$.

When $k=n \leqslant N-1$, the anterior $n+2$ approximate backward differences of fractional partial differential on negative x- and y-coordinate are given by

$$\frac{\partial^v s(x,y)}{\partial x^v} \cong \frac{1}{\Gamma(3-v)}s(x+1,y) + \frac{2^{2-v}-3}{\Gamma(3-v)}s(x,y) + \frac{1}{\Gamma(3-v)}$$
$$\sum_{k=1}^{n-2}\left[-(k-1)^{2-v} + 3k^{2-v} - 3(k+1)^{2-v} + (k+2)^{2-v}\right]s(x-k,y)$$
$$+\frac{1}{\Gamma(3-v)}\left[(2-v)n^{1-v} - 2n^{2-v} + 3(n-1)^{2-v} - (n-2)^{2-v}\right]s(x-n+1,y)$$
$$+\frac{1}{\Gamma(3-v)}\left[(2-3v+v^2)n^{-v} - (2-v)n^{1-v} + n^{2-v} - (n-1)^{2-v}\right]s(x-n,y),$$
$$1 \leqslant v < 2 \qquad (11.111)$$

$$\frac{\partial^v s(x,y)}{\partial x^v} \cong \frac{1}{\Gamma(3-v)}s(x+1,y) + \frac{2^{2-v}-3}{\Gamma(3-v)}s(x,y) + \frac{1}{\Gamma(3-v)}$$
$$\sum_{k=1}^{n-2}\left[-(k-1)^{2-v} + 3k^{2-v} - 3(k+1)^{2-v} + (k+2)^{2-v}\right]s(x,y-k)$$
$$+\frac{1}{\Gamma(3-v)}\left[(2-v)n^{1-v} - 2n^{2-v} + 3(n-1)^{2-v} - (n-2)^{2-v}\right]s(x,y-n+1)$$
$$+\frac{1}{\Gamma(3-v)}\left[(2-3v+v^2)n^{-v} - (2-v)n^{1-v} + n^{2-v} - (n-1)^{2-v}\right]s(x,y-n),$$
$$1 \leqslant v < 2 \qquad (11.112)$$

Similarly we implement the fractional differential mask respectively on the eight symmetric directions, which are notedas *YiFeiPU*-6. In contrast with operator *YiFeiPU*-2, it has same structure and different coefficient, see Fig. 11.41. The coefficient of *YiFeiPU*-6 is given by

$$\begin{cases} C_{s_{-1}} = \dfrac{1}{\Gamma(3-v)} \\ C_{s_0} = \dfrac{2^{2-v}-3}{\Gamma(3-v)} \\ C_{s_1} = \dfrac{3 - 3 \cdot 2^{2-v} + 3^{2-v}}{\Gamma(3-v)} \\ \vdots \\ C_{s_k} = \dfrac{-(k-1)^{2-v} + 3k^{2-v} - 3(k+1)^{2-v} + (k+2)^{2-v}}{\Gamma(3-v)}, 1 \leqslant v < 2 \quad (11.113) \\ \vdots \\ C_{s_{n-2}} = \dfrac{-(n-3)^{2-v} + 3(n-2)^{2-v} - 3(n-1)^{2-v} + n^{2-v}}{\Gamma(3-v)} \\ C_{s_{n-1}} = \dfrac{(2-v)n^{1-v} - 2n^{2-v} + 3(n-1)^{2-v} - (n-2)^{2-v}}{\Gamma(3-v)} \\ C_{s_n} = \dfrac{(2-3v+v^2)n^{-v} - (2-v)n^{1-v} + n^{2-v} - (n-1)^{2-v}}{\Gamma(3-v)} \end{cases}$$

(11.111) and (11.112) show that, when $k \to n = 1$, it has 3×3 mask; when $k \to n = 3$, it has 5×5 mask; and when $k \to n = 2m-1$, it has $(2m+1) \times (2m+1)$ mask. n is usually taken odd number. The algorithm of $YiFeiPU$-6 is the same as $YiFeiPU$-2.

11.2.4 EXPERIMENTS AND ANALYSIS

This section aims at demonstrating that fractional differential masks have better capability in texture-enhancing and segmenting than the traditional approaches for texture-rich image. To this purpose, from the views of telecommunication and signal processing, we analyze the nonlinearly texture-enhancing capability of fractional differential. Then we systematically carry out the relative errors analysis of the proposed 6 fractional different masks and prove that the performance of $YiFeiPU$-2 is the best. Finally, we discuss the capability of texture enhancement of the fractional differential masks and compare them with the traditional approaches.

11.2.4.1 Analyzing the Capability of Non-Linearly Texture Enhancement of Fractional Differential

From the views of signal processing, we analyze the nonlinearly texture-enhancingcapability of fractional differential. It is well known that the Fourier transform of the first-order differential of any quadratic integrable energy signal $s'(t)$ is $Ds(t) \overset{FT}{\Leftrightarrow} (\hat{D}s)(\omega) = (i\omega) \cdot \hat{s}(\omega) = \hat{d}(\omega)\hat{s}(\omega)$. Similarly, from Grümwald-Letnikov based definition, the Fourier transform of fractional differential of signal $D_{G-L}^v s(t) = s^{(v)}(t)$ is given by

$$D_{G-L}^v(s)t = s^{(v)}(t) \overset{FT}{\Leftrightarrow} (\hat{D}_{G-L}^v s)(\omega) = (i\omega)^v \cdot \hat{s}(\omega) - \sum_{k=0}^{n-1} \cdot (i\omega)^k \frac{d^{v-1-k}}{dt^{v-1-k}} s(0) \quad (11.114)$$

where $n-1 < v \leqslant n$. In (11.114), $\sum_{k=0}^{n-1}(i\omega)^k \frac{d^{v-1-k}}{dt^{v-1-k}} s(0)$ is a constant, which will not change the waveform of spectrum density of signal. As for single sideband, it concerns to the initials, and for double sideband, it can be devised. Since we only care about the influence of fractional differential on the waveform of spectrum density, (11.114) can be simplified as

$$D_{G-L}^v S(t) = s^{(v)}(t) \overset{FT}{\Leftrightarrow} (\hat{D}_{G-L}^v s)(\omega) = (i\omega)^v \cdot \hat{s}(\omega) = \hat{d}_v(\omega) \cdot \hat{s}(\omega) \quad (11.115)$$

where, D_{G-L}^v is v-order differential multipliable operator. $\hat{d}_v(\omega) = (i\omega)^v$ is filter function, whose plural exponential and time domain can be respectively expressed as

$$\begin{cases} \hat{d}_v(\omega) = (i\omega)^v = \hat{a}_v(\omega) \cdot \exp(i\theta_v(\omega)) = \hat{a}_v(\omega) \cdot \hat{p}_v(\omega) \\ \hat{a}_v(\omega) = |\omega|^v, \hat{\theta}_v(\omega) = \frac{v\pi}{2}\mathrm{sgn}(\omega) \end{cases} \quad (11.116)$$

$$\hat{d}_v(\omega) \overset{IFT}{\Leftrightarrow} d_v(t) = a_v(t) \cdot v(t) = \frac{1}{2\pi}\int_{-\infty}^{+\infty} (i\omega)^v \cdot e^{i\omega t} d\omega \quad (11.117)$$

$$a_v(t) = \frac{1}{2\pi}\int_{-\infty}^{+\infty} \hat{a}_v \cdot e^{i\omega t} d\omega = \frac{1}{\pi}\int_{-\infty}^{+\infty} |\omega|^v \cdot \cos(i\omega t) d\omega \quad (11.118)$$

$$p_v(t) = \frac{1}{2\pi}\int_{-\infty}^{+\infty} p_v(\omega) \cdot e^{i\omega t} d\omega = \cos\frac{v\pi}{2} \cdot \delta(t) - \sin\frac{v\pi}{2} \cdot \frac{1}{\pi t} \quad (11.119)$$

From (11.116 − 11.119) and in light of the viewpoints of telecommunication modulation, we know that fractional differential is thegeneralized amplitude-and-phase modulation. The amplitude changes with frequency as fractional power exponent, while the phase is generalized Hilbert transform of frequency. The filter function of fractional differential filter is $\hat{d}(\omega) = (i\omega)^v = |\omega|^v \cdot \exp(i\theta_v(\omega))$. The amplitude characteristic is even function and phase characteristic is odd function. Thus we only need to survey the characteristics of fractional differential filter when $\omega > 0$. The frequency response of fractional differential is shown in Fig. 11.42.

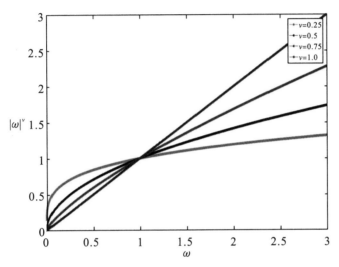

Fig. 11.42 The frequency response of fractional differential.

From Fig. 11.42 we know that, inviewpoints of signal processing, the frequency response of fractional differential is actually a nonlinear filter. In time-domain, fractional differential of signal $s(t)$ can be rewritten as

$$D_{G-L}^v S(t) = d_v(t) \cdot s(t) = a_v(t) \cdot p_v(t) \cdot s(t)$$
$$= \left[\cos\frac{v\pi}{2} \cdot \delta(t) - \sin\frac{v\pi}{2} \cdot \frac{1}{\pi t}\right] \cdot \int_{-\infty}^{+\infty} a_v(t-\tau)s(\tau)d\tau \quad (11.120)$$

When $v=0$, v-order fractional differential is all-pass filter, and its frequency response is $\hat{d}_v(\omega) \equiv 1 \Leftrightarrow d_v(t) = \delta(t)$; When $v<0$, it is fractional integrator and $\hat{d}_v(\omega)$ is singulag low-pass integral filter; When $v>0$, it is fractional derivative operator, and its frequency response is $\lim_{|\omega|\to\infty} |\hat{d}_v(\omega)| \to \infty$. Here, $\hat{d}_v(\omega)$ is singular high-pass differential filter. Note that, while v is increasing, the transmission bands of $\hat{d}_v(\omega)$ become narrower and high-pass characteristic is stronger. In other words, when $v>0$, $\hat{d}_v(\omega)$ nonlinearly enhances high-frequency components of signal $s(t)$ and nonlinearly inhibits its low-frequency components.

Thus, it is naturally to think whether we could design a two-dimensional fractional differential filter for nonlinearly enhancing texture. From Fig. 11.3, we see that, when $0<v<1$, in the section of $\omega>1$, the enhancement of high-frequency components by fractional differential is less than integral one, and the enhancement of high-frequency edge components by fractional differential is inferior to that of first-order one; However, in extremely low-frequency section of $0<\omega<1$, the preservation magnitude of low-frequency contour by fractional differential is superior to that by first-order one. So, fractional differential is a nonlinear attenuator while first-order differential is linear one. That is, in the extremely low-frequency section of $0<\omega<1$, when v becomes smaller, the attenuation of low-frequency components is also less. Similarly, when $v \to 0$, it is an all-pass filter almost keeping signal unchanged. From above-mentioned discussion, we know that as for the image's smooth area whose grey scale does not change intensively, the texture features in smooth area may be greatly attenuating and its differential result is nearly zero (the integral differential of constant is zero), when it is filtered by first-order differential based operator such as Sobel operator, or second-order based one such as Gauss-Laplace operator. For this reason, integral differential linearly attenuates texture features and could not well hold them in such area. On the contrary, fractional differential based operator could nonlinearly preserve textural feature in smooth area to the biggest degree. Thus, as for enhancing the texture in smooth area, fractional differential based operator is superior to integral differential based one.

Then we survey the characteristics of fractional differential and fractional differential mask by experiments. Without losing generality, let $v=0.5$, we do fractional differential for rectangle wave signal and saw-tooth wave signal. The results are shown in Fig. 11.43.

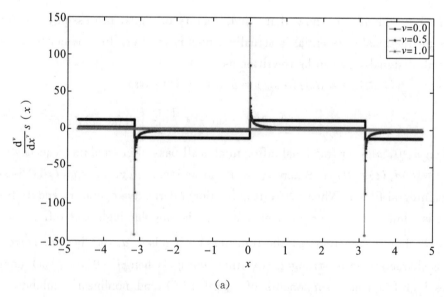

(a)

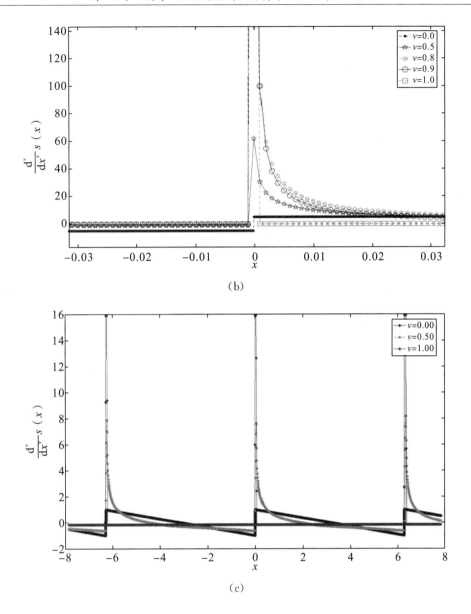

(b)

(c)

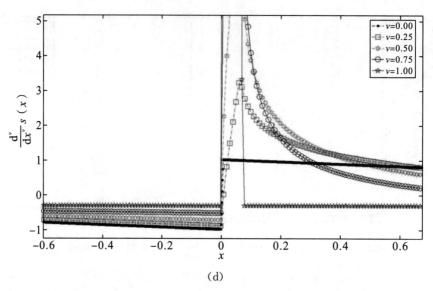

(d)

Fig. 11.43 Characteristics analyzing for fractional differential. (a) Comparison to first-order rectangle wave signal and fractional differential, (b) partial details of rectangle wave signal of fractional differential, (c) First-order and fractional differential of saw-tooth wave signal, (d) partial details of fractional differential of saw-tooth wave signal.

Fig. 11.43 shows that fractional differential have the following characteristics. In the first, fractional differential is from the biggest value in the singular leaping point to zero in smooth area. Note that, by default, any integral differential in smooth area approximately equals to zero (the integral differential of constant is zero), which is the remarkable difference between fractional differential and integral one. In the second, fractional differential in the initial point of grey scale gradient or slope is non-zero, which nonlinearly enhances high-frequency singular information. As we know, when $v=1$, integral differential of high-frequency singular signal is Dirac signal. The classical Sobel operator and Gauss operator is equal to first-order and second-order fractional differential operator respectively. When $0<v<1$, the enhancement of high-frequency singular information is smaller than that when $v=1$. Therefore, integral differential is the special case of fractional differential. Moreover, as its fractional order increased, the enhancement of high-frequency singular information by fractional differential also increases. In the third, the fractional differential of the slope is not zero or constant, it is a nonlinear curve. But, the integral differential of the slope is constant. From the above discussion, we can see that fractional differential could nonlinearly preserve the low-frequency contour feature in the smooth area to the furthest degree, and as well as, nonlinearly enhance high-frequency marginal information in those areas where grey scale changes frequently, and nonlinearly enhance texture details in those areas where grey scale does not change evidently. In brief, fractional differential could nonlinearly enhance the comprehensive texture details. When the image is processed, it requires keeping the original information, improving image quality, enhancing details and texture characteristics, and keeping the marginal details and energy as well. The requirements are difficult to obtain by traditional integral differential based texture-enhancing algorithms, while they are easy to abtain by fractional differential based algorithm.

11.2.4.2 Relative Errors Analyzing for Fractional Differential Mask

The proposed 6 fractional differential masks are all based on approximate analyzing, so errors can not be avoided. Note that relative error is commonly used for measuring errors in physical calculate. As for fractionaldifferential $\frac{d^v s(x)}{dx^v}$, the relative errors of ith algorithm are given by

$$\sigma_{AL-i} = \frac{\frac{d^v s(x)}{dx^v}\big|_{AL-i} - \frac{d^v s(x)}{dx^v}}{\frac{d^v s(x)}{d^v x}} \tag{11.121}$$

where $\frac{d^v s(x)}{dx^v}\big|_{AL-i}$ is the ith algorithmic value and $\frac{d^v s(x)}{dx^v}$ is its analytic value.

As for integral differential, it is known that differentiability means finite orderderivative, and analyticity means infinite order derivative. However, as for fractional differential, the differentiability and analyticity is to say that the function is differintegrable [3], [176] – [177], [230], [429] – [430]. If $s(x)$ is differintegrable function, it equals to differintegrable series that is the finite sum of differintegrable unit function. In fact, most functions are differintegrable in mathematical and physical field.

$$s(x) = \sum_{j=1}^{n} s_j(x) \tag{11.122}$$

$$s_j(x) = (x-a)^q \sum_{r=0}^{\infty} a_r (x-a)^{r/n}, a_0 \neq 0, q > -1 \tag{11.123}$$

Here, $q > -1$ and it is real number (fractional number included), n is positive integer, and $s_j(x)$ is differintegrable unit function. Thus, the differintegrable unit function $s_j(x)$ can be expressed by the form of product of $(x-a)^q$ and $(x-a)^{1/n}$. Without losing generality, set $a=0$ and $n=1$, then we have

$$s_j(x) = \sum_{r=0}^{\infty} a_r x^{q+r} = \sum_{r=0}^{\infty} a_r x^p, a_0 \neq 0, p > -1 \tag{11.124}$$

Therefore, the errors analysis of fractional differential numerical algorithm can be deemed as the corresponding errors analyzing of power function x^p when $p > -1$. If N is big enough, the relative errors of power function x^p can be worked out.

Refer to (11.59), (11.66), (11.82), (11.97), (11.103) and (11.110), we can get thealgorithmic value of power function x^p respectively by the proposed 6 fractional differential algorithms, which are given by

$$\frac{d^v}{dx^v} x^p \bigg|_{AL-1} \simeq \frac{x^{-v} N^v}{\Gamma(-v)} \sum_{k=0}^{N-1} \frac{\Gamma(k-v)}{\Gamma(k+1)} \left(x - \frac{kx}{N}\right)^p \tag{11.125}$$

$$\frac{d^v}{dx^v} x^p \bigg|_{AL-2} \simeq \frac{x^{-v} N^v}{\Gamma(-v)} \sum_{k=0}^{N-1} \frac{\Gamma(k-v)}{\Gamma(k+1)} \left\{ \left(x - \frac{kx}{N}\right)^p + \frac{v}{4} \left[\left(x + \frac{x}{N} - \frac{kx}{N}\right)^p - \left(x - \frac{x}{N} - \frac{kx}{N}\right)^p\right] \right\}$$

$$+\frac{v^2}{8}\Big[\Big(x+\frac{x}{N}-\frac{kx}{N}\Big)^p - 2\Big(x-\frac{kx}{N}\Big)^p + \Big(x-\frac{x}{N}-\frac{kx}{N}\Big)^p\Big]\Big\} \quad (11.126)$$

$$\frac{\mathrm{d}^v}{\mathrm{d}x^v}x^p\Big|_{AL-3} \cong \frac{x^{-v}N^v}{\Gamma(-v)}\sum_{k=0}^{N-1}\frac{\Big(x-\frac{kx}{N}\Big)^p + \Big(x-\frac{x}{N}-\frac{kx}{N}\Big)^p}{-2v}[(k+1)^{-v}-k^{-v}], v<0 \quad (11.127)$$

$$\frac{\mathrm{d}^v}{\mathrm{d}x^v}x^p\Big|_{AL-4} \cong \frac{x^{-v}N^v}{\Gamma(-v)}\sum_{k=0}^{N-1}\Big\{\frac{(k+1)\Big(x-\frac{kx}{N}\Big)^p - k\Big(x-\frac{x}{N}-\frac{kx}{N}\Big)^p}{-v}[(k+1)^{-v}-k^{-v}]$$
$$+\frac{\Big(x-\frac{x}{N}-\frac{kx}{N}\Big)^p - \Big(x-\frac{kx}{N}\Big)^p}{1-v}[(k+1)^{1-v}-k^{1-v}]\Big\}, v<0 \quad (11.128)$$

$$\frac{\mathrm{d}^v}{\mathrm{d}x^v}x^p\Big|_{AL-5} \cong \frac{x^{-v}N^v}{\Gamma(2-v)}\sum_{k=0}^{N-1}\Big[\Big(x-\frac{kx}{N}\Big)^p - \Big(x-\frac{x}{N}-\frac{kx}{N}\Big)^p\Big][(k+1)^{1-v}-k^{1-v}], 0\leqslant v<1 \quad (11.129)$$

$$\frac{\mathrm{d}^v x^p}{\mathrm{d}x^v}\Big|_{AL-6} \cong \frac{x^{-v}N^v}{\Gamma(3-v)}\Big\{\frac{(2-v)\Big(\frac{x}{N}\Big)^p}{N^{v-1}} + \sum_{k=0}^{N-1}\Big[\Big(x+\frac{x}{N}-\frac{kx}{N}\Big)^p - 2\Big(x-\frac{kx}{N}\Big)^p$$
$$+\Big(x-\frac{x}{N}-\frac{kx}{N}\Big)^p\Big][(k+1)^{2-v}-k^{2-v}]\Big\}, 1\leqslant v<2 \quad (11.130)$$

According to Riemann-Liouville based definition in (11.55), power function $\frac{\mathrm{d}^v x^p}{\mathrm{d}x^v}$ can be given by [13].

$$\frac{\mathrm{d}^v x^p}{\mathrm{d}x^v} = \frac{\Gamma(p+1)x^{p-v}}{\Gamma(p-v+1)}, p>-1 \quad (11.131)$$

Then, respectively taking (11.125—11.131) into (11.121), the analytic expression for relative errors σ_{AL-i} of the above 6 algorithms can be gained.

Taking $AL-1$ as example, one can discuss the convergence of fractional differential algorithm. We observe the relationship between relative errors σ_{AL-1} and N & v on the point $x=1.0$ and $p=3/2$, which is shown in Fig. 11.44.

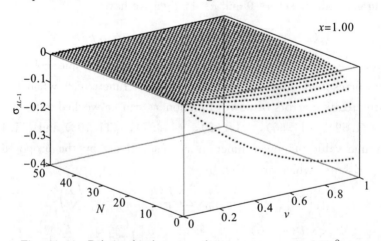

Fig. 11.44 Relationship between relative errors σ_{AL-1} and N & v.

When order v is certain, the relative errors σ_{AL-1} is decreased as N increasing and it is even near to zero; when N is certain, σ_{AL-1} is increased with v increasing. However, when N is the bigger enough, the increase of v has little affection on σ_{AL-1}.

Let $v=0.5$ and $N=150$, the relationship between x and relative errors σ_{AL-1} is shown in Fig. 11.45.

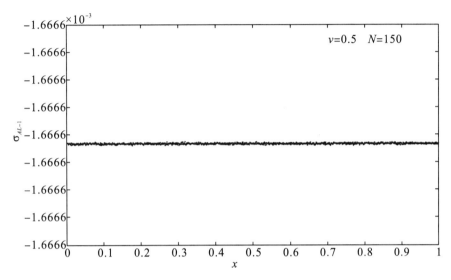

Fig. 11.45 Relationship between x and σ_{AL-1}.

Fig. 11.45 shows that when $N=150$, it has $\sigma_{AL-1} \infty o(10^{-3})$. Thus, relative errors σ_{AL-1} depend on v and N. Moreover, provided N tends to infinite, σ_{AL-i} is near to zero in theoretic analysis. Therefore, one can prove that algorithm $AL-1$ has fine convergence; similarly, we can fatherly prove the other 5 algorithms also have fine convergence.

To ensure the symmetry of fractional differential mask, the dimension of its operator must be odd number and one often take 3×3, 5×5, 7×7 and 9×9 masks. As already noted, the dimension of $YiFeiPU-i$ relates to the item n_{AL-i} of corresponding algorithm $AL-i$. The corresponding relationship is shown in Table 11.1.

Table 11.1 Matching relationship between the dimension of mask and the item n_{AL-i}

$YiFeiPU-i$ \ n_{AL-i}	n_{AL-1}	n_{AL-2}	n_{AL-3}	n_{AL-4}	n_{AL-5}	n_{AL-6}
3×3	3	1	2	2	2	1
5×5	5	3	4	4	4	3
7×7	7	5	6	6	6	5
$\vdots$	$\vdots$	$\vdots$	$\vdots$	$\vdots$	$\vdots$	$\vdots$
$(2m+1) \times (2m+1)$	$2m+1$	$2m-1$	$2m$	$2m$	$2m$	$2m-1$

Table 11.1 shows that, when $n_{AL-1}=3$ and $n_{AL-2}=3$, $YiFePU-1$ in 3×3 and 5×5 mask can be implemented. Let $N = n_{AL-1}$, and take it respectively into (11.121),

(11.125) and (11.131), then we get the relationship between order v and relative error σ_{AL-1}, which is shown in Fig. 11.46.

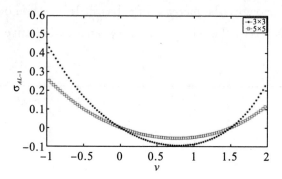

Fig. 11.46 Relationship between order v and relative error σ_{AL-1}.

Fig. 11.46 show that, when v is certain, the corresponding σ_{AL-1} of $N=5$ is smaller than that of $N=3$, and its matching algorithm has better convergence. As for the fractional differential in $0 \leqslant v \leqslant 1$, it has $\sigma_{AL-1} \in (-10^{-1}, 0)$. The relative errors of $YiFeiPU-1$ is the smallest one which can be accepted in engineering.

In general, the often used order of fractional differential for digital image processing is in the range of $-1 \leqslant v < 2$. When $v=0$, it keeps the original image, in other words, there is no differential or integral differential to the image. When $v=1$ or $v=2$, it is the first-order or second-order differential, which do marginal segmentation for digital image. When $-1 < v < 0$, it is the fractional integral, which is fractional linearly denoising. When $0 < v < 1$, it is the fractional differential, which enhances the texture details. When $1 < v < 2$, it is the fractional differential, which segments the texture. And when $v=-1$, it is first-order integral or first-order linearly denoising. In the purpose of determining the best performance of the above six fractional differential masks by theoretic analyzing and experiments, we need compare the relative errors σ_{AL-i} in different masks. Thus, without losing the generality, we compare the relative errors σ_{AL-i} of fractional differential in $-1 \leqslant v \leqslant 2$ at the point $x = \pi$ respectively in 3×3, 5×5 and 7×7 masks, which are shown in Fig. 11.47.

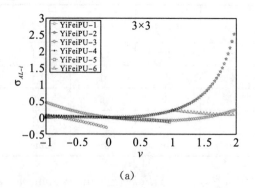

(a)

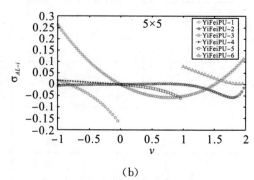

(b)

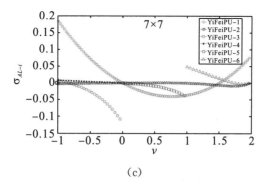

(c)

Fig. 11.47 Relative errors σ_{AL-i} of fractional differential operator $YiFeiPU-i$ (a) 3×3, (b) 5×5, (c) 7×7.

Fig. 11.47 shows that, $YiFeiPU-1$ and $YiFeiPU-2$ has more applications, which can be widely applied for fractional differential and fractional integral. While, $YiFeiPU-3$ and $YiFeiPU-4$ can only be applied to fractional integral, $YiFeiPU-5$ is only for fractional differential in $0 \leqslant v < 1$, and $YiFeiPU-6$ is only for fractional differential in $1 \leqslant v < 2$.

Furthermore, we can find that for the same dimension, $YiFeiPU-2$ has the best convergence and precision. In 3×3 mask, the relative errors of $YiFeiPU-2$ in $-1 \leqslant v < 0.5$ is somewhat like a line, it has $\sigma_{AL-2} \infty o(10^{-2})$ with fine precision; while in $0.5 \leqslant v \leqslant 2$, it is like a power function curve, and the higher the order is, less precision it has. In 5×5 mask, the relative errors of $YiFeiPU-2$ in $-1 \leqslant v < 1.2$ is also like a line and it has $\sigma_{AL-2} \infty o(10^{-2})$ with fine precision; while in $1.2 \leqslant v \leqslant 2$, it is somewhat like a parabola, and the precision of higher order fractional differential is worse than that of lower order one. In 7×7 mask, the relative error of $YiFeiPU-2$ in $-1 \leqslant v \leqslant 2$ is also like a line, and it has $\sigma_{AL-2} \infty o(10^{-2})$ with fine precision. Therefore, with the increase of the dimension of mask, σ_{AL-2} is decreased rapidly. Though the biggest mask of my experiment is only 7×7, $YiFeiPU-2$ has shown its stable and high precision in comparable large range $-1 \leqslant v \leqslant 2$. However, as for $YiFeiPU-1$, $YiFeiPU-3$, $YiFeiPU-4$, $YiFeiPU-5$ and $YiFeiPU-6$, the decrease of σ_{AL-i} is very small as the dimension of mask increases, and algorithm precision only has little improved.

It is because Lagrange 3-point non-linear interpolation has been used in (11.65), the precision of $YiFeiPU-2$ in $-1 \leqslant v \leqslant 2$ is greatly improved with the dimension of mask increasing. In comparison, it is because only 2-point linear interpolation has been used in (11.95), $YiFeiPU-4$ has fine convergence in $-1 \leqslant v < 0$, but with the dimension increasing its precision has little improved.

11.2.4.3 Analyzing for Texture-Enhancing Capability of Multi-Scale Fractional Differential

We have already noted in section 4.2 that $YiFeiPU-2$ has the best convergence and precision. The above characteristics of the operator lead us to further analyzing its capable for enhancing texture details. In my previous works [185] – [186], [434] – [435], we know that when $0 < v < 1$, fractional differential will enhance texture details. Let take $YiFeiPU-2$ in 7×7 mask as example to do fractional differential in $0 \leqslant v \leqslant 1$ on a satellite optic remote sensing image, the experimental results are presented in Fig. 11.48.

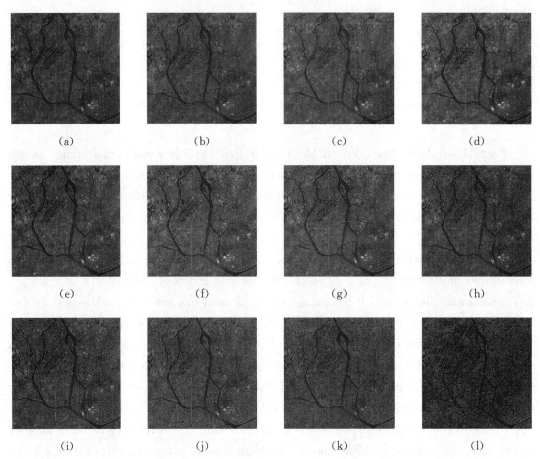

Fig. 11.48 Different order fractional differential of optic remote sensing image. (a) 0 (original image), (b) 0.1, (c) 0.2, (d) 0.25, (e) 0.3, (f) 0.4, (g) 0.5, (h) 0.6, (i) 0.7, (j) 0.8, (k) 0.9, (l) 1.0.

Fig. 11.48 shows that the fractional differential in $0 < v < 1$ is in fact the multi-scale fractional enhancing the image's texture details. Here v is the scale for controlling the enhancing degree. We can see that, from visual angle, firstly, 0 order fractional differential keeps the original image without doing any integral or differential. Secondly, since the integral differential of constant is 0, the first-order differential in smooth area is near to 0, which lost lots of texture details in those area that grey-level does not changing obviously. When doing first-order fractional differential, the image becomes the black in smooth area and the white in high-frequency marginal. It greatly increases marginal information and becomes white verge or black background. Thirdly, we have known in [3], [176] – [177], [230], [430] that the fractional differential of constant is not zero, so when $0<v<1$, the contrast of complex texture details is also enhanced with the increasing of v, and the image becomes clearer. Moreover, the fractional differential of neighboring v is similar, that is to say, it is the continuous interpolation of neighboring order fractional differential. Thus, we can say that fractional differential can well enhance texture details of remote sensing image, even for those textures that only several pixels are known. When $0.9 \leqslant v < 1$, with the increasing of v, the result of fractional differential is closer to that of first-order differential. In the same time, the enhancement for marginal information becomes more visible and verge becomes whiter. Furthermore, with more texture losing, the rich-grained background becomes blacker. When $v=1$, the image after fractional differential becomes white verge and black background.

For quantity analysis, we take grey-level co-occurrence matrix to describe the comprehensive information of the texture details like direction, neighboring distance, changing range and the background. One can pick up 5 classical parameters from co-occurrence matrix, which are angle matrix, contrast, correlation, energy and homogeneity/adverse distance. Then, respectively taking 0° (projection in y-coordinate), 45°, 90° (projection in x-coordinate) and 135°, one can get the corresponding testing values of each parameter in above four angles. So, there are 16 testing values of parameters of grey-level co-occurrence matrix. The parameters of each pixel in 5-pixel distances are shown in Table 11.2.

Table 11.2 Grey-level co-occurrence matrix in four directions (a) 0°, (b) 45°, (c) 90°, (d) 135°

(a)

Fractional differential order / Gray level cooccurrence matrix	Contrast	Correlation	Energy	Honogeneity
0.00	0.5904	0.42582	0.18574	0.77389
0.10	0.67338	0.37279	0.18357	0.75948
0.20	0.93349	0.32579	0.13783	0.71861
0.25	1.2548	0.30072	0.11221	0.68534
0.30	1.3611	0.27144	0.10459	0.67329
0.40	1.9358	0.21702	0.79206	0.62996
0.50	2.3584	0.16444	0.068146	0.60511
0.60	2.9178	0.11996	0.057431	0.57834
0.70	4.6436	0.085754	0.036118	0.51949
0.80	6.5476	0.05973	0.028403	0.48172
0.90	7.9853	0.04732	0.039187	0.4824
1.00	8.6394	0.069256	0.13997	0.57012

(b)

Gray level cooccurrence matrix / Fractional differential order	Contrast	Correlation	Energy	Honogeneity
0.00	0.64379	0.37312	0.18036	0.76051
0.10	0.72483	0.32391	0.17886	0.74712
0.20	0.99141	0.28287	0.13493	0.70765
0.25	1.3246	0.26078	0.10981	0.67469
0.30	1.4291	0.23391	0.10267	0.66362
0.40	2.0126	0.18489	0.077796	0.62802
0.50	2.4307	0.13783	0.066992	0.5966
0.60	2.9904	0.097163	0.056427	0.56978
0.70	4.7371	0.066593	0.035567	0.51167
0.80	6.6604	0.042873	0.028017	0.47387
0.90	8.1278	0.029748	0.038685	0.47498
1.00	8.8962	0.041022	0.13766	0.56421

(c)

Gray level cooccurrence matrix / Fractional differential order	Contrast	Correlation	Energy	Honogeneity
0.00	0.57443	0.44105	0.18813	0.77826
0.10	0.65897	0.3859	0.18577	0.76343
0.20	0.91581	0.33829	0.13941	0.72257
0.25	1.2296	0.31457	0.11342	0.68921
0.30	1.3378	0.28367	0.10571	0.67707
0.40	1.9102	0.22731	0.079928	0.63322
0.50	2.3367	0.17215	0.068676	0.60758
0.60	2.9063	0.12351	0.057725	0.57985
0.70	4.6292	0.088622	0.0363	0.52096
0.80	6.5349	0.061526	0.028499	0.48274
0.90	7.9866	0.047192	0.03913	0.48224
1.00	8.6635	0.066803	0.13955	0.56946

(d)

Gray level cooccurrence matrix / Fractional differential order	Contrast	Correlation	Energy	Honogeneity
0.00	0.64296	0.37394	0.18104	0.75988
0.10	0.72488	0.32387	0.17881	0.74685

(Continued)

Fractional differential order \ Gray level cooccurrence matrix	Contrast	Correlation	Energy	Honogeneity
0.20	0.99048	0.28356	0.13501	0.70778
0.25	1.3239	0.26117	0.10977	0.67471
0.30	1.4258	0.23571	0.10277	0.6643
0.40	2.0077	0.18694	0.077815	0.62141
0.50	2.4249	0.1399	0.06703	0.59721
0.60	2.9845	0.098995	0.56494	0.57057
0.70	4.7236	0.069285	0.35574	0.51189
0.80	6.6341	0.046674	0.28079	0.47488
0.90	8.0942	0.033783	0.038767	0.47587
1.00	8.867	0.044186	0.13847	0.56512

To visual analyze the above data and discuss the relationship between order v and texture-enhancing details, respectively in the directions of 0°, 45°, 90°, 135°, one integrates 5-order nonlinear curve by the above four parameters: contrast, correlation, energy and homogeneity. The results are presented in Fig. 11.49.

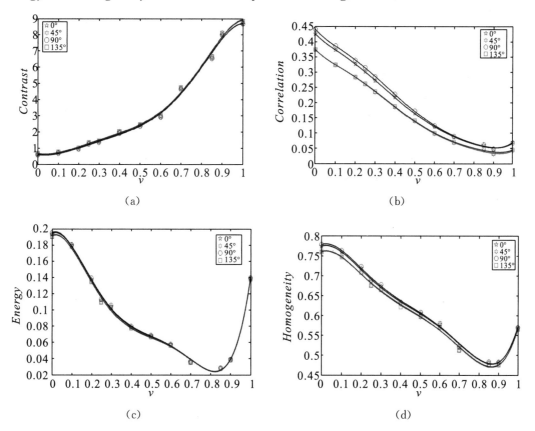

(a)

(b)

(c)

(d)

Fig. 11.49 Integration of 5-order nonlinear curve of parameters of grey-level co-occurrence matrix. (a) contrast, (b) correlation, (c) energy, (d) homogeneity.

Table 11.2 and Fig. 11.49 show that, no matter what the angle is, the change of texture-enhancement has some relations with order v. Firstly, when $v \in [0, 1]$, the contrast of grey-level co-occurrence matrix increases with the increasing of v. Note that when v increases, there are more pixels with great contrast and the texture channel becomes deeper. In general, the image looks clearer. Secondly, when $v \in [0.0, 0.9]$, the correlation of grey-level co-occurrence matrix becomes weaker with the increasing of v. That is to say, when $v \in [0.0, 0.9]$, the partial grey-level correlation is decreased with v increases, and the texture details become clearer. However, when $v \in [0.9, 1.0]$, the correlation of grey-level co-occurrence matrix is increased with v increasing. That is to say, when $v \in (0.9, 1.0]$ the partial grey-level correlation is decreased with v increasing, which leads the image becoming white or black, and the texture details are less visible. Thirdly, when $v \in [0.0, 0.8]$, the energy of grey-level co-occurrence matrix is decreased with the increasing of v. That means when $v \in [0.0, 0.8]$, the partial grey-level correlation is decreased with v increasing, the texture-changing becomes more inhomogeneity and irregularity, and texture details are clearer. When $v \in (0.0, 0.8]$, the energy of grey-level co-occurrence matrix is increased with v increasing. That means when $v \in (0.8, 1.0]$, the partial grey-level correlation is increased with v increasing, the texture-changing is more homogeneity and regularity, and texture details become less visible. Fourthly, when $v \in (0.0, 0.8]$, the homogeneity of grey-level co-occurrence matrix is decreased with v increasing. It means that the texture in different section changes more dramatic and the texture details are clearer. When $v \in (0.8, 1.0]$, the homogeneity is correspondingly increased with v increasing. That is to say, when $v \in (0.8, 1.0]$, with v increasing, it changes little in different sections, the partial is even, and texture details become less visible. In summary, the strongest points of fractional differential for texture-enhancing focus on the range of $v \in (0.8, 0.9]$.

For the purpose of direct observing the changing of grey-scale, without losing generality, we take horizontal projection (y-coordinate) of grey-level of (a), (g) and (l) in Fig. 11.48 as examples, to compare the changing of grey-level values. The results are shown in Fig. 11.50.

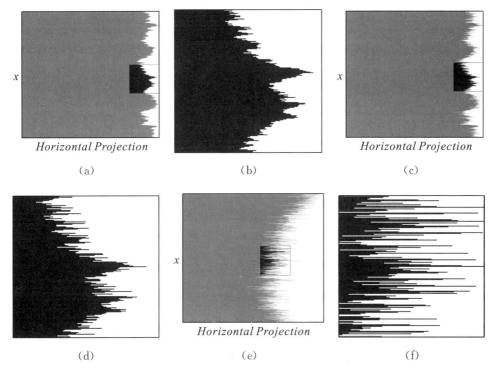

Fig. 11.50 Comparison of horizontal projection of grey-level values between the original satellite optic remote sensing image and its fractional differential. (a) horizontal projection of grey-level in original image, (b) part of horizontal projection of grey-level in original image, (c) horizontal projection of grey-level of $v=0.5$ fractional differential, (d) part of horizontal projection of grey-level of $v=0.5$ fractional differential, (e) horizontal projection of grey-level of $v=1.0$ fractional differential, (f) part of horizontal projection of grey-level of $v=1.0$ fractional differential

Fig. 11.50 shows that, in the first, similar to integral operator, fractional differential operator enhances the high-frequency marginal information in those areas that grey-level changes remarkably, but it is not as strong as integral operator. So it hasn't white-and-black verge. In the second, comparing to integral differential, fractional differential keeps the envelope of grey-level histogram distribution to some degree, and reserves the low-frequency contour information of original image while integrated differential can not do. In the third, both integral and fractional differential will add some burrs to the envelope of grey-level projection of original image, but only fractional differential can be deemed as "adding" burrs, because integral differential has changed its envelope. Note fractional differential can nonlinearly enhance texture details in those areas that grey-levels have little changed, but integral differential is near to zero in those areas. By changing corresponding envelope of grey-level value projection, integral differential has changed its grey-level value diagram distribution, which leads the grey-level value to be too high or too low. Apparently, the image is presented as white or black, but in fact a lot of texture details have lost. Because the grey-level value between neighboring pixels of image is greatly correlated and highly self-similar, the most energy of image converges on low frequency, and the energy of texture and edge information distributes over intermediate and high frequency. Considering the influence of differential based enhancement filter on the mean of the image and its energy, we could see that although integral differential based algorithms can enhance the energy of image on band of high frequency, there is a lot of energy losing on band of low and intermediate frequency. So, the mean of image would change greatly after processed; On the other hand, fractional differential based algorithms can not only maintain the most energy of image on the low frequency, but also nonlinearly enhance its energy over intermediate and high frequency. Then, the mean of image would change slowly after processed. Actually, when images are corrupted with noise, the proposed fractional differential algorithm can also enhance the noise, but its noise enhancement is the interpolation between two neighboring integral differentials ones. Therefore, fractional differential based algorithms can improve of the image quality, while at the same time, restrain the noise enhancement as far as possible. To verify, we do fractional differential to two texture-rich grey images and the results are shown in Fig. 11.51.

(a) (b) (c) (d)

Fig. 11.51 Matching performance comparison between original image and its fractional differential. (a) Amplificatory image of 1/4 top left of 7 (a): original image, (b) amplificatory image of 1/4 top left of 7 (g): 0.5-order, (c) original, (d) 0.6-order.

By respectively doing fractional differential to the three elements of H, S and I in HSI color space, and then revert it to RGB color space, one can gain the fractional differential of color image without distorting. Fig. 11.51 shows that, the backgrounds of original image are those areas with smooth textures, which comprise a great of comprehensive texture details; moreover its grey-level values have little variation. The foreground of original image is those areas with high-frequency verge, and its grey-level values have more variability.

Since scanned films is one of the major fields for the application of texture-enhancement, we also try with the scanned films that has rich grain and high resolution. A film scanner is a device for scanning photographic film directly into a computer without the use of any intermediate printmaking. It provides several benefits by using a flatbed scanner to scan in a print of any size —film scan. By contrast with traditional scanned image, film scan has the feature of smaller variation and higher optical resolution. At the same time, the photographer can directly control over cropping and aspect ratio from the original unmolested image on film, and etc.

The aim of the experiment is to explore texture-enhancement capability of fractional differential based algorithms to scanned films. The images are scanned by different film scanners from a standard sized area of film-arbitrarily chosen to be 0.25 inches by 0.25 inches (0.25" by 0.25"). Film scanners can accept either strips of 35 mm or 120 film, or individual slides. Our digital captures have been normalized to match the dimensions of a 35 mm frame, as explained below. We can get a high-quality JPG of a raw scan snippet representing 0.25" by 0.25" of film. For example, if it's a 4000 dpi film scanner, our snippet should be 1000×1000 pixels. We scanned the same film snippet respectively by three film scanners which are Epson 4990, Leafscan 45 and Nikon LS$-$8000, then did comparative experiments for the above scanned films in the same order between fractional differential based algorithm and an effective nonlinear image enhancement algorithm in frequency space [446]. The experiments show that fractional differential based algorithm has better texture-enhancing results. In order to do comparative experiments on the same pixel-level, we also take scanned film of Epson 4990 film scanner downsampling from 4800 dpi to 2400 dpi (600×600 pixels of 0.25" by 0.25" film at 2400 dpi), and take scanned film of Nikon LS$-$8000 film scanner downsampling from 4000 dpi to 2000 dpi (500×500 pixels of 0.25" by 0.25" film at 2000 dpi). For better visual comparing the effects of scanned films, we particularly give a small overview image of the full frame as well (typically, 300×450 pixels or so). However, the experiment still has some factors that we cannot control, such as scan operator technique and skill, subject Lens quality, film grain, noise, effects of tonality on apparent sharpness, the condition and calibration of your monitor and etc. The experiment results are shown in Fig. 11.52.

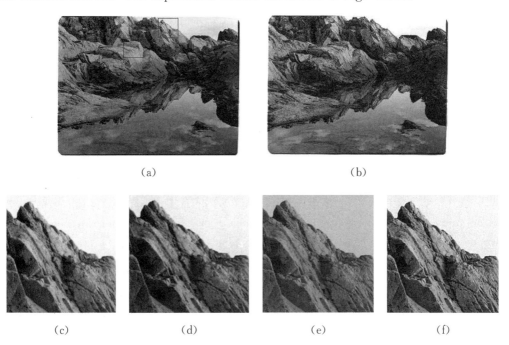

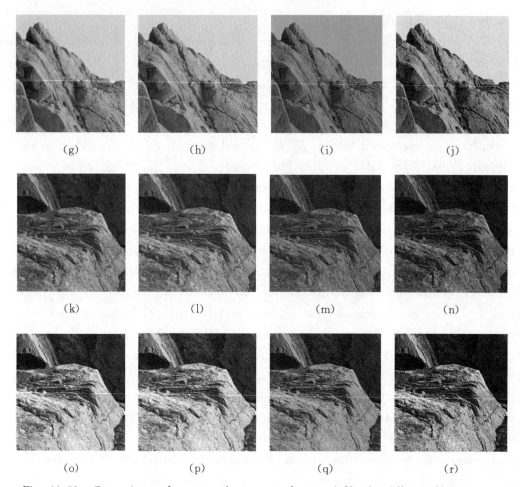

Fig. 11.52 Comparisons of texture-enhancement of scanned film by different film scanners (a) Overview image of the full frame (300×450 pixels), (b) 0.45 order fractional differential of (a), (c) Scanned film snippet from Epson 4990 film scanner (600×600 pixels of 0.25" by 0.25" film at 2400 dpi), (d) non-linearly enhancing the high-frequency components of (b), (e) non-linearly compensation of (b), (f) 0.45 order fractional differential of (b), (g) Scanned film snippet from Leafscan 45 film scanner (645×645 pixels of 0.25" by 0.25" film at 2412 dpi), (h) non-linearly enhancing the high-frequency components of (g), (i) non-linearly compensation of (g), (j) 0.45 order fractional differential of (g), (k) Scanned film snippet from Leafscan 45 film scanner (645×645 pixels of 0.25" by 0.25" film at 2412 dpi), (l) non-linearly enhancing the high-frequency components of (k), (m) non-linearly compensation of (k), (n) 0.45 order fractional differential of (k), (o) Scanned film snippet from Nikon LS−8000 film scanner (500×500 pixels of 0.25" by 0.25" film at 2000 dpi), (p) non-linearly enhancing the high-frequency components of (o), (q) non-linearly compensation of (o), (r) 0.45 order fractional differential of (o).

Fig. 11.52 shows that, no matter which type of film scanners are adopted, we can find that both of nonlinear image enhancement algorithm in frequency space [446] and fractional differential based algorithm can enhance the texture details of scanned films. However, the enhancement to rock and lake water by latter algorithm is obviously better than by the former algorithm both in light and dark areas, and the texture details are also clearer.

For quantity analysis, we take the information entropy [447] − [448] and average gradient [449] − [452] as parameters that are shown in Table 11.3. From Table 11.3 we can see that, in the first, both the nonlinear image enhancement algorithm in frequency space and fractional differential based algorithm have increased the information entropy. However, for all the scanned films getting by all kinds of film scanners, the information entropy gained by fractional differential based algorithm is the biggest one. That is to say, after fractional differential, it keeps the most texture details. In the second, the average gradient has a little increase by nonlinear image enhancement algorithm in frequency space, while it increases more than 2 by fractional differential based algorithm. It shows that fractional differential based algorithm can obviously improve image quality, and increase the contrast of minute texture details and texture transform features.

Table 11.3 quantitative comparison between texture-enhancement algorithms.

Figure \ Quantity	Information entropy	Average gradient	Figure \ Quantity	Information entropy	Average gradient
(a)	7.6492	7.9726	(j)	7.3049	7.2795
(b)	7.7535	12.0214	(k)	7.4095	5.5512
(c)	7.0462	3.4434	(l)	7.4523	6.2602
(d)	7.2552	3.6246	(m)	7.4896	6.6395
(e)	7.3145	3.8003	(n)	7.4682	7.1900
(f)	7.5365	6.4108	(o)	7.8269	6.8229
(g)	6.8985	4.0793	(p)	7.8495	10.0176
(h)	7.1648	4.7475	(q)	7.8965	10.2710
(i)	7.2078	4.9742	(r)	7.9538	13.5023

To further observe the texture-enhancing capability of fractional differential based algorithm, we did a team of comparison experiments of different kind of texture. The results are shown in Fig. 11.53.

(a)

(b)

(c)

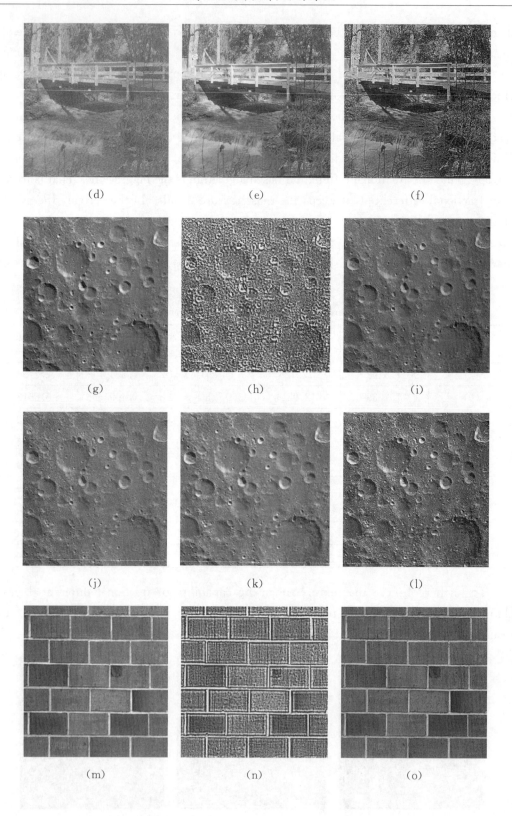

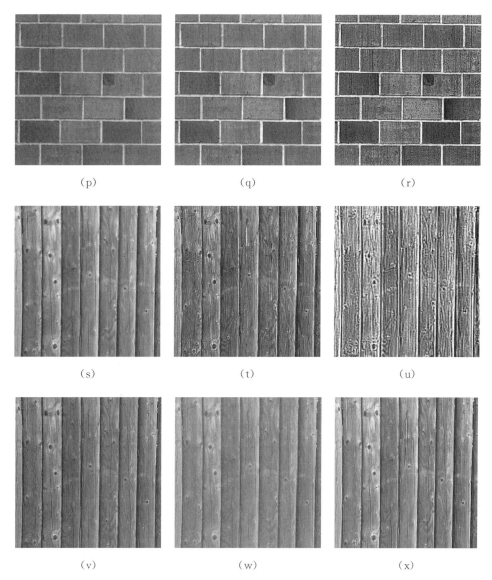

Fig. 11.53 Texture-enhancement capability comparison (a) Original image of bridge 5°, (b) Enhancing result of (a) by Laplacian pyramid based algorithm [452] − [453], (c) Enhancing result of (a) by multiscale wavelet transform based algorithm [453] − [454], (d) Enhancing result of (a) by contourlet based algorithm [455] − [456], (e) Enhancing result of (a) by nonlinear frequency enhancement algorithm [446], (f) 0.5-order fractional differential of (a), (g) Original image of of the surface of moon, (h) Enhancing result of (g) by Laplacian pyramid based algorithm, (i) Enhancing result of (g) by multiscale wavelet transform based algorithm, (j) Enhancing result of (g) by contourlet based algorithm, (k) Enhancing result of (g) by nonlinear frequency enhancement algorithm, (l) 0.55-order fractional differential of (g), (m) Original image of wall, (n) Enhancing result of (m) by Laplacian pyramid based algorithm, (o) Enhancing result of (m) by multiscale wavelet transform based algorithm, (p) Enhancing result of (m) by contourlet based algorithm, (q) Enhancing result of (m) by nonlinear frequency enhancement algorithm, (r) 0.25-order fractional differential of (m), (s) Original image of batten, (t) Enhancing result of (s) by Laplacian pyramid based algorithm, (u) Enhancing result of (s) by

multiscale wavelet transform based algorithm, (v) Enhancing result of (s) by contourlet based algorithm, (w) Enhancing result of (s) by nonlinear frequency enhancement algorithm, (x) 0.45-order fractional differential of (s).

From Fig. 11.53 we could find that Laplacian Pyramid based texture-enhancement algorithm suffers from one serious drawback—the appearance of visible artifacts when large structures are strongly enhanced. By contrast, the fractional differential allows the smooth enhancement of large structures, such visible artifacts can be avoided. For the enhancement of small texture details, the fractional differential based algorithm may have some advantages over the Laplacian Pyramid based algorithm. For the purpose of quantitative evaluating the experimental results, we compare the information entropy [447] – [448] and average gradient [449] – [451]. It shows in Table 11.4.

Table 11.4 quantitative comparison between texture-enhancement algorithms.

Figure \ Quantity	Information entropy	Average gradient	Figure \ Quantity	Information entropy	Average gradient
(a)	7.7152	7.6884	(m)	6.9101	5.2922
(b)	6.7076	31.9033	(n)	6.3146	21.6833
(c)	7.5594	11.1330	(o)	6.9076	7.3686
(d)	7.0845	7.0580	(p)	6.5429	6.0320
(e)	7.3145	3.8003	(q)	7.0799	8.0112
(f)	7.8682	29.7906	(r)	7.4743	18.0816
(g)	6.9658	6.6275	(s)	7.3085	5.3495
(h)	6.9479	32.2919	(t)	7.2392	23.9022
(i)	6.9113	9.2709	(u)	7.1801	7.1961
(j)	6.7693	7.8954	(v)	6.8944	6.0113
(k)	7.1346	9.8197	(w)	7.4677	8.7590
(l)	7.4701	22.3307	(x)	7.5471	19.2621

From Fig. 11.53 and Table 11.4, we can see that, firstly in Fig. 11.53 (a), since there are only 8 neighboring points for each pixel, the fractional mask in the 8 directions has well anti-rotative and it also can well enhance texture of rotated image. Although, Laplacian pyramid based algorithm can greatly increase the average gradient, but it only enhances the margin and losses a lot of texture details, and has the minimum information entropy. While, the average gradient by fractional differential based algorithm is slightly lower than that by Laplacian pyramid based algorithm, however, only fractional differential based algorithm can enhance both the margins and the textures. And it has the biggest information entropy. Secondly, from Fig. 11.53 (g), Fig. 11.53 (m) and Fig. 11.53 (s), it also can get the same conclusion. Therefore, fractional differential could nonlinearly preserve the low-frequency contour feature in the smooth area to the furthest degree, and as well as, nonlinearly enhance high-frequency marginal information in those areas where grey scale changes frequently, and nonlinearly enhance texture details in those areas where grey scale does not change evidently. Fractional differential could nonlinearly enhance the comprehensive texture details.

11.2.4.4　Meanings of Fractional Differential Based Algorithm

At present, fractional differential based algorithm is a novel tends of development. Like the other fractional based mathematical method, fractional differential spans the thinking bridges from integral order to fractional order, from integral dimension to fractional dimension, from Euclid measurement to fractional measurement. Therefore, fractional differential based algorithm is the physical invention of image processing method. Fractional differential based algorithms can well deal with such signals that feature non-linear, non-causal, non-minimum phase system, non-Gaussian, non-even, non integral differential and non-white additive noise. Texture-enhancing is a small part of applications for fractional differential based algorithm. If fractional differential can be wildly applied in modern signal and image processing, we can expect it will produce far-reaching influence.

11.2.5　CONCLUSIONS

How to apply fractional differential to the latest signal analyzing and processing, in particular to digital image processing, is a burgeoning subject branch under discussion. this chapter intends to apply a new mathematic approach—fractional differential to digital processing field. We propose 6 fractional differential masks, then demonstrate the most efficient fractional mask by theoretic and experimental analyzing, and finally discuss the texture enhancement of multiscale fractional mask. Experiments on real data sets show that the fractional differential-based approach is obvious better than traditional integral differential-based algorithms when nonlinearly enhance comprehensive texture details in smooth area of texture-rich digital images.

In recent years, with the advanced requirements for definition and real-time of HDTV, for legible of later bionic medical image (for example: cell image, X image, mammography, CT, MRI, PET and plutonic image), for better quality of bank ticket, for enhancing remote sensing image with comprehensive geographic texture and recognizing the small object, for better quality of bionic intelligent identification as figure print, IRST, and hand print, the fractional differential theory and its numerical implementation and fractional differential filter are expected to be widely used in the above fields.

第12章 分数阶最速下降法：分数阶自适应信号处理[①]

分数阶微积分在信号处理和自适应学习中的应用是一个新兴的研究领域。本章提出了一种利用分数阶微积分的分数阶自适应学习方法，特别地，提出了分数阶最速下降法，研究了分数阶二次能量泛函，并详细分析了该方法的稳定性和收敛性。使用分数阶最速下降法进行了数值实现，并用实验分析了其稳定性。

12.1 Introduction

The integer-order adaptive learning approaches based on the integer-order calculus, such as the classical first-order steepest descent (FOSD) method, produce time-varying parameters relevant for nonlinear systems by generalizing finite quantity of pattern training. The integer-order adaptive learning approaches automatically adapt to the optimal requirements of non-stationary variations in a system. Furthermore, they can produce improvements in system performance even if the characteristics of input signals are unknown or varying with time [457] - [461]. The FOSD method is a widely used gradient-based approach for optimizing the traditional integer-order pattern recognition, adaptive control, and adaptive signal processing problems. Unlike the traditional integer-order Newton optimization approach, the reverse incremental search employed in the FOSD method is in the opposite direction of the first-order gradient of a quadratic energy norm [457] - [459]. For digital analysis, the quadratic energy norm is a typical choice. However, in most of the actual integer-order adaptive systems, the quadratic energy norm of system is unknown; thus, it should be measured and estimated in accordance with the actual random input data of the integer-order adaptive system. The FOSD method utilizes a filtering process that reduces the noise of gradient measurement errors [460] - [461]. The least mean squares (LMS) algorithm based on the FOSD method is simple and effective [460] - [461]. Because the LMS algorithm was initially proposed for non-recursive linear filters, its applications were limited. Later, many improved variants of LMS algorithm have also been proposed, like the LMS-Newton algorithm [464],

---

[①] PU Yifei, ZHOU Jiliu, ZHANG Yi, et al. Fractional Extreme Value Adaptive Training Method: Fractional Steepest Descent Approach [J]. IEEE Transactions on Neural Networks and Learning Systems, 2015, 26 (4): 653-662.

sequential regression algorithm [464], [466], and random searching algorithm [464], [467] etc. However, the integer-order adaptive learning approaches cannot be used for fractional pattern recognition, adaptive control, or adaptive signal processing. In this chapter, a novel fractional adaptive learning (FAL) approach is proposed, which is called the fractional steepest descent approach.

Over the past three hundred years, fractional calculus has become an important branch of mathematical analysis [176], [230] − [231], [429]. Fractional calculus is as old as integral calculus, although, until recently, its utility has been confined to the domain of mathematics only. Fractional calculus is a novel mathematical method that may be useful for physical scientists too. Most of the special functions in mathematical physics involve differintegrable series. Fractional calculus extends and unifies the concepts of difference quotients and Riemann sums [3], [430]. The random variable in a physical process can be deemed as the displacement corresponding to the random movement of particles. Fractional calculus can be used to analyze and process several specific physical problems; in particular, it is useful in fields such as biomedical engineering, diffusion processes, viscoelasticity theory, fractal dynamics, and fractional control [11], [19], [23], [83], [94], [142], [180], [431] − [432]. One main advantage of fractional calculus is that most functions are equivalent to a power series, whereas others are equivalent to the superposition or product of a certain function and a power function [3], [176], [230] − [231], [429] − [430]. Unfortunately, the majority of its usage still lies in describing the transient state of physical change. It is seldom used for the processes involving systemic evolution [11], [19], [23], [83], [89], [94], [126], [142], [180], [184], [431] − [433].

The issue of how to efficiently apply fractional calculus for the purpose of signal analysis and signal processing, especially for the purpose of adaptive learning, is an emerging area of research. Till now, several researchers have concentrated in this problem domain [89], [106], [110], [130] − [131], [157], [185] − [191], [346], [433] − [435], [468]. Fractional calculus has also been hybridized with artificial machine intelligence, mainly because of its inherent strength of long-term memory and non-locality. In the fields of fractional pattern recognition, adaptive control, and adaptive signal processing, fractional partial differential equations, based on the fractional Green formula and fractional Euler-Lagrange equation, must be implemented [189] − [191]. The fractional extreme points of the quadratic energy norm are quite different from the traditional integer-order extreme points such as the first-order stationary points. To determine the fractional extreme points of the quadratic energy norm, we propose generalizing the integer-order steepest descent method to the fractional-order one.

Unlike the classical FOSD method, the FAL approach has two distinct properties. First, the v-order fractional extreme points of the quadratic energy norm E are not coincident with the first-order extreme points of E. In this manner, the iterative search process of the FAL approach can easily pass over the first-order local extreme points of E. Second, if the fractional differential order of the FAL approach satisfies $0<v<1$, then the quadratic energy norm E has a single fractional extreme point or two asymmetric v-order fractional extreme points in a pair. The number of fractional extreme points depends on the value of v. To obtain a single fractional extreme point of E, v must be set to the appropriate value. If $1<v<2$, then the quadratic energy norm E has two asymmetric v-order fractional extreme points in a pair. This asymmetric characteristic is essential to the quadratic energy norm E and the property of nature. If $2 \leqslant k<v<k+1$, where k is a positive integer, the quadratic energy norm E has no fractional extreme point.

12.2 Related Works

This section includes a brief introduction to the basic definitions in the domain of fractional calculus. It is well known that fractal geometry generalizes the Newton-Leibniz derivative. Fractal theory modified the perspective of measure theory; in particular, Euclidean-measure-based fractional calculus is more developed than the Hausdorff-measure-based one. This is why the definition of fractional calculus based on the Euclidean measure is widely used. The commonly-used fractional calculus definitions are the Grünwald-Letnikov definition, Riemann-Liouville definition, and Caputo definition [176], [230] — [231], [429].

The Grünwald-Letnikov fractional calculus of order v is defined by:

$$D_{G-L}^v s(x) = \frac{\mathrm{d}^v}{[\mathrm{d}(x-a)]^v} s(x)|_{G-L}$$

$$= \lim_{N \to \infty} \left\{ \frac{\left(\frac{x-a}{N}\right)^{-v}}{\Gamma(-v)} \sum_{k=0}^{N-1} \frac{\Gamma(k-v)}{\Gamma(k+1)} s\left(x - k\left(\frac{x-a}{N}\right)\right) \right\}, \quad (12.1)$$

where $s(x)$ is the signal under consideration, (a, x) is the duration of $s(x)$, v is a real number (can be fractional in nature too), and Γ is the Gamma function. Here, D_{G-L}^v denotes the Grünwald-Letnikov fractional differential operator. The Grünwald-Letnikov definition of fractional calculus is easy to calculate numerically; its numerical computation only requires the discrete sampling values, $s\left(x - k\left(\frac{x-a}{N}\right)\right)$, of signal $s(x)$. It only carries out the derivative or integral operation for the signal $s(x)$.

The Riemann-Liouville fractional integral of order v is defined by:

$$D_{R-L}^v s(x) = \frac{\mathrm{d}^v}{[\mathrm{d}(x-a)]^v} s(x)|_{R-L}$$

$$= \frac{1}{\Gamma(-v)} \int_a^x (x-\eta)^{-v-1} s(\eta) \mathrm{d}\eta$$

$$= \frac{-1}{\Gamma(-v)} \int_a^x s(\eta) \mathrm{d}(x-\eta)^{-v}, v<0, \quad (12.2)$$

where D_{R-L}^v denotes the Riemann-Liouville fractional differential operator. The values of $v \geqslant 0$, and $0 \leqslant n-1 < v < n$, $n \in \mathbf{R}$ are fixed. The Riemann-Liouville fractional differential of order v is defined by:

$$\begin{aligned} D_{R-L}^v s(x) &= \frac{\mathrm{d}^v}{[\mathrm{d}(x-a)]^v} s(x)\big|_{R-L} \\ &= \frac{\mathrm{d}^n}{\mathrm{d}x^n} \frac{\mathrm{d}^{v-n}}{[\mathrm{d}(x-a)]^{v-n}} s(x)\big|_{R-L} \\ &= \sum_{k=0}^{n-1} \frac{(x-a)^{k-v} s^{(k)}(a)}{\Gamma(k-v+1)} + \frac{1}{\Gamma(n-v)} \int_a^x \frac{s^{(n)}(\eta)}{(x-\eta)^{v-n+1}} \mathrm{d}\eta, 0 \leqslant v < n. \end{aligned}$$

(12.3)

The definition of the Caputo fractional differential of order v is defined by:

$$_a^C D_x^v s(x) = \frac{1}{\Gamma(n-v)} \int_a^x (t-\tau)^{n-v-1} s^{(n)}(\tau) \mathrm{d}\tau, \qquad (12.4)$$

where $_a^C D_x^v$ is the Caputo fractional differential operator. In this manner, the Fourier transform of signal $s(x)$ is defined by:

$$FT[D^v s(x)] = (i\omega)^v FT[s(x)] - \sum_{k=0}^{n-1} (i\omega)^k \frac{\mathrm{d}^{v-1-k}}{\mathrm{d}x^{v-1-k}} s(0), \qquad (12.5)$$

where i is the imaginary unit, and ω is the digital frequency. If $s(x)$ is a causal signal, (12.5) simplifies to $FT[D^v s(x)] = (i\omega)^v FT[s(x)]$.

In this section, a FAL approach is proposed, which has been named as the fractional steepest descent approach and its convergence is described in detail. To better understand the difference between the FAL approach and the integer-order one, the first-order extreme value is taken into account to construct the quadratic energy norm, given as:

$$E = E_{\min}^1 + \eta(s^{1*} - s)^2, \qquad (12.6)$$

where $v > 0$, E is the quadratic energy norm, $\eta \neq 0$ is a constant that controls the degree of convexity-concavity, and $E_{\min}^1$ is the first-order extreme value of E. The performance curve of E is a parabola whose a priori knowledge is often unavailable. The reverse incremental search of the FAL approach is in the negative direction of the v-order fractional gradient of E, which can be given as:

$$s_{k+1} = s_k + \mu(-D_{s_k}^v), \qquad (12.7)$$

where k is the step size or number of iterations, s_k is the current adjusted value of s, s_{k+1}, is the updated adjusted value of s, $D_{s_k}^v$ is the v-order fractional gradient of energy norm E at $s = s_k$, and μ is the constant coefficient that controls the stability and the rate of convergence of the FAL approach. Thus, the iterative search process of the FAL approach can be shown as given in Fig. 12.1.

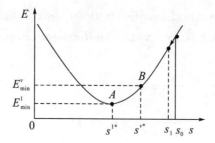

Fig. 12.1 Iterative search process of the fractional adaptive learning approach.

As shown in Fig. 12.1, for the first-order extreme value $E_{\min}^1$, A is the first-order extreme point or stationary point of E, and s^{1*} is the first-order optimal value of s. In contrast, for the fractional extreme value $E_{\min}^v$, B is the v-order fractional extreme point or fractional stationary point of E, and s^{v*} is the fractional optimal value of s. s_0 is the initial value of s, selected at random.

From (12.1), it can be derived that $D_{x-a}^v[0]=0$ and $D_{x-a}^v[c]=c\dfrac{(x-a)^{-v}}{\Gamma(1-v)}$, where $[0]$ denotes the permanent zero constant, and $[c]$ is a non-zero constant. Unlike the integer-order differentials, the fractional differential of a non-zero constant is equal to a non-zero value. Thus, the v-order fractional extreme point B is not coincident with the first-order extreme point A, and $E_{\min}^v$ is not the smallest value of E. If n and N are non-negative integers, $\dfrac{\Gamma(-n)}{\Gamma(-N)}=(-1)^{N-n}\dfrac{N!}{n!}$. From (12.1) and (12.7), and according to properties of fractional calculus, the following is true:

$$D_{s_k}^v = \frac{d^v E}{ds^v}\Big|_{s=s_k} = \frac{[E_{\min}^1 + \eta(s^{1*})^2]s_k^{-v}}{\Gamma(1-v)} - \frac{2\eta s^{1*} s_k^{1-v}}{\Gamma(2-v)} + \frac{2\eta s_k^{2-v}}{\Gamma(3-v)}. \tag{12.8}$$

Equation (12.8) shows that $D_{s_k}^v$ is a nonlinear equation of s_k, involving non-constant coefficients. It is quite difficult to directly derive the analytical expression of $D_{s_k}^v$ using v as a variable. Fortunately, if the order v of the FAL approach is known or specified, the formula described in (12.8) gets transformed to a constant coefficient nonlinear equation. When the iterative search process of the FAL approach converges to a stable solution, $s_{k+1}=s_k=s^{v*}$, $D_{s_k}^v=0$, and $E=E_{\min}^v$. Thus, from (12.6) and (12.8), the following can be obtained:

$$E_{\min}^v = E_{\min}^1 + \eta(s^{1*})^2 - 2\eta s^{1*} s^{v*} + \eta(s^{v*})^2, \tag{12.9}$$

$$(s^{v*})^{-v}\left\{\frac{E_{\min}^1 + \eta(s^{1*})^2}{\Gamma(1-v)} - \frac{2\eta s^{1*}}{\Gamma(2-v)}s^{v*} + \frac{2\eta}{\Gamma(3-v)}(s^{v*})^2\right\} = 0. \tag{12.10}$$

From (12.10), we conclude that $s^{v*} \neq 0$. Thus, (12.10) can be simplified as a constant coefficient quadratic equation, given as:

$$\frac{E_{\min}^1 + \eta(s^{1*})^2}{\Gamma(1-v)} - \frac{2\eta s^{1*}}{\Gamma(2-v)}s^{v*} + \frac{2\eta}{\Gamma(3-v)}(s^{v*})^2 = 0. \tag{12.11}$$

From (12.11), we can derive the relationship between s^{v*} and s^{1*}:

$$s_{1,2}^{v*} = \frac{\Gamma(3-v)}{2\eta}\left\{\frac{\eta s^{1*}}{\Gamma(2-v)} \pm \sqrt{\frac{\eta^2(s^{1*})^2}{\Gamma^2(2-v)} - \frac{2\eta[E_{\min}^1 + \eta(s^{1*})^2]}{\Gamma^2(1-v)\Gamma(3-v)}}\right\}. \quad (12.12)$$

Equation (12.12) implies that the v-order fractional extreme points of E are not always unique; however, they always appear as a pair of points. In general, the v-order fractional extreme points of E in a pair are asymmetric about s^{1*}. If (12.12) has two different solutions, the FAL approach will converge to either s_1^{v*} or s_2^{v*}, respectively. Fig. 12.1 and (12.12) show that the convergence value (s_1^{v*} or s_2^{v*}) of the iterative search process is related to the initial value s_0, selected at random. If (12.12) has two different solutions, the FAL approach is difficult to control and does not converge to the expected stable point. Hence, in order to ensure the convergence to a single v-order fractional extreme point, the two solutions in (12.12) must be equal, i.e., $s^{v*} = s_1^{v*} = s_2^{v*}$. Let us make $s^{v*} \neq 0$. Then, according to the properties of the Gamma function, it can be derived that $\frac{1}{\Gamma(1-v)\Gamma(3-v)} = 0$, if $v = 1, 3$, and $\frac{1}{\Gamma(2-v)} = 0$, if $v = 2$. Thus, from (12.12), the relationship between a single fractional optimal value s^{v*} and the first-order optimal value s^{1*} can be derived as:

$$s^{v*} = \frac{\Gamma(3-v)s^{1*}}{\Gamma(2-v)} \stackrel{v=1}{=} s^{1*}, \text{if } v = 1, 3, \text{and } v \neq 2. \quad (12.13)$$

Equation (12.13) shows that the classical FOSD method is a special case of the FAL approach (called fractional steepest descent approach). When $v \neq 1, 2, 3$ and $\frac{\eta^2(s^{1*})^2}{\Gamma^2(2-v)} - \frac{2\eta[E_{\min}^1 + \eta(s^{1*})^2]}{\Gamma(1-v)\Gamma(3-v)} = 0$, (12.12) has a unique solution $s^{v*} = s_1^{v*} = s_2^{v*}$, and E has a single v-order fractional extreme point. Note that the v-order FAL approach must satisfy:

$$\frac{\Gamma(1-v)\Gamma(3-v)}{\Gamma^2(2-v)} = \frac{2E_{\min}^1}{\eta(s^{1*})^2} + 2, \text{if } v \neq 1, 2, 3. \quad (12.14)$$

Substituting (12.14) into (12.12), the relationship between s^{v*} and s^{1*} can be derived as:

$$s^{v*} = \frac{\Gamma(3-v)s^{1*}}{2\Gamma(2-v)}, \text{if } v \neq 1, 2, 3. \quad (12.15)$$

In this equation, if $v \neq 1, 2, 3$ and v satisfies (12.14), the convergence value of the iterative search process of the FAL approach is s^{v*}. Thus, substituting (12.9) and (12.15) into (12.6), the energy norm E, expressed in terms of the fractional extreme value $E_{\min}^v$ and fractional optimal value s^{v*}, can be derived as:

$$E = E_{\min}^v + \eta\left[\frac{4\Gamma(2-v) - \Gamma(3-v)}{\Gamma(3-v)}(s^{v*})^2 - \frac{4\Gamma(2-v)}{\Gamma(3-v)}s^{v*}s + s^2\right]. \quad (12.16)$$

Substituting (12.8) into (12.7), the transient property of the iterative search process can be derived from s_0 to s^{v*}, for this FAL approach, given as:

$$s_{k+1} = s_k - \frac{\mu[E_{\min}^1 + \eta(s^{1*})^2]s_k^{1-v}}{\Gamma(1-v)} + \frac{2\mu\eta s^{1*}s_k^{1-v}}{\Gamma(2-v)} - \frac{2\mu\eta s_k^{2-v}}{\Gamma(3-v)}. \quad (12.17)$$

Substituting (12.11) into (12.17), one can obtain:

$$s_{k+1} = s_k - \frac{2\mu\eta}{\Gamma(3-v)}\left[s_k^2 - \frac{\Gamma(3-v)}{\Gamma(2-v)}s^{1*}s_k - (s^{v*})^2 + \frac{\Gamma(3-v)}{\Gamma(2-v)}s^{1*}s^{v*}\right]s_k^{-v}. \tag{12.18}$$

Then, substitution of (12.15) into (12.18) results in:

$$s_{k+1} = s_k - \frac{2\mu\eta}{\Gamma(3-v)}(s_k - s^{v*})^2 s_k^{-v}, \text{if } v \neq 1,2,3. \tag{12.19}$$

Equation (12.19) shows that it is a nonlinear, non-constant coefficient, difference equation. It is not possible to directly derive the general term of s_k by mathematical induction. To simplify this nonlinear calculation, let us consider s_k as a discrete sampling of the continuous function $s(t)$ about t. Then, the first-order difference can be used as an approximation for the first-order differential. It means $D_t^1 s(t) \cong s_{k+1} - s_k$. In addition, the power series expansion of s^v about $(s - s^{v*})$ can also be derived. It can be given as: $s^v = \sum_{n=0}^{\infty} \frac{\Gamma(1+v)(s^{v*})^{v-n}(s-s^{v*})^n}{\Gamma(n+1)\Gamma(1-n+v)}$. To further simplify this nonlinear computation, we only take the item summation of the power series expansion of s^v, when $n=0$ and $n=1$, as the approximation of s^v. One can then derive that $s^v \cong v(s^{v*})^{v-1}s$. Thus, (12.19) can be rewritten as:

$$D_t^1 s(t) \cong \frac{-2\mu\eta}{\Gamma(3-v)v(s^{v*})^{v-1}s(t)}[s(t) - s^{v*}]^2, \text{if } v \neq 1,2,3. \tag{12.20}$$

Equation (12.20) is a first-order ordinary differential equation with variables that are separable. Solving (12.20) by separation of variables, one can arrive at the solution $s(t)$. By discretely sampling $s(t)$ about t, the general form of s_k can be obtained:

$$s_k \cong s^{v*} + \exp\left(\frac{-2\mu\eta k}{\Gamma(3-v)v(s^{v*})^{v-1}}\right), \text{if } v \neq 1,2,3. \tag{12.21}$$

For the iterative search process of the FAL approach to converge, it must meet the condition $\lim_{k\to+\infty}\frac{2\mu\eta k}{\Gamma(3-v)v(s^{v*})^{v-1}} = +\infty$, which implies $\frac{2\mu\eta}{\Gamma(3-v)v(s^{v*})^{v-1}} = \chi > 0$, where χ is a positive constant. Thus, the constant coefficient μ satisfies:

$$\mu \cong \frac{\chi\Gamma(3-v)v(s^{v*})^{v-1}}{2\eta}, \text{if } v \neq 1,2,3. \tag{12.22}$$

Variables s_k and k of the classical FOSD method are related by a geometric series [469]. The iterative search process of the FOSD method can be classified as three categories: overdamped oscillation, critically damped oscillation, and underdamped oscillation [469]. Unlike the FOSD method, (12.21) shows that the relationship between s_k and k of the FAL approach is an approximate negative power geometric series. First, when $\chi=1, \mu \cong \frac{\Gamma(3-v)v(s^{v*})^{v-1}}{2\eta}$, if $v \neq 1, 2, 3$. Substituting this approximation into (12.21) yields $s_k \cong s^{v*} + \exp(-k)$, if $v \neq 1, 2, 3$. Thus, when $\chi = 1$, the iterative search process of the FAL approach converges in the form of $\exp(-k)$. Second, when $\chi =$

2, $\mu \cong \dfrac{\Gamma(3-v)v(s^{v*})^{v-1}}{\eta}$, if $v \neq 1$. Substituting this approximation into (12.21) yields $s_k \cong s^{v*} + \exp(-2k)$, if $v \neq 1, 2, 3$. Thus, when $\chi = 2$, the iterative search process of the FAL approach converges rapidly in the form of $\exp(-2k)$. Third, when χ is an integer and $\chi \geqslant 3$, the iterative search process of the FAL approach converges sharply in the form of $\exp(-\chi k)$. Similarly, when $n-1 < \chi = v < n$, the iterative search process of the FAL approach converges rapidly in the form of $\exp(-vk)$. In general, the FAL approach on multi-dimensional performance surface of the quadratic energy norm is a generalization of (12.6). Equation (12.6) shows that the energy norm and reverse incremental search process of the multi-dimensional FAL approach are equivalent to $E = E_{\min}^1 + \eta(s^{1*} - s)^2$ and $s_{k+1} = s_k + \mu(-D_{s_k}^v)$, respectively, where the constant coefficients η and μ are multi-dimensional vectors. Here μ satisfies the condition, given as:

$$\mu \cong \left\{ \dfrac{\chi \Gamma(3-v) v (s^{v*})^{v-1}}{2\eta} \right\}, \text{if } v \neq 1, 2, 3. \tag{12.23}$$

12.3 Experiments and Analysis

12.3.1 Application of the fractional adaptive learning approach

This section briefly describes the application of the FAL approach. Inview of signal processing applications, the fractional differential has the following nonlinear characteristics [185] — [188]. First, the fractional differential of a non-zero constant is not equal to zero. It decreases gradually to zero from the highest value on a singular impulse signal. In contrast, any integer-order differential of a constant is equal to zero. This is a remarkable difference between the fractional differential and the integer-order differential. Second, the fractional differential at the initial point of a ramp function is equal to non-zero, which nonlinearly enhances the high-frequency singular information. If $v = 1$ or $v = 2$, the classical Sobel operator and Gaussian operator coincide with the first-order and second-order differential operators, respectively. In contrast, if $0 < v < 1$, the fractional differential enhancement of high-frequency singular information is smaller than its first-order differential enhancement; moreover, when the order of the fractional differential increases, the enhancement of high-frequency singular information also increases. Third, the fractional differential of a ramp function does not possess a linear relation; instead, it yields a nonlinear curve. In contrast, the integer-order differential of a ramp function does possess a linear relation.

Based on the aforementioned nonlinear characteristics of the fractional differential, the FAL approach can be applied to the image processing problems [185] — [191]. The fractional differential has been demonstrated to preserve, in great detail, the low-frequency contour features in smooth areas. It also enhances the high-frequency marginal information in highly variable greyscale areas and enhances the texture details in areas where greyscale changes are not as obvious. All of this is done in a nonlinear fashion. As such, the FAL approach can be used to perform reverse incremental optimizing searches on the fractional total variation in an image. By using the FAL approach, we can implement a class of multi-scale denoising models for texture images based on the nonlinear fractional partial differential equations that preserve texture details. In this case, the desired objective is achieved in a non-linear fashion [189] − [191].

12.3.2 Stability and convergence of the fractional adaptive learning approach

In this section, the stability and convergence of the FAL approach are discussed. The properties of the v-order fractional differential of the energy norm E are analyzed using fractional calculus. Without loss of generality, we set $E_{\min}^1 = 10$, $\eta = 2$, and $s^{1*} = 5$ in (12.6) and (12.8). Then, as shown in Fig. 12.2, we numerically implement the v-order derivative $D_s^v = \dfrac{d^v E}{ds^v} = \dfrac{[E_{\min}^1 + \eta(s^{1*})^2] s^{-v}}{\Gamma(1-v)} - \dfrac{2\eta s^{1*} s^{1-v}}{\Gamma(2-v)} + \dfrac{2\eta s^{2-v}}{\Gamma(3-v)}$ of the energy norm $E = E_{\min}^1 + \eta(s^{1*} - s)^2$ about s.

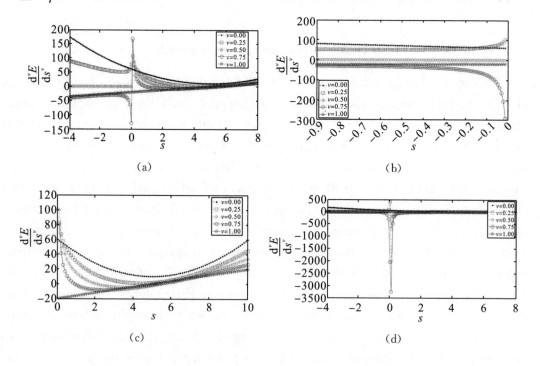

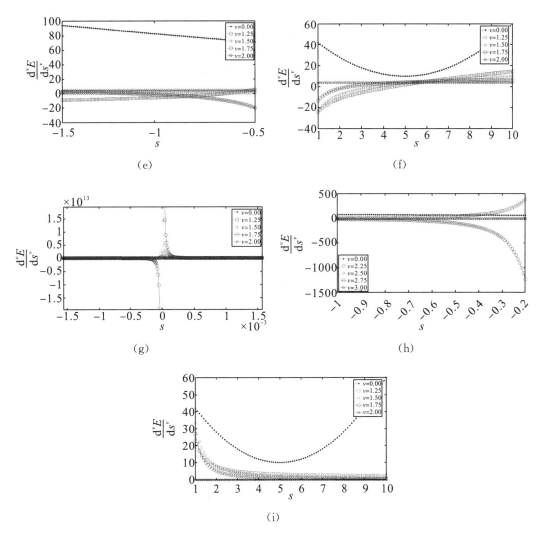

Fig. 12.2 The v-order fractional differential of the energy norm E. (a) $0 \leqslant v \leqslant 1$, (b) partial enlarged details when $-0.9 < s < 0$ and $0 \leqslant v \leqslant 1$, (c) partial enlarged details when $0 < s < 10$ and $0 \leqslant v \leqslant 1$, (d) $1 < v \leqslant 2$, (e) partial enlarged details when $-1.5 < s < -0.5$ and $1 < v \leqslant 2$, (f) partial enlarged details when $1 < s < 10$ and $1 < v \leqslant 2$, (g) $2 < v \leqslant 3$, (h) partial enlarged details when $-1 < s < -0.2$ and $2 < v \leqslant 3$, and (i) partial enlarged details when $1 < s < 10$ and $2 < v \leqslant 3$.

As shown in Fig. 12.2, if $v=0$, D_s^v performs neither the differential nor the integral operations. If $v>0$, D_s^v performs the fractional differential operation. In Figs. 12.2 (a), 12.2 (d), and 12.2 (g), for $D_s^v = \dfrac{\mathrm{d}^v E}{\mathrm{d} s^v} = \dfrac{[E_{\min}^1 + \eta(s^{1*})^2]s^{-v}}{\Gamma(1-v)} - \dfrac{2\eta s^{1*} s^{1-v}}{\Gamma(2-v)} + \dfrac{2\eta s^{2-v}}{\Gamma(3-v)}$, the v-order fractional calculus has two obvious undermentioned features. First, the v-order fractional calculus of most functions is equal to a power function. Second, the v-order fractional calculus of the remaining functions is equal to the sum or product of a certain functions and power function. In this way, the v-order fractional calculus may go on ad infinitum if the denominator of the power function is equal to zero. Therefore, in

Figs. 12.2 (a), 12.2 (d), and 12.2 (g), if $s=0$, $\frac{d^v E}{ds^v}$ has a singular impulse. Figs. 12.2 (a) and 12.2 (b) also show that if $s<0$, $0<v<1$, and $v\neq 1/2$, $\frac{d^v E}{ds^v}$ has a non-zero crossing point. If $s<0$ and $v=1/2$, $\frac{d^v E}{ds^v}\equiv 0$. Equation (12.14) shows that when v_z satisfies $\frac{\Gamma(1-v_z)\Gamma(3-v_z)}{\Gamma^2(2-v_z)} = \frac{2E_{\min}^1}{\eta(s^{1*})^2} + 2$, if $v_z\neq 1$, (12.12) has two identical solutions. There is a single fractional optimal value $s^{v*}=s_1^{v*}=s_2^{v*}$ when the value of v_z is related to $E_{\min}^1$, η, and s^{1*}. As shown in Figs. 12.2 (a) and 12.2 (c), if $0<v<v_z<1$, $D_s^v = \frac{d^v E}{ds^v}$ has a non-zero crossing point. If $0<v=v_z<1$, $D_s^v = \frac{d^v E}{ds^v}$ has a single zero crossing point, and if $0<v_z<v<1$, it has two zero crossing points. In other words, if $0<v<1$, there may not be a single fractional extreme point of E; in general, the v-order fractional extreme points of E occur in pairs, which are asymmetric about s^{1*}. This suggests that if $E_{\min}^1$, s^{1*}, and η are arbitrary, the v-order fractional extreme value s^{v*} of E exists either individually or as a pair. Figs. 12.2 (a), 12.2 (b), and 12.2 (c) show that if $v=1$, $D_s^v = \frac{d^v E}{ds^v}$ has a single zero crossing point. As shown in Figs. 12.2 (d), 12.2 (e), and 12.2 (f), if $1<v<2$, $D_s^v = \frac{d^v E}{ds^v}$ has a pair of zero crossing points asymmetrical about s^{1*}. If $v=2$, $D_s^v = \frac{d^v E}{ds^v} = \eta = 2 \neq 0$. As shown in Figs. 12.2 (g), 12.2 (h), and 12.2 (i), if $2<v<3$, $D_s^v = \frac{d^v E}{ds^v}$ has a non-zero crossing point. If $v=3$, $D_s^v = \frac{d^v E}{ds^v} \equiv 0$. Therefore, if $2 \leq k < v < k+1$, $D_s^v = \frac{d^v E}{ds^v}$ has a non-zero crossing point. If $v=k+1$, $D_s^v = \frac{d^v E}{ds^v} \equiv 0$.

Furthermore, from (12.14), we can further conclude that if $E_{\min}^1 = 10$, $\eta = 2 \neq 0$, and $s^{1*} = 5$, the solution of $\frac{\Gamma(1-v)\Gamma(3-v)}{\Gamma^2(2-v)} = \frac{2E_{\min}^1}{\eta(s^{1*})^2} + 2$, if $v \neq 1$ is equal to $v \cong 0.28577$, which shows that E has a single v-order fractional extreme value. Thus, from (12.8), (12.12), and (12.14), we can numerically implement s_1^{v*}, s_2^{v*}, and the $v \cong 0.28577$ order fractional derivative of E. It is shown in Fig. 12.3.

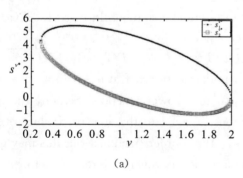

(a)

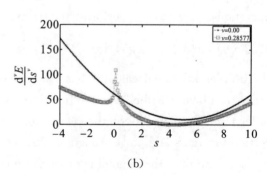

(b)

Fig. 12.3 Relationship between s_1^{v*} and s_2^{v*}, and the $v \cong 0.28577$ order fractional derivative of E: (a) relationship between s_1^{v*} and s_2^{v*}, and (b) the $v \cong 0.28577$ order fractional derivative of E.

From Fig. 12.3 (a), note that if $v \cong 0.28577$, (12.12) has two identical solutions, and there is a single fractional optimal value $s^{v*} = s_1^{v*} = s_2^{v*}$ of E. From Fig. 12.3 (b), if $v \cong 0.28577$, $D_s^v = \dfrac{d^v E}{ds^v}$ has a single zero crossing point of E. The experiment results obtained coincide with the aforementioned theoretical derivation.

Recall that the v-order fractional extreme point of E is not coincident with the first-order extreme point of E. This implies that the iterative search process of the FAL approach easily passes over the first-order local extreme points of E. From (12.14), if v_z satisfies $\dfrac{\Gamma(1-v_z)\Gamma(3-v_z)}{\Gamma^2(2-v_z)} = \dfrac{2E_{\min}^1}{\eta(s^{1*})^2} + 2$, if $v_z \neq 1$, (12.12) has two identical solutions, and there is a single fractional optimal value $s^{v*} = \dfrac{\Gamma(3-v) \; s^{1*}}{2\Gamma(2-v)}$, if $v \neq 1$. Furthermore, if $0 < v_z < v < 1$, (12.12) has two different solutions; specifically, $s_{1,2}^{v*} = \dfrac{\Gamma(3-v)}{2\eta} \left\{ \dfrac{\eta s^{1*}}{\Gamma(2-v)} \pm \sqrt{\dfrac{\eta^2 (s^{1*})^2}{\Gamma^2(2-v)} - \dfrac{2\eta[E_{\min}^1 + \eta(s^{1*})^2]}{\Gamma(1-v)\Gamma(3-v)}} \right\}$. The v-order fractional extreme points of E occur in pairs that are asymmetrical about s^{1*}.

We also consider the stability and convergence rate of the FAL approach. If (12.12) has two different solutions, the FAL approach will converge to s_1^{v*} or s_2^{v*}. The convergence value (s_1^{v*} or s_2^{v*}) of its iterative search process is related to the randomly selected initial value s_0. In Figs. 12.2 (d), 12.2 (e), and 12.2 (f), $v=1.25$ serves as an example for discussion. Equation (12.12) shows that when the iterative search process of the FAL approach converges to a steady state, then $s_{k+1} = s_k = s^{v*}$, $D_{s_k}^v = 0$, and $E = E_{\min}^v$. Specifically, we determined $s_1^{v*} = 4.3906$ and $s_2^{v*} = -0.6406$. For numerical implementation, Let us set $\mu = 0.005$, the total number of iterations $k = 500$, and the initial values $s_{10} = 4.5$ and $s_{20} = -0.25$. From (17), we implement the iterative search process of the FAL approach in one-dimensional space, as shown in Fig. 12.4.

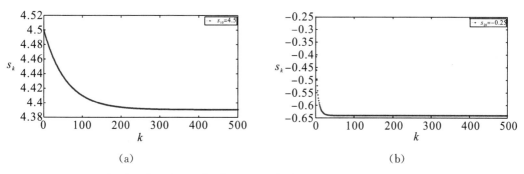

Fig. 12.4 Randomly selected initial value s_0 and the final convergence value of the fractional adaptive learning approach when (a) $s_{10} = 4.5$ and (b) $s_{20} = -0.25$.

As shown in Fig. 12.4, when the iterative search process of the FAL approach converges to a steady state, $s_{1k} = 4.3906 = s_1^{v*}$ and $s_{2k} = -0.6406 = s_2^{v*}$. The reverse incremental search of the FAL approach is in the negative direction of the v-order fractional gradient of E. Thus, if μ is set to an appropriate value, its iterative search process will stop when $D_{s_k}^v = 0$ and converge to the $v = 1.25$ order fractional extreme point of E. The results in Figs. 12.2(d), 12.2(e), and 12.2(f) conform to the theoretical derivations achieved in (12.7), (12.12), and (12.17), respectively. Thus, in order to converge to a single v-order fractional extreme point, the two solutions of (12.12) must be identical, i.e., there is a single fractional optimal value $s^{v*} = s_1^{v*} = s_2^{v*}$.

If (12.14) is satisfied, (12.12) has two identical solutions $s^{v*} = s_1^{v*} = s_2^{v*}$. From (12.14), when $v \cong 0.28577$, it has a single v-order fractional extreme point of E. Thus, from (12.15) and (12.9), we derive $s^{v*} = \dfrac{\Gamma(3-v) \ s^{1*}}{2\Gamma(2-v)} \overset{v \cong 0.28577}{\cong} 4.2856$ and $E_{\min}^v = E_{\min}^1 + \eta(s^{1*})^2 - 2\eta s^{1*}s^{v*} + \eta(s^{v*})^2 \cong 11.0208$. For (12.19), the randomly selected initial value $s_0 = 15$, and the total number of iterations is 200. Furthermore, It is also assumed that $0 < \chi = 0.25 < 2$, $0 < \chi = 1.75 < 2$, and $\chi = 2$ in (12.22). Equations (12.19) and (12.22) are implemented numerically using the iterative search process of the FAL approach in one-dimensional space, as shown in Fig. 12.5.

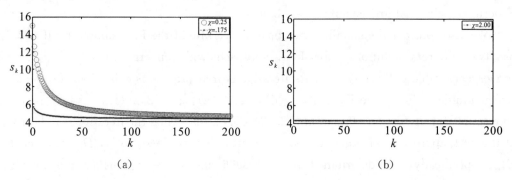

Fig. 12.5 The iterative search process of the fractional adaptive learning approach in one-dimensional space when (a) $0 < \chi = 0.25 < 2$, and $0 < \chi = 1.75 < 2$, and (b) $\chi = 2.00$.

As shown in Fig. 12.5 and (12.21), if $0 < \chi = 0.25 < 2$, $\mu = 0.0099$. The iterative search process of the FAL approach converges in the exponential form of $\exp(-0.25k)$ that depicts that its iterative search process is slower in nature. When $k = 200$, the search process converges to the stable state $s_k = 4.3270$. On the other hand, if $0 < \chi = 1.75 < 2$, $\mu = 0.0691$. The iterative search process of the FAL approach converges in the exponential form of $\exp(-1.75k)$. As χ and μ increase, the iterative search process is accelerated. When $k = 50$, the search process converges to the stable state of $s_k = 4.3270$. Furthermore, if $\chi = 2$, $\mu = 0.0790$. The iterative search process of the FAL approach converges in the exponential form of $\exp(-2k)$ that depicts that its iterative search process is very fast. When $k = 1$, it converges to the stable state of $s_k = 4.2945$, only in a single one step. Note that to simplify the nonlinear calculation in (12.20), only the item summation of the power series expansion of s^v needs to be considered when $n=0$ and $n=1$. This approximation leads to a small deviation. The three extreme values of the aforementioned experiments are approximate to $s_k \cong s^{v*}$, respectively.

Next, the stability and convergence rate of the FAL approach in high-dimensional spaceare examined in depth. In three-dimensional space, we assume $_xE_{\min}^1$, $\eta_x = 2 \neq 0$, $s_x^{1*} = 5$, $_yE_{\min}^1 = {_xE_{\min}^1} = E_{\min}^1$, $\eta_y = 3 \neq 0$, and $s_y^{1*} = 6$. Consequently, $E = E_{\min}^1 + \eta_x(s_x^{1*} - s_x)^2 + \eta_y(s_y^{1*} - s_y)^2$. As shown in (12.14), when $v_x \cong 0.28577$, E has a single v_x-order fractional extreme point in the x coordinate direction. Likewise, when $v_y \cong 0.15625$, E has a single v_y-order fractional extreme value in the y coordinate direction. From (12.9) and (12.15), it follows that $s_x^{v*} = \frac{\Gamma(3-v_x)s_x^{1*}}{2\Gamma(2-v_x)} \stackrel{v_x \cong 0.28577}{\cong} 4.2856$, $_xE_{\min}^v = {_xE_{\min}^1} + \eta_x(s_x^{1*})^2 - 2\eta_x s_x^{1*} s_x^{v*} + \eta_x(s_x^{v*})^2 \cong 11.0208$, $s_y^{v*} = \frac{\Gamma(3-v_y)s_y^{1*}}{2\Gamma(2-v_y)} \stackrel{v_y \cong 0.15625}{\cong} 5.5313$, and $_yE_{\min}^v = {_yE_{\min}^1} + \eta_y(s_y^{1*})^2 - 2\eta_y s_y^{1*} s_y^{v*} + \eta_y(s_y^{v*})^2 \cong 10.6592$. In (12.19), the randomly selected initial values are $_xs_0 = 15$ and $_ys_0 = 12$. In (12.22) and (12.23), we fix the total number of iterations $k = 4000$, and set $\chi = 0.25$, $\chi = 1.75$, and $\chi = 2$. Thus, we determine $\mu = \{\mu_x, \mu_y\} \stackrel{\chi=0.25}{=} \{0.0099, 0.0027\}$, $\mu = \{\mu_x, \mu_y\} \stackrel{\chi=1.75}{=} \{0.0691, 0.0187\}$, $\mu = \{\mu_x, \mu_y\} \stackrel{\chi=2}{=} \{0.079, 0.0214\}$, and $\mu = \max\{\mu_x, \mu_y\} \stackrel{\chi=1.75}{=} \{0.0691, 0.0691\}$. In (12.7) and (12.17), we set $_xs_0 = 15$, $_ys_0 = 12$, $v = 1$, $\mu_x = \mu_y = 0.01$, and $k = 250$. These values are used to implement the three-dimensional iterative search process of the FAL approach as shown in Fig. 12.6.

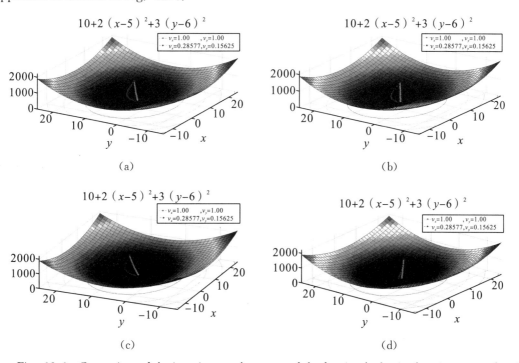

Fig. 12.6 Comparison of the iterative search process of the fractional adaptive learning approach and

classical FOSD method in three-dimensional space when (a) $\chi = 0.25$ and $\mu = \{\mu_x, \mu_y\}^{\chi=0.25} = \{0.0099, 0.0027\}$, (b) $\chi = 1.75$ and $\mu = \{\mu_x, \mu_y\}^{\chi=1.75} = \{0.0691, 0.0187\}$, (c) $\chi = 2.00$ and $\mu = \{\mu_x, \mu_y\}^{\chi=2} = \{0.790, 0.0214\}$, and (d) $\chi = 1.75$ and $\mu = \max\{\mu_x, \mu_y\}^{\chi=1.75} = \{0.0691, 0.0691\}$.

As shown in Fig. 12.6 (a), if $\chi = 0.25$ and $\mu = \{\mu_x, \mu_y\}^{\chi=0.25} = \{0.0099, 0.0027\}$, the iterative search process of the FAL approach converges to $({_x}s_k, {_y}s_k) = (4.3137, 5.6221) \cong (s_x^{v*}, s_y^{v*})$. In Fig. 12.6 (b), if $\chi = 1.75$ and $\mu = \{\mu_x, \mu_y\}^{\chi=1.75} = \{0.0691, 0.0187\}$, the iterative search process converges to $({_x}s_k, {_y}s_k) = (4.2859, 5.5437) \cong (s_x^{v*}, s_y^{v*})$. Similarly, in Fig. 12.6 (c), if $\chi = 2.00$ and $\mu = \{\mu_x, \mu_y\}^{\chi=2} = \{0.0790, 0.0214\}$, the iterative search process of the FAL approach converges to $({_x}s_k, {_y}s_k) = (4.2890, 5.5421) \cong (s_x^{v*}, s_y^{v*})$. Finally, as shown in Fig. 12.6 (d), if $\chi = 1.75$ and $\mu = \max\{\mu_x, \mu_y\}^{\chi=1.75} = \{0.0691, 0.0691\}$, the iterative search process converges to $({_x}s_k, {_y}s_k) = (4.2895, 5.5346) \cong (s_x^{v*}, s_y^{v*})$. These results suggest that if χ and $\mu = \{\mu_x, \mu_y\}$ are different values, the FAL approach converges simultaneously in the x and y coordinate directions.

Note that the nature of the curve surface of quadratic energy norm looks quite similar to a bowl in three-dimensional space. The FAL approach does not converge to the bottom of the bowl of the quadratic energy norm E. The larger the value of χ is, the faster is the convergence rate of FAL approach, and, the sparser is its convergence trajectory. As shown in Figs. 12.6 (a), 12.6 (b), and 12.6 (c), for $\mu_x \neq \mu_y$, the convergence rates in the x and y coordinate directions also vary; the convergence trajectory of the FAL approach is a nonlinear curve. Fig. 12.6 (d) shows that for $\mu_x = \mu_y$, the convergence rates in the x and y coordinate directions are equal; the convergence trajectory of the FAL approach is a straight line. As shown in Figs. 12.6 (a), 12.6 (b), 12.6 (c), and 12.6 (d), the iterative search process of the classical FOSD method converges to $({_x}s_k, {_y}s_k) = (5.0004, 5.0003) \cong (s_x^{1*}, s_y^{1*})$. The classical FOSD method does converge to the bottom bowl of the quadratic energy norm E. As shown in Fig. 12.6, for (s_x^{v*}, s_y^{v*}), there is a small deviation between (s_x^{v*}, s_y^{v*}) and (s_x^{1*}, s_y^{1*}).

12.4 Conclusions

In recent times, the notion of how to apply fractional calculus to perform signal processing, and for the purpose of adaptive learning, has become a potent, emerging research problem and a few research studies have been initiated and reported all over the world. In the fields of fractional pattern recognition, adaptive control, and adaptive signal processing, fractional partial differential equations based on the fractional Green formula and fractional Euler-Lagrange equation have been implemented. The fractional extreme point of the quadratic energy norm is quite different from the traditional integer-order one such as the first-order stationary point. To identify the fractional extreme points of the quadratic energy norm, we generalized the integer-order steepest descent method to a fractional approach. Based on the characteristics of fractional calculus mentioned above, a novel mathematical method is proposed where fractional calculus is used to implement FAL approach, named as fractional steepest descent approach.

Unlike the classical FOSD method, the FAL approach has two different properties. First, the v-order fractional extreme point of quadratic energy norm E is not coincident with the first-order extreme point of E. This means $E_{\min}^v$ is not the smallest value of E. In this manner, the iterative search process of the FAL approach can easily pass over the first-order local extreme points of E. Second, if the fractional differential order of the FAL approach satisfies $0 < v < 1$, the quadratic energy norm E may have a single fractional extreme point, or two asymmetric fractional extreme points in a pair about s^{1*}. The number of fractional extreme points depends on the value of order v. In order to achieve a single fractional extreme point of E, v must be set to an appropriate value. If $1 < v < 2$, the quadratic energy norm E always has two asymmetric v-order fractional extreme points in a pair about s^{1*}. This shows that the asymmetric characteristic is an essential property of the quadratic energy norm E and is also an essential property of nature. If $2 \leqslant k < v < k+1$, where k is a positive integer, the quadratic energy norm E has no fractional extreme point. For $(s_x^{v*}, s_y^{v*}) \neq (s_x^{1*}, s_y^{1*})$, there is a small deviation between (s_x^{v*}, s_y^{v*}) and (s_x^{1*}, s_y^{1*}).

The FAL approach can be suitably applied to implement textural image denoising [189] − [191]. It is well-known that the classical first-order Hopfield neural networks are based on the first-order differential and first-order adaptive learning approach, specifically the FOSD method. Thus, the FAL approach can also be applied to implement fractional Hopfield neural networks and anti-chip cloning attacks for anti-counterfeiting. Some initial discussions in these aspects are presented in [171], [298]. It is intended that in-depth studies in these directions will be undertaken as future course of work.

第 13 章　分数阶变分法中的分数阶欧拉－拉格朗日方程[①]

　　本章讨论了一种新颖概念——基于分数阶变分法的分数阶欧拉－拉格朗日方程，分数阶变分法的反向增量最优搜索是基于分数阶最速下降法的。分数阶微积分具有长时记忆、非局域性和弱奇异性的特征，主要用于信号和图像处理中分数阶确定边界优化问题的研究。第一，为了便于比较，由一阶格林公式推导了分数阶变分方法的一阶欧拉－拉格朗日方程。第二，基于维纳－辛钦定理，推导了分数阶变分方法的欧拉－拉格朗日方程。第三，为了在空间域或时域直接且容易地实现分数阶变分方法，推导了基于分数阶格林公式的分数阶欧拉－拉格朗日方程。第四，推导了分数阶欧拉－拉格朗日方程的求解过程。第五，分析了基于分数阶变分法的分数阶图像修复算法和分数阶去噪算法。实验表明，对于纹理细节丰富的图像，基于分数阶变分法的分数阶图像修复算法的边缘和纹理细节恢复和保持能力优于基于经典一阶变分法的整数阶图像修复算法。

13.1　Introduction

　　It is well known that variational method is an important part of functional analysis, which is firstly introduced by Leonhard Euler in 1744 [470], [471]. In 1755, Joseph-Louis Lagrange proposed a revolutionary technique of variations in his brief letter written to Euler [472]. The classical cases of the integer-order variational method include Fermat's principle of least time, Bernoulli's brachistochrone problem [473], and the isoperimetric problem. Variational method allows us to solve optimization problems using only elementary calculus [474], [475]. The key point of variational method is to determine the function that achieves the extremum seeking algorithm of a functional. The classical first-order Euler-Lagrange equation can be obtained by setting the first-order derivatives of the corresponding functional with regard to each parameter be equal to zero [474], [475]. In addition, the Rayleigh-Ritz method and Galerkin's method led to the formulation of the finite element method for implementing the prediction of a physical process [476], [477]. Because many ordinary and partial differential equations in mathematics, physics, and engineering can be derived as the first-order Euler equation for

---

[①]　PU Yifei. Fractional-Order Euler-Lagrange Equation for Fractional-Order Variational Method: A Necessary Condition for Fractional-Order Fixed Boundary Optimization Problems in Signal Processing and Image Processing [J]. IEEE Access, 2016 (4): 10110-10135.

an appropriate corresponding functional, the integer-order variational method is now widely applied to many scientific fields such as numerical analysis [478], nonconvex problems [479], mechanics [474], [480], chemistry [481], [482], scattering problems [483], electromagnetic field problems [484], control [485], signal processing and image processing [486] — [498], and so on.

Nowadays, fractional calculus has been developed as an important branch of mathematical analyses [3], [176], [178], [230] — [231], which is as old as the integer-order calculus. Fractional calculus extends the concepts of the integer-order difference and Riemann sums. Although until recently, the applications of fractional calculus mainly focused concentration on the field of mathematics and now seem to be a promising mathematical approach for the physical scientists and engineering technicians. More and more scientific researches show that a fractional-order or a fractional dimensional method is now one of the best way to describe many natural phenomena such as fractional diffusion processes [181] — [183], fractional viscoelasticity theory [126], fractal dynamics [184], fractional control [518], [519], fractional neural networks [193], [338], [399], [412], [499] — [500], fractional signal processing [73], [130], [157], [196], [199] — [203], [224], [225], [227], [345], [405], [501], and fractional image processing [186] — [192], [434], [502] — [505], [549], and many other fields in physics [94], [403], [506] — [527].

How to apply fractional calculus to variational method is an emerging scientific field that has been seldom received desired attention. With regard to a signal, the features of its fractional calculus are quite different from those of its integer-order calculus. For example, the fractional differential, except based on the Caputo definition, of a Heaviside function is equal to non-zero, whereas its integer-order differential must be equal to zero [3], [176], [178], [230] — [231]. Fractional calculus has been successfully applied to signal processing and image processing mainly because of its inherent strengths in terms of long-term memory, non-locality, and weak singularity [73], [130], [157], [186] — [192], [196], [199] — [203], [224], [225], [227], [345], [405], [434], [501] — [505]. For instance, a fractional differential mask can nonlinearly maintain the low-frequency contour features in the smooth area of an image and enhance the high-frequency edges and textural details in those areas where the grey level undergoes frequent or unusual variations [186] — [187], [345], [434], [502] — [504], [549]. Therefore, to solve a necessary condition for the fractional-order fixed boundary optimization problems in signal processing and image processing, an interesting theoretical problem emerges naturally: what the fractional-order Euler-Lagrange equation for the fractional-order variational method is. Motivated by this need, in this chapter, we introduce a novel conceptual formulation of the fractional-order Euler-Lagrange equation for the fractional-order variational method, which is based on the fractional-order extremum method. In particular, the reverse incremental optimal search of the fractional-order variational method is based on the fractional-order steepest descent approach [192].

The rest of the manuscript is organized as follows: Section 2 presents in brief the necessary mathematical background of fractional calculus. Section 3 proposes the formulation of the fractional-order Euler-Lagrange equation for the fractional-order variational method. At first, for the convenience of comparison, the first-order Euler-Lagrange equation for the first-order variational method is derived based on the first-order Green formula. Secondly, the fractional-order Euler-Lagrange equation for the fractional-order variational method is derived based on Wiener-Khintchine theorem. Thirdly, in order to directly and easily achieve the fractional-order variational method in spatial domain or time domain, the fractional-order Green formula and the fractional-order Euler-Lagrange equation based on the fractional-order Green formula are derived, respectively. Fourthly, the solution procedure of the fractional-order Euler-Lagrange equation is derived. Section 4 presents the experiment results obtained and the associated analyses carried out. Here, first, a fractional-order inpainting algorithm based on the fractional-order variational method is illustrated. Second, a fractional-order denoising algorithm based on the fractional-order variational method is illustrated. In Section 5, the conclusions of this manuscript are presented.

13.2 Mathematical Background

This section presents a brief introduction to the necessary mathematical background of fractional calculus.

The commonly used fractional calculus definitions in the domain of Euclidean measure are those of Grünwald-Letnikov, Riemann-Liouville, and Caputo [3], [176], [178], [230] − [231]. In addition, there are some other well-known definitions of fractional derivative [528] − [540]. The Grünwald-Letnikov definition of fractional calculus, in a convenient form, for causal signal $f(x)$, is as follows:

$$^{G-L}_{a}D^v_x f(x) = \lim_{N \to \infty} \left\{ \frac{\left(\frac{x-a}{N}\right)^{-v}}{\Gamma(-v)} \sum_{k=0}^{N-1} \frac{\Gamma(k-v)}{\Gamma(k+1)} f\left(x - k\left(\frac{x-a}{N}\right)\right) \right\}, \quad (13.1)$$

where $f(x)$ is a differintegrable function [3], [176], [178], [230] − [231], $[a, x]$ is the duration of $f(x)$, v is a real number, $\Gamma(\alpha) = \int_0^\infty e^{-x} x^{\alpha-1} dx$ is the Gamma function, and $^{G-L}_{a}D^v_x$ denotes the Grünwald-Letnikov defined fractional differential operator.

The Riemann-Liouville definition of the v-order integral, for causal signal $f(x)$, is as follows:

$$^{R-L}_{a}I^v_x f(x) = \frac{1}{\Gamma(v)} \int_a^x \frac{f(x)}{(x-\tau)^{1-v}} d\tau, \quad (13.2)$$

where $v > 0$ and $^{R-L}_{a}I^v_x$ denotes the Riemann-Liouville left-sided fractional integral operator.

The Riemann-Liouville definition of the v-order derivative is as follows:

$$^{R-L}_{a}D^v_x f(x) = \frac{1}{\Gamma(n-v)} \frac{d^n}{dx^n} \int_a^x \frac{f(\tau)}{(x-\tau)^{v-n+1}} d\tau, \tag{13.3}$$

where $n-1 \leqslant v < n$, and $^{R-L}_{a}D^v_x$ denotes the Riemann-Liouville left-handed fractional differential operator. The Laplace transform of the v-order Riemann-Liouville differential operator [78], [541] — [542] are given as:

$$FT[^{R-L}_{0}D^v_x f(x)] = (j\omega)^v FT[f(x)] - \sum_{k=0}^{n-1} (j\omega)^k [^{R-L}_{0}D^{v-1-k}_x f(x)]_{x=0}, \tag{13.4}$$

$$LT[^{R-L}_{0}D^v_x f(x)] = s^v LT[f(x)] - \sum_{k=0}^{n-1} s^k [^{R-L}_{0}D^{v-1-k}_x f(x)]_{x=0}, \tag{13.5}$$

where $FT()$ denotes Fourier transform, $LT()$ denotes Laplace transform, j denotes an imaginary unit, ω denotes an angular frequency, and $s = j\omega$ denotes a Laplace operator. When $f(x)$ is a causal signal, and its fractional primitives are also required to be zero, we can simplify the Laplace transform for $^{R-L}_{0}D^v_x f(x)$ as, respectively:

$$FT[^{R-L}_{0}D^v_x f(x)] = (j\omega)^v FT[f(x)], \tag{13.6}$$

$$LT[^{R-L}_{0}D^v_x f(x)] = s^v LT[f(x)]. \tag{13.7}$$

If signal $f(x)$ is a $(m+1)$-order continuously differentiable function, and $m=[v]=n-1$, where $[v]$ is the round-off number of v, the Grünwald-Letnikov defined fractional calculus is equivalent to the Riemann-Liouville defined one. If the aforementioned condition is not satisfied, the Riemann-Liouville defined fractional calculus is the extension of the Grünwald-Letnikov defined one. Thus, the Riemann-Liouville defined fractional calculus is with more extensive application potential. In this work, we mainly adopt the Riemann-Liouville defined fractional calculus for the following mathematical derivation. We use the equivalent notations $D^v_x = {}^{G-L}_{0}D^v_x = {}^{R-L}_{0}D^v_x$ in an arbitrary, interchangeable manner.

13.3 Fractional-Order Euler-Lagrange Equation for Fractional-Order Variational Method

In this section, with respect to a necessary condition of the fractional-order fixed boundary optimization problems in signal processing and image processing, two different equivalent forms of the fractional-order Euler-Lagrange equation for the fractional-order variational method are derived based on Wiener-Khintchine theorem and the fractional-order Green formula, respectively. For the convenience of illustration, without loss of generality, two-dimensional related issues are only discussed in this section. Likewise, the results of multi-dimensional issues are similar to those of two-dimensional issues.

13.3.1 First-order Euler-Lagrange equation based on fractional-order Green formula

In this subsection, for the convenience of comparison, based on the first-order Green formula, the first-order Euler-Lagrange equation for the first-order variational method [475], [485], [486] need be derived.

Lemma 1: With regard to a first-order continuously differentiable scalar function $u(x,y)$ and a first-order continuously differentiable vector function $\vec{\varphi}$ (φ_x, φ_y), based on the first-order Green formula, an identical equation for the first-order variational method can be derived as:

$$\iint_\Omega u \operatorname{div} \vec{\varphi}\,dx\,dy = \iint_\Omega u(\nabla \cdot \vec{\varphi})\,dx\,dy = -\iint_\Omega \nabla u \cdot \vec{\varphi}\,dx\,dy$$
$$= -\langle \nabla u, \vec{\varphi} \rangle$$
$$= -\langle u, \nabla^* \vec{\varphi} \rangle$$
$$= \iint_\Omega u(-\nabla^* \vec{\varphi})\,dx\,dy, \qquad (13.8)$$

where $u(x, y)$ is an admissible surface (a test function), Ω is an open bounded subsets of real plane R^2 with smooth boundary, Ω bears a rectangular shape for most real applications, div denotes the first-order divergence operator, $dx\,dy$ is the Lebesgue measure on real plane R^2, symbol $\cdot$ denotes the first-order Euclidean inner product, symbol $\langle,\rangle$ denotes the first-order Euclidean inner product, and ∇^* denotes the first-order Hilbert adjoint operator [543] of a Hamilton operator ∇. Note that $\dfrac{\partial \varphi_x}{\partial x} + \dfrac{\partial \varphi_y}{\partial y} = 0$ is the first-order Euler-Lagrange equation of $-\iint_\Omega \nabla u \cdot \vec{\varphi}\,dx\,dy = 0$.

Proof: A Hamiltonian can be defined by:

$$\nabla u = \left(\frac{\partial}{\partial x}, \frac{\partial}{\partial y}\right)u$$
$$= \left(\frac{\partial u}{\partial x}, \frac{\partial u}{\partial y}\right), \qquad (13.9)$$

where $u = u(x,y) \in C_0^1(\Omega)$ denotes a scalar function. Thus, from (13.9), the following can be obtained, respectively:

$$\nabla u \cdot \vec{\varphi} = \left(\frac{\partial u}{\partial x}, \frac{\partial u}{\partial y}\right) \cdot (\varphi_x, \varphi_y)$$
$$= \frac{\partial u}{\partial x}\varphi_x + \frac{\partial u}{\partial y}\varphi_y, \qquad (13.10)$$

$$\operatorname{div}\vec{\varphi} = \frac{\partial \varphi_x}{\partial x} + \frac{\partial \varphi_y}{\partial y}$$
$$= \left(\frac{\partial}{\partial x}, \frac{\partial}{\partial y}\right) \cdot (\varphi_x, \varphi_y)$$
$$= \nabla \cdot \vec{\varphi}, \qquad (13.11)$$

where $\vec{\varphi} = (\varphi_x, \varphi_y)$ denotes a vector function, and div denotes the first-order divergence operator. Thus, from (13.11), one can obtain:

$$u \operatorname{div} \vec{\varphi} = u \frac{\partial \varphi_x}{\partial x} + u \frac{\partial \varphi_y}{\partial y}. \qquad (13.12)$$

Thus, from (13.10) and (13.12), the following can be obtained:

$$\iint_\Omega u \text{ div } \vec{\varphi} \, dx dy + \iint_\Omega \nabla u \cdot \vec{\varphi} \, dx dy$$

$$= \iint_\Omega \left[\left(u \frac{\partial \varphi_x}{\partial x} + u \frac{\partial \varphi_y}{\partial y} \right) + \left(\frac{\partial u}{\partial x} \varphi_x + \frac{\partial u}{\partial y} \varphi_y \right) \right] dx dy$$

$$= \iint_\Omega \left[\left(u \frac{\partial \varphi_x}{\partial x} + \frac{\partial u}{\partial x} \varphi_x \right) + \left(u \frac{\partial \varphi_y}{\partial y} + \frac{\partial u}{\partial y} \varphi_y \right) \right] dx dy$$

$$= \iint_\Omega \left[\left(\frac{\partial (u\varphi_x)}{\partial x} + \frac{\partial (u\varphi_y)}{\partial y} \right) \right] dx dy, \tag{13.13}$$

where Ω denotes a two-dimensional simply connected region. Further, the traditional first-order Green formula can be given as:

$$\oint_C P(x,y) dx + Q(x,y) dy = \iint_\Omega \left(\frac{\partial Q}{\partial x} - \frac{\partial P}{\partial y} \right) dx dy, \tag{13.14}$$

where C denotes the piecewise smooth boundary curve of Ω, $P(x,y)$ and $Q(x,y)$ are continuous both on C and Ω. Thus, from (13.13) and (13.14), it follows that:

$$\iint_\Omega u \text{ div } \vec{\varphi} \, dx dy + \iint_\Omega \nabla u \cdot \vec{\varphi} \, dx dy = \oint_c -u\varphi_y \, dx + u\varphi_x \, dy. \tag{13.15}$$

Without loss of generality, Ω bears a rectangular shape for most real applications, C is a rectangular boundary. Figure 13.1 shows the two-dimensional simply connected region of a rectangular region.

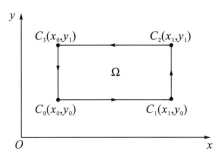

Figure 13.1　Two-dimensional simply connected region of a rectangular region.

Thus, from Figure 13.1 and (13.15), the following can be obtained:

$$\iint_\Omega u \text{ div } \vec{\varphi} \, dx dy + \iint_\Omega \nabla u \cdot \vec{\varphi} \, dx dy$$

$$= \oint_C -u\varphi_y \, dx + u\varphi_x \, dy$$

$$= \int_{x_0}^{x_1} -u\varphi_y \, dx + \int_{y_0}^{y_1} u\varphi_x \, dy + \int_{x_1}^{x_0} -u\varphi_y \, dx + \int_{y_1}^{y_0} u\varphi_x \, dy$$

$$= -\int_{x_0}^{x_1} u\varphi_y \, dx + \int_{y_0}^{y_1} u\varphi_x \, dy + \int_{x_0}^{x_1} u\varphi_y \, dx - \int_{y_0}^{y_1} u\varphi_x \, dy$$

$$= 0. \tag{13.16}$$

From (13.16), it follows that:

$$\iint_\Omega u \,\text{div}\, \vec{\varphi}\,dxdy = \iint_\Omega u(\nabla \cdot \vec{\varphi})\,dxdy$$
$$= -\iint_\Omega \nabla u \cdot \vec{\varphi}\,dxdy$$
$$= -\langle \nabla u, \vec{\varphi}\rangle$$
$$= -\langle u, \nabla^* \vec{\varphi}\rangle$$
$$= \iint_\Omega u(-\nabla^* \vec{\varphi})\,dxdy, \quad (13.17)$$

where symbol $\langle , \rangle$ denotes the first-order inner product and ∇^* denotes the first-order Hilbert adjoint operator of a Hamilton operator ∇.

Note that from (13.17), the following is true:

$$\text{div} = -\nabla^*. \quad (13.18)$$

Equation (13.17) shows that to enable $-\iint_\Omega \nabla u \cdot \vec{\varphi}\,dxdy = 0$ be set up, a necessary condition can be given as:

$$\iint_\Omega u \,\text{div}\, \vec{\varphi}\,dxdy = \iint_\Omega \left(u\frac{\partial \varphi_x}{\partial x} + u\frac{\partial \varphi_y}{\partial y}\right)dxdy = 0. \quad (13.19)$$

Because $u(x, y)$ is a test function, $u(x, y)$ is arbitrary. Therefore, according to the fundamental lemma of variation [475], [485], to enable (13.19) to be set up, a necessary condition can be given as:

$$\frac{\partial \varphi_x}{\partial x} + \frac{\partial \varphi_y}{\partial y} = 0. \quad (13.20)$$

Equation (13.20) shows that $\frac{\partial \varphi_x}{\partial x} + \frac{\partial \varphi_y}{\partial y} = 0$ is the first-order Euler-Lagrange equation of $-\iint_\Omega \nabla u \cdot \vec{\varphi}\,dxdy = 0$. This completes the proof.

Example 1: A first-order variational framework based on the first-order differential is given as:

$$\begin{cases} \text{Minimize:} E[u(x,y)] = \iint_\Omega F\left(x,y,u,\frac{\partial u}{\partial x},\frac{\partial u}{\partial y}\right)dxdy, \\ \text{Subject to:} \langle D^1 u, \vec{n}\rangle = 0 \text{ on } \partial\Omega \end{cases} \quad (13.21)$$

where E denotes an energy functional, $D^1 = \nabla$ denotes the first-order differential operator, $\vec{n}$ denotes the normal to the boundary, $\partial\Omega$ denotes the boundary of Ω, and symbol $\langle , \rangle$ denotes the first-order Euclidean inner product. Equation (13.21) assumes that the function value of $u(x, y)$ continues smoothly as a constant beyond its boundaries. This artificial assumption is required for Neumann boundary condition that would have minor effect on the final results, i.e. the first-order directional derivative $\langle D^1 u, \vec{n}\rangle = \langle \nabla u, \vec{n}\rangle = \nabla u \cdot \vec{n} = \frac{\partial u}{\partial \vec{n}} = 0$. Suppose $u(x, y)$ is a first-order extremal surface of energy functional E, then $\xi(x,y) \in C_0^\infty(\Omega)$ is an admissible surface, a test function, closing to $u(x,y)$. $u(x,y)$ and $\xi(x,y)$ can be merged into a family of surfaces, $u+\beta\xi$, where β is a small parameter. When $\beta=0$, the family of surfaces, $u+\beta\xi$, converts into the first-order extremal surface, $u(x,y)$. The first-order derivative of the extreme points equals to zero. Thus, the anisotropic diffusion of (13.21) can be explained as the first-order dissipation process of energy functional E. To achieve the first-order minimum of (13.21), a first-order energy functional on the family of surfaces, $u+\beta\xi$, is given as:

$$\begin{aligned}
\delta E &= \frac{\partial}{\partial \beta} \iint_\Omega F\left[x,y,u(x,y,\beta),\frac{\partial u(x,y,\beta)}{\partial x},\frac{\partial u(x,y,\beta)}{\partial y}\right] \mathrm{d}x\mathrm{d}y \\
&= \iint_\Omega (F_u \delta u + F_p \delta p + F_q \delta q)\mathrm{d}x\mathrm{d}y \\
&= \iint_\Omega [F_u \delta u + (F_p, F_q) \cdot (\delta p, \delta q)]\mathrm{d}x\mathrm{d}y \\
&= \iint_\Omega [F_u \delta u + \nabla \delta u \cdot (F_p, F_q)]\mathrm{d}x\mathrm{d}y \\
&= 0,
\end{aligned} \qquad (13.22)$$

where $F_u = \frac{\partial F}{\partial u}$, $\delta u = \frac{\partial u(x,y,\beta)}{\partial \beta}$, $p(x,y) = \frac{\partial u(x,y)}{\partial x}$, $q(x,y) = \frac{\partial u(x,y)}{\partial y}$, $\delta p = \frac{\partial p(x,y,\beta)}{\partial \beta}$, $\delta q = \frac{\partial q(x,y,\beta)}{\partial \beta}$, $u(x,y,\beta) = u(x,y) + \beta \delta u = u(x,y) + \beta[\xi(x,y) - u(x,y)]$, $p(x,y,\beta) = \frac{\partial u(x,y,\beta)}{\partial x} = p(x,y) + \beta \delta p$, and $q(x,y,\beta) = \frac{\partial u(x,y,\beta)}{\partial y} = q(x,y) + \beta \delta q$. Thus, from (13.8), (13.20), and (13.22), the following can be obtained:

$$\iint_\Omega [F_u \delta u - \mathrm{div}(F_p, F_q)\delta u]\mathrm{d}x\mathrm{d}y = 0. \qquad (13.23)$$

Because δu is arbitrary, according to the fundamental lemma of variation [475], [485], to enable (13.23) to be set up, a necessary condition can be given as:

$$[F_u - \mathrm{div}(F_p, F_q)] = \left[F_u - \frac{\partial F_p}{\partial x} - \frac{\partial F_q}{\partial y}\right] = 0. \qquad (13.24)$$

Equation (13.24) shows that $\left[F_u - \frac{\partial F_p}{\partial x} - \frac{\partial F_q}{\partial y}\right] = 0$ is the first-order Euler-Lagrange equation of $\frac{\partial}{\partial \beta} \iint_\Omega F\left[x,y,u(x,y,\beta),\frac{\partial u(x,y,\beta)}{\partial x},\frac{\partial u(x,y,\beta)}{\partial y}\right]\mathrm{d}x\mathrm{d}y\bigg|_{\beta=0} = 0$. Thus, $u(x,y)$ is a solution of $\left[F_u - \frac{\partial F_p}{\partial x} - \frac{\partial F_q}{\partial y}\right] = 0$.

13.3.2 Fractional-order Euler-Lagrange equation based on Wiener-Khintchine theorem

In this subsection, based on Wiener-Khintchine theorem, the fractional-order Euler-

Lagrange equation for the fractional-order variational method is derived. J. Bai and J. Zhang claimed that the fractional-order Euler-Lagrange equation for the fractional-order variational method could be derived based on Parseval theorem [346], [544] − [545]. Note that Parseval theorem actually studies the relationship between a self-correlation function and a self-energy-density spectrum or a self-power-density spectrum of an identical continuously differentiable signal $s(t)$, which can be derived as $\frac{1}{2\pi}\int_{-\infty}^{\infty} S(j\omega) \overline{S(j\omega)} d\omega = \int_{-\infty}^{\infty} s^2(t) dt$, where $S(j\omega)$ is the Fourier transforms of $s(t)$ and $\overline{S(j\omega)}$ is the complex conjugation of $S(j\omega)$. However, the derivation of the fractional-order Euler-Lagrange equation for the fractional-order variational method should simultaneously consider two different factors: a fractional-order extremal surface of energy functional and a closing admissible surface. The relationship between a cross-correlation function and a cross energy density spectrum or a cross power density spectrum of two different continuously differentiable signals need be discussed. Therefore, the fractional-order Euler-Lagrange equation for the fractional-order variational method should be derived based on Wiener-Khintchine theorem, rather than Parseval theorem.

Lemma 2: With regard to a v-order continuously differentiable scalar function $u(x, y)$ and a v-order continuously differentiable vector function $\vec{\varphi} = (\varphi_x, \varphi_y)$, based on Wiener-Khintchine theorem, the fractional-order Euler-Lagrange equation of $\iint_{\Omega} D^v u \cdot \vec{\varphi} dx dy = 0$ can be derived as:

$$\frac{1}{4\pi^2} \text{Re}(\overline{D_x^v}\varphi_x + \overline{D_y^v}\varphi_y) = 0, \tag{13.25}$$

where $u(x, y)$ is an admissible surface (a test function), Ω is an open bounded subsets of real plane R^2 with smooth boundary, v is a real number, Re( ) computes the real part of the complex number, symbol $*$ denotes a linear convolution, and $\overline{D_x^v}$ and $\overline{D_y^v}$ denote the conjugate operators of the v-order partial differential operators D_x^v and D_y^v, respectively.

Proof: In view of signal processing, the cross-correlation function between continuous signal $s_1(t)$ and continuous signal $s_2(t)$ can be given as:

$$r_{12}(\tau) = s_1(\tau) * \overline{s_2(-\tau)}$$
$$= \int_{-\infty}^{\infty} s_1(t) \overline{s_2(t-\tau)} dt$$
$$\underset{\text{are real signals}}{\overset{\text{If } s_1(t) \text{ and } s_2(t)}{=}} s_1(\tau) * s_2(-\tau)$$
$$= \int_{-\infty}^{\infty} s_1(t) s_2(t-\tau) dt, \tag{13.26}$$

where $\overline{s_2(-\tau)}$ is the complex conjugation of $s_2(-\tau)$ and symbol $*$ denotes a linear convolution. Thus, whether $s_1(t)$ and $s_2(t)$ are real signals or not, the Fourier transform of (13.26) is as follows:

$$R_{12}(j\omega) = S_1(j\omega) * \overline{S_2(j\omega)}, \tag{13.27}$$

where j denotes an imaginary unit, ω denotes an angular frequency, $\overline{S_2(j\omega)}$ is the complex conjugation of $S_2(j\omega)$, and $R_{12}(j\omega)$, $S_1(j\omega)$, and $S_2(j\omega)$ are the Fourier transforms of $r_{12}(\tau)$, $s_1(t)$, and $s_2(t)$, respectively. Further, if $s_1(t)$ and $s_2(t)$ are real signals, from (13.26) and (13.27), Wiener-Khintchine theorem can be given as:

$$\begin{aligned}
r_{12}(0) &= \frac{1}{2\pi}\int_{-\infty}^{\infty} S_1(j\omega)\,\overline{S_2(j\omega)}\,e^{j\omega\tau}\,d\omega \Big|_{\tau=0} \\
&= \frac{1}{2\pi}\int_{-\infty}^{\infty} S_1(j\omega)\,\overline{S_2(j\omega)}\,d\omega \\
&= s_1(\tau) * s_2(-\tau)\big|_{\tau=0} \\
&= \int_{-\infty}^{\infty} s_1(t) s_2(t)\,dt.
\end{aligned} \tag{13.28}$$

Equation (13.28) shows that first, $r_{12}(0)$ is equal to the cross energy or cross power between $s_1(t)$ and $s_2(t)$. Second, with regard to continuously differentiable signal, the relationship between a cross-correlation function and a cross energy density spectrum or a cross power density spectrum can be derived as $\frac{1}{2\pi}\int_{-\infty}^{\infty} S_1(j\omega)\overline{S_2(j\omega)}\,d\omega = \int_{-\infty}^{\infty} s_1(t)s_2(t)\,dt$. Therefore, from (13.28), $\iint_{\Omega} D^v u \cdot \vec{\varphi}\,dx\,dy$ can be rewritten as:

$$\begin{aligned}
\iint_{\Omega} D^v u \cdot \vec{\varphi}\,dx\,dy &= \iint_{\Omega} \vec{\varphi} \cdot D^v u\,dx\,dy \\
&= \iint_{\Omega} \{\varphi_x(x,y)[D_x^v u(x,y)] + \varphi_y(x,y)[D_y^v u(x,y)]\}\,dx\,dy \\
&= \frac{1}{4\pi^2}\iint_{\Omega_\omega}[\Phi_x(j\omega_x,j\omega_y)\overline{(j\omega_x)^v U(j\omega_x,j\omega_y)} \\
&\quad + \Phi_y(j\omega_x,j\omega_y)\overline{(j\omega_y)^v U(j\omega_x,j\omega_y)}]\,d\omega_x\,d\omega_y \\
&= \frac{1}{4\pi^2}\iint_{\Omega_\omega}[\overline{(j\omega_x)^v}\Phi_x(j\omega_x,j\omega_y)\overline{U(j\omega_x,j\omega_y)} \\
&\quad + \overline{(j\omega_y)^v}\Phi_y(j\omega_x,j\omega_y)\,\overline{U(j\omega_x,j\omega_y)}]\,d\omega_x\,d\omega_y,
\end{aligned} \tag{13.29}$$

where $U(j\omega_x,j\omega_y)$, $\Phi_x(j\omega_x,j\omega_y)$, $\Phi_y(j\omega_x,j\omega_y)$, $(j\omega_x)^v$, and $(j\omega_y)^v$ are the Fourier transforms of $u(x,y)$, $\varphi_x(x,y)$, $\varphi_y(x,y)$, D_x^v, and D_y^v, respectively, Ω_ω is corresponding to Ω in the domain of Fourier transform, and $\overline{(j\omega_x)^v}$, $\overline{(j\omega_y)^v}$, and $\overline{U(j\omega_x,j\omega_y)}$ are the complex conjugations of $(j\omega_x)^v$, $(j\omega_y)^v$, and $U(j\omega_x,j\omega_y)$, respectively. Because $u(x,y)$ is a test function, $\overline{U(j\omega_x,j\omega_y)}$ is arbitrary. Therefore, according to the fundamental lemma of variation, to enable $\iint_{\Omega} D^v u \cdot \vec{\varphi}\,dx\,dy = 0$ to be set up, an essential condition can be given as:

$$\frac{1}{4\pi^2}[\overline{(j\omega_x)^v}\Phi_x(j\omega_x,j\omega_y) + \overline{(j\omega_y)^v}\Phi_y(j\omega_x,j\omega_y)] = 0. \tag{13.30}$$

The inverse Fourier transform of (13.30) is as follows:

$$\frac{1}{4\pi^2}\text{Re}[h_x^v(x,y)*\varphi_x + h_y^v(x,y)*\varphi_y] = \frac{1}{4\pi^2}\text{Re}(\overline{D_x^v}\varphi_x + \overline{D_y^v}\varphi_y)$$
$$= 0, \tag{13.31}$$

where Re( ) computes the real part of the complex number, $h_x^v(x,y) \overset{FT}{\leftrightarrow} \overline{(j\omega_x)^v}$, $h_y^v(x,y) \overset{FT}{\leftrightarrow} \overline{(j\omega_y)^v}$, and $\overline{D_x^v}$ and $\overline{D_y^v}$ denote the conjugate operators of the v-order partial differential operators D_x^v and D_y^v, respectively. The second item on the right side of (13.31), $\frac{1}{4\pi^2}\text{Re}(\overline{D_x^v}\varphi_x + \overline{D_y^v}\varphi_y)$, is expressed in the form of operator. This completes the proof.

Note that first, the aforementioned proof can also be achieved by means of the theory of Hilbert adjoint operator [543]. The Hilbert adjoint operator of a fractional differential operator is a complex conjugate transpose matrix of that of the corresponding fractional differential operator. Second, because the inverse Fourier transforms of $(j\omega_x)^v$ and $(j\omega_y)^v$ belong to Euler integral of the second kind, $\overline{D_x^v}$ and $\overline{D_y^v}$ could not be achieved in spatial domain or time domain directly, which should be implemented in discrete Fourier transform domain. With regard to two-dimensional discrete signal $s(mT, nT)$, according to the period extend theory, its discrete Fourier transform and inverse discrete Fourier transform can be given as, respectively:

$$S(\alpha,\beta) = S\left(\frac{2\pi}{MT}\alpha, \frac{2\pi}{NT}\beta\right)$$
$$= \sum_{m=0}^{M-1}\sum_{n=0}^{N-1}s(mT,nT)e^{-j2\pi(\frac{m\alpha}{M}+\frac{n\beta}{N})}, \tag{13.32}$$

$$s(mT,nT) = \frac{1}{MN}\sum_{u=0}^{M-1}\sum_{v=0}^{N-1}S(\alpha,\beta)e^{j2\pi(\frac{m\alpha}{M}+\frac{n\beta}{N})}, \tag{13.33}$$

where $m = 0, 1, \cdots, M-1$, $n = 0, 1, \cdots, N-1$, T is the interval between the discrete points, $S(\alpha,\beta)$ is the discrete Fourier transform of $s(mT,nT)$, $\alpha = 0, 1, \cdots, M-1$ and $\beta = 0, 1, \cdots, N-1$ in (13.32), and $m = 0, 1, \cdots, M-1$ and $n = 0, 1, \cdots, N-1$ in (13.33). Note that in signal processing and image processing, one usually use the notation of $s(,)$, the lowercase of "s", as a signal in time or space domain, while use the notation of $S(,)$, the uppercase of "s", as the corresponding signal in frequency domain. With regard to digital image processing, we have $T=1$. From (13.32) and (13.33), the first-order difference operators, d_x^1 and d_y^1, of $s(mT, nT)$ can be derived as, respectively:

$$d_x^1 s(mT,nT) = s(mT,nT) - s[(m-1)T,nT]$$
$$= \frac{1}{MN}\sum_{u=0}^{M-1}\sum_{v=0}^{N-1}S(\alpha,\beta)(1-e^{j\frac{2\pi\alpha}{M}})e^{j(\frac{m\alpha}{M}+\frac{n\beta}{N})}, \tag{13.34}$$

$$d_y^1 s(mT, nT) = s(mT, nT) - s[mT, (n-1)T]$$
$$= \frac{1}{MN} \sum_{u=0}^{M-1} \sum_{v=0}^{N-1} S(\alpha, \beta)(1 - e^{j\frac{2\pi\beta}{N}}) e^{j(\frac{m\alpha}{M} + \frac{n\beta}{N})}. \tag{13.35}$$

Thus, by means of mathematical induction, from (13.34) and (13.35), the *nth*-order difference operators, d_x^n and d_y^n, of can be derived as, respectively:

$$d_x^n \overset{DFT}{\leftrightarrow} (1 - e^{j\frac{2\pi\alpha}{M}})^n, \tag{13.36}$$

$$d_y^n \overset{DFT}{\leftrightarrow} (1 - e^{j\frac{2\pi\beta}{N}})^n, \tag{13.37}$$

where *DFT* denotes discrete Fourier transform. Then, extending (13.36) and (13.37) from the integer-order to fractional-order, v-order difference operators, d_x^v and d_y^v, can be derived as:

$$D_x^v \approx d_x^v \overset{DFT}{\leftrightarrow} (1 - e^{j\frac{2\pi\alpha}{M}})^v, \tag{13.38}$$

$$D_y^v \approx d_y^v \overset{DFT}{\leftrightarrow} (1 - e^{j\frac{2\pi\beta}{N}})^v. \tag{13.39}$$

Thus, from (13.38) and (13.39), the following can be obtained:

$$\overline{D_x^v} \approx \overline{d_x^v} \overset{DFT}{\leftrightarrow} \overline{(1 - e^{-j\frac{2\pi\alpha}{M}})^v}, \tag{13.40}$$

$$\overline{D_y^v} \approx \overline{d_y^v} \overset{DFT}{\leftrightarrow} \overline{(1 - e^{-j\frac{2\pi\beta}{N}})^v}, \tag{13.41}$$

where $\alpha = 0, 1, \cdots, M-1$, $\beta = 0, 1, \cdots, N-1$, and $\overline{d_x^v}$ and $\overline{d_y^v}$ are the conjugate operators of d_x^v and d_y^v, respectively. Third, from aforementioned discussion, we can see that the fractional-order Euler-Lagrange equation based on Wiener-Khintchine theorem should be achieved by discrete Fourier transform and inverse discrete Fourier transform successively, which is difficult and complex to implement.

13.3.3 Fractional-order Euler-Lagrange equation based on fractional-order Green formula

In this subsection, in order to directly and easily implement the fractional-order variational method in spatial domain or time domain, the fractional-order Green formula and the fractional-order Euler-Lagrange equation based on the fractional-order Green formula are proposed, respectively.

At first, in order to derive the fractional-order Euler-Lagrange equation, the traditional first-order Green formula should be extended firstly from the first-order one to the fractional-order one.

A two-dimensional simply connected region can be shown as given in Figure 13.2.

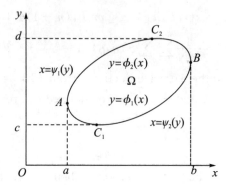

Figure 13.2 A two-dimensional simply connected region.

As shown in Figure 13.2, Ω is a two-dimensional simply connected region, C is the piecewise smooth boundary curve of Ω, and C consists of two piecewise smooth boundary curves, i.e. $y=\Phi_1(x)$, $y=\Phi_2(x)$, $a \leqslant x \leqslant b$ and $x=\Psi_1(y)$, $x=\Psi_2(y)$, $c \leqslant y \leqslant d$. For the convenience of illustration, let us sign the first-order differential operator, the v-order fractional differential operator, the first-order integral operator, the v-order fractional integral operator, the v-order fractional surface integral operator on Ω, the v-order fractional line integral operator on the AC_1B segment of C along the direction of $\overline{AC_1B}$, and the v-order fractional line integral operator on closed curve C in counter-clockwise direction as the symbols of D^1, D^v, $I^1=D^{-1}$, $I^v=D^{-v}$, $\underset{\Omega}{I_x^v I_y^v}$, $\underset{C(AC_1B)}{I^v}$, and $I_{C_1-}^v$, respectively. Further, let $v>0$ in the following derivation.

With regard to a differintegrable function [230] $P(x,y)$, if $P - D^{-v_1}D^{v_1}P \neq 0$, the following can be obtained:

$$D^{v_1}D^{v_2}P = D^{v_1+v_2}P - D^{v_1+v_2}(P - D^{-v_1}D^{v_1}P). \tag{13.42}$$

Then, from (13.42), the following is true:

$$I_x^{v_2}I_y^{v_2}D_y^{v_1}P(x,y) = I_x^{v_2}\{D_y^{v_1-v_2}P(x,y) - D_y^{v_1-v_2}[P(x,y) - D_y^{-v_1}D_y^{v_1}P(x,y)]\} \tag{13.43}$$

Therefore, from Figure 13.2 and (13.43), it follows that:

$$\underset{\Omega}{I_x^{v_2}I_y^{v_2}}D_y^{v_1}P(x,y) = {}_a^b I_x^{v_2} {}_{x\Phi_1(x)}^{\Phi_2(x)} I_y^{v_2}D_y^{v_1}P(x,y)$$

$$= {}_a^b I_x^{v_2}\{D_y^{v_1-v_2}P(x,y) - D_y^{v_1-v_2}[P(x,y) - D_y^{-v_1}D_y^{v_1}P(x,y)]\}\big|_{\Phi_1(x)}^{\Phi_2(x)}$$

$$= - \underset{C(BC_2A)}{I_x^{v_2}}\{D_y^{v_1-v_2}P(x,y) - D_y^{v_1-v_2}[P(x,y) - D_y^{-v_1}D_y^{v_1}P(x,y)]\}$$

$$- \underset{C(AC_1B)}{I_x^{v_2}}\{D_y^{v_1-v_2}P(x,y) - D_y^{v_1-v_2}[P(x,y) - D_y^{-v_1}D_y^{v_1}P(x,y)]\}$$

$$= - \underset{C-}{I_x^{v_2}}\{D_y^{v_1-v_2}P(x,y) - D_y^{v_1-v_2}[P(x,y) - D_y^{-v_1}D_y^{v_1}P(x,y)]\}. \tag{13.44}$$

Likewise, with regard to a differintegrable function [230] $Q(x,y)$, if $Q - D^{-v_1}D^{v_1}Q \neq 0$, the following can be obtained:

$$\underset{\Omega}{I_x^{v_2}I_y^{v_2}}D_x^{v_1}Q(x,y) = \underset{C-}{I_y^{v_2}}\{D_x^{v_1-v_2}Q(x,y) - D_x^{v_1-v_2}[Q(x,y) - D_x^{-v_1}D_x^{v_1}Q(x,y)]\} \tag{13.45}$$

Thus, from (13.44) and (13.45), the fractional-order Green formula can be derived as:

$$I_x^{v_2}I_y^{v_2}[D_x^{v_1}Q(x,y)-D_y^{v_1}P(x,y)]$$
$$=I_x^{v_2}\{D_y^{v_1-v_2}P(x,y)-D_y^{v_1-v_2}[P(x,y)-D_y^{-v_1}D_y^{v_1}P(x,y)]\}$$
$$+I_x^{v_2}\{D_x^{v_1-v_2}Q(x,y)-D_x^{v_1-v_2}[Q(x,y)-D_x^{-v_1}D_x^{v_1}Q(x,y)]\}. \quad (13.46)$$

If D^{-v_1} is the inverse operator of D^{v_1}, (13.42) can be simplified as $D^{v_1}D^{v_2}=D^{v_1+v_2}$. Then, (13.46) can be simplified as:

$$I_x^{v_2}I_y^{v_2}[D_x^{v_1}Q(x,y)-D_y^{v_1}P(x,y)]=I_x^{v_2}D_y^{v_1-v_2}P(x,y)+I_x^{v_2}D_x^{v_1-v_2}Q(x,y). \quad (13.47)$$

Equation (13.47) shows that first, if $v_1=v_2=v$, (13.47) can be further simplified as:

$$I_x^v I_y^v[D_x^v Q(x,y)-D_y^v P(x,y)]=I_x^v P(x,y)+I_y^v Q(x,y). \quad (13.48)$$

Equation (13.48) has the same derived result as that of the literature [546]. Second, if $v_1=v_2=1$, (13.47) can be further simplified as:

$$I_x^1 I_y^1[D_x^1 Q(x,y)-D_y^1 P(x,y)]=I_x^1 P(x,y)+I_y^1 Q(x,y). \quad (13.49)$$

Compared with (13.14) and (13.49), we can see that two formulae are identical. In other words, the traditional first-order Green formula is actually a special case of the fractional-order Green formula.

In addition, let ussign the boundary curve surface of a three-dimensional region Ω as the symbols of S, and suppose the boundary curve C of S is piecewise smooth. In a similar way, if differintegrable function [230] $P(x,y,z)$, $Q(x,y,z)$, and $R(x,y,z)$ are the fractional-order continuously differentiable with respect to x, y, and z in Ω, on S, or on C, the fractional-order Gauss formula and the fractional-order Stokes formula can be derived as, respectively:

$$I_x^{v_2}I_y^{v_2}I_z^{v_2}(D_x^{v_1}P+D_y^{v_1}Q+D_z^{v_1}R)=I_y^{v_2}I_z^{v_2}\{D_x^{v_1-v_2}P-D_x^{v_1-v_2}[P-D_x^{-v_1}D_x^{v_1}P]\}$$
$$+I_x^{v_2}I_z^{v_2}\{D_y^{v_1-v_2}Q-D_y^{v_1-v_2}[Q-D_y^{-v_1}D_y^{v_1}Q]\}$$
$$+I_x^{v_2}I_y^{v_2}\{D_z^{v_1-v_2}R-D_z^{v_1-v_2}[R-D_z^{-v_1}D_z^{v_1}R]\}, \quad (13.50)$$

$$I_y^{v_2}I_z^{v_2}(D_y^{v_1}R-D_z^{v_1}Q)+I_x^{v_2}I_z^{v_2}(D_z^{v_1}P-D_x^{v_1}R)+I_x^{v_2}I_y^{v_2}(D_x^{v_1}Q-D_y^{v_1}P)$$
$$=\frac{1}{2}I_x^{v_2}\{D_y^{v_1-v_2}P-D_y^{v_1-v_2}[P-D_y^{-v_1}D_y^{v_1}P]+D_z^{v_1-v_2}P-D_z^{v_1-v_2}[P-D_z^{-v_1}D_z^{v_1}P]\}$$
$$+\frac{1}{2}I_y^{v_2}\{D_x^{v_1-v_2}Q-D_x^{v_1-v_2}[Q-D_x^{-v_1}D_x^{v_1}Q]+D_z^{v_1-v_2}Q-D_z^{v_1-v_2}[Q-D^{-v_{1z}}D_z^{v_1}Q]\}$$
$$+\frac{1}{2}I_z^{v_2}\{D_x^{v_1-v_2}R-D_x^{v_1-v_2}[R-D_x^{-v_1}D_x^{v_1}R]+D_y^{v_1-v_2}R-D_y^{v_1-v_2}[R-D^{-v_{1y}}D^{v_{1y}}R]\}. \quad (13.51)$$

It can be easily proved that the traditional first-order Gauss formula and the traditional first-order Stokes formula are actually a special case of the fractional-order Gauss formula and that of the fractional-order Stokes formula, respectively.

Secondly, based on the proposed fractional-order Green formula, the fractional-order Euler-Lagrange equation for the fractional-order variational method is proposed.

Lemma 3: With regard to a fractional-order continuously differentiable scalar function $u(x, y)$ and a fractional-order continuous differentiable vector function $\vec{\varphi} = (\varphi_x, \varphi_y)$, based on the fractional-order Green formula, an identical equation for the fractional-order variational method can be derived as:

$$-I_x^{v_2}I_y^{v_2}D^{v_1}u \cdot \vec{\varphi} = -I_x^{v_2}I_y^{v_2}[(D_x^{v_1}u)\varphi_x + (D_y^{v_1}u)\varphi_y]$$
$$= I_x^{v_2}I_y^{v_2}\sum_{n=1}^{\infty}\binom{v_1}{n}[(D_x^{v_1-n}u)D_x^n\varphi_x + (D_y^{v_1-n}u)D_y^n\varphi_y], \quad (13.52)$$

where $u(x, y)$ is an admissible surface (a test function), Ω is an open bounded subsets of real plane R^2 with smooth boundary, Ω bears a rectangular shape for most real applications, v_1 and v_2 are two real numbers, symbol $\cdot$ denotes the first-order Euclidean inner product. Note that $\frac{\Gamma(1+v_1)}{\Gamma(v_1)}[D_x^1\varphi_x + D_y^1\varphi_y] = 0$ is the fractional-order Euler-Lagrange equation of $-I_x^{v_2}I_y^{v_2}D^{v_1}u \cdot \vec{\varphi} = 0$.

Proof: Let us sign the v-order fractional differential operator as a symbol of $D^v = (D_x^v, D_y^v)$. If $v = 0$, D^0 represents an identity operator. Ω bears a rectangular shape for most real applications, C is a rectangular boundary, which can be shown as given in Figure 13.1. With regard to a causal signal, from (13.2), the following can be obtained, respectively:

$$I_x^v s(x,y) = \frac{1}{\Gamma(v)}\int_{a_x}^{x}(x-\eta)^{v-1}s(\eta,y)d\eta, \quad (13.53)$$

$$I_y^v s(x,y) = \frac{1}{\Gamma(v)}\int_{a_y}^{y}(y-\xi)^{v-1}s(x,\xi)d\xi, \quad (13.54)$$

$$I_x^v I_y^v s(x,y) = \frac{1}{\Gamma^2(v)}\int_{a_x}^{x}\int_{a_y}^{y}(x-\eta)^{v-1}(y-\xi)^{v-1}s(\eta,\xi)d\eta d\xi. \quad (13.55)$$

Thus, with regard to a rectangular region Ω, from (13.46), (13.53) − (13.55), and Figure 13.1, it follows that:

$$I_x^{v_2}I_y^{v_2}(D_x^{v_1}Q(x,y) - D_y^{v_1}P(x,y)) = I_{C^-}^{v_2}\{D_y^{v_1-v_2}P(x,y) - D_y^{v_1-v_2}[P(x,y)$$
$$- D_y^{-v_1}D_y^{v_1}P(x,y)]\} + I_{C^-}^{v_2}\{D_x^{v_1-v_2}Q(x,y)$$
$$- D_x^{v_1-v_2}[Q(x,y) - D_x^{-v_1}D_x^{v_1}Q(x,y)]\}$$
$$= {}_{x_0}^{x_1}I_x^{v_2}\{D_y^{v_1-v_2}P - D_y^{v_1-v_2}[P - D_y^{-v_1}D_y^{v_1}P]\}$$
$$+ {}_{x_1}^{x_0}I_x^{v_2}\{D_y^{v_1-v_2}P - D_y^{v_1-v_2}[P - D_y^{-v_1}D_y^{v_1}P]\}$$
$$+ {}_{y_0}^{y_1}I_y^{v_2}\{D_x^{v_1-v_2}Q - D_x^{v_1-v_2}[Q - D_x^{-v_1}D_x^{v_1}Q]\}$$

$$+ {}_{y_1}^{y_0}I_y^{v_2}\{D_x^{v_1-v_2}Q - D_x^{v_1-v_2}[Q - D_x^{v_1-v_2}]\}$$
$$= 0. \tag{13.56}$$

Further, $\sum_{m=0}^{\infty}\sum_{n=0}^{m} \equiv \sum_{n=0}^{\infty}\sum_{m=n}^{\infty}$ and $\binom{v}{r+n}\binom{r+n}{n} \equiv \binom{v}{n}\binom{v-n}{r}$ can be derived, where v is a real number. Thus, from (13.1), it follows that [547]:

$$D_{x-a}^v(fg) = \sum_{n=0}^{\infty}\left[\binom{v}{n}(D_{x-a}^{v-n}f)D_{x-a}^n g\right], \tag{13.57}$$

where $\binom{v}{n} = \frac{\Gamma(1+v)}{\Gamma(1-n+v)\Gamma(1+n)}$. Moreover, according to the homogeneous properties of fractional calculus [230], from (13.56) and (13.57), the following can be obtained:

$$I_x^{v_2}I_y^{x_2}[D_x^{v_1}(u\varphi_x) + D_y^{v_1}(u\varphi_y)] = I_x^{v_2}I_y^{v_2}[D_x^{v_1}(u\varphi_x) - D_y^{v_1}(-u\varphi_y)]$$
$$= I_x^{v_2}I_y^{v_2}\sum_{n=0}^{\infty}\binom{v_1}{n}[(D_x^{v_1-n}u)D_x^n\varphi_x + (D_y^{v_1-n}u)D_y^n\varphi_y]$$
$$= 0. \tag{13.58}$$

Thus, from (13.58), the following is true:

$$-I_x^{v_2}I_y^{v_2}D^{v_1}u \cdot \vec{\varphi} = -I_x^{v_2}I_y^{v_2}[(D_x^{v_1}u)\varphi_x + (D_y^{v_1}u)\varphi_y]$$
$$= I_x^{v_2}I_y^{v_2}\sum_{n=1}^{\infty}\binom{v_1}{n}[(D_x^{v_1-n}u)D_x^n\varphi_x + (D_y^{v_1-n}u)D_y^n\varphi_y], \tag{13.59}$$

where $\binom{v}{0} = \frac{\Gamma(1+v)}{\Gamma(1+v)\Gamma(1)} = 1$. From (13.59), if $-I_x^{v_2}I_y^{v_2}D^{v_1}u \cdot \vec{\varphi} = 0$, one can obtain:

$$-I_x^{v_2}I_y^{v_2}D^{v_1}u \cdot \vec{\varphi} = I_x^{v_2}I_y^{v_2}\sum_{n=1}^{\infty}\binom{v_1}{n}[(D_x^{v_1-n}u)D_x^n\varphi_x + (D_y^{v_1-n}u)D_y^n\varphi_y]$$
$$= 0. \tag{13.60}$$

Since a vector in x direction and a vector in y direction are orthogonal to each other, to enable (13.60) be set up, a necessary condition can be given as:

$$\begin{cases} I_x^{v_2}I_y^{x_2}\sum_{n=1}^{\infty}\binom{v_1}{n}[(D_x^{v_1-n}u)D_x^n\varphi_x] = 0 \\ I_x^{v_2}I_y^{x_2}\sum_{n=1}^{\infty}\binom{v_1}{n}[(D_y^{v_1-n}u)D_y^n\varphi_y] = 0 \end{cases}. \tag{13.61}$$

Because $u(x, y)$ is a test function, $D_x^{v_1-n}u(x, y)$ and $D_y^{v_1-n}u(x, y)$ are arbitrary. Therefore, to enable (13.60) to be set up, from (13.61), an essential condition can be given as:

$$\begin{cases} \binom{v_1}{n}D_x^n\varphi_x \overset{n=1\to\infty}{=} 0 \\ \binom{v_1}{n}D_y^n\varphi_y \overset{n=1\to\infty}{=} 0 \end{cases}. \tag{13.62}$$

For n is a positive integer within $1\to\infty$, to enable $(D_x^n\varphi_x, D_y^n\varphi_y) \overset{n=1\to\infty}{=} (0,0)$ to be set up, only if $(D_x^1\varphi_x, D_y^1\varphi_y) = (0,0)$ is satisfied. Thus, to enable (13.62) to be set up, only if $\begin{cases} \binom{v_1}{1} D_x^1 \varphi_x = 0 \\ \binom{v_1}{1} D_y^1 \varphi_y = 0 \end{cases}$ is satisfied. Further, to enable $\begin{cases} \binom{v_1}{1} D_x^1 \varphi_x = 0 \\ \binom{v_1}{1} D_y^1 \varphi_y = 0 \end{cases}$ to be set up, an essential condition can be given as $\binom{v_1}{1}(D_x^1 \varphi_x + D_y^1 \varphi_y) = 0$. Therefore, to enable (13.60) be set up, an essential condition can be given as:

$$\binom{v_1}{1}[D_x^1 \varphi_x + D_y^1 \varphi_y] = \frac{\Gamma(1+v_1)}{\Gamma(v_1)}[D_x^1 \varphi_x + D_y^1 \varphi_y]$$
$$= 0. \qquad (13.63)$$

Equation (13.63) shows that $\frac{\Gamma(1+v_1)}{\Gamma(v_1)}[D_x^1 \varphi_x + D_y^1 \varphi_y] = 0$ is the fractional-order Euler-Lagrange equation of $-I_{x^2}^{v_2} I_{y^2}^{v_2} D^{v_1} u \cdot \vec{\varphi} = 0$. Equation (13.63) is an equivalent equation of (31), which can be directly and easily implemented in spatial domain or time domain.

Note that first, $\frac{\Gamma(1+v_1)}{\Gamma(v_1)}[D_x^1 \varphi_x + D_y^1 \varphi_y] = 0$ is the fractional-order Euler-Lagrange equation of $-I_{x^2}^{v_2} I_{y^2}^{v_2} D^{v_1} u \cdot \vec{\varphi} = 0$, which is irrelevant to the order v_2 of fractional integral $I_{x^2}^{v_2} I_{y^2}^{v_2}$. Thus, let $v_2 = 1$ for most real applications. Second, if $v_1 = v_2 = 1$, $\frac{\Gamma(1+v_1)^{v_1=1}}{\Gamma(v_1)} = \frac{\Gamma(2)}{\Gamma(1)} = 1$. Thus, (13.63) can be simplified as:

$$D_x^1 \varphi_x + D_y^1 \varphi_y = 0. \qquad (13.64)$$

Equation (13.64) is the first-order Euler-Lagrange equation of $-I_x^1 I_y^1 D^1 u \cdot \vec{\varphi} = 0$. Compared with (20) and (64), we can see that two formulae are identical. In other words, the traditional first-order Euler-Lagrange equation is actually a special case of the fractional-order Euler-Lagrange equation. This completes the proof.

Example 2: Because the traditional first-order Euler-Lagrange equation is actually a special case of the fractional-order Euler-Lagrange equation, the proposed fractional-order Euler-Lagrange equation can also be used to solve the classical issue of for the first-order variational method. The purpose of Example 2 is to verify directly that when we utilize the proposed fractional-order Euler-Lagrange equation based on fractional-order Green formula to deal with the classical first-order image restoration, it has the same results as those obtained by the classical first-order Euler-Lagrange equation, which is a concerned issue in image processing.

Let us suppose that $\Phi_1(D^{v_1} u)$ is the corresponding scalar function of a scalar function u and $\Phi_2(\vec{\varphi})$ is the corresponding scalar function of a vector function $\vec{\varphi}(x, y) = (\varphi_x, \varphi_y)$. Thus, if $-I_x^{v_2} I_y^{v_2} [\Phi_1(D^{v_1} u)\Phi_2(\vec{\varphi}) D^{v_1} u] \cdot \vec{\varphi} = 0$, from (13.59), (13.60), and (13.63), it follows that:

$$-I_x^{v_2} I_y^{v_2} [\Phi_1(D^{v_1} u)\Phi_2(\vec{\varphi}) D^{v_1} u] \cdot \vec{\varphi} = -I_x^{v_2} I_y^{v_2} \Phi_1(D^{v_1} u)\{D^{v_1} u \cdot [\Phi_2(\vec{\varphi}) \vec{\varphi}]\}$$

$$= I_{x\Omega}^{v_2} I_y^{v_2} \sum_{n=1}^{\infty} \binom{v_1}{n} [\Phi_1(D^{v_1}u)D_x^{v_1-n}u]D_x^n\{[\Phi_2(\vec{\varphi})\varphi_x]$$
$$+ [\varphi_1(D^{v_1}u)D_y^{v_1-n}u]D_y^n[\Phi_2(\vec{\varphi})\varphi_y]\}$$
$$= 0. \tag{13.65}$$

Toenable (13.65) be set up, from (13.63), an essential condition can be given as:

$$\binom{v_1}{1}\{D_x^1[\Phi_2(\vec{\varphi})\varphi_x]+D_y^1[\Phi_2(\vec{\varphi})\varphi_y]\}$$
$$= \frac{\Gamma(1+v_1)}{\Gamma(v_1)}\{D_x^1[\Phi_2(\vec{\varphi})\varphi_x]+D_y^1[\Phi_2(\vec{\varphi})\varphi_y]\}$$
$$= 0. \tag{13.66}$$

Equation (13.66) shows that $\frac{\Gamma(1+v_1)}{\Gamma(v_1)}\{D_x^1[\Phi_2(\vec{\varphi})\varphi_x]+D_y^1[\Phi_2(\vec{\varphi})\varphi_y]\} = 0$ is the fractional-order Euler-Lagrange equation of $-I_{x\Omega}^{v_2}I_y^{v_2}[\Phi_1(D^{v_1}u)\Phi_2(\vec{\varphi})D^{v_1}u]\cdot\vec{\varphi}=0$.

Furthermore, in order toutilize the proposed fractional-order Euler-Lagrange equation based on fractional-order Green formula to deal with the classical first-order image restoration, we should construct a fractional-order energy functional in the first place. Without loss of generality, the fractional-order energy functional can be defined by:

$$E(u) = \frac{1}{2} I_{x\Omega}^1 I_y^1 |D^{v_1}u|^{v_2} + \frac{\lambda}{2} I_{x\Omega}^1 I_y^1 (u-u_0)^2$$
$$= \frac{1}{2}\iint_\Omega |D^{v_1}u|^{v_2} dx dy + \frac{\lambda}{2}\iint_\Omega (u-u_0)^2 dx dy, \tag{13.67}$$

where u denotes a clean image, u_0 denotes a noisy image, λ denotes a regularization parameter, and Ω denotes the two-dimensional simply connected region of an image. Suppose that u is a first-order extremal surface of $E(u)$, $\xi(x,y) \in C_0^\infty(\Omega)$ is an admissible surface, a test function, closing to u. u and ξ can be merged into a family of surfaces, $u+\beta\xi$, where β is a small parameter. Then, when $\beta=0$, the family of surfaces, $u+\beta\xi$, converts into the first-order extremal surface, u. This process of seeking the first-order extreme value of a fractional-order norm can be described as a first-order isotropic diffusion. Thus, from (13.67), the functional on a family of surfaces, $u+\beta\xi$ is given as:

$$F(\beta) = I_{x\Omega}^1 I_y^1\left[\frac{1}{2}|D^{v_1}u+\beta D^{v_1}\xi|^{v_2} + \frac{\lambda}{2}(u+\beta\xi-u_0)^2\right]$$
$$= \frac{1}{2}\iint_\Omega |D^{v_1}u+\beta D^{v_1}\xi|^{v_2} dx dy + \frac{\lambda}{2}\iint_\Omega (u+\beta\xi-u_0)^2 dx dy, \tag{13.68}$$

where $\iint_\Omega (u-u_0)^2 dx dy$ is the variance of image noise and $\frac{\lambda}{2}\iint_\Omega (u-u_0)^2 dx dy$ is a fidelity term. For the first item on the right side of (13.68), according to the linearity of the first-order calculus, one can obtain:

$$D_\beta^1 \frac{1}{2} I_{x\Omega}^1 I_y^1 (|D^{v_1}u+\beta D^{v_1}\xi|^{v_2})\bigg|_{\beta=0} = \frac{1}{2} I_{x\Omega}^1 I_y^1 D_\beta^1(|D^{v_1}u+\beta D^{v_1}\xi|^{v_2})\bigg|_{\beta=0}$$
$$= \frac{1}{2}\iint_\Omega (v_2|D^{v_1}u|^{v_2-2}D^{v_1}u)\cdot D^{v_1}\xi dx dy$$

$$= 0. \tag{13.69}$$

In addition, for the second item on the right side of (13.68), according to the linearity of the first-order calculus, we have:

$$D_{\beta}^1 I_x^1 I_y^1 \frac{\lambda}{2}(u+\beta\xi-u_0)^2\bigg|_{\beta=0} = I_x^1 I_y^1 D_{\beta}^1\left[\frac{\lambda}{2}(u+\beta\xi-u_0)^2\right]\bigg|_{\beta=0}$$
$$= \iint_{\Omega} \lambda(u-u_0)\xi \,\mathrm{d}x\mathrm{d}y$$
$$= 0. \tag{13.70}$$

For test function ξ is arbitrary, according to the fundamental lemma of variation [475], [485], from (13.66), (13.69), and (13.70), the first-order Euler-Lagrange formula of (13.67) can be derived as:

$$-\frac{\Gamma(1+v_1)}{2\Gamma(v_1)}v_2\left[D_x^1(|D^{v_1}u|^{v_2-2}D_x^{v_1}u) + D_y^1(|D^{v_1}u|^{v_2-2}D_y^{v_1}u)\right] + \lambda(u-u_0) = 0. \tag{13.71}$$

Equation (13.71) shows that $-\frac{\Gamma(1+v_1)}{2\Gamma(v_1)}v_2\left[D_x^1(|D^{v_1}u|^{v_2-2}D_x^{v_1}u) + D_y^1(|D^{v_1}u|^{v_2-2}D_y^{v_1}u)\right] + \lambda(u-u_0) = 0$ is the first-order Euler-Lagrange formula of $E(u) = \frac{1}{2}\iint_{\Omega}|D^{v_1}u|^{v_2}\mathrm{d}x\mathrm{d}y + \frac{\lambda}{2}\iint_{\Omega}(u-u_0)^2\mathrm{d}x\mathrm{d}y$.

In particular, first, if $v_1=1$ and $v_2=2$ in (13.67), a classical isotropic diffusion denoising model [487] can be given as:

$$E(u) = \frac{1}{2}\iint_{\Omega}|D^1 u|^2 \mathrm{d}x\mathrm{d}y + \frac{\lambda}{2}\iint_{\Omega}(u-u_0)^2 \mathrm{d}x\mathrm{d}y. \tag{13.72}$$

When $v_1=1$ and $v_2=2$, $\frac{\Gamma(1+v_1)}{2\Gamma(v_1)}v_2 \overset{v_1=1,v_2=2}{=} \frac{\Gamma(2)}{\Gamma(1)} = 1$ Thus, in this case, (13.71) can be simplified as:

$$-[D_x^1(D_x^1 u) + D_y^1(D_y^1 u)] + \lambda(u-u_0) = -D^1 \cdot (D^1 u) + \lambda(u-u_0)$$
$$= -\nabla \cdot (\nabla_u) + \lambda(u-u_0)$$
$$= 0. \tag{13.73}$$

Equation (13.73) is identical with the first-order Euler-Lagrange equation for a classical isotropic diffusion denoising model in the literatures of [487]. Second, if $v_1=1$ and $v_2=1$ in (13.67), a classical anisotropy diffusion denoising model, total variation (TV) denoising model [488], [489] can be given as:

$$E(u) = \iint_{\Omega}|D_u^1|\mathrm{d}x\mathrm{d}y + \frac{\lambda}{2}\iint_{\Omega}(u-u_0)^2 \mathrm{d}x\mathrm{d}y. \tag{13.74}$$

When $v_1=1$ and $v_2=1$, $\frac{\Gamma(1+v_1)}{\Gamma(v_1)}v_2 \overset{v_1=1,v_2=2}{=} \frac{\Gamma(2)}{\Gamma(1)} = 1$. Thus, in this case, (13.71) can be simplified as:

$$-[D_x^1(|D^1 u|^{-1}D_x^1 u) + D_y^1(|D^1 u|^{-1}D_y^1 u)] + \lambda(u-u_0)$$
$$= -D^1 \cdot \left(\frac{D^1 u}{|D^1 u|}\right) + \lambda(u-u_0) = -\nabla \cdot \left(\frac{\nabla u}{D^1 u}\right) + \lambda(u-u_0) = 0. \tag{13.75}$$

Equation (13.75) is identical with the first-order Euler-Lagrange equation for a classical anisotropy diffusion denoising model in the literatures of [488], [489]. Third, if $v_1=1$ and $v_2=p$ in (13.67), a classical generalized total variation denoising model [490] can be given as:

$$E(u) = \frac{1}{p}\iint_\Omega |D^1 u|^p \mathrm{d}x\mathrm{d}y + \frac{\lambda}{2}\iint_\Omega (u-u_0)^2 \mathrm{d}x\mathrm{d}y. \quad (13.76)$$

When $v_1=1$ and $v_2=p$, $\dfrac{\Gamma(1+v_1)v_2}{p\Gamma(v_1)} \overset{v_1=1,v_2=p}{=} \dfrac{\Gamma(2)}{\Gamma(1)}=1$. Thus, in this case, (13.71) can be simplified as:

$$\begin{aligned}
&-[D_x^1(|D^1 u|^{p-2} D_x^1 u) + D_y^1(|D^1 u|^{p-2} D_y^1 u)] + \lambda(u-u_0)\\
&= -D^1 \cdot (|D^1 u|^{p-2} D^1 u) + \lambda(u-u_0)\\
&= -\nabla \cdot (|\nabla u|^{p-2} \nabla u) + \lambda(u-u_0)\\
&= 0. \quad (13.77)
\end{aligned}$$

Equation (13.77) is identical with the first-order Euler-Lagrange equation for a classical generalized total variation denoising model in the literatures of [490]. From (13.73), (13.75), and (13.77), we can see that when we utilize the proposed fractional-order Euler-Lagrange equation based on fractional-order Green formula to deal with the classical first-order image restoration problems, it has the same results as those obtained by the classical first-order Euler-Lagrange equation.

13.3.4　Solution procedure of fractional-order Euler-Lagrange equation

In this subsection, the solution procedure of the fractional-order Euler-Lagrange equation is derived. In order to let this chapter self-contained for a reader being not familiar with fractional calculus and the fractional-order gradient descent method, this subsection includes a brief necessary recall on the fractional-order time marching method.

At first, the solution procedure of the first-order Euler-Lagrange equation uses a parabolic equation with time as an evolution parameter (the time marching method), or equivalently, the first-order gradient descent method [487]. Note that, with regard to a two-dimensional signal such as an image, it is treated as a function of time and space, $u(x,y,t)$, and the time marching method is used to search the first-order steady state of (13.20). This means that for $I_x^1 I_y^1 D^1 u \cdot \vec{\varphi} = 0$, we solve:

$$\frac{\partial u(x,y,t)}{\partial t} = \frac{\partial \varphi_x[u(x,y,t)]}{\partial x} + \frac{\partial \varphi_y[u(x,y,t)]}{\partial y}, \quad (13.78)$$

where $\vec{\varphi}(x,y,t) = (\varphi_x, \varphi_y) = \{\varphi_x[u(x,y,t)], \varphi_y[u(x,y,t)]\}$.

Secondly, in a similar way, the solution procedure of the fractional-order Euler-Lagrange equation also uses a parabolic equation with time as an evolution parameter (the time marching method), or equivalently, the fractional-order gradient descent method [192]. Note that, with regard to a two-dimensional signal such as an image, it is treated as a function of time and space, $u(x,y,t)$, and the time marching method is used to search the fractional-order steady state of (13.25) and (13.63), respectively. This means that for $\iint_{\Omega} D^{v_1} u \cdot \vec{\varphi} \mathrm{d}x \mathrm{d}y = 0$ and $I_x^{v_2} I_y^{v_2} D^{v_1} u \cdot \vec{\varphi} = 0$, we solve, respectively:

$$\frac{\partial^{v_2} u(x,y,t)}{\partial t^{v_2}} = -\frac{1}{4\pi^2} \mathrm{Re}\{\overline{D_x^{v_1}}\varphi_x[u(x,y,t)] + \overline{D_y^{v_1}}\varphi_y[u(x,y,t)]\}, \quad (13.79)$$

$$\frac{\partial^{v_3} u(x,y,t)}{\partial t^{v_3}} = \frac{\Gamma(1+v_1)}{\Gamma(v_1)}\{D_x^1\varphi_x[u(x,y,t)] + D_y^1\varphi_y[u(x,y,t)]\}, \quad (13.80)$$

where $\vec{\varphi}(x,y,t) = (\varphi_x, \varphi_y) = \{\varphi_x[u(x,y,t)], \varphi_y[u(x,y,t)]\}$ and v_1, v_2, and v_3 are positive real numbers, respectively.

13.4 Experiment and Analysis

Fractional calculus has been applied to the solution of a necessary condition for the fractional-order fixed boundary optimization problems in signal processing and image processing mainly because of its inherent strengths in terms of long-term memory, non-locality, and weak singularity. Fractional differential of an image can preserve the low-frequency contour feature in the smooth area, and nonlinearly keep high-frequency edge information and texture information in those areas where grey scale changes frequently, and as well as, nonlinearly enhance texture details in those areas where grey scale does not change evidently [186] – [187], [345], [434], [502] – [504], [549]. Thus, it is natural to ponder whether the fractional-order variational method based texture inpainting/denoising could nonlinearly restore/preserve the complex texture details of an image while inpainting/denoising or not.

For the convenience of implementation, the fractional-order Euler-Lagrange equation based on fractional-order Green formula is employed in the following examples. From (13.63), we can see that first, $\frac{\Gamma(1+v_1)}{\Gamma(v_1)}[D_x^1\varphi_x + D_y^1\varphi_y] = 0$ is the fractional-order Euler-Lagrange equation of $-I_x^{v_2}I_y^{v_2}D^{v_1}u \cdot \vec{\varphi} = 0$, which can be directly and easily implemented in spatial domain or time domain. Second, $\frac{\Gamma(1+v_1)}{\Gamma(v_1)}[D_x^1\varphi_x + D_y^1\varphi_y] = 0$ is irrelevant to the order v_2 of fractional integral $-I_x^{v_2}I_y^{v_2}u \cdot \vec{\varphi} = 0$. Thus, let $v_2 = 1$ for most real applications.

13.4.1 Inpainting based on fractional-order variational method

In this subsection, in order to take advantage of the fractional-order Euler-Lagrange equation, a fractional-order inpainting algorithm based on the fractional-order variational method is illustrated.

Example 3: Inpainting is an image restoration problem, which is the process of filling in the missing or desired image information on the unavailable domains. Different from the traditional integer-order variational method based inpainting method [491] − [498], the fractional-order differential of an image and the fraction-power norm of this fractional-order differential are used for a fractional-order variational method based texture inpainting. A fractional-order inpainting algorithm model is to minimize a fractional-order appropriate energy functional:

$$E[u \mid u_0] = I_x^1 I_y^1 \underset{A \cup B}{f(\mid D^{v_1}u \mid^{v_2})} + I_x^1 I_y^1 \underset{B}{\frac{\lambda_B}{2}(u - u_0)^2}$$

$$= \iint_{A \cup B} f(\mid D^{v_1}u \mid^{v_2}) \mathrm{d}x \mathrm{d}y + \iint_B \frac{\lambda_B}{2}(u - u_0)^2 \mathrm{d}x \mathrm{d}y, \qquad (13.81)$$

where B denotes the inpainting domain where original image is missing, A denotes the neighbourhood of B, $u(x, y)$ is an image to be inpainted, $u_0(x, y)$ is the available part of the image u on $\Omega \setminus B$ that is often noisy, Ω is an entire image domain in real plane R^2 with smooth boundary, Ω bears a rectangular shape for most real applications, $f(\ )$ is an analytic function, E denotes a fractional-order multi-variable energy functional, $E[u \mid u_0]$ still means $E[u]$, but with u_0 fixed as known, v_1 and v_2 are positive real numbers, $D^{v_1} = (D_x^{v_1}, D_y^{v_1})$ denotes the v_1-order differential operator, and $\lambda_B(x,y) = \lambda \cdot 1_{\Omega \setminus B}(x, y)$ denotes a regularization parameter. The fractional-order Neumann boundary condition of (13.81) is $\langle D^{v_1}u, \vec{n} \rangle = \frac{\partial^{v_1} u}{\partial \vec{n}^{v_1}} = 0$, where $\vec{n}$ denotes the normal to the boundary of u. It's important to note that at first, we assume that the image to be inpainted, $u(x, y)$, is spatially smooth and rich in textural details. Thus, the corresponding fractional-order regularization term is given by $\iint_{A \cup B} f(\mid D^{v_1}u \mid^{v_2}) \mathrm{d}x \mathrm{d}y$. Secondly, we assume that the inpainting domain is close to its neighbourhood, which means that the difference between $u(x, y)$ and $u_0(x, y)$ should be small. Thus, the penalty term $\iint_B \frac{\lambda_B}{2}(u - u_0)^2 \mathrm{d}x \mathrm{d}y$ is used for the fidelity, which makes the proposed model better conditioned.

Let us assume that $u(x, y)$ is the first-order extremal surface of energy functional E, $\xi(x,y) \in C_0^\infty(\Omega)$ is an admissible surface, a test function, closing to $u(x,y)$. $u(x,y)$ and $\xi(x,y)$ can be merged into a family of surfaces, $u + \alpha \xi$, where α is a small parameter. When $\alpha = 0$, the family of surfaces, $u + \alpha \xi$, converts into the first-order extremal surface, $u(x,y)$. Thus, the first-order anisotropic diffusion of (13.81) can be explained as the first-order dissipation process of a fractional-order multi-variable energy functional E. To achieve the first-order minimum of (13.81), the first-order derivative of a fractional-order energy functional on the family of surfaces, $u + \alpha \xi$, is given as:

$$D_\alpha^1 \delta E \mid_{\alpha=0} = D_\alpha^1 [\delta_1(\alpha) + \delta_2(\alpha)] \mid_{\alpha=0}$$

$$= D_\alpha^1 I_x^1 I_y^1 \underset{A \cup B}{f(\mid D^{v_1}u + \alpha D^{v_1}\xi \mid^{v_2})} \bigg|_{\alpha=0} + D_\alpha^1 I_x^1 I_y^1 \underset{B}{\frac{\lambda_B}{2}(u + \alpha\xi - u_0)^2} \bigg|_{\alpha=0}$$

$$= D_\alpha^1 \iint_{A \cup B} f(\mid D^{v_1}u + \alpha D^{v_1}\xi \mid^{v_2}) \mathrm{d}x \mathrm{d}y \bigg|_{\alpha=0} + D_\alpha^1 \iint_B \frac{\lambda_B}{2}(u + \alpha\xi - u_0)^2 \mathrm{d}x \mathrm{d}y \bigg|_{\alpha=0}$$

$$= 0. \tag{13.82}$$

At first, for the first item on the right side of (13.82), if the first-order differential of $f(|D^{v_1}u + \alpha D^{v_1}\xi|^{v_2})$ with respect to α is existed, $\delta_1(\alpha)$ has a first-order extreme point, a first-order stationary point, when $\alpha = 0$. Thus, from (13.82), the following can be obtained:

$$\begin{aligned}
D_\alpha^1 \delta_1(\alpha)\Big|_{\alpha=0} &= D_\alpha^1 \iint_{A \cup B} f(|\vec{\varphi}|^{v_2}) \mathrm{d}x\mathrm{d}y \Big|_{\alpha=0} \\
&= \iint_{A \cup B} D_\alpha^1 f(|\vec{\varphi}|^{v_2}) \mathrm{d}x\mathrm{d}y \Big|_{\alpha=0} \\
&= \iint_{A \cup B} D^1_{|\vec{\varphi}|^{v_2}} f \Big|_{\alpha=0} v_2 |D^{v_1}u|^{v_2-2} D^{v_1}u \cdot D^{v_1}\xi \mathrm{d}x\mathrm{d}y \\
&= 0, \tag{13.83}
\end{aligned}$$

where with regard to a vector $\vec{\varphi} = D^{v_1}u + \alpha D^{v_1}\xi$, we have $\|\vec{\varphi}\|_2^2 = \sum_{i=1}^{2} |\varphi_i|^2 = (\vec{\varphi})^T \vec{\varphi} = \vec{\varphi} \cdot \vec{\varphi}$, where symbol $\cdot$ denotes the inner product. For the convenience of illustration, in this chapter, the equivalent notations $|\ |=\|\ \|_2$ are used in an arbitrary, interchangeable manner. Equation (13.83) searches for the first-order extreme point of $\delta_1(\alpha)$ with respect to α. Because $\xi(x,y)$ is a test function, $D^{v_1}\xi$ is arbitrary. Therefore, from (13.63), to enable (13.83) to be set up, a necessary condition can be given as:

$$-\frac{\Gamma(1-v_1)}{\Gamma(-v_1)}[D_x^1(D^1_{|\vec{\varphi}|^{v_2}} f|_{\alpha=0} v_2 |D^{v_1}u|^{v_2-2} D_x^{v_1}u) + D_y^1(D^1_{|\vec{\varphi}|^{v_2}} f|_{\alpha=0} v_2 |D^{v_1}u|^{v_2-2} D_y^{v_1}u)] = 0. \tag{13.84}$$

Without loss of generality, let us set $f(|\vec{\varphi}|^{v_2}) = |\vec{\varphi}|^{v_2}$. Thus, $D^1_{|\vec{\varphi}|^{v_2}} f = 1$. Substituting $D^1_{|\vec{\varphi}|^{v_2}} f = 1$ into (13.84), one can obtain:

$$-\frac{\Gamma(1-v_1)}{\Gamma(-v_1)} v_2 [D_x^1(|D^{v_1}u|^{v_2-2} D_x^{v_1}u) + D_y^1(|D^{v_1}u|^{v_2-2} D_y^{v_1}u)] = 0. \tag{13.85}$$

Secondly, for the second item on the right side of (13.82), the following can be obtained:

$$\begin{aligned}
D_\alpha^1 \delta_2(\alpha)|_{\alpha=0} &= D_\alpha^1 I_x^1 I_y^1 \Big|_B \frac{\lambda_B}{2} (u + \alpha\xi - u_0)^2 \Big|_{\alpha=0} \\
&= I_x^1 I_y^1 \Big|_B D_\alpha^1 \Big[\frac{\lambda_B}{2}(u + \alpha\xi - u_0)^2\Big]\Big|_{\alpha=0} \\
&= \iint_B \lambda_B(u - u_0)\xi \mathrm{d}x\mathrm{d}y \\
&= 0. \tag{13.86}
\end{aligned}$$

Since $\xi(x,y)$ is a test function, the corresponding $\xi(x,y)$ is arbitrary. Therefore, according to the fundamental lemma of variation [475], [485], to enable (13.86) to be set up, a necessary condition can be given as:

$$\lambda_B(u - u_0) = 0. \tag{13.87}$$

Thus, from (13.85) and (13.87), we can derive the fractional-order Euler-Lagrange equation of $D_a^1 \delta E(\alpha)|_{\alpha=0} = 0$, given as:

$$-\frac{\Gamma(1-v_1)}{\Gamma(-v_1)}v_2[D_x^1(|D^{v_1}u|^{v_2-2}D_x^{v_1}u) + D_y^1(|D^{v_1}u|^{v_2-2}D_y^{v_1}u)] + \lambda_B(u-u_0) = 0.$$

(13.88)

The solution procedureof (13.88) uses a parabolic equation with time as an evolution parameter (the time marching method), or equivalently, the first-order gradient descent method. Note that, we treat an image as a function of time and space, and use the time marching method to search the fractional-order steady state of (13.88). This means that we solve:

$$\frac{\partial u}{\partial t} = \frac{\Gamma(1-v_1)}{\Gamma(-v_1)}v_2[D_x^1(|D^{v_1}u|^{v_2-2}D_x^{v_1}u) + D_y^1(|D^{v_1}u|^{v_2-2}D_y^{v_1}u)] - \lambda_B(u-u_0).$$

(13.89)

When $\frac{\partial u}{\partial t} = 0$, the time marching method converges to the first-order steady state of (13.88). Furthermore, we must compute $\lambda_B(t)$ in (13.89). Let us suppose that $\iint_A (u - u_0)\mathrm{d}x\mathrm{d}y = 0$ and $\iint_A (u-u_0)^2 \mathrm{d}x\mathrm{d}y = \sigma^2$, where σ^2 is the variance of the added noise of u_0. We multiply by $(u-u_0)$ on the both sides of (13.88) and integrate by parts in the domain of B, the left side of (13.88) vanishes. Thus, the following can be obtained:

$$\lambda_B(t) = \frac{\Gamma(1-v_1)}{\sigma^2 \Gamma(-v_1)}v_2 \iint_A [D_x^1(|D^{v_1}u|^{v_2-2}D_x^{v_1}u) + D_y^1(|D^{v_1}u|^{v_2-2}D_y^{v_1}u)](u-u_0)\mathrm{d}x\mathrm{d}y.$$

(13.90)

Equations (13.89) and (13.90) should benumerically implemented. At first, let us suppose Δt denotes time interval, $n = 0, 1, \cdots$ denotes time (the number of iterations). Thus, $t_n = n\Delta t$. Here $t_0 = 0$ denotes an initial time. In (13.9) and (13.90), $u^n = u(x, y, t_n)$, and $u_0^n(x, y, t_0) = u_0^0(x, y, t_0)$. Secondly, in (13.79), (13.80), (13.89), and (13.90), the fractional-order differential operators, D^v, D_x^v and D_y^v, of a two-dimensional signal such as digital image should be numerically implemented. Equation (13.1) shows that for the Grünwald-Letnikov definition of fractional calculus, the limit symbol may be removed when N is sufficiently large. Thus, the convergence rate and accuracy are improved by introducing a signal value at a non-node into the Grünwald-Letnikov definition, i.e. $^{G-L}D_a^v f(x) \cong \frac{x^{-v}N^v}{\Gamma(-v)}\sum_{k=0}^{N-1}\frac{\Gamma(k-v)}{v(k+1)}f\left(x + \frac{vx}{2N} - \frac{kx}{N}\right)$. Using the Lagrange interpolation polynomial when $n = 3$ points to perform the fractional-order interpolation, we can obtain the fractional-order differential operator in the eight symmetric directions [186] — [187], [345]. Thus, the discrete definitions of D_x^v and D_y^v in two-dimensional space [187] are as follows, respectively:

$$D_x^v u(x,y,t) \approx \left(\frac{v}{4} + \frac{v^2}{8}\right)u(x+1,y,t) + \left(1 - \frac{v^2}{2} - \frac{v^3}{8}\right)u(x,y,t) + \frac{1}{\Gamma(-v)}$$

$$\sum_{k=1}^{n-2}\left[\frac{\Gamma(k-v+1)}{(k+1)!}\cdot\left(\frac{v}{4}+\frac{v^2}{8}\right)+\frac{\Gamma(k-v)}{k!}\cdot\left(1-\frac{v^2}{4}\right)+\frac{\Gamma(k-v-1)}{(k-1)!}\cdot\right.$$
$$\left.\left(-\frac{v}{4}+\frac{v^2}{8}\right)\right]u(x-k,y,t)+\left[\frac{\Gamma(n-v-1)}{(n-1)!\Gamma(-v)}\cdot\left(1-\frac{v^2}{4}\right)+\right.$$
$$\frac{\Gamma(n-v-2)}{(n-2)!\Gamma(-v)}\cdot\left(-\frac{v}{4}+\frac{v^2}{8}\right)\right]u(x-n+1,y,t)+$$
$$\frac{\Gamma(n-v-1)}{(n-1)!\Gamma(-v)}\cdot\left(-\frac{v}{4}+\frac{v^2}{8}\right)u(x-n,y,t), \quad (13.91)$$

$$D_y^v u(x,y,t)\approx\left(\frac{v}{4}+\frac{v^2}{8}\right)u(x,y+1,t)+\left(1-\frac{v^2}{2}-\frac{v^3}{8}\right)u(x,y,t)+\frac{1}{\Gamma(-v)}$$
$$\sum_{k=1}^{n-2}\left[\frac{\Gamma(k-v+1)}{(k+1)!}\cdot\left(\frac{v}{4}+\frac{v^2}{8}\right)+\frac{\Gamma(k-v)}{k!}\cdot\left(1-\frac{v^2}{4}\right)+\frac{\Gamma(k-v-1)}{(k-1)!}\cdot\right.$$
$$\left.\left(-\frac{v}{4}+\frac{v^2}{8}\right)\right]u(x,y-k,t)+\left[\frac{\Gamma(n-v-1)}{(n-1)!\Gamma(-v)}\cdot\left(1-\frac{v^2}{4}\right)+\right.$$
$$\frac{\Gamma(n-v-2)}{(n-2)!\Gamma(-v)}\cdot\left(-\frac{v}{4}+\frac{v^2}{8}\right)\right]u(x-n+1,y,t)+$$
$$\frac{\Gamma(n-v-1)}{(n-1)!\Gamma(-v)}\cdot\left(-\frac{v}{4}+\frac{v^2}{8}\right)u(x,y-n,t). \quad (13.92)$$

In particular, when $v=1$, the fractional-order differential operator is converted to the traditional first-order differential operator. Thirdly, in (13.89), according to the first-order gradient descent method, one can obtain:

$$\frac{\partial u}{\partial t}\approx\frac{u^{n+1}-u^n}{\Delta t}. \quad (13.93)$$

Thus, from (13.89), (13.0), and (13.93), the following can be obtained:

$$u^{n+1}=\frac{\Gamma(1-v_1)}{\Gamma(-v_1)}v_2[D_x^1(|D^{v_1}u^n|^{v_2-2}D_x^{v_1}u^n)+D_y^1(|D^{v_1}u^n|^{v_2-2}D_y^{v_1}u^n)]\Delta t$$
$$-\lambda_B^n\Delta t u_0^n+(1+\lambda_B^n\Delta t)u^n, \quad (13.94)$$

$$\lambda_B^n=\frac{\Gamma(1-v_1)}{\sigma^{n^2}\Gamma(-v_1)}v_2\sum_{x,y}[D_x^1(|D^{v_1}u^n|^{v_2-2}D_x^{v_1}u^n)+D_y^1(|D^{v_1}u^n|^{v_2-2}D_y^{v_1}u^n)](u^n-u^0),$$
$$(13.95)$$

where $\sigma^{n^2}=\sum_{x,y}(u^n-u_0^0)^2$. Fourthly, the numerical implementation of the fractional-order partial differential equation should be restricted by Courant-Friedrichs-Lewy (CFL) condition [548]. Finally, it may be considered that $|D^{v_1}u^n|=0$ during the numerical iteration computation of (13.94) and (13.95). To enable (13.4) and (13.95) to be implemented, if $|D^{v_1}u^n|\leqslant\varepsilon$, let us set $|D^{v_1}u^n|=\varepsilon$, where ε is a small positive constant. Without loss of generality, in the following experiments, let us set $\varepsilon=10^{-5}$.

In order to illustrate the capability of restoring the edges and textural details of the fractional-order inpainting algorithm based on the fractional-order variational method, it was analyzed by considering the integer-order inpainting algorithm based on the classic integer-order variational method [490] − [498] vis-à-vis the proposed fractional-order inpainting algorithm and a suitable texture image i.e. a hair image. We artificially damaged the original hair image at random and used this damaged image as the test image. No *a priori* knowledge regarding the characteristics of the original hair image is either known or required. We set $v_1=2.25$ and $v_2=2.5$ in (13.94) and (13.95). Suppose we set the same parameters, the time interval $\Delta t=0.0001$, the number of iterations $n=150$, for both the fractional-order inpainting algorithm and the integer-order one in this simulation experiment. Thus, the results of the comparative inpainting experiments for the hair image are shown in Fig. 13.3.

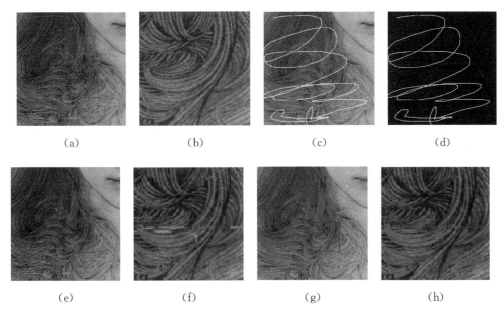

Figure 13.3 Comparative inpainting experiments on a hair image. (a) Original hair image, (b) Enlarged detail of lower right 1/16 of (a), (c) Damaged hair image, (d) Template of inpainting, (e) Integer-order inpainting algorithm, (f) Enlarged detail of lower right 1/16 of (e), (g) Fractional-order inpainting algorithm, (h) Enlarged detail of lower right 1/16 of (g).

To consider the visual effects for the purpose of comparison, from Fig. 13.3 (a), (b), (e), (f), (g), and (h), we can see that the capability of the integer-order inpainting algorithm based on the integer-order variational method to restore the edges and textural details is worse than that of the fractional-order inpainting algorithm based on the fractional-order variational method. In particular, the extent to which the integer-order inpainting algorithm is able to restore the local edges and textural details is relatively weaker than that of the fractional-order inpainting algorithm. The integer-order inpainting algorithm tends to uniformly compensate the damaged image if there are rich local edges and textural details, often resulting in a blocky structure appearance of the output image. In comparison, the result of restoring the edges and textural details of the fractional-order inpainting algorithm is relatively clearer than that of the integer-order inpainting algorithm. The inpainted image regions of the integer-order inpainting algorithm appear relatively smoother than those produced by the fractional-order inpainting algorithm.

Next, with the objective of performing a quantitative analysis, we obtained the peak signal-to-noise ratio (PSNR) between the inpainted image and the original image to evaluate the inpainting effects of each competing algorithm under consideration. The PSNR is defined via the mean squared error (MSE). Given a noise-free $m \times n$ monochrome image I and its noisy approximated image K, the MSE can be defined as $MSE = \frac{1}{mn}\sum_{x=0}^{m-1}\sum_{y=0}^{m-1}[I(x,y) - K(x,y)]^2$. The PSNR (in dB) can be defined as $PSNR = 10 \cdot \log_{10}(MAX_I \mid MSE)$, where, MAX_I is the maximum possible pixel value of the image I. The PSNR of the integer-order inpainting algorithm and that of the fractional-order inpainting algorithm are 16.3560 and 19.8275, respectively. The PSNR of the integer-order inpainting algorithm is much smaller than that of the fractional-order inpainting algorithm, which shows that the high-frequency edges and textural details of the inpainted image of the integer-order inpainting algorithm is sharply diffused and smoothed. This, in turn, indicates that the structural similarity of the edges and textural details between its corresponding inpainted image and the original image is weaker than that of the fractional-order inpainting algorithm. The capability of restoring the edges and textural details of the fractional-order inpainting algorithm is obviously superior to that of the integer-order inpainting algorithm, especially for images rich in textural detail.

Moreover, in order to extend our analysis of the capability of the fractional-order inpainting algorithm to restore edges and textural details, we chose another suitable texture image i.e. a wood grain image to further illustrate. We also randomly damaged the original wood grain image and used this damaged image as the test image. No *a priori* knowledge regarding the characteristics of the original wood grain image is either known or required. We still set $v_1 = 2.25$ and $v_2 = 2.5$ in (13.94) and (13.95). Suppose we set the same parameters, the time interval $\Delta t = 0.0001$, the number of iterations $n = 90$, for both the fractional-order inpainting algorithm and the integer-order one in this simulation experiment. Thus, the results of the comparative inpainting experiments for the wood grain image are shown in Fig. 13.4.

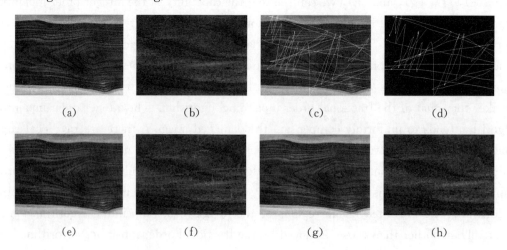

Figure 13.4 Comparative inpainting experiments on a wood grain image. (a) Original wood grain image, (b) Enlarged detail of lower left 1/16 of (a), (c) Damaged wood grain image, (d) Template of inpainting, (e) Integer-order inpainting algorithm, (f) Enlarged detail of lower left 1/16 of (e), (g) Fractional-order inpainting algorithm, (h) Enlarged detail of lower left 1/16 of (g).

From Fig. 13.4 (a), (b), (e), (f), (g), and (h), we can see that the result of restoring the edges and textural details of the fractional-order inpainting algorithm is relatively clearer than that of the integer-order inpainting algorithm. The inpainted image regions of the integer-order inpainting algorithm appear relatively smoother than those produced by the fractional-order inpainting algorithm. Next, the PSNR of the integer-order inpainting algorithm and that of the fractional-order inpainting algorithm are 15.4562 and 16.5723, respectively. The capability of restoring the edges and textural details of the fractional-order inpainting algorithm is obviously superior to that of the integer-order inpainting algorithm, especially for images rich in textural detail.

In addition, as the next experimental step, we then selected another suitable texture image i.e. a moonscape image for the purpose of implementing comparative experiments. We also stochastically damaged the original moonscape image and used this damaged image as the test image. No *a priori* knowledge regarding the characteristics of the original moonscape image is either known or required. We still set $v_1 = 2.25$ and $v_2 = 2.5$ in (13.94) and (13.95). Suppose we set the same parameters, the time interval $\Delta t = 0.0001$, the number of iterations $n = 150$, for both the fractional-order inpainting algorithm and the integer-order one in this simulation experiment. Thus, the results of the comparative inpainting experiments for the moonscape image are shown in Fig. 13.5.

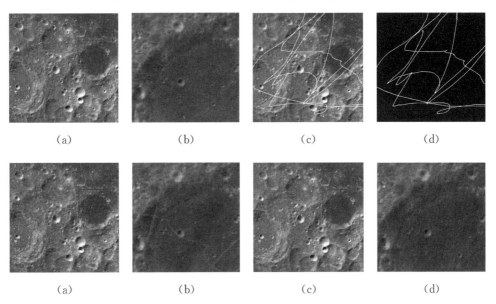

(a)　　　　　　(b)　　　　　　(c)　　　　　　(d)

(a)　　　　　　(b)　　　　　　(c)　　　　　　(d)

Figure 13.5 Comparative inpainting experiments on a moonscape image. (a) Original moonscape image, (b) Enlarged detail of higher right 1/16 of (a), (c) Damaged moonscape image, (d) Template of inpainting, (e) Integer-order inpainting algorithm, (f) Enlarged detail of higher right 1/16 of (e), (g)

Fractional-order inpainting algorithm, (h) Enlarged detail of higher right 1/16 of (g).

From Fig. 13.5 (a), (b), (e), (f), (g), and (h), we can see that the result of restoring the edges and textural details of the fractional-order inpainting algorithm is relatively clearer than that of the integer-order inpainting algorithm. The inpainted image regions of the integer-order inpainting algorithm appear relatively smoother than those produced by the fractional-order inpainting algorithm. Next, the PSNR of the integer-order inpainting algorithm and that of the fractional-order inpainting algorithm are 21.7132 and 23.0138, respectively. The capability of restoring the edges and textural details of the fractional-order inpainting algorithm is obviously superior to that of the integer-order inpainting algorithm, especially for images rich in textural detail.

13.4.2 Image denoising based on fractional-order variational method

In this subsection, in order to take advantage of the fractional-order Euler-Lagrange equation, a fractional-order denoising algorithm based on the fractional-order variational method is illustrated.

Example 4: Image denoising is another image restoration problem. Different from the traditional integer-order variational method based image denoising method [486] – [490], the fractional-order differential of an image, the fraction-power norm of this fractional-order differential, and the fractional-order extreme point of their energy functional are used for the fractional-order variational method based image denoising. A fractional-order denoising algorithm is to minimize a fractional-order appropriate energy functional:

$$E[u] = I_x^1 I_y^1 f(|D^{v_1} u|^{v_2}) + I_x^1 I_y^1 \lambda (u - u_0) u_0$$
$$= \iint_\Omega f(|D^{v_1} u|^{v_2}) \mathrm{d}x \mathrm{d}y + \lambda \iint_\Omega (u - u_0) u_0 \mathrm{d}x \mathrm{d}y, \qquad (13.96)$$

where u denotes a clean image, u_0 denotes a noisy image, λ denotes a regularization parameter, Ω is an the entire image domain in real plane R^2 with smooth boundary, Ω bears a rectangular shape for most real applications, E denotes a fractional-order multi-variable energy functional, $f(\ )$ is an analytic function, v_1 and v_2 are positive real numbers, $D^{v_1} = (D_x^{v_1}, D_y^{v_1})$ denotes the v_1-order differential operator, and λ denotes a regularization parameter. The fractional-order Neumann boundary condition of (13.96) is $\langle D^{v_1} u, \vec{n} \rangle = \dfrac{\partial^{v_1} u}{\partial \vec{n}^{v_1}} = 0$, where $\vec{n}$ denotes the normal to the boundary of u. It's important to note that at first, we assume that the clean image, $u(x, y)$, is spatially smooth and rich in textural details. Thus, the corresponding fractional-order regularization term is given by $\iint_\Omega f(|D^{v_1} u|^{v_2}) \mathrm{d}x \mathrm{d}y$. Secondly, we assume that the clean image is close to the noisy image, which means that the difference between $u(x, y)$ and $u_0(x, y)$ should be small.

Thus, the penalty term $\lambda\iint_{\Omega}(u-u_0)u_0\,dxdy$ is used for the fidelity, which makes the proposed model better conditioned. In order to avoid incomplete beta function being not set up, the fidelity term in (13.96) is set to $\lambda\iint_{\Omega}(u-u_0)u_0\,dxdy$, rather than $\lambda\iint_{\Omega}(u-u_0)^2\,dxdy$, where $\lambda\iint_{\Omega}u_0\,dxdy$ is a cross energy between $(u-u_0)$ and u_0.

Let us assume that $u(x,y)$ is the fractional-order extremal surface of a fractional-order energy functional E, $\xi(x,y)\in C_0^\infty(\Omega)$ is an admissible surface, a test function, closing to $u(x,y)$. Then, $u(x,y)$ and $\xi(x,y)$ can be merged into a family of surfaces, $u+(\alpha-1)\xi$, where α is a small parameter. When $\alpha=1$, the family of surfaces, $u+(\alpha-1)\xi$, converts into the fractional-order extremal surface, $u(x,y)$. Thus, the fractional-order anisotropic diffusion of (13.96) can be explained as the fractional-order dissipation process of a fractional-order multi-variable energy functional E. To achieve the fractional-order minimum of (13.96), the fractional-order derivative of a fractional-order energy functional on the family of surfaces, $u+(\alpha-1)\xi$, is given as:

$$\begin{aligned}D_a^{v_3}\delta E &= D_a^{v_3}[\delta_1(\alpha)+\delta_2(\alpha)]\\ &= D_a^{v_3}I_x^1 I_y^1 f[|D^{v_1}u+(\alpha-1)D^{v_1}\xi|^{v_2}]+D_a^{v_3}I_x^1 I_y^1\lambda[u+(\alpha-1)\xi-u_0]u_0\\ &= D_a^{v_3}\iint_{\Omega}f[|D^{v_1}u+(\alpha-1)D^{v_1}\xi|^{v_2}]dxdy+D_a^{v_3}\lambda\iint_{\Omega}[u+(\alpha-1)\xi-u_0]u_0\,dxdy\\ &= 0,\end{aligned} \qquad (13.97)$$

where v_3 is a positive real number.

At first, for the first item on the right side of (13.97), if the v_3-order differential of $f[|D^{v_1}u+(\alpha-1)D^{v_1}\xi|^{v_2}]$ with respect to α is existed, $\delta_1(\alpha)$ has a v_3-order extreme point, a v_3-order stationary point, when $\alpha=1$. Thus, from (13.97), the following can be obtained:

$$\begin{aligned}D_a^{v_3}\delta_1(\alpha)|_{\alpha=1} &= \frac{\partial^{v_3}}{\partial\alpha^{v_3}}\iint_{\Omega}f(|\vec{\varphi}|^{v_2})dxdy|_{\alpha=1}\\ &= \iint_{\Omega}\frac{\partial^{v_3}}{\partial\alpha^{v_3}}f(|\vec{\varphi}|^{v_2})dxdy|_{\alpha=1}\\ &= 0,\end{aligned}\qquad(13.98)$$

where with regard to a vector $\vec{\varphi}=D^{v_1}u+(\alpha-1)D^{v_1}\zeta$, we have $\|\vec{\varphi}\|_2^2=\sum_{i=1}^{2}|\varphi_i|^2=(\vec{\varphi})^T\vec{\varphi}=\vec{\varphi}\cdot\vec{\varphi}$, where symbol · denotes the inner product. For the convenience of illustration, in this chapter, the equivalent notations $|\ |=\|\ \|_2$ are used in an arbitrary, interchangeable manner. Equation (13.98) searches for the v_3-order extreme point of $\delta_1(\alpha)$ with respect to α. Further, provided v is a fraction, when $n>v$, $\binom{v}{n}=\dfrac{\Gamma(1+v)}{\Gamma(1-n+v)\Gamma(1+n)}=\dfrac{(-1)^n\Gamma(n-v)}{\Gamma(-v)\Gamma(1+n)}\neq 0$. Thus, from (13.57) and Faà di Bruno's formula, it follows that:

$$D_{\alpha-\rho}^{v}f[g(\alpha)] = \frac{(\alpha-\rho)^{-v}}{\Gamma(1-v)}f + \sum_{n=1}^{\infty}\binom{v}{n}\frac{(\alpha-\rho)^{n-v}}{\Gamma(n-v+1)}n!\sum_{m=1}^{n}D_{g}^{m}f\sum\prod_{k=1}^{n}\frac{1}{P_{k}!}\left[\frac{D_{\alpha-\rho}^{k}g}{k!}\right]^{P_{k}},$$
(13.99)

where ρ is an arbitrary constant and $g(\alpha) = |\vec{\varphi}|$. On the right side of (13.99), the item of $n=0$ is separated from summation symbol $\sum_{n=0}^{\infty}$. Equation (13.99) shows that the fractional derivative of a composite function is equal to an infinite sum, in which P_{k} satisfies:

$$\begin{cases} \sum_{k=1}^{n}kP_{k} = n \\ \sum_{k=1}^{n}P_{k} = m \end{cases}.$$
(13.100)

The third summation symbol $\sum$ in (13.99) denotes the sum up of $\left\{\prod_{k=1}^{n}\frac{1}{P_{k}!}\left[\frac{D_{\alpha-\rho}^{k}g}{k!}\right]^{P_{k}}\right\}_{m=1\to n}$ corresponding to the whole combination of $P_{k}|_{m=1\to n}$ that satisfies (13.100). Thus, substituting (13.99) into (13.98), one can obtain:

$$\iint_{\Omega}\frac{\partial^{v_{3}}}{\partial\alpha^{v_{3}}}f(|\vec{\varphi}|^{v_{2}})\mathrm{d}x\mathrm{d}y\bigg|_{\alpha=1} = \iint_{\Omega}\frac{\alpha^{-v_{3}}f(|\vec{\varphi}|^{v_{2}})}{\Gamma(1-v_{3})} + \sum_{n=1}^{\infty}\binom{v_{3}}{n}\frac{\alpha^{n-v_{3}}}{\Gamma(n-v_{3}+1)}n!\sum_{m=1}^{n}D_{|\vec{\varphi}|^{v_{2}}}^{m}$$

$$f\sum\prod_{k=1}^{n}\frac{1}{P_{k}!}\left[\frac{D_{\alpha}^{k}(|\vec{\varphi}|^{v_{2}})}{k!}\right]^{P_{k}}\mathrm{d}x\mathrm{d}y\bigg|_{\alpha=1}$$

$$= \iint_{\Omega}\frac{f(|D^{v_{1}}u|^{v_{2}})}{\Gamma(1-v_{3})} + \sum_{n=1}^{\infty}\frac{(-1)^{n}\Gamma(n-v_{3})}{\Gamma(-v_{3})\Gamma(n-v_{3}+1)}$$

$$\sum_{m=1}^{n}D_{|\vec{\varphi}|^{v_{2}}}^{m}f\bigg|_{\alpha=1}\sum\prod_{k=1}^{n}\frac{1}{P_{k}!}\left[\frac{D_{\alpha}^{k}(|\vec{\varphi}|^{v_{2}})|_{\alpha=1}}{k!}\right]^{P_{k}}\mathrm{d}x\mathrm{d}y$$

$$= 0.$$
(13.101)

In order to simplify the calculation, without loss of generality, let us suppose that $f(\eta)=\eta$. Thus, according to properties of the integer-order calculus, $D_{\eta}^{1}f(\eta)=1$ and $D_{\eta}^{m}f(\eta)\overset{m\geqslant 2}{\equiv}0$. Therefore, (13.101) can be simplified as:

$$\iint_{\Omega}\frac{|D^{v_{1}}u|^{v_{2}}}{\Gamma(1-v_{3})} + \sum_{n=1}^{\infty}\frac{(-1)^{n}\Gamma(n-v_{3})}{\Gamma(-v_{3})\Gamma(n-v_{3}+1)}\left\{\prod_{k=1}^{n}\frac{1}{P_{k}!}\left[\frac{D_{\alpha}^{k}(|\vec{\varphi}|^{v_{2}})|_{\alpha=1}}{k!}\right]^{P_{k}}\right\}_{m=1}\mathrm{d}x\mathrm{d}y = 0.$$
(13.102)

In addition, from (13.100), we can see that if $m=1$, P_{k} satisfies $P_{n}=1$ and $P_{1}=P_{2}=\cdots=P_{n-1}=0$. Thus, (13.102) can be can be further simplified as:

$$\iint_{\Omega}\frac{|D^{v_{1}}u|^{v_{2}}}{\Gamma(1-v_{3})} + \sum_{n=1}^{\infty}\frac{(-1)^{n}\Gamma(n-v_{3})}{\Gamma(-v_{3})\Gamma(n-v_{3}+1)}\frac{D_{\alpha}^{n}(|\vec{\varphi}|^{v_{2}})|_{\alpha=1}}{n!}\mathrm{d}x\mathrm{d}y = 0.$$
(13.103)

Further, the expression of $D_{\alpha}^{n}(|\vec{\varphi}|^{v_{2}})|_{\alpha=1}$ when n is an odd number ($n=2k+1$,

$k=0, 1, 2, \cdots$) is different from that of $D_a^n(\,|\,\vec{\varphi}\,|^{v_2})\,|_{a=1}$ when n is an even number ($n=2k$, $k=1, 2, 3, \cdots$), which can be given as, respectively:

$$D_a^n(\,|\,\vec{\varphi}\,|^{v_2})\,|_{a=1}^{n=2k+1} = \prod_{\tau=1}^{n}(v_2-\tau+1)\,|\,D^{v_1}u\,|^{v_2-n-1}\,|\,D^{v_1}\xi\,|^{n-1}(D^{v_1}\xi)\cdot(D^{v_1}u),$$
(13.104)

$$D_a^n(\,|\,\vec{\varphi}\,|^{v_2})\,|_{a=1}^{n=2k} = \prod_{\tau=1}^{n}(v_2-\tau+1)\,|\,D^{v_1}u\,|^{v_2-n}\,|\,D^{v_1}\xi\,|^{n}. \qquad (13.105)$$

Then, substitution of (13.104) and (13.105) into (13.103) results in:

$$\iint_{\Omega}\sum_{k=0}^{\infty}\frac{\prod_{\tau=1}^{2k}(v_2-\tau+1)\,|\,D^{v_1}u\,|^{v_2-2k-2}\,|\,D^{v_1}\xi\,|^{2k}}{\Gamma(-v_3)(2k)!}(D^{v_1}u)\cdot D^{v_1}$$
$$\left[\frac{\Gamma(2k-v_3)}{\Gamma(2k-v_3+1)}u-\frac{(v_2-2k)\Gamma(2k-v_3+1)}{(2k+1)\Gamma(2k-v_3+2)}\xi\right]dxdy$$

$$=\iint_{\Omega}\sum_{k=0}^{\infty}\frac{\prod_{\tau=1}^{2k}(v_2-\tau+1)}{(2k)!}\left\{\frac{\Gamma(2k-v_3)\,|\,D^{v_1}u\,|^{v_2-2k}}{\Gamma(-v_3)\Gamma(2k-v_3+1)}\,|\,D^{v_1}\xi\,|^{2k}\right.$$
$$\left.-\frac{(v_2-2k)\Gamma(2k-v_3+1)\,|\,D^{v_1}u\,|^{v_2-2k-2}}{(2k+1)\Gamma(-v_3)\Gamma(2k-v_3+2)}\left[(D^{v_1}u)\cdot|\,D^{v_1}\xi\,|^{2k}(D^{v_1}\xi)\right]\right\}dxdy$$
$$=0, \qquad (13.106)$$

where we set $\prod_{\tau=1}^{n}(v_2-\tau+1)\overset{n=0}{=}1$. Because $\xi(x, y)$ is a test function, $|\,D^{v_1}\xi\,|^{2k}$ and $D^{v_1}\xi$ are arbitrary. From (13.63), we can derive that $\frac{\Gamma(1+v_1)}{\Gamma(v_1)}[D_x^1\varphi_x+D_y^1\varphi_y]=0$ is the fractional-order Euler-Lagrange formula of $-\iint_{\Omega}D^{v_1}u\cdot\vec{\varphi}\,dxdy=0$, where $u(x, y)$ denotes an arbitrary scalar function and $\vec{\varphi}=(\varphi_x,\varphi_y)$ denotes a vector function. Therefore, according to the fractional-order Euler-Lagrange formula and the fundamental lemma of variation [475], [485], to enable (13.106) to be set up, a corresponding necessary condition can be given as:

$$\sum_{k=0}^{\infty}\frac{\prod_{\tau=1}^{2k}(v_2-\tau+1)}{(2k)!}\left\{\frac{\Gamma(2k-v_3)\,|\,D^{v_1}u\,|^{v_2-2k}}{\Gamma(-v_3)\Gamma(2k-v_3+1)}\right.$$
$$\left.+\frac{(v_2-2k)\Gamma(1+v_1)\Gamma(2k-v_3+1)\,|\,D^{v_1}u\,|^{v_2-2k-2}}{(2k+1)\Gamma(v_1)\Gamma(-v_3)\Gamma(2k-v_3+2)}\left[D_x^1(D_x^{v_1}u)+D_y^1(D_y^{v_1}u)\right]\right\}=0,$$
(13.107)

where we set $\prod_{\tau=1}^{n}(v_2-\tau+1)\overset{n=0}{=}1$.

Secondly, for the second item on the right side of (13.97), the following can be obtained:

$$D_a^{v_3}\delta_2(\alpha)\,|_{a=1} = D_a^{v_3}I_x^1I_y^1\lambda[u+(\alpha-1)\xi]-u_0]u_0\,|_{a=1}$$

$$= \iint_\Omega \frac{\lambda u_0}{\Gamma(1-v_3)\Gamma(2-v_3)}[\Gamma(2-v_3)(u-u_0-\xi)+\Gamma(1-v_3)\xi]\mathrm{d}x\mathrm{d}y$$
$$= 0. \tag{13.108}$$

Since $\xi(x,y)$ is a test function, the corresponding $[\Gamma(2-v_3)(u-u_0-\xi)+\Gamma(1-v_3)\xi]$ is arbitrary. Therefore, according to the fundamental lemma of variation [475], [485], to enable (13.108) to be set up, a necessary condition can be given as:

$$\frac{\lambda u_0}{\Gamma(1-v_3)\Gamma(2-v_3)} = 0. \tag{13.109}$$

Thus, from (13.107) and (13.109), we can derive the fractional-order Euler-Lagrange equation of $D_a^{v_3}\delta E(\alpha)|_{a=1} = 0$, given as:

$$\sum_{k=0}^\infty \frac{\prod_{\tau=1}^{2k}(v_2-\tau+1)}{(2k)!}\left\{\frac{\Gamma(2k-v_3)\mid D^{v_1}u\mid^{v_2-2k}}{\Gamma(-v_3)\Gamma(2k-v_3+1)}\right.$$
$$+\frac{(v_2-2k)\Gamma(1+v_1)\Gamma(2k-v_3+1)\mid D^{v_1}u\mid^{v_2-2k-2}}{(2k+1)\Gamma(v_1)\Gamma(-v_3)\Gamma(2k-v_3+2)}$$
$$\left.[D_x^1(D_x^{v_1}u)+D_y^1(D_y^{v_1}u)]\right\}+\frac{\lambda u_0}{\Gamma(1-v_3)\Gamma(2-v_3)}=0. \tag{13.110}$$

The solution procedure of (13.110) uses a parabolic equation with time as an evolution parameter (the time marching method), or equivalently, the fractional-order gradient descent method [192]. Note that, we treat an image as a function of time and space, and use the time marching method to search the fractional-order steady state of (13.110). This means that we solve:

$$\frac{\partial^{v_3}u}{\partial_t{}^{v_3}} = -\sum_{k=0}^\infty \frac{\prod_{\tau=1}^{2k}(v_2-\tau+1)}{(2k)!}\left\{\frac{\Gamma(2k-v_3)\mid D^{v_1}u\mid^{v_2-2k}}{\Gamma(-v_3)\Gamma(2k-v_3+1)}\right.$$
$$+\frac{(v_2-2k)\Gamma(1+v_1)\Gamma(2k-v_3+1)\mid D^{v_1}u\mid^{v_2-2k-2}}{(2k+1)\Gamma(v_1)\Gamma(-v_3)\Gamma(2k-v_3+2)}\left.[D_x^1(D_x^{v_1}u)+D_y^1(D_y^{v_1}u)]\right\}$$
$$-\frac{\lambda u_0}{\Gamma(1-v_3)\Gamma(2-v_3)}, \tag{13.111}$$

where we set $\prod_{\tau=1}^n (v_2-\tau+1)\Big|^{n=0}=1$. When $\frac{\partial^{v_3}u}{\partial t^{v_3}}=0$, the time marching method converges to the fractional-order steady state of (13.110). Furthermore, we must compute $\lambda(t)$ in (13.111). Let us suppose that $\iint_\Omega(u-u_0)\mathrm{d}x\mathrm{d}y=0$ and $\iint_\Omega(u-u_0)^2\mathrm{d}x\mathrm{d}y=\sigma^2$, where σ^2 is the variance of the added noise of u_0. We multiply by $(u-u_0)^2/u_0$ on the both sides of (13.110) and integrate by parts in the domain of Ω, the left side of (13.110) vanishes. Thus, the following can be obtained:

$$\lambda = \frac{-\Gamma(1-v_3)\Gamma(2-v_3)}{\sigma^2}\iint_\Omega\sum_{k=0}^\infty\frac{\prod_{\tau=1}^{2k}(v_2-\tau+1)}{(2k)!}\left\{\frac{\Gamma(2k-v_3)\mid D^{v_1}u\mid^{v_2-2k}}{\Gamma(-v_3)\Gamma(2k-v_3+1)}\right.$$

$$+ \frac{(v_2 - 2k)\Gamma(1+v_1)\Gamma(2k-v_3+1)|D^{v_1}u|^{v_2-2k-2}}{(2k+1)\Gamma(v_1)\Gamma(-v_3)\Gamma(2k-v_3+2)}[D_x^1(D_x^{v_1}u) + D_y^1(D_y^{v_1}u)]\Big\} \frac{(u-u_0)^2}{u_0} \mathrm{d}x\mathrm{d}y.$$
(13.112)

Equations (13.111) and (13.112) should be numerically implemented. At first, in (13.111), according to the fractional-order gradient descent method [192], one can obtain:

$$\frac{\partial^{v_3} u}{\partial t^{v_3}} = \Delta t^{-v_3}\left[u^{n+1} - u^n + \frac{2\mu\eta}{\Gamma(3-v_3)}(u^n - u^{v_3*})^2 (u^n)^{-v_3}\right], \quad (13.113)$$

where u^{v_3*} is the v_3-order optimal value of u, $\eta \neq 0$ is a constant that controls the degree of convexity-concavity, and μ is the constant coefficient that controls the stability and the rate of convergence of the fractional-order gradient descent method. u^{v_3*} is unknown beforehand, but every iterative result is an approximate value to u^{v_3*} in the solution procedure, i.e. $u^n \to u^{v_3*}$. Thus, let $(u^n - u^{v_3*})^2 \approx (u^{n-1} - u^n)^2$, $\eta = 1$ and $\mu = 0.005$ in the following example. Thus, from (13.111), (13.112), and (13.113), the following can be obtained:

$$u^{n+1} = I(u^n)\Delta t^{v_3} - \frac{\lambda^n \Delta t^{v_3}}{\Gamma(1-v_3)\Gamma(2-v_3)} u_0^n + u^n - \frac{2\mu}{\Gamma(3-v_3)}(u^{n-1} - u^n)^2 (u^n)^{-v_3},$$
(13.114)

$$\lambda^n = \frac{\Gamma(1-v_3)\Gamma(2-v_3)}{\sigma^{n^2}} \sum_{(x,y)\in\Omega} I(u^n) \frac{(u^n - u_0^n)^2}{u_0^n}, \quad (13.115)$$

$$I(u^n) = -\sum_{k=0}^{1} \frac{\prod_{\tau=1}^{2k}(v_2 - \tau + 1)}{(2k)!} \Big\{\frac{\Gamma(2k-v_3)|D^{v_1}u^n|^{v_2-2k}}{\Gamma(-v_3)\Gamma(2k-v_3+1)}$$
$$+ \frac{(v_2-2k)\Gamma(1+v_1)\Gamma(2k-v_3+1)|D^{v_1}u^n|^{v_2-2k-2}}{(2k+1)\Gamma(v_1)\Gamma(-v_3)\Gamma(2k-v_3+2)}$$
$$[D_x^1(D_x^{v_1}u^n) + D_y^1(D_y^{v_1}u^n)]\Big\}, \quad (13.116)$$

where we set $\prod_{\tau=1}^{n}(v_2 - \tau + 1)\overset{n=0}{=} 1$. Secondly, the numerical implementation of the fractional-order partial differential equation should be restricted by CFL condition [548]. Thirdly, in (13.111) and (13.112), for the convenience of computation, k only takes $k = 0, 1$ as an approximation instead of $k = 0 \to \infty$. Finally, it may be considered that $u^n = 0$ and $|D^{v_1}u^n| = 0$ during the numerical iteration computation of (13.114) and (13.115). To enable (13.114) and (13.115) to be implemented, if $u^n = 0$ and $|D^{v_1}u^n| \leqslant \varepsilon$, let $u^n = \varepsilon$ and $|D^{v_1}u^n| = \varepsilon$, where ε is a small positive constants. Without loss of generality, in the following experiments, let $\varepsilon = 10^{-5}$.

The capability of maintaining the edges and textural details of the fractional-orderdenoising algorithm based on the fractional-order variational method was analyzed by considering the integer-order denoising algorithm based on the classic integer-order variational method [486] — [490] vis-à-vis the proposed fractional-order denoising algorithm and a suitable texture image i. e. a wood grain image. Without loss of generality, let us suppose that the added noise is subject to Gaussian distribution $N(0, 10^{-2})$, with which we randomly degraded the original wood grain image and used this noised image as the test image. No *a priori* knowledge regarding the characteristics of the original wood grain image is either known or required. We set $v_1 = 1.65$, $v_2 = 2.75$, and $v_3 = 1.25$ in (13.114), (13.115), and (13.116). Suppose we set the same parameters, the time interval $\Delta t = 0.000625$, the number of iterations $n = 400$, for both the fractional-order denoising algorithm and the integer-order one in this simulation experiment. Thus, the results of the comparative denoising experiments for the wood grain image are shown in Fig. 13.6.

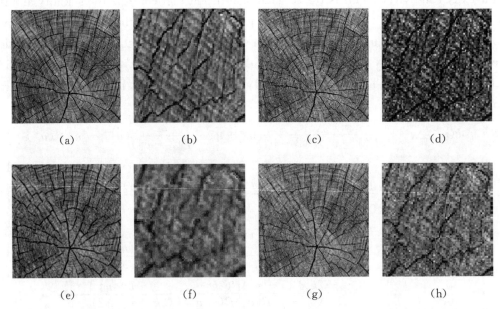

Figure 13.6 Comparative denoising experiments on a wood grain image. (a) Original wood grain image, (b) Enlarged detail of higher right 1/16 of (a), (c) Noised wood grain image, (d) Enlarged detail of higher right 1/16 of (c), (e) Integer-order denoising algorithm, (f) Enlarged detail of higher right 1/16 of (e), (g) Fractional-order denoising algorithm, (h) Enlarged detail of higher right 1/16 of (g).

To consider the visual effects for the purpose of comparison, from Fig. 13.6 (a), (b), (e), (f), (g), and (h), we can see that the capability of the integer-order denoising algorithm based on the integer-order variational method to maintain the edges and textural details is worse than that of the fractional-order denoising algorithm based on the fractional-order variational method. In particular, the extent to which the integer-order denoising algorithm is able to maitain the local edges and textural details is relatively weaker than that of the fractional-order denoising algorithm. The integer-order denoising algorithm tends to overly smooth the noised image if there are rich high-frequency edges and textural details, often resulting in a blurry appearance of the output image. In comparison, the result of maintaining the edges and textural details of the fractional-order denoising algorithm is relatively clearer than that of the integer-order denoising algorithm.

The denoised image of the integer-order denoising algorithm appears relatively smoother and more unclear than that produced by the fractional-order denoising algorithm.

Next, with the objective of performing a quantitative analysis, we obtained the PSNR between the denoised image and the original image to evaluate the denoising effects of each competing algorithm under consideration. The PSNR of the integer-order denoising algorithm and that of the fractional-order denoising algorithm are 20.2832 and 22.4786, respectively. The PSNR of the integer-order denoising algorithm is much smaller than that of the fractional-order denoising algorithm, which shows that the high-frequency edges and textural details of the denoised image of the integer-order denoising algorithm is sharply diffused and smoothed. This, in turn, indicates that the structural similarity of the edges and textural details between its corresponding denoised image and the original image is weaker than that of the fractional-order denoising algorithm. The capability of maintaining the edges and textural details of the fractional-order denoising algorithm is obviously superior to that of the integer-order denoising algorithm, especially for images rich in textural detail.

In addition, as the next experimental step, we then selected another suitable texture image i.e. a radial annulus image for the purpose of achieving comparative experiments. We also stochastically degraded the original radial annulus image with Gaussian distribution noise $N(0, 10^{-2})$ and used this noised image as the test image. No *a priori* knowledge regarding the characteristics of the original radial annulus image is either known or required. We still set set $v_1 = 1.65$, $v_2 = 2.75$, and $v_3 = 1.25$ in (13.114), (13.115), and (13.116). Suppose we set the same parameters, the time interval $\Delta t = 0.000625$, the number of iterations $n = 400$, for both the fractional-order denoising algorithm and the integer-order one in this simulation experiment. Thus, the results of the comparative denoising experiments for the moonscape image are shown in Fig. 13.7.

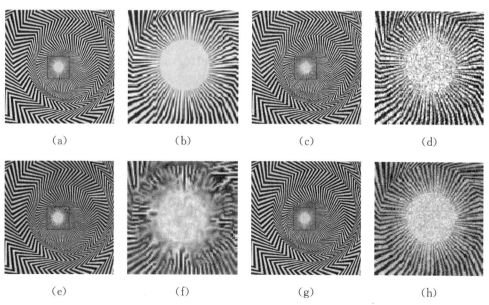

(a)　　　　　　(b)　　　　　　(c)　　　　　　(d)

(e)　　　　　　(f)　　　　　　(g)　　　　　　(h)

Figure 13.7 Comparative denoising experiments on a radial annulus image. (a) Original radial annulus image, (b) Enlarged detail of higher right 1/16 of (a), (c) Noised radial annulus image, (d) Enlarged detail of higher right 1/16 of (c), (e) Integer-order denoising algorithm, (f) Enlarged detail of higher right 1/16 of (e), (g) Fractional-order denoising algorithm, (h) Enlarged detail of higher right 1/16 of (g).

From Fig. 13.7 (a), (b), (e), (f), (g), and (h), we can see that the result of maintaining the edges and textural details of the fractional-order denoising algorithm is relatively clearer than that of the integer-order denoising algorithm. The denoised image of the integer-order denoising algorithm appears relatively smoother than those produced by the fractional-order denoising algorithm. Next, the PSNR of the integer-order denoising algorithm and that of the fractional-order denoising algorithm are 17.1512 and 19.7222, respectively. The capability of maintaining the edges and textural details of the fractional-order denoising algorithm is obviously superior to that of the integer-order denoising algorithm, especially for images rich in textural detail.

13.5 Conclusions

How to apply fractional calculus to variational method is an emerging scientific field that has been seldom received desired attention. Similar to the classical first-order Euler-Lagrange equation for the first-order variational method, in fact, in fractional calculus of variations, the solutions of the fractional-order Euler-Lagrange equation are the functions for which a given fractional-order functional is stationary. In signal processing and image processing, because a differintegrable functional is stationary at its fractional-order local maxima and minima, the fractional-order Euler-Lagrange equation is very useful for solving the fractional-order fixed boundary optimization problems in which, given some fractional-order functional, one seeks the corresponding fractional-order function minimizing (or maximizing) it. The main goal of this chapter is to derive the fractional-order Euler-Lagrange equation for fractional-order variational method, which is a necessary condition for the fractional-order fixed boundary optimization problems in signal processing and image processing. Here, fractional calculus has been proposed to apply to the solution of this problem mainly because of its inherent strengths in terms of long-term memory, non-locality, and weak singularity.

In this chapter, two different equivalent forms of the fractional-order Euler-Lagrange equation for the fractional-order variational method are obtained based on Wiener-Khintchine theorem and the fractional-order Green formula, respectively. At first, because the inverse Fourier transforms of $(j\omega_x)^v$ and $(j\omega_y)^v$ belong to Euler integral of the second kind, $\overline{D_x^v}$ and $\overline{D_y^v}$ could not be achieved in spatial domain or time domain directly, which should be implemented in discrete Fourier transform domain. Thus, the proposed fractional-order Euler-Lagrange equation based on Wiener-Khintchine theorem should be achieved by discrete Fourier transform and inverse discrete Fourier transform successively, which is difficult and complex to implement. Secondly, the proposed fractional-order Euler-Lagrange equation based on fractional-order Green formula can be directly and easily implemented in spatial domain or time domain. It is verified that the traditional first-order Euler-Lagrange equation is actually a special case of the fractional-order Euler-Lagrange equation. When we utilize the proposed fractional-order Euler-Lagrange equation based on fractional-order Green formula to deal with the classical first-order image restoration problems, it has the same results as those obtained by the classical first-order Euler-Lagrange equation. Thirdly, from the comparative experiments of image restoration such as inpainting and denoising, we can see that the capability of restoring and maintaining the edges and textural details of the fractional-order image restoration algorithm based on the fractional-order variational method is superior to that of the integer-order image restoration algorithm based on the classical first-order variational method, especially for images rich in textural details.

The aforementioned discussion has also highlighted additional problems that need to be further studied. For example, further research is required to determine the image-dependent optimal values of v_1, v_2, and v_3. It will be discussed in my future work.

第14章 基于分数阶偏微分方程的退化图像逆处理

为了便于读者整体理解退化图像的复原模型,在其数学公式推导过程中,本章引用了分数阶 E-L 方程和分数阶最速下降法的部分推导。本章仅仅是以退化图像的去噪处理为例进行说明和演示,同理可以运用第13章论述的分数阶变分法对图像修复和图像去模糊等退化图像逆处理。本章的研究目的是构造分数阶偏微分方程,即一类基于分数阶超全变差和分数阶最速下降法的纹理图像多尺度去噪模型。基于分数阶微积分最常用的是 Grümwald-Letnikov 和 Riemann-Liouville 两种定义,首先,分别论述分数阶泛 Green 公式、分数阶泛 Gauss 公式和分数阶泛 Stokes 公式。其次,在此基础上,论述针对二维图像处理的分数阶泛 Euler-Lagrange 方程,论述分数阶最速下降法及其稳定性和收敛率,构造出4种基于分数阶偏微分方程的纹理图像多尺度去噪模型。最后,通过理论推导和对比实验分析论证了分数阶最速下降法的稳定性和收敛性,讨论了该分数阶纹理图像去噪模型的分数阶非线性多尺度去噪能力及其参数的最佳取值。仿真对比实验表明,对于纹理细节信息丰富的图像而言,该分数阶纹理图像去噪模型在去噪的同时,对图像高频边缘和复杂纹理细节的分数阶非线性多尺度保持能力明显优于传统基于整数阶运算的去噪算法。

14.1 引言

基于整数阶偏微分方程的图像处理是图像处理领域中的一个重要分支。一方面,该图像处理方法属于低层图像处理的范畴,其处理结果通常被当作中间结果提供给其他图像处理方法进一步使用;另一方面,随着该图像处理方法的深入研究,人们越来越深刻地挖掘图像和图像处理的本质,并试图用严格的数学理论对现存的传统图像处理方法进行改造,这对于以实用为主的传统图像处理方法是一种挑战。

在基于整数阶偏微分方程的图像处理中,图像去噪是其重要的研究内容。基于整数阶偏微分方程的图像去噪分为两类:基于非线性扩散的方法和基于能量范函最小化的变分法[550-553]。与之对应的两种基本模型是由 Perona 和 Malik 提出的各项异性扩散(PM)模型[554]以及 Rudin, Osher 和 Fatemi 提出的全变分(ROF)模型[487]。PM 模型使用热能的扩散过程来模拟图像的去噪过程,图像去噪的结果就是热能扩散达到平衡时的状态。用全变分来描述上述热能,就是 ROF 模型。在此基础之上,有学者分别将 PM 模型和 ROF 模型推广到彩色图像处理之中[555-556]。有学者研究了模型中的参数选择[557-561],以及如何计算迭代求解过程的最优停止点[562-563]。Rudin 等人提出一种可变时间步长方法来解 Euler-Lagrange 方程[487]。C. R. Vogel 和 M. E. Oman 用不动点迭代方法来提高 ROF 模型的稳定性[564]。D. C. Dobson 和 C. R. Vogel 修改全变分形式来保证 ROF 模型数值计算的收敛性[565]。A. Chambolle 提出一种基于对偶公式的快速算法[566]。J. Darbon 和 M. Sigelle

第 14 章 基于分数阶偏微分方程的退化图像逆处理

利用水平集方法将原始问题分解为相互独立的马尔科夫随机场的优化问题,通过重建得到全局最优解[492,567-568]。有学者提出一种迭代加权范数来求解全变分以提高计算效率[569]。F. Catte 等将原图像先经过一次高斯平滑,使 PM 模型具有适定性[570]。

PM 模型和 ROF 模型都具有容易产生对比信息丢失、纹理信息丢失和阶梯效应等显著缺点[550,571-572]。针对这些缺点,人们提出了许多改进模型。为了保持对比信息和纹理信息,有学者使用 L^1 范数取代 L^2 范数[573-576]。S. Osher 等提出一种迭代正则化方法[577]。G. Gilboa, Y. Y. Zeevi 和 N. Sochen 提出一种随空间变化的自适应数值保真项的方法[578]。S. Esedoglu 和 S. Osher 提出一种保持特定边缘的方向信息[579];为了消除阶梯效应,P. Blomgren 提出一种全变分项随梯度变化的模型[580-581]。有学者将高阶导数引入能量范函中[551,582-586]。有学者将高阶导数和原始 ROF 模型进行结合[587-588]。有学者提出两阶段去噪方法,先平滑图像对应的向量场,再用曲面来匹配被平滑了的向量场[589-590]。上述改进方法对于保持对比信息和纹理信息,以及消除阶梯效应取得了一定的效果。但一方面,这类改进方法会较大地增加计算的复杂性。该类改进的方法能实时保真,但过大的存储和过多的计算让此方法不具实际可行性。另一方面,上述改进方法在本质上仍然都属于基于整数阶微分运算的算法,可能会导致在某些图像边缘区域出现模糊,而且对纹理细节信息的保持效果仍然不是很好。我们致力于将一种新颖的数学方法——分数阶微积分引入到纹理图像的去噪之中,构造分数偏微分方程,即一类基于分数阶超全变差和分数阶最速下降法的纹理图像多尺度去噪模型。该去噪模型在去除图像噪声的同时,既能尽量保留纹理图像平滑区域中的低频轮廓,又能非线性保留灰度值跃变幅度相对较大的高频边缘和纹理细节,还能非线性保留灰度值跃变幅度变化相对不大的纹理细节。

近 300 年以来,分数阶微积分业已成为数学分析的一个重要分支[176,230-231,429],但对于国内外众多数学家乃至工程技术界的物理学家而言,它还鲜为人知。一般而言,欧氏测度下的分数阶微积分是将整数阶微积分的整数步长推广到分数步长的结果,数学上要求必须使用欧氏测度[3,430];同时,由于物理过程中的随机变量为粒子随机运动的位移,所以,在欧氏测度意义下,原则上可以应用分数阶微积分来处理与分析物理状态与过程[11,19,23,83,94,142,180,431-432]。函数的分数阶微积分具有两个明显的特征:大多数函数的分数阶微积分是幂函数;其他函数的分数阶微积分或者是某种函数与幂函数的迭加,或者是某种函数与幂函数的乘积[3,176,230-231,429-430]。这种规律性是否预示着某种自然界的变化规律呢?科学研究已经证实许多分数阶、分数维的数学方法是目前人类对许多自然现象的最佳描述[89,126,401,433]。目前,欧氏测度下的分数阶微积分已经被用于扩散过程、粘弹性理论和随机分形动力学的研究,但绝大多数应用还仅仅停留在处理物理变化过程中的暂态,而对系统的演化过程很少涉及[11,19,23,83,89,94,126,142,180,401,431-433]。

如何将分数阶微积分这一崭新的数学方法应用于现代信号分析与处理[89,106,108,130-131,157,172,185,193,433,440,591-592],特别是数字图像处理之中[110,186-188,434-435,468,593],在国际上是一个研究甚少的新兴学科分支。Guidotti 和 Lambers[544] 以及 BAI 和 FENG[346] 分别对经典的各向异性扩散模型进行了分数阶推广,将其能量范函中的梯度算子从一阶推广到分数阶,并在频域内对得到的分数阶偏微分方程进行数值实现,取得了一定的图像去噪效果。但一方面,该方法仅仅将能量范函中的梯度算子简单地从一阶推广到分数阶,并没有从本质上解决能量范函在各向异性扩散的过程中如何非线性地保留图像纹理细节的难题，图像去噪后对纹理细节信息的保持并不理想；另一方面，该方法没有研究能量范函的分数次幂以及其分数阶极值对图像纹理细节信息进行非线性保留的作用；再一方面，该方法没有运用分数阶微积分的性质来推导相应的分数阶 Euler-Lagrange 方程，而是直接用希尔伯特伴随算子的复共轭转置的性质进行了替换，这样便极大地增加了频域内分数阶偏微分方程数值实现的复杂性；最后，Fourier 变换域内分数阶微积分的传输函数为 $(i\omega)^v$，虽然其形式简单，但 $(i\omega)^v$ 的 Fourier 反变换却是第一类欧拉积分，理论上极难计算。该方法在数值实现时仅仅简单地将频域内的一阶差分从形式上直接推广到分数阶差分，并以此代替频域内的分数阶微分算子，不仅没有解决频域内的分数阶微分算子的 Fourier 反变换是第一类欧拉积分的计算难题，而且在数值计算时引入了较大的误差。针对上述难题，我们首先研究信号分数阶微积分特性，与整数阶微分不同，直流或低频信号的分数阶微分一般不为零。信号的分数阶微分既非线性地加强了信号的高频分量，又在一定程度上非线性地加强了信号的中频分量，同时还极大地非线性保留了信号的低频和直流分量。分数阶微分可以非线性地增强图像信号中近似分形的复杂纹理细节[185-188,434-435,468]。在信号分数阶微分上述特性的基础之上，我们从系统的演化过程的角度，构造一类基于分数偏微分方程的纹理图像多尺度去噪模型（分数阶发展方程）。该去噪模型在去除图像噪声的同时，既能尽量保留纹理图像平滑区域中的低频轮廓，又能非线性保留灰度值跃变幅度相对较大的高频边缘，还能非线性保留灰度值跃变幅度变化相对不大的纹理细节。

14.2 相关工作

众所周知，分形数学理论产生了测度观的转变，分形几何否定了牛顿-莱布尼兹导数的存在性。以 Hausdorff 测度为基础的分形理论，虽然历经了 90 余年的研究，但至今仍然还是一种很不完善的数学理论。Hausdorff 测度下的微积分数学理论的构造至今尚未能完成。目前发展比较成熟的是在欧氏测度下定义的分数阶微积分，它在数学上要求必须使用欧氏测度。在欧氏测度下，分数阶微积分最常用的是 Grümwald-Letnikov 定义和 Riemann-Liouville 定义两种[3,176,230-231,429-430]。

Grümwald-Letnikov 定义信号 $s(x)$ 的 v 阶微积分为

$$D_{G-L}^v s(x) = \frac{d^v}{[d(x-a)]^v} s(x)\Big|_{G-L} = \lim_{N\to\infty}\left\{\frac{\left(\frac{x-a}{N}\right)^{-v}}{\Gamma(-v)} \sum_{k=0}^{N-1} \frac{\Gamma(k-v)}{\Gamma(k+1)} s\left[x - k\left(\frac{x-a}{N}\right)\right]\right\}$$

(14.1)

式中，信号 $s(x)$ 的持续期为 $[a, x]$，v 为任意实数（包括分数），D_{G-L}^v 表示基于 Grümwald-Letnikov 定义的分数阶微分算子，Γ 为 Gamma 函数。由式（14.1）可知，Grümwald-Letnikov 定义在欧氏测度下将整数阶微积分的整数步长推广到分数步长，从而将微积分的整数阶推广到分数阶。分数阶微积分的 Grümwald-Letnikov 定义的计算简便易行，它仅需要与信号 $s(x)$ 自身相关的 $s\left(x - k\left(\frac{x-a}{N}\right)\right)$ 的离散采样值，而不需要信号 $s(x)$ 的导数与积分值。

Riemann-Liouville 定义信号 $s(x)$ 的 v 阶积分$(v<0)$为

$$D_{R-L}^v s(x) = \frac{\mathrm{d}^v}{[\mathrm{d}(x-a)]^v} s(x)\big|_{R-L} = \frac{1}{\Gamma(-v)} \int_a^x (x-\eta)^{-v-1} s(\eta) \mathrm{d}\eta$$

$$= \frac{-1}{\Gamma(-v)} \int_a^x s(\eta) \mathrm{d}(x-\eta)^{-v}, v < 0 \tag{14.2}$$

式中，D_{G-L}^v 表示基于 Riemann-Liouville 定义的分数阶微分算子。对于信号 $s(x)$ 的 v 阶微分（$v \geqslant 0$），n 满足 $n-1 < v \leqslant n$。于是由式（14.2），我们可推导出信号 $s(x)$ 的 v 阶微分的 Riemann-Liouville 定义为

$$D_{R-L}^v s(x) = \frac{\mathrm{d}^v}{[\mathrm{d}(x-a)]^v} s(x)\big|_{R-L} = \frac{\mathrm{d}^n}{\mathrm{d}x^n} \frac{\mathrm{d}^{v-n}}{[\mathrm{d}(x-a)]^{v-n}} s(x)\big|_{R-L}$$

$$= \sum_{k=0}^{n-1} \frac{(x-a)^{k-v} s^{(k)}(a)}{\Gamma(k-v+1)} + \frac{1}{\Gamma(n-v)} \int_a^x \frac{s^{(n)}(\eta)}{(x-\eta)^{v-n+1}} \mathrm{d}\eta \tag{14.3}$$

由式（14.3），我们可以推导信号 $s(x)$ 的 Fourier 变换为

$$FT[D^v s(x)] = (i\omega)^v FT[s(x)] - \sum_{k=0}^{n-1} (i\omega)^k \frac{\mathrm{d}^{v-1-k}}{\mathrm{d}x^{v-1-k}} s(0) \tag{14.4}$$

式中，i 是虚数单位，ω 是数字频率。当信号 $s(x)$ 是因果信号时，式（14.4）可简化为

$$FT[D^v s(x)] = (i\omega)^v FT[s(x)] \tag{14.5}$$

14.3 构造分数阶偏微分方程：一类基于分数阶超全变差和分数阶最速下降法的纹理图像多尺度去噪模型的理论推导和分析

14.3.1 推导分数阶泛 Green 公式、分数阶泛 Gauss 公式和分数阶泛 Stokes 公式

我们要构造适于纹理图像去噪的分数阶偏微分方程，就必须构造分数阶泛 Euler－Lagrange 方程，而构造分数阶泛 Green 公式是构造分数阶泛 Euler－Lagrange 方程的必要前提。因此，我们必须首先将传统的整数阶 Green 公式推广到分数阶，推导并构造分数阶泛 Green 公式。进一步地，为将上述结论推广到三维纹理图像的去噪中，我们推导出分数阶泛 Gauss 公式和分数阶泛 Stokes 公式。

我们令 Ω 是以分段光滑曲线 C 为边界的平面单连通区域，可微积函数 $P(x,y)$ 和 $Q(x,u)$ 在 Ω 和 C 上连续，并存在对 x 和 y 的分数阶连续偏导数；令 D^1 表示 1 阶微分算子，D^v 表示 v 阶分数阶微分算子，$I^1 = D^{-1}$ 表示 1 阶积分算子，$I^v = D^{-v}$ 表示 $v > 0$ 阶分数阶积分算子，$I_x^v I_y^v\big|_\Omega$ 表示在平面 Ω 上的 v 阶分数阶曲面积分算子（将分数阶积分的 Riemann-Liouville 定义式（14.2）从一维推广到二维），$I_{C(AC_1B)}^v$ 表示在曲线 C 的 AC_1B 段上沿 $\overrightarrow{AC_1B}$ 方向的 v 阶分数阶曲线积分算子，I_{C-}^v 表示在闭曲线 C 上沿逆时针方向的 v 阶分数阶闭曲线积分算子；令区域 Ω 的边界 C 是由两曲线 $y = \phi_1(x)$，$y = \phi_2(x)$，$a \leqslant x \leqslant b$ 或 $x = \Psi_1(y)$，$x = \Psi_2(y)$，$c \leqslant y \leqslant d$ 所围成，如图 14.1 所示。

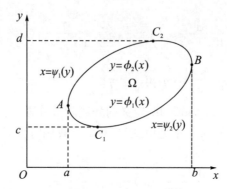

图 14.1 单连通区域 Ω 及其光滑边界曲线 C

对于可微积函数 $P(x,y)$ 而言，当 $P - D^{-v_1}D^{v_1}P \neq 0$ 时，$D^{v_1}D^{v_2}P = D^{v_1+v_2}P - D^{v_1+v_2}(P - D^{-v_1}D^{v_1}P)$。于是，$I_x^{v_2}I_y^{v_2}D_y^{v_1}P(x,y) = I_x^{v_2}\{D_y^{v_1-v_2}P(x,y) - D_y^{-v_2}[P(x,y) - D_y^{-v_1}D_y^{v_1}P(x,y)]\}$，我们可以推导得

$$I_{x\atop\Omega}^{v_2}I_y^{v_2}D_y^{v_1}P(x,y) = {}_a^b I_x^{v_2}{}_{\phi_1(x)}^{\phi_2(x)}I_y^{v_2}D_y^{v_1}P(x,y)$$

$$= {}_a^b I_x^{v_2}\{D_y^{v_1-v_2}P(x,y) - D_y^{-v_2}[P(x,y) - D_y^{-v_1}D_y^{v_1}P(x,y)]\}\big|_{\phi_1(x)}^{\phi_2(x)}$$

$$= -I_{x\atop C(BC_2A)}^{v_2}\{D_y^{v_1-v_2}P(x,y) - D_y^{-v_2}[P(x,y) - D_y^{-v_1}D_y^{v_1}P(x,y)]\}$$

$$- I_{x\atop C(AC_1B)}^{v_2}\{D_y^{v_1-v_2}P(x,y) - D_y^{-v_2}[P(x,y) - D_y^{-v_1}D_y^{v_1}P(x,y)]\}$$

$$= I_{x\atop C^-}^{v_2}\{D_y^{v_1-v_2}P(x,y) - D_y^{-v_2}[P(x,y) - D_y^{-v_1}D_y^{v_1}P(x,y)]\}$$

(14.6)

同理可得

$$I_{x\atop\Omega}^{v_2}I_y^{v_2}D_x^{v_1}Q(x,y) = I_{y\atop C^-}^{v_2}\{D_x^{v_1-v_2}Q(x,y) - D_x^{-v_2}[Q(x,y) - D_x^{-v_1}D_x^{v_1}Q(x,y)]\}$$

(14.7)

由式（14.6）和（14.7），我们可以推导得分数阶泛 Green 公式为

$$I_{x\atop\Omega}^{v_2}I_y^{v_2}(D_x^{v_1}Q(x,y) - D_y^{v_1}P(x,y)) = I_{x\atop C^-}^{v_2}\{D_y^{v_1-v_2}P(x,y) - D_y^{-v_2}[P(x,y) - D_y^{-v_1}D_y^{v_1}P(x,y)]\}$$
$$+ I_{y\atop C^-}^{v_2}\{D_x^{v_1-v_2}Q(x,y) - D_x^{-v_2}[Q(x,y) - D_x^{-v_1}D_x^{v_1}Q(x,y)]\} \quad (14.8)$$

特别地，当 D^{v_1} 和 D^{-v_1} 互逆时（该条件要求较高，一般很难满足），即 $\varphi - D^{-v_1}D^{v_1}\varphi = 0$，则 $D^{v_1}D^{v_2}\varphi = D^{v_1+v_2}\varphi$。由式（14.8）可得分数阶泛 Green 公式在特定条件下的简化表出为

$$I_{x\atop\Omega}^{v_2}I_y^{v_2}(D_x^{v_1}Q(x,y) - D_y^{v_1}P(x,y)) = I_{x\atop C^-}^{v_2}D_y^{v_1-v_2}P(x,y) + I_{y\atop C^-}^{v_2}D_x^{v_1-v_2}Q(x,y)$$

(14.9)

可见，当 $v_1 = v_2 = 1$ 时，由式（14.9）可推得 $I_x^1 I_y^1(D_x^1 Q(x,y) - D_y^1 P(x,y)) = I_{x\atop C^-}^1 P(x,y) + I_{y\atop C^-}^1 Q(x,y)$，传统的整数阶 Green 公式只是分数阶泛 Green 公式的特例；当 $v_1 = v_2 = v$ 时，由式（14.9）可推得 $I_x^v I_y^v (D_x^v Q(x,y) - D_y^v P(x,y)) = I_{x\atop C^-}^v P(x,y) + I_{y\atop C^-}^v Q(x,y)$，传统的分数阶 Green 公式[546]也只是分数阶泛 Green 公式的特例。

同理，令三维空间区域 Ω 的边界曲面 S 以及 S 的边界曲线 C 是光滑的或分片光滑的，可微积函数 $P(x,y,z)$，$Q(x,y,z)$ 和 $R(x,y,z)$ 在 Ω，S 和 C 上存在对 x，y 和 z 的分

数阶连续偏导数，我们还可以推导出分数阶泛 Gauss 公式和分数阶泛 Stokes 公式，即

$$I^{v_2}_{2x}I^{v_2}_{y^1}I^{v_2}_{z^1}(D^{v_1}_x P + D^{v_1}_y Q + D^{v_1}_z R) = I^{v_2}_{y^1}I^{v_2}_{z^1}\{D^{v_1-v_2}_x P - D^{-v_2}_x [P - D^{-v_1}_x D^{v_1}_x P]\}$$
$$+ I^{v_2}_{x^1}I^{v_2}_{z^1}\{D^{v_1-v_2}_y Q - D^{-v_2}_y [Q - D^{-v_1}_y D^{v_1}_y Q]\}$$
$$+ I^{v_2}_{x^1}I^{v_2}_{y^1}\{D^{v_1-v_2}_z R - D^{-v_2}_z [R - D^{-v_1}_z D^{v_1}_z R]\}$$

(14.10)

$$I^{v_2}_{y^1}I^{v_2}_{z^1}(D^{v_1}_y R - D^{v_1}_z Q) + I^{v_2}_{x^1}I^{v_2}_{z^1}(D^{v_1}_z P - D^{v_1}_x R) + I^{v_2}_{x^1}I^{v_2}_{y^1}(D^{v_1}_x Q - D^{v_1}_y P)$$
$$= \frac{1}{2}I^{v_2}_{x^1}\{D^{v_1-v_2}_y P - D^{-v_2}_y [P - D^{-v_1}_y D^{v_1}_y P] + D^{v_1-v_2}_z P - D^{-v_2}_z [P - D^{-v_1}_z D^{v_1}_z P]\}$$
$$+ \frac{1}{2}I^{v_2}_{y^1}\{D^{v_1-v_2}_x Q - D^{-v_2}_x [Q - D^{-v_1}_x D^{v_1}_x Q] + D^{v_1-v_2}_z Q - D^{-v_2}_z [Q - D^{-v_1}_z D^{v_1}_z Q]\}$$
$$+ \frac{1}{2}I^{v_2}_{z^1}\{D^{v_1-v_2}_x R - D^{-v_2}_x [R - D^{-v_1}_x D^{v_1}_x R] + D^{v_1-v_2}_y R - D^{-v_2}_y [R - D^{-v_1}_y D^{v_1}_y R]\}$$

(14.11)

可见，当 D^{v_1} 和 D^{-v_1} 互逆时，若 $v_1=v_2=1$，由式（14.10）和（14.11）可推知传统的整数阶 Gauss 公式和整数阶 Stokes 公式只是分数阶泛 Gauss 公式和分数阶泛 Stokes 公式的特例；若 $v_1=v_2=v$，由式（14.10）和（14.11）可推知传统的分数阶 Gauss 公式[546]和分数阶 Stokes 公式[546]也只是分数阶泛 Gauss 公式和分数阶泛 Stokes 公式的特例。

14.3.2 推导针对二维图像处理的分数阶泛 Euler-Lagrange 方程

基于上述推导所得的分数阶泛 Green 公式，我们可以进一步推导针对二维图像处理的分数阶泛 Euler-Lagrange 方程。令二维空间中的可微积数量函数为 $u(x,y)$ 和可微积矢量函数为 $\vec{\varphi}(x,y)=i\varphi_x+j\varphi_y$，$v$ 阶分数阶微分算子 $D^v=i\frac{\partial^v}{\partial x^v}+j\frac{\partial^v}{\partial y^v}=iD^v_x+jD^v_y=(D^v_x,D^v_y)$，$D^v$ 是一线性算子（当 $v=0$ 时，D^0 表示既不微分也不积分，是一个恒等算子），其中 i 和 j 分别表示在 x 和 y 方向上的单位矢量。一般而言，二维图像区域 Ω 是一个长方形的单连通区域，因此 Ω 的分段光滑边界 C 是一个闭合的长方形曲线，如图 14.2 所示。

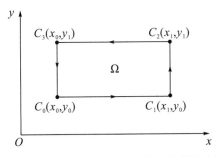

图 14.2 二维单连通图像区域 Ω 及其分段光滑边界曲线 C

由分数阶积分的 Riemann-Liouville 定义式（14.2）可得 $I^v_x s(x,y) = \frac{1}{\Gamma(v)}\int_{a_x}^{x}(x-$

$\eta)^{v-1}s(\eta,y)\mathrm{d}\eta, I_y^v s(x,y) = \dfrac{1}{\Gamma(v)}\int_{a_y}^x (y-\xi)^{v-1}s(x,\xi)\mathrm{d}\xi$ 且 $I_x^v I_y^v s(x,y) = \dfrac{1}{\Gamma^2(v)}\iint_{a_x a_y}^{x\ y}(x-\eta)^{v-1}(y-\zeta)^{v-1}s(\eta,\zeta)\mathrm{d}\eta\mathrm{d}\zeta$. 于是由分数阶泛 Green 公式（14.8）和图 14.2 可推得下式恒成立：

$$\begin{aligned} I_x^{v_2} I_y^{v_2}(D_x^{v_1}Q(x,y) - D_y^{v_1}P(x,y)) &= I_{C^-}^{v_2}\{D_y^{v_1-v_2}P(x,y) - D_y^{v_1-v_2}[P(x,y) - D_y^{-v_1}\\ D_y^{v_1}P(x,y)]\} &+ I_{C^-}^{v_2}\{D_x^{v_1-v_2}Q(x,y) - D_x^{v_1-v_2}[Q(x,y) - D_x^{-v_1}D_x^{v_1}Q(x,y)]\}\\ &= {}_{x_0}^{x_1} I_x^{v_2}\{D_y^{v_1-v_2}P - D_y^{v_1-v_2}[P - D_y^{-v_1}D_y^{v_1}P]\} + {}_{x_1}^{x_0}I_x^{v_2}\{D_y^{v_1-v_2}P - D_y^{v_1-v_2}[P - D_y^{-v_1}D_y^{v_1}P]\}\\ &\quad + {}_{y_0}^{y_1}I_y^{v_2}\{D_x^{v_1-v_2}Q - D_x^{v_1-v_2}[Q - D_x^{-v_1}D_x^{v_1}Q]\}\\ &\quad + {}_{y_1}^{y_0}I_y^{v_2}\{D_x^{v_1-v_2}Q - D_x^{v_1-v_2}[Q - D_x^{-v_1}D_x^{v_1}Q]\} \equiv 0 \end{aligned} \quad (14.12)$$

由于 $\sum_{m=0}^{\infty}\sum_{n=0}^{m} \equiv \sum_{n=0}^{\infty}\sum_{m=n}^{\infty}$ 且 $\binom{v}{r+n}\binom{r+n}{n} \equiv \binom{v}{n}\binom{v-n}{r}$，于是我们可以推导得

$$D_{x-a}^v(fg) = \sum_{m=0}^{\infty}\left[\binom{v}{n}(D_{x-a}^{v-n})D_{x-a}^n g\right] \quad (14.13)$$

由分数阶微积分的齐次性质及式（14.12）可得 $I_\Omega^{v_2}I_y^{v_2}[D_x^{v_1}(u\varphi_x) + D_y^{v_1}(u\varphi_y)] = I_x^{v_2}I_y^{v_2}[D_x^{v_1}(u\varphi_x) - D_y^{v_1}(-u\varphi_y)] = 0$，再由式（14.13），我们可以推导得

$$I_x^{v_2}I_y^{v_2}D^{v_1}u \cdot \vec{\varphi} = I_x^{v_2}I_y^{v_2}[(D_x^{v_1}u)\varphi_x + (D_y^{v_1}u)\varphi_y] =$$
$$-I_x^{v_2}I_y^{v_2}\sum_{n=1}^{\infty}\binom{v_1}{n}[(D_x^{v_1-n}u)D_x^n\varphi_x + (D_y^{v_1-n}u)D_y^n\varphi_y] \quad (14.14)$$

式中，符号·表示内积。与分数阶散度算子 $\mathrm{div}^v\vec{\varphi} = D^v \cdot \vec{\varphi} = D_x^v\varphi_x + D_y^v\varphi_y$ 的定义类似，我们令 v 阶分数阶类微分算子 $P^v = \sum_{n=1}^{\infty}\binom{v}{n}\left[i\dfrac{(D_x^{v-n}u)}{u}D_x^n + j\dfrac{(D_y^{v-n}u)}{u}D_y^n\right]$，$v$ 阶分数阶类散度算子 $\mathrm{div}P^v\vec{\varphi} = P^v \cdot \vec{\varphi} = P_x^v\varphi_x + P_y^v\varphi_y$，$\mathrm{div}P^v$ 和 P^v 都是线性算子，于是根据希尔伯特伴随算子理论[543]，由式（14.14）可以推导得

$$\begin{aligned} I_x^{v_2}I_y^{v_2}D^{v_1}u \cdot \vec{\varphi} &= \langle D^{v_1}u,\vec{\varphi}\rangle^{v_2} = \langle u,(D^{v_1})^*\vec{\varphi}\rangle^{v_2} = I_x^{v_2}I_y^{v_2}u((D^{v_1})^*\vec{\varphi})\\ &= -I_x^{v_2}I_y^{v_2}u(P^{v_1}\cdot\vec{\varphi}) = -I_x^{v_2}I_y^{v_2}u(\mathrm{div}P^{v_1}\vec{\varphi}) \end{aligned} \quad (14.15)$$

式中，$\langle,\rangle^{v_2}$ 表示 v_2 阶分数阶内积的积分形式，$(D^v)^*$ 表示 D^v 的 v 阶分数阶希尔伯特伴随算子。由式（14.15）可得

$$(D^v)^* - \mathrm{div}P^1 \quad (14.16)$$

由式（14.16）可见，分数阶希尔伯特伴随算子 $(D^v)^*$ 是一线性算子。当 $v_1 = v_2 = 1$ 时，由式（14.14）我们可以推导得

$$I_x^1 I_y^1 D^1 u \cdot \vec{\varphi} = \langle D^1 u,\vec{\varphi}\rangle^1 = \langle u,(D^1)^*\vec{\varphi}\rangle^1 = I_x^1 I_y^1 u(D^1)^*\vec{\varphi}) = -I_x^1 I_y^1 u(\mathrm{div}^1\vec{\varphi}) \quad (14.17)$$

式中，$\langle,\rangle^1$ 表示一阶内积的积分形式，$\mathrm{div}^1\vec{\varphi} = D^1 \cdot \vec{\varphi} = D_x^1\vec{\varphi}_x + D_y^1\varphi_y$ 表示一阶散度算子，$(D^1)^*$ 表示 D^1 的一阶希尔伯特伴随算子。针对数字图像而言，由式（14.17）我们可推导得

$$(D^1)^* = -\mathrm{div}^1 \quad (14.18)$$

由式（14.16）和（14.18）可知，一阶希尔伯特伴随算子只是分数阶希尔伯特伴随算

子的特例。当 $I_{x}^{v_2}I_{y}^{v_2}D^{v_1}u \cdot \vec{\varphi} = 0$ 时，由式（14.14）可得

$$I_{x}^{v_2}I_{y}^{v_2}D^{v_1}u \cdot \vec{\varphi} = -I_{x}^{v_2}I_{y}^{v_2}\sum_{n=1}^{\infty}\binom{v_1}{n}[(D_x^{v_1-n}u)D_x^n\varphi_x + (D_y^{v_1-n}u)D_y^n\varphi_y] = 0$$

(14.19)

在式（14.19）中，由于 x 方向和 y 方向的正交性，$(D_x^{v_1-n}u)D_x^n\varphi_x + (D_y^{v_1-n}u)D_y^n\varphi_y = (D_x^{v_1-n}u, D_y^{v_1-n}u) \cdot (D_x^n\varphi_x, D_y^n\varphi_y)$。对于任意二维数量函数 u（测试函数）而言，相应的 $D_x^{v_1-n}u$ 和 $D_y^{v_1-n}u$ 具有随机性，由变分基本引理（fundamental lemma of calculus of variations）[485]，要使式（14.19）成立，只需 $(D_x^n\varphi_x, D_y^n\varphi_y) = (0, 0)$。由于 n 为 $1 \to \infty$ 的正整数，欲使 $(D_x^n\varphi_x, D_y^n\varphi_y) = (0, 0)$，仅需使 $(D_x^1\varphi_x, D_y^1\varphi_y) = (0, 0)$。式（14.19）的等价式为 $-I_{x}^{v_2}I_{y}^{v_2}\binom{v_1}{1}[(D_x^{v_1-1}u)D_x^1\varphi_x + (D_y^{v_1-1}u)D_y^1\varphi_y] = -I_{x}^{v_2}I_{y}^{v_2}\binom{v_1}{1}(D_x^{v_1-1}u, D_y^{v_1-1}u) \cdot (D_x^1\varphi_x, D_y^1\varphi_y) = 0$。于是要使（14.19）成立，当且仅当下式成立：

$$-\binom{v_1}{1}[D_x^1\varphi_x + D_y^1\varphi_y] = \frac{\Gamma(1-v_1)}{\Gamma(-v_1)}[D_x^1\varphi_x + D_y^1\varphi_y] = 0 \quad (14.20)$$

式（14.20）即是与 $I_{x}^{v_2}I_{y}^{v_2}D^{v_1}u \cdot \vec{\varphi} = 0$ 相对应的分数阶泛 Euler-Lagrange 方程。

进一步地，如果 $\Phi_1(D^{v_1}u)$ 是矢量函数 $D^{v_1}u$ 的数量函数，$\Phi_2(\vec{\varphi})$ 是可微积矢量函数 $\vec{\varphi}(x, y) = i\varphi_x + j\varphi_y$ 的数量函数。同理，当 $I_{x}^{v_2}I_{y}^{v_2}\Phi_1(D^{v_1}u)\Phi_2(\vec{\varphi})D^{v_1}u \cdot \vec{\varphi} = 0$ 时，由式（14.14）可得

$$I_{x}^{v_2}I_{y}^{v_2}\Phi_1(D^{v_1}u)\Phi_2(\vec{\varphi})D^{v_1}u \cdot \vec{\varphi} = I_{x}^{v_2}I_{y}^{v_2}\Phi_1(D^{v_1}u)D^{v_1}u \cdot (\Phi_2(\vec{\varphi})\vec{\varphi})$$

$$= -I_{x}^{v_2}I_{y}^{v_2}\sum_{n=1}^{\infty}\binom{v_1}{n}[(\Phi_1(D^{v_1}u)D_x^{v_1-n}u)D_x^n(\Phi_2(\vec{\varphi})\varphi_x) + (\Phi_1(D^{v_1}u)D_y^{v_1-n}u)D_y^n(\Phi_2(\vec{\varphi})\varphi_y)]$$

$$= 0$$

(14.21)

于是要使（14.21）成立，当且仅当下式成立：

$$-\binom{v_1}{1}[D_x^1(\Phi_2(\vec{\varphi})\varphi_x) + D_y^1(\Phi_2(\vec{\varphi})\varphi_y)] = \frac{\Gamma(1-v_1)}{\Gamma(-v_1)}[D_x^1(\Phi_2(\vec{\varphi})\varphi_x) + D_y^1(\Phi_2(\vec{\varphi})\varphi_y)] = 0$$

(14.22)

式（14.22）即是与 $I_{x}^{v_2}I_{y}^{v_2}\Phi_1(D^{v_1}u)\Phi_2(\vec{\varphi})D^{v_1}u \cdot \vec{\varphi} = 0$ 相对应的分数阶泛 Euler-Lagrange 方程。

由于分数阶微积分对于所有的 v 均存在 $D_{x-a}^v[0] = 0$，所以分数阶泛 Euler-Lagrange 方程式（14.20）和（14.22）与分数阶面积分 $I_{x}^{v_2}I_{y}^{v_2}$ 的积分阶次 v_2 无关，因此我们在下面构造适于纹理图像去噪的分数阶偏微分方程模型的能量泛函时均不采用分数阶面积分 $I_{x}^{v}I_{y}^{v}$，而只采用一阶面积分 $I_{x}^{1}I_{y}^{1}$ 的形式。

14.3.3 推导分数阶最速下降法及其稳定性和收敛率

基于上述分数阶泛 Euler-Lagrange 方程，要构造分数偏微分方程，即一类基于分数阶超全变差和分数阶最速下降法的纹理图像多尺度去噪模型，还需要进一步求解基于分数阶超变差（fractional hyper variation）的能量泛函的分数阶极小值，而非传统的整数阶极小值（如一阶极值）。于是，我们还必须将传统的整数阶最速下降法推广到分数阶最速下

降法。

为了便于分析比较分数阶最速下降法与整数阶最速下降法之间的区别，我们仍采用一阶极值来构造能量泛函，即

$$E = E_{\min}^1 + \eta(s^{1*} - s)^2 \tag{14.23}$$

式中，$\eta \neq 0$ 是控制抛物线凹凸程度的常数，$E_{\min}^1$ 是 E 的一阶极值，E 是二次能量泛函，其性能曲线是一条抛物线，一般是未知的。我们令分数阶最速下降法沿着 E 的 $v>0$ 阶分数阶梯度的负值方向进行增量搜索，即

$$s_{k+1} = s_k + \mu(-D_{s_k}^v) \tag{14.24}$$

式中，k 是步长或迭代次数，s_k 是 s 的当前调整值，s_{k+1} 是 s 的新调整值，$D_{s_k}^v$ 是能量泛函 E 在 $s=s_k$ 点的 v 阶分数阶梯度，μ 是一个控制稳定度与收敛率的常系数参数。分数阶最速下降法的迭代搜索过程如图 14.3 所示。

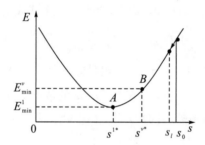

图 14.3　分数阶最速下降法示意图

图 14.3 中，A 点是 E 的一阶极值点或一阶驻点（对应 $E_{\min}^1$），B 点是 E 的 v 阶分数阶极值点或分数阶驻点（对应 $E_{\min}^v$），s^{1*} 是与 $E_{\min}^1$ 相对应的一阶最佳自变量值，s^{v*} 是与 $E_{\min}^v$ 相对应的分数阶最佳自变量值，s_0 是用分数阶最速下降法迭代搜索时 s 的初始猜测值（随机选取）。

由式（14.1）可推证，$D_{x-a}^v[0]=0$ 和 $D_{x-a}^v[c]=c\dfrac{(x-a)^{-v}}{\Gamma(1-v)}$，其中 $[0]$ 表示恒零常数，$[c]$ 表示非零常数。可见，与整数阶微分不同，非零常数的分数阶微分值非零。于是，E 的 v 阶分数阶极值点 B 与 E 的一阶极值点 A 不重合，即 $E_{\min}^v$ 不是 E 的最小值。若 n 和 N 为非负整数，则 $\dfrac{\Gamma(-n)}{\Gamma(-N)}=(-1)^{N-n}\dfrac{N!}{n!}$，于是由式（14.24）、式（14.1）以及分数阶微积分的性质可推得

$$D_{s_k}^v = \dfrac{d^v E}{ds^v}\Big|_{s=s_k} = \dfrac{[E_{\min}^1 + \eta(s^{1*})^2]s_k^{-v}}{\Gamma(1-v)} - \dfrac{2\eta s^{1*} s_k^{1-v}}{\Gamma(2-v)} + \dfrac{2\eta s_k^{2-v}}{\Gamma(3-v)} \tag{14.25}$$

由式（14.25）可知，$D_{s_k}^v$ 是关于 s_k 的一个非常系数非线性方程，很难直接求解出 $D_{s_k}^v$ 关于 v 的解析表达式。但是，当分数阶最速下降法的阶次 v 的取值给定时，式（14.25）转化为一个常系数非线性方程。当分数阶最速下降法的迭代搜索过程收敛到稳态时，即 $s_{k+1}=s_k=s^{v*}$ 时，$D_{s_k}^v=0$ 且 $E=E_{\min}^v$，于是由式（14.23）和（14.25）可推得

$$E_{\min}^v = E_{\min}^1 + \eta(s^{1*})^2 - 2\eta s^{1*} s^{v*} + \eta(s^{v*})^2 \tag{14.26}$$

$$(s^{v*})^{-v}\left\{\dfrac{E_{\min}^1 + \eta(s^{1*})^2}{\Gamma(1-v)} - \dfrac{2\eta s^{1*}}{\Gamma(2-v)}s^{v*} + \dfrac{2\eta}{\Gamma(3-v)}(s^{v*})^2\right\} = 0 \tag{14.27}$$

由式（14.27）知，$s^{v*} \neq 0$。当$s^{v*} \neq 0$时，式（14.27）简化为常系数一元二次方程，即

$$\frac{E_{\min}^1 + \eta(s^{1*})^2}{\Gamma(1-v)} - \frac{2\eta s^{1*}}{\Gamma(2-v)} s^{v*} + \frac{2\eta}{\Gamma(3-v)}(s^{v*})^2 = 0 \qquad (14.28)$$

由式（14.28）可以推导s^{v*}和s^{1*}之间的关系为

$$s_{1,2}^{v*} = \frac{\Gamma(3-v)}{2\eta}\left\{\frac{\eta s^{1*}}{\Gamma(2-v)} \pm \sqrt{\frac{\eta^2(s^{1*})^2}{\Gamma^2(2-v)} - \frac{2\eta[E_{\min}^1 + \eta(s^{1*})^2]}{\Gamma(1-v)\Gamma(3-v)}}\right\} \qquad (14.29)$$

由式（14.29）知，能量泛函E的v阶分数阶极值点不一定唯一，且同阶次的两个不同的分数阶极值点之间一般关于s^{1*}具有不对称性。如果式（14.29）具有两个不同的解，分数阶最速下降法就会收敛于s_1^{v*}或s_2^{v*}之一（与初始猜测值s_0的选取有关）。于是，我们难以控制分数阶最速下降法收敛到所期望的稳定点。一般地，为了保证分数阶最速下降法最后仅收敛到唯一的v阶分数阶极值点，须使式（14.29）具有两个相同的解，即仅存在唯一的分数阶最佳值$s^{v*} = s_1^{v*} = s_2^{v*}$。当$v=1$或$v=3$时，$\frac{1}{\Gamma(1-v)\Gamma(3-v)}=0$，且当$v=2$时，$\frac{1}{\Gamma(2-v)}=0$，又因$s^{v*} \neq 0$，于是由式（14.29）可推得单根$s^{v*}$和$s^{1*}$之间的关系为

$$s^{v*} = \frac{\Gamma(3-v)s^{1*}}{\Gamma(2-v)} \stackrel{v=1}{=} s^{1*}, \text{when } v = 1 \text{ or } 3, \text{and } v \neq 2 \qquad (14.30)$$

由式（14.30）可知，一阶最速下降法只是分数阶最速下降法的一个特例。当$v \neq 1$，2，3，且$\frac{\eta^2(s^{1*})^2}{\Gamma^2(2-v)} - \frac{2\eta[E_{\min}^1 + \eta(s^{1*})^2]}{\Gamma(1-v)\Gamma(3-v)}=0$时，$s^{v*}$仅有单个解，$E$仅具有唯一的$v$阶分数阶极值点。此时，分数阶最速下降法的阶次$v$（$v \neq 1$，2，3）满足如下关系：

$$\frac{\Gamma(1-v)\Gamma(3-v)}{\Gamma^2(2-v)} = \frac{2E_{\min}^1}{\eta(s^{1*})^2} + 2, \text{when } v \neq 1,2 \text{ and } 3 \qquad (14.31)$$

将式（14.31）代入（14.29），可得单根s^{v*}和s^{1*}之间的关系为

$$s^{v*} = \frac{\Gamma(3-v)s^{1*}}{2\Gamma(2-v)}, \text{when } v \neq 1,2 \text{ and } 3 \qquad (14.32)$$

于是，当$v \neq 1$，2，3且满足式（14.31），即s^{v*}仅有单个解时，将式（14.26）和（14.32）代入（14.23），可得用分数阶极值表示一维能量泛函，即

$$E = E_{\min}^v + \eta\left[\frac{4\Gamma(2-v) - \Gamma(3-v)}{\Gamma(3-v)}(s^{v*})^2 - \frac{4\Gamma(2-v)}{\Gamma(3-v)}s^{v*}s + s^2\right] \qquad (14.33)$$

将式（14.25）代入式（14.24），即可分析从初始猜测值s_0到与分数阶最佳值s^{v*}的迭代搜索过程的动态或瞬态性质：

$$s_{k+1} = s_k - \frac{\mu[E_{\min}^1 + \eta(s^{1*})^2]s_k^{-v}}{\Gamma(1-v)} + \frac{2\mu\eta s^{1*}s_k^{1-v}}{\Gamma(2-v)} - \frac{2\mu\eta s_k^{2-v}}{\Gamma(3-v)} \qquad (14.34)$$

将式（14.28）代入式（14.34），可推得

$$s_{k+1} = s_k - \frac{2\mu\eta}{\Gamma(3-v)}\left[s_k^2 - \frac{\Gamma(3-v)}{\Gamma(2-v)}s^{1*}s_k - (s^{v*})^2 + \frac{\Gamma(3-v)}{\Gamma(2-v)}s^{1*}s^{v*}\right]s_k^{-v} \qquad (14.35)$$

将式（14.32）代入式（14.35），即可推得

$$s_{k+1} = s_k - \frac{2\mu\eta}{\Gamma(3-v)}(s_k - s^{v*})^2 s_k^{-v}, \text{when } v \neq 1, 2 \text{ and } 3 \tag{14.36}$$

式（14.36）是一个非线性非常差分方程，不能直接由其前几次迭代通过数学归纳法求解 s_k 的通项式。为了简化非线性计算，在分析分数阶最速下降法的稳定性和收敛率时，我们将 s_k 视为连续函数 $s(t)$ 对时间 t 的离散采样，于是一阶差分近似为一阶微分，即 $D_t^1 s(t) \cong s_{k+1} - s_k$；另外，可推导 s^v 关于 $(s-s^{v*})$ 的幂级数展式为 $s^v = \sum_{n=0}^{\infty} \frac{\Gamma(1+v)(s^{v*})^{v-n}(s-s^{v*})^n}{\Gamma(n+1)\Gamma(1-n+v)}$。为了进一步简化非线性计算，我们仅取 v 乘以 $n=0$ 项与 $n=1$ 项之和来近似 s^v，即 $s^v \cong v(s^{v*})^{v-1}s$，于是式（14.36）演变为

$$D_t^1 s(t) \cong \frac{-2\mu\eta}{\Gamma(3-v)v(s^{v*})^{v-1}s(t)}[s(t)-s^{v*}]^2, \text{when } v \neq 1, 2 \text{ and } 3 \tag{14.37}$$

式（14.37）是可分离变量的一阶常微分方程，用分离变量法求解常微分方程可得 $s(t)$。然后 $s(t)$ 对时间 t 进行离散采样，即得 s_k 的通项式为

$$s_k \cong s^{v*} + \exp\left(\frac{-2\mu\eta k}{\Gamma(3-v)v(s^{v*})^{v-1}}\right), \text{when } v \neq 1, 2 \text{ and } 3 \tag{14.38}$$

由式（14.38）知，要使分数阶最速下降法的迭代搜索过程收敛，仅须使 $\lim_{k\to+\infty}\frac{2\mu\eta k}{\Gamma(3-v)v(s^{v*})^{v-1}}=+\infty$，即 $\frac{2\mu\eta}{\Gamma(3-v)v(s^{v*})^{v-1}}=\chi>0$，其中 χ 为正常数。于是常系数参数 μ 满足：

$$\mu \cong \frac{\chi\Gamma(3-v)v(s^{v*})^{v-1}}{2\eta}, \text{when } v \neq 1, 2 \text{ and } 3 \tag{14.39}$$

众所周知，整数阶最速下降法中 s_k 和 k 的关系是一个几何级数[469]。与之相对应，整数阶最速下降法的迭代搜索过程分为三种类型，即过阻尼振荡、临界阻尼振荡和欠阻尼振荡[469]。与整数阶最速下降法不同，由式（14.38）知，分数阶最速下降法中 s_k 和 k 的关系不是一个几何级数，而是一个近似的负幂级数关系。当 $\chi=1$ 时，$\mu \cong \frac{\Gamma(3-v)v(s^{v*})^{v-1}}{2\eta}$，when $v \neq 1$, 2 and 3，代入式（14.38）可得 $s_k \cong s^{v*}+\exp(-k)$，when $v \neq 1$, 2 and 3，分数阶最速下降法的迭代搜索过程按照 $\exp(-k)$ 的一次指数幂收敛；当 $\chi=2$ 时，$\mu \cong \frac{\Gamma(3-v)v(s^{v*})^{v-1}}{\eta}$，when $v \neq 1$，代入式（14.38）可得 $s_k \cong s^{v*}+\exp(-2k)$，when $v \neq 1$, 2 and 3，其迭代搜索过程按照 $\exp(-k)$ 的二次指数幂迅速收敛；当 χ 取整数且 $\chi \geq 3$ 时，其迭代搜索过程按照 $\exp(-k)$ 的高次整指数幂迅速收敛；当 $n-1<\chi<n$ 时，其迭代搜索过程按照 $\exp(-k)$ 的分数次指数幂的收敛。一般地，多维二次型性能表面上的分数阶最速下降法即将单变量的分数阶最速下降法推广到多维空间中。分数阶最速下降法的多维能量泛函和增量搜索过程分别为 $E=E_{\min}^1+\eta(s^{1*}-s)^2$ 和 $s_{k+1}=s_k+\mu(-D_{s_k}^v)$。此时的常系数参数 μ 为一多维向量，满足下式：

$$\mu \cong \left\{\frac{\chi\Gamma(3-v)v(s^{v*})^{v-1}}{2\eta}\right\}, \text{when } v \neq 1, 2 \text{ and } 3 \tag{14.40}$$

14.3.4 构造第一类基于分数阶偏微分方程的纹理图像多尺度去噪模型

虽然 Guidotti 和 Lambers[544] 以及 BAI 和 FENG[346] 分别对经典的各向异性扩散模型

第14章 基于分数阶偏微分方程的退化图像逆处理

进行了分数阶推广,将其能量范函中的梯度算子从一阶推广到分数阶,并在频域内对得到的分数阶偏微分方程进行数值实现,取得了一定的图像去噪效果。但该方法没有研究能量范函的分数次幂对图像纹理细节信息进行非线性保留的作用;该方法没有运用分数阶微积分的性质来推导相应的分数阶 Euler-Lagrange 方程,而是直接用希尔伯特伴随算子的复共轭转置的性质进行了替换,这样便极大地增加了频域内分数阶偏微分方程数值实现的复杂性;该方法在数值实现时仅仅简单地将频域内的一阶差分从形式上推广到分数阶,并以此代替频域内的分数阶微分算子,不仅没有解决频域内的分数阶微分算子的 Fourier 反变换是第一类欧拉积分在理论上难以计算的难题,而且在数值计算时引入了较大的误差。针对上述难题,我们构造了能量范函的分数次幂,并直接运用分数阶微积分的性质来推导相应的分数阶泛 Euler-Lagrange 方程,于是构造出第一类基于分数阶偏微分方程的纹理图像多尺度去噪模型,该模型在时域内就能进行较为简便的数值实现。

我们令 $s(x,y)$ 表示图像在像素 (x,y) 的灰度值,其中 $\Omega \subset R^2$ 为图像区域,$(x,y) \in \Omega$。令 $s(x,y)$ 表示被噪声污染的退化图像,$s_0(x,y)$ 表示理想的无噪声图像(denote the desired clean image)。由于当待处理的噪声为乘性噪声时,可以利用对数处理将其转换为加性噪声;当待处理的噪声为卷积噪声时,可以利用频域变换和对数处理将其转换为加性噪声。不失一般性,我们令 $n(x,y)$ 表示加性噪声(the additive noise),即

$$s(x,y) = s_0(x,y) + n(x,y) \tag{14.41}$$

与 Hausdorff 测度的分数维 δ-覆盖[594-595]类似,我们令图像 s 的分数阶超变差 (fractional hyper variation) 为 $|D^{v_1}s|^{v_2} = (\sqrt{(D_x^{v_1}s)^2 + (D_y^{v_1}s)^2})^{v_2}$,$v_2$ 是任意实数(包括分数)次方,$|D^{v_1}s|^{v_2}$ 是一个超立方体测度。如上所述,由于分数阶泛 Euler-Lagrange 方程式 (14.20) 和 (14.22) 与分数阶面积分的积分阶次无关,因此我们仍然采用一阶面积分来构造能量泛函。令基于分数阶超变差的能量泛函(分数阶超全变差)为

$$E_{FHTV}(s) = I_x^1 I_y^1 [f(|D^{v_1}s|^{v_2})] = \iint_\Omega f(|D^{v_1}s|^{v_2}) \mathrm{d}x \mathrm{d}y \tag{14.42}$$

式中,$\Omega \subset R^2 ((x,y) \in \Omega)$ 为 $s(x,y)$ 的图像区域。我们令 E_{FHTV} 的一阶极值曲面为 s,试验函数 $\xi(x,y) \in C_0^\infty(\Omega)$ 是与极值曲面接近的容许曲面,将 s 和 ξ 归并为含小参数 β 的曲面族 $s+\beta\xi$,当 $\beta=0$ 时,即为极值曲面。因此,该各向异性扩散(anisotropic diffusion)被解释为求一个分数阶能量泛函的一阶极小化的能量耗散过程。我们令 $\Psi_1(\beta) = E_{FHTV}(s+\beta\xi)$,$\Psi_2(\beta) = E_n(s+\beta\xi) = \iint_\Omega \frac{\lambda}{2}(s+\beta\xi-s_0)^2 \mathrm{d}x \mathrm{d}y$,其中 $\sigma^2 = \iint_\Omega (s-s_0)^2 \mathrm{d}x \mathrm{d}y$ 为图像噪声 $n(x,y)$ 的方差,$E_n(s) = \iint_\Omega \frac{\lambda}{2}(s-s_0)^2 \mathrm{d}x \mathrm{d}y = \frac{\lambda}{2}\sigma^2$ 为保真项,λ 为正则化参数。于是定义在曲面族 $s+\beta\xi$ 上基于分数阶超全变差的分数阶能量泛函为

$$\begin{aligned}\Psi(\beta) &= \Psi_1(\beta) + \Psi_2(\beta) = I_x^1 I_y^1 \left[f(|D^{v_1}s + \beta D^{v_1}\xi|^{v_2}) + \frac{\lambda}{2}(s+\beta\xi-s_0)^2 \right] \\ &= \iint_\Omega \left[f(|D^{v_1}s + \beta D^{v_1}\xi|^{v_2}) + \frac{\lambda}{2}(s+\beta\xi-s_0)^2 \right] \mathrm{d}x \mathrm{d}y\end{aligned} \tag{14.43}$$

对于 $\Psi_1(\beta)$ 而言,如果 $f(|D^{v_1}s + \beta D^{v_1}\xi|^{v_2})$ 对 β 的一阶导数存在,则当 $\beta=0$ 时,

$\Psi(\beta)$ 存在一阶极小值（一阶驻点）。令 $\rho=|D^{v_1}s+\beta D^{v_1}\xi|^{v_2}$，由一阶微积分算子的线性性质可以推导得

$$\begin{aligned}D_\beta^1\Psi_L(\beta)\big|_{\beta=0} &= D_\beta^1 I_x^1 I_y^1 [f(|D^{v_1}s+\beta D^{v_1}\xi|^{v_2})]\big|_{\beta=0} \\ &= I_x^1 I_y^1 D_\beta^1[f(|D^{v_1}s+\beta D^{v_1}\xi|^{v_2})]\big|_{\beta=0} \\ &= \iint_\Omega D_\rho^1 f\big|_{\beta=0} v_2 |D^{v_1}s|^{v_2-2} D^{v_1}s \cdot D^{v_1}\xi \mathrm{d}x\mathrm{d}y = 0 \end{aligned} \quad (14.44)$$

式中，符号 · 表示内积。根据式（14.22），我们可以推导与式（14.44）相对应的分数阶泛 Euler-Lagrange 方程为

$$\frac{\Gamma(1-v_1)}{\Gamma(-v_1)}[D_x^1(D_\rho^1 f\big|_{\beta=0} v_2 |D^{v_1}s|^{v_2-2} D_x^{v_1}s) + D_y^1(D_\rho^1 f\big|_{\beta=0} v_2 |D^{v_1}s|^{v_2-2} D_y^{v_1}s)] = 0$$
(14.45)

式（14.45）中，当 $f(\eta)=\eta$ 时，$D_\eta^1 f=1$，可以推导得

$$\frac{\Gamma(1-v_1)}{\Gamma(-v_1)} v_2 [D_x^1(|D^{v_1}s|^{v_2-2} D_x^{v_1}s) + D_y^1(|D^{v_1}s|^{v_2-2} D_y^{v_1}s)] = 0 \quad (14.46)$$

对于 $\Psi_2(\beta)$ 而言，有

$$\begin{aligned}D_\beta^1\Psi_2(\beta)\big|_{\beta=0} &= D_\beta^1 I_x^1 I_y^1 \frac{\lambda}{2}(s+\beta\xi-s_0)^2\big|_{\beta=0} \\ &= I_x^1 I_y^1 D_\beta^1\Big[\frac{\lambda}{2}(s+\beta\xi-s_0)^2\Big]\big|_{\beta=0} \\ &= \iint_\Omega \lambda(s-s_0)\xi \mathrm{d}x\mathrm{d}y = 0 \end{aligned} \quad (14.47)$$

在式（14.47）中，由于试验函数 ξ 具有随机性，由变分基本引理[485]，要使式（14.47）成立，只需下式成立：

$$\lambda(s-s_0) = 0 \quad (14.48)$$

由于式（14.4）和（14.47）分别是 $\Psi_1(\beta)$ 和 $\Psi_2(\beta)$ 对 β 求一阶极小值，所以我们采用传统的一阶最速下降法[469]来求解式（14.43）对于 β 的一阶极小值，可以推导得

$$\frac{\partial s}{\partial t} = -\frac{\Gamma(1-v_1)}{\Gamma(-v_1)} v_2 [D_x^1(|D^{v_1}s|^{v_2-2} D_x^{v_1}s) + D_y^1(|D^{v_1}s|^{v_2-2} D_y^{v_1}s)] - \lambda(s-s_0)$$
(14.49)

必须计算 $\lambda(t)$。若图像噪声 $n(x,y)$ 为白噪声，则 $\iint_\Omega n(x,y)\mathrm{d}x\mathrm{d}y = \iint_\Omega (s-s_0)\mathrm{d}x\mathrm{d}y = 0$。当 $\frac{\partial s}{\partial t}=0$ 时，式（14.49）收敛于稳定状态。于是在式（14.49）两边同时乘以 $(s-s_0)$ 并在图像区域 Ω 上积分，式（14.49）的左边便消失，可得

$$\lambda(t) = \frac{-\Gamma(1-v_1)}{\sigma^2\Gamma(-v_1)} v_2 \iint_\Omega [D_x^1(|D^{v_1}s|^{v_2-2} D_x^{v_1}s) + D_y^1(|D^{v_1}s|^{v_2-2} D_y^{v_1}s)](s-s_0)\mathrm{d}x\mathrm{d}y$$
(14.50)

我们将式（14.49）和（14.50）所表示的分数阶偏微分方程去噪模型简记为 $YiFeiPU-1$。由于图像分数阶微积分具有特殊性质（它既能尽量保留图像平滑区域中的低频轮廓特征，又能分数阶且多尺度增强图像中灰度值跃变幅度相对较大的高频边缘特

征，还能分数阶且多尺度增强图像中灰度值跃变幅度和频率变化相对不大的高频纹理细节特征)[186-188,434-435,468]，所以在分数阶非线性去噪的同时，尽量保持图像的纹理细节特征与完全滤除在信号的甚低频以及直流部分残留的微弱噪声便形成一个悖论。$YiFeiPU-1$ 虽然能使纹理图像在分数阶非线性去噪的同时尽量分数阶非线性地保留其复杂纹理细节特征，但却很难完全滤除在信号的甚低频以及直流部分残留的微弱噪声。为了使 $YiFeiPU-1$ 能够完全滤除在信号的甚低频和直流部分残留的微弱噪声，最直观的方法就是在纹理图像梯度变化十分缓和的区域减小其凸性。于是我们在数值迭代实现时，还需同时对信号的甚低频和直流部分进行低通滤波。另外，由上述分析可知，式（14.45）中的非线性加权系数 $D_\rho^l f|_{\beta=0}$ 其实并未起作用。在 $YiFeiPU-1$ 中，式（14.42）中的分数阶超全变差的分数阶微分阶次 v_1 和分数次幂的次方数 v_2 都对去噪起了非线性调节的作用。由式（14.42），（14.49）和（14.50）可知，当 $v_1 \neq 1$ 且 $v_2=1$ 或 $v_2=2$ 时，$YiFeiPU-1$ 模型便转化为分数阶各项异形图像处理[346,544]；当 $v_1=1$ 且 $v_2=1$ 时，$YiFeiPU-1$ 模型便转化为传统的 PM 模型和 ROF 模型[487,554]；当 $v_1=1$ 且 $v_2=2$ 或 $v_2=4$ 时，$YiFeiPU-1$ 模型便转化为高阶 PM 模型和 ROF 模型[551,582-588]。

14.3.5 构造第二类基于分数阶偏微分方程的纹理图像多尺度去噪模型

针对工程计算精度要求不高的应用场合，我们运用分数阶微分算子可以用一阶微分算子来线性表出这一数学性质，直接将一阶 Euler-Lagrange 方程在形式上自然推广到分数阶 Euler-Lagrange 方程。于是构造出一种 $YiFeiPU-1$ 模型的近似计算模型，即第二类基于分数阶偏微分方程的纹理图像多尺度去噪模型，该模型以牺牲计算精度为代价，提高数值计算速度。

在 $YiFeiPU-1$ 模型中，我们令 $v_2=1$ 且 $f(\eta)=\eta$，于是图像 s 的分数阶变差为 $|D^{v_1}s|=\sqrt{(D_x^{v_1}s)^2+(D_y^{v_1}s)^2}$，其分数阶全变差（fractional total variation）为 $E_{FTV}(s) = \iint_\Omega |D^{v_1}s| \mathrm{d}x\mathrm{d}y$。按照 Tikhonov 正则化方法[596]，我们令基于分数阶变差的能量泛函为

$$\Psi(s) = \iint_\Omega F(x,y,s,D_x^{v_1}s,D_y^{v_1}s)\mathrm{d}x\mathrm{d}y = \frac{\lambda}{2}(s-s_0)_{L^2(\Omega)}^2 + E_{FTV}(s)$$
$$= \iint_\Omega \left(\frac{\lambda}{2}(s-s_0)^2 + [(D_x^{v_1}s)^2+(D_y^{v_1}s)^2]^{\frac{1}{2}}\right)\mathrm{d}x\mathrm{d}y \tag{14.51}$$

式中，$\sigma^2 = \iint_\Omega (s-s_0)^2 \mathrm{d}x\mathrm{d}y$ 为图像噪声 $n(x,y)$ 的方差，$\frac{\lambda}{2}(s-s_0)_{L^2(\Omega)}^2 = \iint_\Omega \left(\frac{\lambda}{2}\right)(s-s_0)^2 \mathrm{d}x\mathrm{d}y = \frac{\lambda}{2}\sigma^2$ 为保真项，λ 为正则化参数。

由于信号的分数阶微积分是其整数阶微积分的连续内插，分数阶微分算子在数学上可以用一阶微分算子来线性表出[187,230,431]，于是可以证明 $D_{x-a}^{v_1}f = \sum_{n=0}^{\infty} \frac{(-1)^n (x-a)^{n-v_1} D_{x-a}^n f}{\Gamma(-v_1)(n-v_1)n!}$，进而可将 D^{v_1} 视为 D^1 的函数，即 $D^{v_1}s=\Psi(D^1s)$；由于 D^{v_1} 和 D^1 在本质上都是线性算子，故函数 Ψ 存在反函数，即 $D^1s=\Psi^{-1}(D^{v_1}s)$。于是，我

们令

$$F(x,y,s,D_x^{v_1}s,D_y^{v_1}s) = F(x,y,s,\Psi(D_x^1 s),\Psi(D_y^1 s)) = \frac{\lambda}{2}(s-s_0)^2 + |D^{v_1}s|$$

$$= \frac{\lambda}{2}(s-s_0)^2 + |\Psi(D^1 s)| \tag{14.52}$$

对一阶偏微分 $p_x = D_x^1 s = \Psi^{-1}(D_x^{v_1}s)$ 和 $p_y = D_y^1 s = \Psi^{-1}(D_y^{v_1}s)$ 而言，使式（14.52）一阶极小值（一阶驻点）存在的一阶 Euler-Lagrange 方程为 $\frac{\partial F}{\partial s} - \frac{\partial}{\partial x}\left(\frac{\partial F}{\partial p_x}\right) - \frac{\partial}{\partial y}\left(\frac{\partial F}{\partial p_y}\right) = 0$。由于式（14.52）中函数 Ψ 是 D^{v_1} 的阶次 v_1 的函数，其函数形式相当复杂，$\frac{\partial F}{\partial p_x}$ 和 $\frac{\partial F}{\partial p_y}$ 的计算比较困难。为了简化计算，我们放宽条件，直接将一阶 Euler-Lagrange 方程作形式上的分数阶自然推广，从而得到针对 D^{v_1} 的近似分数阶 Euler-Lagrange 方程。虽然这种形式上的分数阶自然推广在数学上不严格相等，但是在实际应用中，这是一个很有效而简便的近似方法。令 $q_x = D_x^{v_1}s = \Psi(D_x^1 s)$，$q_y = D_y^{v_1}s = \Psi(D_y^1 s)$，有 $\frac{\partial F}{\partial s} = \lambda(s-s_0)$，$\frac{\partial F}{\partial q_x} = \frac{D_x^{v_1}s}{|D^{v_1}s|}$ 和 $\frac{\partial F}{\partial q_y} = \frac{D_y^{v_1}s}{|D^{v_1}s|}$ 成立，于是与式（14.52）的一阶极小值（一阶驻点）相对应的近似一阶 Euler-Lagrange 方程为

$$\frac{\partial F}{\partial s} - \frac{\partial}{\partial x}\left(\frac{\partial F}{\partial q_x}\right) - \frac{\partial}{\partial y}\left(\frac{\partial F}{\partial q_y}\right) = \lambda(s-s_0) - \left[D_x^1\left(\frac{D_x^{v_1}s}{|D^{v_1}s|}\right) + D_y^1\left(\frac{D_y^{v_1}s}{|D^{v_1}s|}\right)\right] = 0 \tag{14.53}$$

同理，在式（14.24）和（14.36）中，我们取 $v=1$，用一阶最速下降法来求解式（14.53），可以推导得

$$\frac{\partial s}{\partial t} = \left[D_x^1\left(\frac{D_x^{v_1}s}{|D^{v_1}s|}\right) + D_y^1\left(\frac{D_y^{v_1}s}{|D^{v_1}s|}\right)\right] - \lambda(s-s_0) \tag{14.54}$$

我们必须计算 $\lambda(t)$。若图像噪声 $n(x,y)$ 为白噪声（white noise），则 $\iint_\Omega n(x,y)\mathrm{d}x\mathrm{d}y = \iint_\Omega (s-s_0)\mathrm{d}x\mathrm{d}y = 0$。当 $\frac{\partial s}{\partial t} = 0$ 时，式（14.54）收敛于稳定状态。于是在式（14.54）两边同时乘以（$s-s_0$）并在图像区域 Ω 上积分，式（14.54）的左边便消失，可得

$$\lambda(t) = \frac{1}{\sigma^2}\iint_\Omega \left[D_x^1\left(\frac{D_x^{v_1}s}{|D^{v_1}s|}\right) + D_y^1\left(\frac{D_y^{v_1}s}{|D^{v_1}s|}\right)\right](s-s_0)\mathrm{d}x\mathrm{d}y \tag{14.55}$$

我们将式（14.54）和（14.55）所表示的分数阶偏微分方程去噪模型简记为 $YiFeiPU-2$。另外，与 $YiFeiPU-1$ 的分析同理，为了使 $YiFeiPU-2$ 能够完全滤除在信号的甚低频和直流部分残留的微弱噪声，我们在数值迭代实现时，还需同时对信号的甚低频和直流部分进行低通滤波。

14.3.6 构造第三类基于分数阶偏微分方程的纹理图像多尺度去噪模型

由上述推导可知，除了推导 $YiFeiPU-1$ 模型时提及的缺点外，在 Guidotti 和 Lambers[544]以及 BAI 和 FENG[346]提出的 fractional-order anisotropic diffusion for image denoising 模型中还存在如下缺点：一方面，该方法仅仅将能量范函中的梯度算子简单地从一阶推广到分数阶，并没有从本质上解决能量范函在各向异性扩散的过程中如何非线性地保留图像纹理细节的难题，图像去噪后对纹理细节信息的保持并不理想；另一方面，该方法没有研究能量范函的分数阶极值对图像纹理细节信息进行非线性保留的作用。针对上述难题，我们首先利用图像分数阶微积分的特殊性质，不仅将能量范函中的梯度算子简单地从一阶推广到分数阶，而且采用分数次幂和分数阶极值来构造基于分数阶超全变差的能量泛函；其次，对该能量范函求分数阶极值，而非传统的一阶极值，直接运用分数阶微积分的性质来推导相应的分数阶 Euler-Lagrange 方程，极大地简化了分数阶偏微分方程数值实现的复杂性；最后，运用分数阶最速下降法构造出第三类基于分数阶偏微分方程的纹理图像多尺度去噪模型。

如果直接采用分数阶极值来构造能量泛函，我们可令含有小参数 β 的曲面族 $s+(\beta-1)\xi$，当 $\beta=1$ 时为 v_3 阶极值曲面 s。我们令 $\Psi_1(\beta)=E_{FHTV}[s+(\beta-1)\xi]$，$\Psi_2(\beta)=\iint_\Omega \lambda[s+(\beta-1)\xi-s_0]s_0 \mathrm{d}x\mathrm{d}y$，其中 $\Psi_2(\beta)$ 是噪声 $[s+(\beta-1)\xi-s_0]$ 和无噪信号 s_0 的交叉能量，是对 $[s+(\beta-1)\xi-s_0]$ 和 s_0 相似性的度量。求解 $\Psi_2(\beta)$ 的极小值的过程即是求解噪声和无噪信号的相似性最小的过程。$\Psi_2(\beta)$ 在去噪过程中起非线性保真的作用，λ 为正则化参数。于是定义在曲面族 $s+(\beta-1)\xi$ 上基于分数阶超全变差的分数阶能量泛函为

$$\Psi(\beta)=\Psi_1(\beta)+\Psi_2(\beta)=I_x^1 I_y^1 \{f(|D^{v_1}s+(\beta-1)D^{v_1}\xi|^{v_2})+\lambda[s+(\beta-1)\xi-s_0]s_0\}$$
$$=\iint_\Omega \{f(|D^{v_1}s+(\beta-1)D^{v_1}\xi|^{v_2})+\lambda[s+(\beta-1)\xi-s_0]s_0\}\mathrm{d}x\mathrm{d}y$$

(14.56)

对于 $\Psi_1(\beta)$ 而言，如果 $f(|D^{v_1}s+(\beta-1)D^{v_1}\xi|^{v_2})$ 对 β 的 v_3 阶分数阶导数存在，则当 $\beta=1$ 时，$\Psi_1(\beta)$ 存在 v_3 阶分数阶极小值（分数阶驻点）。由分数阶微分算子的线性性质可得

$$D_\beta^{v_3}\Psi_1(\beta)|_{\beta=1}=\frac{\partial^{v_3}}{\partial \beta^{v_3}}\iint_\Omega f(|\vec{\varphi}|^{v_2})\mathrm{d}x\mathrm{d}y|_{\beta=1}$$
$$=\iint_\Omega \frac{\partial^{v_3}}{\partial \beta^{v_3}}f(|\vec{\varphi}|^{v_2})\mathrm{d}x\mathrm{d}y|_{\beta=1}=0 \quad (14.57)$$

式中，对于向量 $\vec{\varphi}=D^{v_1}s+(\beta-1)D^{v_1}\xi$ 而言，$\vec{\varphi}^2=|\vec{\varphi}|^2=[\sqrt{(\vec{\varphi})^2}]^2=\vec{\varphi}\cdot\vec{\varphi}$，符号 · 表示内积。与基于一阶极值的传统一阶变分法不同，式（14.57）是 $\Psi_1(\beta)$ 关于 β 的 v_3 阶分数阶极值，其目的在于利用图像分数阶微积分的特殊性质（既能尽量保留图像平滑区域中的低频轮廓特征，又能分数阶且多尺度增强图像中灰度值跃变幅度相对较大的高频边缘特征，还能分数阶且多尺度增强图像中灰度值跃变幅度和频率变化相对不大的高频纹理细节特征）[186-188,434-435,468]，使纹理图像在去噪的同时尽量分数阶非线性地保留其复杂纹理细节特征。

只要 v 为分数，可使 $n>v$ 时组合数 $\binom{v}{n}=\frac{(-1)^{-n}\Gamma(n-v)}{\Gamma(-v)\Gamma(n+1)}\neq 0$，则由式（14.13）和 Faà de Bruno 公式 [179]，我们可推得复合函数的分数阶微分法则为

$$D_{\beta-\alpha}^v f[g(\beta)]=\frac{(\beta-\alpha)^{-v}}{\Gamma(1-v)}f+\sum_{n=1}^\infty \binom{v}{n}\frac{(\beta-\alpha)^{n-v}}{\Gamma(n-v+1)}n!\sum_{n=1}^\infty D_g^m f \sum \prod_{k=1}^n \frac{1}{P_k!}\left[\frac{D_{\beta-\alpha}^k g}{k!}\right]^{P_k}$$

(14.58)

式中，$g(\beta) = |\vec{\varphi}|^{v_2}$，$a$ 为任意常数。式（14.58）中的 $n=0$ 这一项已从求和符号中分离出来了。由式（14.58）可知，复合函数的分数阶导数是一个无穷项求和。其中，P_k 满足如下条件：

$$\begin{cases} \sum_{k=1}^{\infty} k P_k = n \\ \sum_{k=1}^{n} P_k = m \end{cases} \tag{14.59}$$

式（14.58）中的第三个求和符号 $\sum$ 表示对满足式（14.59）的 $P_k|_{m=1 \to n}$ 所有组合所对应的 $\left\{ \prod_{k=1}^{n} \frac{1}{P_k!} \left[\frac{D_{\beta-a}^k g}{k!} \right]^{P_k} \right\}|_{m=1 \to n}$ 进行求和。于是由式（14.57），（14.58）以及 Gamma 函数的性质可以推导得

$$\iint_{\Omega} \frac{\partial^{v_3}}{\partial \beta^{v_3}} f(|\vec{\varphi}|^{v_2}) \mathrm{d}x\mathrm{d}y |_{\beta=1} = \iint_{\Omega} \frac{\beta^{-v_3} f(|\vec{\varphi}|^{v_2})}{\Gamma(1-v_3)} + \sum_{n=1}^{\infty} \binom{v_3}{n} \frac{\beta^{n-v_3}}{\Gamma(n-v_3+1)} n! \sum_{m=1}^{n} D^m_{|\vec{\varphi}|^{v_2}} f$$

$$\sum \prod_{k=1}^{n} \frac{1}{P_k!} \left[\frac{D_{\beta}^k (|\vec{\varphi}|^{v_2})}{k!} \right]^{P_k} \mathrm{d}x\mathrm{d}y |_{\beta=1}$$

$$= \iint_{\Omega} \frac{f(|D^{v_1}s|^{v_2})}{\Gamma(1-v_3)} + \sum_{n=1}^{\infty} \frac{(-1)^n \Gamma(n-v_3)}{\Gamma(-v_3)\Gamma(n-v_3+1)}$$

$$\sum_{m=1}^{n} D^m_{|\vec{\varphi}|^{v_2}} f|_{\beta=1} \sum \prod_{k=1}^{n} \frac{1}{P_k!} \left[\frac{D_{\beta}^k (|\vec{\varphi}|^{v_2})|_{\beta=1}}{k!} \right] \mathrm{d}x\mathrm{d}y$$

$$= 0 \tag{14.60}$$

不失一般性，为了简化计算，我们令 $f(\eta) = \eta$，故 $D^1_\eta f(\eta) = 1$，$D^m_\eta f(\eta) \overset{m \geq 2}{=} 0$，于是式（14.60）简化为

$$\iint_{\Omega} \frac{|D^{v_1}s|^{v_2}}{\Gamma(1-v_3)} + \sum_{n=1}^{\infty} \frac{(-1)^n \Gamma(n-v_3)}{\Gamma(-v_3)\Gamma(n-v_3+1)} \left\{ \prod_{k=1}^{n} \frac{1}{P_k!} \left[\frac{D_{\beta}^k (|\vec{\varphi}|^{v_2})|_{\beta=1}}{k!} \right]^{P_k} \right\}_{m=1} \mathrm{d}x\mathrm{d}y = 0 \tag{14.61}$$

由（14.59）可知，当 $m=1$ 时，式（14.61）中 P_k 必须满足 $P_n = 1$ 且 $P_1 = P_2 = \cdots = P_{n-1} = 0$，于是式（14.61）进一步简化为

$$\iint_{\Omega} \frac{|D^{v_1}s|^{v_2}}{\Gamma(1-v_3)} + \sum_{n=1}^{\infty} \frac{(-1)^n \Gamma(n-v_3)}{\Gamma(-v_3)\Gamma(n-v_3+1)} \frac{D_{\beta}^n (|\vec{\varphi}|^{v_2})|_{\beta=1}}{n!} \mathrm{d}x\mathrm{d}y = 0 \tag{14.62}$$

当 n 取奇数（$n = 2k+1$，$k = 0, 1, 2, \cdots$）与当 n 取偶数（$n = 2k$，$k = 1, 2, 3, \cdots$）时，$D_{\beta}^n (|\vec{\varphi}|^{v_2})|_{\beta=1}$ 具有不同的表达式。于是可以分别推导得

$$D_{\beta}^n (|\vec{\varphi}|^{v_2})|_{\beta=1}^{n=2k+1} = \prod_{\tau=1}^{n} (v_2 - \tau + 1) |D^{v_1}s|^{v_2-n-1} |D^{v_1}\xi|^{n-1} (D^{v_1}\xi) \cdot (D^{v_1}s) \tag{14.63}$$

$$D_{\beta}^n (|\vec{\varphi}|^{v_2})|_{\beta=1}^{n=2k+1} = \prod_{\tau=1}^{n} (v_2 - \tau + 1) |D^{v_1}s|^{v_2-n-1} |D^{v_1}\xi|^{n} \tag{14.64}$$

将式（14.63）和（14.64）代入式（14.62），可以推得

$$\iint_{\Omega} \sum_{k=0}^{\infty} \frac{\prod_{\tau=1}^{2k} (v_2 - \tau + 1) |D^{v_1}s|^{v_2-2k-2} |D^{v_1}\xi|^{2k}}{\Gamma(-v_3)(2k)!} (D^{v_1}s) \cdot D^{v_1}$$

$$\left[\frac{\Gamma(2k-v_3)}{\Gamma(2k-v_3+1)}s - \frac{(v_2-2k)\Gamma(2k-v_3+1)}{(2k+1)\Gamma(2k-v_3+2)}\xi\right]\mathrm{d}x\mathrm{d}y = 0 \qquad (14.65)$$

我们令 $\prod_{\tau=1}^{n}(v_2-\tau+1)\overset{n=0}{=}1$。对于任意二维数量函数 ξ（测试函数）而言，相应的 $D^{v_1}\xi$ 具有随机性，于是 $D^{v_1}\left[\frac{\Gamma(2k-v_3)}{\Gamma(2k-v_3+1)}s - \frac{(v_2-2k)\Gamma(2k-v_3+1)}{(2k+1)\Gamma(2k-v_3+2)}\xi\right]$ 亦具有随机性。于是根据式（14.22），我们可以推导与式（14.65）相对应的分数阶泛 Euler-Lagrange 方程为

$$\frac{\Gamma(1-v_1)}{\Gamma(-v_1)\Gamma(-v_3)}\sum_{k=0}^{\infty}\frac{\prod_{\tau=1}^{2k}(v_2-\tau+1)}{(2k!)}\left[D_x^1(\mid D^{v_1}s\mid^{v_2-2k-2}D_x^{v_1}s) + D_y^1(\mid D^{v_1}s\mid^{v_2-2k-2}D_y^{v_1}s)\right] = 0$$
$$(14.66)$$

式中，$\prod_{\tau=1}^{n}(v_2-\tau+1)\overset{n=0}{=}1$。式（14.66）是与式（14.57）相对应的分数阶泛 Euler-Lagrange 方程。

对于 $\Psi_2(\beta)$ 而言，有

$$D_\beta^{v_3}\Psi_2(\beta)\big|_{\beta=1} = D_\beta^{v_3}I_x^1 I_y^1 \lambda[s+(\beta-1)\xi-s_0]s_0\big|_{\beta=1}$$
$$= \iint_\Omega \frac{\lambda s_0}{\Gamma(1-v_3)\Gamma(2-v_3)}[\Gamma(2-v_3)(s-s_0-\xi)+\Gamma(1-v_3)\xi]\mathrm{d}x\mathrm{d}y$$
$$= 0 \qquad (14.67)$$

在式（14.67）中，由于试验函数 ξ 具有随机性，所以 $\Gamma(2-v_3)(s-s_0-\xi)+\Gamma(1-v_3)\xi$ 亦具有随机性，由变分基本引理[485]，要使式（14.67）成立，只需下式成立：

$$\frac{\lambda s_0}{\Gamma(1-v_3)\Gamma(2-v_3)} = 0 \qquad (14.68)$$

由于式（14.57）和（14.67）分别是 $\Psi_1(\beta)$ 和 $\Psi_2(\beta)$ 对 β 求 v_3 阶极小值，所以在式（14.24）和（14.36）中，我们取 $v=v_3\neq 1,2,3$，用 v_3 阶分数阶最速下降法来求解式（14.56）的分数阶极小值可得

$$\frac{\partial^{v_3}s}{\partial t^{v_3}} = \frac{-\Gamma(1-v_1)}{\Gamma(-v_1)\Gamma(-v_3)}\sum_{k=0}^{\infty}\frac{\prod_{\tau=1}^{2k}(v_2-\tau+1)}{(2k)!}\left[D_x^1(\mid D^{v_1}s\mid^{v_2-2k-2}D_x^{v_1}s)\right.$$
$$\left.+D_y^1(\mid D^{v_1}s\mid^{v_2-2k-2}D_y^{v_1}s)\right] - \frac{\lambda s_0}{\Gamma(1-v_3)\Gamma(2-v_3)} \qquad (14.69)$$

式中，$\prod_{\tau=1}^{n}(v_2-\tau+1)\overset{n=0}{=}1$。$\frac{\partial^{v_3}s}{\partial t^{v_3}}$ 按照式（14.36）所示的分数阶差分方法实现，当 v_3 满足式（14.31）时，仅收敛到唯一 v_3 阶分数阶极值点。我们必须计算 $\lambda(t)$。若图像噪声 $n(x,y)$ 为白噪声（white noise），则 $\iint_\Omega n(x,y)\mathrm{d}x\mathrm{d}y = \iint_\Omega (s-s_0)\mathrm{d}x\mathrm{d}y = 0$。当 $\frac{\partial^{v_3}s}{\partial t^{v_3}}$ 时，式（14.69）收敛于稳定状态。于是在式（14.69）两边同时乘以 $(s-s_0)^2$，并在图像区域 Ω 上积分，式（14.69）的左边便消失，可得

$$\lambda(t) = \frac{-\Gamma(1-v_1)\Gamma(1-v_3)\Gamma(2-v_3)}{\sigma^2 \Gamma(-v_1)\Gamma(-v_3) s_0} \iint_\Omega \sum_{k=0}^{\infty} \frac{\prod_{\tau=1}^{2k}(v_2-\tau+1)}{(2k)!}$$
$$[D_x^1(|D^{v_1}s|^{v_2-2k-2}D_x^{v_1}s) + D_y^1(|D^{v_1}s||D^{v_2-2k-2}D_x^{v_1}s)](s-s_0)^2 \,\mathrm{d}x\,\mathrm{d}y$$
(14.70)

我们将式（14.69）和（14.70）所表示的分数阶偏微分方程去噪模型简记为 $YiFeiPU$-3。与 $YiFeiPU$-1 的分析同理，为了使 $YiFeiPU$-3 能够完全滤除在信号的甚低频和直流部分残留的微弱噪声，我们在数值迭代实现时，还需同时对信号的甚低频和直流部分进行低通滤波。另外，对比式（14.69），（14.70）与式（14.49），（14.50）可知，与 $YiFeiPU$-1 模型相比，$YiFeiPU$-3 模型不仅通过连乘函数 $\prod_{\tau=1}^{2k}(v_2-\tau+1)$ 以及 $|D^{v_1}s|$ 的幂次方 v_2-2k-2 的形式较大增强了分数阶微分阶次 v_2 对去噪的非线性调节作用，而且通过在分母中增加 $\Gamma(-v_3)$ 的形式增加了分数阶微分阶次 v_3 对去噪的非线性调节作用。另外，由式（14.69）知，当 $v_3=0$ 时，$YiFeiPU$-3 模型即为传统的位势方程（椭圆型方程）；当 $v_3=1$ 时，$YiFeiPU$-3 模型即为传统的热传导方程（抛物型方程）；当 $v_3=2$ 时，$YiFeiPU$-3 模型即为传统的波动方程（双曲型方程）；当 $0<v_3<1$ 时，$YiFeiPU$-3 模型即为传统的位势方程和传统的热传导方程之间的连续内插；当 $1<v_3<2$ 时，$YiFeiPU$-3 模型即为传统的热传导方程和波动方程之间的连续内插。可见，在数学和物理意义上，$YiFeiPU$-3 模型将传统的基于偏微分方程的图像处理从基于传统的热传导方程的各向异性扩散的普遍基本处理方法推广到了更广阔的领域。

14.3.7 构造第四类基于分数阶偏微分方程的纹理图像多尺度去噪模型

由上述推导可知，$YiFeiPU$-3 模型在式（14.56）中直接采用分数阶极值来构造基于分数阶超全变差的能量泛函，相当于 $YiFeiPU$-3 模型首先将参考坐标从一阶极值平移到分数阶极值处，然后构造相应的能量范函。这其实与我们在前面推导分数阶最速下降法时仍采用一阶极值来构造能量泛函不同。针对这一问题，我们根据分数阶最速下降法，仍采用如式（14.23）所示的一阶极值来构造基于分数阶超全变差的能量泛函，与 $YiFeiPU$-3 模型的推导同理，可构造出第四类基于分数阶偏微分方程的纹理图像多尺度去噪模型。

第一，我们改造式（14.56）中的 $\Psi_1(\beta)$。我们仍采用式（14.23）所示的一阶极值来构造能量泛函，根据分数阶最速下降法，由式（14.33），我们令 $\Psi_1(\beta) = I_x^1 I_y^1 [f(\|\vec{\phi}\|^{v_2})] = \iint_\Omega f(\|\vec{\phi}\|^{v_2})\,\mathrm{d}x\,\mathrm{d}y$。其中，当 v_3 满足式（14.26），（14.31）和（14.32）时，我们令向量 $\vec{\phi}[D^{v_1}s, (\beta-1)D^{v_1}\xi] = (\beta-1)D^{v_1}\xi - \frac{2\Gamma(2-v_3)}{\Gamma(3-v_3)}D^{v_1}s$，并根据式（14.33）特别定义的向量的 $\vec{\phi}$ 范数为 $\|\vec{\phi}\| = \sqrt{\frac{4\Gamma(2-v_3)-\Gamma(3-v_3)}{\Gamma(3-v_3)}(D^{v_1}s)^2 - \frac{4(\beta-1)\Gamma(2-v_3)}{\Gamma(3-v_3)}D^{v_1}s \cdot D^{v_1}\xi + (\beta-1)^2(D^{v_1}\xi)^2}$ = $\sqrt{(\vec{\phi})^2 + C}$，其中 $C = \frac{4\Gamma(2-v_3)\Gamma(3-v_3)-\Gamma^2(3-v_3)-4\Gamma^2(2-v_3)}{\Gamma^2(3-v_3)}(D^{v_1}s)^2$，符号·表

示内积。对于向量$\vec{\phi}$，$D^{v_1}s$ 和 $D^{v_1}\xi$ 而言，$(\vec{\phi})^2=|\vec{\phi}|^2=(\sqrt{(\vec{\phi})^2})^2=\vec{\phi}\cdot\vec{\phi}$，$(D^{v_1}s)^2=|D^{v_1}s|^2=(\sqrt{(D^{v_1}s)^2})^2=D^{v_1}s\cdot D^{v_1}s$，且 $(D^{v_1}\xi)^2=|D^{v_1}\xi|^2=(\sqrt{(D^{v_1}\xi)^2})^2=D^{v_1}\xi\cdot D^{v_1}\xi$。

第二，我们保持式（14.56）中的 $\Psi_2(\beta)=\iint_\Omega\lambda[s+(\beta-1)\xi-s_0]s_0\mathrm{d}x\mathrm{d}y$ 不变。于是定义在曲面族 $s+(\beta-1)\xi$ 上基于分数阶超全变差的分数阶能量泛函为

$$\Psi(\beta)=\Psi_1(\beta)+\Psi_2(\beta)=I_x^1I_y^1\{f(\|\vec{\phi}\|^{v_2})+\lambda[s+(\beta-1)\xi-s_0](s-s_0)\}$$

$$=\iint_\Omega\{f(\|\vec{\phi}\|^{v_2})+\lambda[s+(\beta-1)\xi-s_0](s-s_0)\}\mathrm{d}x\mathrm{d}y \tag{14.71}$$

于是，我们可将式（14.57）改写为

$$D_\beta^{v_3}\Psi_1(\beta)\big|_{\beta=1}=\frac{\partial^{v_3}}{\partial\beta^{v_3}}\iint_\Omega f(\|\vec{\phi}\|^{v_2})\mathrm{d}x\mathrm{d}y\bigg|_{\beta=1}=\iint_\Omega\frac{\partial^{v_3}}{\partial\beta^{v_3}}f(\|\vec{\phi}\|^{v_2})\mathrm{d}x\mathrm{d}y\bigg|_{\beta=1}=0$$
(14.72)

于是，与式（14.62）的推导同理可以推得

$$\iint_\Omega\frac{\left[\sqrt{\left|\frac{4\Gamma(2-v_3)-\Gamma(3-v_3)}{\Gamma(3-v_3)}\right|}\right]^{v_2}|D^{v_1}s|^{v_2}}{\Gamma(1-v_3)}$$

$$+\sum_{n=1}^\infty\frac{(-1)^n\Gamma(n-v_3)}{\Gamma(-v_3)\Gamma(n-v_3+1)}\frac{D_\beta^n(\|\vec{\phi}\|^{v_2})|_{\beta=1}}{n!}\mathrm{d}x\mathrm{d}y=0 \tag{14.73}$$

为了保证 $\frac{4\Gamma(2-v_3)-\Gamma(3-v_3)}{\Gamma(3-v_3)}$ 始终为正数，以使 $\sqrt{\frac{4\Gamma(2-v_3)-\Gamma(3-v_3)}{\Gamma(3-v_3)}}$ 有意义，我们取其绝对值 $\left|\frac{4\Gamma(2-v_3)-\Gamma(3-v_3)}{\Gamma(3-v_3)}\right|$。当 n 取奇数（$n=2k+1$，$k=0,1,2,\cdots$）与当 n 取偶数（$n=2k$，$k=1,2,3,\cdots$）时，$D_\beta^n(\|\vec{\phi}\|^{v_2})|_{\beta=1}$ 具有不同的表达式，于是可以推导得

$$D_\beta^n(\|\vec{\phi}\|^{v_2})\big|_{\beta=1}^{n=2k+1}=\frac{-2\Gamma(2-v_3)\prod_{\tau=1}^n(v_2-\tau+1)}{\Gamma(3-v_3)}\left[\sqrt{\left|\frac{4\Gamma(2-v_3)-\Gamma(3-v_3)}{\Gamma(3-v_3)}\right|}\right]^{v_2-n-1}$$

$$|D^{v_1}s|^{v_2-n-1}|D^{v_1}\xi|^{n-1}(D^{v_1}\xi)\cdot(D^{v_1}s) \tag{14.74}$$

$$D_\beta^n(\|\vec{\phi}\|^{v_2})\big|_{\beta=1}^{n=2k}=\prod_{\tau=1}^n(v_2-\tau+1)\left[\sqrt{\left|\frac{4\Gamma(2-v_3)-\Gamma(3-v_3)}{\Gamma(3-v_3)}\right|}\right]^{v_2-n}|D^{v_1}s|^{v_2-n}|D^{v_1}\xi|^n$$
(14.75)

将式（14.74）和（14.75）代入式（14.73），可以推得

$$\iint_\Omega\sum_{k=0}^\infty\frac{\prod_{\tau=1}^{2k}(v_2-\tau+1)\left[\sqrt{\left|\frac{4\Gamma(2-v_3)-\Gamma(3-v_3)}{\Gamma(3-v_3)}\right|}|D^{v_1}s|\right]^{v_2-2k-2}|D^{v_1}\xi|^{2k}}{\Gamma(-v_3)\Gamma(3-v_3)(2k)!}(D^{v_1}s)\cdot$$

$$D^{v_1}\left[\frac{\Gamma(2k-v_3)[4\Gamma(2-v_3)-\Gamma(3-v_3)]}{\Gamma(2k-v_3+1)}s+\frac{2(v_2-2k)\Gamma(2k-v_3+1)\Gamma(2-v_3)}{(2k+1)\Gamma(2k-v_3+2)}\xi\right]\mathrm{d}x\mathrm{d}y=0$$
(14.76)

我们令 $\prod_{\tau=1}^{n}(v_2-\tau+1)\overset{n=0}{=}1$。对于任意二维数量函数 ξ（测试函数）而言，相应的 $D^{v_1}\xi$ 具有随机性，于是 $D^{v_1}\left[\dfrac{\Gamma(2k-v_3)[4\Gamma(2-v_3)-\Gamma(3-v_3)]}{\Gamma(2k-v_3+1)}s+\dfrac{2(v_2-2k)\Gamma(2k-v_3+1)\Gamma(2-v_3)}{(2k+1)\Gamma(2k-v_3+2)}\xi\right]$ 亦具有随机性。根据式（14.22），我们可以推导与式（14.76）相对应的分数阶泛 Euler–Lagrange 方程为

$$\dfrac{\Gamma(1-v_1)}{\Gamma(-v_1)\Gamma(-v_3)\Gamma(3-v_3)}\sum_{k=0}^{\infty}\dfrac{\prod_{\tau=1}^{2k}(v_2-\tau+1)\left[\sqrt{\left|\dfrac{4\Gamma(2-v_3)-\Gamma(3-v_3)}{\Gamma(3-v_3)}\right|}\right]^{v_2-2k-2}}{(2k)!}$$
$$[D_x^1(|D^{v_1}s|^{v_2-2k-2}D_x^{v_1}s)+D_y^1(|D^{v_1}s|^{v_2-2k-2}D_y^{v_1}s)]=0 \quad (14.77)$$

式中，$\prod_{\tau=1}^{2k}(v_2-\tau+1)\overset{n=0}{=}1$。式（14.77）是与式（14.72）相对应的分数阶泛 Euler–Lagrange 方程。同理，由于式（14.72）和（14.67）分别是 $\Psi_1(\beta)$ 和 $\Psi_2(\beta)$ 对 β 求 v_3 阶极小值，所以在式（14.24）和（14.36）中，我们取 $v=v_3\neq 1,2,3$，用 v_3 阶分数阶最速下降法来求解式（14.71）的分数阶极小值，可得

$$\dfrac{\partial^{v_3}s}{\partial t^{v_3}}=\dfrac{-\Gamma(1-v_1)}{\Gamma(-v_1)\Gamma(-v_3)\Gamma(3-v_3)}\sum_{k=0}^{\infty}\dfrac{\prod_{\tau=1}^{2k}(v_2-\tau+1)\left[\sqrt{\left|\dfrac{4\Gamma(2-v_3)-\Gamma(3-v_3)}{\Gamma(3-v_3)}\right|}\right]^{v_2-2k-2}}{(2k)!}$$
$$[D_x^1(|D^{v_1}s|^{v_2-2k-2}D_x^{v_1}s)+D_y^1(|D^{v_1}s|^{v_2-2k-2}D_y^{v_1}s)]\dfrac{\lambda s_0}{\Gamma(1-v_3)\Gamma(2-v_3)}$$
$$(14.78)$$

式中，$\prod_{\tau=1}^{2k}(v_2-\tau+1)\overset{n=0}{=}1$。$\dfrac{\partial^{v_3}s}{\partial t^{v_3}}$ 按照式（14.36）所示的分数阶差分方法实现，当 v_3 满足式（14.31）时，仅收敛到唯一 v_3 阶分数阶极值点。与式（14.70）的推导同理，我们可以推得

$$\lambda(t)=\dfrac{-\Gamma(1-v_1)\Gamma(1-v_3)\Gamma(2-v_3)}{\sigma^2\Gamma(-v_1)\Gamma(-v_3)\Gamma(3-v_3)s_0}\iint_{\Omega}\sum_{k=0}^{\infty}$$
$$\dfrac{\prod_{\tau=1}^{2k}(v_2-\tau+1)\left[\sqrt{\left|\dfrac{4\Gamma(2-v_3)-\Gamma(3-v_3)}{\Gamma(3-v_3)}\right|}\right]^{v_2-2k-2}}{(2k)!}$$
$$[D_x^1(|D^{v_1}s|^{v_2-2k-2}D_x^{v_1}s)+D_y^1(|D^{v_1}s|^{v_2-2k-2}D_y^{v_1}s)](s-s_0)^2\mathrm{d}x\mathrm{d}y \quad (14.79)$$

我们将式（14.78）和（14.79）所表示的分数阶偏微分方程去噪模型简记为 YiFeiPU-4。与 YiFeiPU-1 的分析同理，为了使 YiFeiPU-4 能够完全滤除在信号的甚低频和直流部分残留的微弱噪声，我们在数值迭代实现时，还需同时对信号的甚低频和直流部分进行低通滤波。另外，对比式（14.78），（14.79）与式（14.69），（14.70）可知，与 YiFeiPU-3 模型相比，YiFeiPU-4 模型通过在分子中增加 $\sqrt{\dfrac{4\Gamma(2-v_3)-\Gamma(3-v_3)}{\Gamma(3-v_3)}}$ 以及在分母中增加 $\Gamma(3-v_3)$ 的形式，增强了分数阶微分阶次 v_3 对去噪的非线性调节作用。另外，由式（14.78）知，当 $v_3=0$ 时，YiFeiPU-4 模型即为传统的位势方程（椭圆型方程）；当 $v_3=1$ 时，YiFeiPU-4 模型即为传统的热传导方程（抛物型方程）；当 $v_3=2$ 时，

$YiFeiPU$-4 模型即为传统的波动方程（双曲型方程）；当 $0<v_3<1$ 时，$YiFeiPU$-4 模型即为传统的位势方程和传统的热传导方程之间的连续内插；当 $1<v_3<2$ 时，$YiFeiPU$-4 模型即为传统的热传导方程和波动方程之间的连续内插。可见，在数学和物理意义上，同 $YiFeiPU$-3 模型一样，$YiFeiPU$-4 模型也将传统的基于偏微分方程的图像处理从基于传统的热传导方程的各向异性扩散的普遍基本处理方法推广到了更广阔的领域。

14.4 实验仿真及其理论分析

14.4.1 分数阶最速下降法的稳定性与收敛率分析

我们知道，信号分数阶微分具有如下特性[157,185]：第一，常数的分数阶微分值不为零，常数的整数阶微分值必为零。在信号幅值不变或变化不大的平滑区域内，其分数阶微分值是从对应奇异跳变处的极大值渐趋于零。而信号幅值平坦段的任意整数阶微分必为零。这是分数阶微分相对于整数阶微分的显著区别之一。第二，在信号幅值阶梯或斜坡的起始点处的分数阶微分值非零，起到了加强高频奇异信息的作用。随着分数阶微分阶次的增加，其对高频奇异信号的加强幅度逐渐增大。当 $0<v<1$ 时，分数阶微分对高频奇异信息的增强幅度比 $v=1$ 阶微分的增强幅度小。信号的整数阶微分只是分数阶微分的特例。第三，沿信号幅值斜坡的分数阶微分值非零，亦非常数，是非线性曲线，而沿信号幅值斜坡的整数阶微分值为常数。因此在数字图像处理之中，分数阶微分能够非线性增强图像的复杂纹理细节特征。图像分数阶微分既能尽量非线性保留图像平滑区域中的低频轮廓信息，又能非线性增强图像中灰度值跃变幅度相对较大的高频边缘信息，还能非线性增强图像中灰度值跃变幅度和频率变化相对不大的高频纹理细节信息[185−188,434−435,468]。此部分内容请参见第 12 章的实验部分（图 12.2～图 12.6）。

14.4.2 基于分数阶偏微分方程的纹理图像多尺度去噪模型的数值实现

如式（14.49），（14.50），（14.54），（14.55），（14.69），（14.70），（14.78）和（14.79）所示，上述四类基于分数阶偏微分方程的纹理图像多尺度去噪模型均需要数值实现二维数字图像的分数阶微分算子。对于分数阶微积分的 Grümwald−Letnikov 定义式（14.1），当 N 足够大时，可以去掉极限符号。为了提高收敛速度和收敛精度，我们在 Grümwald−Letnikov 定义式中引入信号 $s(x)$ 在非节点处的信号值，即 $\left.\dfrac{\mathrm{d}^v}{\mathrm{d}x^v}s(x)\right|_{G-L} \cong \dfrac{x^{-v}N^v}{\Gamma(-v)}\sum_{k=0}^{N-1}\dfrac{\Gamma(k-v)}{\Gamma(k+1)}s\left(x+\dfrac{vx}{2N}-\dfrac{kx}{N}\right)$。于是当 $v\neq 1$ 时，应用拉格朗日三点插值公式对信号 $s(x)$ 进行分数插值，可分别构造出在 8 个对称方向上的数字图像的 $YiFeiPU$-2 分数阶微分算子[187,486]，如图 14.4 所示。

		⋯	0	C_{s_n}	0	⋯		
		⋯	0	$C_{s_{n-1}}$	0	⋯		
		⋯	0	$C_{s_{n-2}}$	0	⋯		
			⋮	⋮	⋮			
		⋯	0	C_{s_k}	0	⋯		
			⋮	⋮	⋮			
		⋯	0	C_{s_1}	0	⋯		
		⋯	0	C_{s_0}	0	⋯		
		⋯	0	$C_{s_{s-1}}$	0	⋯		

(a)

⋮	⋮	⋮		⋮		⋮	⋮	⋮
0	0	0	⋯	0	⋯	0	0	0
C_{s_n}	$C_{s_{n-1}}$	$C_{s_{n-2}}$	⋯	C_{s_k}	⋯	C_{s_1}	C_{s_0}	$C_{s_{-1}}$
0	0	0	⋯	0	⋯	0	0	0
⋮	⋮	⋮		⋮		⋮	⋮	⋮

(b)

		⋯	0	C_{s_n}	0	⋯		
		⋯	0	$C_{s_{n-1}}$	0	⋯		
		⋯	0	C_{s_0}	0	⋯		
			⋮	⋮	⋮			
		⋯	0	C_{s_k}	0	⋯		
			⋮	⋮	⋮			
		⋯	0	$C_{s_{n-2}}$	0	⋯		
		⋯	0	$C_{s_{n-1}}$	0	⋯		
		⋯	0	C_{s_n}	0	⋯		

(c)

⋮	⋮	⋮		⋮		⋮	⋮	⋮
0	0	0	⋯	0	⋯	0	0	0
$C_{s_{-1}}$	C_{s_0}	C_{s_1}	⋯	C_{s_k}	⋯	$C_{s_{n-2}}$	$C_{s_{n-1}}$	C_{s_n}
0	0	0	⋯	0	⋯	0	0	0
⋮	⋮	⋮		⋮		⋮	⋮	⋮

(d)

						⋯	0	$C_{s_{-1}}$
					⋯	0	C_{s_0}	0
				⋯	0	C_{s_1}	0	⋮
				⋰	⋰	0	⋮	
		⋯	0	C_{s_k}	0	⋯		
	⋮	0	⋰	⋰				
⋮	0	$C_{s_{n-2}}$	0	⋯				
0	$C_{s_{n-1}}$	0	⋯					
C_{s_n}	0	⋯						

(e)

						⋯	0	C_{s_n}
					⋯	0	$C_{s_{n-1}}$	0
				⋯	0	$C_{s_{n-2}}$	0	⋮
				⋰	⋰	0	⋮	
		⋯	0	C_{s_k}	0	⋯		
	⋮	0	⋰	⋰				
⋮	0	C_{s_1}	0	⋯				
0	C_{s_0}	0	⋯					
$C_{s_{-1}}$	0	⋯						

(f)

(g)

$$\begin{pmatrix} C_{s_n} & 0 & \cdots & & & & & & \\ 0 & C_{s_{n-1}} & 0 & \cdots & & & & & \\ \vdots & 0 & C_{s_{n-2}} & 0 & \cdots & & & & \\ & \vdots & 0 & \ddots & \ddots & & & & \\ & & \vdots & & 0 & C_{s_k} & 0 & \cdots & \\ & & & \vdots & & \ddots & \ddots & 0 & \vdots \\ & & & & \cdots & 0 & C_{s_1} & 0 & \vdots \\ & & & & \cdots & & 0 & C_{s_0} & 0 \\ & & & & & & \cdots & 0 & C_{s_{-1}} \end{pmatrix}$$

(h)

$$\begin{pmatrix} C_{s_{-1}} & 0 & \cdots & & & & & & \\ 0 & C_{s_0} & 0 & \cdots & & & & & \\ \vdots & 0 & C_{s_1} & 0 & \cdots & & & & \\ & \vdots & 0 & \ddots & \ddots & & & & \\ & & \vdots & & 0 & C_{s_k} & 0 & \cdots & \\ & & & \vdots & & \ddots & \ddots & 0 & \vdots \\ & & & & \cdots & 0 & C_{s_{n-2}} & 0 & \vdots \\ & & & & \cdots & & 0 & C_{s_{n-1}} & 0 \\ & & & & & & \cdots & 0 & C_{s_n} \end{pmatrix}$$

图 14.4 8 个方向上的分数阶微分掩模 D^v。(a) x 轴负方向上的分数阶微分算子（用 D_{x-}^v 表示），(b) y 轴负方向上的分数阶微分算子（用 D_{y-}^v 表示），(c) x 轴正方向上的分数阶微分算子（用 D_{x+}^v 表示），(d) y 轴正方向上的分数阶微分算子（用 D_{y+}^v 表示），(e) 左下对角线方向上的分数阶微分算子（用 D_{ldd}^v 表示），(f) 右上对角线方向上的分数阶微分算子（用 D_{rud}^v 表示），(g) 左上对角线方向上的分数阶微分算子（用 D_{lud}^v 表示），(h) 右下对角线方向上的分数阶微分算子（用 D_{rdd}^v 表示）

在图 14.4 中，$C_{s_{-1}}$ 是覆盖在非因果像素点上 $s_{-1} = s(x + x/N)$ 的算子系数值。C_{s_0} 是覆盖在感兴趣点（pixel of interest）$s_0 = s(x)$ 上的算子系数值。n 一般取奇数。当 $k \to n = 2m - 1$ 时，我们构造出 $(2m+1) \times (2m+1)$ 的分数微分算子，其算子系数分别为[187,468]：

$$C_{s_{-1}} = \frac{v}{4} + \frac{v^2}{8}, \quad C_{s_0} = 1 - \frac{v^2}{2} - \frac{v^3}{8}, \quad C_{s_1} = -\frac{5v}{4} + \frac{5v^3}{16} + \frac{v^4}{16}, \cdots, C_{s_k} = \frac{1}{\Gamma(-v)}$$

$$\left[\frac{\Gamma(k-v+1)}{(k+1)!} \cdot \left(\frac{v}{4} + \frac{v^2}{8} \right) + \frac{\Gamma(k-v)}{k!} \cdot \left(1 - \frac{v^2}{4} \right) + \frac{\Gamma(k-v-1)}{(k-1)!} \cdot \left(-\frac{v}{4} + \frac{v^2}{8} \right) \right], \cdots, C_{s_{n-2}}$$

$$= \frac{1}{\Gamma(-v)} \left[\frac{\Gamma(n-v-1)}{(n-1)!} \cdot \left(\frac{v}{4} + \frac{v^2}{8} \right) + \frac{\Gamma(n-v-2)}{(n-2)!} \cdot \left(1 - \frac{v^2}{4} \right) + \frac{\Gamma(n-v-3)}{(n-3)!} \cdot \left(-\frac{v}{4} + \frac{v^2}{8} \right) \right],$$

$$C_{s_{n-1}} = \frac{\Gamma(n-v-1)}{(n-1)! \Gamma(-v)} \cdot \left(1 - \frac{v^2}{4} \right) + \frac{\Gamma(n-v-2)}{(n-2)! \Gamma(-v)} \cdot \left(-\frac{v}{4} + \frac{v^2}{8} \right), \quad C_{s_n} = \frac{\Gamma(n-v-1)}{(n-1)! \Gamma(-v)} \cdot$$

$\left(-\frac{v}{4} + \frac{v^2}{8} \right)$。对于数字灰度图像而言，分数阶微分算子的数值运算规则采用算子卷积的空域滤波方案。我们选取上述 8 个方向上模值最大的分数阶偏微分值作为该像素点的分数阶微分值。对于数字图像的分数阶偏微分方程中的 $v \neq 1$ 阶分数阶微分算子 $D^v = i_1 D_{x-}^v + i_2 D_{y-}^v + i_3 D_{x+}^v + i_4 D_{y+}^v + i_5 D_{ldd}^v + i_6 D_{rud}^v + i_7 D_{lud}^v + i_8 D_{rdd}^v$ 而言，其中 D_{x-}^v，D_{y-}^v，D_{x+}^v，D_{y+}^v，D_{ldd}^v，D_{rud}^v，D_{lud}^v，D_{rdd}^v 可根据图 14.9 及其相关系数进行数值计算[187,486]；当 $v = 1$ 时，为了保持数值计算的稳定性，我们选取了一种特殊的差分形式来近似一阶微分：$D_x^1 s(x,y) = \frac{2[s(x+1,y) - s(x-1,y)] + s(x+1,y+1) - s(x-1,y+1) + s(x+1,y-1) - s(x-1,y-1)}{4}$，$D_y^1 s(x,y) = \frac{2[s(x,y+1) - s(x,y-1)] + s(x+1,y+1) - s(x+1,y-1) + s(x-1,y+1) - s(x-1,y-1)}{4}$。

若时间等分间隔为 Δt，n 时刻为 $t_n = n \Delta t$，$n = 0, 1, \cdots$（$t_0 = 0$ 表示初始时刻），则

式 (14.41) 中 n 时刻的数字图像为 $s_{x,y}^n = s(x,y,t_n)$, $s_{x,y}^0 = s_0(x,y,t_0) + n(x,y,t_0)$, $s_{x,y}^0$ 为待去噪的原始图像，s_0 为理想的无噪声图像，它是一个恒定值，故 $s_0(x,y,t_0) = s_0(x,y,t_n)$。于是，我们可用一阶差分来近似一阶微分，即

$$\frac{\partial s}{\partial t} \cong \frac{s_{x,y}^{n+1} - s_{x,y}^n}{\Delta t} \quad (14.80)$$

同理，利用式 (14.1) 和 (14.36)，我们可用分数阶最速下降法来近似实现关于时间的分数阶微分，即

$$\frac{\partial^v s}{\partial t^v} = \Delta t^{-v} \left[s_{x,y}^{n+1} - s_{x,y}^n + \frac{2\mu\eta}{\Gamma(3-v)} (s_{x,y}^n - s_{x,y}^{v*})^2 (s_{x,y}^n)^{-v} \right], \text{when } v \neq 1 \quad (14.81)$$

另外，由于理想的无噪声图像 $s_0(x,y,t_0)$ 事先不知道，但是每次数值迭代的去噪中间结果 $s_{x,y}^n$ 都是对理想的无噪声图像 $s_0(x,y,t_0)$ 一次逼近，即 $s_{x,y}^n \to s_0(x,y,t_0) = s_0(x,y,t_n)$。故为了在数值迭代时尽量逼近 $s - s_0$，我们令 $(s - s_0)_{x,y} \cong s_{x,y}^0 - s_{x,y}^n$。为了提高算法的计算精度和抗旋转能力，在实际计算时我们需要同时考虑上述 8 个方向上的分数阶微分值。于是由式 (14.80)，可得式 (14.49) 和 (14.50) 的数值实现方程分别为

$$s_{x,y}^{n+1} = -\frac{\Gamma(1-v_1)}{\Gamma(-v_1)} v_2 \left[D_x^1(|D^{v_1} s_{x,y}^n|^{v_2-2} D_x^{v_1} s_{x,y}^n) + D_y^1(|D^{v_1} s_{x,y}^n|^{v_2-2} D_y^{v_1} s_{x,y}^n) \right] \Delta t$$
$$- \lambda^n \Delta t s_{x,y}^0 + (1 + \lambda^n \Delta t) s_{x,y}^n \quad (14.82)$$

$$\lambda^n = -\frac{\Gamma(1-v_1)}{\sigma^{n^2} \Gamma(-v_1)} v_2 \sum_{x,y} \left[D_x^1(|D^{v_1} s_{x,y}^n|^{v_2-2} D_x^{v_1} s_{x,y}^n) + D_y^1(|D^{v_1} s_{x,y}^n|^{v_2-2} D_y^{v_1} s_{x,y}^n) \right]$$
$$(s_{x,y}^0 - s_{x,y}^n) \quad (14.83)$$

式中，$\sigma^{n^2} = \sum_{x,y} (s_{x,y}^0 - s_{x,y}^n)^2$。在数值迭代计算时，一方面，我们不需要预先获知或估计噪声的方差，而只需要令第一次数值迭代时的 σ^{1^2} 为一个较小的正数。在下述实验中我们均令 $\sigma^{1^2} = 0.01$。将 σ^{1^2} 代入式 (14.83) 以启动数值迭代计算的过程，于是每一次迭代所得的 σ^{n^2} 都不一样，但每一 σ^{n^2} 都是对噪声真正方差的一次逼近。另一方面，在数值迭代计算的过程中，可能出现 $|D^{v_1} s_{x,y}^n| = 0$ 的情况，为不使式 (14.82) 和 (14.83) 无意义，此时只需要令 $|D^{v_1} s_{x,y}^n|$ 为一个很小的正数。在下述实验中，当出现 $|D^{v_1} s_{x,y}^n| = 0$ 的情况时，我们均令 $|D^{v_1} s_{x,y}^n| = 10^{-6}$。同理，可得式 (14.54) 和 (14.55) 的数值实现方程分别为

$$s_{x,y}^{n+1} = \left[D_x^1 \left(\frac{D_x^{v_1} s_{x,y}^n}{|D^{v_1} s_{x,y}^n|} \right) + D_y^1 \left(\frac{D_y^{v_1} s_{x,y}^n}{|D^{v_1} s_{x,y}^n|} \right) \right] \Delta t - \lambda^n \Delta t s_{x,y}^0 + (1 + \lambda^n \Delta t) s_{x,y}^n$$
$$(14.84)$$

$$\lambda^n = \frac{1}{\sigma^{n^2}} \sum_{x,y} \left[D_x^1 \left(\frac{D_x^{v_1} s_{x,y}^n}{|D^{v_1} s_{x,y}^n|} \right) + D_y^1 \left(\frac{D_y^{v_1} s_{x,y}^n}{|D^{v_1} s_{x,y}^n|} \right) \right] (s_{x,y}^0 - s_{x,y}^n) \quad (14.85)$$

在数值迭代计算时，对于式 (14.81) 而言，由于分数阶最速下降法的最佳图像 $s_{x,y}^{v*}$ 事先不知道，但是每次数值迭代的去噪中间结果 $s_{x,y}^n$ 都是对 $s_{x,y}^{v*}$ 的一次逼近，即 $s_{x,y}^n \to s_{x,y}^{v*}$。故为了在数值迭代时尽量逼近 $(s_{x,y}^n - s_{x,y}^{v*})^2$，我们令 $(s_{x,y}^n - s_{x,y}^{v*})^2 \cong (s_{x,y}^{n-1} - s_{x,y}^n)^2$。对于分数阶超全变差式 (14.56) 而言，式 (14.81) 中 $\eta = 1$。在下述数值计算的实验中，对于式 (14.69) 和 (14.70) 中的 k 而言，我们均只取 $k = 0, 1$ 进行近似计算。于是由式 (14.81)，可得式 (14.69) 和 (14.70) 的数值实现方程分别为

$$s_{x,y}^{n+1} = \mathrm{P}(s_{x,y}^n)\Delta t^{v_3} - \frac{\lambda^n \Delta t^{v_3}}{\Gamma(1-v_3)\Gamma(2-v_3)}s_{x,y}^n + s_{x,y}^n - \frac{2\mu}{\Gamma(3-v_3)}(s_{x,y}^{n-1}-s_{x,y}^n)^2(s_{x,y}^n)^{-v_3},$$
$$\text{when } v_3 \neq 1,2 \text{ and } 3 \quad (14.86)$$

$$\lambda^n = \frac{\Gamma(1-v_3)\Gamma(2-v_3)}{\sigma^{n^2} s_{x,y}^n}\sum_{x,y}\mathrm{P}(s_{x,y}^n)(s_{x,y}^0 - s_{x,y}^n)^2 \quad (14.87)$$

式中，$\prod_{\tau=1}^{n}(v_2-\tau+1)\overset{n=0}{=}1$，$\mathrm{P}(s_{x,y}^n) = \frac{-\Gamma(1-v_1)}{\Gamma(-v_1)\Gamma(-v_3)}\sum_{k=0}^{1}\frac{\prod_{\tau=1}^{2k}(v_2-\tau+1)}{(2k)!}$ $[D_x^1(|D^{v_1}s_{x,y}^n|^{v_2-2k-2}D_x^{v_1}s_{x,y}^n) + D_y^1(|D^{v_1}s_{x,y}^n|^{v_2-2k-2}D_y^{v_1}s_{x,y}^n)]$。同理，可得式（14.78）和（14.79）的数值实现方程分别为

$$s_{x,y}^{n+1} = \mathrm{Q}(s_{x,y}^n)\Delta t^{v_3} - \frac{\lambda^n \Delta t^{v_3}}{\Gamma(1-v_3)\Gamma(2-v_3)}s_{x,y}^n + s_{x,y}^n - \frac{2\mu}{\Gamma(3-v_3)}(s_{x,y}^{n-1}-s_{x,y}^n)^2(s_{x,y}^n)^{-v_3},$$
$$\text{when } v_3 \neq 1,2 \text{ and } 3 \quad (14.88)$$

$$\lambda^n = \frac{\Gamma(1-v_3)\Gamma(2-v_3)}{\sigma^{n^2} s_{x,y}^n}\sum_{x,y}\mathrm{Q}(s_{x,y}^n)(s_{x,y}^0 - s_{x,y}^n)^2 \quad (14.89)$$

式中，$\prod_{\tau=1}^{n}(v_2-\tau+1)\overset{n=0}{=}1$，$\mathrm{Q}(s_{x,y}^n) = \frac{-\Gamma(1-v_1)}{\Gamma(-v_1)\Gamma(-v_3)\Gamma(3-v_3)}\sum_{k=0}^{1}$ $\frac{\prod_{\tau=1}^{2k}(v_2-\tau+1)\left[\sqrt{\left|\frac{4\Gamma(2-v_3)-\Gamma(3-v_3)}{\Gamma(3-v_3)}\right|}\right]^{v_2-2k-2}}{(2k)!}[D_x^1(|D^{v_1}s_{x,y}^n|^{v_2-2k-2}D_x^{v_1}s_{x,y}^n) +$ $D_y^1(|D^{v_1}s_{x,y}^n|^{v_2-2k-2}D_y^{v_1}s_{x,y}^n)]$。在下述数值迭代计算时，为了保证分数阶最速下降法收敛，我们均取 μ 为一个较小的数，$\mu = 0.005$。

另外，在下述数值迭代计算时，第一，式（14.69）和（14.78）中的 k 不用取 $0 \to \infty$，仅取 $k=0,1$ 进行近似计算即可。第二，为了使 $|D^{v_1}s_{x,y}^n|^{v_2-2k-2}$ 有意义，当 $|D^{v_1}s_{x,y}^n| \leqslant 0.0689$ 时，我们均取 $|D^{v_1}s_{x,y}^n| = 0.0689$。第三，为了使 $(s_{x,y}^n)^{-v_3}$ 有意义，当 $s_{x,y}^n = 0$ 时，我们均取 $s_{x,y}^n = 0.00001$。第四，为了保证分数阶最速下降法收敛，我们均取 μ 为一个较小的数，$\mu = 0.005$。第五，如前所述，为了使 YiFeiPU-1，YiFeiPU-2，YiFeiPU-3 和 YiFeiPU-4 能够完全滤除在信号的甚低频和直流部分残留的微弱噪声，最直观的方法就是在纹理图像梯度变化十分缓和的区域减小其凸性。于是我们在数值迭代实现时，还需同时对信号的甚低频和直流部分进行低通滤波。在式（14.82）～式（14.89）中的具体做法是：对于一维信号而言，当信号受噪声干扰不严重时，为了保证去噪的质量，当 $|D^{v_1}s_x^n| < \alpha_A$ 且 $\alpha_{NA} < \alpha_A$ 时，我们令 $s_x^{n+1} = \frac{s_{x-1}^{n+1}+2s_x^{n+1}+s_{x+1}^{n+1}}{4}$，否则 $s_x^{n+1} = s_x^{n+1}$。其中，$\alpha_A = \frac{1}{N_x}\sum_x^{N_x}|D^{v_1}s_x^n|$，$\alpha_{NA} = \frac{|D^{v_1}s_{x-1}^n|+2|D^{v_1}s_x^n|+|D^{v_1}s_{x+1}^n|}{4}$。当信号受噪声干扰十分严重时，为了加快去噪的速度，我们直接令 $s_x^{n+1} = \frac{s_{x-1}^{n+1}+2s_x^{n+1}+s_{x+1}^{n+1}}{4}$；对于二维图像信号而言，我们令 $s_{x,y}^{n+1} = \frac{{}^x s_{x,y}^{n+1}+{}^y s_{x,y}^{n+1}+{}^r s_{x,y}^{n+1}+{}^l s_{x,y}^{n+1}}{4}$。其中，当 $\alpha_{NA}^x = \min(\alpha_{NA}^x, \alpha_{NA}^y, \alpha_{NA}^r, \alpha_{NA}^l)$ 时，${}^x s_{x,y}^{n+1} = \frac{s_{x-1,y}^{n+1}+2s_{x,y}^{n+1}+s_{x+1,y}^{n+1}}{4}$，${}^y s_{x,y}^{n+1} = {}^r s_{x,y}^{n+1} = {}^l s_{x,y}^{n+1} = s_{x,y}^{n+1}$。当 $\alpha_{NA}^y = \min(\alpha_{NA}^x, \alpha_{NA}^y, \alpha_{NA}^r,$

α_{NA}^l)时,$^y s_{x,y}^{n+1} = \dfrac{s_{x,y-1}^{n+1}+2s_{x,y}^{n+1}+s_{x,y+1}^{n+1}}{4}$, $^x s_{x,y}^{n+1}=^r s_{x,y}^{n+1}=^l s_{x,y}^{n+1}=s_{x,y}^{n+1}$。当 $\alpha_{NA}^r = \min$ (α_{NA}^x, α_{NA}^y, α_{NA}^r, α_{NA}^l)时,$^r s_{x,y}^{n+1}=\dfrac{s_{x-1,y+1}^{n+1}+2s_{x,y}^{n+1}+s^{n+1}x+1,y-1}{4}$, $^x s_{x,y}^{n+1}=^y s_{x,y}^{n+1}=^l s_{x,y}^{n+1}=s_{x,y}^{n+1}$。当 $\alpha_{NA}^l = \min$ (α_{NA}^x, α_{NA}^y, α_{NA}^r, α_{NA}^l)时,$^l s_{x,y}^{n+1}=\dfrac{s_{x-1,y-1}^{n+1}+2s_{x,y}^{n+1}+s_{x+1,y+1}^{n+1}}{4}$, $^x s_{x,y}^{n+1}=^y s_{x,y}^{n+1}=^r s_{x,y}^{n+1}=s_{x,y}^{n+1}$。否则 $^x s_{x,y}^{n+1}=^y s_{x,y}^{n+1}=^r s_{x,y}^{n+1}=^l s_{x,y+4}^{n+1}=s_{x,y}^{n+1}$。其中,$\alpha_{NA}^x = \dfrac{|D^{v_1}s_{x-1,y}^n|+2|D^{v_1}s_{x,y}^n|+|D^{v_1}s_{x+1,y}^n|}{4}$, $\alpha_{NA}^y = \dfrac{|D^{v_1}s_{x,y-1}^n|+2|D^{v_1}s_{x,y}^n|+|D^{v_1}s_{x,y+1}^n|}{4}$, $\alpha_{NA}^r = \dfrac{|D^{v_1}s_{x-1,y+1}^n|+2|D^{v_1}s_{x,y}^n|+|D^{v_1}s_{x+1,y-1}^n|}{4}$, $\alpha_{NA}^l = \dfrac{|D^{v_1}s_{x-1,y-1}^n|+2|D^{v_1}s_{x,y}^n|+|D^{v_1}s_{x+1,y+1}^n|}{4}$, x 表示 x 轴方向, y 表示 y 轴方向, r 表示右对角线方向, l 表示左对角线方向。第六,为了处理图像边界像素,我们需要对图像向外进行一个像素的扩展。第七,由于在式(14.82)~式(14.89)中加入了低通滤波,为了消除方程在数值迭代求解过程中有可能出现不稳定的发散点,对于一维信号而言,当 $|s_x^{n+1}|>6|s_x^n|$ 时,我们令 $s_x^{n+1}=s_x^n$。对于二维图像信号而言,当 $|s_{x,y}^{n+1}|>6|s_{x,y}^n|$ 时,我们令 $s_{x,y}^{n+1}=s_{x,y}^n$。

14.4.3 基于分数阶偏微分方程的纹理图像多尺度去噪模型的分数阶非线性多尺度去噪能力分析

为了更为直观地分析和说明基于分数阶偏微分方程的纹理图像多尺度去噪模型在去除图像噪声的同时,既能尽量保留纹理图像平滑区域中的低频轮廓,又能非线性保留灰度值跃变幅度相对较大的高频边缘和纹理细节,还能非线性保留灰度值跃变幅度变化相对不大的纹理细节的分数阶非线性多尺度去噪能力,我们首先用由矩形波信号、正弦波信号和锯齿波信号所合成的一维信号进行对比实验。在本实验中,我们在基于分数阶偏微分方程的纹理图像多尺度去噪模型中所选用的分数阶阶次 v_1 和 v_3 以及分数阶幂次 v_2 并不是最佳值,而只是根据经验确定的一组相对较好的值。v_1, v_2 和 v_3 的最佳取值将在本章最后部分进行分析和讨论。另外,在本实验中,数值实迭代过程均以峰值信噪比 PSNR(Peak Signal-to-Noise Ratio)达到最高值时作为迭代的终止条件。本实验如图 14.5 所示。

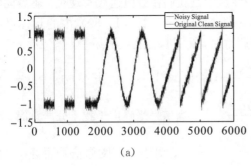

(a)

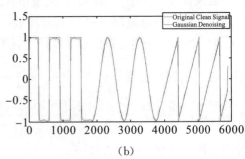

(b)

第 14 章 基于分数阶偏微分方程的退化图像逆处理

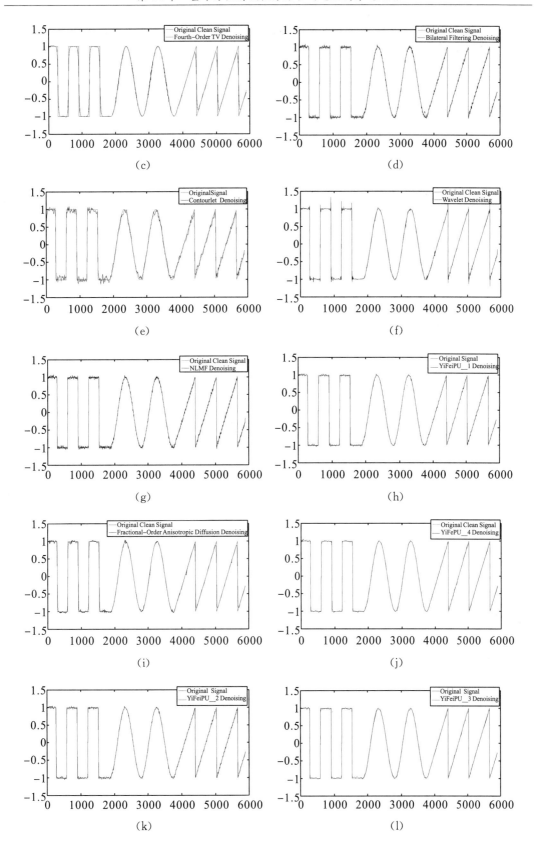

图 14.5 对由矩形波信号、正弦波信号和锯齿波信号合成的一维信号进行去噪处理。(a) Original Clean Signal and Noisy Signal (Adds White Gaussian Noise to the Original Clean Signal, PSNR (Peak Signal-to-Noise Ratio) =25.2486), (b) Gaussian Denoising, (c) Fourth-Order TV Denoising [582] — [584], (d) Bilateral Filtering Denoising [597] — [598], [601], (e) Contourlet Denoising [456], [599], (f) Wavelet Denoising [600] — [601], (g) NLMF (Non-Local Means noise Filtering) Denoising [602] — [603], (h) Fractional-Order Anisotropic Diffusion Denoising [346], (i) $YiFeiPU$-1 Denoising ($v_1=1.75$, $v_2=1.95$, $\Delta t=0.0296$), (j) $YiFeiPU$-2 Denoising ($v_1=1.75$, $\Delta t=0.0296$), (k) $YiFeiPU$-3 Denoising ($v_1=1.025$, $v_2=2.25$, $v_3=1.05$, $\Delta t=0.0296$), (l) $YiFeiPU$-4 Denoising ($v_1=1.025$, $v_2=2.25$, $v_3=1.05$, $\Delta t=0.0296$)

从主观视觉效果看，分析图 14.5 可知，第一，Gaussian Denoising 与 Fourth-Order TV Denoising 的去噪效果相对较差，对信号的高频奇异成分有较大的扩散和平滑作用。在图 14.5 (b) (c) 中，矩形波信号和锯齿波信号的高频边缘突变被明显地平滑，其高频奇异成分的能量在邻域内被明显地扩散。第二，Fractional-Order Anisotropic Diffusion Denoising，$YiFeiPU$-1 Denoising 和 $YiFeiPU$-2 Denoising 的去噪能力适中，虽然保留信号高频奇异成分的能力明显比 Gaussian Denoisin, Fourth-Order TV Denoising 和 Contourlet Denoising 强，但却比 Bilateral Filtering Denoising, Wavelet Denoising 和 NLMF Denoising 弱，而且滤除噪声不干净。在图 14.5 (h) ~ (j) 中，矩形波信号和锯齿波信号的高频边缘突变被较弱地平滑，高频奇异成分的能量在邻域内被较弱地扩散。另外，denoised signal 中残留微小毛刺。第三，Bilateral Filtering Denoising, Contourlet Denoising, Wavelet Denoising 和 NLMF Denoising 的去噪能力较好，能够较好地保留信号的高频奇异成分，但滤除噪声仍不干净。在图 14.5 (d) ~ (g) 中，虽然矩形波信号和锯齿波信号的高频边缘突变被保持得较好，但是 denoised signal 中残留着较多小的毛刺。第四，$YiFeiPU$-3 Denoising 和 $YiFeiPU$-4 Denoising 的去噪能力相对最好，不仅能较好地保留信号的高频奇异成分，而且对噪声滤除也比较干净。在图 14.5 (k) ~ (l) 中，不仅矩形波信号和锯齿波信号的高频边缘突变被保持得较好，而且 denoised signal 中基本没有小的毛刺残留。

从客观定量分析看，我们不仅采用 PSNR，而且采用 the correlation coefficients between noisy signal or denoised signal and original clean signal 来综合度量图 14.5 中的上述各种算法的去噪效果，见表 14.1。

表 14.1 对由矩形波信号、正弦波信号和锯齿波信号合成的一维信号的综合去噪效果

Denoising Algorithm \ Denoising Effect	PSNR	Correlation Coefficients
Noisy Signal	25.2486	0.9951
Gaussian Denoising	26.9611	0.9953
Fourth-Order TV Denoising	28.2589	0.9959
Bilateral Filtering Denoising	35.0247	0.9990
Contourlet Denoising	28.2945	0.9959
Wavelet Denoising	32.1565	0.9975

续表

Denoising Algorithm \ Denoising Effect	PSNR	Correlation Coefficients
NLMF Denoising	33.3088	0.9976
Fractional-Order Anisotropic Diffusion Denoising	29.8692	0.9975
$YiFeiPU$-1 Denoising	29.6963	0.9978
$YiFeiPU$-2 Denoising	29.2516	0.9977
$YiFeiPU$-3 Denoising	38.1902	0.9996
$YiFeiPU$-4 Denoising	39.0434	0.9996

分析表 14.1 可知，在上述几种去噪方法中，第一，Gaussian Denoising、Fourth-Order TV Denoising 与 Contourlet Denoising 的 $26.9611 \leqslant PSNR \leqslant 28.2945$，相对较小，其 $0.9953 \leqslant$ Correlation Coefficients $\leqslant 0.9959$，也相对较小，说明其去噪能力相对较差，其 denoised signal 和 original clean signal 之间的相似度相对较小。第二，Fractional-Order Anisotropic Diffusion Denoising、$YiFeiPU$-1 Denoising 和 $YiFeiPU$-2 Denoising 的 $29.2516 \leqslant PSNR \leqslant 29.8692$，相对中等，其 $0.9975 \leqslant$ Correlation Coefficients $\leqslant 0.9978$，也相对中等，说明其去噪能力相对适中，其 denoised signal 和 original clean signal 之间的相似度相对适中。第三，Bilateral Filtering Denoising、Wavelet Denoising、NLMF Denoising、NLMF Denoising、$YiFeiPU$-3 和 Denoising 的 $33.3088 \leqslant PSNR \leqslant 39.0434$，相对较大，其 $0.9975 \leqslant$ Correlatio Coefficients $\leqslant 0.9996$，也相对较大，说明其去噪能力相对较好，其 denoised signal 和 original clean signal 之间的相似度相对较大。其中，$YiFeiPU$-3 Denoising 和 $YiFeiPU$-4 Denoising 的 PSNR 和相关系数最高，说明其去噪能力相对最好，其 denoised signal 和 original clean signal 之间的相似度相对最高。

当 Gaussian Noise 干扰十分强烈时，特别是原图被噪声彻底淹没时，为了进一步分析和说明基于分数阶偏微分方程的纹理图像多尺度去噪模型对顽健噪声的分数阶非线性多尺度去噪能力，我们选取上述实验中相对较好的 Bilateral Filtering Denoising、Wavelet Denoising、NLMF Denoising、$YiFeiPU$-3 Denoising 和 $YiFeiPU$-4 Denoising 算法，对随机产生的一维信号进行对比实验。在本实验中，我们在基于分数阶偏微分方程的纹理图像多尺度去噪模型中所选用的分数阶阶次 v_1 和 v_3 以及分数阶幂次 v_2 并不是最佳值，而只是根据经验确定的一组相对较好的值。v_1、v_2 和 v_3 的最佳取值将在本章最后部分进行分析和讨论。另外，在本实验中，数值实迭代过程均以峰值信噪比 PSNR（Peak Signal-to-Noise Ratio）达到最高值时作为迭代的终止条件。本实验如图 14.6 所示。

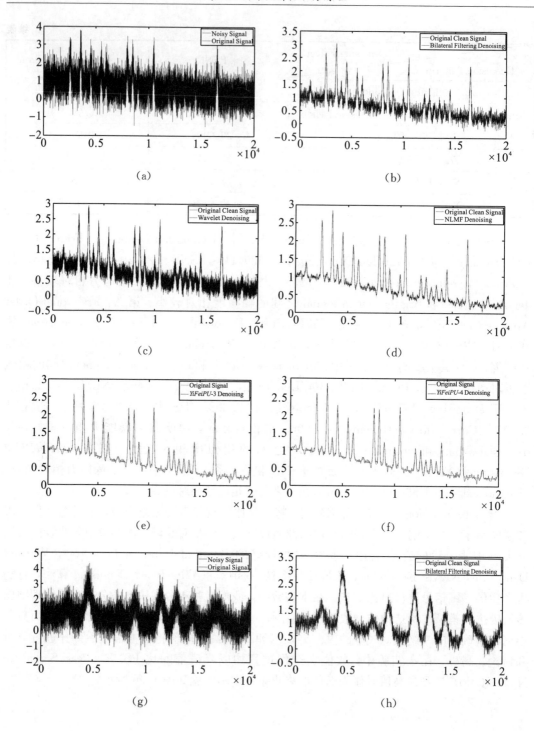

第 14 章 基于分数阶偏微分方程的退化图像逆处理

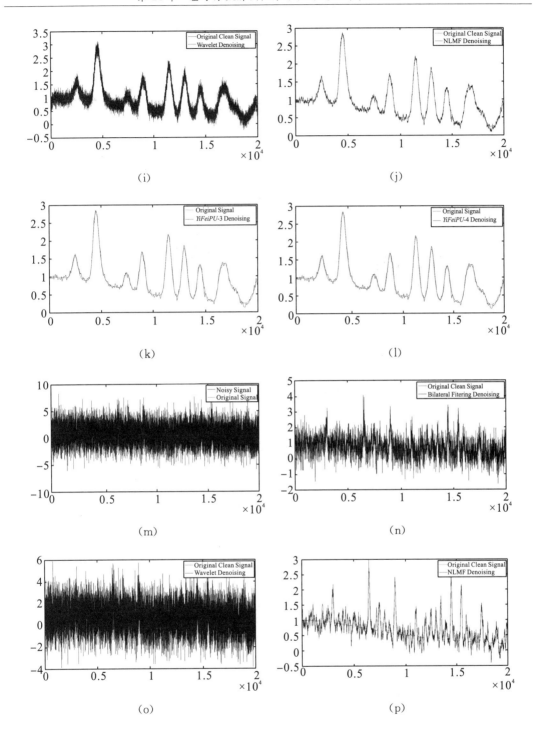

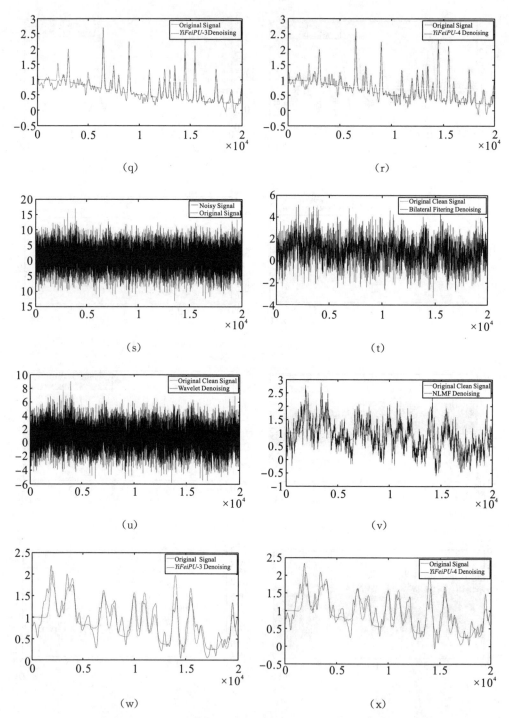

图 14.6 当噪声干扰十分强烈时,对随机产生的一维信号进行去噪处理。(a) Original Clean Signal and Noisy Signal (Adds White Gaussian Noise to the Original Clean Signal, PSNR=14.4294), (b) Bilateral Filtering Denoising [597] − [598], [601], (c) Wavelet Denoising [600] − [601], (d) NLMF Denoising [602] − [603], (e) $YiFeiPU$-3 Denoising (v_1=1.75, v_2=2.25, v_3=1.05, Δt=0.145), (f) $YiFeiPU$-4 Denoising (v_1=1.75, v_2=2.25, v_3=1.05, Δt=0.125), (g) Original Clean Signal and Noisy Signal (Adds White Gaussian Noise to the Original Clean Signal, PSNR=14.2479), (h) Bilateral

Filtering Denoising [597] — [598], [601], (i) Wavelet Denoising [600] — [601], (j) NLMF Denoising [602] — [603], (k) $YiFeiPU$-3 Denoising ($v_1=1.75$, $v_2=2.25$, $v_3=1.05$, $\Delta t=0.145$), (l) $YiFeiPU$-4 Denoising ($v_1=1.75$, $v_2=2.25$, $v_3=1.05$, $\Delta t=0.125$), (m) Original Clean Signal and Noisy Signal (Adds White Gaussian Noise to the Original Clean Signal, PSNR=2.8808), (n) Bilateral Filtering Denoising [597] — [598], [601], (o) Wavelet Denoising [600] — [601], (p) NLMF Denoising [602] — [603], (q) $YiFeiPU$-3 Denoising ($v_1=1.75$, $v_2=2.25$, $v_3=1.05$, $\Delta t=0.145$), (r) $YiFeiPU$-4 Denoising ($v_1=1.75$, $v_2=2.25$, $v_3=1.05$, $\Delta t=0.125$), (s) Original Clean Signal and Noisy Signal (Adds White Gaussian Noise to the Original Clean Signal, PSNR=−5.6540), (t) Bilateral Filtering Denoising [597] — [598], [601], (u) Wavelet Denoising [600] — [601], (v) NLMF Denoising [602] — [603], (w) $YiFeiPU$-3 Denoising ($v_1=1.75$, $v_2=2.25$, $v_3=1.05$, $\Delta t=0.145$), (x) $YiFeiPU$-4 Denoising ($v_1=1.75$, $v_2=2.25$, $v_3=1.05$, $\Delta t=0.125$)

从主观视觉效果看，分析图 14.6 可知，第一，在图 14.6 (a) ~ (f) 中，original clean signal 是随机产生的脉冲尖波，其高频奇异成分很大。所加的高斯白噪声较强 (PSNR=14.4294)，original clean signal 已经被噪声彻底淹没，但是 original clean signal 的轮廓还依稀可见。相比而言，Bilateral Filtering Denoising 和 Wavelet Denoising 的去噪效果相对较差。在图 14.6 (b)、(c) 中，虽然其 denoised signal 基本具备 original clean signal 的主要波形和信息特征，其脉冲尖波的高频奇异部分保持较好，但是在 original clean signal 的低频部分，其 denoised signal 还残留了大量的毛刺，噪声滤除不干净。NLMF Denoising, $YiFeiPU$-3 Denoising 和 $YiFeiPU$-4 Denoising 的去噪效果相对较好。在图 14.6 (d) ~ (f) 中，其 denoised signal 不仅在脉冲尖波的高频奇异部分与 original clean signal 吻合较好，而且在 original clean signal 的低频部分噪声滤除较干净，其 denoised signal 残留的毛刺较少。其中，$YiFeiPU$-3 Denoising 和 $YiFeiPU$-4 Denoising 的 denoised signal 与 original clean signal 的整体吻合相对最好，其去噪效果相对最好。第二，在图 14.6 (g) ~ (l) 中，original clean signal 是随机波动信号，其高频奇异成分相对较小，其中频成分相对较大。所加的高斯白噪声较强 (PSNR=14.2479)，original clean signal 亦已被噪声彻底淹没，但是 original clean signal 的轮廓仍依稀可见。相比而言，Bilateral Filtering Denoising 和 Wavelet Denoising 的去噪效果相对较差。在图 14.6 (h) (i) 中，虽然其 denoised signal 基本具备 original clean signal 的主要波形和信息特征，其随机波动信号的高频和中频部分保持较好，但是在 original clean signal 的整个频段，其 denoised signal 还残留了大量的毛刺，噪声滤除不干净。NLMF Denoising, $YiFeiPU$-3 Denoising 和 $YiFeiPU$-4 Denoising 的去噪效果相对较好。在图 14.6 (j) ~ (l) 中，其 denoised signal 不仅在随机波动信号的高频和中频部分与 original clean signal 吻合较好，而且在 original clean signal 的整个频段其噪声滤除较干净，其 denoised signal 残留的毛刺较少。其中，$YiFeiPU$-3 Denoising 和 $YiFeiPU$-4 Denoising 的 denoised signal 与 original clean signal 的整体吻合相对最好，其去噪效果相对最好。第三，在图 14.6 (m) ~ (r) 中，original clean signal 是随机产生的脉冲尖波，其高频奇异成分很大。所加的高斯白噪声很强 (PSNR=2.8808)，不仅 original clean signal 已被噪声彻底淹没，而且 original clean signal 的轮廓完全不可见。相比而言，Wavelet Denoising 的去噪效果相对最差。在图 14.6 (o) 中，其 denoised signal 仍完全淹没于噪声之中，基本不具备 original clean signal 的主要波形和信息特征。Bilateral Filtering Denoising 的去噪效果相对较差。在图

14.6（n）中，虽然 denoised signal 基本具备 original clean signal 的主要波形和信息特征，但是在 original clean signal 的整个频段，其 denoised signal 还残留了大量的毛刺，噪声滤除不干净。NLMF Denoising，$YiFeiPU$-3 Denoising 和 $YiFeiPU$-4 Denoising 的去噪效果相对较好。在图 14.6（p）～（r）中，其 denoised signal 不仅在脉冲尖波的高频奇异部分与 original clean signal 吻合较好，而且在 original clean signal 的低频部分噪声滤除较干净，其 denoised signal 残留的毛刺较少。其中，$YiFeiPU$-3 Denoising 和 $YiFeiPU$-4 Denoising 的 denoised signal 与 original clean signal 的整体吻合相对最好，其去噪效果相对最好。第四，在图 14.6（s）～（x）中，original clean signal 是随机波动信号，其高频奇异成分相对较小，其中频成分相对较大。所加的高斯白噪声非常强（PSNR＝－5.6540），不仅 original clean signal 已被噪声彻底淹没，而且 original clean signal 的轮廓完全不可见。相比而言，Wavelet Denoising 的去噪效果相对最差。在图 14.6（u）中，其 denoised signal 仍完全淹没于噪声之中，基本不具备 original clean signal 的主要波形和信息特征。Bilateral Filtering Denoising 和 NLMF Denoising 的去噪效果相对较差。在图 14.6（t）、(v) 中，虽然 denoised signal 基本具备 original clean signal 的主要波形和信息特征，但是在 original clean signal 的整个频段，其 denoised signal 还残留了大量的毛刺，噪声滤除不干净。$YiFeiPU$-3 Denoising 和 $YiFeiPU$-4 Denoising 的去噪效果相对最好。在图 14.6（w）、(x) 中，其 denoised signal 不仅在随机波动信号的高频和中频部分与 original clean signal 吻合相对最好，而且在 original clean signal 的低频部分噪声滤除较干净，其 denoised signal 残留的毛刺较少。

从客观定量分析看，我们不仅采用 PSNR，而且采用 the correlation coefficients between noisy signal or denoised signal and original clean signal 来综合度量图 14.6 中的上述各种算法的去噪效果，见表 14.2。

表 14.2 当 Gaussian Noise 干扰十分强烈时，对随机产生的一维信号的综合去噪效果

(a) 图 14.6 (a) ～ (f)

Denoising Algorithm / Denoising Effect	PSNR	Correlation Coefficients
Noisy Signal	14.4294	0.6658
Bilateral Filtering Denoising	24.8495	0.9475
Wavelet Denoising	25.6361	0.9544
NLMF Denoising	32.5325	0.9920
$YiFeiPU$-3 Denoising	33.1761	0.9928
$YiFeiPU$-4 Denoising	33.7535	0.9931

(b) 图 14.6 (g) ～ (l)

Denoising Algorithm / Denoising Effect	PSNR	Correlation Coefficients
Noisy Signal	14.2479	0.6993
Bilateral Filtering Denoising	24.7166	0.9763
Wavelet Denoising	25.9757	0.9664
NLMF Denoising	33.3324	0.9937
$YiFeiPU$-3 Denoising	34.8902	0.9955
$YiFeiPU$-4 Denoising	35.1737	0.9962

(c) 图 14.6 (m) ~ (r)

Denoising Algorithm \ Denoising Effect	PSNR	Correlation Coefficients
Noisy Signal	2.8808	0.2171
Bilateral Filtering Denoising	13.1951	0.5891
Wavelet Denoising	6.2909	0.2954
NLMF Denoising	21.9354	0.8889
$YiFeiPU$-3 Denoising	24.6000	0.9297
$YiFeiPU$-4 Denoising	24.6557	0.9316

(d) 图 14.6 (s) ~ (x)

Denoising Algorithm \ Denoising Effect	PSNR	Correlation Coefficients
Noisy Signal	-5.6540	0.1143
Bilateral Filtering Denoising	4.5176	0.3481
Wavelet Denoising	0.7087	0.2173
NLMF Denoising	13.4951	0.7230
$YiFeiPU$-3 Denoising	16.3783	0.8557
$YiFeiPU$-4 Denoising	17.2040	0.8475

分析表 14.2 可知，在上述几种去噪方法中，当噪声干扰十分强烈时，特别是信号被噪声彻底淹没时，第一，Bilateral Filtering Denoising 和 Wavelet Denoising 的去噪能力相对较差。在表 14.2 (a) ~ (d) 中，其 PSNR 和相关系数始终相对较小。特别地，当 noisy signal 的 PSNR 进一步大幅下降时，Wavelet Denoising 的去噪能力急剧下降。在表 14.2 (c)、(d) 中，其 PSNR 和相关系数均远小于其他去噪模型。第二，NLMF Denoising 的去噪能力相对较好，其 PSNR 和相关系数始终相对较高。第三，$YiFeiPU$-3 Denoising 和 $YiFeiPU$-4 Denoising 的去噪能力相对最好，其 PSNR 和相关系数始终相对最高。

综合上述两个对比实验（图 14.5 和图 14.6）的主观视觉效果与客观定量分析可知，第一，无论噪声的强弱，基于分数阶偏微分方程的纹理图像多尺度去噪模型无论是在信号的高频，还是在信号的中频和低频段都有很好的去噪能力。其噪声信号不仅与原图信号的高频边缘突变和中频轮廓吻合相对最好，而且在原始信号的低频部分噪声滤除较干净，其噪声信号残留的毛刺较少。第二，当噪声干扰十分强烈时，特别是信号被噪声彻底淹没时，基于分数阶偏微分方程的纹理图像多尺度去噪模型对顽健噪声具有相对最好的去噪能力，所以该分数阶去噪模型可以有效用于扩频通信的信号和噪声的分离。第三，众所周知，图 14.5 中的矩形波信号和锯齿波信号的高频边缘突变与图 14.6 中的脉冲尖波，都具有很强的高频奇异成分，对应着二维图像的高频边缘和纹理细节信息。图 14.5 中的正弦波信号和锯齿波信号的爬坡部分与图 14.6 中的随机波动信号，其高频奇异成分相对较小，其中频成分相对较大，对应着二维图像中的中频纹理细节信息。图 14.5 中的矩形波信号的常值直流部分，对应着二维图像中的低频和直流轮廓和背景信息。如上所述，基于分数阶偏微分方程的纹理图像多尺度去噪模型对上述一维信号具有良好的去噪能力。于是可以很自然地想到，该分数阶去噪模型对纹理图像亦应具有良好的去噪能力。

为了更为直观地分析和说明基于分数阶偏微分方程的纹理图像多尺度去噪模型对纹理图像具有良好的分数阶非线性多尺度去噪能力，我们仍然选取上述实验中相对较好的 Bilateral Filtering Denoising, Wavelet Denoising, NLMF Denoising, $YiFeiPU$-3 Denoising 和 $YiFeiPU$-4 Denoising 算法，对富于纹理细节信息的球铁金相图像进行对比实验。在本实验中，我们在基于分数阶偏微分方程的纹理图像多尺度去噪模型中所选用的分数阶阶次 v_1 和 v_3 以及分数阶幂次 v_2 并不是最佳值，而只是根据经验确定的一组相对较好的值。v_1、v_2 和 v_3 的最佳取值将在本章最后部分进行分析和讨论。另外，在本实验中，数值实迭代过程均以峰值信噪比 PSNR（Peak Signal-to-Noise Ratio）达到最高值时作为迭代的终止条件。本实验如图 14.7 所示。

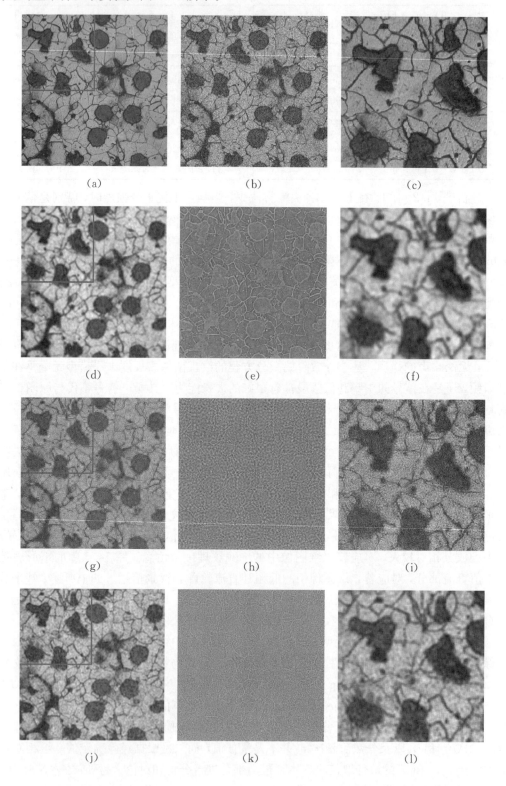

(a) (b) (c)
(d) (e) (f)
(g) (h) (i)
(j) (k) (l)

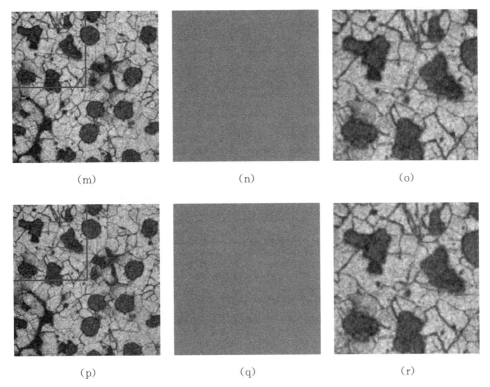

图 14.7 对富于纹理细节信息的球铁金相图像进行去噪处理。(a) Original Clean Image, (b) Noisy Image (Adds White Gaussian Noise to the Original Clean Image, PSNR=13.6872), (c) 局部放大显示 (a) 红框中的 1/4 部分, (d) Denoised Image of Bilateral Filtering Denoising [597] — [598], [601], (e) Bilateral Filtering Denoising 的残差图像, (f) 局部放大显示 (d) 红框中的 1/4 部分, (g) Denoised Image of Wavelet Denoising [600] — [601], (h) Wavelet Denoising 的残差图像, (i) 局部放大显示 (g) 红框中的 1/4 部分, (j) Denoised Image of NLMF Denoising [602] — [603], (k) NLMF Denoising 的残差图像, (l) 局部放大显示 (j) 红框中的 1/4 部分, (m) Denoised Image of $YiFeiPU$-3 Denoising ($v_1=1.75$, $v_2=2.25$, $v_3=1.05$, $\Delta t=10^{-10}$), (n) $YiFeiPU$-3 Denoising 的残差图像, (o) 局部放大显示 (m) 红框中的 1/4 部分, (p) Denoised Image of $YiFeiPU$-4 Denoising ($v_1=1.75$, $v_2=2.25$, $v_3=1.05$, $\Delta t=10^{-10}$), (q) $YiFeiPU$-4 Denoising 的残差图像, (r) 局部放大显示 (p) 红框中的 1/4 部分

从主观视觉效果看，分析图 14.7 可知，第一，Bilateral Filtering Denoising 与 Wavelet Denoising 对图像的高频边缘和纹理细节有明显的扩散和平滑作用，其去噪能力相对较差。在图 14.7（e）、(h) 中，其残差图像中的图像边缘和纹理细节相对比较清晰，说明 Bilateral Filtering Denoising 与 Wavelet Denoising 从噪声图像中去除的噪声与在原始图像中所加入的噪声有一定差距。在图 14.7（f）、(i) 中，其去噪图像相对比较模糊，在图 14.7（i）中，其噪声滤除不干净，说明 Bilateral Filtering Denoising 与 Wavelet Denoising 对图像边缘和纹理细节信息的保持能力相对比较差。第二，NLMF Denoising 对图像边缘和纹理细节信息的保持能力相对较好，但对图像边缘和纹理附近的去噪能力略差。在图 14.7（k）中，其残差图像中的图像边缘和纹理细节虽然比图 14.7（e）、(h) 中的相对明显减弱，但仍然略微可见，说明 NLMF Denoising 从 noisy image 中去除的噪声与在原始图像中所加入的噪声比较接近。在图 14.7（l）中，其 denoised image 中的边缘和纹理细节

模糊程度较小，相对比较清晰，说明 NLMF Denoising 能够相对较好地保持图像边缘和纹理细节信息。另外，在图 14.7 (k) 中，其残差图像中的图像边缘和纹理细节的邻域附近比较平滑，在图 14.7 (l) 中，其 denoised image 中的边缘和纹理细节邻域附近的残留噪声的强度要高于其他区域，说明 NLMF Denoising 对边缘和纹理附近的去噪能力略差。第三，$YiFeiPU$-3 Denoising 和 $YiFeiPU$-4 Denoising 不仅能相对最好地保持图像边缘和纹理细节信息，而且对噪声滤除也比较干净，其去噪能力相对最好。在图 14.7 (n)、(q) 中，其残差图像中的图像边缘和纹理细节仅隐约可见，说明 $YiFeiPU$-3 Denoising 和 $YiFeiPU$-4 Denoising 从 noisy image 中去除的噪声与在原始图像中所加入的噪声最接近。在图 14.7 (o)、(r) 中，其 denoised image 中的边缘和纹理细节模糊程度很小，相对最清晰，说明 $YiFeiPU$-3 Denoising 和 $YiFeiPU$-4 Denoising 对图像边缘和纹理细节信息的保持能力相对最好。

从客观定量分析看，我们不仅采用 PSNR, the correlation coefficients between noisy image or denoised image and original clean image[101]，而且采用平均灰度共生矩阵来综合度量图 14.7 中的上述各种算法的去噪效果。我们对图 14.7 中各图的每一像素在 5 像素距离（pixel distance）内计算其灰度共生矩阵参数，并在角度矩取 $0°$（y 正方向上投影），$45°$，$90°$（x 方向上投影）和 $135°$ 四个方向上分别从灰度共生矩阵中导出反映矩阵状况的典型参数，即对比度、相关性、能量和同质性/逆差距，并将其 4 个方向上的值分别进行平均。其测量值见表 14.3。

表 14.3 对富于纹理细节信息的球铁金相图像的综合去噪效果

Denoising Effect / Denoising Algorithm	PSNR	Correlation Coefficients	Contrast	Correlation	Energy	Homogeneity
Noisy Image	13.6872	0.9991	5.9118	0.2491	0.0197	0.4580
Bilateral Filtering Denoising	20.4845	0.9995	1.1200	0.8038	0.0787	0.7092
Wavelet Denoising	16.5708	0.9994	2.1387	0.3910	0.0473	0.5833
NLMF Denoising	21.0494	0.9996	1.2524	0.7150	0.0728	0.6822
$YiFeiPU$-3 Denoising	21.4826	0.9996	1.4365	0.6876	0.0688	0.6646
$YiFeiPU$-4 Denoising	21.5482	0.9997	1.4371	0.6646	0.0688	0.6608

分析表 14.3 可知，在上述几种去噪方法中，第一，Bilateral Filtering Denoising 和 Wavelet Denoising 的去噪能力相对较差。其 PSNR 和相关系数均相对较小，说明其噪声图像中的高频边缘和纹理细节的扩散和平滑作用较大，噪声去除不干净，其噪声图像与原图的相似度较小。另外，Bilateral Filtering Denoising 的平均灰度共生矩阵中的对比度较小，说明其噪声图像中对比度大的像素对较少，其纹理的沟纹越浅，视觉效果越模糊。Wavelet Denoising 的平均灰度共生矩阵中的对比度（Contrast）最大，虽然说明其噪声图像中对比度大的像素对较多，但这是由于噪声去除不干净造成的，并不能说明其纹理的沟纹越深且视觉效果较清晰。第二，NLMF Denoising，$YiFeiPU$-3 Denoising 和 $YiFeiPU$-4 Denoising 的去噪能力相对较好。其 PSNR 和相关系数相对较高，说明其噪声图像中的高频边缘和纹理细节保持得较好，噪声去除较干净，其噪声图像与原始图像的相似度较大。

其中，$YiFeiPU$-4 Denoising 的 PSNR 和相关系数均最高，说明其去噪最干净，其噪声图像和原始图像的相似度最大。$YiFeiPU$-4 Denoising 的平均灰度共生矩阵中的对比度相对较大，说明其噪声图像中对比度大的像素对较多，说明其纹理的沟纹较深，视觉效果较清晰。其相关性相对较小，说明其噪声图像中的局部灰度相关性较弱，其纹理细节的较明显。其能量相对较小，说明其噪声图像中的纹理模式变化较不均一和不规则，其纹理细节的较明显。其同质性/逆差距相对较小，说明其噪声图像中的图像纹理的不同区域间变化较剧烈，纹理细节较明显。综上可见，$YiFeiPU$-4 Denoising 是其中相对最好的去噪算法。

当 Gaussian Noise 干扰十分强烈时，特别是当原始图像被噪声彻底淹没时，为了进一步分析和说明基于分数阶偏微分方程的纹理图像多尺度去噪模型对顽健噪声的分数阶非线性多尺度去噪能力，我们仍选取 Bilateral Filtering Denoising，Wavelet Denoising，NLMF Denoising，$YiFeiPU$-3 Denoising 和 $YiFeiPU$-4 Denoising 算法，对富于内脏器官纹理细节的腹部 MRI 进行对比实验。在本实验中，我们在基于分数阶偏微分方程的纹理图像多尺度去噪模型中所选用的分数阶阶次 v_1 和 v_3 以及分数阶幂次 v_2 并不是最佳值，而只是根据经验确定的一组相对较好的值。v_1，v_2 和 v_3 的最佳取值将在本章最后部分进行分析和讨论。另外，在本实验中，数值实迭代过程均以峰值信噪比 PSNR（Peak Signal-to-Noise Ratio）达到最高值时作为迭代的终止条件。本实验如图 14.8 所示。

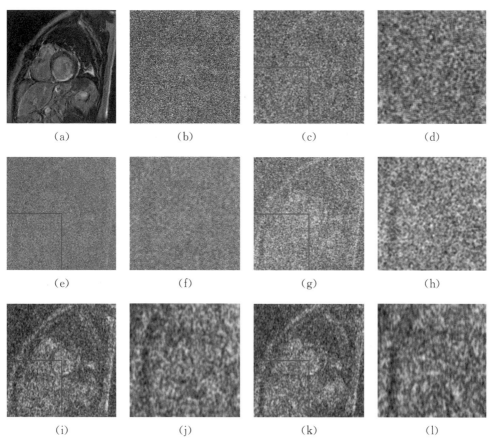

图 14.8 当 Gaussian Noise 干扰十分强烈时，对富于内脏器官纹理细节的腹部 MRI 进行去噪处理。(a) Original Clean MRI，(b) Noisy Image（Adds White Gaussian Noise to the Original Clean MRI，PSNR

$=5.4389$),(c) Denoised Image of Bilateral Filtering Denoising [597]-[598],[601],(d) 局部放大显示(c) 红框中的 1/4 部分,(e) Denoised Image of Wavelet Denoising [600]-[601],(f) 局部放大显示(e) 红框中的 1/4 部分,(g) Denoised Image of NLMF Denoising [602]-[603],(h) 局部放大显示(g) 红框中的 1/4 部分,(i) Denoised Image of $YiFeiPU$-3 Denoising ($v_1=1.75$,$v_2=2.75$,$v_3=1.05$,$\Delta t=5\times10^{-6}$),(j) 局部放大显示 (i) 红框中的 1/4 部分,(k) Denoised Image of $YiFeiPU$-4 Denoising ($v_1=1.75$,$v_2=2.5$,$v_3=1.05$,$\Delta t=5\times10^{-6}$),(l) 局部放大显示 (i) 红框中的 1/4 部分

从主观视觉效果看,分析图 14.8 可知,当噪声干扰十分强烈时,特别是 MRI 被噪声彻底淹没时,第一,Bilateral Filtering Denoising 与 Wavelet Denoising 的去噪能力相对较差。在图 14.8(c)~(f) 中,其噪声图像仅仅轮廓隐约可见,其内脏器官的边缘和纹理细节特征几乎难以辨识。第二,NLMF Denoising 的去噪能力相对较好。在图 14.8(g)、(h) 中,其噪声图像的图像轮廓更清晰,但其内脏器官的边缘和纹理细节特征仍模糊不清。第三,$YiFeiPU$-3 Denoising 和 $YiFeiPU$-4 Denoising 的去噪能力相对最好。在图 14.8(i)~(l) 中,不仅其噪声图像的图像轮廓相对最清晰,而且其内脏器官的边缘和纹理细节特征已能够大致辨识。

从客观定量分析看,我们仍然采用 PSNR,the correlation coefficients between noisy image or denoised image and original clean MRI[101],平均灰度共生矩阵来综合度量图 14.8 中的上述各种算法的去噪效果,见表 14.4。

表 14.4 当 Gaussian Noise 干扰十分强烈时,对富于内脏器官纹理细节的腹部 MRI 的综合去噪效果

Denoising Effect Denoising Algorithm	PSNR	Correlation Coefficients	Contrast	Correlation	Energy	Homogeneity
Noisy Image	5.4389	0.9857	21.2697	0.0013	0.0844	0.4811
Bilateral Filtering Denoising	11.4039	0.9894	1.1790	0.1796	0.1113	0.6562
Wavelet Denoising	9.9246	0.9892	1.8443	0.0175	0.0806	0.5941
NLMF Denoising	12.5855	0.9921	1.1795	0.1247	0.1083	0.6406
$YiFeiPU$-3 Denoising	17.4317	0.9975	1.1847	0.0836	0.0913	0.5737
$YiFeiPU$-4 Denoising	17.4572	0.9976	1.1853	0.0830	0.0893	0.5715

分析表 14.4 可知,当噪声干扰十分强烈时,特别是当 original clean MRI 被噪声彻底淹没时,在上述几种去噪方法中,第一,Bilateral Filtering Denoising 和 Wavelet Denoising 的去噪能力相对较差。其 PSNR 和相关系数均相对较小,说明其噪声去除不干净,其噪声图像与 original clean MRI 的相似度较小。另外,Wavelet Denoising 的平均灰度共生矩阵中的对比度(Contrast)最大,虽然说明其噪声图像中对比度大的像素对较多,但这是由于噪声去除不干净造成的,并不能说明其纹理的沟纹越深且视觉效果较清晰。第二,NLMF Denoising,$YiFeiPU$-3 Denoising 和 $YiFeiPU$-4 Denoising 的去噪能力相对较好。其 PSNR 和相关系数相对较高,说明其噪声去除相对较干净,其噪声图像与 original clean MRI 的相似度较大。其中,$YiFeiPU$-4 Denoising 的 PSNR 和相关系数均最高,且其平均灰度共生矩阵中的对比度(Contrast)相对较大,其相关性相对较小,其能量相对较小,其同质性/逆差距相对较小,说明 $YiFeiPU$-4 Denoising 是其中相对最好的去噪算法。

当同时被各种噪声(本实验同时加入 White Gaussian Noise,Salt & Pepper Noise and

Speckle Noise)强烈干扰时,特别是当原始图像几乎被这些噪声彻底淹没时,为了进一步分析和说明基于分数阶偏微分方程的纹理图像多尺度去噪模型对顽健噪声的分数阶非线性多尺度去噪能力,我们仍选取 Bilateral Filtering Denoising, Wavelet Denoising, NLMF Denoising, $YiFeiPU$-3 Denoising 和 $YiFeiPU$-4 Denoising 算法,对富于陨石坑纹理细节的月球卫星遥感图像进行对比实验。在本实验中,我们在基于分数阶偏微分方程的纹理图像多尺度去噪模型中所选用的分数阶阶次 v_1 和 v_3 以及分数阶幂次 v_2 并不是最佳值,而只是根据经验确定的一组相对较好的值。v_1,v_2 和 v_3 的最佳取值将在本章最后部分进行分析和讨论。另外,在本实验中,数值实迭代过程均以峰值信噪比 PSNR(Peak Signal-to-Noise Ratio)达到最高值时作为迭代的终止条件。本实验如图 14.9 所示。

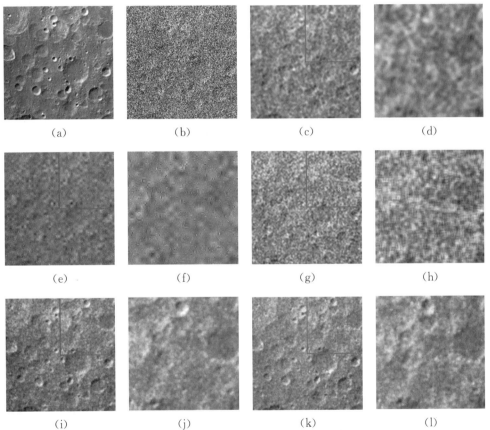

图 14.9 当同时被 White Gaussian Noise, Salt & Pepper Noise and Speckle Noise 三种噪声强烈干扰时,对富于陨石坑纹理细节的月球卫星遥感图像进行去噪处理。(a) Original Clean Image,(b) Noisy Image(Adds White Gaussian Noise(Its Standard Variance=0.02),Salt & Pepper Noise(Its Noise Density =0.2) and Speckle Noise(Its Standard Variance=0.1) to the Original Clean Image, PSNR=8.8564),(c) Denoised Image of Bilateral Filtering Denoising[597]-[598],[601],(d) 局部放大显示(c)红框中的 1/4 部分,(e) Denoised Image of Wavelet Denoising[600]-[601],(f) 局部放大显示(e)红框中的 1/4 部分,(g) Denoised Image of NLMF Denoising[602]-[603],(h) 局部放大显示(g)红框中的 1/4 部分,(i) Denoised Image of $YiFeiPU$-3 Denoising(v_1=1.75,v_2=2.75,v_3=1.05,Δt=10^{-10}),(j) 局部放大显示(i)红框中的 1/4 部分,(k) Denoised Image of $YiFeiPU$-4 Denoising(v_1=1.75,v_2=2.75,v_3=1.05,Δt=10^{-10}),(l) 局部放大显示(i)红框中的 1/4 部分

从主观视觉效果看，分析图 14.9 可知，当同时被 White Gaussian Noise，Salt & Pepper Noise and Speckle Noise 强烈干扰时，特别是当原始月球卫星遥感图像几乎被这些噪声彻底淹没时，第一，Bilateral Filtering Denoising，Wavelet Denoising 和 NLMF Denoising 的去噪能力相对较差。在图 14.9（c）～（h）中，其噪声图像仅仅轮廓隐约可见，其月球卫星遥感图像的陨石坑纹理细节特征几乎难以辨识。第二，$YiFeiPU$-3 Denoising 和 $YiFeiPU$-4 Denoising 的去噪能力相对最好。在图 14.9（i）～（l）中，不仅其噪声图像的图像轮廓相对最清晰，而且其月球卫星遥感图像的陨石坑纹理细节特征已能够较为清楚地辨识。

从客观定量分析看，我们仍然采用 PSNR，the correlation coefficients between noisy image or denoised image and original clean image[101]，平均灰度共生矩阵来综合度量图 14.9 中的上述各种算法的去噪效果，见表 14.5。

表 14.5 当同时被 White Gaussian Noise，Salt & Pepper Noise and Speckle Noise 三种噪声强烈干扰时，对富于陨石坑纹理细节的月球卫星遥感图像的综合去噪效果

Denoising Algorithm \ Denoising Effect	PSNR	Correlation Coefficients	Contrast	Correlation	Energy	Homogeneity
Noisy Image	8.8564	0.9968	12.5054	0.0159	0.0230	0.3930
Bilateral Filtering Denoising	19.3530	0.9991	0.4934	0.6371	0.1601	0.7841
Wavelet Denoising	18.3278	0.9990	0.5736	0.7302	0.2348	0.7765
NLMF Denoising	19.9585	0.9989	1.0104	0.4319	0.0999	0.6797
$YiFeiPU$-3 Denoising	21.1448	0.9995	0.61033	0.5881	0.1476	0.7598
$YiFeiPU$-4 Denoising	21.1455	0.9996	0.6609	0.5809	0.1363	0.7495

分析表 14.5 可知，当同时被 White Gaussian Noise，Salt & Pepper Noise and Speckle Noise 强烈干扰时，特别是当原始月球卫星遥感图像几乎被这些噪声彻底淹没时，在上述几种去噪方法中，第一，Bilateral Filtering Denoising，Wavelet Denoising 和 NLMF Denoising 的去噪能力相对较差。其 PSNR 和相关系数均相对较小，说明其噪声去除不干净，其噪声图像与原始月球卫星遥感图像的相似度较小。另外，NLMF Denoising 的平均灰度共生矩阵中的对比度（Contrast）最大，虽然说明其噪声图像中对比度大的像素对较多，但这是由于噪声去除不干净造成的，并不能说明其纹理的沟纹越深且视觉效果较清晰。第二，$YiFeiPU$-3 Denoising 和 $YiFeiPU$-4 Denoising 的去噪能力相对较好。其 PSNR 和相关系数相对较高，说明其噪声去除相对较干净，其噪声图像与原始月球卫星遥感图像的相似度较大。其中，$YiFeiPU$-4 Denoising 的 PSNR 和相关系数均最高，且其平均灰度共生矩阵中的对比度相对较大，其相关性相对较小，其能量相对较小，其同质性/逆差距相对较小，说明 $YiFeiPU$-4 Denoising 是其中相对最好的去噪算法。

综合上述三个对比实验（图 14.7、图 14.8 和图 14.9）的主观视觉效果与客观定量分析可知，第一，无论噪声的强弱和类型，基于分数阶偏微分方程的纹理图像多尺度去噪模型均具有很好的去噪能力。不仅其 PSNR 相对最高，其噪声图像的噪声去除相对最干净，而且其相关系数相对最高，其噪声图像与原始图像吻合得相对最好。第二，基于分数阶偏微分方程的纹理图像多尺度去噪模型在去除图像噪声的同时，既能尽量保留纹理图像平滑

区域中的低频轮廓，又能非线性保留灰度值跃变幅度相对较大的高频边缘和纹理细节，还能非线性保留灰度值跃变幅度变化相对不大的纹理细节的分数阶非线性多尺度去噪能力。之所以该分数阶去噪模型具有上述分数阶非线性多尺度去噪能力，是因为信号的分数阶微积分具有独特的特性所造成的。信号分数阶微分具有如下特性[157,185]：①常数的分数阶微分值不为零，常数的整数阶微分值必为零。在信号幅值不变或变化不大的平滑区域内，其分数阶微分值是从对应奇异跳变处的极大值渐趋于零，而信号幅值平坦段的任意整数阶微分必为零。这是分数阶微分相对于整数阶微分的显著区别之一。②在信号幅值阶梯或斜坡的起始点处的分数阶微分值非零，起到了加强高频奇异信息的作用。随着分数阶微分阶次的增加，其对高频奇异信号的加强幅度逐渐增大。当 $0<v<1$ 时，分数阶微分对高频奇异信息的增强幅度比 $v=1$ 阶微分的增强幅度小。信号的整数阶微分只是分数阶微分的特例。③沿信号幅值斜坡的分数阶微分值非零，亦非常数，是非线性曲线，而沿信号幅值斜坡的整数阶微分值为常数。因此在数字图像处理之中，分数阶微分能够非线性增强图像的复杂纹理细节特征。图像分数阶微分既能尽量非线性保留图像平滑区域中的低频轮廓信息，又能非线性增强图像中灰度值跃变幅度相对较大的高频边缘信息，还能非线性增强图像中灰度值跃变幅度和频率变化相对不大的高频纹理细节信息[185−188,434−435,468]。

在上述的分析和处理之中，我们在基于分数阶偏微分方程的纹理图像多尺度去噪模型中所选用的分数阶阶次 v_1 和 v_3 以及分数阶幂次 v_2 并不是最佳值，而只是根据经验确定的一组相对较好的值。那么我们很自然地要问：v_1，v_2 和 v_3 对该分数阶去噪模型的分数阶非线性多尺度去噪能力有何贡献？其值何时最佳？为此，我们对分数阶微分阶次 v_1 的最佳值进行分析和测试。我们选取上述实验中相对最好的 $YiFeiPU$-4 Denoising，对富于纹理细节信息的 Bridge 图像（该图具有 5°旋转角，可同时测试该分数阶去噪模型的抗旋转性能）对 v_1 的最佳值进行测试。在本实验中，我们取 $v_1 \in (0, 3]$，并固定 $v_2=2.25$，$v_3=1.05$ 和 $\Delta t=10^{-10}$ 不变，如图 14.10 所示。

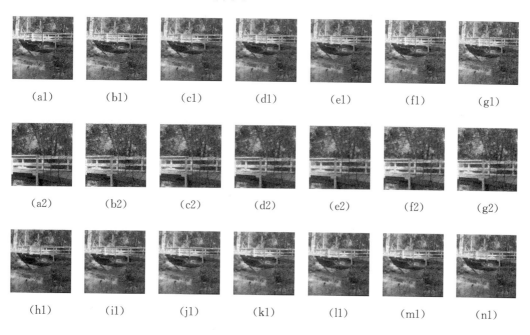

| (h2) | (i2) | (j2) | (k2) | (l2) | (m2) | (n2) |

图 14.10 对 $YiFeiPU$-4 Denoising 中 v_1 的最佳值进行测试，其中，$v_2=2.25$，$v_3=1.05$ 且 $\Delta t=10^{-10}$。(a1) Original Clean Image，(a2) 局部放大显示（a1）红框中的 1/4 部分，(b1) Noisy Image (Adds White Gaussian Noise to the Original Clean Image, PSNR=18.4877)，(b2) 局部放大显示（b1）红框中的 1/4 部分，(c1) $v_1=0.25$，(c2) 局部放大显示（c1）红框中的 1/4 部分，(d1) $v_1=0.50$，(d2) 局部放大显示（d1）红框中的 1/4 部分，(e1) $v_1=0.75$，(e2) 局部放大显示（e1）红框中的 1/4 部分，(f1) $v_1=1.00$，(f2) 局部放大显示（f1）红框中的 1/4 部分，(g1) $v_1=1.25$，(g2) 局部放大显示（g1）红框中的 1/4 部分，(h1) $v_1=1.50$，(h2) 局部放大显示（h1）红框中的 1/4 部分，(i1) $v_1=1.75$，(i2) 局部放大显示（i1）红框中的 1/4 部分，(j1) $v_1=2.00$，(j2) 局部放大显示（j1）红框中的 1/4 部分，(k1) $v_1=2.25$，(k2) 局部放大显示（k1）红框中的 1/4 部分，(l1) $v_1=2.50$，(l2) 局部放大显示（l1）红框中的 1/4 部分，(m1) $v_1=2.75$，(m2) 局部放大显示（m1）红框中的 1/4 部分，(n1) $v_1=3.00$，(n2) 局部放大显示（n1）红框中的 1/4 部分。

从主观视觉效果看，分析图 14.10 可知，第一，随着 v_1 的逐渐增大，其噪声图像中的噪声颗粒越少，其去除的噪声越多，其噪声图像中的树枝、桥和水纹等的边缘和纹理细节与原始图像的吻合越好，相似度越高；第二，随着 v_1 的逐渐增大，虽然其噪声图像的对比度随之逐渐缓缓减弱，图像的清晰度随之逐渐缓缓降低，但其边缘和纹理细节特征仍能够清楚地辨识。

从客观定量分析看，我们仍然采用 PSNR, the correlation coefficients between noisy image or denoised image and original clean image [101]，平均灰度共生矩阵来综合度量图 14.10 中 v_1 各阶次的去噪效果，见表 14.6。

表 14.6 对 $YiFeiPU$-4 Denoising 中 v_1 的最佳值进行测试的综合去噪效果

Denoising Effect $YiFeiPU$-4 Denoising	PSNR	Correlation Coefficients	Contrast	Correlation	Energy	Homogeneity
Noisy Image	18.4877	0.99969	2.5999	0.5696	0.0318	0.5634
$v_1=0.25$	21.0500	0.99947	1.3042	0.7310	0.0519	0.6628
$v_1=0.50$	21.4689	0.99953	1.1488	0.7488	0.0578	0.6811
$v_1=0.75$	22.2654	0.99965	0.9781	0.7665	0.0658	0.7067
$v_1=1.00$	22.3331	0.99965	0.9604	0.7860	0.0668	0.7138
$v_1=1.25$	22.8218	0.99979	0.9516	0.7920	0.0690	0.7163
$v_1=1.50$	22.8607	0.99975	0.8020	0.8054	0.0725	0.7344
$v_1=1.75$	23.6697	0.99984	0.6839	0.8331	0.0789	0.7580
$v_1=2.00$	23.6859	0.99984	0.7826	0.8181	0.0716	0.7384
$v_1=2.25$	23.8916	0.99986	0.7757	0.8214	0.0720	0.7414
$v_1=2.50$	23.8488	0.99986	0.7680	0.8238	0.0727	0.7437

续表

Denoising Effect YiFeiPU−4 Denoising	PSNR	Correlation Coefficients	Contrast	Correlation	Energy	Homogeneity
$v_1=2.75$	23.7615	0.99986	0.7523	0.8336	0.0759	0.7536
$v_1=3.00$	23.6467	0.99985	0.7834	0.8194	0.0716	0.7404

为了直观分析表14.6中的数据,从而说明分数阶微分阶次 v_1 与 YiFeiPU-4 Denoising 的综合去噪效果之间的对应变化关系,我们将表14.6中的PSNR、相关系数、对比度、相关性、能量和同质性/逆差距6个参数分别进行5至9阶的非线性曲线拟合,如图14.11所示。

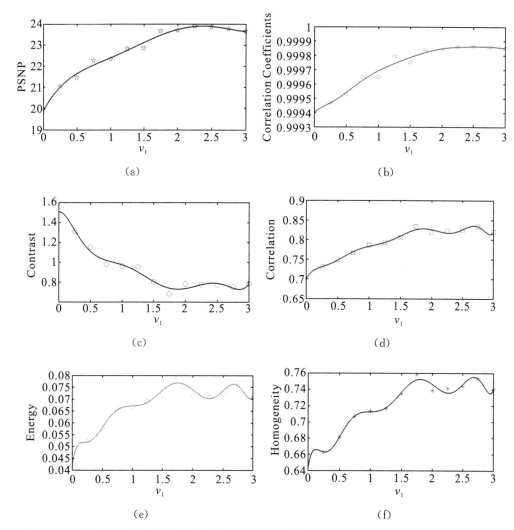

图14.11 对表14.6中的参数分别进行5至9阶的非线性曲线拟合。(a) PSNR,(b) Correlation Coefficients,(c) Contrast,(d) Correlation,(e) Energy,(f) Homogeneity

分析表14.6和图14.11可知,分数阶微分阶次 v_1 与 YiFeiPU-4 Denoising 的综合去噪效果之间的对应变化关系是有规律的。第一,由表14.6和图14.11(a)、(b)可知,随

着 v_1 的逐渐增大，其 PSNR 和相关系数先是随之逐渐增大，并在 $v_1=2.25$ 附近均增大到最大值。然后，随着 v_1 的进一步逐渐增大，其 PSNR 和相关系数随之逐渐减小。说明在 $v_1=2.25$ 左右时，$YiFeiPU$-4 Denoising 的去噪程度最强，其噪声图像和原始图像的吻合最好，相似度最高。第二，由表 14.6 和图 14.11（c）可知，随着 v_1 的逐渐增大，其平均灰度共生矩阵中的对比度（Contrast）先是随之逐渐减小，并在 $v_1=1.75$ 左右减小到最小值。然后，随着 v_1 的进一步逐渐增大，其对比度（Contrast）的减小趋势得到了遏制，基本趋于稳定，并在 $v_1=2.25$ 左右出现了一个局部极大值。结合表 14.6 和图 14.11（a）、(b) 可知，当 $v_1<1.75$ 时，虽然其对比度（Contrast）相对较大，但其 PSNR 和相关系数相对较小。虽然说明在其噪声图像中对比度大的像素对相对较多，但这是由于其噪声去除不干净造成的，并不能说明其纹理的沟纹越深且视觉效果较清晰。另外，在 $v_1=2.25$ 左右时，其对比度（Contrast）出现极大值，说明其噪声图像中对比度大的像素对相对较多，其纹理的沟纹相对较深，视觉效果相对更清晰。第三，由表 14.6 和图 14.11（d）可知，随着 v_1 的逐渐增大，其平均灰度共生矩阵中的相关性（Correlation）先是随之逐渐增大，并在 $v_1=1.75$ 左右增大到最大值。然后，随着 v_1 的进一步逐渐增大，其相关性（Correlation）的增大趋势得到了遏制，基本趋于稳定，并在 $v_1=2.25$ 左右出现了一个局部极小值。结合表 14.6 和图 14.11（a）、(b) 可知，当 $v_1<1.75$ 时，虽然其相关性（Correlation）相对较小，但其 PSNR 和相关系数较小。虽然说明在其噪声图像中局部灰度相关性相对较弱，但这是由于其噪声去除不干净造成的，并不能说明其纹理细节的相对较明显。另外，在 $v_1=2.25$ 左右时，其相关性（Correlation）出现极小值，说明其噪声图像中局部灰度相关性相对较弱，其纹理细节相对较明显。第四，由表 14.6 和图 14.11（e）可知，随着 v_1 的逐渐增大，其平均灰度共生矩阵中的能量（Energy）先是随之逐渐增大，并在 $v_1=1.75$ 左右增大到最大值。然后，随着 v_1 的进一步逐渐增大，其能量（Energy）的增大趋势得到了遏制，基本趋于稳定，并在 $v_1=2.25$ 左右出现了一个局部极小值。结合表 14.6 和图 14.11（a）、(b) 可知，当 $v_1<1.75$ 时，虽然其能量（Energy）相对较小，但其 PSNR 和相关系数较小。虽然说明在其噪声图像中的纹理模式变化相对较不均一和较不规则，但这是由于其噪声去除不干净造成的，并不能说明其纹理细节的相对较明显。另外，在 $v_1=2.25$ 左右时，其能量（Energy）出现极小值，说明其噪声图像中的纹理模式变化相对较不均一和较不规则，其纹理细节相对较明显。第五，由表 14.6 和图 14.11（f）可知，随着 v_1 的逐渐增大，其平均灰度共生矩阵中的同质性/逆差距（Homogeneity）先是随之逐渐增大，并在 $v_1=1.75$ 左右增大到最大值。然后，随着 v_1 的进一步逐渐增大，其同质性/逆差距（Homogeneity）的增大趋势得到了遏制，基本趋于稳定，并在 $v_1=2.25$ 左右出现了一个局部极小值。结合表 14.6 和图 14.11（a）、(b) 可知，当 $v_1<1.75$ 时，虽然其同质性/逆差距（Homogeneity）相对较小，但其 PSNR 和相关系数较小。虽然说明在其噪声图像中的图像纹理的不同区域间变化相对较剧烈，但这是由于其噪声去除不干净造成的，并不能说明其纹理细节的相对较明显。另外，在 $v_1=2.25$ 左右时，其同质性/逆差距（Homogeneity）出现极小值，说明其噪声图像中的图像纹理的不同区域间变化相对较剧烈，其纹理细节相对较明显。综上所述，当分数阶微分阶次 $v_1=2.25$ 左右时，$YiFeiPU$-4 Denoising 的综合去噪效果相对最佳，其噪声图像中的纹理细节相对最明显。

进一步地，我们着手对分数阶幂次方 v_2 的最佳值进行分析和测试。我们仍然选取上述

实验中相对最好的 $YiFeiPU$-4 Denoising，将富于纹理细节信息的 Goldhill 图像对 v_2 的最佳值进行测试。在本实验中，我们取 $v_2 \in (0, 6]$，令 v_1 取其在上述实验中的最佳值，$v_1 = 2.25$，并固定 $v_3 = 1.05$ 和 $\Delta t = 10^{-10}$ 不变，如图 14.12 所示。

图 14.12 对 $YiFeiPU$-4 Denoising 中 v_2 的最佳值进行测试，其中，$v_1 = 2.25$，$v_3 = 1.05$ 且 $\Delta t = 10^{-10}$。(a1) Original Clean Image, (a2) 局部放大显示 (a1) 红框中的 1/4 部分, (b1) Noisy Image (Adds White Gaussian Noise to the Original Clean Image, PSNR=17.2084), (b2) 局部放大显示 (b1) 红框中的 1/4 部分, (c1) $v_2 = 0.50$, (c2) 局部放大显示 (c1) 红框中的 1/4 部分, (d1) $v_2 = 1.00$, (d2) 局部放大显示 (d1) 红框中的 1/4 部分, (e1) $v_2 = 1.50$, (e2) 局部放大显示 (e1) 红框中的 1/4 部分, (f1) $v_2 = 2.00$, (f2) 局部放大显示 (f1) 红框中的 1/4 部分, (g1) $v_2 = 2.50$, (g2) 局部放大显示 (g1) 红框中的 1/4 部分, (h1) $v_2 = 3.00$, (h2) 局部放大显示 (h1) 红框中的 1/4 部分, (i1) $v_2 = 3.50$, (i2) 局部放大显示 (i1) 红框中的 1/4 部分, (j1) $v_2 = 4.00$, (j2) 局部放大显示 (j1) 红框中的 1/4 部分, (k1) $v_2 = 4.50$, (k2) 局部放大显示 (k1) 红框中的 1/4 部分, (l1) $v_2 = 5.00$, (l2) 局部放大显示 (l1) 红框中的 1/4 部分, (m1) $v_2 = 5.50$, (m2) 局部放大显示 (m1) 红框中的 1/4 部分, (n1) $v_2 = 6.00$, (n2) 局部放大显示 (n1) 红框中的 1/4 部分

从主观视觉效果看，分析图 14.12 可知，第一，随着 v_2 的逐渐增大，其噪声图像中的噪声颗粒越少，其去除的噪声越多，其噪声图像中的房屋、窗框和树木等的边缘和纹理细节与 original clean image 的吻合得相对更好，相似度更高；第二，当 v_2 逐渐增大到 $v_2 = 4.5$ 左右之后，其噪声图像开始随之略微逐渐变模糊；第三，随着 v_2 的逐渐增大，其噪声图像的对比度，除了在 $v_2 = 1$ 之外，几乎没有随之发生变化，图像的清晰度也几乎没有随之发生变化。

从客观定量分析看，我们仍然采用 PSNR, the correlation coefficients between noisy image or denoised image and original clean image[101]，平均灰度共生矩阵来综合度量图 14.12 中 v_2 各阶次的去噪效果，见表 14.7。

表 14.7 对 $YiFeiPU$-4 Denoising 中 v_2 的最佳值进行测试的综合去噪效果

Denoising Effect / $YiFeiPU$-4 Denoising	PSNR	Correlation Coefficients	Contrast	Correlation	Energy	Homogeneity
Noisy Image	17.2084	0.99969	2.0930	0.6021	0.0385	0.5856
$v_2=0.50$	25.5938	0.99994	0.3866	0.9040	0.1127	0.8314
$v_2=1.00$	25.5943	0.99995	0.3869	0.9039	0.1126	0.8314
$v_2=1.50$	25.5943	0.99995	0.3866	0.9040	0.1127	0.8314
$v_2=2.00$	25.5947	0.99995	0.3866	0.9040	0.1127	0.8314
$v_2=2.50$	25.5949	0.99995	0.3866	0.9040	0.1127	0.8314
$v_2=3.00$	25.5949	0.99995	0.3866	0.9040	0.1127	0.8314
$v_2=3.50$	25.5948	0.99995	0.3866	0.9040	0.1127	0.8314
$v_2=4.00$	25.5946	0.99995	0.3866	0.9040	0.1127	0.8314
$v_2=4.50$	25.5941	0.99995	0.3866	0.9040	0.1127	0.8314
$v_2=5.00$	25.5935	0.99994	0.3866	0.9040	0.1127	0.8314
$v_2=5.50$	25.5926	0.99993	0.3866	0.9040	0.1127	0.8314
$v_2=6.00$	25.5916	0.99992	0.3866	0.9040	0.1127	0.8314

为了直观分析表 14.7 中的数据，从而说明分数阶幂次方 v_2 与 $YiFeiPU$-4 Denoising 的综合去噪效果之间的对应变化关系，我们将表 14.7 中的 PSNR、相关系数、对比度、相关性、能量和同质性/逆差距 6 个参数分别进行 1 至 9 阶的非线性曲线拟合，如图 14.13 所示。

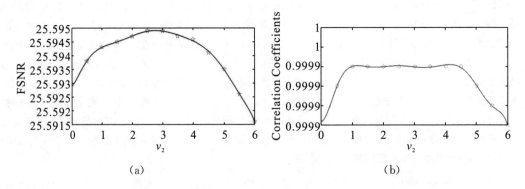

(a)   (b)

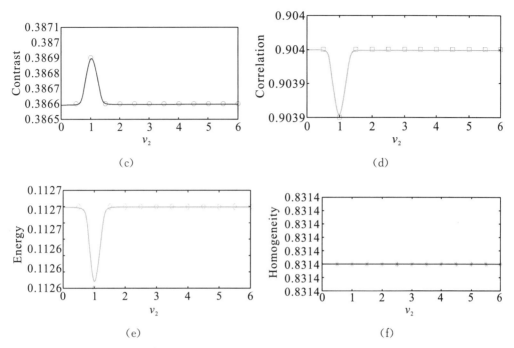

图 14.13 对表 14.7 中的参数分别进行 1 至 9 阶的非线性曲线拟合。(a) PSNR，(b) Correlation Coefficients，(c) Contrast，(d) Correlation，(e) Energy，(f) Homogeneity

分析表 14.7 和图 14.13 可知，分数阶幂次方 v_2 与 $YiFeiPU$-4 Denoising 的综合去噪效果之间的对应变化关系也是有规律的。第一，由表 14.7 和图 14.13 (a)、(b) 可知，随着 v_2 的逐渐增大，其 PSNR 和相关系数先是随之逐渐增大。其 PSNR 在 $v_2 \in [2.50,3.50]$ 之间时，增大到最大值。其相关系数在 $v_2 = 1.00$ 时，增大到最大值。然后，随着 v_2 进一步逐渐增大到一定值之后，其 PSNR 和相关系数又随之逐渐减小。其 PSNR 从 $v_2 = 3.50$ 开始随之逐渐减小，其相关系数从 $v_2 = 4.50$ 开始随之逐渐减小。这说明在 $v_2 \in [2.50,3.50]$ 左右时，$YiFeiPU$-4 Denoising 的去噪程度最强，其噪声图像和原始图像的吻合最好，相似度最高。第二，由表 14.7 和图 14.13 (c)～(f) 可知，随着 v_2 的逐渐增大，其平均灰度共生矩阵中的对比度 (Contrast) 仅在 $v_2 = 1.00$ 时出现了一个正的奇异跳变，其平均灰度共生矩阵中的相关性 (Correlation) 和能量 (Energy) 均仅在 $v_2 = 1.00$ 时出现了一个负的奇异跳变。除此之外，其相关性 (Correlation) 和能量 (Energy) 均不随 v_2 的逐渐增大而发生变化，均保持常数。其同质性/逆差距 (Homogeneity) 始终保持常数。结合表 14.7 和图 14.13 (a)、(b) 可知，当 $v_2 = 1.00$ 时，虽然其对比度 (Contrast) 相对较大，其相关性 (Correlation) 和能量 (Energy) 均相对较小，但其 PSNR 和相关系数相对较小。虽然说明在其噪声图像中对比度大的像素对相对较多，局部灰度相关性相对较弱，纹理模式变化相对较不均一和较不规则，但这是由于其噪声去除不干净造成的，并不能说明其纹理的沟纹越深且视觉效果较清晰。综上所述，分数阶幂次方 v_2 除了 $v_2 = 1.00$ 之外，均不能对噪声图像中纹理细节的清晰度造成影响。在 $v_2 \in [2.50,3.50]$ 时，$YiFeiPU$-4 Denoising 的 PSNR 和相关系数相对最高，其综合去噪效果相对最佳。

最后，我们着手对分数阶微分阶次 v_3 的最佳值进行分析和测试。我们仍然选取上述实验中相对最好的 $YiFeiPU$-4 Denoising，将富于纹理细节信息的 Barb 图像对 v_3 的最佳值进行测试。在本实验中，我们取 $v_3 \in (0, 3)$。因为在式（14.78），（14.79），（14.88）和（14.89）中，$v_3 \neq 1, 2, 3$，所以我们分别取 $v_3 = 1.0001 \cong 1.00$，$v_3 = 2.0001 \cong 2.00$ 和 $v_3 = 2.9999 \cong 3.00$ 予以近似。我们令 v_1 取其在上述实验中的最佳值，$v_1 = 2.25$，v_2 取其在上述实验中的最佳值，$v_2 = 2.50 \in [2.50, 3.50]$，并固定 $\Delta t = 10^{-10}$ 不变，如图 14.14 所示。

图 14.14 对 $YiFeiPU$-4 Denoising 中 v_3 的最佳值进行测试，其中，$v_1 = 2.25$，$v_2 = 2.50$ 且 $\Delta t = 10^{-10}$。(a1) Original Clean Image，(a2) 局部放大显示（a1）红框中的 1/4 部分，(b1) Noisy Image (Adds White Gaussian Noise to the Original Clean Image, PSNR=17.9199)，(b2) 局部放大显示（b1）红框中的 1/4 部分，(c1) $v_3 = 0.25$，(c2) 局部放大显示（c1）红框中的 1/4 部分，(d1) $v_3 = 0.50$，(d2) 局部放大显示（d1）红框中的 1/4 部分，(e1) $v_3 = 0.75$，(e2) 局部放大显示（e1）红框中的 1/4 部分，(f1) $v_3 = 1.0001$，(f2) 局部放大显示（f1）红框中的 1/4 部分，(g1) $v_3 = 1.25$，(g2) 局部放大显示（g1）红框中的 1/4 部分，(h1) $v_3 = 1.50$，(h2) 局部放大显示（h1）红框中的 1/4 部分，(i1) $v_3 = 1.75$，(i2) 局部放大显示（i1）红框中的 1/4 部分，(j1) $v_3 = 2.0001$，(j2) 局部放大显示（j1）红框中的 1/4 部分，(k1) $v_3 = 2.25$，(k2) 局部放大显示（k1）红框中的 1/4 部分，(l1) $v_3 = 2.50$，(l2) 局部放大显示（l1）红框中的 1/4 部分，(m1) $v_3 = 2.75$，(m2) 局部放大显示（m1）红框中的 1/4 部分，(n1) $v_3 = 2.9999$，(n2) 局部放大显示（n1）红框中的 1/4 部分

从主观视觉效果看，分析图 14.14 可知，第一，当 v_3 逐渐增大到 $v_3 = 0.25$ 左右时，其噪声图像中的头巾、藤椅和书本等的边缘和纹理细节与原始图像的吻合相对最好，相似度相对最高；第二，当 v_3 逐渐增大到 $v_3 = 1.0001$ 左右时，其噪声图像中的头巾、藤椅和书本等的边缘和纹理细节与原始图像的吻合相对最差，相似度最低；第三，随着 v_3 的逐渐增大，其噪声图像的对比度，除了在 $v_3 = 1.0001$ 之外，随之发生的略微变化。

从客观定量分析看，我们仍然采用 PSNR，the correlation coefficients between noisy image or denoised image and original clean image[101]，平均灰度共生矩阵来综合度量图 14.14 中 v_3 各阶次的去噪效果，见表 14.8。

表 14.8　对 $YeFeiPU$-4 Denoising 中 v_3 的最佳值进行测试的综合去噪效果

Denoising Effect $YiFeiPU$-4 Denoising	PSNR	Correlation Coefficients	Contrast	Correlation	Energy	Homogeneity
Noisy Image	17.9199	0.99969	2.3560	0.5328	0.0356	0.5731
$v_3=0.25$	23.3446	0.99987	0.5716	0.8535	0.0825	0.7844
$v_3=0.50$	23.3319	0.99987	0.5714	0.8514	0.0818	0.7831
$v_3=0.75$	23.3323	0.99987	0.5736	0.8529	0.0823	0.7838
$v_3=1.0001$	20.5971	0.99935	0.5985	0.8455	0.0797	0.7762
$v_3=1.25$	23.3324	0.99987	0.5665	0.8560	0.0825	0.7834
$v_3=1.50$	23.3283	0.99987	0.5736	0.8516	0.0819	0.7822
$v_3=1.75$	23.3313	0.99987	0.5645	0.8562	0.0831	0.7857
$v_3=2.0001$	23.3156	0.99987	0.5689	0.8510	0.0822	0.7841
$v_3=2.25$	23.3260	0.99987	0.5178	0.8542	0.0834	0.7856
$v_3=2.50$	23.2961	0.99987	0.5364	0.8529	0.0823	0.7849
$v_3=2.75$	23.3012	0.99987	0.5059	0.8550	0.0852	0.7867
$v_3=2.9999$	23.2978	0.99987	0.5160	0.8541	0.0839	0.7857

为了直观分析表 14.8 中的数据，从而说明分数阶微分阶次 v_3 与 $YiFeiPU$-4 Denoising 的综合去噪效果之间的对应变化关系，我们将表 14.8 中的 PSNR、相关系数、对比度、相关性、能量和同质性/逆差距 6 个参数分别进行 1 至 12 阶的非线性曲线拟合，如图 14.15 所示。

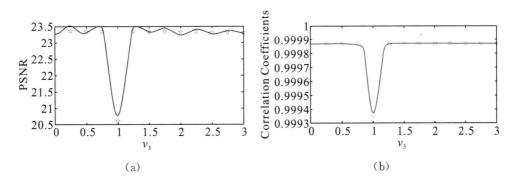

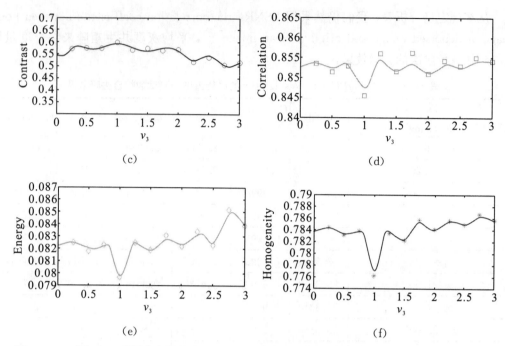

图 14.15 对表 14.8 中的参数分别进行 1 至 12 阶的非线性曲线拟合。(a) PSNR, (b) Correlation Coefficients, (c) Contrast, (d) Correlation, (e) Energy, (f) Homogeneity

分析表 14.8 和图 14.15 可知, 分数阶微分阶次 v_3 与 $YiFeiPU$-4 Denoising 的综合去噪效果之间的对应变化关系也是有规律的。第一, 由表 14.8 和图 14.15 (a) 可知, 当 v_3 逐渐增大到 $v_3=0.25$ 左右时, 其 PSNR 增大到最大值, 当 v_3 逐渐增大到 $v_3=1.0001$ 左右时, 其 PSNR 锐减到最小值。随着 v_3 的进一步逐渐增大, 其 PSNR 呈现略微下降的趋势, 并且 PSNR 均在 $v_3=0.25 \times p$, $p \neq 1$ 为奇数的地方出现局部极大值。说明当 $v_3=0.25$ 时, 以及 $v_3=0.25 \times p$, $p \neq 1$ 为奇数时, $YiFeiPU$-4 Denoising 的去噪程度相对较强, 其噪声图像和原始图像的吻合得相对较好, 相似度相对较高。同时, 亦说明基于一阶最速下降法的一阶偏微分热传导方程的综合去噪效果最差。这是由于图像的一阶微分, 其图像边缘和纹理要么是黑, 要么是白。一阶微分对图像的纹理细节的损失最大的原因所致[185-188,434-435,468]。第二, 由表 14.8 和图 14.15 (b) 可知, 随着 v_3 的逐渐增大, 其相关系数除了在 $v_3=1.0001$ 左右时出现锐减外, 其余地方均保持恒定值。说明 $v_3=1.0001$ 左右时, $YiFeiPU$-4 Denoising 的去噪效果最差, 其噪声图像和原始图像的吻合得相对最差, 相似度相对最低。第三, 由表 14.8 和图 14.15 (c) ~ (f) 可知, 随着 v_3 的逐渐增大, 其平均灰度共生矩阵中的对比度 (Contrast) 在 $v_3=1.0001$ 时出现最大值, 其平均灰度共生矩阵中的相关性 (Correlation)、能量 (Energy) 和同质性/逆差距 (Homogeneity) 均在 $v_3=1.0001$ 时出现最小值。除此之外, 它们的值均随 v_3 的逐渐增大而发生小的波动, 且均在 $v_3=0.25 \times p$, $p \neq 1$ 为奇数的地方出现局部极大值或极小值。结合表 14.8 和图 14.15 (a)、(b) 可知, 当 $v_3=1.0001$ 左右时, 虽然其对比度 (Contrast) 相对最大, 其相关性 (Correlation)、能量 (Energy) 和同质性/逆差距 (Homogeneity) 均相对最小, 但其 PSNR 和相关系数相对最小。虽然说明在其噪声图像中对比度大的像素对相对最多, 局部灰度相关性相对最弱, 纹理模式变化相对最不均一和最不规则, 图像纹理的不同区域间变化相对最剧烈,但这是由于其噪声去除不干净造成的,并不能说明其纹理的沟纹最深且视觉效果最清晰。另外,说明当$v_3=0.25$,以及$v_3=0.25\times p$,$p\neq 1$为奇数时,其PSNR和相关系数相对较大,其噪声图像的平均灰度共生矩阵中的相关性(Correlation)、能量(Energy)和同质性/逆差距(Homogeneity)均出现局部极大值或极小值。在其噪声图像中对比度大的像素对相对较多,局部灰度相关性相对较弱,纹理模式变化相对较不均一和较不规则,图像纹理的不同区域间变化相对较剧烈,其纹理的沟纹相对较深且视觉效果相对较清晰。结合上述式(14.31)和(14.71)以及关于分数阶最速下降法的实验可知,这其实本质上是由于当$v_3=0.25$左右时,分数阶最速下降法的分数阶能量范函$\Psi(\beta)$仅存在唯一分数阶极值点的原因所致。综上所述,当分数阶微分阶次$v_3=0.25$左右时,YiFeiPU-4 Denoising的综合去噪效果相对最佳,其噪声图像中的纹理细节相对最明显。当$v_3=0.25\times p$,$p\neq 1$为奇数时,YiFeiPU-4 Denoising的综合去噪效果仅次之。当$v_3=1.0001$左右时,YiFeiPU-4 Denoising的综合去噪效果相对最差。

14.5 结论

近300年以来,分数阶微积分业已成为数学分析的一个重要分支,但对于国内外众多数学家乃至工程技术界的物理学家而言,它还鲜为人知。如何将分数阶微积分这一崭新的数学方法应用于现代信号分析与处理,特别是数字图像处理之中,在国际上是一个研究甚少的新兴学科分支。我们致力于将一种新颖的数学方法——分数阶微积分引入到纹理图像的去噪之中,构造分数偏微分方程,即一类基于分数阶超全变差和分数阶最速下降法的纹理图像多尺度去噪模型。该去噪模型在去除图像噪声的同时,既能尽量保留纹理图像平滑区域中的低频轮廓,又能非线性保留灰度值跃变幅度相对较大的高频边缘和纹理细节,还能非线性保留灰度值跃变幅度变化相对不大的纹理细节。

我们的前期研究成果表明,信号的分数阶微积分特性与其整数阶微分不同,直流或低频信号的分数阶微分一般不为零。信号的分数阶微分既非线性地加强了信号的高频分量,又在一定程度上非线性地加强了信号的中频分量,同时还极大地非线性保留了信号的低频和直流分量。分数阶微分可以非线性地增强图像信号中近似分形的复杂纹理细节。在信号分数阶微分上述特性的基础之上,我们从系统的演化过程的角度,首先,分别论述分数阶泛Green公式、分数阶泛Gauss公式和分数阶泛Stokes公式。其次,在此基础上,论述针对二维图像处理的分数阶泛Euler-Lagrange方程,又进一步论述分数阶最速下降法及其稳定性和收敛率,构造出4种基于分数阶偏微分方程的纹理图像多尺度去噪模型。最后,通过理论推导和对比实验分析论证了分数阶最速下降法的稳定性和收敛性,讨论了该分数阶纹理图像去噪模型的分数阶非线性多尺度去噪能力及其参数的最佳取值。仿真对比实验表明,对于纹理细节信息丰富的图像而言,该分数阶纹理图像去噪模型在去噪的同时,对图像高频边缘和复杂纹理细节的分数阶非线性多尺度保持能力明显优于传统基于整数阶运算的去噪算法。

近年来,随着下一代高清晰数字电视对图像清晰度要求的不断提高,生物医学图像(如细胞图像、X光片、乳腺钼靶片、CT图像、MRI、PET图像、超声图像等)对后期图像处理清晰度要求的不断提高,银行票据智能识别系统对票据复杂纹理细节特征预处理质量要求的不断提高，卫星遥感图像对其复杂地理纹理细节的清晰度要求以及对仅有几个像素的小目标识别能力要求的不断提高，指纹、虹膜、掌纹等生物特征智能识别系统对复杂生物纹理特征预处理质量要求的不断提高，分数偏微分方程即一类基于分数阶超全变差和分数阶最速下降法的纹理图像多尺度去噪模型有望在这些领域中得到广泛的应用。

第15章 图像的分数阶对比度增强：一种Retinex的分数阶变分框架[①]

本章讨论了一种新颖的方法Retinex分数阶变分法，这是一种用于纹理图像的多尺度非局部对比度增强Retinex分数阶偏微分方程（FPDE）。传统的整数阶纹理增强算法在处理振铃、伪影现象等问题上有一些缺陷，这是研究中亟待解决的问题。分数阶微积分具有长期记忆、非局域性、弱奇点等优点，且分数阶微积分能非线性增强图像中复杂纹理细节，在信号处理和图像处理领域的应用中占据突出地位。因此，为了解决传统的整数阶纹理增强算法中存在的上述问题，我们在这里提出了一个有趣的理论研究，是否能用分数阶微积分既保留边缘细节，又能多尺度非局域性增强图像纹理。我们引入了Retinex分数阶变分法，首先用分数阶最速下降法实现分数阶偏微分方程，讨论了限制性分数阶优化算法的实现，并用实验来分析了分数阶偏微分方程在增强对比度的同时保持边缘和纹理细节的能力。分数阶偏微分方程具有保留图像边缘和纹理细节的能力，这使该算法优于传统的整数阶对比度增强算法，特别是对于富含纹理细节的图像，其处理效果尤为明显。

15.1 Introduction

Recent advances in digital image processing have led to an increased interest in research activities related to the domain of the contrast enhancement of texture images. Numerous methods ranging from the simple linear contrast stretch to highly sophisticated adaptive algorithms are available in the literature for image contrast enhancement. Generally, such contrast enhancement algorithms can be classified into two categories: global techniques and adaptive techniques. For the global methods, a single transformation is applied to all the pixels of an image, whereas, the adaptive methods generally involve an input-output mapping procedure that undergoes adaptive changes with the local characteristics of an image. It is well known that histogram equalization (HE) is a widely used global technique for contrast enhancement, whose input-output mapping is determined by the cumulative distribution function of the histogram of an image. Another popular alternative is multi-channel filtering that allows the entire spatial frequency range of the input image to be divided into several spatial frequency bands, and, subsequently,

---

[①] PU Yifei, Patrick Siarry, Amitava Chatterjee, et al. A Fractional-Order Variational Framework for Retinex: Fractional-Order Partial Differential Equation Based Formulation for Multi-scale Nonlocal Contrast Enhancement with Texture Preserving [J]. IEEE Transactions on Image Processing, 2018, 27 (3): 1214-1229.

the mid to high spatial frequency components are suitably amplified relative to the low frequency components, in spatial domain [605], [606]. Some adaptive techniques are based on unsharp masking in which the contrast gain can be adaptively varied as a function of some local characteristics of an image [607] — [610]. In another class of approach, adaptive histogram equalization (AHE) enables the use of regional histograms to create locally varying grey scale transformations, in a bid to improve the image contrast in smaller regions [611] — [615]. To overcome the specific problem of over enhancement in ordinary histogram equalization, some algorithms, such as contrast-limited histogram equalization (CLHE) and contrast-limited adaptive histogram equalization (CLAHE), have been developed, in which the local contrast gain is limited by restricting the height of local histograms [616]. In addition, recovering the illumination from a given image is known to be a mathematically ill-posed problem, and various algorithms have so far been proposed in the literature for this purpose. These algorithms essentially attempt to propose different alternate means of overcoming this limitation. The retinex algorithm was motivated by Land's seminal research on the human visual system [617]. The retinex theory shows that our visual system is able to practically recognize and match colors under a wide range of different illuminations, a property that is commonly referred to as the color constancy phenomenon. As a matter of fact, Land's findings indicated that even when retinal sensory signals originating from different color patches under different illumination levels were identical, subjects were able to correctly identify the color of the surface reflectance. This initial work acted as an inspiration for many different retinex implementations developed later, which can be divided into two major groups. One group of such retinex algorithms use homomorphic filters, namely single scale retinex (SSR) [618] — [624] and multi-scale retinex (MSR) [625] — [627] algorithms, which explore the relative luminance of the image using a variety of image paths or by comparing the current pixel color to a set of random pixels. The other group of these retinex algorithms uses a convolution mask [628] — [632], or variational techniques [633] — [635] to determine a local contrast result. The traditional first-order variational framework for retinex (VFR) adopts a Bayesian view point of the estimation problem, which leads to an algebraic regularization term that contributes to better conditioning of the reconstruction problem. The aforementioned contrast enhancement algorithms are all essentially based on integer-order computations, which have some existing common shortcomings that are needed to be overcome. For example, ringing artifacts and staircase effects often appear around the sharp transitions within a contrast-enhanced texture image and constitute a major source of problem that requires extensive research attention.

In recent times, fractional calculus has evolved as an important, contemporary branch of mathematical analyses [142], [176], [178], [230] — [231], which can be potentially employed in a variety of engineering and other problems. Fractional calculus is as old as the integer-order calculus, although until recently, its applications were exclusively in the domain of mathematics and now seems to be gaining its acceptance as a novel promising mathematical method among the physical scientists and engineering technicians. The basic characteristic feature of fractional calculus is that it extends the concepts of the integer-order difference and Riemann sums. The fractional differential, except based on the Caputo definition, of a Heaviside function is non-zero, whereas its integer-order differential must be zero [142], [178]. The random variable in a physical process can be deemed as the displacement of the random movement of particles. The fractional calculus of various functions possesses one obvious feature: the fractional calculus of most functions is equal to a power series, whereas that of the remaining functions is equal to the superposition or product of a certain function and a power function [142], [176], [178], [230] − [231]. Several scientific studies have demonstrated that a fractional-order or a fractional dimensional approach is now the best way that many natural phenomena can be described. This approach has obtained promising results and ideas demonstrating that fractional calculus can be an interesting and useful tool in many scientific fields such as diffusion processes [181] − [183], viscoelasticity theory [126], fractal dynamics [184], fractional control [89] − [90], signal processing [130], [196], [200] − [203], [225], [227], [345], [399], [405], and image processing [186] − [190], [192], [503] − [505].

The application of fractional calculus to signal analysis and signal processing [130], [196], [200] − [203], [225], [227], [345], [399], [405], especially to image processing [186] − [190], [192], [503] − [505], is an emerging field of research which has seldom received desired attention. The properties of the fractional calculus of a signal are quite different from those of its integer order calculus. Fractional calculus has been hybridized with signal processing and image processing mainly because of its inherent strengths in terms of long-term memory, nonlocality, and weak singularity [142], [176], [178], [230] − [231]. Some progress in the studies on fractional-order image processing not only validates the fractional-order partial differential equations, but also provides interesting and practical suggestions for future research. For instance, a fractional differential provides the flexibility of enhancing the complex textural details of an image in a nonlinear manner [186] − [187], [345], [503] − [504], it can maintain the low-frequency contour features in the smooth area of an image in a nonlinear fashion, and creates the possibility of enhancing the high-frequency edges and textural details, in a nonlinear manner, in those areas where the grey level undergoes frequent or unusual variations [186] − [187], [345], [503] − [504]. The fractional-order steepest descent method can determine the fractional-order extreme points of the energy norm, which are quite different from the traditional integer-order extreme points, such as the first-order stationary points [192]. Therefore, to solve the aforementioned common problems of the traditional integer-order contrast enhancement algorithms, an interesting theoretical problem emerges naturally: would it be possible to hybridise the capability to preserve the edges and textural details of fractional calculus with texture image multi-scale nonlocal contrast enhancement, what is the fractional-order variational framework for retinex, and how can VFR [633] — [635] be generalised to a fractional-order framework. Motivated by this need, in this chapter, we introduce a novel conceptual formulation of the fractional-order variational framework for retinex, which is a fractional-order partial differential equation (FPDE) formulationof retinex for the multi-scale nonlocal contrast enhancement of texture images. In particular, the reverse incremental optimal search of the fractional-order variational framework for retinex is based on the fractional-order steepest descent method [192]. In comparison with the method of image contrast enhancement proposed in [187], the fractional differential-based approach utilizes directly fractional differential masks, fractional differential operators, for the purpose of processing the grey values of an image [187], while the fractional-order variational framework for retinex proposed by this work uses the FPDE to implement a restrictive fractional-order optimization algorithm. The numerical implementation of the FPDE makes use of the fractional differential masks proposed by [187]. The capability of preserving the edges and textural details of the FPDE is an important advantage that leads to the superiority of the proposed approach compared to the traditional integer-order computation-based contrast enhancement algorithms, especially for images rich in textural details.

The rest of the manuscript is organized as follows: Section 2 presents in brief the necessary mathematical background of fractional calculus, required for subsequent presentation of the work carried out. Section 3 proposes the formulation of a fractional-order variational framework for retinex, which is a FPDE formulationof retinex for the multi-scale nonlocal contrast enhancement of texture images. At first, we implement the FPDE by means of the fractional-order steepest descent method. This is followed by the formulation and implementation of the restrictive fractional-order optimization algorithm. Section 4 presents the experiment results obtained and the associated analyses carried out. Here, at first, we discuss the numerical implementation of the FPDE. Second, we analyze the capability of preserving the edges and textural details of the FPDE, while enhancing the contrast. Third, we further analyze the influence of variations of the parameters of the FPDE in achieving the final result. In Section 5, the conclusions of this manuscript are presented.

15.2 Mathematical Background

This section presents a brief introduction to the necessary mathematical background of fractional calculus.

The commonly used fractional calculus definitions are those of Grünwald-Letnikov, Riemann-Liouville, and Caputo [142], [176], [178], [230] — [231]. The Grünwald-

Letnikov definition of fractional calculus, in a convenient form, for causal signal $f(x)$, is as follows:

$$^{GL}_{a}D^v_x f(x) = \lim_{N \to \infty} \left\{ \frac{\left(\frac{x-a}{N}\right)^{-v}}{\Gamma(-v)} \sum_{k=0}^{N-1} \frac{\Gamma(k-v)}{\Gamma(k+1)} f\left(x - k\left(\frac{x-a}{N}\right)\right) \right\}, \quad (15.1)$$

where $f(x)$ is a differintegrable function [142], [176], [178], [230] − [231], $[a, x]$ is the duration of $f(x)$, v is a non-integer, $\Gamma(\alpha) = \int_0^\infty e^{-x} x^{a-1} dx$ is the Gamma function, and $^{GL}_{a}D^v_x$ denotes the Grünwald-Letnikov defined fractional differential operator. In this work, we mainly adopt the Grünwald-Letnikov defined fractional calculus for the following mathematical derivation. We use the equivalent notations $D^v_x = {}^{GL}_{a}D^v_x$ in an arbitrary, interchangeable manner.

15.3 Fractional-Order Variational Framework for Retinex Based on FPDE

15.3.1 Implementation of FPDE

In this subsection, we implement the FPDE for texture image multi-scale nonlocal contrast enhancement by means of the fractional-order steepest descent method.

It is well known that E. H. Land formulated the retinex theory [617], i.e. the first attempt to simulate and explain how the human visual system perceives color, given as follows:

$$S(x,y) = R(x,y) \cdot L(x,y), \quad (15.2)$$

where $S(x, y)$ denotes the reflected image, $R(x, y)$ denotes the surface reflection characteristics, and $L(x, y)$ denotes the incident light. The physical property of reflecting objects is such that they reflect only part of incident light. Here we further assume that reflectivity $0 < R \leq 1$ and illumination effect $0 < L < \infty$. Since R is restricted to be within the unit interval, from (15.2), we have $L \geq S > 0$. To handle the product form, we first implement the logarithmic transformation of (15.2). Since the logarithmic function is monotonous, we have $l \geq s$. Thus, it follows that:

$$s(x,y) = r(x,y) + l(x,y), \quad (15.3)$$

where $s(x,y) = Log[S(x,y)]$, $r(x,y) = Log[R(x,y)]$, $l(x,y) = Log[L(x,y)]$, and $l \geq s$. Fractional calculus has potentially received prominent applications in the domain of signal processing and image processing mainly because of its strong points like long-term memory, nonlocality, weak singularity [142], [176], [178], [230] − [231], and because of the ability of a fractional differential to enhance the complex textural details of an image in a nonlinear manner [186] − [187], [345], [503] − [504]. Therefore, to solve the aforementioned common problems of the traditional integer-order contrast enhancement algorithms, it is natural to consider whether it will be possible to hybridize the capabilities of fractional calculus to preserve the edges and textural details with texture image multi-scale nonlocal contrast enhancement, to consider the nature of the fractional-order variational framework for retinex, and to consider the generalisation of the VFR [633] - [635] to a fractional-order framework. Further, we can assume that the illumination continues smoothly as a constant beyond the image boundaries. This artificial assumption is required for Neumann boundary condition that would have minor effect on the final results. The fractional-order variational framework for retinex based on a state-of-the-art mathematical method, fractional calculus, is given as:

$$\begin{cases} \text{Minimize}: E(l) = \iint_\Omega [\|D^{v_1} l\|^{v_2} + \alpha_1 |l-s|^{v_2} + \alpha_2 \|D^{v_1}(l-s)\|^{v_2}] dx dy \\ \text{Subject to}: l \geqslant s, \text{ and} \langle D^{v_1} l, \vec{n} \rangle = 0 \text{ on } \partial\Omega \end{cases},$$

(15.4)

where E denotes an energy norm, $\Omega \subset R^2$ denotes an image region, $D^{v_1} = (D_x^{v_1}, D_y^{v_1})$, $(x, y) \in \Omega$, v_1 and v_2 are two fractions, $\partial\Omega$ denotes the boundary of Ω, $\vec{n}$ is the normal to the boundary, and a_1 and a_2 are non-negative real coefficients, respectively. In particular, with regard to a vector $\vec{\Psi} = (\Psi_1, \Psi_2, \cdots, \Psi_n)$, we have $\|\vec{\Psi}\|_2^2 = \sum_{i=1}^n |\Psi_i|^2 = (\vec{\Psi})^T \vec{\Psi} = \vec{\Psi} \cdot \vec{\Psi}$, where symbol • denotes the inner product. For the convenience of illustration, in inner product space, we use the equivalent notations $\| \| = \| \|_2$ in an arbitrary, interchangeable manner. We use the fractional-order norm to model the fidelity term instead of the L2 norm and L1 norm for the purpose of fidelity in (15.4). The fractional-power-exponent v_2 is, in fact, the multi-scale fractional-order power of the energy norm for the edges and the textural details of an image. Here v_2 is the scale factor responsible for controlling the fractional-order power. The capability of the multi-scale filtering to preserve edges and textural details of the fractional-order norm is an important advantage over the conventional L2 norm and L1 norm, especially for images rich in textural details [189] - [190], [505]. It is important to note that, at first, we assume that the incident light is spatially smooth and rich in textural details. Thus, the corresponding fractional-order regularization term is given by $\iint_\Omega \|D^{v_1} l\|^{v_2} dx dy$. Further, we assume that the incident light is close to the reflected image, which means that the fractional-order norm of $r(x, y)$ should be small (i.e. $R(x, y)$ tends towards Black). Thus, the penalty term $\iint_\Omega \alpha_1 |l-s|^{v_2} dx dy$ is used for the purpose of checking fidelity, which makes the proposed model better conditioned. In addition, we assume that the surface reflection characteristics are piecewise constant and rich in textural details. Thus, the corresponding fractional-order total variation is used as the regularization term $\iint_\Omega \alpha_2 |D^{v_1}(l-s)|^{v_2} dx dy$, which represents a fractional-order Bayesian penalty expression. It forces the surface reflection characteristics $r(x, y)$ to be a visually pleasing image, rich in textural details. Therefore, in order to obtain better visual effect for the processed image using the proposed model, we set $\alpha_2 > \alpha_1$ in (15.4). Secondly, v_1 and v_2 are included as only two model parameters in (15.4). As we know, introducing more free parameters in a model and to properly tune them can actually make the attainment of the desired values of these parameters and their convergence more difficult and potentially will increase the associated computational complexity. This may limit the applicability of such a model in real problems. Thus, we choose the same parameters v_1 and v_2 in three different terms in (15.4).

Suppose l is a fractional-order extremal surface of E, then $\xi(x, y) \in C_0^\infty(\Omega)$ is an admissible surface, a test function, closing to l. l and ξ can be merged into a family of surfaces, $s + (\beta-1)\xi$, where β is a small parameter. When $\beta=1$, the family of surfaces, $s + (\beta-1)\xi$, converts into the fractional-order extremal surface, l. The fractional-order derivative of the extreme points equals to zero. The fractional-order fractional extreme point of energy norm is not always coincident with the first-order extreme point [192]. Thus, the anisotropic diffusion of (15.4) can be explained as the fractional-order dissipation process of energy. To achieve the fractional-order minimum of (15.4), a fractional-order energy norm on the family of surfaces, $s + (\beta-1)\xi$, is given as:

$$F(\beta) = \iint_\Omega g(\beta) \mathrm{d}x \mathrm{d}y = \iint_\Omega \{ \|D^{v_1} l + (\beta-1) D^{v_1} \xi\|^{v_2} + \alpha_1 |l - s + (\beta-1)\xi|^{v_2}$$
$$+ \alpha_2 \|D^{v_1}(l-s) + (\beta-1) D^{v_1} \xi\|^{v_2} \} \mathrm{d}x \mathrm{d}y, \qquad (15.5)$$

where $g(\beta)$ is an analytical function. Thus, when $\beta=1$, $D_\beta^{v_3} F|_{\beta=1} = 0$. According to the chain rule of the fractional derivatives of a composite function [230], from (15.1), it follows that:

$$D_\beta^{v_3} g(\beta) = \frac{\beta^{-v_3} g(\beta)}{\Gamma(1-v_3)} + \sum_{n=1}^\infty \frac{(-1)^n \beta^{n-v_3} (D_\beta^n g)}{\Gamma(-v_3)(n-v_3)\Gamma(n+1)}, \qquad (15.6)$$

where v_3 is a fraction and $D_\beta^{v_3} g(\beta)$ is the v_3-order derivative respect to β. In particular, the value of $D_\beta^n g(\beta)|_{\beta=1}$ becomes different, depending on whether n is odd or even. Thus, from (15.5) and (15.6), one obtains (15.7) and (15.8) depending on whether n is odd or even, respectively, given as follows:

$$D_\beta^n g(\beta)|_{\beta=1}^{n=2k+1} = \prod_{\tau=1}^n (v_2 - \tau + 1) \{ \|D^{v_1} l\|^{v_2-n-1} \|D^{v_1} \xi\|^{n-1} (D^{v_1} \xi) \cdot (D^{v_1} l)$$
$$+ \alpha_1 |l-s|^{v_2-n-1} |\xi|^{n-1} \xi(l-s) + \alpha_2 \|D^{v_1}(l-s)\|^{v_2-n-1} \|D^{v_1} \xi\|^{n-1} (D^{v_1} \xi) \cdot [D^{v_1}(l-s)] \},$$
$$(15.7)$$

$$\{D_\beta^n g(\beta)|_{\beta=1}^{n=2k} = \prod_{\tau=1}^n (v_2 - \tau + 1) \{ \|D^{v_1} l\|^{v_2-n} \|D^{v_1} \xi\|^n + \alpha_1 |l-s|^{v_2-n} |\xi|^n$$
$$+ \alpha_2 \|D^{v_1}(l-s)\|^{v_2-n} \|D^{v_1} \xi\|^n] \}, \qquad (15.8)$$

where $k = 0, 1, 2, \cdots$. From (15.5), (15.6), (15.7), and (15.8), when $\beta=1$, it follows that:

$$D_{\beta^3}^{v_2} F|_{\beta=1} = \iint_\Omega \sum_{k=0}^\infty \frac{\prod_{\tau=1}^{2k}(v_2 - \tau + 1)}{\Gamma(-v_3)(2k)!} \{\|D^{v_1}l\|^{v_2-2k-2}\|D^{v_1}\xi\|^{2k}(D^{v_1}l) \cdot D^{v_1}\chi_1$$

$$+\alpha_1|l-s|^{v_2-2k-2}|\xi|^{2k}(l-s)\chi_2 + \alpha^2\|D^{v_1}(l-s)\|^{v_2-2k-2}\|D^{v_1}\xi\|^{2k}[D^{v_1}(l-s)] \cdot D^{v_1}\chi_2\}dxdy$$

$$= \iint_\Omega \sum_{k=0}^\infty \frac{\prod_{\tau=1}^{2k}(v_2 - \tau + 1)}{\Gamma(-v_3)(2k)!}$$

$$\left\{\begin{array}{l} \frac{\Gamma(2k-v_3)}{\Gamma(2k-v_3+1)}[\|D^{v_1}l\|^{v_2-2k}]\|D^{v_1}\xi\|^{2k} + \alpha_1|l-s|^{v_2-2k}|\xi|^{2k} \\ +\alpha_2\|D^{v_1}(l-s)\|^{v_2-2k}\|D^{v_1}\xi\|^{2k} \\ -\frac{(v_2-2k)\Gamma(2k-v_3+1)}{(2k+1)\Gamma(2k-v_3+2)}(\alpha_1|l-s|^{v_2-2k-2}(l-s))[|\xi|^{2k}\xi] \\ -\frac{(v_2-2k)\Gamma(2k-v_3+1)}{(2k+1)\Gamma(2k-v_3+2)}[(\|D^{v_1}l\|^{v_2-2k-2}(D^{v_1}l) \\ +\alpha_2\|D^{v_1}(l-s)\|^{v_2-2k-2}D^{v_1}(l-s) \cdot (\|D^{v_1}\xi\|^{2k}D^{v_1}\xi)] \end{array}\right\} dxdy = 0, \quad (15.9)$$

where χ_1 and χ_2 are given as:

$$\chi_1 = \frac{\Gamma(2k-v_3)}{\Gamma(2k-v_3+1)}l - \frac{(v_2-2k)\Gamma(2k-v_3+1)}{(2k+1)\Gamma(2k-v_3+2)}\xi, \quad (15.10)$$

$$\chi_2 = \frac{\Gamma(2k-v_3)}{\Gamma(2k-v_3+1)}(l-s) - \frac{(v_2-2k)\Gamma(2k-v_3+1)}{(2k+1)\Gamma(2k-v_3+2)}\xi, \quad (15.11)$$

and $\prod_{\tau=1}^n (v_2 - \tau + 1) \stackrel{n=0}{=} 1$. Because the test function ξ is arbitrary, the corresponding $D^{v_1}\xi$ is also arbitrary. According to the fractional-order Euler-Lagrange formula [189] – [190], [505], we can derive that $\frac{\Gamma(1+v_1)}{\Gamma(v_1)}[D_x^1\varphi_x + D_y^1\varphi_y] = 0$ is the fractional-order Euler-Lagrange formula of $\iint_\Omega D^{v_1}u \cdot \vec{\varphi}\,dxdy = 0$, where $u(x,y)$ denotes an arbitrary scalar function and $\vec{\varphi} = (\varphi_x, \varphi_y)$ denotes a vector function. We can consider that to enable $\iint_\Omega [A(x,y)z(x,y) + \vec{B} \cdot D^{v_1}z]dxdy = 0$ be set up, a necessary condition can be given as $\begin{cases} \iint_\Omega [A(x,y)z(x,y)]dxdy = 0 \\ \iint_\Omega \vec{B} \cdot D^{v_1}z\,dxdy = 0 \end{cases}$, where $A(x,y)$ is a scalar function and $\vec{B} = (B_x, B_y)$. Thus, we consider that if $z(x,y) \in C_0^\infty(\Omega)$ is arbitrary, a necessary condition for $\iint_\Omega [A(x,y)z(x,y) + \vec{B} \cdot D^{v_1}z]dxdy = 0$ to be satisfied is $A(x,y) - \frac{\Gamma(1+v_1)}{\Gamma(v_1)}[D_x^1 B_x + D_y^1 B_y] = 0$. Thus, inspired by the fractional-order Euler-Lagrange formula [189] – [190], [505], and the fundamental lemma of variation [485], [536], from (15.9), it follows that:

$$\sum_{k=0}^{\infty} \frac{\prod_{\tau=1}^{2k}(v_2-\tau+1)}{\Gamma(-v_3)(2k)!} \left\{ \begin{array}{l} \frac{\Gamma(2k-v_3)}{\Gamma(2k-v_3+1)}[\|D^{v_1}l\|^{v_2-2k}+\alpha_1|l-s|^{v_2-2k}+\alpha_2\|D^{v_1}(l-s)\|^{v_2-2k}] \\ -\frac{(v_2-2k)\Gamma(2k-v_3+1)}{(2k+1)\Gamma(2k-v_3+2)}\alpha_1|l-s|^{v_2-2k-2}(l-s) \\ +\frac{\Gamma(1+v_1)(v_2-2k)\Gamma(2k-v_3+1)}{\Gamma(v_1)(2k+1)\Gamma(2k-v_3+2)} \begin{bmatrix} D_x^1(\|D^{v_1}l\|^{v_2-2k-2}(D_x^{v_1}l) \\ +\alpha_2\|D^{v_1}(l-s)\|^{v_2-2k-2}D_x^{v_1}(l-s)) \\ +D_y^1(\|D^{v_1}l\|^{v_2-2k-2}(D_y^{v_1}l) \\ +\alpha_2\|D^{v_1}(l-s)\|^{v_2-2k-2}D_y^{v_1}(l-s)) \end{bmatrix} \end{array} \right\} = 0,$$

(15.12)

where $\prod_{\tau=1}^{n}(v_2-\tau+1)\overset{n=0}{=}1$. To enable (15.9) to be implemented, (15.12) is a necessary condition. The nonlinear solution procedure of (15.12) uses a parabolic equation with time as an evolution parameter (the time marching method), or equivalently, the fractional-order gradient descent method [192], [487]. Note that, we treat an image as a function of time and space, and use the time marching method to search the fractional-order steady state of (15.12). This means that we solve:

$$\frac{\partial^{v_3}l}{\partial t^{v_3}} = -\sum_{k=0}^{\infty} \frac{\prod_{\tau=1}^{2k}(v_2-\tau+1)}{\Gamma(-v_3)(2k)!} \left\{ \begin{array}{l} \frac{\Gamma(2k-v_3)}{\Gamma(2k-v_3+1)}[\|D^{v_1}l\|^{v_2-2k}+\alpha_1|l-s|^{v_2-2k}+\alpha_2\|D^{v_1}(l-s)\|^{v_2-2k}] \\ -\frac{(v_2-2k)\Gamma(2k-v_3+1)}{(2k+1)\Gamma(2k-v_3+2)}\alpha_1|l-s|^{v_2-2k-2}(l-s) \\ +\frac{\Gamma(1+v_1)(v_2-2k)\Gamma(2k-v_3+1)}{\Gamma(v_1)(2k+1)\Gamma(2k-v_3+2)} \begin{bmatrix} D_x^1(\|D^{v_1}l\|^{v_2-2k-2}(D_x^{v_1}l) \\ +\alpha_2\|D^{v_1}(l-s)\|^{v_2-2k-2}D_x^{v_1}(l-s)) \\ +D_y^1(\|D^{v_1}l\|^{v_2-2k-2}(D_y^{v_1}l) \\ +\alpha_2\|D^{v_1}(l-s)\|^{v_2-2k-2}D_y^{v_1}(l-s)) \end{bmatrix} \end{array} \right\},$$

(15.13)

where $\prod_{\tau=1}^{n}(v_2-\tau+1)\overset{n=0}{=}1$. According to the fractional-order steepest descent method [192], (15.13) converges to the v_3-order minimum of (15.4). Equation (15.13) presents a restrictive fractional-order optimization algorithm of (15.4). From (15.4), (15.5), and (15.9), we can see that, compared to the traditional first-order total-variation-based VFR, the FPDE is essentially the fractional-order variational framework for retinex.

15.3.2 Implementation of restrictive fractional-order optimization algorithm

In this subsection, from (15.13), we propose how the restrictive fractional-order optimization algorithm for texture image multi-scale nonlocal contrast enhancement can be formulated and implemented.

With respect to color image processing, the restrictive fractional-order optimization algorithm for texture image multi-scale nonlocal contrast enhancement is given as:

Step 1: Convert the reflected image $S(x, y)$ from RGB color space to HSV one. Only use its V value in the HSV color model, $V(x, y)$, for further processing. To prevent the logarithm of $V(x, y)$ from being negative infinity, from (15.2) and (15.3), let $S(x,y) = Log[V(x,y)+1]$.

Step 2: Make a suitable assumption for the non-negative real coefficients α_1 and α_2 of (15.13). Suppose the initial iterative value of $l(x,y)$ is equal to $s(x,y)$. According to the fractional-order steepest descent method [192], (15.13) iteratively converges to the v_3-order estimated optimal value of $l(x, y)$.

Step 3: From (15.3), we obtain $r(x,y)=s(x,y)-l(x,y)$. The intensity value of the surface reflection characteristics of the original image is $R_V(x,y)=Log^{-1}[r(x,y)]-1$. To prevent $R_V(x,y)$ from excessive texture enhancing by (15.13) and to restrict $R_V(x,y) \in [0,255]$, apply the Gamma correction to $R_V(x,y)$. The Gamma correction of $R_V(x,y)$ with an adjustable parameter γ is defined as $R_V^{Gc}=W \cdot (R/W)^{1/\gamma}$, where R_V^{Gc} is the Gamma correction value of R_V, W is the white value (it is equal to 255 in a 8 bit image and it is also equal to 255 in the value channel of an HSV image). The usual adjustable parameter is set to be 2.2 in the tests [635].

Step 4: Let $V(x,y)=R_V(x,y)$. Combine $V(x,y)$ with the H and S values of $S(x,y)$ to obtain the surface reflection characteristics of the original image, $R(x,y)$, in the HSV color space. Convert $R(x,y)$ from an HSV to an RGB colour space. Then, we obtain the contrast-enhanced texture image, $R(x,y)$.

15.4　Experiment and Analysis

15.4.1　Numerical implementation of FPDE

In thissubsection, we discuss the numerical implementation of the FPDE.

At first, we should numerically implement the fractional-order differential operators, D^v, D_x^v, and D_y^v, of a two-dimensional digital image in (15.13). Equation (15.1) shows that for the Grünwald-Letnikov definition of fractional calculus, the limit symbol may be removed when N is sufficiently large. Thus, the convergence rate and accuracy are improved by introducing a signal value at a non-node into the Grünwald-Letnikov definition, i.e. ${}_a^{GL}D_x^v f(x) \cong \frac{x^{-v} N^v}{\Gamma(-v)} \sum_{k=0}^{N-1} \frac{\Gamma(k-v)}{\Gamma(k+1)} f(x+\frac{vx}{2N}-\frac{kx}{N})$. Using the Lagrange interpolation polynomial when three points to perform the fractional-order interpolation, we can obtain the fractional-order differential operator in the eight symmetric directions [186] − [187], [345]. Thus, the discrete definitions of D_x^v and D_y^v in two-dimensional space [187] are as follows:

$$D_x^v(x,y) \cong (\frac{v}{4}+\frac{v^2}{8})s(x+1,y) + (1-\frac{v^2}{2}-\frac{v^3}{8})s(x,y) + \frac{1}{\Gamma(-v)}\sum_{k=1}^{N-2}\left[\frac{\Gamma(k-v+1)}{(k+1)!}\cdot\right.$$
$$\left.\left(\frac{v}{4}+\frac{v^2}{8}\right)+\frac{\Gamma(k-v)}{k!}\cdot(1-\frac{v^2}{4})+\frac{\Gamma(k-v-1)}{(k-1)!}\cdot\left(-\frac{v}{4}+\frac{v^2}{8}\right)\right]s(x-k,y)$$
$$+\left[\frac{\Gamma(N-v-1)}{(N-1)!\Gamma(-v)}\cdot\left(1-\frac{v^2}{4}\right)+\frac{\Gamma(N-v-2)}{(N-2)!\Gamma(-v)}\cdot\left(-\frac{v}{4}+\frac{v^2}{8}\right)\right]s(x-N+1,y)$$

第15章 图像的分数阶对比度增强：一种Retinex的分数阶变分框架

$$+ \frac{\Gamma(N-v-1)}{(N-1)!\Gamma(-v)} \cdot \left(-\frac{v}{4}+\frac{v^2}{8}\right)s(x-N,y), \quad (15.14)$$

$$D_y^v(x,y) \cong \left(\frac{v}{4}+\frac{v^2}{8}\right)s(x,y+1) + \left(1-\frac{v^2}{2}-\frac{v^3}{8}\right)s(x,y) + \frac{1}{\Gamma(-v)}\sum_{k=1}^{N-2}\left[\frac{\Gamma(k-v+1)}{(k+1)!} \cdot \right.$$

$$\left.\left(\frac{v}{4}+\frac{v^2}{8}\right) + \frac{\Gamma(k-v)}{k!} \cdot \left(1-\frac{v^2}{4}\right) + \frac{\Gamma(k-v-1)}{(k-1)!} \cdot \left(-\frac{v}{4}+\frac{v^2}{8}\right)\right]s(x,y-k)$$

$$+ \left[\frac{\Gamma(N-v-1)}{(N-1)!\Gamma(-v)} \cdot \left(1-\frac{v^2}{4}\right) + \frac{\Gamma(N-v-2)}{(N-2)!\Gamma(-v)} \cdot \left(-\frac{v}{4}+\frac{v^2}{8}\right)\right]s(x,y-N+1)$$

$$+ \frac{\Gamma(N-v-1)}{(N-1)!\Gamma(-v)} \cdot \left(-\frac{v}{4}+\frac{v^2}{8}\right)s(x,y-N). \quad (15.15)$$

In particular, when $v=1$, the fractional-order differential operator is converted to the traditional first-order differential operator.

Secondly, we should numerically implement (15.13) using the fractional-order steepest descent method [192]. Suppose that Δt denotes the equal interval of time, $t_n = n\Delta t$, where n denotes the sampling time instant (or the iteration number), and $t_0 = 0$ denotes the initial time, where $n = 0, 1, \cdots$. With respect to (15.13), $l_{x,y}^n = l(x, y, t_n)$ denotes a $l(x, y)$ at time n, $s_{x,y}^0 = s(x, y, t_0)$ denotes the logarithmic transformed value of a reflected image, the input image, at the initial time, and $l_{x,y}^0 = l(x, y, t_0)$ denotes the logarithmic transformed value of the incident light at the initial time. Because $l_{x,y}^0$ is an unknown condition, let us assume $l_{x,y}^0 = 1.05 \cdot s_{x,y}^0$. To simplify the calculation, we only consider $k = 0, 1$ in (15.13). Therefore, according to the fractional-order steepest descent method [192], from (15.13), it follows that:

$$l_{x,y}^{n+1} = P(l_{x,y}^n)\Delta t^{v_3} + l_{x,y}^n - \frac{2\mu}{\Gamma(3-v_3)}(l_{x,y}^{n-1} - l_{x,y}^n)^2 (l_{x,y}^n)^{-v_3}, \quad (15.16)$$

where μ is the convergence rate of the fractional-order steepest descent method, and $P(l_{x,y}^n)$ satisfies:

$$P(l_{x,y}^n) = -\sum_{k=0}^{\infty} \frac{1}{\Gamma(-v_3)(2k)!} \prod_{\tau=1}^{2k}(v_2 - \tau + 1) \left\{ \begin{array}{l} \frac{\Gamma(2k-v_3)}{\Gamma(2k-v_3+1)}\left[\|D^{v_1}l_{x,y}^n\|^{v_2-2k} + \alpha_1|l_{x,y}^n - s_{x,y}^0|^{v_2-2k} \right. \\ \left. + \alpha_2\|D^{v_1}(l_{x,y}^n - s_{x,y}^0)\|^{v_2-2k}\right] - \frac{(v_2-2k)\Gamma(2k-v_3+1)}{(2k+1)\Gamma(2k-v_3+2)} \\ \left[\alpha_1|l_{x,y}^n - s_{x,y}^0|^{v_2-2k-2}(l_{x,y}^n - s_{x,y}^0)\right] \\ + \frac{\Gamma(1+v_1)(v_2-2k)\Gamma(2k-v_3+1)}{\Gamma(v_1)(2k+1)\Gamma(2k-v_3+2)} \left[\begin{array}{l} D_x^1(\|D^{v_1}l_{x,y}^n\|^{v_2-2k-2}(D_x^{v_1}l_{x,y}^n) \\ +\alpha_2\|D^{v_1}(l_{x,y}^n - s_{x,y}^0)\|^{v_2-2k-2}D_x^{v_1}(l_{x,y}^n - s_{x,y}^0)) \\ + D_y^1\left(\begin{array}{l}\|D^{v_1}l_{x,y}^n\|^{v_2-2k-2}(D_y^{v_1}l_{x,y}^n) \\ +\alpha_2\|D^{v_1}(l_{x,y}^n - s_{x,y}^0)\|^{v_2-2k-2}D_y^{v_1}(l_{x,y}^n - s_{x,y}^0)\end{array}\right)\end{array}\right] \end{array}\right\},$$

$$(15.17)$$

where $\prod_{\tau=1}^{n}(v_2 - \tau + 1) \overset{n=0}{=} 1$. In (15.16) and (15.17), let us assume $\mu = 0.1$, $\alpha_1 = 0.05$, and $\alpha_2 = 0.1$ in the following experiments. For the stopping criteria of the numerical implementation of the FPDE, we continue its iteration until $\frac{\|l_{x,y}^{n+1} - l_{x,y}^n\|}{\|l_{x,y}^{n+1}\|} \leqslant \varepsilon_l$ and $\frac{\|(l_{x,y}^{n+1} - s_{x,y}^{n+1}) - (l_{x,y}^n - s_{x,y}^n)\|}{\|l_{x,y}^{n+1} - s_{x,y}^{n+1}\|} \leqslant \varepsilon_{l-s}$. Without loss of generality, let us assume $\varepsilon_l = \varepsilon_{l-s}$

$=10^{-3}$ in the following experiments. Note that, at first, in order to keep stability, the numerical implementation of the FPDE is restricted by Courant-Friedrichs-Lewy (CFL) condition [66]. In particular, with respect to the CFL condition for the fractional-order parabolic equations, we should extend the CFL condition to the fractional-order one. We begin by defining the numerical domain of dependence of a point for a particular solution scheme, for the problem of (15.13). From (15.14) and (15.15), the numerical approximation of the v-order partial differential of $s(x,t)$ can be rewritten as, respectively:

$$D_x^v s(x,t) \cong \frac{1}{(\Delta x)^v \Gamma(-v)} \sum_{k=0}^{N-1} \frac{\Gamma(k-v)}{\Gamma(k+1)} \left[s_k^b + \frac{v}{4}(s_{k-1}^b - s_{k+1}^b) + \frac{v^2}{8}(s_{k-1}^b - 2s_k^b + s_{k+1}^b) \right],$$
(15.18)

$$D_t^v s(x,t) \cong \frac{1}{(\Delta t)^v \Gamma(-v)} \sum_{m=0}^{N-1} \frac{\Gamma(m-v)}{\Gamma(m+1)} \left[s_a^m + \frac{v}{4}(s_a^{m-1} - s_a^{m+1}) + \frac{v^2}{8}(s_a^{m-1} - 2s_a^m + s_a^{m+1}) \right],$$
(15.19)

where $\Delta x = x/N$, $\Delta t = t/N$, $s_a^b = s(s\Delta x, b\Delta t)$, $N \geqslant 2$, is the number of the selected discrete sampling points of a signal for the numerical implementation of its fractional differential, and a and b are two positive integers. Note that when $N=2$, (15.18) and (15.19) are converted into the classical first-order differential. For a digital image, the lattice distance $\Delta x = 1$. Thus, from (15.18) and (15.19), the v-order difference scheme of (15.16) and (15.17) can be derived as:

$$s_a^b = \frac{1}{\left(1 - \frac{v^2}{2} - \frac{v^3}{8}\right)} \left\{ \begin{array}{l} -\left(\frac{v}{4} + \frac{v^2}{8}\right) s_a^{b-1} - \frac{1}{\Gamma(-v)} \sum_{k=1}^{N-2} \left[\begin{array}{l} \frac{\Gamma(k-v+1)}{(k+1)!} \cdot \left(\frac{v}{4} + \frac{v^2}{8}\right) + \frac{\Gamma(k-v)}{k!} \cdot \\ (1 - \frac{v^2}{4}) + \frac{\Gamma(k-v-1)}{(k-1)!} \cdot \left(-\frac{v}{4} + \frac{v^2}{8}\right) \end{array} \right] s_a^{b+k} \\ -\left[\frac{\Gamma(N-v-1)}{(N-1)!\Gamma(-v)} \cdot \left(1 - \frac{v^2}{4}\right) + \frac{\Gamma(N-v-2)}{(N-2)!\Gamma(-v)} \cdot \left(-\frac{v}{4} + \frac{v^2}{8}\right) \right] s_a^{b+N-1} \\ -\frac{\Gamma(N-v-1)}{(N-1)!\Gamma(-v)} \cdot \left(-\frac{v}{4} + \frac{v^2}{8}\right) s_a^{b+N} \\ +\left(\frac{c\Delta t}{\Delta x}\right)^v \sum_{k=0}^{N-1} \frac{\Gamma(k-v)}{\Gamma(k+1)} \left[s_k^b + \frac{v}{4}(s_{k-1}^b - s_{k+1}^b) + \frac{v^2}{8}(s_{k-1}^b - 2s_k^b + s_{k+1}^b) \right] \end{array} \right\},$$
(15.20)

where c is a positive constant. We set $R = (c\Delta t/\Delta x)^v$. Thus, from (15.20), it is then clear that the solution at a point $(b\Delta x, a\Delta t)$ depends on the values at the nine lattice point, as illustrated in Fig. 15.1 (a).

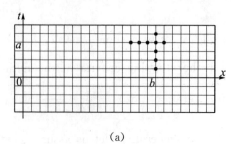

(a)

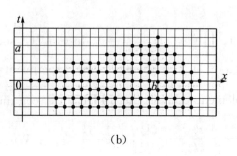

(b)

Fig. 15.1 Numerical domain of dependence of point ($b\Delta x$, $a\Delta t$) for v-order difference scheme (20) where $R=1(\Delta t/\Delta x)^v$. (a) Dependence at ath time level of solution at point ($b\Delta x$, $a\Delta t$), (b) Numerical domain of dependence of point ($b\Delta x$, $a\Delta t$).

In Fig. 15.1, for theconvenience of illustration, without loss of generality, we set $N=5$. Likewise, when N is another positive integer, it has the similar results to those of $N=5$. From Fig. 15.1 (b), we can see that if we continue this process down to the zero time level, the solution depends on the points $(((b-(N-2)(a+1))\Delta x,0),\cdots,(b\Delta x,0),((b+a+1)\Delta x,0))$. For this reason, in this situation we define the numerical domain of dependence of the point ($b\Delta x$, $a\Delta t$) for the fractional-order difference scheme (15.20) to be the interval:

$$D_a = [(b-(N-2)(a+1))\Delta x, (b+a+1)\Delta x]. \tag{15.21}$$

It should be emphasized thatall the points on which the solution at the point ($b\Delta x$, $a\Delta t$) depend for all possible refinements satisfying $R = c(\Delta t/\Delta x)^v$ for a fixed R are contained in the interval D_a. Moreover, the analytic domain of dependence of the point (x, t) for the fractional-order partial differential equation (15.13) is $x_0 = x - ct$. If we consider the point $(x,t)=(b\Delta x, a\Delta t)$, then:

$$x_0 = x - ct = b\Delta x - ca\Delta t = (b - R^{1/v}a)\Delta x. \tag{15.22}$$

Then, from (15.21) and (15.22), we can see that $x_0 \in [(b-(N-2)(a+1))\Delta x, (b+a+1)\Delta x]$, if and only if $(b-(N-2)(a+1))\Delta x \leqslant (b-R^{1/v}a)\Delta x \leqslant (b+a+1)\Delta x$. Therefore, we can derive the following equivalent inequalities:

$$-(a+1)/a < -1 \leqslant R^{1/v} \leqslant N-2 < (N-2)(a+1)/a. \tag{15.23}$$

Equation (15.23) shows that theanalytical domain of dependence is contained in the numerical domain of dependence if and only if $-1 \leqslant R^{1/v} \leqslant N-2$. In particular, when $N=2$, (15.18) and (15.19) are converted into the classical first-order differential, i.e. $v=1$. Thus, we have $-1 \leqslant R = c\Delta t/\Delta x \leqslant 0$, which is the classical first-order CFL condition [548]. The classical first-order CFL condition is a special case of the fractional-order CFL condition. Therefore, we can see that the fractional-order CFL condition for the v-order partial differential equation (15.13) and the v-order difference scheme (15.20) is equivalent to the inequality $-1 \leqslant R^{1/v} \leqslant N-2$. This is a necessary condition for the stability of the numerical implementation of (15.16) and (15.17). Secondly, projecting onto the constraint $l \geqslant s$ is done by $l_{x,y}^{n+1} = \max(l_{x,y}^{n+1}, s_{x,y}^0)$, where max( ) is a maximum function. Thirdly, it may be considered that $l_{x,y}^n = 0$, $\|D^{v_1}l_{x,y}^n\| = 0$, or $\|D^{v_1}(l_{x,y}^n - s_{x,y}^0)\| = 0$ during the numerical iteration computation of (15.16) and (15.17). To enable (15.16) and (15.17) to be implemented, if $l_{x,y}^n = 0$, $\|D^{v_1}l_{x,y}^n\| \leqslant \varepsilon_1$, or $\|D^{v_1}(l_{x,y}^n - s_{x,y}^0)\| \leqslant \varepsilon_1$, let us set $l_{x,y}^n = \varepsilon_2$, $\|D^{v_1}l_{x,y}^n\| = \varepsilon_1$, or $\|D^{v_1}(l_{x,y}^n - s_{x,y}^0)\| = \varepsilon_1$, where ε_1 and ε_2 are two small positive constants, respectively. Without loss of generality, in the following experiments, let us set $\varepsilon_1 = 0.006$ and $\varepsilon_2 = 10^{-5}$, respectively.

15.4.2 Capability of preserving edges and textural details

In thissubsection, we analyze the capability of preserving the edges and textural details of the FPDE.

Example 1: The capability of preserving the edges and textural details of the FPDE was analysed by considering several well established contrast enhancement algorithms for the purpose of comparison i. e. HE, CLHE, AHE, CLAHE, SSR, MSR, and VFR, vis-à-vis our proposed FPDE algorithm and a suitable, benchmark texture image i. e. the boat image was considered from the digital image processing standard image library of the University of Southern California. Note that, the VFR model of Ng et al. [635] is different from that of Kimmel et al. [633], because the former model also considers the reflection function and hence is more appropriate and reasonable for the decomposition. Furthermore, the VFR model of Ng et al. [635] is also different from that of Morel et al. [634], because the former model additionally incorporates some constraints and a fidelity term in the its energy functional. These constraints and the fidelity term guarantee that some theoretical analysis can be performed and established. Thus, we choose the VFR model of Ng et al. [635] to implement the following experiment and perform the analysis in this chapter. With reference to the VFR model of Ng et al. [635], we fix the regularization parameter for the L2 norm of the illumination function, for the fidelity, and for the theoretical setting to be equal to 1, 0.1, and 10^{-5}, and iterations to be equal to 20 for the following tests, respectively. We artificially reduced the contrast of the original image of the boat to 6%, and used this low visibility image as the test image. No *a priori* knowledge regarding the characteristics of the original image is either known or required. Without loss of generality, we set $v_1 = 1.25$, $v_2 = 0.85$, and $v_3 = 0.90$ in (15.17). With regard to (15.4) and (15.17), we choose the energy model of
$$\min_{l} \iint_{\Omega} [\|D^{v_1} l\|^{v_2} + \alpha_1 |l - s|^{v_2} + \alpha_2 \|D^{v_1}(l - s)\|^{v_2}] dx dy.$$
We set $\alpha_1 = 0.05$ and $\alpha_2 = 0.1$ in (15.17). Then, the results of the comparative contrast enhancement experiments for the boat image are shown in Fig. 15.2.

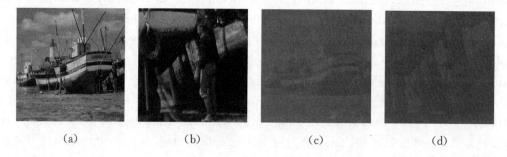

(a)　　　　　　(b)　　　　　　(c)　　　　　　(d)

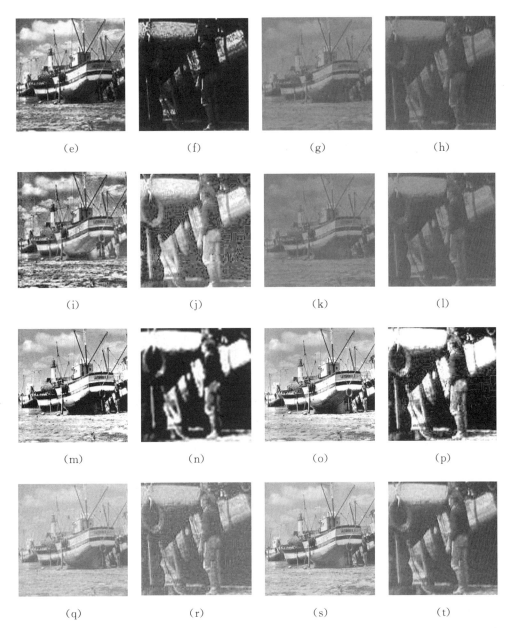

Fig. 15.2 Comparative contrast enhancement experiments for boat. (a) Original image, (b) Enlarged detail of lower left 1/16 of (a), (c) Low visibility test image (contrast=6%), (d) Enlarged detail of lower left 1/16 of (c), (e) HE of (c), (f) Enlarged detail of lower left 1/16 of (e), (g) CLHE of (c), (h) Enlarged detail of lower left 1/16 of (g), (i) AHE of (c), (j) Enlarged detail of lower left 1/16 of (i), (k) CLAHE of (c), (l) Enlarged detail of lower left 1/16 of (k), (m) SSR of (c), (n) Enlarged detail of lower left 1/16 of (m), (o) MSR of (c), (p) Enlarged detail of lower left 1/16 of (o), (q) VFR of (c), (r) Enlarged detail of lower left 1/16 of (q), (s) FPDE of (c), (t) Enlarged detail of lower left 1/16 of (s).

To consider the visual effects for the purpose of comparison, at first, from Fig. 15.2 (e), (f), (i), (j), (m), and (n), we can see that the capability of HE, AHE, and SSR to preserve the edges and textural details are worse than that of the other contrast enhancement algorithms because they can hardly enhance all parts of the image simultaneously. In particular, the extent to which HE and SSR are able to enhance the local shadows is relatively weaker than that of the other contrast enhancement algorithms. AHE tends to over enhance the image contrast if there are high peaks in the histogram, often resulting in a harsh, noisy appearance of the output image. It shows that when contrast enhancement is performed, the capabilities of preserving the edges and textural details of HE, AHE, and SSR are the worst. Secondly, from Fig. 15.2 (g), (h), (o), and (p), we can see that the capabilities of CLHE and MSR to preserve the edges and textural details are worse than that of CLAHE, VFR, and FPDE. The results of their contrast enhancement are relatively clearer than that of HE, AHE, and SSR, but the degree of loss around the edges and of textural details is more extensive than that of CLAHE, VFR, and FPDE algorithms. In particular, the use of CLHE for contrast enhancement produces a slightly halo artifact. The MSR obviously allows the bright local region to become too bright; hence, the contrast formed by the edges and the textural details of the bright local region are largely removed by MSR. Thirdly, from Fig. 15.2 (k), (l), and (q) − (t) we can further see that the capabilities of CLAHE, VFR, and FPDE algorithms to preserve the edges and the textural details are superior to that of the other contrast enhancement algorithms. In particular, the edges and the textural details in Fig. 15.2 (s), and (t) appear to be the most well defined ones. The capabilities of CLAHE and VFR algorithms to preserve the edges and the textural details are worse than that of the FPDE algorithm. Although these two algorithms produce a relatively cleaner contrast enhancement, their ability to maintain edges and textural details is not as good as that of FPDE. The contrast-enhanced images of CLAHE and VFR appear relatively smoother than those produced by FPDE. In comparison, the ability of FPDE to preserve edges and textural details of textured images is better than those of other contrast enhancement algorithms, while the algorithm is simultaneously capable of performing contrast enhancement to satisfaction.

Next, with the objective of performing a quantitative analysis, we obtained the information entropy, contrast, peak signal-to-noise ratio (PSNR), and structural similarity (SSIM) between the contrast-enhanced image and the original image to comprehensively evaluate the contrast enhancement effects of each competing algorithm under consideration. In this regard, SSIM is used as a significant performance index that can be used to measure image degradation as a function of the perceived change in structural information, which is based on the human visual system. Table 15.1 presents a comparative evaluation of several of these performance indices for competing algorithms under consideration.

Table 15.1 Quantitative evaluation of contrast enhancement performances for competing algorithms under consideration.

Algorithms \ Enhancement Effects	Information Entropy	Contrast	PSNR	SSIM
HE	0.2446	2.1510	13.7822	0.8513
CLHE	0.0229	0.1729	15.0260	0.9525
AHE	0.2136	0.2215	13.6392	0.8474
CLAHE	0.0248	0.1788	15.0905	0.9782
SSR	0.2306	2.1880	14.0982	0.8956
MSR	0.4143	3.8083	14.2494	0.9050
VFR	0.0252	0.1867	15.1028	0.9801
FPDE	0.0250	0.1868	15.6931	0.9959

Table 15.1 shows that, firstly, the capabilities of HE, AHE, and SSR algorithms to preserve the edges and textural details are worse than that of the other contrast enhancement algorithms. The PSNR and SSIM of HE, AHE, and SSR are the smallest in comparison. The results in the table show that the high-frequency edges and textural details of the contrast-enhanced image of HE, AHE, and SSR are sharply diffused and smoothed. This, in turn, indicates that the structural similarity between their corresponding contrast-enhanced image versions and the original image is the smallest. However, although HE, AHE, and SSR algorithms produce the largest values for information entropy and contrast, these are the results of their excessive enhancement of bright local regions and noise in the test image. Secondly, the capabilities of CLHE and MSR to preserve edges and textural details are worse than those of CLAHE, VFR, and FPDE algorithms. The PSNR and SSIM of CLHE and MSR are relatively smaller than those of CLAHE, VFR, and FPDE algorithms. The information entropy and contrast values of CLHE and MSR algorithms are relatively smaller than those of HE, AHE, and SSR, but are relatively larger than those of CLAHE, VFR, and FPDE algorithms. The effects of their contrast enhancement are relatively clearer than those of HE, AHE, and SSR algorithms, but the degree of loss around the edges and the effect on textural details are more severe than for CLAHE, VFR, and FPDE algorithms. Thirdly, the capability of FPDE to preserve edges and textural details is obviously superior to those of the other competing contrast enhancement algorithms, especially for images rich in textural detail. In comparison, the PSNR and SSIM of FPDE are larger than those of the other contrast enhancement algorithms, thereby indicating that the structural similarity between its contrast-enhanced image version and the original image is the largest. The information entropy and contrast values obtained with FPDE are comparable to those of CLAHE and VFR algorithms, showing that the effect of its contrast enhancement on bright local regions and noise in the test image are comparable to those of CLAHE and VFR

algorithms. The contrast-enhanced image produced by FPDE is the clearest compared to those produced by the other contrast enhancement algorithms.

Example 2: To extend our analysis of the capability of FPDE algorithm to preserve edges and textural details, we selected those algorithms which proved to be efficient in the aforementioned comparative experiments, namely CLAHE, VFR, and FPDE, to perform comparative experiments on a sequence of photos of a building captured by a Nikon D40 camera using different exposure levels. In order to extend our analysis of the capability of preserving edges and textural details of the same FPDE algorithm, we additionally perform experimental comparisons with identical energy model and its parameter values for both Example 2 and Example 1. The aperture value, ISO speed, and focal distance of the camera were chosen as f/8, ISO-200, and 35 mm respectively. The exposure times of the sequence of different exposure photos of the building were 1/500 s, 1/320 s, 1/200 s, 1/125 s, 1/160 s, 1/50 s, 1/30 s, 1/20 s, 1/13 s, and 1/16 s, respectively. To consider the visual effects for the purpose of comparison, the optimal time of exposure of the optimal exposure photo was chosen as 1/20 s. We chose the minimum exposure photo of this sequence of different exposure photos as the test image, of which the exposure time was 1/500 s. The results of the comparative contrast-enhancement experiments for the building are shown in Fig. 15.3.

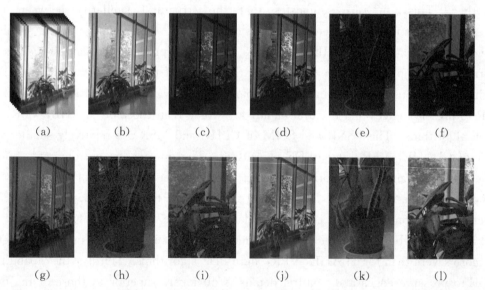

Fig. 15.3 Comparative contrast enhancement experiments of building. (a) Sequence of different exposure photos, (b) Optimal exposure photo of (a), (c) Test image (Minimum exposure photo of (a)), (d) CLAHE, (e) Enlarged detail of lower left 1/16 of (d), (f) Enlarged detail of upper left 1/16 of (d), (g) VFR, (h) Enlarged detail of lower left 1/16 of (g), (i) Enlarged detail of upper left 1/16 of (g), (j) FPDE, (k) Enlarged detail of lower left 1/16 of (j), (l) Enlarged detail of upper left 1/16 of (j).

To consider the visual effects for the purpose of comparison, Fig. 15.3 shows that the similar contrast-enhancement effects were achieved as mentioned above for the comparative experiments. At first, from Fig. 15.3 (d) — (i), we can see that the capabilities of CLAHE and VFR algorithms to preserve edges and textural details are worse than those of the FPDE algorithm. In particular, Fig. 15.3 (d) — (i) show that the contrast enhancement ability of CLAHE is relatively weaker than that of VFR and FPDE. Moreover, the appearance of the contrast-enhanced photo obtained using CLAHE algorithm is significantly darker (Fig. 15.3 (d) — (f)). The edges and the textural details of its stem leaf and flowerpot are significantly weaker than those of FPDE in Fig. 15.3 (j) — (l). Figure 15.3 (g) — (i) show that the degree to which VFR enhances local shadows is relatively weaker than that of FPDE in Fig. 15.3 (j) — (l). VFR can significantly enhance the noise in the photo simultaneously. Secondly, from Fig. 15.3 (j) — (l), one can easily see that the edges and the textural details produced by FPDE are the most distinct ones and this algorithm could preserve the edges and the textural details of its stem leaf and flowerpot better than CLAHE and VFR, while, simultaneously enhancing the contrast as desired. Fig. 15.3 (j) and (k) show that the red colour of the flowerpot could only be restored by FPDE, whereas CLAHE and VFR were unable to do so. The contrast-enhanced image of the FPDE is the clearest compared to those obtained using CLAHE and VFR.

The results of the corresponding quantitative analyses are listed in Table 15.2, where the identical performance indices were considered, as in Table 15.1, to compare the contrast enhancement performances.

Table 15.2 Quantitative evaluation of contrast enhancement performances for competing algorithms under consideration.

Algorithms \ Enhancement Effects	Information Entropy	Contrast	PSNR	SSIM
CLAHE	0.0098	0.1182	13.2810	0.9034
VFR	0.0130	0.1249	14.0451	0.9312
FPDE	0.0126	0.1285	15.2640	0.9963

Table 15.2 shows that with regard to the textured image, the capability of FPDE algorithm to preserve edges and textural details is the best with the simultaneous achievement of contrast enhancement. Hence, the information entropy, contrast, PSNR, and SSIM obtained with FPDE are all relatively higher than those respectively obtained with CLAHE and VFR algorithms. This shows that the high-frequency edges and textural details of the contrast-enhanced image produced by FPDE are well preserved compared to those of CLAHE and VFR. Overall it can be seen that the edges and the textural details of the contrast-enhanced-image of FPDE are the most distinct ones.

In addition, since it is hard to judge the image quality utilizing the entropy and contrast index, we presentthe distortion identification-based image verity and integrity evaluation (DIIVINE) index [67], a non-reference image quality assessment index, to judge the image quality. The DIIVINE index assesses the quality of a distorted image without needing the existence of a reference image. DIIVINE is based on a 2-stage framework involving distortion identification followed by distortion-specific quality assessment. DIIVINE is based on natural scene statistics which govern the behavior of natural images. The DIVINE index values of Fig. 15.3 (b) − (d), (g), and (j) are 0.9640, 0.8253, 0.8650, 0.9082, and 0.9593, respectively. Hence, the DIVINE index obtained with FPDE is relatively higher than those respectively obtained with CLAHE and VFR algorithms. This shows that the high-frequency edges and textural details of the contrast-enhanced image produced by FPDE are well preserved compared to those of CLAHE and VFR.

Example 3: As our next experimental step, we then selected CLAHE, VFR, and FPDE algorithms for the purpose of conducting comparative experiments, on a sea beach image of scanned films. In order to further analyze the capability of preserving edges and textural details of the same FPDE algorithm, we also perform experimental comparisons with identical energy model and its parameter values for both Example 3 and Example 1. The objective of this experiment was to explore the contrast enhancement of scanned films by the FPDE algorithm. These films were scanned by a film scanner from a standard sized area of film which was arbitrarily chosen to be 0.25 inches by 0.25 inches (0.25" by 0.25"). Film scanners can accept either strips of 35 mm or 120 films, or individual slides. Thus, we could obtain a high-quality JPG of a raw scan snippet representing 0.25" by 0.25" of film; e.g. in the case of a 4000 dpi film scanner, our snippet would contain 1000 ×1000 pixels. We used an Epson 4990 film scanner to film scan the film snippet of the sea beach image. The results of the comparative contrast-enhancement experiments carried out for the sea beach image (800×600 pixels) are shown in Fig. 15.4.

Fig. 15.4 Comparative contrast enhancement experiments on a sea beach image of scanned films. (a) Original image, (b) Enlarged detail of lower right 1/16 of (a), (c) CLAHE, (d) Enlarged detail of lower right 1/16 of (c), (e) VFR, (f) Enlarged detail of lower right 1/16 of (e), (g) FPDE, (h) Enlarged

detail of lower right 1/16 of (g).

Once more, considering the visual effects for the purpose of comparison, at first, Fig. 15.4 shows that the contrast-enhancement effects of CLAHE and VFR algorithms are worse than those of FPDE algorithm. Figure 15.4 (c) and (d) show that the contrast-enhanced image produced by CLAHE algorithm has little colour aberration. Figure 15.4 (e) and (f) also show that the contrast-enhanced image obtained using VFR algorithm has little excessive contrast enhancement. On the contrary, Fig. 15.4 (g) and (h) show that the contrast-enhanced image produced by FPDE algorithm does not show any colour aberration. In addition, Fig. 15.4 shows that the capability of FPDE algorithm to preserve the edges and the textural details is the best. Figure 15.4 (g) and (h) also show that the contrast-enhanced image obtained using FPDE algorithm is comparatively clear, compared to other competing algorithms considered. Furthermore, the edges and the textural details of the contrast-enhanced image of FPDE are more distinct than those of CLAHE and VFR algorithms. In particular, Fig. 15.4 (h) shows that the contrast-enhanced edges and textural details of the shadow of the board produced by FPDE algorithm are more distinct than those enhanced by CLAHE and VFR algorithms. Thus, FPDE algorithm has a good ability to preserve the edges and the textural details of an image, while enhancing the contrast at the same time.

The results of the corresponding quantitative analyses are listed in Table 15.3, considering the identical performance indices that were utilized in Table 15.1 and Table 15.2 to quantitatively compare contrast enhancement performances of the competing algorithms at hand.

Table 15.3 Quantitative evaluation of contrast enhancement performances for competing algorithms under consideration.

Algorithms	Enhancement Effects	Information Entropy	Contrast	PSNR	SSIM
CLAHE		0.0015	0.0106	15.7915	0.9950
VFR		0.0010	0.0109	21.5361	0.9973
FPDE		0.0018	0.0115	23.8730	0.9997

The results in Table 15.3 show that with regard to the textured image, the capability of FPDE to preserve the edges and the textural details is the best while enhancing the contrast at the same time. The information entropy, contrast, PSNR, and SSIM of FPDE are all higher than those obtained with CLAHE and VFR.

In addition, the corresponding computational time in terms of CPU time required for processing Fig. 15.4 by FPDE, CLAHE, and VFR algorithms are computed as 25.70, 0.49, and 6.35 seconds, respectively, under MATLAB R2009a environment, with 3.00 GB RAM, and Intel (R) Core (TM) i5 CPU 2.40 GHz. From (15.16) and (15.17), we can see that because the fractional differential method is a nonlinear computation model, the computational complexity of FPDE algorithm is obviously higher than those of CLAHE and VFR algorithms. The time-consuming computation of FPDE algorithm is a major disadvantage compared to the traditional integer-order computation-based contrast enhancement algorithms.

Example 4: Next, we used six under-exposed real images to analyze the capability of FPDE to preserve edges and textural details. In order to achieve uniform comparative conclusions, we perform analysis with different energy models and the associated parameter values of Example 4 with those of Example 1. With regard to (15.4) and (15.17), we choose the energy model of $\min\limits_{l}\iint\limits_{\Omega}[\|D^{v_1}l\|^{v_2}+\alpha_2\|D^{v_1}(l-s)\|^{v_2}]\mathrm{d}x\mathrm{d}y$. We set $\alpha_1=0$ and $\alpha_2=0.1$ in (15.17). The results of the contrast-enhancement experiments for these six under-exposed textured images are shown in Fig. 15.5.

Fig. 15.5 Contrast enhancement experiments of FPDE for six under-exposed real images. (a) Original image1, (b) Contrast enhancement of (a), (c) Original image2, (d) Contrast enhancement of (c), (e) Original image3, (f) Contrast enhancement of (e), (g) Original image4, (h) Contrast enhancement of (g), (i) Original image5, (j) Contrast enhancement of (i), (k) Original image1, (l) Contrast enhancement of (k).

Once again, let us consider the visual effects for the purpose of comparison. From Fig. 15.5, we can see that the FPDE algorithm is capable of preserving the edges and the textural details of the image, while simultaneously achieving the desired contrast enhancement. In addition, the high-frequency edges and the textural details of the contrast-enhanced images produced by FPDE are more well-preserved. Thus, the contrast-enhanced images obtained by FPDE are significantly more distinct than those of the corresponding original images.

15.4.3 Influence of Parameter Variations on Performances of FPDE

In this subsection, we further analyze the influences on the overall contrast enhancement performances that one can achieve by varying the following parameters of FPDE: the fractional-order v_1, fractional-power-exponent v_2, and fractional-order v_3. In this work we have proposed the application of a novel, state-of-the-art mathematical method, fractional calculus, to the contrast-enhanced textured image produced by the FPDE method. From (15.4), (15.5), and (15.9), we can see that, compared with the traditional first-order total-variation-based VFR algorithm, FPDE is, in essence, the fractional-order variational framework for retinex. To facilitate the comparison and analyses, we chose VFR and FPDE algorithms to analyze the influences of parameter variations, v_1, v_2, and v_3, of FPDE. Moreover, for the convenience of analysis, we perform experimental comparisons with the same energy model and its parameter values as those chosen for Example 1. The results of the contrast-enhancement experiments for the three under-exposed real textured images are shown in Fig. 15.6.

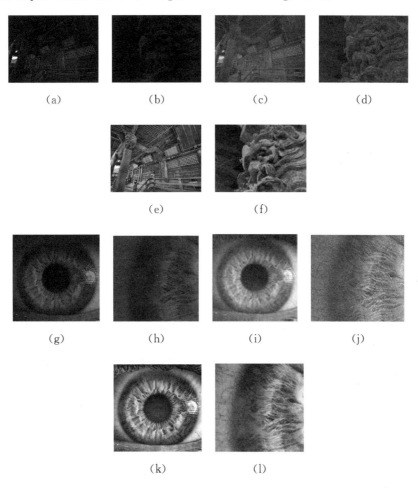

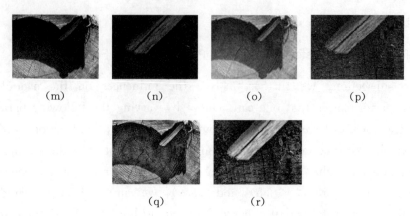

Fig. 15.6 Contrast enhancement experiments of VFR and FPDE. (a) Original image1, (b) Enlarged detail of upper left 1/64 of (a), (c) Contrast enhancement of (a) by VFR, (d) Enlarged detail of upper left 1/64 of (c), (e) Contrast enhancement of (a) by FPDE, (f) Enlarged detail of upper left 1/64 of (e), (g) Original image2, (h) Enlarged detail of upper left 1/64 of (g), (i) Contrast enhancement of (g) by VFR, (j) Enlarged detail of upper left 1/64 of (i), (k) Contrast enhancement of (g) by FPDE, (l) Enlarged detail of upper left 1/64 of (k), (m) Original image3, (n) Enlarged detail of upper left 1/64 of (m), (o) Contrast enhancement of (m) by VFR, (p) Enlarged detail of upper left 1/64 of (o), (q) Contrast enhancement of (m) by FPDE, (r) Enlarged detail of upper left 1/64 of (q).

Furthermore, the comparative quantitative evaluations are listed in Table 15.4.

Table 15.4 Quantitative evaluation of contrast enhancement performances.

Algorithms / Enhancement Effects	Information Entropy	Contrast
VFR of original image 1	0.0031	0.0017
FPDE of original image 1	0.0038	0.0022
VFR of original image 2	0.0023	0.0069
FPDE of original image 2	0.0029	0.0078
VFR of original image 3	0.0011	0.0055
FPDE of original image 3	0.0016	0.0068

From Fig. 15.6 and Table 15.4 we can see that the contrast-enhanced edges and textural details of the wood carving, pupil, and wood grain produced by FPDE algorithm are more distinct than the corresponding features of contrast enhancement achieved using VFR algorithm. The true reason for the achievement of this superior capability of the fractional-order total-variation-based FPDE algorithm in preserving the edges and the textural details compared to the traditional first-order total-variation-based VFR algorithm, while simultaneously enhancing the contrast, can be explained as follows:

At first, from (15.4), (15.5), and (15.9), we can see that when $v_1 = 1$, $v_2 = 2$, and $v_3 = 1$, the FPDE algorithm actually converts to the VFR algorithm. Thus it can be seen that the VFR algorithm is a special case of the FPDE algorithm. Comparing fractional calculus with the traditional integer-order calculus, fractional calculus extends the concepts of the integer-order difference and Riemann sums. The fractional differential of a Heaviside function is non-zero, whereas its integer-order differential must be zero [142], [176], [178], [230] - [231]. Thus, the traditional first-order total-variation-based image processing algorithms are essentially based on the integer-order calculation, and have a common shortcoming, i.e. the ability to easily lose contrast information. These results in the ringing artifacts and staircase effects and a heavier loss may be incurred in terms of edges and textural details. However, previous studies have shown that one of the main reasons why fractional calculus could be successfully incorporated into domain of signal processing and image processing was primarily because of its long-term memory, nonlocality, and weak singularity. As mentioned before, a fractional differential can enhance the complex textural details of an image in a nonlinear manner [186] - [187], [189] - [190], [345], [505]. In addition, a fractional differential can nonlinearly maintain the low-frequency contour features in the smooth area of an image, and nonlinearly enhance the high-frequency edges and textural details in those areas with frequent and unusual changes in the grey level occur [186] - [187], [189] - [190], [345], [505]. From (15.4), (15.5), and (15.9), we can see that compared with the traditional first-order total-variation-based VFR, FPDE is essentially the fractional-order variational framework for retinex. It is primarily because FPDE possesses the properties of fractional calculus, this equips FPDE with the capability of preserving the edges and the textural details while enhancing the contrast is superior to that of the other traditional integer-order computation-based contrast-enhancement algorithms.

Secondly, from (15.4), (15.5), and (15.9), it can also be seen that the capability of FPDE to preserve edges and textural details is the result of the comprehensive functionality of multiple factors, i.e. v_1, v_2, and v_3. The function of v_1 is to extract the fractional-order gradient information of the edges and the textural details of an image. In fact, the v_1-order differential is the multi-scale fractional-order enhancement for the edges and the textural details of an image. Here v_1 is the scale factor responsible for controlling the degree of enhancement. The function of v_2 is to construct the fractional-order total variation of the v_1-order gradient information. The v_2-order power is in fact the multi-scale fractional-order power of the energy norm for the edges and the textural details of an image. Here v_2 is the scale factor responsible for controlling the fractional-order power. Furthermore, the function of v_3 is to determine the fractional-order extreme points of the v_2-order total variation. The fractional-order extreme points of the energy norm are significantly different from the traditional integer-order extreme points such as the first-order stationary points [192]. The reverse incremental search of the fractional-order steepest descent method is in the negative direction of the v_3-order fractional derivative of the v_2-order total variation. The optimal values of v_1, v_2, and v_3 are image-dependent.

15.5 Conclusions

The application of fractional calculus to signal analysis and signal processing, especially in the domain of image processing, is an emerging field of study which has very recently started to receive prominence. The properties of the fractional calculus of a signal are quite different from those of its integer-order calculus. Fractional calculus has been hybridized with signal processing and image processing mainly because of its inherent strength in terms of long-term memory, nonlocality, and weak singularity. The fractional differential can nonlinearly maintain the low-frequency contour features in smooth areas of an image, and nonlinearly enhance the high-frequency edges and textural details in those areas in which frequent and less obvious changes in the grey level occur. Therefore, to solve the aforementioned common problems associated with the traditional integer-order contrast enhancement algorithms, it is natural to consider whether it would theoretically be possible to hybridise the capabilities of fractional calculus to preserve edges and textural details while, simultaneously, enhancing the contrast of a textured image. Moreover, it would also be interesting to consider what the fractional-order variational framework for retinex is, and how to generalize the VFR algorithm to a fractional-order algorithm. Motivated by this need, in this work, it has been proposed to introduce a novel conceptual formulation of the fractional-order variational framework for retinex, which is a FPDE formulationof retinex for the multi-scale nonlocal contrast enhancement of texture images. In particular, the reverse incremental optimal search of the fractional-order variational framework for retinex is based on the fractional-order steepest descent method. The capability of FPDE to preserve edges and textural details is an important advantage which is superior to the abilities of the traditional integer-order computation-based contrast enhancement algorithms, especially for images rich in textural details.

The aforementioned discussion has also highlighted additional problems that need to be further studied. For example, further research is required to determine the image-dependent optimal values of v_1, v_2, and v_3. The authors wish to focus their future scope of research work in this direction.

参考文献

[1] Taylor M E. Partial Differential Equations. II: Qualitative studies of linear equations, Part of the Applied Mathematical Sciences book series (AMS, volume 116) [M]. New York: Springer Verlag, 1996.

[2] Audounet J, Matignon D, Montseny G. Diffusive representations of fractional and pseudo-differential operators (Published Conference Proceedings style), in Nonlinear control in the Year 2000, Part of the Lecture Notes in Control and Information Sciences book series (LNCIS, volume 259) [M]. New York: Springer Verlag, 2000: 163-182.

[3] Samko S G, Kilbas A A, Marichev O I. Fractional Integrals and Derivatives [M]. Yverdon, Switzerland: Gordon and Breach, 1993.

[4] Tatom F B. The Relationship Between Fractional Calculus and Fractals [J]. Fractals-complex Geometry Patterns & Scaling in Nature & Society, 1995, 3 (1): 217-229.

[5] Mandelbrot B B, Van Ness J W. Fractional Brownian Motions, Fractional Noise and Applications [J]. SIAM Review, 1968, 10 (3): 422-437.

[6] YAO Kui, ZHANG Xia. Research Announcements on the Fractional Calculus of a Type of Weierstrass Functions [J]. Advance in Mathematics, 2002, 31 (5): 483-484.

[7] Nigmatullin R R, Mehaute A L. Is there geometrical/physical meaning of the fractional integral with complex exponent? [J]. Journal of Non-Crystalline Solids, 2005, 351 (33-36): 2888-2899.

[8] Meerschaert M M, Mortensen J, Scheffler H P. Vector Grünwald formula for fractional derivatives [J]. Fractional Calculus & Applied Analysis, 2004, 7 (1): 61-81.

[9] Podlubny I. Derivácie neceloíselného rádu: história, teória, aplikácie. Plenárna prednáka na konferencii Slovenskej matematickej spolonosti, Jasná, 23. novembra 2003. (Fractional-order derivatives: History, theory, and applications. Plenary lecture at the Conference of the Slovak Mathematical Society, Jasná, November 23, 2002, in Slovak.)

[10] Ver la referencia bibliográfica. Fractional differential equations, PODLUBNY, Igor, 1999.

[11] Kempfle S, Schaefer I, Beyer H R. Fractional Calculus via Functional Calculus:

Theory and Applications [J]. Nonlinear Dynamics, 2002, 29 (1-4): 99-127.
[12] 吴俊霖. 正交函数运算矩阵及其在微分方程中之应用 [D]. 台北: 成功大学, 2003.
[13] Gorenflo R, Mainardi F. Fractional calculus: Integral and Differential Equations of Fractional Order, in: Carpinteri A, Mainardi F (Eds.), Fractals and Fractional Calculus in Continuum Mechanics [M]. New York: Springer Verlag, 1997: 223-276.
[14] CHEN W. A new definition of the fractional Laplacian, Simula Research Laboratory [J]. Arxiv Cornell University Library, 2002, 69 (11): 30-36.
[15] Chen Y Q, Moore K L. Discretization Schemes for Fractional-Order Differentiators and Integrators [J]. IEEE transactions on circuits and systems—I: fundamental theory and applications, 2002, 49 (3): 363-367.
[16] Ortigueira M D, Machado J A T, da Costa J S. Which differintegration? [J]. IEE Proc. -Vis. Image Signal Process, 2005, 152 (6): 846-849.
[17] Duarte F B M, Machado J A T. Pseudoinverse Trajectory Control of Redundant Manipulators: A Fractional Calculus Perspective [C]. Proceeding of IEEE International Conference on Robotics & Automation, Washington DC, 2002: 2406-2411.
[18] Engheta N. Fractionalization methods and their applications to radiation and scattering problems [J]. International Conference on Mathematical Methods in Electromagnetic Theory, 2000, 1: 34-40.
[19] Engheta N. On Fractional Calculus and Fractional Multipoles in Electromagnetism [J]. IEEE Trans on Antennas and Propagation, 1996, 44 (4): 554-566.
[20] Ortigueira M. Fractional discrete-time linear systems [C]. Proceeding of IEEE International Conference on Acoustics, Munich, Germany, 1997: 2241-2245.
[21] Pei S C, Tseng C C. A comb filter design using fractional-sample delay [C]. IEEE Transactions on Circuits & Systems II Analog & Digital Signal Processing, Hong Kong, 1997: 2228-2231.
[22] Akay O, Boudreaux-Bartels G F. Fractional Convolution and Correlation via Operator Methods and an Application to Detection of Linear FM Signals [J]. IEEE transactions on signal processing, 2001, 49 (5): 979-994.
[23] Nader Engheta. On the Role of Fractional Calculus in Electromagnetic Theory [J]. IEEE Antennas and Propagation Magazine, 1997, 39 (4): 35-46.
[24] Matignon D, Montseny G. Analysis and numerical simulation of long-memory viscoelastic systems by means of diffusive representations [C]. International Conference on Scientific Computations, Beirut, Lebanon, 1999: 421-430.
[25] Bender C, Elliott R J. A Note on the Clark-Ocone Theorem for Fractional Brownian Motions with Hurst Parameter bigger than a Half [J]. Stochasticsan International Journal of Probability & Stochastic Processes, 2003, 75 (6): 391-405.
[26] Baeumer B, Meerschaert M M. Stochastic solutions for fractional cauchy problems

[J]. Fractional Calculus & Applied Analysis, 2001, 4 (4): 481-500.

[27] Liu F, Anh V, Turner I, et al. Numerical simulation for solute transport in fractal porous media [J]. Anziam Journal, 2004, 217 (1): 159-165.

[28] Herrick M G, Benson D A, Meerschaert M M, et al. Hydraulic conductivity, velocity, and the order of the fractional dispersion derivative in a highly heterogeneous system [J]. Water Resources Research, 2001, 38 (11): 1227-1239.

[29] Meerschaert M M, Tadjeran C. Finite difference approximations for fractional advection-dispersion &ow equations [J]. Journal of Computational and Applied Mathematics, 2004, 172: 65-77.

[30] Chen W, Holm S. Physical interpretation of fractional diffusion-wave equation via lossy media obeying frequency power law [A]. Arxiv Cornell University Library, 2003, 15.

[31] Reyes-Melo M E, Martinez-Vega J J, Guerrero-Salazar C A, et al. Application of fractional calculus to modelling of relaxation phenomena of organic dielectric materials [C]. Proceedings of IEEE International Conference on Solid Dielectrics, Toulouse, France, 2004, 2: 530-533.

[32] Paulin M G, Hoffman L F, Assad C. Dynamics and the Single Spike [J]. IEEE Transactions on neural networks, 2004, 15 (5): 987-994.

[33] Engheta N. On fractional calculus and fractional multipoles in electromagnetism [J]. IEEE Transactions on antennas and propagation, 1996, 44 (4): 554-566.

[34] Batista A G, Ortigueira M D, Rodrigues M. Time-Frequency and Time-Scale characterisation of the beat-by-beat High Resolution-Electrocardiogram. https://www.researchgate.net/profile/Manuel_Ortigueira/publication/237835204_Time-Frequency_and_TimeScale_Characterisation_of_the_Beat-by-Beat_High-Resolution_Electrocardiogram/links/55dc51f008ae9d6594940fee.pdf?origin=publication_detail.

[35] Feeny B F, Lin G, Das T. Reconstructing the Phase Space With Fractional Derivatives [C]. Proceedings of DETC'03 Proceedings of DETC03 ASME 2003 Design Engineering Technical Conferences and Computers and Information in Engineering Conference Chicago, Illinois, USA, September 2-6, 2003: 1-11.

[36] Dorák U, Petrá I, Terpák J, et al. Comparison of the methods for discrete approximation of the fractional-order operator [J]. Acta Montanistica Slovaca, 2003, 8 (4): 179-188.

[37] Poty A, Melchior P, Orsoni B, et al. ZV and ZVD shapers for explicit fractional derivative systems [C]. Proceedings of ICAR 2003 The 11th International Conference on Advanced Robotics Coimbra, Portugal, June 30-July 3, 2003: 399-344.

[38] Matignon D, Montseny G, eds. Fractional Differential Systems: models, methods

and applications, vol. 5 of ESAIM: Proceedings, December 1998, SMAI. URL: http://www.emath.fr/Maths/Proc/Vol.5/index.htm.

[39] Audounet J, Matignon D, Montseny G. Opérateurs différentiels fractionnaires. opérateurs pseudo-différentiels. représentations diffusives. Journées Thématiques Opérateurs pseudo-différentiels et représentations diffusives en modélisation, contrôle et signal, November 1999. URL: http://www.laas.fr/gt-opd/.

[40] Hwang C, Leu J F, Tsay S Y. Technical Notes and Correspondence A Note on Time-Domain Simulation of Feedback Fractional-Order Systems [J]. IEEE Transactions on Automatic Control, 2002, 47 (4): 625-631.

[41] Ortigueira M D. Introduction to fractional linear systems. Part 2: Discrete-time case [J]. IEE proceedings on vision, image and signal processing, 2000, 147 (1): 71-78.

[42] Unser M, Horbelt S, Blu T. Fractional derivatives, splines and tomography [C]. Proc. European Signal Processing Conference (EUSIPCO'2000), Tampere, Finland, September 5-8, 2000: 1-4.

[43] Chen W. A note on fractional derivative modeling of broadband frequency-dependent absorption: Model III [A]. Arxiv Cornell University Library, 2002.

[44] Farid H. Discrete-Time Fractional Differentiation from Integer Derivatives. https://www.researchgate.net/publication/228584180_Discrete-Time_Fractional_Differentiation_from_Integer_Derivatives, 2004.

[45] Tseng C C. Improved Design of Fractional Order Differentiator Using Fractional Sample Delay [J]. IEEE International Symposium on Circuits & Systems, 2005, 4: 3713-3716.

[46] WEI Y H, YUAN X, TENG X D, et al. Generalized Hilbert Transform and Digital Realization [J]. Journal of University of Electronic Science & Technology of China, 2005, 34 (2): 175-178.

[47] Anh V V. Mcvinish R. Fractional Differential Equations Driven by Levy noise [J]. Journal of Applied Mathematics and Stochastic Analysis, 2003, 16 (2): 97-119.

[48] Jensen M J. Ordinary Least Squares Estimate of the Fractional Dierencing Parameter Using Wavelets as Derived from Smoothing Kernels [D]. Carbondale: Southern Illinois University, 2012.

[49] Onufrienko V. New Description of Sfatial Harmqnics of Surface Waves [C]. MMET'98 Proceedings, 1998: 219-221.

[50] Vinagre B M. Applications of Fractional Calculus in Control and Signal Processing [C]. Escuela de Ingenier'ýas Industriales, Departamento de Electr'onica e Ingenier'ýa Electromec'anica Universidad de Extremadura, 5-7/06/2001.

[51] Veliev E I, Onufrienko V M. Fractal electrical and magnetical radiators [C]. Physics & Engineering of Millimeter & Submillimeter Waves, Msmw 98 Third International Kharkov Symposium, Kharkov, Ukraine, 1998, 1 (1): 357-360.

[52] Vesma J, Saramaki T. Design and properties of polynomial-based fractional delay filters [C]. IEEE International Symposium on Circuits and Systems, Geneva, Switzerland, 2000, 1 (1): 104-107.

[53] Pullia A, Riboldi S. Time-Domain Simulation of Electronic Noises [J]. IEEE Transactions on Nuclear Science, 2004, 51 (4): 1817-1823.

[54] Pullia A, Maderna M. Computer simulation of the electronic noise of solidstate detectors. 2004: 566-570, http://ieeexplore.ieee.org/stamp/stamp.jsp?arnumber=1352106.

[55] Zaborovski V, Podgurski Y, Yegorov S. New traffic model on the base of fractional calculus. http://www.neva.ru/conf/art/art8.html.

[56] Zaborovsky V, Meylanov R. Peer-to-peer fractal models: a new approach to describe multiscale network processes. http://www.neva.ru/conf/art3/.

[57] Ste$\acute{\text{p}}$hane Mallat. 信号处理的小波导论 [M]. 杨力华, 戴道清, 黄文良, 等, 译. 北京: 机械工业出版社, 2002.

[58] 奥本海姆. 信号与系统（英文版）[M]. 2 版. 北京: 电子工业出版社, 2002.

[59] 崔锦泰. 小波分析导论 [M]. 程正兴, 译. 白居宪, 审校. 西安: 西安交通大学出版社, 1995.

[60] 陈元亨. 信息与信号理论基础 [M]. 北京: 高等教育出版社, 1989.

[61] 科恩. 时频分析: 理论与应用 [M]. 白居宪, 译. 西安: 西安交通大学出版社, 1994.

[62] 张恭庆, 林源渠. 泛函分析讲义 [M]. 北京: 北京大学出版社, 2003.

[63] 张贤达. 现代信号处理 [M]. 2 版. 北京: 清华大学出版社, 2002.

[64] 迈耶. 小波与算子（第一卷）[M]. 尤众, 译. 北京: 世界图书出版公司, 1992.

[65] 袁晓. 一类新的复解析子波构造及其性质研究 [J]. 电子学报, 2000, 28 (4): 123-126.

[66] 袁晓, 虞厥邦, 陈向东, 等. 超高斯谱函数及其时—频局域化特征 [J]. 电子学报, 2001, 29 (1): 80-83.

[67] 陶德元, 袁晓, 何小海. 一类复子波的时—频局域化特征分析 [J]. 电子科技大学学报, 2001, 30 (1): 21-25.

[68] 袁晓, 虞厥邦. 复解析小波变换与语音信号包络提取和分析 [J]. 电子学报, 1999, 27 (5): 142-144.

[69] 袁晓, 虞厥邦. Bubble 小波的正交条件研究 [J]. 电子科技大学学报, 1998, 27 (1): 25-28.

[70] 哈雷特, 克莱逊. 微积分 [M]. 胡乃冏, 邵勇, 徐可, 等, 译. 北京: 高等教育出版社, 2002.

[71] 维德罗. 自适应信号处理 [M]. 王永德, 龙宪惠, 译. 成都: 四川大学出版社, 1989.

[72] 袁晓, 陈向东, 王俊波. 经典规范正交子波的一种简单广义化方法及其应用 [J]. 电子与信息学报, 2002, 24 (12): 1870-1878.

[73] 蒲亦非，袁晓，廖科，等. 连续子波变换数值实现中尺度采样间隔的确定 [J]. 四川大学学报（工程科学版），2004, 36 (6): 111-116.

[74] PU Yifei, YUAN Xiao, LIAO Ke, et al. The Ascertainment of Scale Sampling Step for Numerical Realization of The Continuous Wavelet Transform [C]. Proceedings of the 2004 International Conference on Intelligent Mechatronics and Automation, Chengdu, China, 2004: 842-846.

[75] PU Yifei, LIAO Ke, ZHOU Jiliu, et al. The ascertainment of scale sampling step for numerical realization adopting binary pick sampling of the continuous wavelet transform [C]. Proceedings of the Third International Conference on Machine Learning and Cybernetics, 2004: 2063-2068.

[76] PU Yifei, YUAN Xiao, LIAO Ke, et al. The ascertainment of scale sampling step for numerical realization adopting even dot-and-grid sampling of the continuous wavelet transform [C]. Proceedings of Seventh International Conference on Signal Processing, 2004: 824-829.

[77] Vinagre B M, Chen Y Q. Fractional calculus applications in automatic control and robotics [C]. 41st IEEE CDC, Tutorial workshop 2, Las Vegas, 2002.

[78] Ozaktas H M, Zalevsky Z, Kutay M A. The Fractional Fourier Transform: with Applications in Optics and Signal Processing [M]. England: John Wiley & Sons Ltd., 2001.

[79] McBride A C, Kerr F H. On Namias's fractional Fourier transforms [J]. Ima Journal of Applied Mathematics, 1987, 39 (2): 159-175.

[80] Rocco Andrea, West Bruce. Fractional calculus and the evolution of fractal phenomena [J]. Physica A, 1999, 265 (3-4): 535-546.

[81] Kalia R N, Srivastava H M. Fractional Calculus and Its Applications Involving Functions of Several Variables [J]. Applied Mathematics Letters, 1999, 12 (5): 19-23.

[82] Adda F B. The differentiability in the fractional calculus [J]. Nonlinear Analysis Theory Methods & Applications, 2001, 47 (8): 5423-5428.

[83] Chen M P, Srivastava H M. Fractional Calculus Operators and their Applications Involving Power Functions and Summation of Series [J]. Applied Mathematics and Computation, 1997, 81 (2-3): 287-304.

[84] Ahmad W M, Sprott J C. Chaos in fractional-order autonomous nonlinear systems [J]. Chaos Solitons & Fractals, 2003 (16): 339-351.

[85] Ahmad W M, Harb A M. On nonlinear control design for autonomous chaotic systems of integer and fractional orders [J]. Chaos Solitons & Fractals, 2003, 18 (4): 693-701.

[86] Bárcena R, Delasen M, Garrido A J. Auototuning of fractional order hold circuits for digital control systems [C]. Proceedings of the 2001 IEEE International Conference on Control Applications, Mexico, 2001: 7-12.

[87] Bárcena R, Mdl Sen, Sagastabeitia I, et al. Discrete control for a computer hard disk by using a fractional order hold device [J]. IEE Proceeding on Control Theory Applation, 2001, 148 (2): 117-124.

[88] Vinagre B M, Petráš I, Podlubny I, et al. Using Fractional Order Adjustment Rules and Fractional Order Reference Models in Model-Reference Adaptive Control [J]. Nonlinear Dynamics, 2002, 29 (1-4): 269-279.

[89] Shunji Manabe. A Suggestion of Fractional-Order Controller for Flexible Spacecraft Attitude Control [J]. Nonlinear Dynamic, 2002, 29 (1): 251-268.

[90] Podlubny I, Petráš I, Vinagre B M, et al. Analogue Realizations of Fractional-Order Controllers [J]. Nonlinear Dynamics, 2002, 29 (1-4): 281-296.

[91] Barbosa R S, Machado J A T. Describing Function Analysis of Systems with Impacts and Backlash [J]. Nonlinear Dynamics, 2002, 29 (1-4): 235-250.

[92] OrsoniB, Melchior P, Oustaloup A, et al. Fractional Motion Control: Application to an XY Cutting Table [J]. Nonlinear Dynamics, 2002, 29 (1): 297-314.

[93] Kempfle S, Schäfer I, Beyer H. Fractional Calculus via Functional Calculus: Theory and Applications [J]. Nonlinear Dynamics, 2002, 29 (1-4): 99-127.

[94] Magin R L. Fractional Calculus in Bioengineering: Part1, Part 2 and Part 3, Critical Reviews in Biomedical Engineering [J]. Begell House Inc., 2004, 32 (1-4): 1-378.

[95] DeduranJ A, Kalla S L, Srivastava H M. Fractional calculus and the sums of certain families of infinite series [J]. Journal of Mathematical Analysis and Applications, 1995, 190 (3): 738-754.

[96] Altintas O, Irmak H, Srivastava H M. Fractional calculus and certain starlike functions with negative coefficients [J]. Computers & Mathematics with Applications, 1995, 30 (2): 9-15.

[97] Alcoutlabi M, Martinez-Vega J J. Aplication of fractional calculusto viscoelastic behaviour modeling and to the physical ageing phenomenon in glassy amorphous polymers [J]. Polymer, 1998, 39 (25): 6269-6277.

[98] Chen M P, Irmak H, Srivastava H M. A certain subclass of analytic functions involving operators of fractional calculus [J]. Computers & Mathematics with Applications, 1998, 35 (5): 83-91.

[99] Enelund M, Lesieutre G A. Time domain modeling of damping using anelastic displacement fields and fractional calculus [J]. International Journal of Solids and Structures, 1999, 36 (29): 4447-4472.

[100] Enelund M, Lesieutre G A. An Application of fractional calculus to a new class of multivalent functions with negative coefficients [J]. Computers and Mathematics with Applications, 1999, 38 (5-6): 169-182.

[101] Scalas E, Gorenflo R, Mainardi F. Fractional calculusto and continuous-time finance [J]. Physica A: Statistical Mechanics and its Applications, 2000, 284 (1-

4): 376-384.

[102] Mainardi F, Raberto M, Gorenflo R, et al. Fractional calculusto and continuous-time finance II: the waiting-time distribution [J]. Physica A: Statistical Mechanics and its Applications, 2000, 287: 468-481.

[103] TU S T, WU T C, Srivastava H M, et al. Commutativity of the Leibniz rules in fractional calculus [J]. Computers and Mathematics with Applications, 2000, 40 (2): 303-312.

[104] Carpinteri A, Cornetti P. A fractional calculus approach to the description of stress and strain localization in fractal media [J]. Chaos Solitons & Fractals, 2002, 13 (1): 85-94.

[105] LIN S D, TU S T, Srivastava H M, et al. Certain operators of fractional calculus and their applications to differential equations [J]. Computers and Mathematics with Applications, 2002, 44 (12): 1557-1565.

[106] Perrin E, Harba R, Berzin-Joseph C, et al. nth-order fractional Brownian motion and fractional Gaussian noises [J]. IEEE Trans on Signal Processing, 2001, 49 (5): 1049-1059.

[107] Perez A, D'Attellis C E, Rapacioli M, et al. Analyzing blood cell concentration as a stochastic process [J]. Engineering in Medicine & Biology Magazine IEEE, 2001, 20 (6): 170-175.

[108] Ninness B. Estimation of 1/f noise [J]. IEEE Transactions on Information Theory, 1998, 44 (1): 32-46.

[109] Yazici B, Kashyap R L. Affine stationary processes with applications to fractional Brownian motion [J]. IEEE International Conference on Acoustics, 1997 (5): 3669-3672.

[110] Szu-Chu Liu, Shyang Chang. Dimension estimation of discrete-time fractional Brownian motion with applications to image texture classification [J]. IEEE Trans on Image Processing, 1997, 6 (8): 1176-1184.

[111] Kahng A B. Exploiting fractalness of error surfaces: New methods for neural network learning [J]. IEEE International Symposium on Circuits & Systems, 1992 (1): 41-44.

[112] YANG S, LI Z S, WANG X L. Vessel radiated noise recognition with fractal features [J]. Electronics Letters, 2000, 36 (10): 923-925.

[113] YaziciB, Kashyap R L. A class of second-order stationary self-similar processes for 1/f phenomena [J]. IEEE Transactions on Signal Processing, 1997, 45 (2): 396-410.

[114] Reed I S, Lee P C, Truong T K. Spectral representation of fractional Brownian motion in n dimensions and its properties [J]. IEEE Transactions on Information Theory, 1995, 41 (5): 1439-1451.

[115] LIU J C, Hwang W L, CHEN M S. Estimation of 2-D noisy fractional Brownian

motion and its applications using wavelets [J]. IEEE Transactions on Image Processing, 2000, 9 (8): 1407-1419.

[116] Kawasaki S, Morita H. Evaluation for convergence of wavelet-based estimators on fractional Brownian motion [J]. IEEE International Symposium on Information Theory, 2000: 470.

[117] CHANG Y C, CHANG S Y. A fast estimation algorithm on the Hurst parameter of discrete-time fractional Brownian motion [J]. IEEE Transactions on Signal Processing, 2002, 50 (3): 554-559.

[118] LIU J C, Hwang W L, CHEN M S. Estimation of 2-D noisy fractional Brownian motion and its applications using wavelets [J]. IEEE Transactions on Image Processing, 2000, 9 (8): 1407-1419.

[119] 张贤达. 时间序列分析 [M]. 北京: 清华大学出版社, 1996.

[120] 张贤达. 现代信号处理 [M]. 北京: 清华大学出版社, 1995.

[121] Lamber-Torres G. Application of rough sets in power system control center data mining [J]. Power Engineering Society Winter Meeting, 2002 (1): 627-631.

[122] Planka L, Mrozek A. Rule-based stabilization of the inverted pendulum [J]. Computational Intelligence, 1995, 11 (2): 348-356.

[123] Czogala E. Idea of a rough fuzzy controller and its application to the stabilization of a pendulum-car system [J]. Fuzzy Sets and Systems, 1995, 72 (1): 61-73.

[124] Takagi T, Sugeno M. Fuzzy identification of systems and its applications to modeling and control [J]. IEEE tranctions on systems, man, and cybernetics, 1985 (15): 116-132.

[125] Boss B A. Brief History and Exposition of the Fundamental Theory of Fractional Calculus [R]. Fractional Calculus and its applications, Springer Lecture Notes in Mathematics, New York: Springer-Verlag, 1975, 457: 1-36.

[126] Koeller R C. Applications of the fractional calculus to the theory of viscoelasticity [J]. J. Appl Mech, 1984, 51 (2): 294-298.

[127] Argyris J. Chaotic vibrations of a nonlinear viscoelastic beam [J]. Chaos Solitons Fractals, 1996, 7 (1): 151-163.

[128] 袁晓, 陈向东, 李齐良, 等. 微分算子与子波构造 [J]. 电子学报, 2002, 30 (5): 769-773.

[129] 袁晓, 虞厥邦. 分数导数与数字微分器设计 [J]. 电子学报, 2004, 32 (10): 1658-1665.

[130] Tseng C C. Design of Fractional Order Digital FIR Differentiators [J]. IEEE Signal Processing Letters, 2001, 8 (3): 77-79.

[131] CHEN Y Q, Vinagre B M. A New IIR-Type Digital Fractional Order Dierentiator [J]. Elsevier Science, 2003: 1-12.

[132] Regalia P A. Comments on A Weighted Least-Squares Method for the Design of Stable 1-D and 2-D IIR Digital Filters [J]. IEEE Transactions on Signal

Processing, 1999, 47 (7): 2063-2065.

［133］Samadi S, Igarashi Y, Iwakura H. Design and Multiplierless Realization of Maximally Flat FIR Digital Hilbert Transformers ［J］. IEEE Transactions on Signal Processing, 1999, 47 (7): 1946-1953.

［134］Komodromos M Z, Russell S F, Tang P T P. Design of FIR Hilbert Transformers and Differentiators in the Complex Domain ［J］. IEEE Transactions on Circuits & Systems I Fundamental Theory & Applications, 1998, 45 (1): 64-67.

［135］陈遵德，陈富贵. 非整数阶微积分的滤波特性及数值算法 ［J］. 数值计算与计算机应用, 1999 (1).

［136］朱正佑，李根国，程昌钧. 分数积分的一种数值计算方法及其应用 ［J］. 应用数学和力学, 2003, 24 (4): 331-341.

［137］柯朗，约翰. 微积分和数学分析引论（第二卷，第二分册）［M］. 张恭庆，廖可人，邓东皋，等，译. 叶其孝，校. 北京：科学出版社, 1989.

［138］Oldham K B, Spanier J. The Fractional Calculus ［M］. New York and London: Academic Press, 1974.

［139］林孔容. 关于分数阶导数的几种不同的定义的分析与比较 ［J］. 闽江学院学报, 2003, 24 (5): 3-6.

［140］高安秀树. 分数维 ［M］. 北京：地震出版社, 1989.

［141］Podlubny I. Fractional Differential Equations ［M］. New York: Academic Press, 1999.

［142］Samko S G, Kilbas A A, Marichev O I. Fractional integrals and derivatives: theory and applications ［M］. Yverdon: Gordon & Breach, 1993.

［143］Petráš I, Podlubny I, O'Leary P, et al. Analogue realization of fractional order controllers ［J］. Nonlinear Dynamics, 2002 (29): 281-296.

［144］薛定宇，陈阳泉. 高等应用数学问题的MATLAB求解 ［M］. 北京：清华大学出版社, 2004.

［145］白亿同，罗玉芳，胡永旭. 分数微分积分及其级数展开式 ［J］. 武汉测绘科技大学学报, 1993 (1): 66-75.

［146］罗玉芳，白亿同. 关于处处没有导数的连续函数求分数阶导数的例子 ［J］. 武汉测绘科技大学学报, 1993 (12): 72-75.

［147］倪致祥. 从阶乘的推广到分数阶导数 ［J］. 阜阳师范学院学报（自然科学版）, 2001, 18 (1): 40-43.

［150］Hilfer R. Applications of fractional calculus in physics ［M］. Singapore: World Scientific, 2000.

［151］Podlubny I. Fractional differential equations ［M］. San Diago: Academic Press, 1999.

［152］陆善镇，王昆扬. 实分析 ［M］. 北京：北京师范大学出版社, 1997.

［153］四川大学数学系高等数学教研室. 高等数学（第三版，第一册）［M］. 北京：高等教育出版社 1995.

[154] 赵元英，袁晓，滕旭东，等. 常用周期信号的分数微分运算 [J]. 四川大学学报（工程科学版），2004，36（2）：94-97.

[155] 王国胤. Rough 集理论与知识获取 [M]. 西安：西安交通大学出版社，2001：117-152.

[156] 盛骤，谢式千，潘承毅. 概率论与数理统计 [M]. 北京：高等教育出版社，1993.

[157] 蒲亦非，袁晓，廖科，等. 现代信号分析与处理中分数阶微积分的五种数值实现算法 [J]. 四川大学学报（工程科学版），2005，37（5）：118-124.

[158] PU Yifei, YUAN Xiao, LIAO Ke, et al. AN Efficient Fractional Order Wavelet-based Numerical Engineering Algorithms for Signal's Fractional Calculus [C]. Proceedings of 6th International Progress on Wavelet Analysis and Active Media Technology, Singapore, 2005：683-689.

[159] PU Yifei, YUAN Xiao, LIAO Ke, et al. Theory and Efficient Numerical Algorithms for Signal's Fractional Calculus Based on Fractional Order Wavelet Transform [C]. Proceedings of 6th International Progress on Wavelet Analysis and Active Media Technology, Singapore, 2005：717-724.

[160] Ortigueira M D. Introduction to fractional linear systems. Part 1. Continuous-time case [J]. IEE Proceedings-Vision, Image and Signal Processing, 2000, 147 (1)：62-70.

[161] CHEN Y Q, Moore K L. Discretization Schemes for Fractional-Order Differentiators and Integrators [J]. IEEE Transactions on Circuits & Systems I Fundamental Theory & Applications, 2002, 49 (3)：363-367.

[162] Barcena, Sen D L. On the sufficiently small sampling period for the convenient tuning of fractional-order hold circuits [J]. IEE Proceedings-Control Theory and Applications, 2003, 150 (2)：183-188.

[163] Paliouras V, Dagres J, Tsakalides P, et al. VLSI Architectures for Blind Equalization Based on Fractional-Order Statistics [J]. IEEE International Conference on Electronics, 2002 (2)：799-802.

[164] Ahmad W, EI-khazali R, Eiwakil A S. Fractional-order Wien-bridge oscillator [J]. Electronics Letters, 2002, 37 (18)：1110-1112.

[165] Sugi M, Saito K. Non-integer exponents in electronic circuits: F-matrix representation of the power-law conductivity [J]. Ieice Transactions on Fundamentals of Electronics Communications & Computeres, 1992 (6)：720-725.

[166] Samavati H, Hajimiri A, Shahani A R, et al. Fractal capacitors [J]. IEEE Journal of Solid-State Circuits, 1998, 33 (12)：2053-2041.

[167] Oustaloup A. Fractional Order Sinusoidal Oscillators: Optimization and Their Use in Highly Linear FM Modulation [J]. IEEE Transactions on circuits and system, 1981, 28 (10)：1007-1009.

[168] Kobayashi K, Nemoto Y, Sato R. Equivalent Circuits of Binomial Form

Nonuniform Coupled Transmission Lines [J]. IEEE Transactions on Microwave Theory & Techniques, 1981, 29 (8): 817-824.

[169] 廖科, 袁晓, 蒲亦非, 等. 1/2 阶分数演算的模拟 OTA 电路实现 [J]. 四川大学学报（工程科学版）, 2005, 37 (6): 150-154.

[170] LIAO K, YUAN X, PU Y F, et al. Fractance analog realization using one order Newton method [C]. IEEE 2005 International Symposium on Microwave, Antenna, Propagation and EMC Technologies For Wireless Communications, MAPE 2005: 467-472.

[171] PU Y F, YUAN X, LIAO K, et al. A Recursive Net-Grid-Type Analog Fractance Circuit for Any Order Fractional Calculus [C]. Proceedings of the IEEE Conference on Mechatronics & Automation, Niagara Falls, July 2005: 1375-1380.

[172] PU Y F, YUAN X, LIAO K, et al. Structuring analog fractance circuit for 1/2 order fractional calculus [C]. IEEE Proceedings of 6th International Conference on ASIC, Shanghai, 2005: 1039-1042.

[173] PU Y F, YUAN X, LIAO K, et al. A Recursive Two-Circuits Series Analog Fractance Circuit for Any Order Fractional Calculus [C]. Proceedings of the 20th Conference on SPIE, Changchun, August 2005: 509-519.

[174] ZHOU Jiliu, PU Yifei, YUAN Xiao, et al. Any Fractional Order H Type Analog Fractance Circuit [C]. Proceedings of ASICON 2005, October 2005: 1074-1077.

[175] Oldham K B, Spanier J. The Fractional Calculus: Integrations and Differentiations of Arbitrary-order [M]. New York: Academic Press, 1974: 1-234.

[176] McBride A C. Fractional Calculus [M]. New York: Halsted Press, 1986.

[177] Nishimoto K. Fractional Calculus [M]. New Haven: University of New Haven Press, 1989.

[178] Podlubny I. Fractional Differential Equations: An Introduction to Fractional Derivatives, Fractional Differential Equations, Some Methods of Their Solution and Some of Their Applications [M]. San Diego: Academic Press, 1998: 1-340.

[179] Abramowitz M, Stegun I. Handbook of Mathematical Functions with Formulas, Graphs and Mathematical Tables [M]. Washington D. C.: U. S. Department of Commerce Press, 1964: 1-1046.

[180] Butzer PL, Westphal U. An Introduction to Fractional Calculus [C]. Ch. 1 in Applications of Fractional Calculus in Physics, Singapore: World Scientific, 2000: 1-85.

[181] Özdemir N, Karadeniz D. Fractional Diffusion-Wave Problem in Cylindrical Coordinates [J]. Physics Letters A, 2008, 372 (38): 5968-5972.

[182] Özdemir N, Agrawal O P, Karadeniz D, et al. Analysis of an Axis-Symmetric Fractional Diffusion-Wave Problem [J]. Journal of Physics A: Mathematical and Theoretical, 2009, 42 (35): 1-10.

[183] Povstenko Y. Solutions to the Fractional Diffusion-Wave Equation in A Wedge [J]. Fractional Calculus and Applied Analysis, 2014, 17 (1): 122-135.

[184] Rossikhin Y A, Shitikova M V. Applications of fractional calculus to dynamic problems of linear and nonlinear heredi-tary mechanics of solids [J]. Appl. Mech. Rev., 1997, 50: 15-67.

[185] PU Y F. Research on application of fractional calculus to latest signal analysis and processing [D]. Chengdu: Sichuan University, 2006.

[186] PU Y F, WANG W X, ZHOU J L, et al. Fractional differential approach to detecting textural features of digital image and its fractional differential filter implementation [J]. Science in China Series F: Information Sciences, 2008, 51 (9): 1319-1339.

[187] PU Y F, ZHOU J L, YUAN X. Fractional Differential Mask: A Fractional Differential Based Approach for Multiscale Texture Enhancement [J]. IEEE Trans. Image Processing, 2010, 19 (2): 491-511.

[188] PU Y F, ZHOU J L. A Novel Approach for Multi-scale Texture Segmentation Based on Fractional Differential [J]. International Journal of Computer Mathematics, 2011, 88 (1): 58-78.

[189] PU Y F, Siarry P, ZHOU J L, et al. Fractional Partial Differential Equation Denoising Models for Texture Image [J]. Science China Information Sciences, 2014, 57 (7): 1-19.

[190] PU Y F, Siarry P, ZHOU J L, et al. A Fractional Partial Differential Equation Based Multi-Scale Denoising Model for Texture Image [J]. Mathematical Methods in Applied Sciences, 2014, 37 (12): 1784-1806.

[191] PU Y F, ZHOU J L, Siarry P, et al. Fractional Partial Differential Equation: Fractional Total Variation and Fractional Steepest Descent Approach Based Multi-Scale Denoising Model for Texture Image [J]. Abstract and Applied Analysis, 2013: 1-19.

[192] PU Y F, ZHOU J L, ZHANG Y, et al. Fractional Extreme Value Adaptive Training Method: Fractional Steepest Descent Approach [J]. IEEE Trans. Neural Networks and Learning Systems, 2015, 26 (4): 653-662.

[193] PU Y F, YUAN X, LIAO K, et al. Implement any Fractional Order Neural-Type Pulse Oscillator with Net-Grid-Type Analog Fractance Circuit [J]. Journal of Sichuan University (Engineering Science Edition), 2006, 38 (1): 128-132.

[194] PU Y F. Implement Any Fractional Order Multilayer Dynamic Associative Neural Network [C]. Proc. 6th IEEE Conf. on 'ASIC', Shanghai, 2005: 635-638.

[195] Carlson G E, Halijak C A. Approximation of fractional-order capacitors (1/s) 1/n by a regular Newton process [J]. IEEE Trans. on Circuit Theory, 1964, 11 (2): 210-213.

[196] Roy S C D. On the Realization of A Constant-Argument Immitance of Fractional

Operator [J]. IEEE Trans. on Circuit Theory, 1967, 14 (3): 264-374.

[197] Jones H E, Shenoi B A. Maximally Flat Lumped-Element Approximation to Fractional Operator Immitance Function [J]. IEEE Trans. on Circuit and Systems, 1970, 17 (1): 125-128.

[198] Nakagawa M, Sorimachi K. Basic Characteristics of a Fractance Device [J]. IEICE Trans Fundamentals of Electronics, Communications and Computer Sciences, 1992, E75-A (12): 1814-1819.

[199] Oustaloup A, Levron F, Nanot F, et al. Frequency Band Complex Noninteger Differentiator: Characterization and Synthesis [J]. IEEE Trans. on Circ. and Syst., 2000, 47 (1): 25-40.

[200] CHEN Y Q, Vinagre B M. A New IIR-Type Digital Fractional Order Differentiator [J]. Signal Processing, 2003, 83 (11): 2359-2365.

[201] Elwakil A S. Fractional-Order Circuits and Systems: An Emerging Interdisciplinary Research Area [J]. IEEE Circuits Syst. Mag., 2010, 10 (4): 40-50.

[202] Freeborn T J. A Survey of Fractional-Order Circuit Models for Biology and Biomedicine [J]. IEEE Journal on Emerging and Selected Topics in Circuits and Systems, 2013, 3 (3): 416-424.

[203] Tembulkar T, Darade S, Jadhav S R, et al. Design of Fractional Order Differentiator & Integrator Circuit Using RC Cross Ladder Network [J]. International Journal of Emerging Engineering Research and Technology, 2014, 2 (7): 127-135.

[204] PU Y F. Material Performance Measurement of A Promising Circuit Element: Fractor-Part I: Driving-Point Impedance Function of the Arbitrary-Order Fractor in Its Natural Implementation [J]. Materials Research Innovations, 2015, 19 (S10): 176-182.

[205] PU Y F. Material Performance Measurement of A Promising Circuit Element: Fractor-Part II: Measurement Units and Physical Dimensions of Fractance and Rules for Fractors in Series and Parallel [J]. Materials Research Innovations, 2015, 19 (S10): 183-189.

[206] Dorčák L, Terpák J, Petráš I, et al. Comparison of the Electronic Realization of the Fractional-Order System and Its Model [C]. Proc. 13th International Carpathian Control Conference, High Tatras, 2012: 119-124.

[207] Dorčák L, Terpák J, Petráš I, et al. Electronic Realization of Fractional-Order System [C]. Proc. 2012 International Multidisciplinary Scientific GeoConference, Albena Resort, 2012: 17-23.

[208] Dorčák L, Valsa J, Gonzalez E A, et al. Analogue Realization of Fractional-Order Dynamical Systems [J]. Entropy, 2013, 15 (10): 4199-4214.

[209] Valsa J, Dvorak P, Friedl M. Network Model of the CPE [J]. Radioengineering,

2011, 20 (3): 619-626.

[210] Gonzalez E A, Petráš I. Advances in Fractional Calculus: Control and Signal Processing Applications [C]. Proc. 16th International Carpathian Control Conference, Szilvasvarad, 2015: 147-152.

[211] Gonzalez E A, Dorčák L, Petráš I, et al. On the Mathematical Properties of Generalized Fractional-Order Two-Port Networks Using Hybrid Parameters [C]. Proc. 14th International Carpathian Control Conference, Rytro, 2013: 88-93.

[212] Abulencia G L, Abad A C. Analog Realization of A Low-Voltage Two-Order Selectable Fractional-Order Differentiator in A 0.35um CMOS Technology [C]. Proc. 2015 International Conference on Humanoid, Nanotechnology, Information Technology, Communication and Control, Environment and Management, Cebu City, 2015: 1-6.

[213] Tepljakov A, Petlenkov E, Belikov J. Application of Newton's Method to Analog and Digital Realization of Fractional-order Controllers [J]. International Journal of Microelectronics and Computer Science, 2012, 3 (2): 45-52.

[214] Tepljakov A, Petlenkov E, Belikov J. Efficient Analog Implementations of Fractional-Order Controllers [C]. Proc. 14th International Carpathian Control Conference, Rytro, 2013: 377-382.

[215] Gonzalez E A, Dorčák L, Monje C A, et al. Conceptual Design of A Selectable Fractional-Order Differentiator for Industrial Applications [J]. Fractional Calculus and Applied Analysis, 2014, 17 (3): 697-716.

[216] Chua L O. Memristor-The Missing Circuit Element [J]. IEEE Transactions on Circuit Theory, 1971, CT-18 (5): 507-519.

[217] Chua L O, KANG S M. Memristive Devices and Systems [J]. Proceedings of the IEEE, 1976, 64 (2): 209-223.

[218] Chua L O. Device Modeling via Basic Nonlinear Circuit Elements [J]. IEEE Transactions on Circuit Systems, 1980, CAS-27 (11): 1014-1044.

[219] Chua L O. Nonlinear Circuit Foundations for Nanodevices, Part I: The Four-Element Torus [J]. Proceedings of the IEEE, 2003, 91 (11): 1830-1859.

[220] Chua L O. Resistance Switching Memories are Memristors [J]. Applied Physics A, 2011, 102 (4): 765-783.

[221] Chua L O. The Fourth Element [J]. Proceedings of the IEEE, 2012, 100 (6): 1920-1927.

[222] Prodromakis T, Toumazou C, Chua L O. Two Centuries of Memristors [J]. Nature Materials, 2012 (11): 478-481.

[223] Adhikari S P, Sah M P, Hyongsuk K, et al. Three Fingerprints of Memristor [J]. IEEE Transactions on Circuits and Systems I, 2013, 60 (11): 3008-3021.

[224] PU Y F, YUAN X. Fracmemristor: Fractional-Order Memristor [J]. IEEE Access, 2016 (4): 1872-1888.

[225] PU Y F. Measurement Units and Physical Dimensions of Fractance-Part I: Position of Purely Ideal Fractor in Chua's Axiomatic Circuit Element System and Fractional-Order Reactance of Fractor in Its Natural Implementation [J]. IEEE Access, accepted to be published, 2016.

[226] Moreau X, Khemane F, Malti R, et al. Approximation of a Fractance by a Network of Four Identical RC Cells Arranged in Gamma and a Purely Capacitive Cell [R]. New Trends in Nanotechnology and Fractional Calculus Applications, Springer Science + Business Media, Berlin, 2010.

[227] Carlson G E, Halijak C A. Approximation of fractional capacitors $(1/s)^{1/n}$ by a regular Newton process [J]. IEEE Trans. on Circuit Theory, 1964, 11 (2): 210-213.

[228] Jones H E, Shenoi B A. Maximally Flat Lumped-Element Approximation to A Fractional Operator Immitance Function [J]. IEEE Trans. on Circuit and Systems, 1970, 17 (1): 125-128.

[229] Clauset A, Shalizi C R, Newman M E J. Power-Law Distributions in Empirical Data [J]. SIAM Review, 2009, 51 (4): 661-703.

[230] Oldham K B, Spanier J. The Fractional Calculus: Integrations and Differentiations of Arbitrary Order [M]. New York: Academic, 1974.

[231] Nishimoto K. Fractional Calculus: Integrations and Differentiations of Arbitrary Order [M]. New Haven: University of New Haven Press, 1989.

[232] Steiglitz K. An RC Impedance Approximant to $s^{-1/2}$ [J]. IEEE Trans. on Circuit Theory, 1964, 11 (1): 494-495.

[233] Ferdi Y. Computation of Fractional Order Derivative and Integral via Power Series Expansion and Signal Modeling [J]. Nonlinear dynamics, 2006, 46 (1-2): 1-15.

[234] Ferdi Y. Some Applications of Fractional Order Calculus to Design Digital Filters for Biomedical Signal Processing [J]. Journal of Mechanics in Medicine and Biology, 2012, 12 (2): 501-509.

[235] Machado J A T. Fractional Generalization of Memristor and Higher Order Elements [J]. Communications in Nonlinear Science and Numerical Simulation, 2013, 18 (12): 264-275.

[236] Abdelhouahad M S, Lozi R, Chua L O. Memfractance: A Mathematical Paradigm for Circuit Elements with Memory [J]. International Journal of Bifurcation and Chaos, 2014, 24 (9): 29.

[237] Strukov D B, Snider G S, Stewart D R, et al. The Missing Memristor Found [J]. Nature, 2008, 453 (7191): 80-83.

[238] Borghetti J, Snider G S, Kuekes P J, et al. 'Memristive' Switches Enable 'Stateful' Logic Operations via Material Implication [J]. Nature, 2010, 464 (7290): 873-878.

[239] Valov I, Linn E, Tappertzhofen S, et al. Nanobatteries in Redox-Based Resistive

Switches Require Extension of Memristor Theory [J]. Nature Communications 4, Article number: 1771, 2013.

[240] Meuffels P, Schroeder H. Comment on "Exponential Ionic Drift: Fast Switching and Low Volatility of Thin-Film Memristors" by Strukov D B and Williams R S in Appl. Phys. A, 2009, 94: 515-519 [J]. Applied Physics A, 2011, 105 (1): 65-67.

[241] Meuffels P, Soni R. Fundamental Issues and Problems in the Realization of Memristors [A]. Cornell University Library, 2012: 14. Available: http://arxiv.org/abs/1207.7319.

[242] Di Ventra M, Pershin Y V. On the Physical Properties of Memristive, Memcapacitive, and Meminductive Systems [J]. Nanotechnology, 2013, 24 (25): 12.

[243] Slipko V A, Pershin Y V, Di Ventra M. Changing the State of A Memristive System with White Noise [J]. Physical Review E, 2013 (87): 9 pages, Article ID 042103.

[244] Hashem N, Das S. Switching-Time Analysis of Binary-Oxide Memristors via A Non-Linear Model [J]. Applied Physics Letters, 2012, 100 (26): 3 pages, Article ID 262106.

[245] Linn E, Siemon A, Waser R, et al. Applicability of Well-Established Memristive Models for Simulations of Resistive Switching Devices [J]. IEEE Transactions on Circuits and Systems I, 2014, 61 (8): 2402-2410.

[246] Pershin Y V, Di Ventra M. Memory Effects in Complex Materials and Nanoscale Systems [J]. Advances in Physics, 2011, 60 (2): 145-227.

[247] Biolek D, Biolek Z, Biolkova V. Pinched Hysteresis Loops of Ideal Memristors, Memcapacitors, and Meminductors Must be 'Self-Crossing' [J]. Electronics Letters, 2011, 47 (25): 1385-1387.

[248] Terabe K, Hasegawa T, Liang C, et al. Control of Local Ion Transport to Create Unique Functional Nanodevices based on Ionic Conductors [J]. Science and Technology of Advanced Materials, 2007, 8 (6): 536-542.

[249] Beck A, Bednorz J G, Gerber C, et al. Reproducible Switching Effect in Thin Oxide Films for Memory Applications [J]. Applied Physics Letters, 2000, 77 (1), Article ID 139.

[250] Krieger J H, Spitzer S M. Non-traditional, Non-volatile Memory Based on Switching and Retention Phenomena in Polymeric Thin Films [C]. Proceedings of the IEEE Non-Volatile Memory Technology Symposium, Orlando, 2004: 121-124.

[251] Crupi M, Pradhan L, Tozer S. Modelling Neural Plasticity with Memristors [J]. IEEE Canadian Review, 2012 (68): 10-14.

[252] Bessonov A A, Kirikova M N, Petukhov D I, et al. Layered Memristive and Memcapacitive Switches for Printable Electronics [J]. Nature Materials, 2014

(4): 199-204.

[253] Chanthbouala A, Garcia V, Cherifi R O, et al. A Ferroelectric Memristor [J]. Nature Materials, 2012, 11 (10): 860-864.

[254] WANG X, CHEN Y, XI H, et al. Spintronic Memristor through Spin Torque Induced Magnetization Motion [J]. IEEE Electron Device Letters, 2009, 30 (3): 294-297.

[255] Chanthbouala A, Matsumoto R, Grollier J, et al. Vertical-Current-Induced Domain-Wall Motion in MgO-based Magnetic Tunnel Junctions with Low Current Densities [J]. Nature Physics, 2011, 7 (8): 626-630.

[256] Krzysteczko P, Günter R, Thomas A. Memristive Switching of MgO based Magnetic Tunnel unctions [J]. Applied Physics Letters, 2009, 95 (11): 3 pages, Article ID 112508.

[257] Krzysteczko P, Münchenberger J, Schäfers M, et al. The Memristive Magnetic Tunnel Junction as a Nanoscopic Synapse-Neuron System [J]. Advanced Materials, 2012, 24 (6): 762-766.

[258] Pershin Y V, Di Ventra M. Spin Memristive Systems: Spin Memory Effects in Semiconductor Spintronics [J]. Physical Review B, 2008, 78 (11): 14 pages, Article ID 113309.

[259] Pershin Y V, Di Ventra M. Current-Voltage Characteristics of Semiconductor/ Ferromagnet Junctions in the Spin-Blockade Regime [J]. Physical Review B, 2008, 77 (7): 3 pages, Article ID 073301.

[260] Lehtonen E. Two Memristors Suffice to Compute All Boolean Functions [J]. Electronics Letters, 2010, 46 (3): 239-240.

[261] Pershin Y V, La Fontaine S, Di Ventra M. Memristive Model of Amoeba Learning [J]. Physical Review E, 2009, 80 (2): 6 pages, Article ID 021926.

[262] Saigusa T, Tero A, Nakagaki T, et al. Amoebae Anticipate Periodic Events [J]. Physical Review Letters, 2008, 100 (1), Article ID 018101.

[263] Versace M, Chandler B. MoNETA: A Mind Made from Memristors [J]. IEEE Spectrum, 2010, 12 (12): 30-37.

[264] Snider G, Amerson R, Carter D, et al. From Synapses to Circuitry: Using Memristive Memory to Explore the Electronic Brain [J]. IEEE Computer, 2011, 44 (2): 21-28.

[265] Merrikh-Bayat F, Bagheri-Shouraki S, Rohani A. Memristor Crossbar-based Hardware Implementation of IDS Method [J]. IEEE Transactions on Fuzzy Systems, 2011, 19 (6): 1083-1096.

[266] Chua L O. Memristor, Hodgkin-Huxley, and Edge of Chaos [J]. Nanotechnology, 2013, 24 (38), Article ID 383001.

[267] Di Ventra M, Pershin Y V, Chua L O. Circuit Elements with Memory: Memristors, Memcapacitors and Meminductors [J]. Proceedings of the IEEE,

2009, 97 (10): 1717-1723.

[268] Eshraghian K, Rok Cho K R, Kavehei O, et al. Memristor MOS Content Addressable Memory (MCAM): Hybrid Architecture for Future High Performance Search Engines [J]. IEEE Transactions on Very Large Scale Integration (VLSI) Systems, 2011, 19 (8): 1407-1417.

[269] Tse Nga Ng, Russo B, Arias A C. Solution-Processed Memristive Junctions Used in a Threshold Indicator [J]. IEEE Transactions on Electron Devices, 2011, 58 (10): 3435-3443.

[270] Corinto F, Ascoli A. Memristive Diode Bridge with LCR Filter [J]. Electronics Letters, 2012, 48 (14): 824-825.

[271] Petráš I. Fractional-Order Memristor-Based Chua's Circuit [J]. IEEE Transactions on Circuits and Systems-II, 2010, 57 (12): 975-979.

[272] Radwan A G, Moaddy K, Hashim I. Amplitude Modulation and Synchronization of Fractional-Order Memristor-Based Chua's Circuit [J]. Abstract and Applied Analysis, 2013: 10 pages, Article ID 758676.

[273] Tenreiro Machado J A. Fractional Generalization of Memristor and Higher Order Elements [J]. Communications in Nonlinear Science and Numerical Simulation, 2013, 18 (12): 264-275.

[274] Ortigueira M D. An Introduction to the Fractional Continues-Time Linear Systems: The 21st Century Systems [J]. IEEE Circuits and Systems Magazine, 2008, 8 (3): 19-26.

[275] Krishna B T. Studies on Fractional Order Differentiators and Integrators: A Survey [J]. Signal Processing, 2011, 91 (3): 386-426.

[276] Ahmed S, Radwan A G, Soliman A M. Fractional-Order Mutual Inductance: Analysis and Design [J]. International Journal of Circuit Theory & Applications, 2016, 44 (1): 85-97.

[277] Hopfield J J. Neurons with Graded Response Have Collective Computational Properties Like Those of Two-State Neurons [J]. Proceedings of the National Academy of Sciences, 1984, 81 (10): 3088-3092.

[278] Newcomb R W. Neural-type microsystem: some circuits and consideration [C]. Proc. of the 1980 IEEE conference on Circuits and Computers, New York, 1980.

[279] Barranco B L. A novel CMOS analog neural oscillator cell [J]. IEEE trans, 1989, CAS-36 (5): 756-760.

[280] Xiao Yuan. On the models of a class of programmable neural oscillator cell [C]. Proc. of CHINA 1991 ICCAS, 1991 (1): 279-281.

[281] 蒲亦非，袁晓，廖科，等. 一种实现任意分数阶神经型脉冲振荡器的格形模拟分抗电路 [J]. 四川大学学报（工程科学版），2006, 38 (1): 128-132.

[282] 波色. 滤波器 [M]. 杜锡钰，等，译. 北京：人民邮电出版社，1958.

[283] 龙建忠，马代兴，何其超，等. 电路与系统理论 [M]. 成都：四川大学出版

社，1995.

[284] 特默斯，米特纳. 现代滤波器理论与设计 [M]. 王志杰，译. 刘宜伦，校. 北京：人民邮电出版社，1984.

[285] 拉姆. 模拟和数字滤波器设计与实现 [M]. 冯鹬云，应启衍，陆延丰，等，译. 北京：人民邮电出版社，1985.

[286] Eric Bogatin. 信号完整性分析 [M]. 李玉山，李丽平，等，译. 北京：电子工业出版社，2005.

[287] Alberto Isidori. 非线性控制系统 [M]. 王奔，庄圣贤，译. 北京：电子工业出版社，2005.

[288] Richard S. Muller, Theodore I. Kamins, Mansun Chan. 集成电路器件电子学 [M]. 王燕，张莉，译. 许军，校. 北京：电子工业出版社，2004.

[289] 希林，彼罗菲. 电子电路分立与集成 [M]，华中工学院工业电子学教研室，译. 陈婉儿，校. 北京：中国农业机械出版社，1984.

[290] Ash W R. Design for a brain [M]. 2nd ed. New York: Wiley, 1960.

[291] Siegelmann H T, Sontag E D. Turing computability with neural nets [J]. Applied Mathematics letters, 1991 (4): 77-80.

[292] Siegelmann H T, Horne B G, Giles C L. Computational capabilities of recurrent NARX neural netwoks [J]. Systems, man, and Cybernetics, 1997, 27: 208-215.

[293] Giles C L. Dynamically driven recurrent neural networks: Models, learning algorithms, and applications [C]. Tutorial 4, International Conference on Neural Networks, Washington, DC, 1996.

[294] Simon Haykin. 神经网络原理 [M]. 叶世伟，史忠植，译. 北京：机械工业出版社，2004.

[295] Hopfield J J. Neural networks and physical systems with emergent collective computational properties [C]. Proceedings of the National Academy of Sciences, 1982, 79: 2554-2558.

[296] Hopfield J J, Tank D W. Neural computation of decisions in optimization problems [J]. Biological Cybernetics, 1985, 52 (5): 141-154.

[297] PU Yifei, LIAO Ke, ZHOU Jiliu. A learning algorithm of multilayer dynamics associative neural network based on generalized Hebb rule [C]. Proceedings 2004 International Conference on Intelligent Mechatronics and Automation, IEEE, 2004: 836-841.

[298] PU Yifei. Implement any fractional order multilayer dynamics associative neural network [C]. Proc. 6th IEEE Conf. ASIC, Shanghai, China, 2005: 635-638.

[299] 奥本海姆. 信号与系统（英文版）[M]. 2版. 北京：电子工业出版社，2002.

[300] Hagan M T, Demuth H B, Beale M H. 神经网络设计 [M]. 戴葵，译. 李伯民，审校. 北京：机械工业出版社，2002.

[301] 加卢什金. 神经网络原理 [M]. 阎平凡，译. 北京：清华大学出版社，2002.

[302] 张乃尧，阎平凡. 神经网络与模糊控制［M］. 北京：清华大学出版社，2000.

[303] Puskorius G V, Feldkamp L A, Davis L I. Jr. Dynamic neural network methods applied to onvehicle idle speed control［J］. Proceedings of the IEEE, 1996（84）：1407-1420.

[304] Hebb D O. The Organization of Behavior［M］. New York：Wiley, 1949.

[305] 徐宗本，张讲社，郑亚林. 计算智能中的仿生学：理论与算法［M］. 北京：科学出版社，2003.

[306] Rafael C. Gonzalez, Richard E. Woods. 数字图象处理［M］. 阮秋琦，阮宇智，等，译. 北京：电子工业出版社，2003.

[307] Castleman K R. 数字图象处理［M］. 朱志刚，等，译. 北京：电子工业出版社，1998.

[308] 杨凯，孙家抦，卢健，等. 遥感图象处理原理和方法［M］. 北京：测绘出版社，1988.

[309] Milan Sonka, Vaclav Hlavac, Roger Boyle. 图象处理、分析与机器视觉［M］. 艾海舟，武勃，等，译. 北京：人民邮电出版社，2003.

[310] 普拉特. 数字图象处理学［M］. 高荣坤，王贻良，等，译. 北京：科学出版社，1984.

[311] Mlillesand T, Kiefer R W. 遥感与图象判读［M］. 黎勇奇，吴振鑫，晓岸，译. 杨廷槐，校. 北京：高等教育出版社，1986.

[312] 袁晓，虞厥邦. Bubble 小波的正交条件研究［J］. 成都：电子科技大学学报，1998，27（1）：25-28.

[313] PU Yifei, YUAN Xiao, LIAO Ke, et al. Fractional Calculus of Two-Dimensional Digital Image and Its Numerical Implementation. Waiting for being published.

[314] Regan D D. Human perception of Objects：Early Visual Processing of Spatial Form Defined by Luminance, Color, Texture, Motion, and Binocular Disparity, Sinauer Associates, Sunderland, Mass, 2000.

[315] Atchison D A, Smith G. Optics of the Human Eye. Butterwoh-Heinemann, Boston, Mass, 2000.

[316] Oyster C W. The Human Eye：Structure and Function, Sinauer Associates. Sunderland, Mass, 1999.

[317] Gordon I E. Theory of Visual Perception［M］. 2nd ed. New York：John Wiley & Sons, 1997.

[318] Hubel D H. Eye, Brain, and Vision, Scientific Amer. Library. W. H. Freeman, New York, 1988

[319] Cornsweet T N. Visual Perception［M］. New York：Academic Press, 1970.

[320] Blouke M M, Sampat N, Canosa J. Sensors and Camera Systems for Scientific, Industrial, and Digital Photography Applications-II［M］. Bellingham, Wash：SPIE Press, 2001.

[321] Levine M D. Vision in Man and Machine［M］. New York：McGraw-Hill, 1985.

[322] 陶然，齐林，王越. 分数阶 Fourier 变换的原理与应用 [M]. 北京：清华大学出版社，2004.

[323] Tank D W, Hopfield J J. Simple 'Neural' Optimization Networks: an A/D Converter, Signal Decision Circuit and a Linear Programming Circuit [J]. IEEE Trans. Circuits and System, 1986, 33 (5): 533-541.

[324] Li J, Michel A N, Porod W. Analysis and Synthesis of a Class of Neural Networks: Linear Systems Operating on a Closed Hypercube [J]. IEEE Trans. Circuits and System, 1989, 36 (11): 1405-1422.

[325] Michel A N, Farrell J A. Associative Memories via Artificial Neural Networks [J]. IEEE Control Systems Magazine, 1990, 10 (3): 6-17.

[326] Kosko B. Adaptive Bidirectional Associative Memories [J]. Applied Optics, 1987, 26 (23): 4910-4918.

[327] Kosko B. Bidirectional Associative Memories [J]. IEEE Trans. Systems, Man, and Cybernetics, 1988, 18 (1): 49-60.

[328] Samad T, Harper P. High-Order Hopfield and Tank Optimization Networks [J]. Parallel Computing, 1990, 16 (2-3): 287-292.

[329] Kosmatopoulos E B, Christodoulou M A. Structural Properties of Gradient Recurrent High-Order Neural Networks [J]. IEEE Trans. Circuits and Systems II: Analog and Digital Signal Processing, 1995, 42 (9): 592-603.

[330] LIU X, Teo K L, XU B. Exponential Stability of Impulsive High-Order Hopfield-Type Neural Networks with Time-Varying Delays [J]. IEEE Trans. Neural Networks, 2005, 16 (6): 1329-1339.

[331] LIU X, WANG Q. Impulsive Stabilization of High-Order Hopfield-Type Neural Networks With Time-Varying Delays [J]. IEEE Trans. Neural Networks, 2008, 19 (1): 71-79.

[332] HUANG H, Ho D W C, Lam J. Stochastic Stability Analysis of Fuzzy Hopfield Neural Networks with Time-Varying Delays [J]. IEEE Trans. Circuits and Systems II: Express Briefs, 2005, 52 (5): 251-255.

[333] LI H, CHEN B, ZHOU Q, et al. Robust Stability for Uncertain Delayed Fuzzy Hopfield Neural Networks with Markovian Jumping Parameters [J]. IEEE Trans. Systems, Man, and Cybernetics, Part B: Cybernetics, 2009, 39 (1): 94-102.

[334] YANG R, GAO H, SHI P. Novel Robust Stability Criteria for Stochastic Hopfield Neural Networks with Time Delays [J]. IEEE Trans. Systems, Man, and Cybernetics, Part B: Cybernetics, 2009, 39 (2): 467-474.

[335] ZHANG B, XU S, ZONG G, et al. Delay-Dependent Exponential Stability for Uncertain Stochastic Hopfield Neural Networks with Time-Varying Delays [J]. IEEE Trans. Circuits and Systems, 2009, 56 (6): 1241-1247.

[336] Özdemir N, İskender B B, Özgür N Y. Complex Valued Neural Network with Möbius Activation Function [J]. Commun. Nonlinear Sci. Numer. Simul., 2011,

16 (12): 4698-4703.

[337] Alofi A, Cao J, Elaiw A, et al. Delay-Dependent Stability Criterion of Caputo Fractional Neural Networks with Distributed Delay [J]. Discrete Dynamics in Nature and Society, 2014: 1-6, Article ID 529358.

[338] Kaslik E. Dynamics of Fractional-Order Neural Networks [C]. Proceedings of the 2011 International Joint Conference on Neural Networks, San Jose, 2011: 611-618.

[339] ZHANG R, QI D, WANG Y. Dynamics analysis of fractional-order three-dimensional Hopfield neural network [C]. Proceedings of the Sixth International Conference on Natural Computation, Yantai, 2010: 3037-3039.

[340] Raja M A Z, Khan J A, Qureshi I M. A New Stochastic Approach for Solution of Riccati Differential Equation of Fractional-Order [J]. Ann Math Artif Intell., 2010, 60 (3-4): 229-250.

[341] Raja M A Z, Khan J A, Qureshi I M. Swarm Intelligence Optimized Neural Network for Solving Fractional-order Systems of Bagley-Torvik Equation [J]. Engineering Intelligent Systems, 2011, 19 (1): 41-51.

[342] Raja M A Z, Khan J A, Qureshi I M. Evolutionary Computation Technique for Solution of Riccati Differential Equation of Arbitrary Order [J]. World Academy Science Engineering Technology, 2009, 3 (10): 303-309.

[343] Raja M A Z, Khan J A, Qureshi I M. Solution of Fractional-order System of Bagley-Torvik Equation Using Evolutionary Computational Intelligence [J]. Mathematical Problems in Engineering, 2011: 1-18, Article ID 765075.

[344] CHEN Y Q, Vinagre B M. A New IIR-Type Digital Fractional-order Differentiator [J]. Signal Processing, 2003, 83 (11): 2359-2365.

[345] PU Y F. Research on Application of Fractional Calculus to Latest Signal Analysis and Processing [D]. Chengdu: Sichuan University, 2006.

[346] BAI J, FENG X C. Fractional-Order Anisotropic Diffusion for Image Denoising [J]. IEEE Trans. Image Processing, 2007, 16: 2492-2502.

[347] Ham F M, Kostanic I. Principles of Neurocomputing for Science and Engineering [M]. New York: McGraw-Hill Companies, Inc., 2001.

[348] Storkey A J. Increasing the Capacity of a Hopfield Network Without Sacrificing Functionality [C]. Proceedings of the 7th International Conference on Artificial Neural Networks, Lausanne, 1997: 451-456.

[349] Storkey A J, Valabregue R. The basins of attraction of a new Hopfield learning rule [J]. Neural Networks, 1999, 12 (6): 869-876.

[350] Hopfield J J, Tank D W. Computing with Neural Circuits: A Model [J]. Science, 1986, 233 (4764): 625-633.

[351] Lach J, Mangione W H, Potkonjak M. Fingerprinting Techniques for Field-Programmable Gate Array Intellectual Property Protection [J]. IEEE Trans.

Computer-Aided Design of Integrated Circuits and Systems, 2001, 20 (10): 1253-1261.

[352] Kirovski D, Potkonjak M. Local Watermarks: Methodology and Application to Behavioral Synthesis [J]. IEEE Trans. Computer-Aided Design of Integrated Circuits and Systems, 2003, 22 (9): 1277-1283.

[353] Guneysu T, Moller B, Paar C. Dynamic Intellectual Property Protection for Reconfigurable Devices [C]. Proceedings of IEEE International Conference on Field-Programmable Technology, eds. Amano H, Ye A, Ikenaga T (Kitakyushu, Japan, 2007), 2007: 169-176.

[354] Roy J A, Koushanfar F, Markov I L. Ending Piracy of Integrated Circuits [J]. Computer, 2010, 43 (10): 30-38.

[355] Marsh, Kean T. A Security Tagging Scheme for ASIC Designs and Intellectual Property Cores. Design&Reuse, 2007. Available: http://www.design-reuse.com/articles/15105/a-security-tagging-scheme-for-asic-designs-and-intellectual-property-cores.html.

[356] Guajardo J, Kumar S S, Schrijen G J, et al. FPGA Intrinsic PUFs and Their Use for IP Protection, Cryptographic Hardware and Embedded Systems-CHES 2007, vol. 4727 of the series Lecture Notes in Computer Science, 2007: 63-80.

[357] Suh G E, Clarke D, Gassend B, et al. AEGIS: Architecture for Tamper-Evident and Tamper-Resistant Processing [C]. Proceedings of the 17th annual international conference on Supercomputing, eds. Banerjee U, Gallivan K, González A (San Francisco, USA, 2003), 2003: 160-171.

[358] Berger S, Cáceres R, Goldman K A, et al. vTPM: Virtualizing the Trusted Platform Module [C]. Proceedings of the 15th conference on USENIX Security Symposium-Volume 15, eds. Keromytis A D (Vancouver, CA, 2006), 2006: 305-320.

[359] Nithyanand R, Solis J. A Theoretical Analysis: Physical Unclonable Functions and the Software Protection Problem [C]. Proceedings of the 2012 IEEE Symposium on Security and Privacy Workshops, eds. Sven Dietrich (San Francisco, USA, 2012), 2012: 1-11.

[360] Pappu R, Recht B, Taylor J, et al. Physical One-Way Functions [J]. Science, 2002, 297 (5589): 2026-2030.

[361] Gassend, Clarke D, van Dijk M, et al. Silicon Physical Random Functions [C]. Proceedings of the 9th ACM Computer and Communications Security Conference, eds. Atluri V (Washington D. C., USA, 2002), 2002: 148-160.

[362] Lim, Lee J W, Gassend B, et al. Extracting Secret Keys from Integrated Circuits [J]. IEEE Transactions on Very Large Scale Integration (VLSI) Systems, 2005, 13 (10): 1200-1205.

[363] Böhm C, Hofer M. Physical Unclonable Functions in Theory and Practice [M].

Berlin: Springer Press, 2013.

[364] Böhm C, Hofer M, Pribyl W. A Microcontroller SRAM-PUF [C]. Proceedings of the 5th International Conference on Network and System Security, eds. XU L, Bertino E, MU Y (Milan, Italy, 2011), 2011: 269-273.

[365] Claes M, van der Leest V, Braeken A. Comparison of SRAM and FF PUF in 65nm Technology [C]. Proceedings of the 16th Nordic Conference on Secure IT-Systems, eds. Laud P (Tallinn, Estonia, 2011), 2011: 47-64.

[366] Kardas S, Kiraz M S, Bingöl M A, et al. A Novel RFID Distance Bounding Protocol Based on Physically Unclonable Functions [C]. Proceedings of the 7th International Workshop, RFIDSec, eds. Juels A, Paar C (Amherst, USA, 2011), 2011: 78-93.

[367] Selimis G, Konijnenburg M, Ashouei M, et al. Evaluation of 90nm 6T-SRAM as Physical Unclonable Function for Secure Key Generation in Wireless Sensor Nodes [C]. Proceedings of the 2011 IEEE International Symposium on Circuits and Systems, eds. Paulo S. R. Diniz and Ricardo L. de Queiroz (Rio de Janeiro, Brazil, 2011), 2011: 567-570.

[368] Kardaş S, Celik S, Yıldız M, et al. PUF-Enhanced Offline RFID Security and Privacy [J]. Journal of Network and Computer Applications, 2012, 35 (6): 2059-2067.

[369] Fyrbiak M, Kison C, Jeske M, et al. Combined HW-SW Adaptive Clone-Resistant Functions as Physical Security Anchors [C]. Proceedings of the 2013 NASA/ESA Conference on Adaptive Hardware and Systems, eds. Benkrid K (Torino, Italy, 2013), 2013: 130-137.

[370] Maiti A, Kim I, Schaumont P. A Robust Physical Unclonable Function with Enhanced Challenge-Response Set [J]. IEEE Trans. Information Forensics and Security, 2012, 7 (1): 333-345.

[371] Lee J W, Lim D, Gassend B, et al. A Technique to Build a Secret Key in Integrated Circuits for Identification and Authentication Applications [C]. Proceedings of the 2004 Symposium on VLSI Circuits. Digest of Technical Papers (Honolulu, USA, 2004), 2004: 176-179.

[372] Gassend B, Clarke D, Van Dijk M, et al. Delay-based Circuit Authentication and Applications [C]. Proceedings of the 18th Annual ACM symposium on Applied computing, eds. Lamont G, Menezes R (Melbourne, Australia, 2003), 2003: 294-301.

[373] Suh G E, Devadas S. Physical Unclonable Functions for Device Authentication and Secret Key Generation [C]. Proceedings of the 44th annual Design Automation Conference, eds. Levitan S P (San Diego, USA, 2007), 2007: 9-14.

[374] Morozov S, Maiti A, Schaumont P. An Analysis of Delay Based PUF Implementations on FPGA. Reconfigurable Computing: Architectures, Tools and

Applications, vol. 5992 of the series Lecture Notes in Computer Science, 2010: 382-387.

[375] Su Y, Holleman J, Otis B P. A Digital 1. 6 pJ/bit Chip Identification Circuit Using Process Variations [J]. IEEE Journal of Solid-State Circuits, 2008, 43 (1): 69-77.

[376] Tuyls P, Schrijen G J, Škorić B, et al. Read-Proof Hardware from Protective Coatings. Cryptographic Hardware and Embedded Systems-CHES 2006, vol. 4249 of the series Lecture Notes in Computer Science, 2006: 369-383.

[377] Majzoobi M, Koushanfar F, Potkonjak M. Techniques for Design and Implementation of Secure Reconfigurable PUFs [J]. ACM Trans. Reconfigurable Technol. Syst. , 2009, 2 (1): 1-33.

[378] Helfmeier C, Boit C, Nedospasov D, et al. Cloning Physically Unclonable Functions [C]. Proceedings of the IEEE International Symposium on Hardware Oriented Security and Trust, eds. Karri R, Koushanfar F, Hsiao M (Austin, USA, 2013), 2013: 1-6.

[379] Schusterr. Side-Channel Analysis of Physical Unclonable Functions (PUFs) [D]. Germany: Technische Universität München, 2010.

[380] Ruhrmair U, van Dijk M. PUFs in Security Protocols: Attack Models and Security Evaluations [C]. Proceedings of the 2013 IEEE International Symposium on Security and Privacy, eds. Sommer R (Berkeley, CA, 2013), 2013: 19-22.

[381] Merli D. Hardware Attacks on PUFs [C]. Proceedings of the 2012 NASA/ESA Conference on Adaptive Hardware and Systems, eds. Patel U (Erlangen, Germany, 2012), 2012: 25-28.

[382] Delvaux J, Verbauwhede I. Fault Injection Modeling Attacks on 65 nm Arbiter and RO Sum PUFs via Environmental Changes [J]. IEEE Trans. Circuits and Systems I, 2014, 61 (6): 1701-1713.

[383] LIAO X F, WANG J, CAO J D. Global and Robust Stability of Interval Hopfield Neural Networks with Time-Varying Delays [J]. International Journal of Neural Systems, 2003: 13 (3): 171-182.

[384] Al-Alawi R. FPGA Implementation of a Pyramidal Weightless Neural Networks Learning System [J]. International Journal of Neural Systems, 2003, 13 (4): 225-237.

[385] Bénédic Y, Wira P, MerckléJ. A New Method for the Re-Implementation of Threshold Logic Functions with Cellular Neural Networks [J]. International Journal of Neural Systems, 2008, 18 (4): 293-303.

[386] Strack B, Jacobs K M, Cios K J. Simulating Vertical and Horizontal Inhibition with Short-Term Dynamics in A Multi-column Multi-Layer Model of Neocortex [J]. International Journal of Neural Systems, 2014, 24 (5): 19 pages.

[387] WANG Z, GUO L, Adjouadi M. A Generalized Leaky Integrate-and-Fire Neuron

Model with Fast Implementation Method [J]. International Journal of Neural Systems, 2014, 24 (5): 15pages.

[388] Murray F. Analogue VLSI for Probabilistic Networks and Spike-Time Computation [J]. International Journal of Neural Systems, 2001, 11 (1): 23-32.

[389] Johnston S P, Prasad G, Maguire L, et al. An FPGA Hardware/Software Co-Design Towards Evolvable Spiking Neural Networks For Robotics Application [J]. International Journal of Neural Systems, 2010, 20 (6): 447-461.

[390] Rosselló J L, Canals V, Morro A, et al. Hardware Implementation of Stochastic Spiking Neural Networks [J]. International Journal of Neural Systems, 2012, 22 (4): 11 pages.

[391] Friedrich J, Urbanczik R, Senn W. Code-Specific Learning Rules Improve Action Selection by Populations of Spiking Neurons [J]. International Journal of Neural Systems, 2014, 24 (5): 16 pages.

[392] Shapero S, Zhu M, Hasler J, et al. Optimal Sparse Approximation with Integrate and Fire Neurons [J]. International Journal of Neural Systems, 2014, 24 (5): 16 pages.

[393] Rosselló J L, Canals V, Oliver A, et al. Studying the Role of Synchronized and Chaotic Spiking Neural Ensembles in Neural Information Processing [J]. International Journal of Neural Systems, 2014, 24 (5): 11 pages.

[394] Zhang G, Rong H, Neri F, et al. An Optimization Spiking Neural P System for Approximately Solving Combinatorial Optimization Problems [J]. International Journal of Neural Systems, 2014, 24 (5): 16 pages.

[395] Ghosh-Dastidar S, Adeli H. Spiking Neural Networks [J]. International Journal of Neural Systems, 2009, 19 (4): 295-308.

[396] Ghosh-Dastidar S, Adeli H. A New Supervised Learning Algorithm for Multiple Spiking Neural Networks with Application in Epilepsy and Seizure Detection [J]. Neural Networks, 2009, 22 (10): 1419-1431.

[397] Ghosh-Dastidar S, Adeli H. Improved Spiking Neural Networks for EEG Classification and Epilepsy and Seizure Detection [J]. Integrated Computer-Aided Engineering, 2007, 14 (3): 187-212.

[398] Adeli, Ghosh-Dastidar S. Automated EEG-based Diagnosis of Neurological Disorders-Inventing the Future of Neurology (CRC Press, Taylor & Francis, Boca Raton, Florida, 2010).

[399] PU Y F, ZHANG Y, ZHOU J L. Fractional Hopfield Neural Networks: Fractional Dynamic Associative Recurrent Neural Networks [J]. IEEE Trans. Neural Networks and Learning Systems, 2017, 28 (10): 2319-2333.

[400] Petráš. Fractional-Order Nonlinear Systems: Modeling, Analysis and Simulation [M]. Berlin: Springer Berlin Heidelberg, 2011.

[401] Rossikhin Y A, Shitikova M V. Applications of Fractional Calculus to Dynamic Problems of Linear and Nonlinear Hereditary Mechanics of Solids [J]. Applied Mechanics Reviews, 1997, 50 (1): 15-67.

[402] Nakagawa M, Sorimachi K. Basic Characteristics of a Fractance Device [J]. IEICE Trans. Fundamentals of Electronics, Communications and Computer Sciences, 1992, E75-A (12): 1814-1819.

[403] Butzer P L. An Introduction to Fractional Calculus [M] // Butzer P L, Westphal U. Applications of Fractional Calculus in Physics. Singapore: World Scientific Press, 2000.

[404] Heymans N, Podlubny I. Physical Interpretation of Initial Conditions for Fractional Differential Equations with Riemann-Liouville Fractional Derivatives [J]. Rheolgica Acta, 2006, 45 (5): 765-772.

[405] Pu Y F. Measurement Units and Physical Dimensions of Fractance-Part II: Fractional-Order Measurement Units and Physical Dimensions of Fractance and Rules for Fractors in Series and Parallel [J]. IEEE Access, 2016 (4): 3398-3416.

[406] Liu S H. Fractal Model for the AC Response of Rough Interface [J]. Physical Review Letters, 1985, 55 (5): 529-532.

[407] Kaplan T, Gray L J. Effect of Disorder on A Fractal Model for the AC Response of A Rough Interface [J]. Physical review B Condensed Matter, 1985, 33 (11): 7360-7366.

[408] Kaplan T, Liu S H, Gray L J. Inverse-Cantor-Bar Model for the AC Response of A Rough Interface [J]. Physical review B Condensed Matter, 1986, 34 (7): 4870-4873.

[409] Kaplan T, Gray L J, Liu S H. Self-Affine Fractal Model for a Metal-Electrolyte Interface [J]. Physical review B Condensed Matter, 1987, 35 (10): 5379-5381.

[410] PU Y F. Fractional-Order Circuit Gene Based Anti-Counterfeiting Detector for Defense against Chip Cloning Attacks [P]. China Patent for Invention 201410247717. 6, June 2014.

[411] HUANG X, WANG Z, LI Y X. Nonlinear Dynamics and Chaos in Fractional Hopfield Neural Networks with Delay [J]. Advances in Mathematical Physics, Article ID 657245, 2013: 9 pages.

[412] SONG C, CAO J D. Dynamics in Fractional-Order Neural Networks [J]. Neurocomputing, 2014 (142): 494-498.

[413] Choon Ki Ahn, Peng Shi, Michael V. Basin. Two-Dimensional Dissipative Control and Filtering for Roesser Model [J]. IEEE Transactions on Automatic Control, 2015, 60 (7): 1745-1759.

[414] Choon Ki Ahn, Ligang Wu, Peng Shi. Stochastic Stability Analysis for 2-D Roesser Systems with Multiplicative Noise [J]. Automatica, 2016 (69):

356-363.

[415] Choon Ki Ahn, Peng Shi, Ligang Wu. Receding Horizon Stabilization and Disturbance Attenuation for Neural Networks with Time-Varying Delay [J]. IEEE Transactions on Cybernetics, 2015, 45 (12): 2680-2692.

[416] Alexander C K, Sadiku M N O. Fundamentals of Electric Circuits [M]. Boston: McGraw-Hill Higher Education, 2000.

[417] Mandelbrot B B. The Fractal Geometry of Nature [M]. United States: W. H. Freeman and Company, 1982.

[418] Gordon D K, Philipson W R. A texture-enhancement procedure for separating orchard from forest in Thematic Mapper data [J]. International Journal of Remote Sensing, 1986, 7 (2): 301-304.

[419] Dutra L V, Mascarenhas N D D. Texture enhancement of Synthetic Aperture Radar (SAR) images with speckle noise reduction filters [J]. NASA STI/Recon Technical Report N, the SAO/NASA Astrophysics Data System, 1991 (93): 19833-19840.

[420] Joachim Weickert. Multiscale texture enhancement [J]. Computer Analysis of Images and Patterns; Lecture Notes in Computer Science, Springer, Berlin, 1995, 970: 230-237.

[421] Scheermesser T, Bryngdahl O. Control of texture in image halftoning [J]. J. Optical Society of America A, 1996, 13 (8): 1645-1652.

[422] Taponecco F, Urness T, Interrante V. Directional enhancement in texture-based vector field visualization [C]. Proceedings of 4th International Conference on Computer Graphics and Interactive Techniques in Australasia and South East Asia, 2006: 197-204.

[423] Alhinai K G, Khan M A, Canas A A. Enhancement of sand dune texture from landsat imagery using difference of Gaussian filter [J]. International Journal of Remote Sensing, 2008 (12): 1063-1069.

[424] Ofek E, Shilat E, Rappopport A, et al. Highlight and Reflection Independent Multiresolution Textures from Image Sequences [J]. IEEE Computer Graphics and Applications, 1997, 17 (2): 141-165.

[425] Irani M, Peleg S. Super resolution from image sequences [C]. Proc. International Conference on Pattern Recognition, Atlantic City, NJ, 1990: 115-120.

[426] Reinhard Koch, Marc Pollefeys, Luc Van Gool. Automatic 3D Model Acquisition from Uncalibrated Image Sequences [J]. IEEE Proceedings of International Conference on Computer Graphics, 1998: 597-604.

[427] Schröder P, Sweldens Wim. Spherical wavelets: Texture processing [J]. Rendering Techniques, 1995, 13 (8): 252-263.

[428] Larsen R, Stegmann M B, Darkner S, et al. Texture enhanced appearance models

[J]. Comput Vision Image Understanding, 2007, 106 (1): 20-30.

[429] Love E R. Fractional Derivatives of Imaginary Order [J]. J. London Math. Soc. 1971, 3: 241-259.

[430] Miller K S. Derivatives of Noninteger Order [J]. Math. Mag., 1995, 68: 183-192.

[431] Kilbas A A, Srivastava H M, Trujiilo J J. Theory and Applications of Fractional Differential Equations [M]. Amsterdam, Netherlands: Elsevier, 2006.

[432] Sabatier J, Agrawal O P, Tenreiro Machado J A. Advances in Fractional Calculus: Theoretical Developments and Applications in Physics and Engineering [M]. Springer, 2007.

[433] Chen W, Holm S. Fractional Laplacian time-space models for linear and nonlinear lossy media exhibiting arbitrary frequency dependency [J]. J. Acoustic Society of America, 2004, 115 (4): 1424-1430.

[434] PU Yifei. Fractional Calculus Approach to Texture of Digital Image [C]. IEEE Proceedings of 8th International Conference on Signal Processing, 2006: 1002-1006.

[435] PU Yifei. Fractional Differential Filter of Digital Image [P]. Invention Patent of China, No. 200610021702. 3, 2006.

[436] Dennis M, Healy Jr. Modern Signal Processing [M]. Cambridge University Press, 2004.

[437] Almeida L. The Fractional Fourier Transform and Time-Frequency Representations [J]. IEEE Transaction on Signal Processing, 1994, 42 (11).

[438] Ying Huang, Bruce Suter. The Fractional Wave Packet Transform [J]. Multidimensional Systems and Signal Processing, Springer Netherlands, 1998, 9 (4): 399-402.

[439] CHEN Linfei, ZHAO Daomu. Image encryption with fractional wavelet packet method [J]. International Journal for Light and Electron Optics, 2006, 119 (6): 286-291.

[440] Unser M, Blu T. Fractional Splines and Wavelets [J]. SIAM Review, 2000, 42 (1): 43-67.

[441] Beran J. Statistics for Long-Memory Processes [M]. Chapman & Hall, 1994.

[442] Tatom FB. The Relationship between Fractional Calculus and Fractals [J]. Fractals, 1995, 3 (1): 217-229.

[443] Jean-Bernard Martens. The Hermite Transform-Theory [J]. IEEE Trans. on Acoustics, Speech and Signal Processing, 1990, 38 (9): 1595-1606.

[444] Jean-Bernard Martens. The Hermite Transform-Applications [J]. IEEE Trans. on Acoustics, Speech and Signal Processing, 1990, 38 (9): 1607-1618.

[445] Jean-Bernard Martens. Local Orientation Analysis in Images by Means of the Hermite Transform [J]. IEEE Transactions on Image Processing, 1997, 6 (8):

1103-1116.

[446] Hayit Greenspan, Charles H. Anderson, Sofia Akber. Image Enhancement by Nonlinear Extrapolation in Frequency Space [J]. IEEE Transactions on Image Processing, 2000, 9 (6): 1035-1048.

[447] Shannon C E. A Mathematical Theory of Communication [J]. The Bell System Technical J., 1948 (27): 379-423, 623-656.

[448] Shannon C E, Weaver W. Mathematical Theory of Communication [M]. Urbana, IL: University of Illinois Press, 1963.

[449] Weszka J S, Rosenfeld A. Histogram modification for threshold selection [J]. IEEE Transactions on Systems, Man, And Cybernet, 1979, 9 (1): 38-52.

[450] Zhang Y J, Gerbrands J J. Transition region determination based thresholding [J]. Pattern Recognition Letters, 1991, 12 (1): 13-23.

[451] Groenewald A M, Barnard E, Botha E C. Related approaches to gradient-based thresholding [J]. Pattern Recognition Letters, 1993, 14 (7): 567-572.

[452] Burt P J, Adelson E H. The Laplacian Pyramid as a Compact Image Code [J]. IEEE Transactions on Communications, 1983, COM-31 (4): 532-540.

[453] Dippel S, Stahl M, Wiemker R, et al. Multiscale Contrast Enhancement for Radiographies: Laplacian Pyramid Versus Fast Wavelet Transform [J]. IEEE Transactions on Medical Imaging, 2002, 21 (4): 343-353.

[454] Laine A F, Fan J, Yang W. Wavelets for contrast enhancement of digital mammography [J]. IEEE Engineering in Medicine and Biology Society Magazine, 1995, 14 (5): 536-550.

[455] Starck J L, Murtagh F, Candes E, et al. Gray and Color Image Contrast Enhancement by the Curvelet Transform [J]. IEEE Transactions on Image Processing, 2003, 12 (6): 706-717.

[456] Do M N, Vetterli M. The Contourlet transform: an efficient directional multiresolution image representation [J]. IEEE Transactions on Image Processing, 2005, 14 (12): 2091-2106.

[457] Haykin S. Adaptive Filter Theory [M]. New Jersey: Prentice-Hall, 2002.

[458] Alexander S T. Adaptive Signal Processing: Theory and Applications [M]. New York: Springer-Verlag, 1986.

[459] Belfiore M G. Adaptive Filters and Signal Analysis [M]. New York: Dekker, 1988.

[460] Dinniz P S R. Adaptive Filtering: Algorithms and Practical Implementation [M]. Boston: Kluwer, 1997.

[461] Bitmead R P. Performance of adaptive estimation algorithms in dependent random environments [J]. IEEE Trans. Autom. Control., 1980, 25 (4): 788-794.

[462] Bitmead R P. Convergence properties of LMS adaptive estimators with unbounded dependent inputs [J]. IEEE Trans. Autom. Control., 1984, 29 (5): 477-497.

[463] Douglas S C. A family of normalized LMS algorithms [J]. IEEE Signal Process. Lett., 1994, 1 (3): 49-51.

[464] Dinniz P S R, Biscainho L W P. Optimal variable step size for the LMS/Newton algorithm with application to subband adaptive filtering [J]. IEEE Trans. Signal Process., 1992, 40 (11): 2825-2829.

[465] Graupe D. Identification of System [M]. New York: Van Nostrand Reinhold, 1972.

[466] Ahmed N, Soldan D L, Hummels D R, et al. Sequential regression considerations of adaptive filtering [J]. IET Electron. Lett., 1977, 13 (15): 446-448.

[467] Stearns S D. Digital Signal Analysis [M]. New York: Hayden, 1975.

[468] PU Yifei. High Precision Fractional Calculus Filter of Digital Image [P]. Invention Patent of China, No. ZL201010138742. 2, 2010.

[469] Snyman J A. Practical Mathematical Optimization: An Introduction to Basic Optimization Theory and Classical and New Gradient-Based Algorithms [M]. Beijing: Springer-Verlag, 2005.

[470] Euler L. Methodus Inveniendi Lineas Curvas Maximi Minimive Proprietate Gaudentes sive Solutio Problematis Isoperimetrici Latissimo Sensu Accepti [M]. Laussanne and Geneva, 1744.

[471] Euler L. Methodus Inveniendi Lineas Curvas, Additamentum I: De Curvis Elasticis (1744) [C]. Opera Omnia Ser. Prima, vol. XXIV, Orell Füssli, Bern, 1952: 231-297.

[472] Goldstine H H. A History of the Calculus of Variations from the 17th Through the 19th Century [M]. New York: Springer-Verlag, 1980.

[473] Erlichson H. Johann Bernoulli's Brachistochrone Solution Using Fermat's Principle of Least Time [J]. Eur. J. Phys., 1999, 20 (5): 299-304.

[474] Lanczos C. The Variational Principles of Mechanics [M]. New York: Dover Publications, 1986.

[475] Lao D Z. Fundamental of the Calculus of Variations (in Chinese) [M]. Beijing: National Defense Industry Press, 2015.

[476] Courant R. Variational Method for the Solution of Problems Equilibrium and Vibrations [J]. Bulletin of the American Mathematical Society, 1943, 49: 1-23.

[477] Turner M J, Clough R W, Martin H C, et al. Stiffness and Deflection Analysis of Complex Structures [J]. Journal of the Aeronautical Sciences, 1956, 23 (9): 805-823.

[478] Elsgolc L E. Calculus of variations [M]. London: Pergamon Press, 1961.

[479] Ekeland I. Nonconvex minimization problems [J]. Bulletin of the American Mathematical Society, New Series. 1, 1979 (3): 443-474.

[480] Hanc J, Tuleja S, Hancova M. Simple Derivation of Newtonian Mechanics from the Principle of Least Action [J]. Am. J. Phys., 2003, 71 (4): 386-391.

[481] Epstein S T. The Variation Method in Quantum Chemistry [M]. New York: Academic Press, 1974.

[482] Nesbet R K. Variational Principles and Methods in Theoretical Physics and Chemistry [M]. New York: Cambridge U. P., 2003.

[483] Adhikari S K. Variational Principles for the Numerical Solution of Scattering Problems [M]. New York: Wiley, 1998.

[484] Chung S J, Chen C H. Partial Variational Principle for Electromagnetic Field Problems: Theory and Applications [J]. IEEE Transactions on Microwave Theory and Techniques, 1988, 36 (3): 473-479.

[485] Leitmann G. The Calculus of Variations and Optimal Control: An Introduction [M]. Berlin: Springer, 1981.

[486] Chan T F, Shen J H. A Good Image Model Eases Restoration—on the Contribution of Rudin-Osher-Fatemi's BV Image Model [R]. IMA Technical Report 1829, 2002: 1-19. Available: https://www.researchgate.net/publication/237434927.

[487] Rudin L I, Osher S, Fatemi E. Nonlinear Total Variation based Noise Removal Algorithms [J]. Physica D, 1992, 60 (1-4): 259-268.

[488] Rudin L I, Osher S. Total Variation based Image Restoration with Free Local Constraints [C]. Proc. 1st IEEE Int. Conf. on Image Processing, Austin, TX, 1994: 31-35.

[489] Chan T F, Osher S, Shen J H. The Digital TV Filter and Nonlinear Denoising [J]. IEEE Trans. on Image Processing, 2001, 10 (2): 231-241.

[490] Song B. Topics in Variational PDE Image Segmentation, Inpainting and Denosing [D]. Los Angeles: University of California Los Angeles, 2003.

[491] Esedoglu S, Shen J. Digital Inpainting based on the Mumford-Shah-Euler Image Model [J]. European Journal of Applied Mathematics, 2002, 13 (4): 353-370.

[492] Darbon J, Sigelle M. Image Restoration with Discrete Constrained Total Variation Part I: Fast and Exact Optimization [J]. Journal of Mathematical Imaging and Vision, 2006, 26 (3): 261-276.

[493] Liu C, Shum H Y, Freeman W T. Face Hallucination: Theory and Practice [J]. International Journal of Computer Vision, 2007, 75 (1): 115-134.

[494] Datsenko D, Elad M. Example-based Single Document Image Super-Resolution: A Global MAP Approach with Outlier Rejection [J]. Multidimensional Systems and Signal Processing, 2007, 18 (2): 103-121.

[495] Marquina A, Osher S. Image Super-Resolution by TV-Regularization and Bregman Iteration [J]. Journal of Scientific Computing, 2008, 37 (3): 367-382.

[496] Lieu L H, Vese L A. Image Restoration and Decomposition via Bounded Total Variation and Negative Hilbert-Sobolev Spaces [J]. Applied Mathematics and Optimization, 2008, 58 (2): 167-193.

[497] Roth S, Black M J. Fields of Experts [J]. International Journal of Computer Vision, 2009, 82 (2): 103-111.

[498] Dahl J, Hansen P C, Jensen S H, et al. Algorithms and Software for Total Variation Image Reconstruction via First-Order Methods [J]. Numerical Algorithms, 2010, 53 (1): 67-92.

[499] PU Y F, ZHANG Y, ZHOU J L. Defense against Chip Cloning Attacks Based on Fractional Hopfield Neural Networks [J]. International Journal of Neural Systems, 2017, 27 (4).

[500] PU Y F. Analog Circuit Realization of Arbitrary-Order Fractional Hopfield Neural Networks: A Novel Application of Fractor to Defense against Chip Cloning Attacks [J]. IEEE Access, 2016 (4): 5417-5435.

[501] PU Y F, ZHANG N, WANG H, et al. Order-Frequency Characteristics of a Promising Circuit Element: Fractor [J]. Journal of Circuits, Systems, and Computers, 2016, 25 (12): 17 pages, Article ID 1650156.

[502] PU Y F, WANG W X. Fractional differential masks of digital image and their numerical implementation algorithms [J]. Acta Automatica Sinica, 2007, 33 (11): 1128-1135.

[503] XU M J, YANG J Z, ZHAO D Z, et al. An Image-Enhancement Method Based on Variable-Order Fractional Differential Operators [J]. Bio-Medical Materials and Engineering, 2015, 26 (s1): S1325-S1333.

[504] CHEN D L, ZHENG C R, XUE D Y, et al. Non-Local Fractional Differential-Based Approach for Image Enhancement [J]. Research Journal of Applied Sciences, Engineering and Technology, 2013, 6 (17): 3244-3250.

[505] PU Y F, ZHANG N, ZHANG Y, et al. A Texture Image Denoising Approach Based on Fractional Developmental Mathematics [J]. Pattern Analysis and Applications, 2016, 19 (2): 427-445.

[506] Ortigueira M D. Fractional Calculus for Scientists and Engineers [M]. Netherlands: Springer, 2011.

[507] Herrmann R. Fractional Calculus: An Introduction for Physicist [M]. River Edge: World Scientific, 2011.

[508] Herrmann R. On the Origin of Space [J]. Central European Journal of Physics, 2013, 11 (10): 1212-1220.

[509] El-Nabulsi R A. The Fractional Boltzmann Transport Equation [J]. Computers and Mathematics with Applications, 2011, 62 (3): 1568-1575.

[510] Herrmann R. Infrared Spectroscopy of Diatomic Molecules—A Fractional Calculus Approach [J]. International Journal of Modern Physics B, 2013, 27 (6): 17 pages, Article ID 1350019.

[511] Herrmann R. Gauge Invariance in Fractional Field Theories [J]. Physics Letters A, 2008, 372 (34): 5515-5522.

[512] El-Nabulsi R A, Wu G C. Fractional Complexified Field Theory from Saxena-Kumbhat Fraction Integral, Fractional Derivative of Order (α, β) and Dynamical Fractional Integral Exponent [J]. African Diaspora Journal of Mathematics, 2012, 13 (2): 45-61.

[513] El-Nabulsi R A. The fractional White Dwarf Hydrodynamical Nonlinear Differential Equation and Emergence of Quark Stars [J]. Applied Mathematics and Computation, 2011, 218 (6): 2837-2849.

[514] Riascos A P, Mateos J L. Fractional Dynamics on Networks: Emergence of Anomalous Diffusion and Lévy Flights [J]. Physical Review E, 2014, 90 (3), Article ID 032809.

[515] El-Nabulsi R A. A Cosmology Governed by A Fractional Differential Equation and the Generalized Kilbas-Saigo-Mittag-Leffler Function [J]. International Journal of Theoretical Physics, 2016, 55 (2): 625-635.

[516] El-Nabulsi R A. Implications of the Ornstein-Uhlenbeck-Like Fractional Differential Equation in Cosmology [J]. Revista Mexicana De Física, 2016, 62 (3): 240-250.

[517] El-Nabulsi R A. Fractional Variational Symmetries of Lagrangians, the Fractional Galilean Transformation and the Modified Schrödinger Equation [J]. Nonlinear Dynamics, 2015, 81 (1): 939-948.

[518] Freed A, Diethelm K, Luchko Y. Fractional-Order Viscoelasticity (FOV): Constitutive Development Using the Fractional Calculus [M]. Ohio: NASA's Glenn Research Center, 2002.

[519] El-Nabulsi R A. Fractional Functional with Two Occurrences of Integrals and Asymptotic Optimal Change of Drift in the Black-Scholes Model [J]. Acta Mathematica Vietnamica, 2015, 40 (4): 689-703.

[520] El-Nabulsi R A. The Fractional Kinetic Einstein-Vlasov System and Its Implications in Bianchi Spacetimes Geometry [J]. Int. Journal of Theoretical Physics, 2014, 53 (8): 2712-2726.

[521] Baumann G. Fractional Calculus and Symbolic Solution of Fractional Differential Equations, in Part of the series Mathematics and Biosciences in Interaction, Fractals in Biology and Medicine [M]. Switzerland: Birkhäuser Basel, 2005: 287-298.

[522] Cafagna D. Fractional Calculus: A Mathematical Tool from the Past for Present Engineering [J]. IEEE Industrial Electronics Magazine, 2007, 1 (2): 35-40.

[523] Marazzato R, Sparavigna A C. Astronomical image processing based on fractional calculus: the AstroFracTool [M]. Eprint Arxiv, 2009: 9 pages. Available: http://xueshu.baidu.com/s? wd = paperuri％ 3A％ 280b528e1ff6aa66ba6c1420f776954816％ 29&filter = sc _ long _ sign&tn = SE _ xueshusource _ 2kduw22v&sc _ vurl = http％ 3A％ 2F％ 2Fwww.oalib.com％

2Fpaper%2F3549008&ie=utf-8&sc _ us=12382112217299432113.

[524] Ahmed E, Elgazzar A S. On Fractional Order Differential Equations Model for Nonlocal Epidemics [J]. Physica A: Statistical Mechanics and its Applications, 2007, 379 (2): 607-614.

[525] Tejado I, Valério D, Valério N. Fractional Calculus in Economic Growth Modelling: The Spanish Case, in CONTROLO'2014—Proceedings of the 11th Portuguese Conference on Automatic Control [R]. Volume 321 of the series Lecture Notes in Electrical Engineering, Porto, 2014: 449-458.

[526] El-Nabulsi R A. Fractional Derivatives Generalization of Einstein's Field Equations [J]. Indian Journal of Physics, 2013, 87 (2): 195-200.

[527] El-Nabulsi R A. Fractional Variational Approach with Non-Standard Power-Law Degenerate Lagrangians and A Generalized Derivative Operator [J]. Tbilisi Mathematical Journal, 2016, 9 (1): 279-294.

[528] Gosh U, Sarkar S, Das S. Solutions of Linear Fractional Non-Homogeneous Differential Equations with Jumarie Fractional Derivative and Evaluation of Particular Integrals [J]. American Journal of Mathematical Analysis, 2015, 3 (3): 54-64.

[529] Luchko Y, Trujillo J. Caputo-Type modification of the Erdélyi-Kober Fractional Derivative [J]. Fractional Calculus and Applied Analysis, 2007, 10 (3): 249-267.

[530] El-Nabulsi R A. Fractional Elliptic Operators from a Generalized Glaeske-Kilbas-Saigo-Mellin Transform [J]. Functional Analysis, Approximation and Computation, 2015, 7 (1): 29-33.

[531] El-Nabulsi R A. The Fractional Calculus of Variations from Extended Erdélyi-Kober Operator [J]. International Journal of Modern Physics B, 2009, 23 (16): 3349-3361.

[532] El-Nabulsi R A. Fractional Variational Problems from Extended Exponentially Fractional Integral [J]. Applied Mathematics and Computation, 2011, 217 (22): 9492 - 9496.

[533] Sneddon I N. The Use in Mathematical Physics of Erdélyi-Kober Operators and Some of Their Generalizations [C]. Proceedings of the International Conference Held at the University of New Haven, New Haven, 1974: 37-79.

[534] El-Nabulsi R A, Torres D F M. Fractional Action-Like Variational Approach [J]. Journal of Mathematical Physics, 2008, 49, Article ID 053521.

[535] El-Nabulsi R A. Calculus of Variations with Hyperdifferential Operators from Tabasaki-Takebe-Toda Lattice Arguments [J]. Revista de la Real Academia de Ciencias Exactas, Fisicas y Naturales. Serie A. Matematicas, 2013, 107 (2): 419-436.

[536] El-Nabulsi R A, Torres D F M. Necessary Optimality Conditions for Fractional

Action-Like Integrals of Variational Calculus with Riemann-Liouville Derivatives of Order ($α$, $β$) [J]. Mathematical Methods in the Applied Sciences, 2007, 30 (15): 1931-1939.

[537] Jumarie G. Modified Riemann-Liouville Derivative and Fractional Taylor Series of Non-Differentiable Functions Further Results [J]. Computers & Mathematics with Applications, 2006, 51 (9-10): 1367-1376.

[538] El-Nabulsi R A. Glaeske-Kilbas-Saigo Fractional Integration and Fractional Dixmier Trace [J]. Acta Mathematica Vietnamica, 2012, 37 (2): 149-160.

[539] Katugampola U N. A New Approach to Generalized Fractional Derivatives [J]. Bulletin of Mathematical Analysis and Applications, 2014, 6 (4): 1-15.

[540] El-Nabulsi R A. Universal Fractional Euler-Lagrange Equation from a Generalized Fractional Derivative Operator [J]. Central European Journal of Physics, 2011, 9 (1): 250-256.

[541] Li Y, Chen Y Q, Podlubny I. Stability of Fractional-Order Nonlinear Dynamic Systems: Lyapunov Direct Method and Generalized Mittag-Leffler Stability [J]. Computers & Mathematics with Applications, 2010, 59 (5): 1810-1821.

[542] Li Y, Chen Y Q, Podlubny I. Mittag-Leffler Stability of Fractional Order Nonlinear Dynamic Systems [J]. Automatica, 2009, 45 (8): 1965-1969.

[543] Rudin W. Functional Analysis [M]. New York: McGraw Hill, 1991.

[544] Guidotti P, Lambers J V. Two New Nonlinear Nonlocal Diffusions for Noise Reduction [J]. J. Math. Imaging Vis., 2009, 33 (1): 25-37.

[545] Zhang J. Fractional-Order Variational PDE Based Image Modeling and Denoising Algorithms [D]. Nanjing: Nanjing University of Science and Technology, 2010.

[546] Tarasov V E. Fractional Vector Calculus and Fractional Maxwell's Equations [J]. Annals of Physics, 2008, 323 (11): 2756-2778.

[547] Osler T J. Leibniz Rule for Fractional Derivatives Generalized and an Application to Infinite Series [J]. SIAM Journal on Mathematical Analysis, 1970, 18 (3): 658-674.

[548] Thomas J W. Numerical Partial Differential Equations: Finite Difference Methods [M]. New York: Springer-Verlag, 1995.

[549] PU Y F. Application of Fractional Differential Approach to Digital Image Processing [J]. Journal of Sichuan University (Engineering Science Edition), 2007, 39 (3): 58-78.

[550] Chan T, Esedoglu S, Park F, et al. Recent Developments in Total Variation Image Restoration, in Mathematical Models of Computer Vision [M]. Springer Verlag, 2005.

[551] Buades A, Coll B, Morel J M. A Review of Image Denoising Algorithms, with a new one [J]. Multiscale Model. Simul., 2005, 4 (2): 490-530.

[552] Weickert J. Anisotropic Diffusion in Image Processing, ECMI Series [M].

Stuttgart: Teubner, 1998.

[553] Aubert G, Kornprobst P. Mathematical Problems in Image Processing: Partial Differential Equationsand the Calculus of Variations, Applied Mathematical Sciences [M]. New York: Springer-Verlag, 2006: 147.

[554] Perona P, Malik J. Scale-Space and Edge Detecting Using Anisotropic Diffusion [J]. IEEE Trans. Pattern Anal. Mach. Intell., 1990, 12 (7): 629-639.

[555] Sapiro G, Ringach D. Anisotropic Diffusion of Multivalued Images with Applications to Color Filtering [J]. IEEE Trans. Image Process., 1996, 5 (11): 1582-1586.

[556] Blomgren P, Chan T. Color TV: Total Variation Methods for Restoration of Vector-Valued Images [J]. IEEE Trans. Image Process., 1998, 7 (3): 304-309.

[557] Galatsanos N P, Katsaggelos A K. Methods for Choosing the Regularization Parameter and Estimating the Noise Variance in Image Restoration and Their Relation [J]. IEEE Trans. Image Process., 1992, 1 (3): 322-336.

[558] LI S Z. Close-Form Solution and Parameter Selection for Convex Minimization-Based Edge-Preserving Smoothing [J]. IEEE Trans. Pattern Anal. Mach. Intell., 1998, 20 (9): 916-932.

[559] Nguyen N, Milanfar P, Golub G. Efficient generalized cross-validation with applications to parametric image restoration and resolution enhancement [J]. IEEE Trans. Image Process., 2001, 10 (9): 1299-1308.

[560] Strong D M, Aujol J F, Chan T F. Scale Recognition, Regularization Parameter Selection, and Meyer's G Norm in Total Variation Regularization [J]. Multiscale Model. Simul., 2006, 5 (1): 273-303.

[561] Thompson A M, Brown J C, Kay J W, et al. A Study of Methods of Choosing the Smoothing Parameter in Image Restoration by Regularization [J]. IEEE Trans. Pattern Anal. Mach. Intell., 1991, 13 (4): 326-339.

[562] Mrazek P, Navara M. Selection of Optimal Stopping Time for Nonlinear Diffusion Filtering [J]. Int. J. Comput. Vis., 2003, 52 (2-3): 189-203.

[563] Gilboa G, Sochen N, Zeevi Y Y. Estimation of Optimal PDE-Based Denoising in the SNR Sense [J]. IEEE Trans. Image Process., 2006, 15 (8): 2269-2280.

[564] Vogel C R, Oman M E. Iterative Methods for Total Variation Denoising [J]. SIAM J. Sci. Comput., 1996, 17 (1): 227-238.

[565] Dobson D C, Vogel C R. Convergence of an Iterative Method for Total Variation Denoising [J]. SIAM J. Numer. Anal., 1997, 34 (5): 1779-1791.

[566] Chambolle A. An Algorithm for Total Variation Minimization and Applications [J]. J. Math. Imaging Vis., 2004, 20 (1-2): 89-97.

[567] Darbon J, Sigelle M. Exact Optimization of Discrete Constrained Total Variation Minimization Problems [C]. Proc. IWCIA, 2004: 548-557.

[568] Darbon J, Sigelle M. Image Restoration with Discrete Constrained Total Variation Part II: Levelable Functions, Convex Priors and Non-Convex Cases [J]. J. Math. Imaging Vis., 2006, 26 (3): 277-291.

[569] Wohlberg B, Rodriguez P. An Iteratively Reweighted Norm Algorithm for Minimization of Total Variation Functionals [J]. IEEE Trans. Signal Process. Lett., 2007, 14 (12).

[570] Catte F, Lions P L, Morel J M, et al. Image Selective Smoothing and Edge Detection by Nonlinear Diffusion [J]. SIAM J. Numer. Anal., 1992, 29 (1): 182-193.

[571] Meyer Y. Oscillating Patterns in Image Processing and in Some Nonlinear Evolution Equations [M]. AMS, 2001.

[572] Strong D, Chan T. Edge-Preserving and Scale-Dependent Properties of Total Variation Regularization [J]. Inv. Probl., 2003, 19 (6): 165-187.

[573] Alliney S. A Property of the Minimum Vectors of a Regularizing Functional Defined by Means of the Absolute Norm [J]. IEEE Trans. Signal Process., 1997, 45 (4): 913-917.

[574] Nikolova M. A Variational Approach to Remove Outliers and Impulse Noise [J]. J. Math. Imaging Vis., 2004, 20 (12): 99-120.

[575] Chan T, Esedoglu S. Aspects of Total Variation Regularized L^1 Function Approximation [J]. SIAM J. Appl. Math., 2005, 65 (5): 1817-1837.

[576] Nikolova M. Minimizers of Cost-Functions Involving Nonsmooth Data-Fidelity Terms [J]. SIAM J. Numer. Anal., 2002, 40 (3): 965-994.

[577] Osher S, Burger M, Goldfarb D, et al. An Iterative Regularization Method for Total Variation Based on Image Restoration [J]. Multiscale Model. Simul., 2005, 4 (2): 460-489.

[578] Gilboa G, Zeevi Y Y, Sochen N. Texture Preserving Variational Denoising Using an Adaptive Fidelity Term [C]. Proc. VLSM, Nice, France, 2003: 137-144.

[579] Esedoglu S, Osher S. Decomposition of Images by the Anisotropic Rudin-Osher-Fatemi Model [J]. Comm. Pure Appl. Math., 2004, 57 (12): 1609-1626.

[580] Blomgren P, Chan T, Mulet P. Extensions to Total Variation Denoising [C]. Proc. SPIE, San Diego, 1997, 3162.

[581] Blomgren P, Mulet P, Chan T, et al. Total Variation Image Restoration: Numerical Methods and Extensions [C]. ICIP, Santa Barbara, 1997: 384-387.

[582] Chan T, Marquina A, Mulet P. High-Order Total Variation-Based Image Restoration [J]. SIAM J. Sci. Comput., 2000, 22 (2): 503-516.

[583] You Y L, Kaveh M. Fourth-Order Partial Differential Equation for Noise Removal [J]. IEEE Trans. Image Process, 2000, 9 (10): 1723-1730.

[584] Lysaker M, Lundervold A, Tai X C. Noise Removal Using Fourth Order Partial Differential Equation with Applications to Medical Magnetic Resonance Images in

Space and Time [J]. IEEE Trans. Image Process, 2003, 12 (12): 1579-1590.
[585] Gilboa G, Sochen N, Zeevi Y Y. Image Enhancement and Denoising by Complex Diffusion Processes [J]. IEEE Trans. Pattern Anal. Mach. Int., 2004, 26 (8): 1020-1036.
[586] Chambolle A, Lions P. Image Recovery via Total Variation Minimization and Related Problem [J]. Numer. Math., 1997, 76 (2): 167-188.
[587] Lysaker M, Tai X C. Iterative Images Restoration Combining a Total Variation Minimization and a Second-Order Functional [J]. Int. J. Comput. Vis., 2006, 66 (1): 5-18.
[588] LI F, SHEN C M, FAN J S, et al. Image Restoration Combining a Total Variational Filter and a Fourth-Order Filter [J]. J. Vis. Commun. Image Represent., 2007, 18 (4): 322-330.
[589] Lysaker M, Osher M, Tai X C. Noise Removal Using Smoothed Normals and Surface Fitting [J]. IEEE Trans. Image Process., 2004, 13 (10): 1345-1357.
[590] DONG F F, LIU Z, KONG D X, et al. An Improved LOT Model for Image Restoration [J]. J. Math. Imaging Vis., 2009, 34 (1): 89-97.
[591] Duits R, Felsberg M, Florack L, et al. α scale spaces on a bounded domain [C]. Proc. 4th Int. Conf. Scale Spaces, 2003: 494-510.
[592] Didas S, Burgeth B, Imiya A, et al. Regularity and Scale Space Properties of Fractional High Order Linear Filtering [J]. Scale Spaces PDE Meth. Comput. Vis., 2005, 3459: 13-25.
[593] Mathieu B, Melchior P, Oustaloup A, et al. Fractional differentiation for edge detection [J]. Signal Processing, 2003, 83: 2421-2432.
[594] Halmos P. Measure theory [M]. D. van Nostrand and Co., 1950.
[595] Munroe M E. Introduction to Measure and Integration [M]. Addison Wesley, 1953.
[596] Tychonoff A N, Arsenin V Y. Solution of Ill-posed Problems [M]. Washington: Winston & Sons, 1977.
[597] Tomasi C, Manduchi R. Bilateral Filtering for Gray and Color Images [C]. Proceedings of the IEEE International Conference on Computer Vision, 1998: 839-846.
[598] Zhang M, Gunturk B K. Multiresolution Bilateral Filtering for Image Denoising [J]. IEEE Transactions on Image Processing, 2008, 17 (12): 2324-2333.
[599] Duncan D Y Po, Minh N Do. Directional Multiscale Modeling of Images using the Contourlet Transform [J]. IEEE Transactions on Image Processing, 2006, 15 (6): 1610-1620.
[600] Chen G Y, Bui T D. Multiwavelets Denoising using Neighboring Coefficients [J]. IEEE Signal Processing Letters, 2003, 10 (7): 211-214.
[601] YU H, ZHAO L, WANG H. Image Denoising using Trivariate Shrinkage Filter in

the Wavelet Domain and Joint Bilateral Filter in the Spatial Domain [J]. IEEE Transactions on Image Processing, 2009, 18 (10): 2364-2369.

[602] Antoni Buades, Bartomeu Coll, Jean-Michel Morel. A Non-Local Algorithm for Image Denoising [J]. IEEE Computer Society Conference on Computer Vision and Pattern Recognition, 2005, 2: 60-65.

[603] Buades A, Coll B, Morel J M. Nonlocal image and Movie Denoising [J]. International Journal of Computer Vision, 2008, 76 (2): 123-139.

[604] Wang Z, Bovik A C, Sheikh H R, et al. Image quality assessment: From error measurement to structural similarity [J]. IEEE Transactions on Image Processing, 2004, 13 (4): 600-612.

[605] Levi L. Unsharp Masking and Related Image Enhancement Techniques [J]. Computer Graphics and Image Processing, 1974, 3 (2): 163-177.

[606] Tahoces P G, Correa J, Souto M, et al. Enhancement of Chest and Breast Radiographs by Automatic Spatial Filtering [J]. IEEE Trans. Medical Imaging, 1991, 10 (3): 330-335.

[607] Peli T, Lim J S. Adaptive Filtering for Image Enhancement [J]. Optical Eng., 1982, 21 (1): 108-112.

[608] Sezan M I, Tekalp A M, Schaetzing R. Automatic Anatomically Selective Image Enhancement in Digital Chest Radiography [J]. IEEE Trans. Medical Imaging, 1989, 8 (2): 154-162.

[609] Fahnestock J D, Schowengerdt R A. Spatially Variant Contrast Enhancement using Local Range Modification [J]. Optical Eng., 1983, 22 (3): 378-381.

[610] Kim J Y, Kim L S, Hwang S H. An Advanced Contrast Enhancement Using Partially Overlapped Sub-Block Histogram Equalization [J]. IEEE Trans. Circuits and Systems for Video Technology, 2001, 11 (4): 475-484.

[611] Hummel R A. Image Enhancement by Histogram Transformation [J]. Computer Graphics and Image Processing, 1977, 6 (2): 184-195.

[612] Pizer S M, Zimmerman J B, Staab E V. Adaptive Gray Level Assignment in CT Scan Display [J]. J. Comput. Assisted Tomography, 1984, 8 (2): 300-305.

[613] Sherrier R H, Johnson G A. Regionally Adaptive Histogram Equalization of the Chest [J]. IEEE Trans. Medical Imaging, 1987, 6 (1): 1-7.

[614] Zimmerman J B. Effectiveness of Adaptive Contrast Enhancement [D]. University of North Carolina, 1985.

[615] Ji T L, Sundareshan M K, Roehrig H. Adaptive Image Contrast Enhancement Based on Human Visual Properties [J]. IEEE Trans. Medical Imaging, 1994, 13 (4): 573-58.

[616] Pizer S M, Amburn E P, Austin J D, et al. Adaptive Histogram Equalization and Its Variations [J]. Computer Vision, Graphics and Image Processing, 1987, 39 (3): 355-368.

[617] Land E H, McCann J J. Lightness and Retinex Theory [J]. J. Opt. Soc. Am., 1971, 61 (1): 1-11.

[618] Land E H. Recent Advances in the Retinex Theory and Some Implications for Cortical Computations: Color Vision and the Natural Image [J]. Proc. Nat. Acad. Sci. USA, 1983, 80 (16): 5163-5169.

[619] Frankle J, McCann J. Method and Apparatus for Lightness Imaging [P]. U. S. Patent 4 384 336, May 17, 1983.

[620] Jobson D J, Rahman Z, Woodell G A. Properties and Performance of the Center/Surround Retinex [J]. IEEE Trans. Image Processing, 1997, 6 (3): 451-462.

[621] Jobson D J, Rahman Z U, Woodell G A. A Multiscale Retinex for Bridging the Gap between Color Images and the Human Observation of Scenes [J]. IEEE Trans. Image Processing, 1997, 6 (7): 965-976.

[622] Land E H. An Alternative Technique for the Computation of the Designator in the Retinex Theory of Color Vision [J]. Proc. Nat. Acad. Sci. USA, 1986, 83 (10): 3078-3080.

[623] Marini D, Rizzi A. A Computational Approach to Color Adaptation Effects [J]. Image and Vision Computing, 2000, 18 (13): 1005-1014.

[624] Provenzi E, Fierro M, Rizzi A, et al. Random Spray Retinex: A New Retinex Implementation to Investigate the Local Properties of the Model [J]. IEEE Trans. on Image Processing, 2007, 16 (1): 162-171.

[625] Rahman Z, Jobson D J, Woodell G A. Multi-Scale Retinex for Color Image Enhancement [C]. Proc. IEEE International Conference on Image Processing, Lausanne, 1996: 1003-1006.

[626] Barnard K, Funt B. Investigations into Multi-Scale Retinex, Colour Imaging: Vision and Technology [M]. CIteseer, New York: Wiley Press, 1999: 9-17.

[627] Petro A B, Sbert C, Morel J M. Multiscale Retinex [J]. Image Processing On Line, 2014, 4: 71-88.

[628] Blake A. Boundary Conditions for Lightness Computation in Mondrian World [J]. Computer Vision Graphics and Image Processing, 1985, 32 (3): 314-327.

[629] Horn B K P. Determining Lightness from An Image [J]. Computer Graphics and Image Processing, 1974, 3 (4): 277-299.

[630] Terzopoulos D. Image Analysis Using Multigrid Relaxation Methods [J]. IEEE Trans. Pattern Analysis and Machine Intelligence, 1986, 8 (2): 129-139.

[631] Funt B V, Ciurea F, McCann J. Retinex in Matlab™ [J]. Journal of Electronic Imaging, 2004, 13 (1): 48-57.

[632] Bertalmio M, Caselles V, Provenzi E. Issues about Retinex Theory and Contrast Enhancement [J]. International Journal of Computer Vision, 2009, 83 (1): 101-119.

[633] Kimmel R, Elad M, Shaked D, et al. A Variational framework for Retinex [J].

International Journal of Computer Vision, 2003, 52 (1): 7-23.

[634] Morel J M, Petro A B, Sbert C. A PDE Formulationof Retinex Theory [J]. IEEE Trans. Image Processing, 2010, 19 (11): 2825-2837.

[635] Ng M K, Wang W. A Total Variation Model for Retinex [J]. SIAM Journal on Imaging Sciences, 2011, 4 (1): 345-365.

[636] CHEN W G. Variational Principles of Mechanics (in Chinese) [M]. Shanghai: Tongji University Press, 1989.

[637] Moorthy A K, Bovik A C. Blind Image Quality Assessment: From Natural Scene Statistics to Perceptual Quality [J]. IEEE Trans. on Image Process, 2011, 20 (12): 3350-3364.

[638] YUAN X. Mathematical Principles of Fractance Approximation Circuits [M]. Beijing: Science Press, 2015.

[639] Fouda M E, Radwan A G. On the Fractional-Order Memristor Model [J]. Journal of Fractional Calculus and Applications, 2013, 4 (1): 1-7.

[640] Fouda M E, Radwan A G. Fractional-order Memristor Response under DC and Periodic Signals [J]. Circuits Systems and Signal Processing, 2015, 34 (3): 961-970.

[641] YU Y J, BAO B C, KANG H Y, et al. Calculating area of Fractional-Order Memristor Pinched Hysteresis Loop [J]. The Journal of Engineering, DOI: 10.1049/joe. 2015. 0154, 2015: 3 pages.

[642] YU Y J, WANG Z H. A fractional-Order Memristor Model and the Fingerprint of the Simple Series Circuits including a Fractional-Order Memristor [J]. Acta Physica Sinica, 2015, 64 (23): 9 pages, Article ID 238401.

[643] SHI M, HU S L. Pinched Hysteresis Loop Characteristics of a Fractional-Order HP TiO2 Memristor [C]. Proc. International Conference on Life System Modeling and Simulation and International Conference on Intelligent Computing for Sustainable Energy and Environment, Nanjing, 2017: 705-713.

[644] Chua L O. Introduction to Nonlinear Network Theory, Volume 1, Foundations of Nonlinear Network Theory [M]. New York: Robert E. Krieger Publishing Company, 1978.

[645] Chua L O. Introduction to Nonlinear Network Theory, Volume 2, Resistive Nonlinear Networks [M]. New York: Robert E. Krieger Publishing Company, 1978.

[646] Chua L O. Dynamic Nonlinear Network: State-of-the-art [J]. IEEE Transactions on Circuit Systems, 1980, CAS-27 (11): 1059-1087.

[647] Wyatt J L, Chua L O, Oster G F. Nonlinear n-port Decomposition via the Laplace Operator [J]. IEEE Transactions on Circuit Systems, 1978, 25 (9): 741-754.

[648] Velmurugan G, Rakkiyappan R, Cao J. Finite-Time Synchronization of Fractional-Order Memristor-Based Neural Networks with Time Delays [J]. Neural

Networks, 2016, 73 (1-2): 36-46.

[649] Riana R. First Order Mem-Circuit: Modeling, Nonlinear Oscillations and Bifurcations [J]. IEEE Transactions on Circuits and Systems I: Regular Papers, 2013, 60 (6): 1570-1583.

[650] Kolka Z, Biolek D, Biolkova V. Frequency-Domain Steady-State Analysis of Circuits with Mem-Elements [J]. Analog Integrated Circuits and Signal Processing, 2013, 74 (1): 79-89.

[651] Liu S H. Fractal Model for the AC Response of Rough Interface [J]. Physical Review Letters, 1985, 55 (5): 529-532.

[652] Kim H, Sah M P, Yang C J, et al. Memristor Emulator for Memristor Circuit Applications [J]. IEEE Transactions on Circuits and Systems I: Regular Papers, 2012, 59 (10): 2422-2431.

[653] Budhathoki R K, Sah M P, Adhikari S P, et al. Composite Behavior of Multiple Memristor Circuits [J]. IEEE Transactions on Circuits and Systems I: Regular Papers, 2013, 60 (10): 2688-2700.

[654] Corinto F, Ascoli A, Gilli M. Nonlinear Dynamics of Memristor Oscillators [J]. IEEE Transactions on Circuits and Systems I: Regular Papers, 2011, 58 (6): 1323-1336.